The Chemical Reactor from Laboratory to Industrial Plant

AF708128

Riccardo Tesser • Elio Santacesaria

The Chemical Reactor from Laboratory to Industrial Plant

A Modern Approach to Chemical Reaction Engineering

Second Edition

Springer

Riccardo Tesser
Dipartimento di Scienze Chimiche,
Complesso di Monte Sant'Angelo
University of Naples Federico II
Naples, Italy

Elio Santacesaria (deceased)
Eurochem Engineering s.r.l.
Milan, Italy

ISBN 978-3-031-82504-0 ISBN 978-3-031-82502-6 (eBook)
https://doi.org/10.1007/978-3-031-82502-6

© The Editor(s) (if applicable) and The Author(s), under exclusive license to Springer Nature Switzerland AG 2018, 2025

This work is subject to copyright. All rights are solely and exclusively licensed by the Publisher, whether the whole or part of the material is concerned, specifically the rights of translation, reprinting, reuse of illustrations, recitation, broadcasting, reproduction on microfilms or in any other physical way, and transmission or information storage and retrieval, electronic adaptation, computer software, or by similar or dissimilar methodology now known or hereafter developed.
The use of general descriptive names, registered names, trademarks, service marks, etc. in this publication does not imply, even in the absence of a specific statement, that such names are exempt from the relevant protective laws and regulations and therefore free for general use.
The publisher, the authors and the editors are safe to assume that the advice and information in this book are believed to be true and accurate at the date of publication. Neither the publisher nor the authors or the editors give a warranty, expressed or implied, with respect to the material contained herein or for any errors or omissions that may have been made. The publisher remains neutral with regard to jurisdictional claims in published maps and institutional affiliations.

This Springer imprint is published by the registered company Springer Nature Switzerland AG
The registered company address is: Gewerbestrasse 11, 6330 Cham, Switzerland

If disposing of this product, please recycle the paper.

Preface

The First Edition of this book has received a good reception from students and chemical engineering researchers, therefore, the authors have been encouraged by Springer to publish a Second Edition. In this new edition, the structure of the book, well described in the Preface of the First Edition, has not been changed. Still, all the seven chapters of the First Edition have been carefully and critically re-examined, some errors found have been corrected, some arguments have been integrated or updated, and many different new exercises have been inserted. In particular, the MATLAB codes of all the exercises have rendered more accessible to the readers. Moreover, two new chapters, Chaps. 8 and 9, have been added to the book, one devoted to Polymerization and another one dealing with Bioreactors.

Chapter 8, dedicated to Polymerization, first describes briefly the historical evolution of the industrial production of polymers. Then, the polymers' average molecular weight and molecular weight distribution are defined together with the description of the techniques commonly employed for their experimental determination. The different polymerization mechanisms and the related polymerization kinetics are successively described as they represent the basic approach to polymerization reactors modelling. For this purpose, polycondensations, polyadditions, radical polymerizations, and stereospecific polymerizations are presented to define their mechanism, the kinetic approach, and reactor modelling strategy. Different examples and solved exercises have been elaborated by focusing on the previously defined aspects of the polymerization reactions

For what concerns Chap. 9, devoted to Bioreactors, a list of the microorganisms that can be cultivated for obtaining chemical products or biomass is reported in the introduction. Then, fermentation processes and microorganism kinetic growth are described. The historical evolution of the kinetic models, for the interpretation of the fermentation processes, is successively reported, considering both unstructured and structured models with their respective variants of segregated and unsegregated models. Starting from the unstructured basic Monod kinetic model and related modifications, the compartmental structured models are considered, in particular metabolic and cybernetic models. These models have been applied to the production of baker yeast from Saccharomyces cerevisiae, as an example of microorganism kinetic growth in aerobic conditions. Then, it has been considered the production of bioethanol, also employing the same microorganism but in anaerobic conditions. A brief review of the non-mechanistic models such as Neural Network mathematical models and Fuzzy models is reported at the end of the chapter. Different examples and solved exercises, related to both unstructured and structured models, are reported in this chapter.

By writing the two new chapters the authors were conscious that the topics treated therein could be the basis for a deeper description in independent books devoted to the specific subjects. This is true also for the previously mentioned seven chapters and effectively many reviews and books of, respectively, Thermodynamics, Kinetics, Catalysis, Mass and Heat Transfer, Multiphasic reactions and reactors, Reactor Modelling, Unit operations and Scale-up, have been published in the literature. Clearly, it is difficult to report in a short chapter all the fundamental information for each of the mentioned topics, but the scope of this book is mainly to furnish the readers with a method for solving Chemical Engineering problems starting from the laboratory experimental approach towards the industrial plant realization. Normally, this has been done following an approach that considers all the steps to be faced, such as: (1) the theoretical physical and chemical definition of the problem; (2) the laboratory work necessary to identify in what conditions is opportune to operate; (3) the definition of the algorithms and models for interpreting the laboratory experimental data; (4) the modelling and optimizing industrial size chemical engineering operations. The exercises of each chapter have been chosen always taking in mind both the scope of the book and the just described working scheme.

Naples, Italy Riccardo Tesser
Milan, Italy Elio Santacesaria

Preface to the First Edition

The contents of this book derive, first of all, from lessons lectured by both of the authors to students of the course, "Principles of Industrial Chemistry and Related Exercises," taken for the Master's Degree of Industrial Chemistry at the University Federico II of Naples, Italy. The content is also the result of many years of experimental and theoretical research work conducted by the authors in the field of chemical-reaction engineering. During their 30 years of research and teaching, science and technology have made an enormous progress, in particular after the advent of very powerful personal computers. In the old textbooks, many problems of industrial chemistry had approximated analytical or graphical solutions to avoid long and tedious hand-made calculations. The first computers—which were slow, expensive, and bulky—allowed for more rigorous calculations but were accessible only to specialists using a rigid software language, such as FORTRAN. The large diffusion of personal computers that are fast, cheap, and small—combined with a flexible and powerful software, such as MATLAB—allow everybody to solve many complicated problems with a numerical approach that is simpler, faster, and more satisfactory as to precision. Therefore, in this book, together with the theoretical approach to different topics, which is necessary to know for understanding the chemical reactor behaviour (e.g., thermodynamics of physical and chemical transformation, catalysis, kinetics, and mass transfer), many exercises are proposed inside the chapters devoted to the mentioned topics. The solutions to the exercises are described in detail inside the text, but the reader can find (at the Springer website) the MATLAB codes related to any single exercise and can interact directly with the proposed mathematical models and related solutions. The solutions are based on a numerical approach. A brief description of the main algorithm used, along with simple examples, is also reported on the Springer website with the MATLAB codes for the exercises.

The book is organized into seven chapters. Chapter 1 is a brief introduction describing what is important to know for developing industrial processes. Chapter 2 is devoted to the thermodynamics of chemical and physical equilibrium. Chapter 3 deals with the role of catalysis in promoting chemical reactions. Chapters 4 and 5 are related to the kinetics in, respectively, the homogeneous and heterogeneous phase and the relation between kinetics and the reaction mechanism. The last two chapters describe the effect of mass and heat transfer in, respectively, gas–solid and multi-phase reactors. In our opinion, this book could be useful for master's and doctoral students of chemical-engineering science and industrial chemistry but also for researchers working in the field of catalysis, kinetics, reactor design, and simulation.

Naples, Italy Riccardo Tesser
Milan, Italy Elio Santacesaria

Contents

Notation

Lowercase Letters

a, b	Attractive and repulsive constants in equations of state
a_i	Activity of component *i*
a_L	Gas–liquid specific inter-phase area
b_i	Adsorption equilibrium constant of *i* component
d_p	Particle diameter
e	Charge of the electron
f_i	Fugacity of component *i*
f_R	Reference fugacity of component *i*
g^E	Excess Gibbs energy
h	Heat-transfer coefficient
k_B	Boltzmann constant
k	Thermal conductivity
k_{eff}, k_T	Effective thermal conductivity
k_n	Kinetic constant of an elementary reaction of *n* order
k_c	Mass-transfer coefficient for a concentration gradient
k_g	Mass-transfer coefficient for a pressure gradient
k_L	Gas–liquid (liquid side) mass-transfer coefficient
k_{app}	Apparent kinetic constant
k°	Pre-exponential factor
n_i	Number of *i* moles
n_i^o	Initial mole number of *i*
p_i	Partial pressure of component *i*
p_{ij}	Probability of interaction between components *i* and *j*
p_i^o	Vapour pressure of *i* component
r_j	Reaction rate for *j*th reaction
q_e	Amount of solute adsorbed on the solid
q_{max}	Maximum adsorption capacity
r	Particle radius

s	Standard error
t	Time
t_{inv}	Inverse of Student distribution
u_{ij}	Energy of interaction
u_x	Velocity in x direction
v_i	Volume of component i
x_i	Liquid-phase mole fraction of component i
y_i	Vapour-phase mole fraction of component i

Uppercase Letters

A	Helmoltz free energy
A_{p}	Activity of the poisoned catalyst
A_{o}	Activity of the fresh catalyst
$B(T)$, $C(T)$, $D(T)$	Functions in virial
B_{M}	Virial coefficient for mixture
B_{ij}	Cross virial coefficient
C_{P}	Specific heat at constant pressure
C_{V}	Specific heat at constant volume
C_i	Concentration of component i
D	Distillate molar flow rate
D_i	Diffusion coefficient of component i
D_{eff} or D_{e}	Effective diffusion coefficient
D_{be}	Bulk-diffusion coefficient
D_{ke}	Knudsen diffusion coefficient
D_{ij}	Diffusion coefficient of component i in j
D_{23}	Sauter diameter
Ea	Activation energy
E	Enhancement factor
E_{L}	Back-mixing coefficient
F	Volumetric or molar flow rate
G	Gibbs free energy
G_{m}	Mass velocity
H	Enthalpy
H_{o}	Hammett function
H_{f}	Height of fluidized bed
H_i	Henry solubility parameter of i component
H_{A}	Hatta number
K_{P}	Equilibrium constant in terms of partial pressures
K_{C}	Equilibrium constant in terms of concentrations
K_{L}	Equilibrium constant in liquid phase
K_{n}	Equilibrium constant in terms of number of moles
$K_{x(\text{or } y)}$	Equilibrium constant in terms of mole fractions
K_{f}	Equilibrium constant in terms of fugacites
K_{ϕ}	Equilibrium constant in terms of fugacity coefficients

K_Γ	Equilibrium constant in terms of activities
K_e	Generic equilibrium constant
K	Vapour–liquid equilibrium constant
J	Jacobian matrix of partial derivatives
J_D, J_H	Analogy factors, respectively, related to mass and heat transfer
J_G, J_L	Mass transfer rates for gas–liquid system
L	Liquid molar flow rate
M	Molecular weight
M_W	Weisz modulus
N	Number of molecules
N_{AV}	Avogadro number
Nu	Nusselt number
^{o}	Initial conditions of the variable preceding the symbol
P	Total pressure
Pr	Prandtl number
P_C	Critical pressure
P_{Cm}	Pseudo-critical pressure
P_R	Reduced pressure
Q	Heat
Q_i	Volumetric flow rate of i
Q_{ads}	Heat of adsorption
Q_v	Heat exchanged by a mole of ideal gas at constant volume
Q_T	Heat exchanged by a mole of ideal gas at constant temperature
R	Gas constant
Re	Reynolds number
R^2	Correlation coefficient
R°	Initial molar ratio between reactants
S_i	Fractional selectivity related to i component
S	Entropy
Sh	Sherwood number
Sc	Schmidt number
T	Temperature
T_B	Boiling temperature
T_C	Critical temperature
T_{Cm}	Pseudo-critical temperature
T_R	Reduced temperature
U	Internal energy
V_i	Volume of i
V	Vapour molar flow rate
V_R	Reactor volume
V_C	Critical volume
V_R	Reduced volume
W	Mechanical work or mass of catalyst
We	Weber number

$\overline{X}$	Partial molar quantity
Z_i, z_i	Compressibility factor of*i*component
Z_C	Critical compressibility factor
Z	Solvent dielectric constant
Z_{AB}	Number of collision between *A* and *B*

Greek Letters

α, β, γ, δ	Stoichiometric coefficients
β_j	Adjustable parameters
β	Prater number
α	Vaporization ratio (V/F)
γ	Adiabatic exponent
γ_i	Activity coefficient of component *i*
δ	Film thickness
δ_i	Solubility parameters
φ	Fugacity coefficient
φ_{c}, φ_{max}	Gas absorption, maximum gas absorption
ϕ	Fugacity coefficient
ϕ_{L}	Thiele modulus
ϕ_i	Volumetric fraction
$\phi(\beta)$	Objective function
λ	Fractional conversion
λ_{e}	Equilibrium fractional conversion
λ_{ev_i}	Cohesion energy of component *i*
μ	Chemical potential
μ_{p}	Dipole moment
μ_{r}	Reduced dipole moment
ξ	Extent of reaction or coordinate of reaction
ω	Acentric factor
ω_{m}	Acentric factor for the mixture
ν_i	Stoichiometric coefficient of component *i*
Γ_i	Activity of component *i*
ΔH	Enthalpy exchange
Λ	Interaction parameter in Wilson equation
τ	Interaction parameter in NRTL equation
ϑ	Surface fraction in UNIQUAC equation
Θ	Volume fraction in UNIQUAC equation
Θ_i	Fraction of site occupied by component *i*
σ	Site on solid surface
σ_{ij}	Lennard–Jones parameter
σ_i	Collision integral of component *i*
ΔE	Activation energy
θ	Porosity
δ_{s}	Density of solid in fluidized bed

δ_f	Density of fluid in fluidized bed
ε_f^o	Void degree in fluidized bed
ε	Lennard–Jones parameter
τ	Tortuosity
μ	Viscosity
Ω_μ, Ω_D	Collision integrals
ρ	Density
η	Catalyst effectiveness factor
η_j	Murfree efficiency on plate j
η_i^j	Fractional yield of i with respect to j

Math Symbols

$\oint$	Cyclic integral
∂	Partial derivative

Chapter 1
Introduction to the Study of Chemical Industrial Processes

Abstract This chapter starts with a brief introduction describing the structure and characteristics of chemical industrial plants considered as an ensemble of unit operations. Next, the roles of thermodynamics, catalysis, kinetics, transport phenomena, and mass and energy balance in modeling and conducting chemical industrial processes are briefly described to justify the focused approach to these topics in successive chapters. Finally, consideration is given to the power of modern computers, the use of numerical methods for solving modeling problems, and the simulation of chemical industrial processes. Some examples of the most employed mathematical algorithms are mentioned, while some examples and exercises are given as electronic supplementary material (ESM) accessible to readers online.

Keywords Unit operations · Modeling and conducting industrial processes · Computer simulation, Algorithms and numerical methods

1.1 Structure and Characteristics of Chemical Industrial Plants

The heart of any chemical plant is represented by the chemical reactor, a vessel in which a chemical reaction occurs yielding a desired product accompanied by eventual by-products formed as a consequence of side reactions. The reactor's shape is conditioned by several factors, such as the physical state of the reagents and products, the heat released by the reaction, the adopted operative conditions, etc. The materials used for building the reactors can be different, but generally, the most economical materials available from industry (1) that are compatible with the corrosiveness of the reactants and (2) that are resistant to the temperature and pressure adopted for performing the reaction are the ones chosen. However, we can distinguish between the following types of reactors.

Supplementary Information The online version contains supplementary material available at https://doi.org/10.1007/978-3-031-82502-6_1.

© The Author(s), under exclusive license to Springer Nature Switzerland AG 2025
R. Tesser, E. Santacesaria, *The Chemical Reactor from Laboratory to Industrial Plant*, https://doi.org/10.1007/978-3-031-82502-6_1

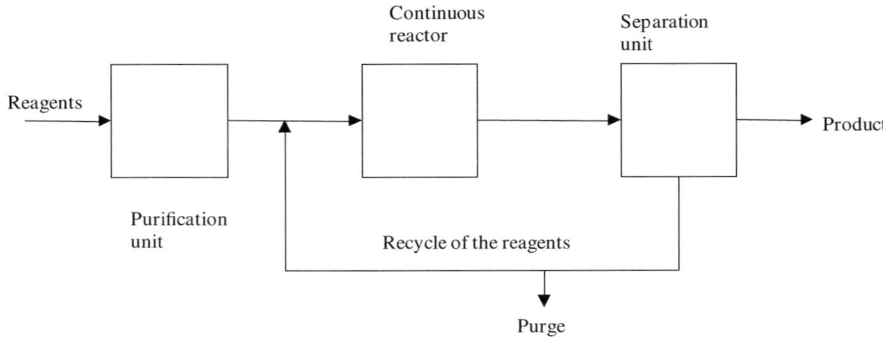

Fig. 1.1 Scheme of a continuous process operating under steady-state conditions

Batch Reactors

All of the reactants are initially loaded in the reactor and kept at a determined temperature until the reaction is completed. At the end of the reaction, the reaction mixture is discharged, and the obtained products are separated from the unreacted reagents and the eventual by-products. The production is characterized in this case by a cyclic operation as follows: (1) loading the reactor with the reactants; (2) starting the reaction by heating and mixing; (3) discharging the reactor; (4) separating the product from the other components in the reaction mixture; (5) recovering and recycling the employed catalyst; and (6) reloading the reactor for a new cycle.

Continuous Reactors

In a continuous reactor, the reactants are opportunely purified, preheated, pressurized, and fed to the reactor. Normally, inside the reactor, a catalyst promotes the reaction; therefore, at the reactor outlet, the product mixed with both unreacted reagents and by-products can be recovered. This mixture is then submitted to separation in other continuous units. When the reaction equilibrium is not favorable, the conversion of the reagents is only partial, and the abundant unreacted reagents accompanying the product must be continuously recycled to the reactor after separation. Then, if an inert component is present inside the reagents (e.g., when nitrogen from the air is used as a reagent, argon component gradually accumulates inside the reactor), it must be purged by discharging a flow stream of an opportune entity to avoid a frequent plant stopping. Obviously, part of the reagents is normally lost in this operation. A characteristic of the continuous plant is the achievement of a **steady-state condition**, in which the composition at, respectively, the inlet and outlet of the reactor is rigorously constant over time. The concentrations and temperature profiles inside the reactor are also constant. A continuous plant can be described with a series of operations of the type (Fig. 1.1).

Semi-batch Reactors

In semi-batch reactors, one or more reagents are preliminarily put into the reactor and heated at the desired temperature. One or more reagents are then fed continuously to produce the reaction. This type of reactor is frequently employed in the case of gas–liquid reactions because it is simpler to feed a stream of gas kept at constant pressure instead of loading all the gas necessary for the reaction compressed at increased pressure. Another case for which this type of reactor is preferred is liquid–liquid reactions, in which the reaction is highly exothermic. By opportunely adjusting the flow rate of one of the reagents, the heat release and consequently the temperature inside the reactor can be kept under control.

Chemical Industrial Plants and the Unit Operation Strategy

Any industrial chemical plant, even if it is very complex, can be regarded as an ensemble of different unit operations that can be considered separately for the scope of modeling and optimization. The "unit operation" strategy, as applied to the industrial chemical plants, was introduced by the chemical engineer, Arthur D. Little, of MIT (1915) (see Servos (1980)) and is still universally used. A large variety of chemical reactions are performed in industrial processes under different operative conditions, that is, we have thousands of different processes, but all of them have in common some basic operations—such as mixing, heat exchange, distillation, filtration, etc.—that are similar in each process. Therefore, whatever process we consider, a limited number of unit operations must be considered to completely describe the plant by independently modeling or simulating every single unit.

An attempt also has been made to classify chemical reactions involved in industrial processes by their similarity (polymerization, esterification, transesterification, sulfonation, nitration, etc.), but the success of this classification was limited to the didactic scope. By drawing the sequence of unit operations occurring in a plant, we have a scheme of the plant (flow sheet) that can be more or less detailed. In these schemes, the devices used for each unit operation are drawn in a standardized way and are normally easily recognizable. An example of a simplified flow sheet is shown in Fig. 1.2.

As mentioned previously, the heart of the plant is the **chemical reactor** in which the reaction occurs, thus yielding the desired product. The study of the chemical reactor is the subject of this book. Let us consider, first of all, some variables that must be absolutely known to follow the evolution of a reaction inside a reactor.

Definition of Conversion, Yield, and Selectivity

Consider a generic reaction:

$$\alpha \mathrm{A} + \beta \mathrm{B} \rightarrow \gamma \mathrm{C} + \delta \mathrm{D} \tag{1.1}$$

The yield of C with respect to A is given by:

$$\eta_{\mathrm{A}}^{\mathrm{C}} = \frac{n_{\mathrm{C}}}{n_{\mathrm{C_{tot}}}} \tag{1.2}$$

being n_{C} is the number of C moles that have actually formed, and $n_{\mathrm{C_{tot}}}$ is the number of C moles that will be formed if the reaction goes to completion. If the initial moles of A are $\overset{\circ}{n}_{\mathrm{A}}$, we can write that:

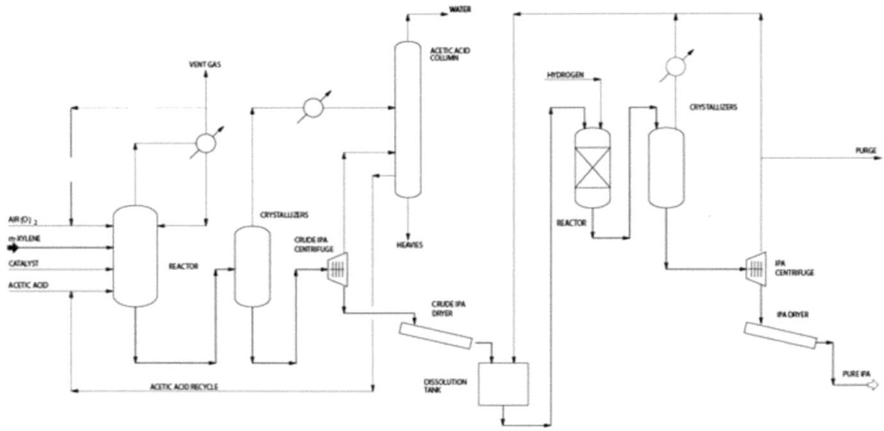

Fig. 1.2 Simplified scheme (flow sheet) of the process for the production of isophthalic acid from *m*-xylene

$$\frac{\overset{\circ}{n}_{\mathrm{A}}}{\alpha} = \frac{n_{\mathrm{C_{tot}}}}{\gamma} \tag{1.3}$$

Hence,

$$n_{\mathrm{C_{tot}}} = \frac{\overset{\circ}{n}_{\mathrm{A}}}{\alpha}\gamma \tag{1.4}$$

Consequently, the yield will be:

$$\eta_{\mathrm{A}}^{\mathrm{C}} = \frac{\alpha n_{\mathrm{C}}}{\gamma \overset{\circ}{n}_{\mathrm{A}}} \tag{1.5}$$

In the same way the yield of D with respect to A:

$$\eta_{\mathrm{A}}^{\mathrm{D}} = \frac{\alpha n_{\mathrm{D}}}{\delta \overset{\circ}{n}_{\mathrm{A}}} \tag{1.6}$$

The yield of a process can be limited by three factors:

(1) The reaction rate
(2) The chemical equilibrium
(3) The presence of side reactions

In some cases, to limit the formation of by-products originated by a long reaction time, it could be convenient to stop the reaction at a relatively early stage to recover the product and recycle the reactant. By operating in this way, the overall yield remains high because the conversion per passage is kept low but the reaction time is

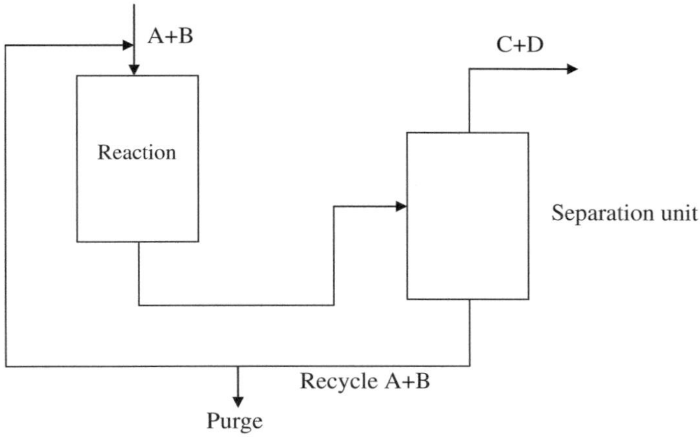

Fig. 1.3 Scheme of a process working at low conversion of the reactants

decreased. The conversion of a reactant is the ratio between the amount reacted and the initial amount of A, that is,

$$\lambda = \frac{\overset{\circ}{n}_A - n_A}{\overset{\circ}{n}_A} = \frac{\text{Reacted moles of A}}{\text{Initial moles of A}} \tag{1.7}$$

By operating at a low conversion level, unreacted A must be continuously recycled, and the process is shown in Fig. 1.3.

In this case, we can evaluate a conversion per passage (in some cases <10%) and the corresponding yield per passage given by the amount of C + D produced with respect to A + B fed to the reactor; the yield per passage must be distinguished by the yield of product that is the ratio between C + D produced and A + B reacted.

The ratio between the yield and the conversion is called selectivity, a very important factor indicating how much of a product is privileged in production with respect to other possible products. Knowing selectivity in the production of different products means knowing the product distribution. The following example, based on the reaction of benzene hydrogenation, better explains the concept of selectivity:

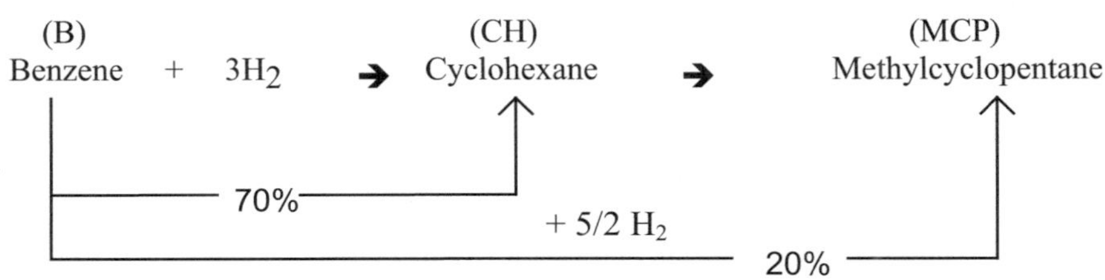

Starting from 1 mol of benzene, we will have:

Benzene residual (10%) B	0.1 mol
Benzene reacted to give CH	0.7 Mol
Benzene reacted to give MCP	0.2 Mol

Conversion of Benzene

$$\lambda = \frac{\overset{\circ}{n}_B - n_B}{\overset{\circ}{n}_B} = \frac{1 - 0.1}{1} = 0.9 \tag{1.8}$$

Yield

$$\eta_B^{CH} = \frac{n_{CH}}{\overset{\circ}{n}_B} = \frac{0.7}{1} = 0.7 \quad \eta_B^{MCP} = \frac{n_{MCP}}{\overset{\circ}{n}_B} = \frac{0.2}{1} = 0.2 \tag{1.9}$$

Selectivity

$$S_{CH} = \frac{\eta_B^{CH}}{\lambda} = \frac{0.7}{0.9} = 0.777 \quad S_{MCP} = \frac{\eta_B^{MCP}}{\lambda} = \frac{0.2}{0.9} = 0.222 \tag{1.10}$$

In conclusion, by the hydrogenation of benzene, we obtain 77.7% cyclohexane and 22.2% methyl-cyclopentane.

These preliminary remarks make clear the importance of obtaining the highest possible yield and selectivity in the industrial processes, but this goal can be achieved only by studying all of the reactions occurring in a process: (a) the reaction scheme and the stoichiometry of each reaction; (b) the chemical equilibria; (c) the mass and heat balance; (d) the chemical reaction rate; (e) the role of catalysis; and (f) the heat- and mass-transfer rates. Moreover, the most convenient operative conditions must be individuated.

1.2 Thermodynamics, Catalysis, Kinetics, and Transport Phenomena: Their Role in Modeling and Conducting Chemical Industrial Processes

The first step in the realization of a chemical industrial plant is the definition of the sequence of operations (unit operations) that must be made to obtain a product with the purity level required by the market. Then, it is necessary to design each apparatus of the plant by defining its size and the operative modalities. This requires knowing all of the chemical and physical phenomena occurring in any single apparatus. Thermodynamics allows us to solve all of the equilibrium and energetic problems (both chemical and physical) of any possible transformation (chemical reactions, changes of physical state). Kinetics studies the rate of reactions and its dependence on temperature and composition. The reaction rate strongly affects the size and productivity of the reactor. The catalysts have a key role in industrial processes because they can greatly increase the rate of the desired reaction, therefore also positively affecting selectivity. Transport phenomena are often important in limiting the reaction rate or determining problems of heat removal. All of these aspects must be preliminarily studied in the laboratory in plants of small size before passing to the scaling-up project: It is wise to initially build a pilot plant and finally the industrial plant. In the next chapters, all of the mentioned aspects will be treated in detail, although it should be noted that each of the mentioned topics could be the subject of a dedicated book.

1.3 Material and Energy Balances

To model and simulate the behavior of an apparatus constituting a unit operation of a chemical plant, it is necessary to solve a set of equations of material and energy balance. The same also must be done for elaborating experimental data collected in a laboratory device in the context of plant design. In both cases, we will apply the laws of mass and energy conservation (i.e., matter and energy cannot be created or destroyed but just transformed). The material balance can be made in moles or in mass and can be related to a single component or to the overall mass. In the most general way, we can write the balance in moles for a generic component "*i*" as follows:

$$\begin{aligned}\text{Moles of } i \text{ entering/time} = {} & (\text{moles of } i \text{ outgoing} \\ & + \text{moles of } i \text{ formed or reacted} \\ & + \text{accumulation of } i)/\text{time}\end{aligned}$$

In a continuous reactor operating under a steady-state regime, the accumulation term is equal to 0, while in a batch reactor, the flow rates at the inlet and outlet are null. Last, in a semi-batch reactor, the outlet term or, more seldom, the inlet one can be null. In a device in which only physical transformations occur, the generation term will be null. The material balance is most conveniently expressed in mass because this balance is not affected by the stoichiometry. We can write for the component "*i*" as follows:

$$\begin{aligned}\text{Mass of } i \text{ entering/time} = {} & (\text{mass of } i \text{ outgoing} \\ & + \text{mass of } i \text{ formed or reacted} \\ & + \text{accumulation of } i)/\text{time}\end{aligned}$$

The material balance related to the overall mass can be written as follows:

$$\text{Total mass entering/time} = (\text{total mass outgoing} + \text{accumulation})/\text{time}$$

The energy balance is strictly coupled with the mass balance because any component goes in and out with its enthalpy of formation. In energy balance, the reaction enthalpy change must be considered with accuracy. Some reactions are exothermic and produce heat, while others are endothermic and absorb heat. Moreover, the eventual change of physical state (e.g., from liquid to vapor) also must be considered.

1.4 Introduction to the Numerical Solution of the Most Frequently Employed Algorithms with Examples Developed in MATLAB

In the following chapters, many different exercises will be reported inside the text describing how to solve them. The related calculations were developed in MATLAB, and the corresponding MATLAB codes of the solutions will be reported for each chapter as Electronic Supplementary Material (ESM), Data 1.1 accessible by this Springer Link:

link.springer.com/chapter/10.1007/978-3-319-97439-2_n#SupplementaryMaterial where "*n*" is the chapter number.

For example, it is possible to find the ESM of this chapter at:

https://link.springer.com/chapter/10.1007/978-3-319-97439-2_1 #SupplementaryMaterial

You will find a brief description of the mathematical algorithms that will be used in many exercises reported in successive chapters. Each of them has simple examples of numerical solutions using the MATLAB software. In particular, the following systems are considered:

- System of linear equations
- Single nonlinear algebraic equation
- System of nonlinear algebraic equations
- System of ordinary differential equations
- Nonlinear least squares and optimization
- Numerical integration for definite integrals

Simple examples of the mentioned systems will be proposed and solved, and the corresponding MATLAB codes can be found in a file accessible at the previously mentioned Web link. This ESM is preparatory to face the more complex exercises reported in the next chapters.

Reference

Servos, J.W.: The industrial relations of science: chemical engineering at MIT, 1900-1939. JSTORE- J. Storage. **71**(4), 530–549 (1980)

Chapter 2
Thermodynamics of Physical and Chemical Transformations

Abstract In this chapter, physical and chemical transformations are considered as a route from a less stable to a more stable equilibrium state. Initially, only physical transformations are considered describing the first and second thermodynamic laws and thermodynamic properties like internal energy, enthalpy, and entropy. Next, thermodynamic equilibrium in chemically reacting systems is considered, defining "chemical potential" as the driving force for reactions and examining in detail chemical equilibrium conditions. Equilibrium reactions between gases at different pressures are considered and the fugacity concept to describe the behavior of real gases is introduced. Different methods for determining pressure and fugacity are described. Equilibrium reactions between reactants in the liquid phase are also considered, introducing the concept of activity. Different methods for determining activity coefficients are described. Several examples of equilibrium calculations are reported. MATLAB code associated with these examples is available online. Finally, vapor–liquid equilibrium is examined in detail considering its applications in a flash-unit and a tray-distillation column.

Keywords Chemical equilibrium · Real gas · Fugacity · Vapor-liquid equilibria (VLE) · Conversion

2.1 Introduction to Physical and Chemical Equilibrium

A system can be considered in equilibrium when no change occurs inside it. The equilibrium state for a system can be more or less stable, and consequently, a system can evolve spontaneously from an unstable to a more stable state as shown in Fig. 2.1.

The transition from a less stable to a more stable equilibrium state occurs through a transformation that can be accompanied by the following:

Supplementary Information The online version contains supplementary material available at https://doi.org/10.1007/978-3-031-82502-6_2.

© The Author(s), under exclusive license to Springer Nature Switzerland AG 2025
R. Tesser, E. Santacesaria, *The Chemical Reactor from Laboratory to Industrial Plant*, https://doi.org/10.1007/978-3-031-82502-6_2

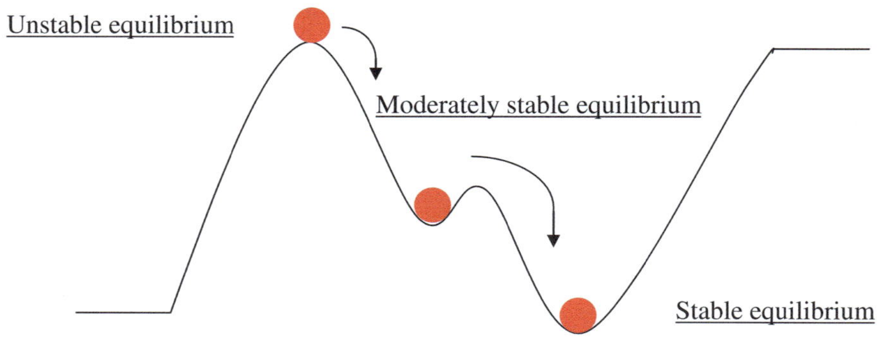

Fig. 2.1 An exemplification of the equilibrium states for a mechanical system

(a) A flux of energy in the form of heat (by conduction, convection, radiation) or mechanical work
(b) A flux of matter (by convective motion, diffusion)
(c) A chemical reaction

The rates with which these transformations occur are not relevant for thermodynamics that consider only the equilibrium states both in the absence (physical equilibria) and in the presence of a chemical reaction (chemical equilibria). Time is not considered a variable in thermodynamics. Transformation rate laws are studied in a different ambit called "kinetics," which will be treated later in a dedicated chapter.

Considering a reactive system, thermodynamics provide three important pieces of information:

(1) Allow to establish whether or not a certain reaction under some defined conditions of temperature and pressure is favored and therefore can spontaneously occur
(2) Allow to evaluate the amount of heat released or absorbed by the reaction under different operative conditions of pressure and temperature
(3) Allow to evaluate, on a theoretical basis, what could be the maximum yields obtainable under different operative conditions and hence suggest what would be the best operative conditions without recurring to extensive experimentation

For transformations occurring in the absence of a chemical reaction, thermodynamics allow to evaluate and foresee different physical equilibrium states useful for the industrial processes such as:

(1) PVT behavior of ideal and real gases
(2) Vapor–liquid equilibria of pure components
(3) Vapor–liquid equilibria of liquid mixtures
(4) Solubility of gases in liquids
(5) Partition of one or more components between two immiscible liquids
(6) Liquid–solid phase equilibria

2.2 Thermodynamic Properties and Equilibrium Conditions of Physical Transformations

2.2.1 The First Thermodynamic Law

We can recognize in nature three different thermodynamic systems:

(1) Isolated systems: systems characterized by the absence of any exchange of matter and energy with the external environment
(2) Closed systems: systems that can exchange energy, but not matter, with the environment
(3) Open systems: systems that can exchange both matter and energy with the environment

It is possible to ideally, or practically, build all the mentioned systems using walls with opportune properties. If, for example, the walls of a recipient are rigid, we have a system characterized by a constant volume; insulating walls do not exchange heat and give place to an adiabatic box. Another property of the walls can be high conductivity, which allows operating under isothermal conditions. Consider then two systems, separated by a wall, characterized by two different equilibrium states. The wall is a constraint allowing the existence of different equilibria in the two separated systems. If the wall is removed, a transformation occurs, and a new unique equilibrium state is reached. The physical transformations in a closed system containing an ideal gas are governed by the first law of thermodynamics, that is, the energy conservation principle. According to this principle, we can write the energy balance equation:

$$\mathrm{d}U = Q + W \tag{2.1}$$

where U is the overall internal energy of the system; and Q and W are, respectively, the exchanged heat and the mechanical work. U is a function of state; that is, it depends exclusively on the initial and final state of the system and is represented here as an exact differential, while the contribution of Q and W in determining the final U value can be defined only if we know the conditions under which a "reversible" transformation occurs (e.g., at constant temperature [isotherm], at constant volume [isochoric], at constant pressure [isobar], and without heat exchange [adiabatic]). By imposing a condition for the occurrence of the transformation, it is possible to evaluate both Q and W in each case. We consider here reversible transformations involving a so-called ideal or perfect gas. A **reversible transformation**, involving a perfect gas, is an ideal process, occurring very slowly and passing from a virtual equilibrium state to a successive one, and each transition can go in one direction or in the opposite as a consequence of a very small change of the applied conditions. An **ideal or perfect gas** is a gas constituted by spherical molecules having negligible volume and not interacting with each other. The motion of the ideal gas molecules occurs in all the directions with the same probability, and the collisions between the molecules and the walls of the container are perfectly elastic.

As mentioned previously, dU is an exact differential: This means that $U = f(T, P, V)$ where T, P, and V are, respectively, the physical variables temperature, pressure, and volume. Considering that only two of these three variables are independent, it is enough to write $U = f(T, P)$ or $U = f(T, V)$ or $U = f(P, V)$; by considering the first function, we have:

$$U_2 - U_1 = f(T_2, P_2) - f(T_1, P_1) \tag{2.2}$$

If we consider then an infinitesimal variation of the variable U, we can write:

$$\mathrm{d}U = \left(\frac{\delta U}{\delta T}\right)_P \mathrm{d}T + \left(\frac{\delta U}{\delta P}\right)_T \mathrm{d}P \tag{2.3}$$

The choice of the two variables, T and P, is arbitrary, and we can write other two similar equations considering as variables (T, V) or (P, V).

Equation (2.3) allows calculating the change of U between two different thermodynamic states as:

$$\int_1^2 \mathrm{d}U = U_2(T_2, P_2) - U_1(T_1, P_1) \tag{2.4}$$

If we make a transformation that returns exactly to the initial state, we will have:

$$\oint \mathrm{d}U = 0 \tag{2.5}$$

Q and W are not exact differentials, and there are infinite possible transformations for going from state 1 to state 2 of U. In conclusion, to have the possibility to integrate Eq. (2.3), we must impose the type of occurring transformation, as will be shown in the next sections.

2.2.2 *Transformations at Constant Pressure*

W in Eq. (2.1) represents the mechanical work. Considering a gas kept at constant pressure (isobaric transformation), we can have a work originated by an expansion in which $V_2 > V_1$ or, alternatively, a work originated by a compression in which

$V_2 < V_1$. In any case, by considering an infinitesimal variation of the volume, we can write:

$$dW = -PdV \quad \text{and} \quad \int_1^2 dW = -P\int_1^2 dV \tag{2.6}$$

The result of the integration is Eq. (2.7).

$$\Delta W = -P(V_2 - V_1) \tag{2.7}$$

that is, the compression work is positive and, on the contrary, the expansion work is negative.

Q is the heat exchanged by the system, and also in this case, we can have a positive or a negative term if the heat is acquired or released by the system. By imposing, as in the previous case, the pressure constant, we can write:

$$\Delta U_P = Q_P - P\Delta V \tag{2.8}$$

From this equation, we can deduce that:

$$Q_P = U_2 - U_1 + PV_2 - PV_1 \tag{2.9}$$

By rearranging the expression:

$$Q_P = (U_2 + PV_2) - (U_1 + PV_1) = H_2 - H_1 \tag{2.10}$$

By observing this expression, with being P constant, it results that Q_P depends only on the initial and final state of the system. In other words, H is a function of state such as U and is named enthalpy:

$$H = \text{Enthalpy} = U + PV, \text{Heat exchanged } (P = \text{constant}) \text{ normally referred to 1 mol} \tag{2.11}$$

As it will be seen later, this thermodynamic property is extensively used more than U because it can easily be measured experimentally. It is then possible to evaluate both the specific and molar heat as:

$$C_P = \left(\frac{\delta H}{\delta T}\right)_P \tag{2.12}$$

where H refers either to a unit of mass or to one mole of ideal gas.

2.2.3 Transformations at Constant Volume

Because V is constant, $\Delta V = 0$; consequently, the mechanical work is null, $W = PdV = 0$. In this case, Eq. (2.1) reduces to:

$$\Delta U = Q_V \tag{2.13}$$

Again, Q_V becomes a function of state denoting the heat exchanged by a mole of ideal gas in a system kept at constant volume. It is now possible to evaluate both the specific and molar heat at constant volume, that is:

$$C_v = \left(\frac{\delta Q_V}{\delta T}\right)_V \tag{2.14}$$

where Q_V refers either to a unit of mass or to one mole of ideal gas.

2.2.4 Transformations at Constant Temperature

In this case, $\Delta T = 0$, and for an ideal gas $dU = 0$. As a consequence:

$$Q_T + W = 0 \quad \text{and} \quad Q_T = -W = PdV \tag{2.15}$$

Remembering that for one mole of an ideal gas, the general gas law $PV = RT$ is valid, we can write:

$$Q_T = -W = RT \int_1^2 \frac{dV}{V} = RT \ln \frac{V_2}{V_1} \tag{2.16}$$

As under isothermal conditions the Boyle law is also valid (i.e., $PV =$ constant), some different expressions equivalent to Eq. (2.16) can be written as:

$$\Delta Q_T = RT \ln \frac{P_1}{P_2} = P_1 V_1 \ln \frac{V_2}{V_1} = P_1 V_1 \ln \frac{P_1}{P_2} = P_2 V_2 \ln \frac{V_2}{V_1} = P_2 V_2 \ln \frac{P_1}{P_2} \tag{2.17}$$

The mechanical work, under isothermal conditions, is a function of state depending only on the initial and final conditions. The isothermal transformation occurs at a prefixed temperature, $T1 - T2 - T3$ or $T4$, as shown in Fig. 2.2, and the surface area in the same plot represents the work of the transformation occurring, for example, between A and B.

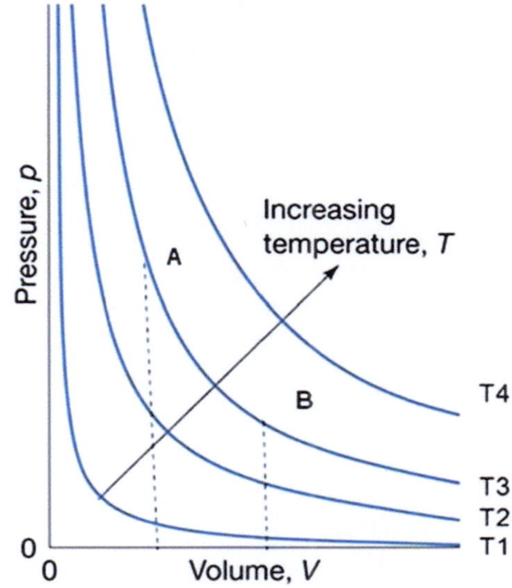

Fig. 2.2 Isothermal transformations of an ideal gas obeying to Boyle's law

2.2.5 *Adiabatic Transformations*

Adiabatic transformations are characterized by no heat exchange, that is, $Q = 0$. Consequently, for the first thermodynamic principle:

$$dU = -PdV \tag{2.18}$$

where U is an exact differential and in analogy with Eq. (2.3), we can write

$$dU = \left(\frac{\delta U}{\delta V}\right)_T dV + \left(\frac{\delta U}{\delta T}\right)_V dT \tag{2.19}$$

For an ideal gas $\left(\frac{\delta U}{\delta V}\right)_T = 0$, therefore, considering Eqs. (2.13) and (2.14), it is possible to write:

$$dU = \left(\frac{\delta U}{\delta T}\right)_V dT = C_V dT \tag{2.20}$$

and also:

$$C_V dT + PdV = 0 \tag{2.21}$$

If C_V is independent of the temperature, it is possible to integrate Eq. (2.21) after the substitution of $P = RT/V$ related to one mole of ideal gas:

$$C_V dT + P dV = C_V \frac{dT}{T} - R \frac{dV}{V} = 0 \tag{2.22}$$

By integrating between states 1 and 2, we obtain:

$$C_V \ln \frac{T_2}{T_1} + R \ln \frac{V_2}{V_1} = 0 \tag{2.23}$$

It is easy to show that $C_p - C_v = R$ and also that $C_p/C_v = \gamma$ where $\gamma = 1.666$ for a monoatomic ideal gas, 1.44 for a diatomic gas, and 1.33 for a polyatomic gas. Equation (2.23) can therefore be rewritten as:

$$\ln \frac{T_2}{T_1} + \left(\frac{C_P}{C_V} - 1\right) \ln \frac{V_2}{V_1} = \ln \frac{T_2}{T_1} + (\gamma - 1) \ln \frac{V_2}{V_1} = 0 \tag{2.24}$$

which means: $\frac{T_1}{T_2} = \left(\frac{V_2}{V_1}\right)^{\gamma - 1}$

and, with few steps, considering the state equation for an ideal gas, it is possible to write the relation between pressure and volume for an adiabatic transformation:

$$P_1 V_1^\gamma = P_2 V_2^\gamma, \quad \text{that is,} \quad PV^\gamma = \text{constant} \tag{2.25}$$

The mechanical work for an adiabatic transformation can be calculated by integrating Eq. (2.18). Other expressions can be obtained by converting the difference of temperature in a (*PV*) difference or opportunely involving C_p and C_v. Figure 2.3 shows a comparison between isothermal and adiabatic transformations in a *PV* plot and the corresponding works.

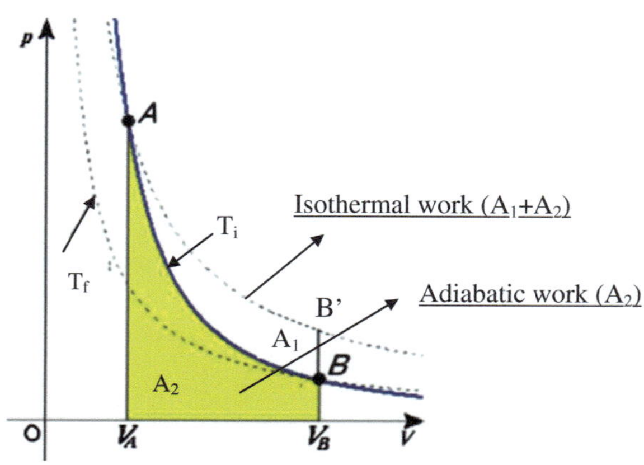

Fig. 2.3 Comparison between isothermal and adiabatic transformation in a *P–V* plot and the corresponding mechanical work

2.2.6 *The Second Thermodynamic Law*

The first thermodynamic law is important for individuating the correlation between heat and work in well-established conditions that are far from practice (considering the reversible transformations of an ideal gas). However, it is not adequate to evaluate, for example, whether or not a reaction is possible, in what direction the equilibrium is shifted by a change of pressure or temperature, the criterion for recognizing an equilibrium state, or the nature of the equilibrium that can be reached (more or less stable). For giving an answer to all the abovementioned questions, the second law of thermodynamics must be introduced. The second law derives from the everyday observations of some natural events and can be expressed in different ways. We observe, for example, in nature that, if there is no hindrance, water spontaneously flows from a higher to a lower level in the same way that heat always spontaneously flows from an object at higher temperature to another at lower temperature. In a more general way, we can say that thermodynamic systems give place to spontaneous transformations passing from a higher energetic level to a lower one or, equivalently, all the systems spontaneously change to reach a condition of more stable equilibrium. These spontaneous processes are not reversible, and to return from the final to the initial condition, we must spend an amount of work greater than the energy change of the direct process. In conclusion, in all the irreversible transformations, some energy is lost. All these observations have been elaborated in decades of activity by three well-known scientists: Carnot, Clausius, and Lord Kelvin.

A new thermodynamic function of state, called "entropy," was introduced for better describing the irreversible transformations. This function can be defined as the ratio between the heat exchanged and the temperature at which the exchange occurs:

$$S = \frac{Q}{T} \tag{2.26}$$

S is a perfect differential, although both Q and T are not. As a matter of fact, starting from the expression:

$$Q = C_V \mathrm{d}T + P\mathrm{d}V \tag{2.27}$$

dividing all the terms by T:

$$\mathrm{d}S = \frac{Q}{T} = C_V \frac{\mathrm{d}T}{T} + R\frac{\mathrm{d}V}{V} \tag{2.28}$$

if C_V can be assumed independent of the temperature, we can integrate the expression between states 1 and 2 giving a value of ΔS dependent only on the final and initial conditions, that is, the entropy is a thermodynamic function of state and has the advantage of being easily measurable. For a reversible process, Eq. (2.28) is valid, and for an adiabatic process $\mathrm{d}S = 0$, while for a spontaneous irreversible process we can write:

$$\mathrm{d}S > \frac{Q}{T} \tag{2.29}$$

It is possible to conclude that for any irreversible process, the entropy always increases. This is simply a different statement of the second thermodynamic law. Moreover, it is clear that the product, *TS*, in an irreversible process corresponds to the lost and therefore not usable thermal energy. At this point, it is possible to introduce two new functions of state, respectively, called Gibbs free energy "*G*" (i.e., free energy at constant pressure) and Helmholtz free energy "*A*" (free energy at constant volume). For this purpose, we can write:

$$G = \text{Gibbs free energy} = H - TS \tag{2.30}$$

$$A = \text{Helmholtz free energy} = U - TS \tag{2.31}$$

A correlation exists between the different variables of state. We can write, for example:

$$U = U(S, V) \tag{2.32}$$

from which:

$$\mathrm{d}U = \left(\frac{\partial U}{\partial S}\right)_V \mathrm{d}S + \left(\frac{\partial U}{\partial V}\right)_S \mathrm{d}V = T\mathrm{d}S - P\mathrm{d}V \tag{2.33}$$

being:

$$T = \left(\frac{\partial U}{\partial S}\right)_V \text{ and } P = -\left(\frac{\partial U}{\partial V}\right)_S \quad \text{(intensive variables)} \tag{2.34}$$

It can be observed that the energetic terms appearing in Eq. (2.33) are the products of an **intensive variable** (*T* and *P*), which can change in any point of the system, and an **extensive variable** (*S* and *V*) having the property of the additivity of mass or volume. The intensive variables constitute the driving force for any possible transformation.

All the previously described functions of state are, on the contrary, extensive variables.

It is possible to write relations similar to Eq. (2.33) for all the functions of state, such as:

$$dU = TdS + PdV \tag{2.35}$$

$$dH = TdS + VdP \tag{2.36}$$

$$dG = -SdT + VdP \tag{2.37}$$

$$dA = -SdT - PdV \tag{2.38}$$

These equations are the basis for interpreting all the physical equilibrium states. However, before proceeding in this respect, it is necessary to define the criteria for considering a system in equilibrium. These criteria are related to both intensive and extensive variables.

2.2.7 Criteria for Defining the Thermodynamic Equilibrium of Physical Transformations

For a thermodynamic system to be in equilibrium, all intensive (temperature, pressure) and extensive thermodynamic properties (U, G, A, H, and S) must be constant. Hence, the total change in any of those properties must be zero at equilibrium. These criteria derive from the postulates of the second law of thermodynamics. Considering the intensive variables, we can write:

$$\text{Thermal equilibrium is characterized by } \Delta T = 0 \tag{2.39}$$

$$\text{Mechanical equilibrium is characterized by } \Delta P = 0 \tag{2.40}$$

Then, according to the second law of thermodynamics, these criteria also suggest how the system goes toward the equilibrium condition:

$$T_{\text{higher}} \xrightarrow{\text{Heat flow}} T_{\text{lower}} \quad \text{until } \Delta T = 0 \tag{2.41}$$

$$P_{\text{greater}} \xrightarrow{\text{flow of mass or energy}} P_{\text{smaller}} \quad \text{until } \Delta P = 0 \tag{2.42}$$

Considering first of all the entropy, the conditions of equilibrium for this variable of state are:

$$dS_{V,U} = 0 \quad d^2S_{V,U} < 0 \tag{2.43}$$

That is, at equilibrium, the entropy does not change, and the first differential suggests that an equilibrium condition is reached for both a minimum or a maximum value of the entropy, while the second differential suggests that a stable equilibrium is reached only for a maximum value of the entropy. Similarly, if we consider the free energy of Gibbs, the equilibrium condition is:

$$dG_{\mathrm{T,P}} = 0 \quad d^2G_{\mathrm{T,P}} > 0 \tag{2.44}$$

The Gibbs free energy in a transformation will diminish until reaching a minimum value corresponding to the equilibrium condition. A similar behavior can be foreseen for the Helmholtz free energy A:

$$dA_{\mathrm{V,T}} = 0 \quad d^2A_{\mathrm{V,T}} > 0 \tag{2.45}$$

2.3 Thermodynamic Equilibrium in Chemical-Reacting Systems

2.3.1 *Introduction to the Thermodynamics of Chemical Equilibrium*

If, together with a physical transformation, we have also the occurrence of a chemical reaction, it is possible to write:

$$U = U(S, V, n_1, n_2, n_3 \ldots n_k) \tag{2.46}$$

That is, the internal energy in this case is also a function of the chemical composition being n_1, n_2, n_3 … n_k the number of moles of the components 1, 2, 3 … k. Obviously, the composition can change as a consequence of the reaction. The exact differential of Eq. (2.46) will be:

$$dU = \left(\frac{\partial U}{\partial S}\right)_{V,n_i} dS + \left(\frac{\partial U}{\partial V}\right)_{S,n_i} dV + \sum_{i=1}^{k} \left(\frac{\partial U}{\partial n_i}\right)_{S,V,n_j} dn_i \tag{2.47}$$

The third term of this expression is again characterized by the product of an intensive variable time and extensive one. The intensive variable is called "**chemical potential,**" μ, and is the driving force pushing the chemical reaction toward the equilibrium condition. For a generic "*i*" component, we can write:

$$\mu_i = \left(\frac{\partial U}{\partial n_i}\right)_{S,V,n_j} \tag{2.48}$$

By rearranging Eq. (2.47), we have:

$$dU = TdS - PdV + \sum_i \mu_i dn_i \tag{2.49}$$

This relation is fundamental for deriving other useful relationships that do not contain new information but are easier to be used:

$$dH = TdS + VdP + \sum_i \mu_i dn_i \tag{2.50}$$

$$dG = -SdT + VdP + \sum_i \mu_i dn_i \tag{2.51}$$

$$dA = -SdT - PdV + \sum_i \mu_i dn_i \tag{2.52}$$

As a consequence, the following identities are valid:

$$\mu_i = \left(\frac{\partial U}{\partial n_i}\right)_{S,V,n_j} = \left(\frac{\partial G}{\partial n_i}\right)_{T,P,n_j} = \left(\frac{\partial H}{\partial n_i}\right)_{S,P,n_j} = \left(\frac{\partial A}{\partial n_i}\right)_{T,V,n_j} \tag{2.53}$$

Being the chemical potential the driving force for a reaction, we must add a new specific criterion for the chemical equilibrium to the already mentioned criteria for defining the physical equilibrium:

$$\mu_{\text{greater}} \xrightarrow{\text{Flow of mass}} \mu_{\text{lower}} \quad \text{until} \quad \Delta\mu = 0 \tag{2.54}$$

In conclusion, equilibrium is reached when $\Delta\mu = 0$.

For what concerns the extensive variable n_i, equilibrium is reached when the composition of the reacting system does not change, that is, $\Delta n_i = 0$.

Let us consider now a generic chemical reaction occurring in homogeneous phase:

$$\alpha_A A + \alpha_B B \leftrightarrow \alpha_M M + \alpha_N N \tag{2.55}$$

More concisely, we can write $\Sigma_i\ \alpha_i A_i = 0$, where α_i are the stoichiometric coefficients assumed positive for the products and negative for the reagents. As the reaction proceeds, a defined amount of formed M and N correspond to an amount of reacted A or B. For expressing this interdependence between reagents and products, we can write:

$$-\frac{dn_A}{\alpha_A} = -\frac{dn_B}{\alpha_B} = \frac{dn_M}{\alpha_M} = \frac{dn_N}{\alpha_N} = d\xi \tag{2.56}$$

In conclusion, only one variable must be defined to follow the reaction extent, for example, dn_A, but it is more convenient to define the "advancement degree" or "coordinate" of the reaction ξ. Thanks to ξ, the composition of the system is always known because:

$$n_i = n_i^o + \alpha_i \xi \quad \text{and} \quad dn_i = \alpha_i d\xi \tag{2.57}$$

We previously saw that the chemical equilibrium conditions are:

$$dG_{T,P} = 0 \quad \text{and} \quad d^2 G_{T,P} > 0 \tag{2.58}$$

If T and P are constants, G will depend only on the number of moles of the components, that is:

$$dG_{T,P} = \sum_i \left(\frac{\partial G}{\partial n_i}\right)_{T,P,n_j j \neq i} dn_i = \sum_i \mu_i dn_i = 0 \tag{2.59}$$

and hence:

$$dG_{T,P} = \sum_i \alpha_i \mu_i d\xi = 0 \tag{2.60}$$

By plotting $G_{\mathrm{T,P}}$ as a function of the reaction coordinate, ξ, the equilibrium according to the first conditions reported in Eq. (2.59) is obtained for a minimum value of $G_{\mathrm{T,P}}$. From this plot, shown in Fig. 2.4, we can derive the value of ξ at the equilibrium, that is, the equilibrium composition.

The slopes of the curve give information about the direction of the reaction:

- if $\Delta G < 0$ reaction is favored and goes from left to the right;
- if $\Delta G > 0$ reaction is not favored and goes in the opposite direction; and
- if $\Delta G = 0$ the reaction is at equilibrium.

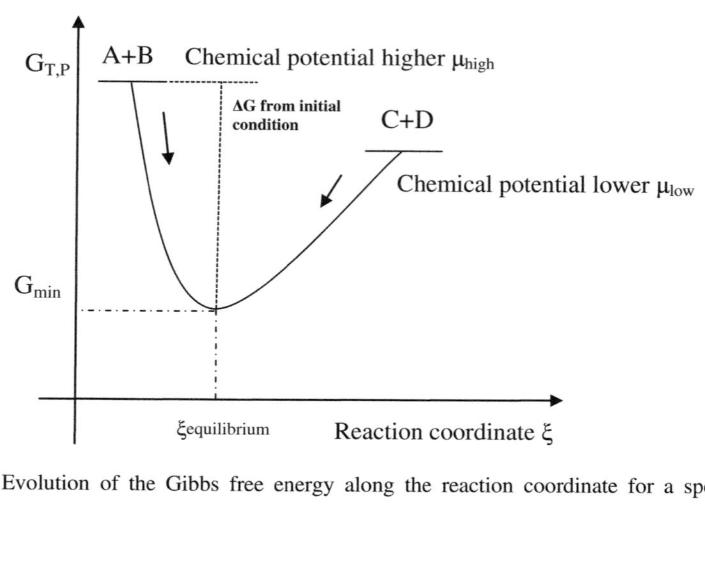

Fig. 2.4 Evolution of the Gibbs free energy along the reaction coordinate for a spontaneous reaction

2.3.2 Equilibrium of Reactions Between Ideal Gases

Considering Eq. (2.55), there are four different equilibrium possibilities:

(1) Equilibrium in homogeneous phase between ideal gases
(2) Equilibrium in homogeneous phase between real gases
(3) Equilibrium in liquid homogeneous phase
(4) Equilibrium in heterogeneous phases

For all these systems, a unique relation is valid, which is:

$$\sum_i \alpha_i \mu_i = 0 \tag{2.61}$$

but clearly, the application of this relation is conditioned by the knowledge of the dependence of the chemical potential of each component on the composition of the system. For an ideal gas, for example, we can write:

$$\mu_i = \left(\frac{\partial G}{\partial n_i}\right)_{T,P,n_j} = \overline{G_i} = \text{Partial molar} \quad \text{Gibbs free energy} \tag{2.62}$$

Differentiating $\overline{G}_i$ with respect to the pressure:

$$\left(\frac{\partial \overline{G}_i}{\partial P}\right)_{T,n_j} = \frac{\partial}{\partial P}\left(\frac{\partial G}{\partial n_i}\right)_{T,P,n_j} = \frac{\partial}{\partial n_i}\left(\frac{\partial G}{\partial P}\right)_{T,n_j} = \left(\frac{\partial V}{\partial n_i}\right)_{P,T,n_j} = \overline{V}_i \tag{2.63}$$

with $\overline{V}_i$ being the partial molar volume.

For an ideal gas, we can write that $V = \left(\frac{RT}{P}\right)\sum_i n_i$

$$\overline{V}_i = \left(\frac{\partial V}{\partial n_i}\right)_{T,P,n_j} = \frac{RT}{P}\left(\frac{\partial n_T}{\partial n_i}\right)_{T,P,n_j} = \frac{RT}{P} \tag{2.64}$$

Moreover, $p_i = y_i P$ and $\mathrm{d}p_i = y_i \mathrm{d}P$; therefore,

$$\mathrm{d}\overline{G}_i = \mathrm{d}\mu_i = \overline{V}_i \mathrm{d}P = (RT/p_i) y_i \mathrm{d}P = RT \mathrm{d} \ln p_i \tag{2.65}$$

By integrating, it results in:

$$\mu_i = \mu_i^o + RT \ln\left(\frac{p_i}{p_R}\right) \quad \text{and assuming} \quad p_R = 1 \tag{2.66}$$

$$\mu_i = \mu_i^o + RT \ln p_i \tag{2.67}$$

where μ_i^o is the standard chemical potential, that is, the chemical potential for which $p_i = 1$. p_R is the reference state arbitrarily assumed equal to 1 atm and corresponding to the standard state having chemical potential μ_i^o. Then, remembering again the equilibrium condition:

$$\begin{aligned}
&\Sigma_i \alpha_i \mu_i = 0 \\
&\Sigma_i \alpha_i \mu_i^o + RT\,\Sigma_i \alpha_i \ln p_i = 0 \\
&\underbrace{\Sigma_i \alpha_i \mu_i^o}_{\boxed{\Delta G^o_{\mathrm{T,P}}}} + RT \ln \Pi_i p_i^{\alpha_i} = 0 \\
&\Delta G^o_{\mathrm{T,P}} = -RT \ln \Pi_i p_i^{\alpha_i} \\
&K_{\mathrm{P}} = \Pi_i p_i^{\alpha_i} = \frac{p_M^{\alpha_M} p_N^{\alpha_N}}{p_A^{\alpha_A} p_B^{\alpha_B}} = e^{-\Delta G^o/RT}
\end{aligned} \tag{2.68}$$

Remember that for any term of pressure we have considered, a reference pressure $p_R = 1$ is at the denominator.

Then we can also write:

$$\Delta G = G_{\text{products}} - G_{\text{reagents}} = \alpha_M G_M + \alpha_N G_N - \alpha_A G_A - \alpha_B G_B \tag{2.69}$$

$$\Delta G = \Delta G^\circ + RT \ln \frac{(p_M)^{\alpha_m} (p_N)^{\alpha_N}}{(p_A)^{\alpha_A} (p_B)^{\alpha_B}} \tag{2.70}$$

At the equilibrium $\Delta G = 0$; therefore, $\Delta \overset{\circ}{G}_{\mathrm{T,P}} = -RT \ln K_{\mathrm{P}}$ and, consequently:

$$\Delta \overset{\circ}{G}_{\mathrm{T,P}} = RT \ln \prod_i p_i^{\alpha_i} - RT \ln K_{\mathrm{P}} \text{"Reaction isotherm"} \tag{2.71}$$

ΔG^o is the change of standard free energy and corresponds to ΔG when $K_p = 1$. Finally, we can write the equilibrium constant for a reaction between ideal gases as:

$$K_{\mathrm{P}} = \frac{p_M^{\alpha_M} p_N^{\alpha_N}}{p_A^{\alpha_A} p_B^{\alpha_B}} = \exp\left(\frac{-\Delta G^\circ}{RT}\right) \tag{2.72}$$

ΔG^o for all the chemical elements is assumed equal to 0 in order to have a reference point for calculating the standard free energy of the compound formation.

Some useful examples of a practical approach to chemical-equilibria calculations for reactions between ideal gases.

A practical example of a spontaneous equilibrium reaction occurring without changing the number of moles is the synthesis of HI studied by Bodenstein (1897) and reported in many textbooks as a classic example of equilibrium reaction proceedings without a change in the number of molecules. The reaction is:

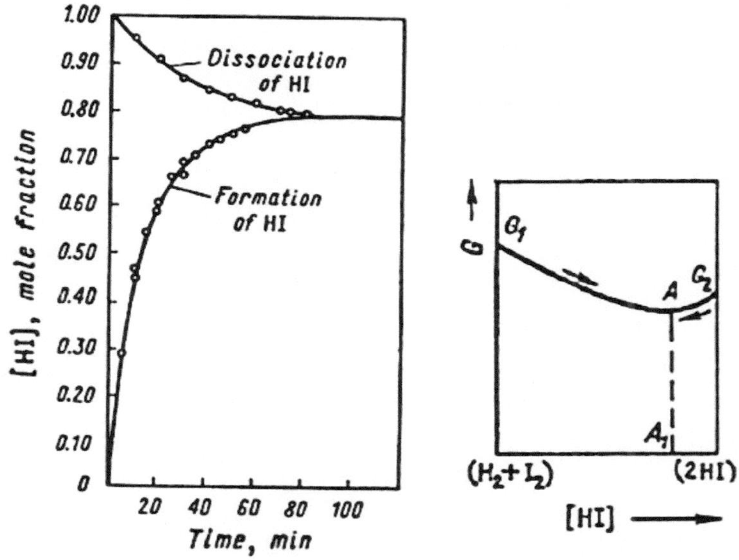

Fig. 2.5 (Left panel) approach to the equilibrium in the reaction of formation and decomposition of HI. Reaction performed at 448 °C [data from Bodenstein and re-elaborated by Gerasimov et al. (1974)] and (right panel) isothermal evolution of the Gibbs free energy for the reaction $H_2 + I_2 \leftrightarrow 2HI$

$$H_2(\text{gas}) + I_2(\text{gas}) \longrightarrow 2HI(\text{gas}) \tag{2.73}$$

Bodenstein studied both the synthesis and the reverse decomposition reaction and obtained the following results, which were successively confirmed by other authors (Fig. 2.5).

Both of the reactions, at 448 °C, tend to a limit that is equal to approximately 21.4% of HI dissociation and 78.6% of HI conversion from the elements. This means that equilibrium is somewhat shifted toward synthesis. However, thermodynamic data for this reaction can be found, for example, in the book published by Stull et al. (1969), and some of these data are reported in Table 2.1 for different temperatures. Performing the reaction at low pressure and high temperature, we can consider both reactants and products as ideal gases.

From Table 2.1, we can see, first of all, that ΔG_f° for the reaction:

$$1/2H_2(\text{gas}) + 1/2I_2(\text{gas}) \rightleftarrows HI(\text{gas}) \tag{2.74}$$

is always negative, meaning that in the range of the chosen temperatures, the reaction of HI formation is thermodynamically favored and spontaneously occurs but never completely. On the basis of the thermodynamic parameters, we can evaluate what is the equilibrium constant at any temperature, calculating the corresponding value from the relation:

Table 2.1 Some thermodynamic properties of HI evaluated at different temperatures

T (K)	$C_p°$ (cal/mol K)	$S°$ (cal/mol K)	$\Delta H_f^°$	$\Delta G_f^°$
500	7.11	52.98	–1.35	–2.41
600	7.25	54.29	–1.43	–2.62
700	7.42	55.42	–1.49	–2.81
800	7.60	56.42	–1.55	–3.00
900	7.77	57.33	–1.58	–3.17

$$K_P = \frac{p_{HI}}{p_{H_2}^{(1/2)} p_I^{(1/2)}} = \exp\left(\frac{-\Delta G°}{RT}\right) \tag{2.75}$$

Instead of using the partial pressure for expressing the equilibrium constant, we can use the concentrations or the number of moles, remembering that:

$$p_i = n_i \frac{RT}{V} = C_i RT \tag{2.76}$$

Clearly, introducing the concentrations C_i or the number of moles n_i, in this case, the constant remains the same because the terms RT or RT/V are eliminated, that is, $K_p = K_c = K_n$. Bodenstein found an equilibrium constant at 444.5 °C of approximately 6.76 ± 0.2. Starting from the thermodynamic data of Table 2.1, we can calculate with Eq. (2.75) a value of $K_p = \exp(-\Delta G°/RT) \simeq 7.22$, which agrees with the experimental value.

Another observation is that $\Delta H_f^°$, the enthalpy change of HI formation, is also negative, that is, the reaction is moderately exothermic, while decomposition is on the contrary endothermic. It is important to point out that if instead of considering Eq. (2.74) we consider Eq. (2.73), the equilibrium constant becomes $K_P' = K_P^2 \simeq 4 \pm 2$.

Exercise 2.1. Equilibrium Calculation for the Synthesis of HI from the Elements Imagine putting in a vessel of 1 l, 1 mol of H_2, and 3 mol of I_2, bring the temperature at 717.66 K, and wait enough time to reach equilibrium. As seen, the equilibrium constant of Eq. (2.73) at that temperature is 46. Calculate the equilibrium composition. Consider the conversion of hydrogen as unknown:

λ = Reacted moles/initial moles $= \frac{n_{H_2}^o - n_{H_2}}{n_{H_2}^o}$

$R°$ = Initial molar ratio between the reactants $= \frac{n_{I_2}^o}{n_{H_2}^o} = 3$ (Table 2.2)

$$K_P' = \frac{\left[2n_{H_2}^o \lambda\right]^2}{\left[n_{H_2}^o * (1-\lambda)\right] * \left[n_{H_2}^o * (R° - \lambda)\right]} = 46 \tag{2.77}$$

Simplifying

Table 2.2 Initial and equilibrium composition in the synthesis of HI

Initial concentration	H_2	I_2	HI
	$n^o_{H_2} = 1.00$	$n^o_{I_2} = 3.00$	$n^o_{HI} = 0.00$
Composition at equilibrium	$n_{H_2} = n^o_{H_2} {}^*(1-\lambda)$	$n_{I_2} = n^o_{H_2} {}^*(R^\circ - \lambda)$	$2n^o_{H_2}\lambda$

$$K'_P = \frac{[2\lambda]^2}{[1-\lambda] * [R^\circ - \lambda]} = \frac{4\lambda^2}{\lambda^2 - \lambda^*(R^\circ + 1) + R^\circ} \tag{2.78}$$

$$42\lambda^2 - 184\lambda + 138 = 0$$

$$\lambda = \frac{184 \pm \sqrt{184^2 - 4 * 42 * 138}}{84} = \begin{cases} \nearrow 0.96 \text{ Correct result} \\ \searrow 3.42 \text{ Impossible result} \end{cases} \tag{2.79}$$

In conclusion, at equilibrium, 96% of H_2 reacts and the residual moles are 0.04, 32.00% of I_2 reacts and the residual I_2 mole are 2.04, and 1.92 mol of HI have been formed. These results can be obtained with a MATLAB calculation program that can be found as Electronic Supplementary Material, Data 2.1.

Results

Moles at equilibrium ($H_2 : I_2 = 1{:}3$)		
H_2	I_2	HI
0.0393	2.0393	1.9213
H_2 conversion = 0.9607		

With the synthesis of HI from the elements H_2 and I_2, we have seen an example of equilibrium reaction occurring without a change in the overall number of moles as a consequence of the reaction. Other examples of this type are of greater industrial interest. For example, the water–gas shift reaction is useful for producing H_2 from water and CO according to the following reaction:

$$H_2(\text{gas}) + CO_2(\text{gas}) \rightleftarrows CO(\text{gas}) + H_2O(\text{gas}) \tag{2.80}$$

Some equilibrium constants for this reaction have been determined by Neumann and Kohler (1928) and are listed in Table 2.3.

As can be seen, if we want to produce hydrogen, the equilibrium of Eq. (2.80) must be shifted as much as possible to the left; this means to operate at low temperature in the presence of excess water (law of mass action).

Table 2.3 Some equilibrium constants experimentally evaluated by Neumann and Kohler (1928) related to the water–gas shift reaction

T (K)	K_P
673	0.080
873	0.410
1073	0.93
1273	1.66

Exercise 2.2. Equilibrium of the Gas Shift and Yield in Hydrogen Production Calculate the hydrogen yields starting from an equimolecular mixture of CO and H_2O and from a mixture containing 10 mol of H_2O and 1 of CO. The temperature of reaction in both cases is 673 K. Produce a plot of equilibrium CO conversion by varying the ratio $R^\circ = n_{H2O}{}^\circ / n_{CO}{}^\circ$ from 0.5 to 10 at the same temperature. Considering the reverse of reaction (2.80), at 673 K, the corresponding equilibrium constant will be 1/0.080 = 12.5.

Defining:

λ = Conversion of CO = Reacted moles of CO/initial moles of CO $= \frac{\overset{\circ}{n}_{CO} - n_{CO}}{\overset{\circ}{n}_{CO}}$

R° = Initial molar ratio between H_2O and

$$CO = \frac{n^o_{H_2O}}{n^o_{CO}} \begin{matrix} \nearrow 1 \\ \searrow 10 \end{matrix}$$

$$K'_p = \frac{n_{CO_2} n_{H_2}}{n_{CO} n_{H_2O}} = \frac{\left(\overset{\circ}{n}_{CO}\lambda\right)^2}{\overset{\circ}{n}_{CO}{}^*(1-\lambda) * \overset{\circ}{n}_{CO}{}^*(R^\circ - \lambda)} = \frac{\lambda^2}{(1-\lambda) * (R^\circ - \lambda)} = 12.5 \tag{2.81}$$

$$11.5\lambda^2 - 12.5\lambda(R^\circ + 1) + 12.5R^\circ = 0 \tag{2.82}$$

The correct solution for $R^\circ = 1$ is $\lambda = 0.779$, while for $R^\circ = 10$ $\lambda = 0.991$, that is, by keeping high the ratio $\frac{\overset{\circ}{n}_{H_2O}}{\overset{\circ}{n}_{CO}}$ all CO is transformed in H_2, that is useful for producing ammonia. The results reported below can be obtained using a MATLAB calculation program found as Electronic Supplementary Material, Data 2.1.

Results

CO	H_2O	H_2	CO_2
Case no. 1 Moles at equilibrium (CO:H_2O = 1:1)			
0.2205	0.2205	0.7795	0.7795
CO conversion = 0.7795 (–)			
Case no. 2 Moles at equilibrium (CO:H_2O = 1:10)			
0.0087	9.0087	0.9913	0.9913
CO conversion = 0.9913 (–)			

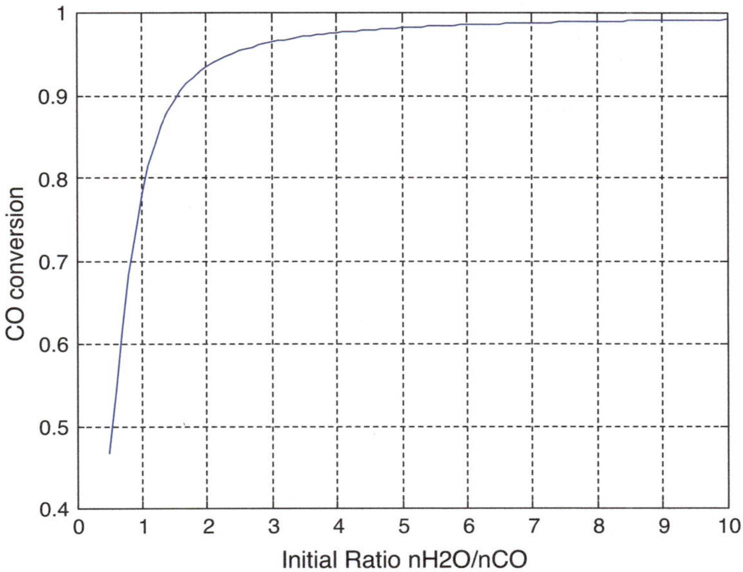

Fig. 2.6 Conversion of CO at equilibrium in the gas shift reaction for different initial mole ratio n_{H_2O}/n_{CO°

As can be seen, by dosing the amount of water, the thermodynamic allows to evaluate the extent of the reaction (see Fig. 2.6). Clearly, if the reaction occurs at high pressure, we cannot consider reactants and products as ideal gases; this aspect will be examined in a next section. However, many reactions exist in which the number of moles passing from reactants to the products changes by increasing or decreasing. For example:

$$NO + O_2 \rightleftarrows NO_2 \tag{2.83}$$

$$N_2O_4 \rightleftarrows 2NO_2 \tag{2.84}$$

$$SO_2 + 1/2\,O_2 \rightleftarrows SO_3 \tag{2.85}$$

$$N_2 + 3\,H_2 \rightleftarrows 2NH_3 \tag{2.86}$$

In these cases, we cannot write (as previously) $K_p = K_c = K_n$, but remembering Eq. (2.76) and considering the synthesis of ammonia, we can write:

$$K_p = \frac{P^2_{NH_3}}{P_{N_2}P^3_{H_2}} = \frac{y^2_{NH_3}}{y_{N_2}y^3_{H_2}}(P)^{-2} = \frac{n^2_{NH_3}}{n_{N_2}n^3_{H_2}}\left(\frac{RT}{V}\right)^{-2} = \frac{C^2_{NH_3}}{C_{N_2}C^3_{H_2}}(RT)^{-2} \tag{2.87}$$

That is:

$$K_{\mathrm{p}} = K_x(P)^{-2} = K_n\left(\frac{RT}{V}\right)^{-2} = K_n(n_{\mathrm{tot}}P)^{-2} = K_{\mathrm{c}}(RT)^{-2} \tag{2.88}$$

It is again convenient to consider, as a reference, the conversion of one single component fixing its stoichiometric coefficient to 1. Consider, for example, the conversion of N_2. We have an initial condition defined by $n^o_{N_2}$ and $n^o_{H_2}$ being the moles of NH_3 initially equal 0. The reacted moles will be $n^o_{N_2}\lambda$ and the residual moles of $N_2 = n^o_{N_2}(1-\lambda)$. For each mole of reacted N_2, 3 mol of H_2 disappear; the moles disappearing are therefore $3n^o_{N_2}\lambda$ and the residual moles of hydrogen are $n^o_{H_2}\left(1-3n^o_{N_2}\lambda\right)$. The moles of formed NH_3 will be $2n^o_{N_2}\lambda$. The total moles initially are $\left(n^o_{N_2}+n^o_{H_2}\right)$, while at equilibrium: $n_{\mathrm{tot}} = n^o_{N_2}(1-\lambda) + n^o_{H_2}\left(1-3n^o_{N_2}\lambda\right) + 3n^o_{N_2}\lambda$.

$$K_{\mathrm{p}} = K_n(n_{\mathrm{tot}}P)^{-2} = \frac{\left(2n^o_{N_2}\lambda\right)^2}{n^o_{N_2}(1-\lambda)n^o_{H_2}\left(1-3n^o_{N_2}\lambda\right).}(n_{\mathrm{tot}}P)^{-2} \tag{2.89}$$

If we start with the reactants in a stoichiometric ratio, that is, $\frac{n^o_{H_2}}{n^o_{N_2}} = 3$

$$K_{\mathrm{p}} = \frac{(2\lambda)^2}{3(1-\lambda)\left(1-3n^o_{N_2}\lambda\right)}(n_{\mathrm{tot}}P)^{-2} \tag{2.90}$$

This reaction is thermodynamically favored by the pressure because for obtaining a constant value of K_{p}, by increasing the total pressure P, the value of K_n must be increased, and therefore the yields in the formation of NH_3 will also increase.

Exercise 2.3. Thermodynamic Equilibrium of Ammonia Synthesis at Low Pressure (Reaction Between Ideal Gases)
Stoichiometric amounts of N_2 and H_2 are reacted at 350 °C ($K_{\mathrm{p}} = 7.07 \times 10^{-4}$) at 1 and 10 atm and at 500 °C ($K_{\mathrm{p}} = 1.45 \times 10^{-5}$). Determine the equilibrium composition expressed in molar fractions. Consider a reaction starting with 1 mol of nitrogen.

Initial reacting moles: $n^o_{N_2}$ $\quad n^o_{H_2} = 3n^o_{N_2}$
Reacted moles: Nitrogen $= n^o_{N_2}\lambda$ $\quad$ Hydrogen $= 3n^o_{N_2}\lambda$
Residual moles: Nitrogen $= n^o_{N_2}(1-\lambda)$ $\quad$ Hydrogen $= n^o_{H_2} - 3n^o_{N_2}\lambda = 3n^o_{N_2}(1-\lambda)$
Moles of NH_3 obtained: $2n^o_{N_2}\lambda$
Assuming $n^o_{N_2} = 1$ and $P = 1$ atm, it results in:

$$K_{\mathrm{P}} = K_n(n_{\mathrm{tot}}P)^{-2} = \frac{4\lambda^2}{27(1-\lambda)^4}(n_{\mathrm{tot}}P)^{-2} = 7.07 \times 10^{-4} \quad n_{\mathrm{tot}} = (4-2\lambda) \tag{2.91}$$

$$K_P = \frac{4\lambda^2}{27(1-\lambda)^4(4-2\lambda)^2} = 7.07 \times 10^{-4} \tag{2.92}$$

At $P = 1$ bar, it results $\lambda = 0.17$. As a consequence, the equilibrium composition is:

$$y_{H_2} = \frac{3(1-\lambda)}{n_{tot}} = \frac{3(1-0.17)}{(4-2\lambda)} = \frac{2.49}{3.66} = 0.680 \tag{2.93}$$

$$y_{N_2} = \frac{(1-\lambda)}{3.66} = 0.226 \tag{2.94}$$

$$y_{NH_3} = \frac{2\lambda}{3.66} = 0.094 \tag{2.95}$$

At $P = 10$ atm, we have:

$$K_P = \frac{4\lambda^2}{2700(1-\lambda)^4(4-2\lambda)^2} = 7.07 \times 10^{-4} \tag{2.96}$$

The conversion at equilibrium results in $\lambda = 0.505$, and the equilibrium composition will be:

$$y_{H_2} = 0.497 \quad y_{N_2} = 0.165 \quad y_{NH_3} = 0.338$$

The conclusion is that the pressure favors the ammonia yield. If we consider now a temperature of 500 °C, K_p in this case is 1.45×10^{-5}; therefore:

$$K_P = \frac{4\lambda^2}{2700(1-\lambda)^4(4-2\lambda)^2} = 1.45 \times 10^{-5} \tag{2.97}$$

and $\lambda = 0.215$, and the new equilibrium composition will be:

$$y_{H_2} = 0.659 \quad y_{N_2} = 0.219 \quad y_{NH_3} = 0.121 \tag{2.98}$$

that is, increasing the temperature, the yield of ammonia strongly decreases so the problem is:

(1) to find a catalyst promoting the reaction at low temperature or
(2) to operate at high pressure.

As can be seen, thermodynamics suggests the best strategy to follow for obtaining the optimal performance for a reaction. Last, it is opportune to mention that we have considered an ideal system; however, clearly by increasing the pressure, we must consider for this reaction, occurring in gas phase, the effect of nonideality, as will be seen in the next sections. The described results were obtained with a MATLAB calculation program available as Electronic Supplementary Material, Data 2.1.

Results

	N_2	H_2	NH_3
Case no. 1 Mole fractions at equilibrium (*P* = 1 atm, *T* = 350 °C)			
	0.2264	0.6791	0.0945
Nitrogen conversion = 0.1728 (–)			
Case no. 2 Mole fractions at equilibrium (*P* = 10 atm, *T* = 350 °C)			
	0.1655	0.4964	0.3381
Nitrogen conversion = 0.5053 (–)			
Case no. 1 Mole fractions at equilibrium (*P* = 10 atm, *T* = 500 °C)			
	0.2196	0.6589	0.1214
Nitrogen conversion = 0.2166 (–)			

2.3.3 Generalities About Chemical Equilibrium for Reactions Between Real Gases

Real gases have a rather different behavior with respect to the perfect gas. As a matter of fact, the molecules of a perfect gas are considered like points of negligible volume without any reciprocal interaction. In contrast, the molecules of a real gas are characterized by a small but not negligible volume, and the molecules interact with each other more or less. Pressure, as a consequence, is attenuated by the reciprocal attraction of the molecules, while there is a portion of volume (co-volume) that cannot be compressed corresponding to the overall incompressible volume of the molecules.

Therefore, instead of the general equation of state (EOS) $PV = RT$, which is valid for one mole of ideal gas, other equations must be used. Van der Waals (1873), for example, in a pioneering work suggested the relation for real gases:

$$\left(P + \frac{a}{V^2}\right)(V - b) = RT \tag{2.99}$$

where "a" is an attractive term meaning that molecules are attracted each other, thus affecting the total pressure, while "b," the co-volume, is a repulsive term that considers that molecules have their own volume and when molecules are very close, strong repulsive forces hinder any further compression.

The isotherms corresponding to the Van der Waals equation (Fig. 2.7) are somewhat different from the isotherms of ideal gases shown in Fig. 2.2.

As can be seen, real gases approximate the behavior of ideal gases only at high temperature over the critical temperature, T_c. Moreover, C is the critical point

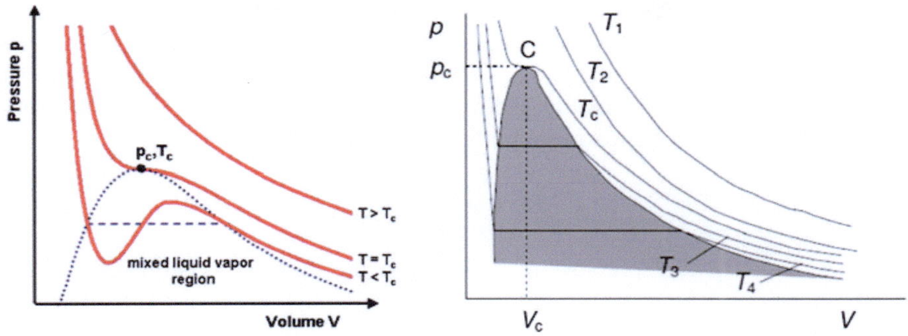

Fig. 2.7 Isotherms of real gases according to Van der Waals EOS

individuated by the values of P_c = critical pressure and V_c = critical volume at the critical temperature T_c. The critical point, C, is a characteristic point corresponding to a flex in the isotherm. Over these points, only the gaseous phase exists, while below this point, liquid and vapor phase coexist. The Van der Waals equation has been a pioneering breakthrough for describing real gases but Eq. (2.99) is inaccurate, in particular at high pressures, and normally other equations of state are currently used for obtaining more correct results.

For determining the equilibrium of the reactions between real gases, as previously mentioned, the condition:

$$\sum_i \alpha_i \mu_i = 0 \tag{2.100}$$

is still valid, and the formalism developed for expressing the chemical potential of Eq. (2.66) is conserved by writing, as suggested by Lewis (1901):

$$\mu_i = \mu_i^o + RT \ln\left(\frac{f_i}{f_R}\right) \tag{2.101}$$

where the term f_i, named "fugacity," has been placed in the expression instead of the pressure. The fugacity can be identified with the pressure, as for an ideal gas, only at very low pressure (p_i near to 0). By comparing f_i with the pressure, we have that:

$$f_i/p_i \rightarrow 1 \text{ for } p_i \rightarrow 0$$

The most convenient reference state is $f_R = 1$ atm = reference fugacity of the component "i," which arbitrarily assumed equal to 1. Hence, in analogy with Eq. (2.67), we can also write:

$$\mu_i = \mu_i^o + RT \ln f_i \tag{2.102}$$

According to the Lewis and Randall (1961) approximation, the fugacity of a component, *i*, in an ideal mixture of real gases can be regarded as the fugacity of the pure component, f_i°, at the same temperature and pressure of the mixture times its molar fraction. An ideal mixture has a null mixing heat and does not show any change of volume for the mixing. In this case, we can write:

$$f_i = f_i^{\circ} y_i = p_i \varphi_i = P y_i \varphi_i \tag{2.103}$$

where p_i is the partial pressure of i and φ_i is the "fugacity coefficient" of the i component.

Calculation of the fugacity of a mixture is realized, in this case, through calculation of the fugacity of the pure components. The chemical equilibrium between real gases will be made as follows:

$$\begin{gathered}
\sum_i \alpha_i \mu_i = 0 \\
\sum_i \alpha_i \mu_i^o + RT \sum_i \alpha_i \ln f_i = 0 \\
\underbrace{\sum_i \alpha_i \mu_i^o}_{\boxed{\Delta G^o_{\mathrm{T,P}}}} + RT \ln \prod_i f_i^{\alpha_i} = 0 \\
\Delta G^o_{\mathrm{T,P}} = -RT \ln \prod_i f_i^{\alpha_i} \\
\prod_i f_i^{\alpha_i} = \frac{f_M^{\alpha_M} f_N^{\alpha_N}}{f_A^{\alpha_A} f_B^{\alpha_B}} = \frac{p_M^{\alpha_M} p_N^{\alpha_N}}{p_A^{\alpha_A} p_B^{\alpha_B}} \frac{\varphi_M^{\alpha_M} \varphi_N^{\alpha_N}}{\varphi_A^{\alpha_A} \varphi_B^{\alpha_B}} = K_{\mathrm{P}} K_{\varphi}
\end{gathered} \tag{2.104}$$

where K_{p} is the thermodynamic constant assuming the reacting gases as ideal. K_{φ} grouping all the fugacity coefficients is not a constant because is dependent on the composition.

2.3.4 *The Van der Waals EOS and the Corresponding State Law*

We have seen that the calculation of fugacity is important for a correct evaluation of the equilibrium constants in reactions occurring between real gases; however, as will be seen later, the same calculation is also useful for describing vapor–liquid equilibrium (VLE) of both pure compounds and their mixtures. Considering the fugacity of an i component in a mixture and remembering Eq. (2.63):

$$\left(\frac{\partial \mu_i}{\partial P}\right)_T = \overline{V}_i \quad \text{Partial molar volume} \tag{2.105}$$

$$\left[\mathrm{d}\mu_i = \overline{V}_i \mathrm{d}P = RT \mathrm{d} \ln f_i\right]_T \tag{2.106}$$

By integrating:

$$RT\ln\varphi_i = RT\ln\frac{f_i}{Py_i} = \int_0^P \left(\overline{V}_i - \frac{RT}{P}\right)\mathrm{d}P \tag{2.107}$$

$\varphi_i = \frac{f_i}{P_i} = \frac{f_i}{Py_i} =$ **coefficient of fugacity of *i* in a gas mixture**

For a pure component $\overline{V}_i = V_i =$ molar volume of i, then by assuming $Z_i = \frac{PV_i}{RT}$ we obtain:

$$RT\ln\left(\frac{f_i^\circ}{P}\right)_{\text{pure } i} = \int_0^P \left(V_i - \frac{RT}{P}\right)\mathrm{d}P = \int_0^P \frac{(Z_i - 1)}{P}\mathrm{d}P \tag{2.108}$$

Z_i = Compressibility factor of the i component

As seen, the fugacity of the i component in a mixture, according to the Lewis–Randall rule, is equal to the fugacity of a pure component times its molar fraction $f_i = f_i^\circ y_i$ according to Eq. (2.103). The molar volume of a gas mixture normally follows the Amagat law (1880) (see also Wisniak 2005), that is, at a fixed temperature and pressure, the total volume of the mixture is equal to the sum of the volumes of the components $V = \sum_i n_i v_i$, where $n_i =$ number of i moles. According to the Amagat law, if there is no volume change by mixing different gases, the partial molar volume of each component is equal to the molar volume of the pure component.

The integration of Eq. (2.108) can be made if the dependence of the volume V_i (or of the factor, Z_i) on the pressure is known, that is, we need an EOS as the Van der Walls or a more accurate one. However, by rearranging the Van der Waals equation (Eq. 2.99), we can write, for example:

$$P = \frac{RT}{V - b} - \frac{a}{V^2} \tag{2.109}$$

by differentiating:

$$\mathrm{d}P = \left(-\frac{RT}{(V - b)^2} + \frac{2a}{V^3}\right)\mathrm{d}V \tag{2.110}$$

hence,

$$RT\mathrm{d}\ln f = \left(-\frac{VRT}{(V - b)^2} + \frac{2a}{V^2}\right)\mathrm{d}V \tag{2.111}$$

Integrating in the range $V^* \to \infty$ and V:

$$\ln\frac{f}{f^*}=\frac{b}{V-b}-\frac{b}{V^*-b}-\ln(V-b)+\ln(V^*-b)-\frac{2a}{RTV}+\frac{2a}{RTV^*} \tag{2.112}$$

Considering that when $V^* \to \infty$, $P \to 0$ and $V^* - b = RT/P^*$, $P^* = f^* 1/V^* = 0$, we have:

$$\ln f=\ln\frac{RT}{V-b}+\frac{b}{V-b}-\frac{2a}{RTV} \tag{2.113}$$

As mentioned before, the Van der Waals equation is not accurate in describing the behavior of real gases because the constants "a" and "b" are characteristic of any single compound. Its performance can be greatly improved by determining the constants "a" and "b" as a function of the critical variables. This can be performed by applying the relation at the critical point (see Fig. 2.7), that is, in the point at which we have, for any compound, known values of P_C, V_C, and T_C. By equating to zero the first and second derivative of Eq. (2.109) and writing Eq. (2.109) with the critical variables, it is possible to determine "a" and "b" of Van der Waals equation as a function of the critical variables resulting in:

$$a=3P_C V_C^2=\frac{27R^2T_C^2}{64P_C} \tag{2.114}$$

$$b=\frac{V_C}{3}=\frac{RT}{8P_C} \tag{2.115}$$

Because T_C and P_C experimental data are more precise than V_C data, normally the relations of a and b containing P_C and T_C are preferred. The Van der Waals equation, with a and b constants determinable from critical variables, has a more general validity but it is still not sufficient. In fact, the "compressibility factor" Z_C at the critical point becomes a constant equal to:

$$Z_C=\frac{P_C V_C}{RT_C}=\frac{3}{8}=0.375 \tag{2.116}$$

Considering that the compressibility factor Z for one mole of an ideal gas is equal to 1, Z_C is an index of the deviation of real gases from ideality. In practice, the experimentally observed values of Z_C are in the range of 0.23–0.30, and many substances show a value of 0.27. The conclusion is that other refinements of the Van der Waals equation are necessary. A further improvement can be obtained by considering the reduced critical parameters, which are:

$$P_R=\frac{P}{P_C}\quad V_R=\frac{V}{V_C}\quad T_R=\frac{T}{T_C}\quad \text{and remembering that } R=\frac{8P_C V_C}{3T_C} \tag{2.117}$$

By substituting the reduced variables and R in Eq. (2.109), we obtain, at last:

$$\left(P_R + \frac{3}{V_R^2}\right)(3V_R - 1) = 8T_R \tag{2.118}$$

This is an equation of universal validity because we do not care which fluids we are considering: We just need to know the reduced conditions. Equation (2.118) is the basis of the principle of corresponding states, according to which different substances behave alike at the same reduced state. Therefore, real gases under the same conditions of reduced pressure and temperature would have the same reduced volume.

It is important to point out that hydrogen, helium, and neon (quantum gases) deviate from the common behaviors. Calculations for these gases must be made by introducing pseudo critical reduced parameters (Lowry 1923):

$$P_R = \frac{P}{P_c + 8} \quad \text{and} \quad T_R = \frac{T}{T_c + 8} \tag{2.119}$$

However, Standing and Katz (1942) constructed a generalized plot (*Z*-chart) to obtain *Z* values from the reduced variables P_R and T_R. In Fig. 2.8, it is possible to appreciate the agreement obtained by the corresponding state law for many different compounds in an intermediate range of pressure (see Su 1946). As seen in Fig. 2.8, the value of *Z* tends toward 1 as the gas pressure approaches 0. However, all gases

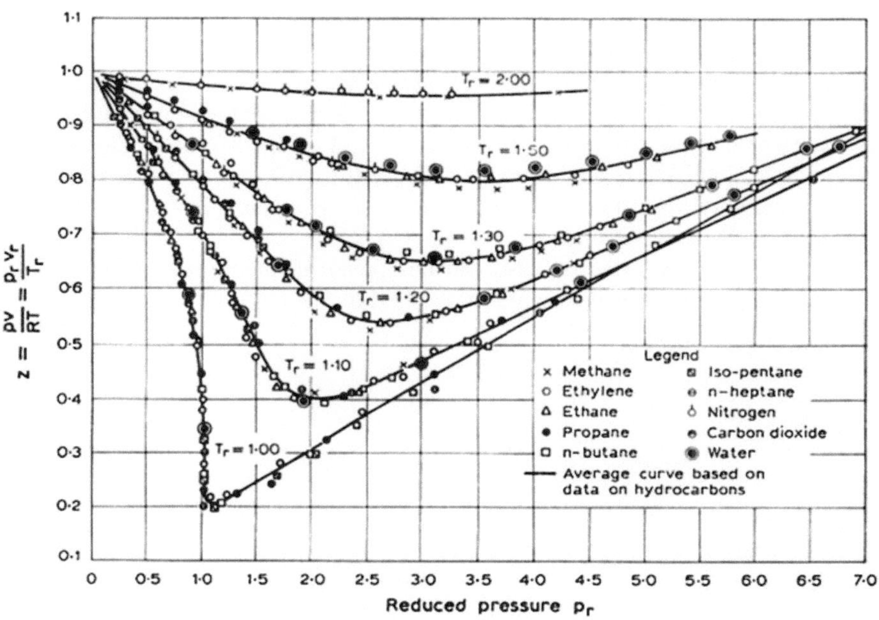

Fig. 2.8 Compressibility factor as a function of the reduced variables for different substances. Reprinted with permission from Su (1946) Copyright (1946) American Chemical Society. See also Wallace and Linning (1970)

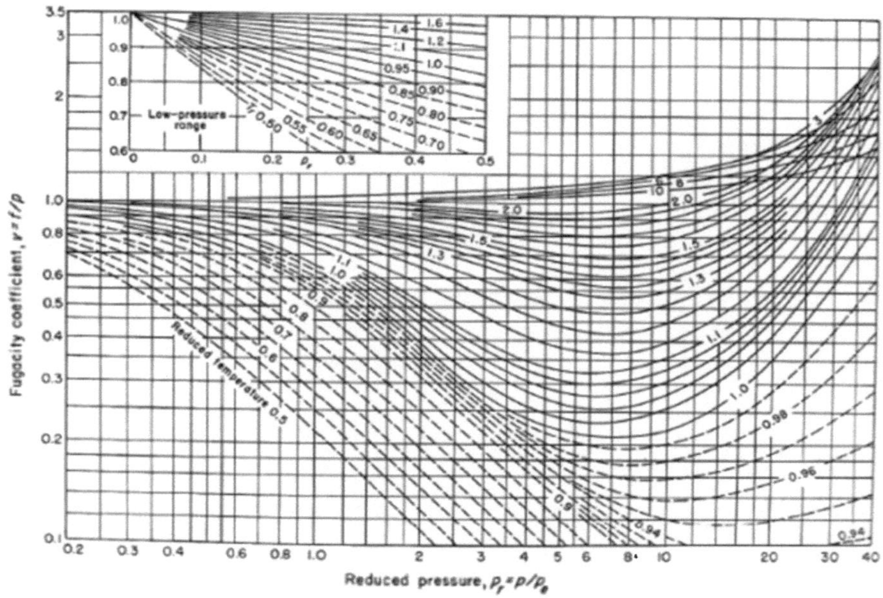

Fig. 2.9 Calculated fugacity coefficients as a function of the reduced pressure at different reduced temperatures

tend toward ideal behavior (<1) at intermediate pressures because the intermolecular forces of attraction between the molecules cause the actual volumes to be less than the ideal values and toward less ideal behavior (>1) at very high pressures because the intermolecular repulsive forces cause the actual volumes to be greater than the ideal values. In Fig. 2.9, the generalized fugacity coefficient as a function of the reduced pressure for different pressure ranges (see Hougen and Watson 1947; see also Lee and Kesler 1975).

All the developed concepts also can be applied to mixtures of real gases by determining pseudo critical variables with the Kay rule (1936):

$$P_{\mathrm{Cm}} = \sum_i y_i P_{Ci} \quad P_{\mathrm{R}} = \frac{P}{P_{\mathrm{Cm}}} \tag{2.120}$$

$$T_{\mathrm{Cm}} = \sum_i y_i T_{Ci} \quad T_{\mathrm{R}} = \frac{T}{T_{\mathrm{Cm}}} \tag{2.121}$$

where P_{Cm} and T_{Cm} are, respectively, the pseudo critical pressure and temperature of the mixture.

Not all the fluids obey the law of the corresponding states, in particular fluids constituted by nonspherical molecules or strongly asymmetric molecules having great polarity. However, in both cases, the introduction of one or two new parameters allows to expand the validity of the corresponding state law. In particular, one of these parameters is the **acentric factor**, ω, introduced by Kenneth Pitzer (1955). This parameter is related to the non-sphericity of a molecule. For spherical molecules, ω is almost zero (noble gases). Nonspherical molecules have values >0, but only the most severely nonspherical molecules have values approaching unity. The acentric factor is defined as:

$$\omega = -1 - \log_{10} \frac{P^{\circ}}{P_{C}} \tag{2.122}$$

where P° is the vapor pressure of a liquid at the reduced temperature $T_{R} = 0.7$. If ω is <0.2, the departures from corresponding states are approximately linear in ω. In this case, the compressibility factor, Z, can be written as:

$$Z(P_R, T_R) = Z_0(P_R, T_R) + \omega Z_1(P_R, T_R) \tag{2.123}$$

where Z_0 is the compressibility factor of a substance comprising spherical molecules and Z_1 is the correction for non-sphericity. Values of Z_0 and Z_1 have been tabulated or plotted in different publications and books (see, e.g., Reid et al. 1987).

In a mixture, the acentric factor can be determined from the values of the pure components as:

$$\omega_{m} = \sum_{i} y_i \omega_i \tag{2.124}$$

In the case of polar molecules, the corresponding state law can be applied only after the introduction of at least one additional parameter.

Exercise 2.4. Thermodynamic Equilibrium in Ammonia Synthesis at High Pressure (Reaction Between Real Gases): The Graphical Approach of the Principle of the Corresponding States
Repeat Exercise 2.3 related to the ammonia synthesis by assuming a temperature of 400 °C and a pressure of 300 atm; then also calculate the composition at 450 °C and 300 atm. In this case, we begin collecting the critical parameters of the substances involved in the reaction, calculate the reduced pressure and temperature, and individuate the fugacity coefficients on the plots of Fig. 2.9. Remember that the reduced pressure and temperature for hydrogen must be calculated with Eq. (2.119). All these data are listed in Table 2.4.

Table 2.4 Reduced variables at 673.16 K and 300 atm[a]

Gas	T_c (K)	P_c (atm)	T_r	P_r	f_i°/P	f_i° (atm)
H_2	33.2	18.8	16.33	14.42	1.10	330
N_2	126	33.5	5.34	8.95	1.15	345
NH_3	405.5	111.5	1.66	2.69	0.88	264

[a]Fugacities of pure components and related fugacity coefficients were determined by graphical approach

Consider now the following stoichiometry:

$$1/2\,N_2 + 3/2\,H_2 \rightleftarrows NH_3 \tag{2.125}$$

$$K_f = \frac{f_{NH_3}}{\left(f_{H_2}\right)^{3/2}\left(f_{N_2}\right)^{1/2}} = \frac{f^\circ_{NH_3}}{\left(f^\circ_{H_2}\right)^{3/2}\left(f^\circ_{N_2}\right)^{1/2}}\,\frac{y_{NH_3}}{\left(y_{H_2}\right)^{3/2}\left(y_{N_2}\right)^{1/2}} = K^\circ_f\,K_y \tag{2.126}$$

$$K^\circ_f = \frac{f^\circ_{NH_3}}{\left(f^\circ_{H_2}\right)^{3/2}\left(f^\circ_{N_2}\right)^{1/2}} = \frac{264}{(345)^{1/2}(330)^{3/2}} = 2.371\times 10^{-3} \tag{2.127}$$

According to Glasstone (1947):

$$\Delta G^\circ = -9130 + 7.46T\ln T - 3.69\times 10^{-3}T^2 + 2.35^{*}10^{-7}T^3 - 12.07RT \tag{2.128}$$

$$\text{At } T = 400 + 273.16 = 673.16, \text{ it results in } \Delta G^\circ = 5831 \text{ cal/mol} \tag{2.129}$$

$$\Delta G^\circ_{673.16} = -RT\ln K_f = 5831 \tag{2.130}$$

$$K_f = \exp\left(-\frac{5831}{RT}\right) = \exp\left(-\frac{5831}{1.987^{*}673.16}\right) = 0.0128 \tag{2.131}$$

$$K_f = K^\circ_f\,K_y = 2.371\times 10^{-3}K_y = 0.0128 \tag{2.132}$$

$$K_y = \frac{y_{NH3}}{\left(y_{H_2}\right)^{3/2}\left(y_{N_2}\right)^{1/2}} = \frac{0.0128}{2.371\times 10^{-3}} = 5.39 \tag{2.133}$$

Remembering the mass balance reported in Exercise 2.3, we have:

$$n_{tot} = n^o_{N_2}(4 - 2\lambda) \tag{2.134}$$

where λ is the conversion = reacted moles/initial moles of nitrogen. The molar fraction will be:

$$y_{NH_3} = \frac{\text{Moles obtained of } NH_3}{\text{Total moles}} = \frac{2n^o_{N_2}\lambda}{n^o_{N_2}(4-2\lambda)} = \frac{2\lambda}{(4-2\lambda)} \tag{2.135}$$

$$y_{H_2} = \frac{\text{Residual moles of } H_2}{\text{Total moles}} = \frac{3(1-\lambda)}{(4-2\lambda)} \tag{2.136}$$

$$y_{N_2} = \frac{\text{Residual moles of } N_2}{\text{Total moles}} = \frac{(1-\lambda)}{(4-2\lambda)} \tag{2.137}$$

$$K_y = \frac{y_{NH3}}{(y_{H_2})^{3/2}(y_{N_2})^{1/2}} = \frac{\frac{2\lambda}{(4-2\lambda)}}{\left(\frac{3(1-\lambda)}{(4-2\lambda)}\right)^{3/2}\left(\frac{(1-\lambda)}{(4-2\lambda)}\right)^{1/2}} = 0.385\frac{\lambda(4-2\lambda)}{(1-\lambda)^2}$$
$$= 5.39 \tag{2.138}$$

$$14 = \frac{\lambda(4-2\lambda)}{(1-\lambda)^2} \quad \text{developing : } 16\lambda^2 - 32\lambda + 14 = 0 \quad \lambda = 0.64 \tag{2.139}$$

The equilibrium composition will therefore be:

$$y_{NH_3} = 0.478 \quad y_{H_2} = 0.391 \quad y_{N_2} = 0.130 \tag{2.140}$$

The obtained value of ammonia yield is in good agreement with the value of 0.48 found by Larson (1924) and by Gillespie and Beattie (1930) as well as reported by Vancini (1971), as can be seen in Table 2.5.

Let us consider now what happens for a temperature >450 °C leaving all the other conditions unchanged. In this case, some data of Table 2.4 must be recalculated as shown in Table 2.6.

$$\Delta G^{\circ}_{723.16} = -RT \ln K_f = 7212 \tag{2.141}$$

$$K_f = \exp\left(-\frac{7212}{RT}\right) = 0.00661 \tag{2.142}$$

$$K_f = K_f^{\circ} K_y = 0.00661 \tag{2.143}$$

Table 2.5 Ammonia yields at equilibrium in percent of ammonia

Pressure (atm)					
T (°C)	10	100	300	600	1000
300	14.73	52.04	70.96	84.21	92.55
400	3.85	25.37	48.18	66.17	79.82
450	2.11	16.40	35.87	54.00	69.69
500	1.21	10.51	25.80	42.32	57.47
550	0.76	6.82	18.23	32.18	41.16
600	0.49	4.53	12.84	24.04	31.43
700	0.23	2.18	7.28	12.60	12.87

Table 2.6 Reduced variables at 723.16 K and 300 atm[a]

Gas	T_c (K)	P_c (atm)	T_r	P_r	f_i°/P	f_i° (atm)
H_2	33.2	18.8	17.6	14.42	1.09	327
N_2	126	33.5	5.7	8.90	1.14	342
NH_3	405.5	111.5	1.78	2.69	0.91	273

[a]Fugacities of pure components and related fugacity coefficients were determined by graphical approach

$$K_{\rm f}^{\circ}=\frac{f_{\rm NH_3}^{\circ}}{\left(f_{\rm H_2}^{\circ}\right)^{3/2}\left(f_{\rm N_2}^{\circ}\right)^{1/2}}=\frac{273}{(342)^{1/2}(327)^{3/2}}=2.497\times 10^{-3} \tag{2.144}$$

$$K_y=\frac{y_{\rm NH3}}{\left(y_{\rm H_2}\right)^{3/2}\left(y_{\rm N_2}\right)^{1/2}}=0.385 \quad \frac{\lambda(4-2\lambda)}{(1-\lambda)^2}=\frac{0.00661}{2.497\times 10^{-3}}=2.647 \tag{2.145}$$

$$\text{Developing}: 8.86\lambda^2-17.72\lambda+6.86=0 \quad \lambda=0.524 \tag{2.146}$$

The equilibrium composition will therefore be:

$$y_{\rm NH_3}=0.357 \quad y_{\rm H_2}=0.482 \quad y_{\rm N_2}=0.161 \tag{2.147}$$

This value of ammonia yield agrees with the experimental value (see Table 2.5). Again, we can conclude that the production of ammonia is favored by low temperature and by high pressure: The problem is just to find a catalyst able to promote the reaction at a lower temperature at a reasonable rate. All the described results can be obtained using a MATLAB calculation program available as Electronic Supplementary Material, Data 2.1.

Results

	N_2	H_2	NH_3
Case no. 1 Mole fractions at equilibrium (*P* = 300 atm, *T* = 450°C)			
	0.1304	0.3913	0.4783
Nitrogen conversion = 0.6471 (–)			
Case no. 2 Mole fractions at equilibrium (*P* = 300 atm, *T* = 450°C)			
	0.1606	0.4819	0.3574
Nitrogen conversion = 0.5266 (–)			

Exercise 2.5. Calculation of the Compressibility Factor of a Real Gas with an EOS

Calculate the compressibility factor Z of ammonia using the Redlich–Kwong EOS.

$$P=\frac{RT}{(V-b)}-\frac{a}{V(V+b)\sqrt{T}} \tag{2.148}$$

In this expression, a and b can be calculated from the critical constants as:

$$a=\frac{0.42748R^2T_{\rm C}^{2.5}}{P_{\rm C}} \quad b=\frac{0.0866RT_{\rm C}}{P_{\rm C}} \tag{2.149}$$

The critical constants for ammonia are $T_c=405.6$ K and $P_c=112$ atm.

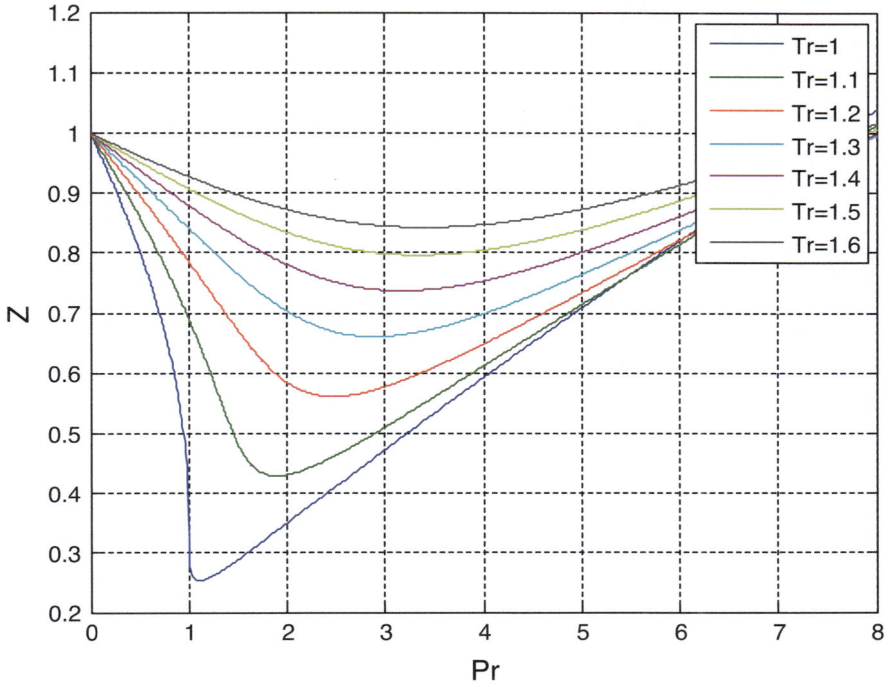

Fig. 2.10 Compressibility factor as a function of the reduced pressure calculated using the RK-EOS for ammonia

Make the calculation for different reduced temperature, T_R, from 1 to 1.6 with a step of 0.1, and construct a plot of Z as a function of the reduced pressure, P_R, for each considered T_R in a range of P_R from 0 to 8. To facilitate calculation, the RK-EOS equation can be expressed in the following form:

$$(V-b)PV(V+b)\sqrt{T} = RTV(V+b)\sqrt{T} - a(V-b) \tag{2.150}$$

without the denominator.

The result is the plot shown Fig. 2.10 obtained using the MATLAB program reported as Electronic Supplementary Material, Data 2.1.

2.3.5 *Alternative Equations of State*

Equations of state are useful for determining different thermodynamic properties, such as heat capacities, enthalpies, entropies, etc., and for studying thermodynamic equilibrium properties. For these reasons, hundreds of equations of state (EOS) have been proposed to improve and substitute the Van der Waals equation. Redlich and Kwong (1949), for example, proposed a modification of the Van der Waals equation that resulted more accurate in determining the fugacities at higher pressures:

$$\underset{\text{Van der Waals (1873)}}{\underline{P=\frac{RT}{V-b}-\frac{a}{V^2}}} \qquad \underset{\text{Redlich and Kwong (1949)}}{\underline{P=\frac{RT}{V-b}-\frac{a}{V(V+b)T^{0.5}}}} \tag{2.151}$$

The values of a and b for the Redlich and Kwong equation can be determined as a function of the critical variables as was previously made for the Van der Waals equation:

$$a=0.42748\frac{R^2T_C^{2.5}}{P_C} \qquad b=0.08664\frac{RT_C}{P_C} \tag{2.152}$$

Many other improvements have been successively introduced, creating a class of "cubic equations of state," being equations of the third order with respect to the volume. In particular, the Redlich and Kwong equation was modified by Soave by substituting the terms $(a/T^{3/2})$ with the term $a=a(T)$, thus obtaining the popular Redlich–Kwong–Soave (RKS) equation (Soave 1972):

$$P=\frac{RT}{V-b}-\frac{a(T)}{V(V+b)} \tag{2.153}$$

where:

$$a(T)=0.42748\left(\frac{R^2T_C^2}{P_C}\right)\left\{1+m\left[1-\left(\frac{T}{T_C}\right)^{0.5}\right]\right\}^2 \tag{2.154}$$

$$m=0.480+1.57\omega-0.176\omega^2 \text{ and } b=0.08664\frac{RT_C}{P_C} \tag{2.155}$$

where ω is the already seen acentric factor.

Another interesting modification of the Redlich and Kwong equation is the one proposed by Peng and Robinson (PR) (1976):

$$P=\frac{RT}{V-b}-\frac{a(T)}{V(V+b)+b(V-b)} \tag{2.156}$$

PR redefined $a(T)$ as:

$$a(T) = 0.45724\left(\frac{R^2T_C^2}{P_C}\right)\left\{1 + k\left[1 - \left(\frac{T}{T_C}\right)^{0.5}\right]\right\}^2 \tag{2.157}$$

$$k = 0.37464 + 1.5422\omega - 0.26922\omega^2 \tag{2.158}$$

$$b = 0.07780\frac{RT_C}{P_C} \tag{2.159}$$

The RKS and PR equations of state are widely used in industry because they are easy to use and represent satisfactorily the P, T and phase composition in both binary and multicomponent systems. The only necessary information is the critical properties and the values of the acentric factor of the pure components. These equations well reproduce the equilibrium phase pressure, but they normally fail in calculating the saturated liquid volume. Cubic equations of state well reproduce VLEs, but they have scarce accuracy in reproducing the volumetric properties of pure fluids, particularly under supercritical conditions. For this purpose, Peneloux et al. (1982) improved the volumetric behavior of cubic equations of state by introducing a volume-shift parameter. This modification was applied by Soave et al. (1993) to the original RKS-EOS to obtain a relation able to predict high-pressure fugacity coefficients with a great accuracy:

$$P = \frac{RT}{V + c - b} - \frac{a(T)}{(V + c)(V + c + b)} \tag{2.160}$$

This modification has the great advantage that the values of the parameters $a(T)$ and b need not be changed. This allows use of the RKS equation for calculating a and b parameters, while the correction, c, is useful just for evaluating the density of the fluid and to evaluate more correct equilibrium compositions (see Bertucco et al. 1995).

Another important EOS is the virial equation because it can be derived directly from statistical thermodynamic. It was introduced by Kammerlingh Onnes (1901) as a generalization of the general law of ideal gas. According to Kammerlingh Onnes, for a gas containing N molecules, we can write:

$$\frac{P}{k_B T} = \delta + B_2(T)\delta^2 + B_3(T)\delta^3 + \cdots \tag{2.161}$$

where P is the pressure, k_B is the Boltzmann constant, T is the absolute temperature, and δ is the density of the gas expressed as N/V = number of molecules for volume unit. It must be pointed out that if we consider only the first term of the virial expansion, assuming Na = Avogadro number (number of molecules/mole), we obtain $pV = nN_Ak_BT = nRT$, that is, the law of ideal gas. Therefore, the terms after the first describe the deviation of real gases from the ideal one, and each virial coefficient interprets the deviation as a consequence of a particular type of interaction between the molecules and depends only on temperature. Clearly, the accuracy of the equation increases by increasing the number of coefficients introduced in the expression. From Eq. (2.161), it is possible to obtain the expression related to one mole of gas:

$$\frac{PV}{RT} = 1 + \frac{B(T)}{V} + \frac{C(T)}{V^2} + \frac{D(T)}{V^3} \cdots \tag{2.162}$$

where B, C, and D are, respectively, named "first," "second," and "third virial coefficients." B describes the interactions between two molecules, C the interactions between three molecules, and so on. An advantage of the virial equation is that the coefficients have physical meaning and can be derived theoretically from the intermolecular potential function. Often, the virial equation is truncated at the second term, that is:

$$Z = \frac{PV}{RT} = 1 + \frac{B(T)}{V} \tag{2.163}$$

To evaluate $B(T)$, different methods have been proposed based in particular on the integration of the expression of the intermolecular energy to the distance between the molecules. Unfortunately, our knowledge of the intermolecular energies is limited; therefore, the estimation of B is more usually made by employing the corresponding state relations. According to Van Ness and Abbott (1982), for non-polar molecules, we can write:

$$\frac{BP_c}{RT_c} = B^{(0)} + \omega B^{(1)} \tag{2.164}$$

$$B^{(0)} = 0.083 - \frac{0.422}{T_r^{1.6}} \quad B^{(1)} = 0.139 - \frac{0.172}{T_r^{4.2}} \tag{2.165}$$

where ω is an acentric factor (see Eq. 2.122). In Fig. 2.11, the agreement obtained for the second virial coefficient by Van Ness and Abbott (see Reid et al. 1987) can be appreciated.

For describing the behavior of polar molecules according to Tsonopoulos (1974), another term must be added to Eq. (2.164):

$$B^{(2)} = \frac{\alpha}{T_r^6} - \frac{\beta}{T_r^8} \tag{2.166}$$

with $\beta = 0$ when the molecules do not form hydrogen bonds. α is described by equations that are functions of the reduced dipole moment:

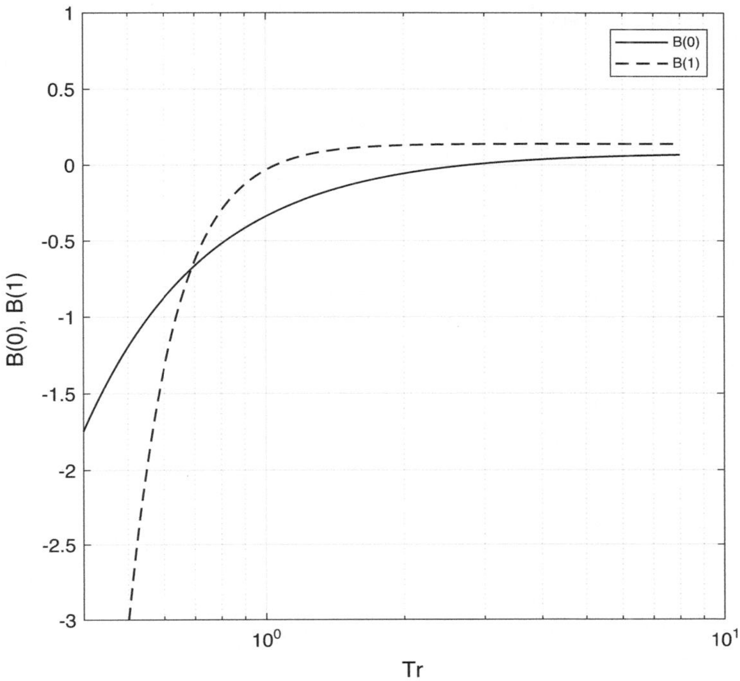

Fig. 2.11 Second virial coefficients calculated with Eq. (2.165) fitting the behavior of 14 different compounds

$$\mu_r = \frac{10^5 \mu_p^2 P_C}{T_C^2} \tag{2.167}$$

where μ_p is the dipole moment (debyes), P_c is the critical pressure (bars), and T_c is the critical temperature (K).

An example of α relation that is valid for ketones, aldehydes, nitriles, ethers, and esters is:

$$\alpha = -2.112 \times 10^{-4} \mu_r - 3.877 \times 10^{-21} \mu_r^8 \tag{2.168}$$

However, a limit of this approach is that α and β are expressed with different relations for different classes of compounds.

McCann and Danner (1984) developed a method based on the group contributions method for determining the second virial coefficient. The method has the same accuracy of the one suggested by Tsonoupolos, but it also can be applied to compounds for which little information is available.

2.3.6 Fugacity Evaluation from an EOS

As has been seen, when a gaseous system is kept at pressure >10–20 bars, it cannot be approximated to an ideal system, and for describing correctly the gaseous phase, we must evaluate the fugacity of all the components of the system. The fugacity of a pure "*i*" component can be determined starting from Eq. (2.107):

$$\ln \varphi_i = \ln \frac{f_i}{P} = \int_0^P \left(\frac{V_i}{RT} - \frac{1}{P}\right) \mathrm{d}P \tag{2.169}$$

where f_i is the fugacity and φ_i is the fugacity coefficient for the *i*th component.

The fugacity of a component "*i*" in a mixture, by applying the Lewis–Randall approximation, can be determined as:

$$\ln \varphi_{i,m} = \ln \frac{f_i}{y_i P} = \int_0^P \left(\frac{\overline{V}_i}{RT} - \frac{1}{P}\right) \mathrm{d}P \tag{2.170}$$

where y_i is the molar fraction of *i* and $\overline{V}_i$ is the partial molar volume of *i*.

Remembering that for one mole of an ideal gas we can write:

$PV_{i\text{-ideal}} = RT$, that is,

$$\frac{1}{P} = \frac{V_{i-\text{ideal}}}{RT} \tag{2.171}$$

Equation (2.169) becomes:

$$\ln \varphi_i = \ln \frac{f_i}{P} = \frac{1}{RT} \int_0^P (V_i - V_{i-\text{ideal}}) \mathrm{d}P \tag{2.172}$$

This equation allows to evaluate the fugacity of a real gas or a nonideal vapor for a given temperature and pressure. To solve this equation, we need an EOS that allows evaluating V_i as a function of P. In the simplest case, we can use the virial equation:

$$PV = RT + BP + CP^2 \ldots \tag{2.173}$$

$$V = \frac{RT}{P} + B + CP + \cdots \tag{2.174}$$

In this case, Eq. (2.172) can be solved analytically, and we will have:

$$\ln \varphi_i = \ln \frac{f_i}{P} = \frac{1}{RT} \int_0^P (B + CP + \cdots)\mathrm{d}P \tag{2.175}$$

$$\ln \varphi_i = \ln \frac{f_i}{P} = \frac{BP}{RT} + \frac{CP^2}{2RT} + \cdots \tag{2.176}$$

Because the virial equation is not accurate at higher pressures, the problem can be solved by using another EOS coupled with a numerical approach or the principle of the corresponding states with a graphical approach.

To apply the corresponding state principle, we can write Eq. (2.172) in the following form:

$$\ln \varphi_i = \ln \frac{f_i}{P} = \frac{1}{RT} \int_0^P \left(Z\frac{RT}{P} - \frac{RT}{P}\right)\mathrm{d}P = \int_0^P \left(\frac{Z-1}{P}\right)\mathrm{d}P = \int_0^P \times (Z-1)\mathrm{d}\ln P \tag{2.177}$$

If the critical variables of the component "i" are known, from the plots of Fig. 2.8, giving Z as a function of the reduced variables, T_R and P_R, we can evaluate how Z changes in a given range of P_R, at a certain T_R value. When the fugacity of the pure components is known, it is possible to evaluate the fugacity of a mixture by applying Eq. (2.170).

An alternative approach is to apply directly a more accurate EOS. The most employed equations of state in industry for the calculation of gas-phase nonideality are the RKS-EOS and the PR-EOS equations because they are easier to employ, accurate in their results, and also can be applied to the liquid phase. Considering again the RKS-EOS equation:

$$P = \frac{RT}{V-b} - \frac{a(T)}{V(V+b)} \tag{2.178}$$

Defining the dimensionless factors A and B:

$$A = \frac{aP}{R^2T^2} \tag{2.179}$$

$$B = \frac{bP}{RT} \tag{2.180}$$

and remembering that:

$$V = Z\frac{RT}{P} \tag{2.181}$$

Equation (2.141) also can be written as:

$$Z^3 - Z^2 + Z\left(A - B - B^2\right) - AB = 0 \tag{2.182}$$

Considering then a pure component "*i*", we have:

$$a_i = 0.42748\left(\frac{R^2 T_{Ci}^2}{P_{Ci}}\right)\left\{1 + m_i\left[1 - \left(\frac{T}{T_{Ci}}\right)^{0.5}\right]\right\}^2 \tag{2.183}$$

$$m_i = 0.480 + 1.57\omega_i - 0.176\omega_i^2 \quad \text{and} \quad b_i = 0.08664\frac{RT_{Ci}}{P_{Ci}} \tag{2.184}$$

$$\omega = -\text{Log}_{10}\left(P_{Ri}^{sat}\right)_{T_{Ri}} - 1 \tag{2.185}$$

$P_{Ri}^{sat} = \frac{P_{Si}}{P_{Ci}}$ = reduced saturation vapor pressure
P_{Si} = vapor pressure of *i* at the reduced temperature T_R
P_{Ci} = critical pressure
The dimensionless terms *A* and *B* for the pure component "*i*" become:

$$A_i = 0.42748\frac{P/P_{Ci}}{\left(T/T_{Ci}\right)^2}\left\{1 + m\left[1 - \left(\frac{T}{T_{Ci}}\right)^{0.5}\right]\right\}^2 \tag{2.186}$$

$$B_i = 0.08664\frac{P/P_{Ci}}{T/T_{Ci}} \tag{2.187}$$

The fugacity coefficient for the pure component will be calculated as:

$$\ln\frac{f_i}{P_i} = Z - 1 - \ln(Z - B_i) - \frac{A_i}{B_i}\ln\left(1 + \frac{B_i}{Z}\right) \tag{2.188}$$

The compressibility factor *Z* to be used in Eq. (2.188) is obtained by solving Eq. (2.182) taking the highest root that is related to the vapor phase, while the smallest one is related to the liquid phase when we have a VLE.

For a multicomponent mixture, we must adopt appropriate mixing rules to the parameters *a* and *b* such as:

$$a = \sum_i\sum_j a_{ij}y_iy_j \quad a_{ij} = \left(1 - k_{ij}\right)\sqrt{a_i a_j} \tag{2.189}$$

where k_{ij} is a binary interaction coefficient considering the deviation from the geometric mean. For *b*, we can write:

$$b = \sum_i \sum_j b_{ij} y_i y_j \quad b_{ij} = \frac{b_i + b_j}{2} \tag{2.190}$$

The calculation of fugacity is laborious, but by using the RKS-EOS equation, the only required parameters are the critical constants and the acentric factor for each component plus k_{ij} for all possible binary components. k_{ij} for nonpolar components can be assumed equal to zero as a first approximation.

Exercise 2.6. Determination of the Fugacity Coefficient of a Pure Component and a Mixture

- **Part A: Fugacity coefficient of a pure component**

Calculate the fugacity coefficient of pure gaseous ammonia at $T = 600$ K and at P between 10 and 600 atm using the RKS-EOS. Compare the obtained results with the correlation proposed by Dyson and Simon (1986). Data for ammonia are listed in Table 2.7.

Correlation of Dyson and Simon (1986):

$$\phi_{NH_3} = 0.1439 + 0.002029T - 0.0004488P - 0.1143 \times 10^{-5}T^2 + 0.2761 \times 10^{-6}P^2 \tag{2.191}$$

- **Solution**

The RKS-EOS and related parameters for pure components are the following expressions:

$$P = \frac{RT}{V - b} - \frac{\alpha a}{V(V + b)} \tag{2.192}$$

$$a = 0.42748 \frac{\alpha (RT_C)^2}{P_C} \tag{2.193}$$

$$b = 0.8664 \frac{(RT_C)}{P_C} \tag{2.194}$$

$$m = 0.480 + 1.574\omega - 0.17\omega^2 \tag{2.195}$$

$$\alpha = \left[1 + m\left(1 - T_R^{0.5}\right)\right]^2 \tag{2.196}$$

The EOS can be rearranged in the form of compressibility factor as:

Table 2.7 Some thermodynamic data of pure ammonia

Properties	Value
Critical temperature (K)	405
Critical pressure (atm)	115.3
Acentric factor	0.253

$$z^3 - z^2 + (A - B - B^2)z - AB = 0 \tag{2.197}$$

with A and B defined as:

$$A = \frac{aP}{(RT)^2} \quad B = \frac{bP}{RT} \tag{2.198}$$

This polynomial of the third degree can be solved numerically obtaining three roots. The highest root corresponds to the compressibility factor related to gas phase. Finally, with the value of the compressibility factor, the fugacity coefficient can be calculated with the expression:

$$\ln \phi = (z - 1) - \ln(z - b) - \frac{A}{B} \ln\left(1 + \frac{B}{z}\right) \tag{2.199}$$

The results obtained by calculations, respectively, at constant temperature (A) and at constant pressure (B) are shown in Fig. 2.12a, b.

The described results were obtained with a MATLAB program reported as Electronic Supplementary Material, Data 2.1.

- **Part B: Fugacity Coefficient of a Mixture of Real Gases**

Calculate the fugacity coefficient of a mixture of ammonia, hydrogen, and nitrogen at $T = 600$ K and at P between 10 and 600 atm using the RKS-EOS. The molar composition of the mixture is 0.333, 0.555, and 0.167 for, respectively, ammonia, hydrogen, and nitrogen. Data of the pure components are listed in Table 2.8.

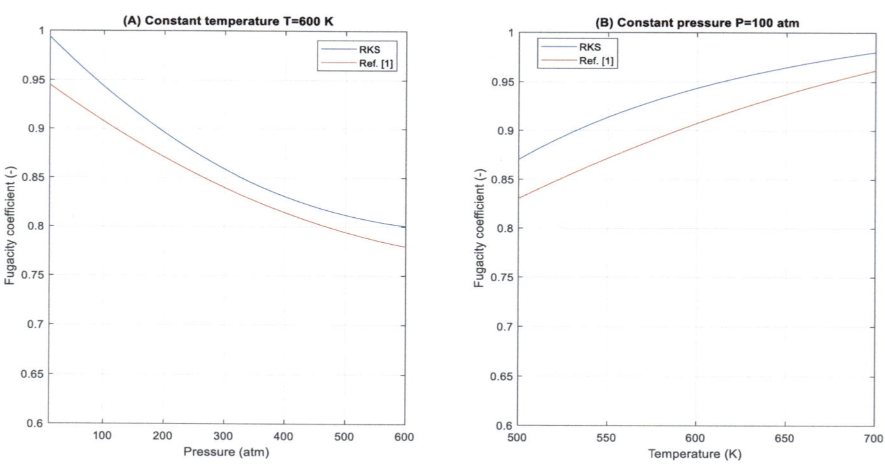

Fig. 2.12 Fugacity coefficients calculated for pure ammonia using, respectively, the RKS-EOS and the Dyson and Simon (1986) correlation. (**a**) calculation at constant temperature; (**b**) calculation at constant pressure

Table 2.8 Some thermodynamic properties of the components of a mixture

Properties	NH_3	H_2	N_2
Critical temperature (K)	405	33	126
Critical pressure (atm)	115.3	13.2	34.6
Acentric factor	0.253	−0.21	0.037

- **Solution**

The RKS-EOS and related parameters for a mixture will be:

$$P=\frac{RT}{V-b_{\mathrm{m}}}-\frac{\alpha a_{\mathrm{m}}}{V(V+b_{\mathrm{m}})} \tag{2.200}$$

$$a_i=0.42748\frac{\alpha_i(RT_{\mathrm{Ci}})^2}{P_{\mathrm{Ci}}} \tag{2.201}$$

$$b_i=0.8664\frac{(RT_{\mathrm{Ci}})}{P_{\mathrm{Ci}}} \tag{2.202}$$

$$m_i=0.480+1.574\omega_i-0.176\omega_i^2 \tag{2.203}$$

$$\alpha_i=\left[1+m_i\left(1-T_{\mathrm{Ri}}^{0.5}\right)\right]^2 \tag{2.204}$$

The parameters for a mixture can be evaluated through suitable mixing rules, such as, for example:

$$a_{\mathrm{m}}=\sum_{i=1}^{\mathrm{NC}}\sum_{j=1}^{\mathrm{NC}}y_iy_j\sqrt{a_ia_j} \tag{2.205}$$

$$b_{\mathrm{m}}=\sum_{i=1}^{\mathrm{NC}}y_ib_i \tag{2.206}$$

The EOS can be rearranged in the form of compressibility factor as:

$$z_{\mathrm{m}}^3-z_{\mathrm{m}}^2+\left(A_{\mathrm{m}}-B_{\mathrm{m}}-B_{\mathrm{m}}^2\right)z_{\mathrm{m}}-A_{\mathrm{m}}B_{\mathrm{m}}=0 \tag{2.207}$$

with A_{m} and B_{m} defined as:

$$A_{\mathrm{m}}=\frac{a_{\mathrm{m}}P}{(RT)^2} \quad B_{\mathrm{m}}=\frac{b_{\mathrm{m}}P}{RT} \tag{2.208}$$

This polynomial of the third degree in z_{m} can be solved numerically obtaining three roots. The highest root corresponds to the mixture compressibility factor related to the gas phase. Finally, with the value of the compressibility factor, the fugacity coefficient for each component in the gaseous mixture can be calculated as:

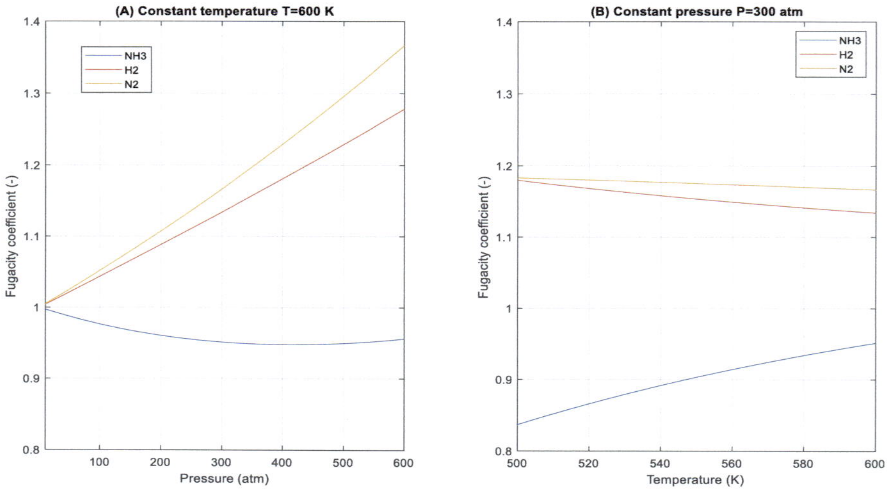

Fig. 2.13 Fugacity coefficients of the components of a gaseous mixture. (**a**) Calculation at constant temperature; (**b**) calculation at constant pressure

$$\ln \phi_i = \frac{b_i}{b_m}(z_m - 1) - \ln(z_m - B_m) - \frac{A_m}{B_m}\left(2\frac{a_i^{0.5}}{a_m^{0.5}} - \frac{b_i}{b_m}\right)\ln\left(1 + \frac{B_m}{z_m}\right) \quad (2.209)$$

The obtained results are reported in Fig. 2.13a, b.

The described results were obtained using a MATLAB program reported as Electronic Supplementary Material, Data 2.1.

Exercise 2.7. Determination of the Equilibrium Composition of Ammonia Synthesis Calculated Using the RKS-EOS

As previously seen, ammonia synthesis occurs through the following equilibrium reaction:

$$\frac{1}{2}N_2 + \frac{3}{2}H_2 \leftrightarrows NH_3 \quad (2.210)$$

The experimental data for the equilibrium constants at different temperatures are listed in Table 2.9:

For this reaction, the variation of Gibbs free energy can be expressed as a function of temperature by Eq. (2.211):

$$\begin{aligned} \Delta G_f = & -31.035^*X^*(1 - \ln(X)) - 25.341^*X^2/2 \\ & - 13.512^*X^3/6 + 13.148^*X^4/12 - 37.904 + 144.635^*X \end{aligned} \quad (2.211)$$

where $X = T/1000$.

Data for pure components are listed in Table 2.10.

Table 2.9 Equilibrium constants for ammonia synthesis from experimental data

T (K)	298	300	400	500	600	700	800	900	1000
$\ln K_p$	2.831	2.783	0.749	–0.522	–1.400	–2.044	–2.537	–2.928	–3.245

Table 2.10 Some thermodynamic properties of the pure components

Properties	NH_3	H_2	N_2
Critical temperature (K)	405	33	126
Critical pressure (atm)	115.3	13.2	34.6
Acentric factor	0.253	−0.21	0.037

Table 2.11 Experimental conversion data for ammonia synthesis under different conditions of pressure and temperature

Data set	P (atm)	101.32	202.65	303.97	405.3	506.62	607.95	709.27	810.6
T=617.15 K	Equilibrium conversion	56.71	70.28	77.47	82.15	85.51	88.06	90.04	91.63
T=713.15 K	Equilibrium conversion	30.39	45.51	55.26	62.32	67.77	72.15	75.77	78.81

Experimental data of the equilibrium conversion for this reaction are also available at different temperatures and pressure. In Table 2.11, two data sets, respectively, T=617.15 and 713.15 K, are reported as a function of pressure.

Build two plots: (1) one in which K_p is reported as a function of absolute temperature and (2) one, in the usual way, in which $\ln K_p$ is reported as a function of the inverse of temperature (between 300 and 1200 K). In both of these plots, also put in the experimental data from Table 2.9. Then build two other additional plots (T=617.15 K and T=713.5 K), in which the equilibrium conversion is reported as a function of the total pressure between 50 and 900 bar. In these plots, consider both the cases of ideal- and real-gas behavior described by the RKS-EOS. The first part of the exercise is easy to solve. The only relation that we need is the expression between ΔG_f, absolute temperature, and equilibrium constant, K_f:

$$\ln K_f = -\frac{\Delta G_f}{RT} \tag{2.212}$$

The other tasks of the first part consist only of simple data calculation and plotting.

For each of the required temperatures (617.15 and 713.15 K), the procedure is the same. Considering the initial and the final (equilibrium) state, the relation between the moles of each component and the reaction extent is the following:

$$n_i^{eq} = n_i^{\circ} + \nu_i \xi \quad \nu_i = \text{stoichiometric coefficient} \;\; \xi = \text{reaction extent} \tag{2.213}$$

The equilibrium relation to be solved for reaction extent is then:

$$K_{\mathrm{f}} = K_{\mathrm{y}} K_{\phi} P^{\sum \nu_i} \quad K_{\mathrm{y}} = \frac{y_{\mathrm{NH_3}}}{y_{\mathrm{N_2}}^{0.5} y_{\mathrm{H_2}}^{1.5}} \quad K_{\phi} = 1(\text{ideal gas}) \quad K_{\phi} = \prod_j \phi_j^{\nu_j}(\text{real gas}) \tag{2.214}$$

Fugacity coefficients for describing real-gas behavior can be evaluated by using the RKS-EOS. The RKS-EOS, and related parameters for the pure components, are summarized in the following expressions:

$$P = \frac{RT}{V - b_i} - \frac{a_i}{V(V + b_i)} \tag{2.215}$$

$$a_i = 0.42748 \frac{\alpha_i (RT_{Ci})^2}{P_{Ci}} \tag{2.216}$$

$$b_i = 0.8664 \frac{(RT_{\mathrm{Ci}})}{P_{\mathrm{Ci}}} \tag{2.217}$$

$$m_i = 0.480 + 1.574\omega_i - 0.17\omega_i^2 \tag{2.218}$$

$$\alpha_i = \left[1 + m_i\left(1 - T_{\mathrm{Ri}}^{0.5}\right)\right]^2 \tag{2.219}$$

The parameters for the mixture can be evaluated through suitable mixing rules, such as, for example:

$$a_{\mathrm{m}} = \sum_{i=1}^{\mathrm{NC}} \sum_{j=1}^{\mathrm{NC}} y_i y_j \sqrt{a_i a_j} \tag{2.220}$$

$$b_{\mathrm{m}} = \sum_{i=1}^{\mathrm{NC}} y_i b_i \tag{2.221}$$

The EOS can be rearranged in the form of compressibility factor as:

$$z_{\mathrm{m}}^3 - z_{\mathrm{m}}^2 + \left(A_{\mathrm{m}} - B_{\mathrm{m}} - B_{\mathrm{m}}^2\right) z_{\mathrm{m}} - A_{\mathrm{m}} B_{\mathrm{m}} = 0 \tag{2.222}$$

with A_{m} and B_{m} defined as:

$$A_{\mathrm{m}} = \frac{a_{\mathrm{m}} P}{(RT)^2} \quad B_{\mathrm{m}} = \frac{b_{\mathrm{m}} P}{RT} \tag{2.223}$$

This polynomial of the third degree in z_{m} can be solved numerically obtaining three roots. The highest root corresponds to the mixture compressibility factor related to the gas phase. Finally, with the value of the compressibility factor, the fugacity coefficient of each component in the gaseous mixture can be calculated using the expression:

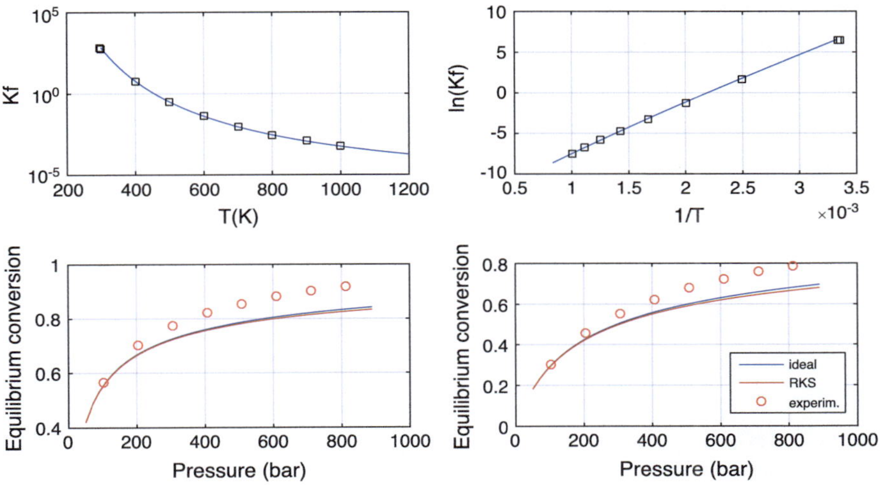

Fig. 2.14 Plots obtained with the calculations described in Exercise 2.7

$$\ln \phi_i = \frac{b_i}{b_\mathrm{m}}(z_\mathrm{m} - 1) - \ln(z_\mathrm{m} - B_\mathrm{m}) - \frac{A_\mathrm{m}}{B_\mathrm{m}}\left(2\frac{a_i^{0.5}}{a_\mathrm{m}^{0.5}} - \frac{b_i}{b_\mathrm{m}}\right)\ln\left(1 + \frac{B_\mathrm{m}}{z_\mathrm{m}}\right) \quad (2.224)$$

The plots obtained with the calculations are shown in Fig. 2.14.

It is interesting to observe that the RKS-EOS, which is successfully employed for describing VLEs under subcritical conditions, is not accurate in this type of calculation because it does not reproduce well the molar volume. A third parameter, as suggested by Peneloux (1982) and by Bertucco et al. (1995), is necessary (Eq. 2.159) to obtain greater accuracy.

The described results were obtained using a MATLAB program reported Electronic Supplementary Material, Data 2.1.

2.3.7 *Evaluation of Critical Parameters with Semiempirical Methods*

When the critical parameters for a given substance are not known, it is possible to estimate them using some empirical methods. The most popular ones are the methods based on the groups contribution, such as the ones suggested, respectively, by Lydersen (1955), Ambrose (1978, 1980), Joback and Reid (1987), and Constantinou and Gani (1994).

Ambrose Method

In the Ambrose method, T_C, P_C, and V_C are estimated using the following relations:

$$T_C = T_B\left[1 + \left(1.242 + \sum \Delta_T\right)^{-1}\right] \tag{2.225}$$

$$P_C = M\left(0.339 + \sum \Delta_P\right)^{-2} \tag{2.226}$$

$$V_C = 40 + \sum \Delta_V \tag{2.227}$$

where T_C is expressed in Kelvin, P_C in bars, and V_C in cm^3/mol. T_B is the boiling point at 1 atm, and M is the molecular weight. Δ values are determined by adding the tabulated group contributions.

Joback Method

The equations, in this case, are:

$$T_C = T_B\left[0.584 + 0.965\sum \Delta_T - \left(\sum \Delta_T\right)^2\right]^{-1} \tag{2.228}$$

$$P_C = \left(0.113 + 0.0032 n_A - \sum \Delta_P\right)^{-2} \tag{2.229}$$

$$V_C = 17.5 + \sum \Delta_V \tag{2.230}$$

where T_C is expressed in Kelvin, P_C in bars, and V_C in cm^3/mol. T_B is the boiling point at 1 atm, and n_A is the number of atoms in the molecule. Δ values are determined by adding the tabulated group contributions.

Lydersen Method

In this case, the equations are:

$$T_C = T_B\left[0.567 + \sum \Delta_T - \left(\sum \Delta_T\right)^2\right]^{-1} \tag{2.231}$$

$$P_C = M\left(0.34 + \sum \Delta_P\right)^{-2} \tag{2.232}$$

$$V_C = 40 + \sum \Delta V \tag{2.233}$$

where T_C is expressed in Kelvin, P_C in bars, and V_C in cm^3/mol. T_B is the boiling point at 1 atm, and n_A is the number of atoms in the molecule. Δ values are determined by adding the tabulated group contributions reported in Appendix 1.

2.3.8 Chemical Equilibrium in Liquid Phase

For defining the equilibrium of the reactions in liquid phase, the following condition is always valid:

$$\sum_i \alpha_i \mu_i = 0 \tag{2.234}$$

and again, the formalism developed for expressing the chemical potential of Eqs. (2.66) and (2.88) is conserved by writing:

$$\mu_i = \mu_i^\circ + RT \ln \frac{f_i}{f_R} \tag{2.235}$$

However, it is convenient to choose as standard state for the liquids that of a pure substance at an opportune pressure, normally the atmospheric pressure or the vapor pressure of the pure component, because in this case the standard chemical potential depends only on the temperature. The fugacity for a liquid can be written as:

$$f_i = \gamma_i x_i f_i^0 \tag{2.236}$$

where γ_i is the "activity coefficient." It is easy to show that the following relationship can be written:

$$\gamma_i = \gamma_{ir} \exp \int_{P_r}^{P} \frac{\overline{V}_i}{RT} \mathrm{d}P \tag{2.237}$$

where γ_{ir} is independent of the pressure. Then we can write:

$$a_i = \frac{f_i}{f_i^0} = x_i \gamma_{ir} \Gamma_i = \text{activity} \tag{2.238}$$

where Γ_i reflects the importance of the pressure in affecting the activity; however, as commonly occurs, $\Gamma_i \simeq 1$; therefore:

$$a_i = x_i \gamma_{ir} \tag{2.239}$$

where x_i, the molar fraction, can be substituted more conveniently by the molarity (mol/L) or by the molality (mol/1000 g of solvent). In any case, the chemical potential is more usually written as:

$$\mu_i = \mu_i^o + RT \ln a_i \tag{2.240}$$

And by following the same procedure used before, we can write the equilibrium constant in liquid phase as:

$$K = \frac{a_M^m \; a_N^n}{a_A^a \; a_B^b} = \frac{x_M^m \; x_N^n}{x_A^a \; x_B^b} \cdot \frac{\gamma_M^m \; \gamma_N^n}{\gamma_A^a \; \gamma_B^b} \cdot \frac{\Gamma_M^m \; \Gamma_N^n}{\Gamma_A^a \; \Gamma_B^b} = K_x \; K_\gamma \; K_\Gamma = e^{-\Delta G^\circ / RT} \tag{2.241}$$

Considering that normally $K_\Gamma \simeq 1$, a rigorous determination of the equilibrium constant requires evaluation of the activity coefficients of each reactant and product γ_I^a. The activity coefficients express the deviation from the ideality of a component in a liquid mixture, that is, they suggest degree to which the activity is different from the concentration. The procedure for calculating the activity coefficients in liquid solutions will be described in more detail later in the text in a dedicated section.

2.3.9 Equilibrium Constants and the Reference Systems

The condition $\sum_i \alpha_i \; \mu_i = 0$, describing the chemical equilibrium, is independent of the chosen standard states. It is possible, therefore, to write an equilibrium constant also using nonhomogeneous concentration units. This concept can be better explained with an example. Consider the reaction of the synthesis of urea, starting from ammonia and carbon dioxide, dissolved in aqueous phase:

$$2\,NH_3(g) + CO_2(g) \leftrightarrow CO(NH_2)_2(sol) + H_2O(1) \tag{2.242}$$

The equilibrium conditions at 25 °C will be:

$$\begin{aligned}
\mu_{NH_3} &= \mu_{NH_3}^o + RT \ln \bar{f}_{NH_3} \\
\mu_{CO_2} &= \mu_{CO_2}^o + RT \ln \bar{f}_{CO_2} \\
\mu_{H_2O} &= \mu_{H_2O}^o + RT \ln \gamma_{H_2O} x_{H_2O} \\
\mu_{CO(NH_2)_2} &= \mu_{CO(NH_2)_2}^o + RT \ln \gamma_{urea} x_{urea} \\
\Delta G^\circ &= \mu_{urea}^o(m) + \mu_{H_2O}^o(x) - \mu_{CO_2}^o - 2\mu_{NH_3}^o = -3.200 \\
K_e &= \exp\left(-\frac{\Delta G^\circ}{RT}\right) = 222 \\
\mu_{urea} &+ \mu_{H_2O} - 2\mu_{NH_3} - \mu_{CO_2} = 0 \quad \text{Equilibrium condition} \\
K_e &= \frac{\gamma_{urea} m_{urea} \gamma_{H_2O} x_{H_2O}}{\bar{f}_{CO_2} \bar{f}_{NH_3}^2} \\
&= \exp\left(-\frac{\Delta G^\circ}{RT}\right) \quad \text{Equilibrium constant with non} \\
&\quad - \text{homogeneous concentration units}
\end{aligned}$$

Standard states	μ_i^o (cal/mol)
NH_3 (g) fugacity = 1	−3.976
CO_2 (g) fugacity = 1	−94.260
H_2O(l) pure liquid at 1 atm	−56.690
$CO(NH_2)_2$ molar solution ideal at 1 atm	−48.720

2.3.10 Heterogeneous Equilibrium

As stated previously, for defining a reached chemical equilibrium, we can write:

$$\text{Reaction} \quad \sum_{i=1}^{N} \alpha_1 M_i = 0 \tag{2.243}$$

$$\text{Equilibrium condition} \quad \sum_{i=1}^{N} \alpha_1 \ \mu_i \ = 0 \tag{2.244}$$

Considering the possibility that the substances from 1 to n are gaseous and from $n + 1$ to N are in a condensed phase, Eq. (2.173) can be rewritten as:

$$\sum_{i=1}^{n} \alpha_i \ \mu_i \quad + \quad \sum_{n=n+1}^{N} \alpha_i \ \mu_i \quad = \ 0 \tag{2.245}$$

$$RT \sum_{i=1}^{n} \ln p_i^{\alpha_i} + \sum_{i=1}^{n} \alpha_i \mu_i^o + \sum_{i=n+1}^{N} \alpha_i \mu_i = 0 \tag{2.246}$$

Define a partial equilibrium constant exclusively related to the gaseous phase K_p' for $i = 1, n$

$$-RT \ln K_\text{p}' \sum_{i=1}^{n} \alpha_i \mu_i^o + \sum_{i=n+1}^{N} \alpha_i \mu_i \tag{2.247}$$

The components in the condensed phase can be considered as pure components—in this case $\mu_\text{i} = \mu^\circ{}_i$ (chemical potential of a pure solid or a liquid at 1 atm.)—and then we have:

$$-RT \ln K_\text{p}' = \sum_{i=1}^{N} \alpha_i \mu_i^o = \Delta G_\text{T}^\circ \tag{2.248}$$

Therefore,

$$K_{\mathrm{p}}' = \prod_{i=1}^{n} p_i^{\alpha_i} \tag{2.249}$$

In conclusion, in the presence of solids, the equilibrium constant is determined by considering only the partial pressures of the gaseous components.

2.3.11 Dependence of the Chemical Equilibrium Constant on Temperature

We have seen that the chemical equilibrium constant is strictly related to $\Delta G°$, that is, the standard free energy of Gibbs. Starting from the following equation:

$$\Delta G° = \Delta H° - T\Delta S° \tag{2.250}$$

It is possible to derive from this the Gibbs–Helmholtz equation:

$$\frac{\mathrm{d}(\Delta G°/T)}{\mathrm{d}T} = -\frac{\Delta H°}{T^2} \tag{2.251}$$

and remembering that $K_{\mathrm{e}} = e^{-\Delta G°/RT}$, we can obtain the Vant' Hoff equation for the most general case:

$$\frac{\mathrm{d}\ln K_{\mathrm{e}}}{\mathrm{d}T} = -\frac{\Delta H°}{RT^2} \tag{2.252}$$

which gives the dependence of the equilibrium constant on the temperature provided that $\Delta H°$ and $\Delta S°$ are constant in the considered range of temperature. In fact, this relation can be used for small intervals of temperature where both $\Delta H°$ and $\Delta S°$ can be considered approximately constants.

More rigorously, we can start from the following relationship:

$$\left(\frac{\partial H_i}{\partial T}\right)_P = C_{\mathrm{P}_i} = \text{Molar thermal capacity} \tag{2.253}$$

where H_i is the molar enthalpy of i pure at temperature T.

By considering the standard conditions, it possible to write the following equation:

$$\frac{\mathrm{d}\Delta H°}{\mathrm{d}T} = \frac{\mathrm{d}\sum \alpha_i H_i°}{\mathrm{d}T} = \sum \alpha_i C_{\mathrm{P}_i}° \tag{2.254}$$

The dependence of the thermal molar capacity on temperature is normally expressed with empirical polynomial relations of the following type:

$$C^{\circ}_{\mathrm{p}_i} = a_i + b_i T + c_i T^2 + \cdots \tag{2.255}$$

with parameters a_i, b_i, c_i … being tabulated for each compound. It follows that:

$$\frac{\mathrm{d}\Delta H^{\circ}}{\mathrm{d}T} = \sum \alpha_i \left(a_i + b_i T + c_i T^2 + \cdots\right) \tag{2.256}$$

By integrating:

$$(\Delta H^{\circ})_{\mathrm{T}} = (\Delta H^{\circ})_{\mathrm{T}_0} + \left[\sum \alpha_i \left(a_i T + \frac{b_i T^2}{2} + \frac{c_i T^3}{3} \cdots\right)\right] \tag{2.257}$$

The dependence of the reaction heat on the pressure is null for ideal gases and very small for solids and liquids, but it cannot be neglected for real gases. Then, remembering that:

$$\frac{\mathrm{d}\ln K_{\mathrm{p}}}{\mathrm{d}T} = \frac{\Delta H^{\circ}}{RT^2} \tag{2.258}$$

with K_{p} being the thermodynamic equilibrium constant calculated by considering the partial pressures of the components involved in the reaction, we obtain:

$$\frac{R\mathrm{d}\ln K_{\mathrm{P}}}{\mathrm{d}T} = \frac{(\Delta H^{\circ})_{T_0}}{T^2} + \left[\sum \alpha_i \left(\frac{a_i}{T} + \frac{b_i}{2} + \frac{c_i}{3} T \ldots\right)\right] \tag{2.259}$$

The integration of this relation gives the dependence of K_{p} on temperature, for any temperature range, considering the integration constant C.

$$R\ln k_{\mathrm{P}} = C - \frac{(\Delta H^{\circ})_{T_0}}{T} + \sum \alpha_i \left(a_i \ln T + \frac{b_i T}{2} + \frac{c_i T^2}{6}\right) \tag{2.260}$$

Then, by integrating the relation $\frac{\mathrm{d}(\Delta G^{\circ}/T)}{\mathrm{d}T} = -\frac{\Delta H^{\circ}}{T^2}$ or remembering that: $-\mathrm{RT}\ln K_{\mathrm{P}} = \Delta G^{\circ}_{\mathrm{T}}$ we obtain:

$$\Delta G^{\circ}_{\mathrm{T}} = \sum \alpha_i \mu_i^0 = -CT + \Delta H_0 - \sum \alpha_i \left(a_i T \ln T + \frac{b_i T^2}{2} + \frac{c_i T^3}{6}\right) \tag{2.261}$$

The standard free energy of formation, as well as the corresponding standard enthalpy and entropy at 25 °C and 1 atm of pressure for several compounds, have been tabulated in many textbooks and data banks. The values of these thermodynamic properties for the chemical elements are conventionally assumed to be equal to 0 in their physical state (fundamental state) at 25 °C and 1 atm.

As mentioned previously, the constants *a*, *b*, *c*, etc., for calculating the values of C_P are also tabulated for many compounds, but when the values are not available, there are calculation procedures for an approximate estimation of this parameter, which we will see in the next section.

2.3.12 Estimation of Thermodynamic Properties Starting from Molecule Structure

Two very important data can be derived from thermodynamic calculations: (1) the equilibrium condition, which means to know the equilibrium composition at a given temperature, and (2) the pressure and the heat required or released by the reaction from the initial to the equilibrium condition. These calculations can be made provided that the thermodynamic data of each component of the reaction are known. For many organic and inorganic compounds, few thermodynamic data (free energies of formation, formation enthalpies, heat capacities, etc.) are available and tabulated. In addition, in some cases, data are not available at all. Equilibrium and enthalpy change calculations can be made, in these cases, by estimating the unknown thermodynamic properties with the help of empirical or semiempirical methods. In particular, two class of methods are used:

(1) Methods based on the bond energy of the molecules
(2) Methods based on group contributions

The first method consists of evaluating the formation heat of the compounds as the sum of the energy of the bonds appearing in the molecular structure. The method, if applied in the described way, is simple but not precise. Attempts to obtain greater precision require complicated and tedious calculations. For this reason, the methods based on the group contributions are usually the most popular. Different methods based on group contributions have been proposed by, respectively, Anderson et al. (1944), by Franklin (1949), and by Van Krevelen and Chermin (1951). The best approach is to use of a combination of some of the available methods to have the possibility of evaluating all the properties necessary for the calculations.

2.3.13 Heat of Formation

The standard heats of formation, ΔH°_f, are the most commonly available data reported in many textbooks. The standard condition is normally referred to 298 K (25 °C). Different group contribution methods have been proposed for the calculation of $\Delta H^\circ_{f\ 298K}$. A good compromise between complexity and precision can be found in the method suggested by Verma and Doraiswamy (1965), the values of which are reported in Appendix 2. $\Delta H^{\circ}_{\mathrm{f}\ 298\mathrm{K}}$ determined with this method is expressed in kcal/mol. The standard enthalpy change, at a temperature different from 25 °C, can be calculated with the following relation:

$$\Delta H^{\circ}_{\mathrm{fT}} = \Delta H^{\circ}_{\mathrm{f298K}} + \int_{298}^{T} C^{\circ}_{p}\, dT \tag{2.262}$$

C°_{p} will be calculated as described in the next chapter.

2.3.14 Heat Capacity Calculation

The thermal capacity of ideal gases is a function of the temperature and different relations were proposed to describe this relation. The most commonly used is a polynomial of the following type:

$$C^{\circ}_{\mathrm{p}} = a + bT + cT^{2} + dT^{3} \ldots \tag{2.263}$$

The parameters—*a*, *b*, *c*, and *d*—are tabulated for many substances. Rihani and Doraiswamy (1965) gave the possibility to calculate these parameters by the additive group contributions, which is also reported in Appendix 2. The temperature is expressed in Kelvin, and the value of C°_{P} is calculated in calories/mol K. The calculation can be made for $T > 270$ K.

2.3.15 Gibbs Free Energy

Gibbs free energy, useful for evaluating equilibrium constants, can be estimated with the relationship:

$$\Delta G^{\circ}_{\mathrm{f}} = \Delta H^{\circ}_{\mathrm{f}} - \Delta S^{\circ}_{\mathrm{f}}\, T \tag{2.264}$$

An estimation method is the one suggested by Van Krevelen and Chermin (1951)—which considered $\Delta H^{\circ}_{\mathrm{f}}$ and $\Delta S^{\circ}_{\mathrm{f}}$ to be constant inside the two ranges of temperature—which are, respectively, 300–600 and 600–1500 K; therefore:

$$\Delta G^{\circ}_{\mathrm{f}} = A - BT \tag{2.265}$$

where *A* and *B* are calculated with the group contributions reported in Appendix 2. *T* is expressed in K and the result in kcal/mol. Some corrections must be made that are related to the symmetry of the compounds. The correction $R\ln\sigma$ must be added to the constant B, for example, $\sigma = 4$ for methane and $\sigma = 2$ for acetone. A further correction, $-R\ln\eta$, which is always to be added to B, must be made in the case of optical isomers, and η is the number of such isomers. The error in the evaluation of ΔG°_{f} with this method is $<\cong 5$ kcal/mol.

Exercise 2.8. Mass and Heat Balance for the Reaction $SO_2 + ½\ O_2 \rightarrow SO_3$
Consider the equilibrium of the following reaction:

$$SO_2 + \frac{1}{2}O_2 \leftrightarrows SO_3 \tag{2.266}$$

The equilibrium constant, K_p, is a function of the temperature according to the following equation:

$$\log K_p = \frac{4956}{T} - 4.678 \quad (T\,\text{in}\,\text{K}) \tag{2.267}$$

Part 1—Build a diagram in which the equilibrium conversion is reported as a function of the temperature between 300 and 1200 °C assuming an initial gas composition of 12 mol of SO_2, 10 mol of O_2, and 78 mol of N_2. Assume a total pressure of 1 atm.

Calculate then the equilibrium conversion and the adiabatic temperature reached, in a first stage, when a mixture with the previously reported initial composition is held at a temperature of 400 °C. Repeat the calculation for a second adiabatic stage in which the feed is represented by the outlet stream of the first stage cooled to 400 °C. Represent the calculation results as straight lines on the same previously constructed plot. Table 2.12 lists some thermodynamic properties of the components involved in the reaction.

Part 2—Build a plot in which the effect of total pressure on equilibrium curve is reported. Draw equilibrium curves correspondent to a total pressure of, respectively, 1, 5, and 10 atm.

For the material balance, Table 2.13 can be written, related to initial and equilibrium state, introducing the extent of reaction ξ.

Mole fractions and partial pressure can be calculated according to the following relations:

Table 2.12 Constants for determining specific heat of the involved components

Constant for specific heat	A ($\times 10^3$)	B ($\times 10^5$)	C ($\times 10^8$)	D ($\times 10^{12}$)
SO_2	38.91	3.904	−3.1040	8.606
O_2	29.10	1.158	−0.6076	1.311
SO_3	48.50	9.188	−8.540	32.40
N_2	29.00	0.2199	0.5723	−2.871

Heat of reaction: −100 kJ/mol
Note: Specific heat is expressed as $C_P = A + BT^2 + CT^3 + DT^3$ cal/(mol K)

Table 2.13 Initial moles and equilibrium composition as a function of the reaction extent

Components	Initial moles	Equilibrium
SO_2	12	12–ξ
O_2	10	10–0.5 ξ
SO_3	0	ξ
N_2	78	78
Total	100	100–0.5 ξ

$$y_i = n_i^{\mathrm{eq}}/n_{\mathrm{tot}}^{\mathrm{eq}} \quad P_i = y_i P_{\mathrm{tot}} \tag{2.268}$$

For the solution of part 1 of the exercise, it is sufficient to write a MATLAB code in which a loop of temperature from 300 to 1200 °C is implemented. At each temperature, a nonlinear equation, with ξ as unknown, must be solved:

$$f(\xi) = \frac{P_{\mathrm{SO_3}}}{P_{\mathrm{O_2}}^{0.5} P_{\mathrm{SO_2}}} - K_{\mathrm{P}} = 0 \tag{2.269}$$

Conversion of SO_2 can be calculated as:

$$x_{\mathrm{SO_2}} = \frac{\xi}{n_{\mathrm{SO_2}}^{o}} \tag{2.270}$$

For the solution of the second part of the exercise, material and energy balances must be solved simultaneously. These balances are represented by two equations in two unknowns as follows:

$$f_1(\xi, T_{\mathrm{eq}}) = \frac{P_{\mathrm{SO_3}}}{P_{\mathrm{O_2}}^{0.5} P_{\mathrm{SO_2}}} - K_{\mathrm{P}}(T_{\mathrm{eq}}) = 0 \tag{2.271}$$

$$f_2(\xi, T_{\mathrm{eq}}) = \xi \Delta H_{\mathrm{R}} + \sum_{i=1}^{N} n_i^{\mathrm{eq}} H_i^{\mathrm{eq}} - \sum_{i=1}^{N} n_i^{o} H_i^{o} = 0 \tag{2.272}$$

where H are the enthalpies of various components evaluated, respectively, at 400 °C and at T_{eq} as follows:

$$H_i^{o} = \int_{25}^{400} C_{\mathrm{p}}^{i}(T)\mathrm{d}T \quad H_i^{\mathrm{eq}} = \int_{25}^{T_{\mathrm{eq}}} C_{\mathrm{p}}^{i}(T)\mathrm{d}T \tag{2.273}$$

The results of the calculations for the first stage are:

$$\xi = 7.7564 \quad x_{\mathrm{eq}} = 0.6464 \quad T_{\mathrm{eq}} = 621.9257 \quad {}^{\circ}\mathrm{C} \tag{2.274}$$

and the following plot (Fig. 2.15) is obtained.

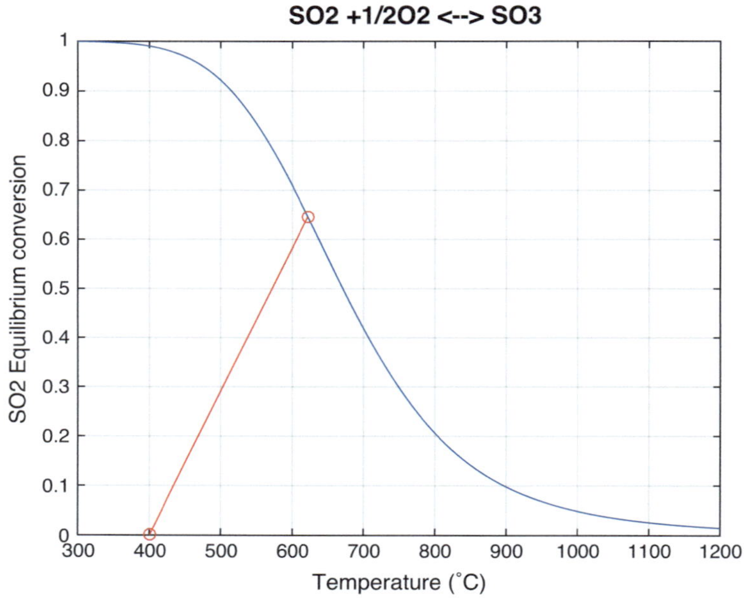

Fig. 2.15 Equilibrium conversion of SO_2 as a function of temperature. The diagonal straight lines correspond to the increase of temperature occurring in the two adiabatic stages starting from 400 °C and a fixed initial composition. The horizontal line corresponds to the cooling between the stages under adiabatic conditions. Blue curve is equilibrium behavior

Results

Stage	Extent of reaction (mol)	Conversion (–)	Equil. T (°C)
1	7.7564	0.6464	621.93
2	3.3559	0.9620	496.66

The plot required by part 2 of the exercise is depicted in Fig. 2.16. As expected, an increase in pressure, at a fixed temperature, corresponds to an increase in equilibrium conversion of sulfur dioxide.

The described results were obtained using a MATLAB program reported as Electronic Supplementary Material, Data 2.1.

2.3.16 Simultaneous Chemical Equilibria

If different reversible reactions occur simultaneously, determination of the equilibrium composition requires, first of all, to individuate what reactions can be considered thermodynamically independent. Then a system of algebraic equations with a

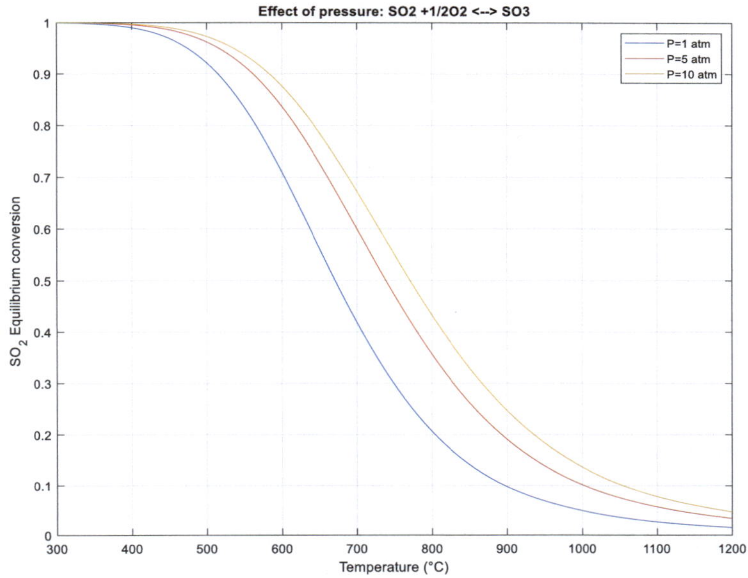

Fig. 2.16 Equilibrium conversion of SO_2 as a function of temperature. Effect of total pressure

number of unknowns equal to the number of the components involved in the equilibrium reactions must be solved.

To individuate, in a complex reaction system, the independent reactions, we can follow this procedure:

(1) Write the reaction of formation of each involved chemical species starting from the elements in their atomic state.
(2) Combine the reactions in such a way as to eliminate all the atomic species.
(3) Rewrite the obtained equations that are thermodynamically independent.

Example

Methane dehydrogenation.

In the equilibrium mixture, we have H_2, CH_4, C_2H_6, and C_3H_8.

According to the previously illustrated procedure, it is possible to write:

$$\begin{aligned} &2H \rightarrow H_2 \\ &C + 4H \rightarrow CH_4 \\ &2C + 6H \rightarrow C_2H_6 \\ &3C + 8H \rightarrow C_3H_8 \end{aligned} \tag{2.275}$$

As in our system, there are no elements in the atomic state, so we eliminate first the atomic hydrogen (e.g., 4H means $2H_2$) writing:

$$\begin{aligned} &C + 2H_2 \rightarrow CH_4 \\ &2C + 3H_2 \rightarrow C_2H_6 \\ &3C + 4H_2 \rightarrow C_3H_8 \end{aligned} \tag{2.276}$$

Then we eliminate the atomic carbon by multiplying all the components of the first reaction by 2 and subtracting it from the second. The same operation is performed by multiplying all the components of the first reaction by 3 and subtracting it from the third. Finally, we obtain:

$$2CH_4 \rightarrow C_2H_6 + H_2 \tag{2.277}$$

$$3CH_4 \rightarrow C_3H_8 + 2H_2 \tag{2.278}$$

These reactions are thermodynamically independent, thus giving the stoichiometry of the system. It is opportune to point out that these reactions allow to calculate the equilibrium composition, which is a thermodynamic constraint, but they give no information about the mechanisms and the rates with which methane is dehydrogenated because this is the domain of kinetics.

2.3.17 *An Example Calculation of Equilibrium Composition in a Complex System Characterized by the Presence of Multiple Reactions*

After determination of the thermodynamically independent reactions, it is possible to calculate the equilibrium composition by solving the corresponding mass balance equation system. This calculation seems simple to sketch out; however, in the presence of two or more simultaneous equilibria, the systems obtainable are highly nonlinear and therefore soluble only by adopting a numerical procedure. Let us consider, for example, the production of synthesis gas (a mixture of $CO + H_2$) from methane. Assuming that under the reaction conditions the formation of elemental carbon is negligible, the thermodynamically independent reactions are:

$$CH_4 + H_2O \rightleftarrows CO + 3H_2 \tag{2.279}$$

$$CO + H_2O \rightleftarrows CO_2 + H_2 \tag{2.280}$$

The determination of the equilibrium composition requires to know the number of moles of, respectively, $n_{CH_4}, n_{H_2O}, n_{CO}, n_{H_2}, n_{CO_2}$, that is, five components.

If we start from 1 mol of CH_4 and 4 mol of H_2O, the following balance can be written:

$n_{CH_4} = 1 - x \quad x = \text{moles of } CH_4 \text{ reacted in the first reaction}$

$n_{H_2O} = 4 - x - y \quad y = \text{moles of } CO_2 \text{ reacted in the reaction 2}$

$n_{CO} = x - y$

$n_{H_2} = 3x + y$

$n_{CO_2} = y$

The two equilibrium expressions are:

$$K_{P1} = \frac{p_{CO} p_{H_2}^3}{p_{CH_4} p_{H_2O}} = \frac{n_{CO} n_{H_2}^3}{n_{CH_4} n_{H_2O}} \left(\frac{P_{TOT}}{\sum n_i} \right)^2 \tag{2.281}$$

$$K_{P2} = \frac{p_{CO_2} p_{H_2}}{p_{CO} p_{H_2O}} = \frac{n_{CO_2} n_{H_2}}{n_{CO} n_{H_2O}} \tag{2.282}$$

$$\text{Total moles} = \sum n_i = 1 - x + 4 - x - y + x - y + 3x + y + y = 5 + 2x \tag{2.283}$$

By substituting:

$$K_{P1} = \frac{(x - y)(3x + y)^3}{(1 - x)(4 - x - y)} \left(\frac{P_{TOT}}{5 + 2x} \right)^2 \tag{2.284}$$

$$K_{P2} = \frac{y(3x + y)}{(x - y)(4 - x - y)} \tag{2.285}$$

As can be seen, we have only two equations with two unknowns x and y, but the first equation is of the fourth order in y and can be easily solved only by a numerical procedure.

Exercise 2.9. Calculation of the Composition for Multiple Equilibria
Methane steam reformation and water–gas shift equilibrium are represented by the following reactions:

$$CH_4 + H_2O \rightleftarrows CO + 3H_2 \tag{2.286}$$

$$CO + H_2O \rightleftarrows H_2 + CO_2 \tag{2.287}$$

The equilibrium constants for the two reactions can be calculated by the following two relations as a function of temperature:

$$\ln K_{p1} = 30.420 - \frac{27,106}{T} \tag{2.288}$$

$$\ln K_{p2} = 34.218 - \frac{31,266}{T} \tag{2.289}$$

- Part 1: Calculate the conversion of reactions 1 and 2 starting from 1 mol of methane and 4 mol of water. Assume a temperature of 900 K.

- Part 2: Repeat the calculation as in part 1 but varying the temperature between 600 and 1200 K and producing a plot of the conversion of the two reactions as a function of the temperature in the assigned range.

For the solution of the first part, we must evaluate the partial pressures of each component under equilibrium conditions:

$$\begin{aligned}
&n_{\mathrm{CO}}^{\mathrm{eq}} = n_{\mathrm{CO}}^{o} - x \quad n_{\mathrm{H_2O}}^{\mathrm{eq}} = n_{\mathrm{H_2O}}^{o} - x \quad n_{\mathrm{H_2}}^{\mathrm{eq}} = n_{H_2}^{o} + x n_{\mathrm{CO_2}}^{\mathrm{eq}} = n_{\mathrm{CO_2}}^{o} + x \\
&n_{\mathrm{T}}^{\mathrm{eq}} = n_{\mathrm{CO}}^{\mathrm{eq}} + n_{\mathrm{H_2O}}^{\mathrm{eq}} + n_{\mathrm{H_2}}^{\mathrm{eq}} + n_{\mathrm{CO_2}}^{\mathrm{eq}} \\
&y_i = n_i^{\mathrm{eq}} / n_{\mathrm{T}}^{\mathrm{eq}} \\
&P_i = P_{\mathrm{tot}} y_i
\end{aligned} \tag{2.290}$$

The equilibrium relation, that is, the function of the only unknown x, becomes:

$$\frac{P_{\mathrm{H_2}} P_{\mathrm{CO_2}}}{P_{\mathrm{CO}} P_{\mathrm{H_2O}}} - K_{\mathrm{p}} = 0 \tag{2.291}$$

This function can be solved numerically obtaining the value of x.

- **Solutions**

First reaction = 0.4533
Second reaction = 0.2488
The solution of the second part allows to obtain the plot shown in Fig. 2.17.
The described results were obtained using a MATLAB program reported as Electronic Supplementary Material, Data 2.1.

2.3.18 *Influence of Operative Conditions on the Yields of a Process: A Qualitative Approach*

The yields of a process can be affected by the following factors:

(a) Temperature
(b) Pressure
(c) Presence of an inert gas
(d) The law of mass action

- ***Effect of temperature on the yields of a process***

The equilibrium constant depends only on temperature through the enthalpy of the reaction. If we increase the temperature, the formation of the product is favored if the reaction is endothermic; the opposite occurs if the reaction is

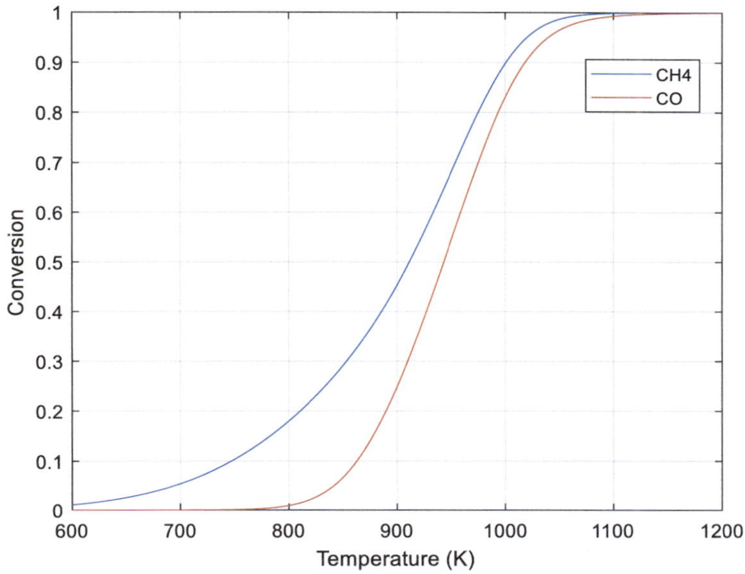

Fig. 2.17 Evolution of the reaction extent with the temperature for, respectively, CH_4 and CO

exothermic. The obtainable composition for any temperature can rigorously be obtained by applying the previously described equations.

- ***Effect of pressure on the yields of a process***

  Pressure influences yields only in the case of reactions occurring between reactants in gaseous phase when a change in the overall number of moles occurs and the system is far from the behavior of ideal gas. In this case, the equilibrium constant can be written as:

$$K_e = K_f K_y \left(\frac{P_{TOT}}{\sum n_i} \right)^{\Delta n} \tag{2.292}$$

where $K_f = \prod_i \varphi_i^{\alpha_i}$ is the ratio of the fugacity coefficients elevated at their stoichiometric coefficients and $K_y = \prod_i y_i^{\alpha_i}$ is the ratio of the molar fractions elevated at their stoichiometric coefficients. Pressure can exert influence on both the value of K_f and the term $P_{TOT}^{\Delta n}$ in the case of $\Delta n \neq 0$. Clearly, for the remaining constant K_e, the composition must change because K_y changes when K_f and/or $P_{TOT}^{\Delta n}$ change. The reactions occurring with a volume contraction will be favored by pressure; the opposite occurs with an increase of the overall number of moles.

- ***Effect of an inert gas***

  An inert gas changes the number of total moles, and—as in the previous case—the equilibrium composition can be affected only for reactions occurring between reactants in gaseous phase when a change in the overall number of moles occurs and the system is far from the behavior of ideal gas. By considering Eq. (2.292), it can be seen that the effect of the inert gases is that of modifying the value of $\sum n_i$ and the composition changes only when $\Delta n \neq 0$.
- ***Effect of components on the equilibrium composition: the law of mass action***

  An excess of one of the reactants shifts the equilibrium toward the formation of products (law of mass action) . A similar effect can be obtained by subtracting a product from the equilibrium.

2.3.19 *Thermodynamics of Some Hydrocarbons Transformations*

We have already seen the importance of thermodynamics for evaluating the best operative conditions for many different reactions of industrial interest, such as ammonia and methanol synthesis or the production of synthesis gas. However, thermodynamics is fundamental for some chemical processes. This occurs when the kinetics and eventual mass transfer for the occurring reaction are very fast. In this case, the reaction goes rapidly toward thermodynamic equilibrium, and the knowledge of the thermodynamic data is just what occurs for modeling the reactor. For example, many hydrocarbon transformations occur at very high temperatures, and the reactions are ruled by the thermodynamic properties of the involved molecules. In these cases, for having a general idea of the subject, it is useful to view a diagram introduced by Francis (1928) giving the standard free energy of formation referred to any g-atom of carbon for different temperatures (see Rossini et al. 1953; Guéret et al. 1997), as reported in Fig. 2.18. By observing this figure, it is possible to note, first of all, that almost all the hydrocarbons are less stable than the elements, carbon and hydrogen, whose standard free energy of formation is conventionally assumed to be equal to zero. This means that any hydrocarbon transformation has the problem of forming carbon black, which can be prevented by limiting as much as possible the rate of this thermodynamically favored reaction.

Another observation is that alkenes have slopes more minor than the corresponding alkanes: This means that alkenes are more stable than alkanes at higher temperature. This suggests a way to produce ethylene, propylene, butylenes, and butadiene by the cracking of alkanes.

Looking at the plots shown in Fig. 2.18, if we want to produce, for example, propylene from propane, we must operate at temperature much >1100 K, which is the point at which the two curves intersect. We must operate at temperature much >1100 K to achieve a high reaction rate that rapidly reaches equilibrium. Then the

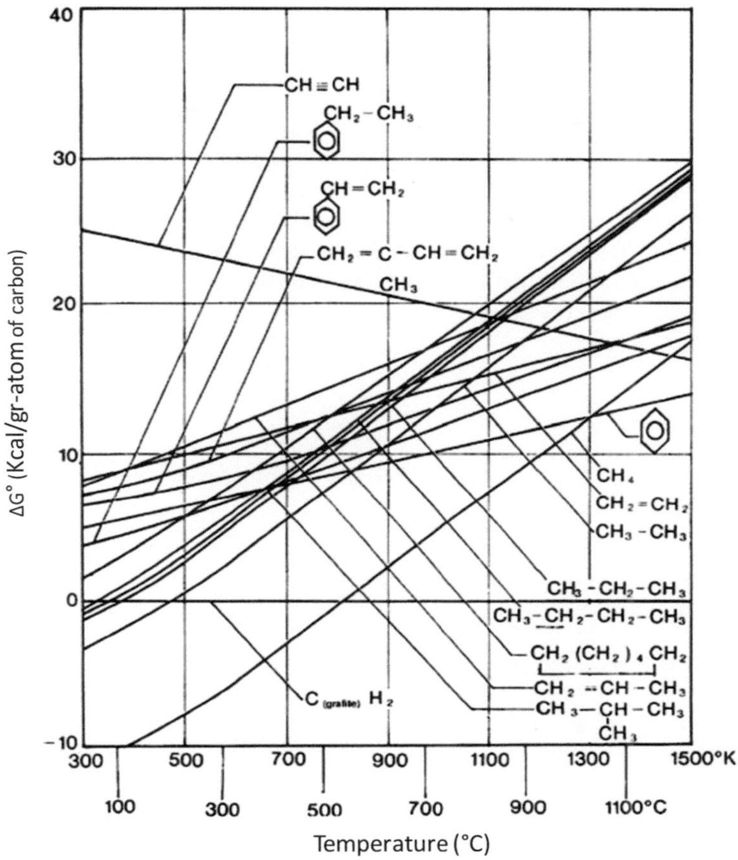

Fig. 2.18 Variation with temperature of the formation of standard free energy for different hydrocarbons expressed as kcal/g-atom of carbon (Francis diagram). Re-elaborated with permission from Guéret et al. (1997). Copyright Elsevier (1997)

products are rapidly quenched to limit the formation of carbon black. The procedure also is the same for producing ethylene from alkanes of higher molecular weight or for the production of acetylene. Acetylene, in particular, is more stable than other hydrocarbons at very high temperatures. If we want to produce acetylene from methane, we must operate at a temperature >1450 K, while starting from butane, the temperature of stability inversion is 1120 K. However, the higher the temperature, the higher the equilibrium constant and the rate of acetylene formation. In fact, the temperatures employed in industry are >1500 °C, and the reaction occurs in a few milliseconds.

2.3.20 Procedures for Calculating the Components Activities of a Liquid-Phase Mixture and Related Coefficients

The activity coefficients give information about the deviation from ideality of the concentration of a component in a liquid mixture as a consequence of the interaction between the molecules. To find a correlation between the activity coefficient and the molecular interactions, Hildebrand and, independently, Scatchard (1949) studied the behavior of the "regular solutions" (see Hildebrand et al. 1970). Regular solutions are constituted by liquid mixtures of noionic, nonpolar compounds and have comparable dimensions. For these solutions, Hildebrand suggested the following relation:

$$RT \ln \gamma_i = V_i \left(\delta_i - \overline{\delta}\right)^2 \tag{2.293}$$

where V_i is the molar volume of i and δ_i is the solubility parameter (characteristic of each component). It can be determined as:

$$\delta_i = \sqrt{\frac{\lambda_{\mathrm{ev}_i}}{V_i}} \tag{2.294}$$

where $\lambda_{\mathrm{ev}i}$ is the cohesion energy of $i \simeq$ heat of vaporization. Then, we also can write:

$$\overline{\delta} = \sum_i \phi_i \delta_i \quad \phi_i = \frac{x_i V_i}{\sum_i x_i V_i} = \text{volumetric fraction} \tag{2.295}$$

Another useful example of activity coefficient calculation is the one developed by Debye and Hückel (1923) for ion solution. The activity, in this case, is usually expressed in "molality" (moles of solute/in 1000 g of solvent). Molality has the advantage of being independent of temperature. Hence:

$$a_i = \gamma_i m_i \tag{2.296}$$

Debye–Huckel derived the following expression:

$$\ln \gamma_1 = \frac{Z_i^2 e^2 \beta}{2\varepsilon kT(1 + ka)} \tag{2.297}$$

where a is the diameter of the ion, Z is the dielectric constant of the solvent, e is the charge of the electron, and k is the Boltzmann constant. Then:

$$\beta = \sqrt{\frac{8\pi e^2}{\varepsilon kT}} \quad \text{with } I = \text{ionic strength} = \frac{1}{2}\sum_j n_j Z_j^2 \tag{2.298}$$

where n = ionic density = number of ions/volume.

More frequently, γ_i is calculated starting from the "free energy of excess" g^E. The significance of g^E can be argued from the following expression:

$$G_\text{M} = G_\text{id} + n_\text{T}\, g^E \tag{2.299}$$

that is, the free energy of the mixture, G_M, is the sum of the free energy of the ideal solution plus a value of excess. The relation between the activity coefficient and the excess free energy is:

$$\ln \gamma_i = \frac{1}{RT}\left(\frac{\partial n_\text{T} g^E}{\partial n_i}\right)_{T,P,n_j} \tag{2.300}$$

The excess free energy, g^E, can be considered the result of two different contributions, one enthalpic and the other entropic:

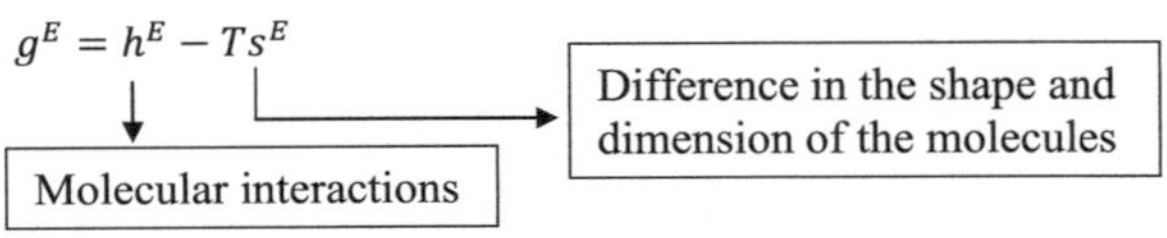

(2.301)

The methods of calculation for determining g^E were developed starting from two different limit approaches: One was originated by Hildebrand and which considers $s^E = 0$ (solution of components having molecules of similar size with nonspecific interactions), and the other was suggested by Flory (1942) and by Huggins (1941) and assumes $h^E = 0$ (athermic solutions).

In the first case, the excess free energy can be expressed as:

$$g^E = (x_1 V_1 + x_2 V_2)\,\phi_1\,\phi_2(\delta_1 - \delta_2)^2 \tag{2.302}$$

with:

$$\delta_i = \sqrt{\frac{\Delta H_{vi} - RT}{V_i}} \qquad \phi_i = \frac{x_i V_i}{V} = \frac{x_i V_i}{\sum_j x_j V_j} \tag{2.303}$$

where

δ_i	is the solubility parameter of i,
ΔH_{vi}	vaporization enthalpy,
ϕ_i	volume fraction of i,
x_i	molar fraction of i, and
V_i	molar volume.

It derives that:

$$\ln \gamma_{11} = \frac{V_1 \phi_2^2}{RT} (\delta_1 - \delta_2)^2 \quad \ln \gamma_{21} = \frac{V_2 \phi_1^2}{RT} (\delta_1 - \delta_2)^2 \tag{2.304}$$

The expression can be extended to a multicomponent solution as follows:

$$\ln \gamma_{i,m} = \frac{V_i}{RT} (\delta_1 - \delta_m)^2 \quad \delta_m = \sum_j \phi_j \delta_j \tag{2.305}$$

Margules (1895) and Van Laar (1910) introduced some semiempirical parameters with respect to the Scatchard–Hildebrand theory to obtain g^E, γ_1, and γ_2 through the expressions listed in Table 2.14.

In the approach followed by Flory–Huggins, we have:

$$g^E \cong -TS^E \quad S^E = -R\left(x_1 \ln \frac{\phi_1}{x_1} + x_2 \ln \frac{\phi_2}{x_2} \right) \tag{2.306}$$

Therefore:

$$g_E = RT\left(x_1 \ln \frac{\phi_1}{x_1} + x_2 \ln \frac{\phi_2}{x_2} \right) \tag{2.307}$$

For multicomponent systems:

$$g^E = \left(\sum_i x_i \ln \frac{\phi_i}{x_i} \right) RT \tag{2.308}$$

Wilson (1964) further developed the Flory–Huggins theory by introducing the concept of "**local composition.**" If we consider a binary mixture, for example, type 1 molecules could be surrounded mainly by type 2 molecules, the contrary could occur for the type 1 molecule. This determines a modification at a local level of the composition of the mixture (see Fig. 2.19).

Wilson interpreted any possible behavior of this type by introducing a probability factor as can be seen in the following relation:

$$\frac{p_{12}}{p_{11}} = \frac{x_2 e^{-\lambda_{12}/RT}}{x_1 e^{-\lambda_{11}/RT}} = \frac{\text{probability of interaction } 1-2}{\text{probability of interaction } 1-1} \tag{2.309}$$

where λ_{11} and λ_{12} are, respectively, the averaged potential energies of interaction of type 1–1 and type 1–2. The approach developed by Wilson, assuming implicitly the existence of the molecular interactions, clearly deviates from the Flory–Huggins

Table 2.14 Some different expressions derived from the Scatchard–Hildebrand model for determining activity coefficients in a binary mixture[a]

Author	Equations
Hildebrand	$RT\ \ln \gamma_1 = \widetilde{V}_1^1 \Phi_2^2 (\delta_1 - \delta_2)^2$ $\delta_1 = \sqrt{\frac{\Delta \widetilde{H}_{ev,i} - RT}{\widetilde{V}_i}}$ $RT\ \ln \gamma_2 = \widetilde{V}_2^1 \Phi_1^2 (\delta_1 - \delta_2)^2$
Van Laar	$\ln \gamma_1 = A_1 \Phi_2^2 = \frac{A_1 x_2^2}{(x_1 A_1/A_2 + x_2)^2}$ $\ln \gamma_2 = A_2 \Phi_1^2 = \frac{A_2 x_1^2}{(x_1 + x_2 A_2/A_1)^2}$
Margules	$\ln \gamma_1 = x_2^2 [A_1 + 2x_1 (A_2 - A_1)]$ $\ln \gamma_2 = x_1^2 [A_2 + 2x_2 (A_1 - A_2)]$

[a] A_1 and A_2 are empirical parameters. $\widetilde{V}_i^1$ is the molar volume at normal boiling point, and A_{ij} is an averaged value of the interaction energy of the molecules i and j. $\Phi_i = \frac{x_i \widetilde{V}_i}{\sum_j x_j \widetilde{V}_i}$

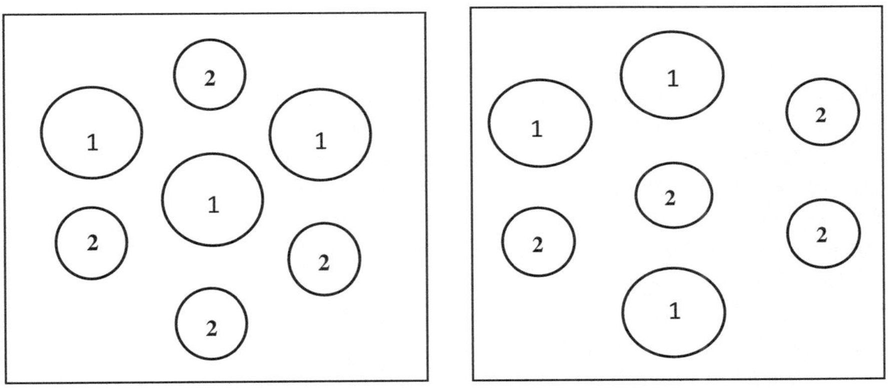

Fig. 2.19 Example of altered composition at a local level for a binary mixture

model and can be considered an intermediate model between the Flory–Huggins and Scatchard–Hildebrand models.

The volume fractions are calculated as local values:

$$\phi_1 = \frac{p_{11} V_1}{p_{11} V_1 + p_{12} V_2} \quad \phi_2 = \frac{p_{22} V_2}{p_{22} V_2 + p_{21} V_1} \tag{2.310}$$

and—according to Wilson—the excess free energy can be written as:

$$\frac{g^E}{RT} = -x_1\ \ln(x_1 + \Lambda_{12} x_2) - x_2\ \ln(\Lambda_{12} x_1 + x_2) \tag{2.311}$$

with:

$$\Lambda_{12} = \frac{V_2}{V_1} e^{-(\lambda_{12} - \lambda_{22})/RT} \quad \Lambda_{21} = \frac{V_1}{V_2} e^{-(\lambda_{12} - \lambda_{22})/RT} \tag{2.312}$$

For a multicomponent system, we can write:

$$\frac{g^E}{RT} = -\sum_i x_i \ln\left(\sum_j x_i \Lambda_{ij}\right) \tag{2.313}$$

with

$$\Lambda_{\text{ij}} = \frac{V_j}{V_i} e^{-\left(\lambda_{ij} - \lambda_{jj}\right)/RT} \tag{2.314}$$

Normally, $\lambda_{ij} = \lambda_{ji}$, but

$$\Lambda_{ij} \neq \Lambda_{ji} \tag{2.315}$$

The Wilson model improves the description of the vapor–liquid equilibria of binary and multicomponent mixtures with respect to the previously mentioned methods, but it is not able to foresee the occurrence of liquid–liquid separation.

Another theory, developed by Renon and Prausnitz (1968), called "nonrandom two-liquid theory" (NRTL), gives place to performances comparable with the Wilson model for what concerns the description of the VLE, but it is also able to predict the possibility of liquid–liquid separation for a given composition (miscibility gap). The NRTL method starts from the two-liquid theory developed by Scott (1956), according to which the property of a binary mixture can be described through the properties of two hypothetical fluids of particular characteristics. Some properties defined as "residual" must be defined first. These properties depend exclusively on the relative position of the molecules. A residual property, Y, for the two liquids theory is determinable by:

$$Y = x_1 Y^{(1)} + x_2 Y^{(2)} \tag{2.316}$$

If the molecules are positioned in the space in a preferred way, as seen, for example, in Fig. 2.19, the Gibbs free energy for type 1 ($g^{(1)}$) and type 2 cells ($g^{(2)}$) can be evaluated as:

$$g^{(1)} = x_{11} g_{11} + x_{21} g_{21} \tag{2.317}$$

$$g^{(2)} = x_{12} g_{12} + x_{22} g_{22} \tag{2.318}$$

where g_{11}, g_{22}, and g_{12} represent the interaction of, respectively, the type 1-1, 2-2, and 1-2, while x_{ij} is a local molar fraction. Obviously, it holds:

$$x_{12} + x_{22} = 1 \tag{2.319}$$

Moreover, for pure components, $x_{12} = x_{21} = 0$; therefore:

$$g_{\text{pure}}^{(1)} = g_{11} \quad \text{and} \quad g_{\text{pure}}^{(2)} = g_{22} \tag{2.320}$$

Then it is possible to write:

$$g^E = x_1\left(g^{(1)} - g_{\text{puro}}^{(1)}\right) + x_2\left(g^{(2)} - g_{\text{pure}}^{(2)}\right) \tag{2.321}$$

By substituting opportunely:

$$g^E = x_1 x_{12}(g_{21} - g_{11}) + x_2 x_{12}(g_{12} - g_{22}) \tag{2.322}$$

It is possible then to express the local molar fractions in a way similar to the one suggested by Wilson, that is,

$$\frac{x_{21}}{x_{11}} = \frac{x_2\, e^{-\alpha_{12}\, g_{21}/RT}}{x_1\, e^{-\alpha_{12}\, g_{11}/RT}} \tag{2.323}$$

$$\frac{x_{12}}{x_{22}} = \frac{x_1\, e^{-\alpha_{21}\, g_{12}/RT}}{x_2\, e^{-\alpha_{21}\, g_{22}/RT}} \tag{2.324}$$

where α_{12} and α_{21} represent the tendency of the mixture components to assume privileged configurations, not random ones; then it is generally assumed that $\alpha_{12} = \alpha_{21}$. If α_{12} and α_{21} are equal to zero, the mixing is random, and the local composition is equal to the overall composition. The expression for determining the excess free energy is, in this case:

$$g^E = RT\, x_1\, x_2\left(\frac{\tau_{21}\, G_{21}}{x_1 + x_2\, G_{21}} + \frac{\tau_{12}\, G_{12}}{x_2 + x_1\, G_{12}}\right) \tag{2.325}$$

where:

$$\tau_{12} = (g_{12} - g_{22})/RT \quad \tau_{21} = (g_{21} - g_{11})/RT \tag{2.326}$$

$$G_{12} = e^{-\alpha_{12}\, \tau_{12}} \quad G_{21} = e^{-\alpha_{21}\, \tau_{21}} \tag{2.327}$$

It can be assumed that $g_{12} = g_{21}$ and $\alpha_{12} = \alpha_{21}$.

The α parameter gives more flexibility to the NRTL method with respect to the Wilson model. For this reason, the NRTL method is also able to describe the occurrence and the extension of an eventual miscibility gap.

The equations, respectively, related to the Wilson and NRTL methods are listed in Table 2.15.

Table 2.15 Some different expressions derived from the Flory–Huggins model for determining the activity coefficients in binary mixtures

Wilson
$\ln\gamma_1 = -\ln(x_1 + \Lambda_{12}x_2) + x_2\left[\frac{\Lambda_{12}}{x_1+\Lambda_{12}x_2} - \frac{\Lambda_{21}}{\Lambda_{21}x_1+x_2}\right]$
$\ln\gamma_2 = -\ln(x_2 + \Lambda_{21}x_1) + x_1\left[\frac{\Lambda_{12}}{x_1+\Lambda_{21}x_2} - \frac{\Lambda_{21}}{\Lambda_{21}x_1+x_2}\right]$
$\Lambda_{ij} = \frac{\tilde{V}_j}{\tilde{V}_i} e^{-(\lambda_{ij}-\lambda_{ii})/RT}$
N.R.T.L.
$\frac{g^E}{RT} = x_1x_2\left(\frac{\tau_{21}G_{21}}{x_1+x_2G_{21}} + \frac{\tau_{12}G_{12}}{x_2+x_1G_{12}}\right)$
$\ln\gamma_1 = x_2^2\left[\tau_{21}\left(\frac{G_{21}}{x_2+x_1G_{21}}\right)^2 + \frac{\tau_{12}G_{12}}{(x_2+x_1G_{12})^2}\right]$
$\ln\gamma_2 = x_1^2\left[\tau_{12}\left(\frac{G_{12}}{x_2+x_1G_{12}}\right)^2 + \frac{\tau_{21}G_{21}}{(x_1+x_2G_{21})^2}\right]$

Another method for determining the activity coefficients, γ_i, always based on the determination of the local composition and on the Scott theory of two liquids, was developed by Abrams and Prausnitz (1975) and is called the "universal quasi chemical theory" (UNIQUAC). This theory considers ln γ_i to be the result of two contributions, one combinatorial dependent on the shape and size of the molecules and another residual depending on the energies of interaction. Therefore, it is possible to write:

$$\ln\gamma_i = \ln\gamma_i^C + \ln\gamma_i^R \tag{2.328}$$

The combinatorial term corresponds to:

$$\ln\gamma_i^e = \ln\frac{\vartheta_i}{x_i} + \frac{z}{2}q_i\ln\frac{\Theta_i}{\vartheta_i} + l_i - \frac{\vartheta_i}{x_i}\sum\nolimits_j x_j l_j \tag{2.329}$$

where:

$$l_i = \frac{z}{2}(r_i - q_i) - (r_i - 1) \tag{2.330}$$

$$\Theta_\mathrm{i} = \frac{q_i x_i}{\sum_j q_j x_j} = \text{volumetric fraction} \tag{2.331}$$

$$\vartheta_i = \frac{r_i x_i}{\sum_j r_j x_j} = \text{surface fraction} \tag{2.332}$$

z = the coordination number usually kept = 10.

The parameters related to the pure components r_i and q_i depend on the volume and molecular surface of Van der Waals. The residual contribution to γ_i can be calculated as:

$$\ln \gamma_i^R = q_i \left[1 - \ln \sum_j \Theta_j \tau_{ji} - \sum_j \frac{\Theta_j \tau_{ij}}{\sum_k \Theta_k \tau_{kj}} \right] \tag{2.333}$$

with

$$\tau_{jj} = \exp\left(u_{ij} - u_{ii}\right)/RT \tag{2.334}$$

$$u_{ij} = u_{ji} \quad \tau_{ji} \neq \tau_{ij} \tag{2.335}$$

$$u_{ij} - u_{ii} = \text{energy of interaction.} \tag{2.336}$$

In this model, τ_{ij} is an adjustable parameter that can be determined from the experimental data collected for binary mixture. The UNIQUAC method well reproduces the binary systems and also the miscibility gap. Moreover, starting from this method, another method—called "*uni*versal *f*unctional group *a*ctivity *c*oefficient model (UNIFAC) and based on group contribution—was developed by Fredenslund et al. (1975); it was later extended by Fredenslund et al. (1977). UNIFAC is a predictive method based on group contributions. The values of the group contributions were determined by mathematical regression analysis performed on thousands of experimental published data available in different data banks. Another predictive method, called "analytical solution of groups" (ASOG), was developed by Kojima and Tochigi (1979).

2.4 Calculations Related to Physical Equilibria

2.4.1 Physical Equilibria

In the previous sections, we examined the methods for calculating the fugacity and activity coefficients. These methods are useful, as seen before, for correct evaluation of the chemical equilibrium composition, but they are also useful—as it will be seen in the next sections—for the determination of physical equilibria, such as the vapor–liquid equilibria of single components or of mixtures containing two or more components. These equilibria are fundamental for describing separation units of the industrial plants, but they are also important for evaluating the partition of a component between liquid and vapor phase in a chemical reactor.

2.4.2 VLE of a Single Pure Component

Let us first consider the thermodynamic equilibrium for systems comprising just one component. In this case, the equilibrium is not affected by the chemical composition; therefore, the chemical potential can be neglected. Examples of this type of equilibrium are those considering phase changes, such as evaporation, condensation, melting, crystallization, sublimation, and allotropic change. However, we will consider here, as an example, only the vapor–liquid equilibrium.

If we consider that a pure component is partitioned between two different phases, α and β, as at equilibrium $\mathrm{d}G = 0$, we can write:

$$\Delta G = G_\beta - G_\alpha = 0 \text{ and } G_\beta = G_\alpha \tag{2.337}$$

By differentiating $\mathrm{d}G_\beta = \mathrm{d}G_\alpha$ and remembering that $\mathrm{d}G = V\mathrm{d}P - S\mathrm{d}T$, we can write:

$$V_\alpha \mathrm{d}P - S_\alpha \mathrm{d}T = V_\beta \mathrm{d}P - S_\beta \mathrm{d}T \tag{2.338}$$

$$\text{or}: \quad \frac{\mathrm{d}P}{\mathrm{d}T} = \frac{S_\beta - S_\alpha}{V_\beta - V_\alpha} = \frac{\Delta S}{\Delta V} \tag{2.339}$$

However, because $\Delta S = \frac{\Delta H}{T}$, and with ΔH as the heat absorbed or released for the change of state (heat of melting, of evaporation, of sublimation, etc.), we finally obtain the Clapeyron equation:

$$\frac{\mathrm{d}P}{\mathrm{d}T} = \frac{\Delta H}{T\Delta V} \tag{2.340}$$

In the case of vapor–liquid equilibrium, we can write, in particular:

$$\frac{\mathrm{d}p_i^\circ}{\mathrm{d}T} = \frac{\widetilde{\Delta H}_{\mathrm{ev}}}{T\left(\widetilde{V}^v - \widetilde{V}^l\right)}. \tag{2.341}$$

where $p^\circ{}_i$ is the vapor pressure. The liquid volume of a mole is negligible with respect to the corresponding volume of vapor, and we remember that for a perfect gas:

$$\widetilde{V}^v = \frac{RT}{P} \tag{2.342}$$

By introducing these relations in Eq. (2.341), the Clausius–Clapeyron equation is obtained:

$$\frac{\mathrm{d}\ln p_i^\circ}{\mathrm{d}T} = \frac{\widetilde{\Delta H}_{\mathrm{ev}}}{RT^2} \tag{2.343}$$

By assuming, as a first approximation, that the evaporation-enthalpy change is independent of temperature, by integrating, we obtain:

$$\log_{10} p_i^\circ = A - \frac{\widetilde{\Delta H}_{\text{ev}}}{2.303RT} \tag{2.344}$$

This relation is similar to the empirical relation known as the Antoine equation, which is largely applied in industry to evaluate the vapor pressure of many liquids at different temperatures:

$$\log_{10} p_i^\circ = A - \frac{B}{t(^\circ C) + C} \tag{2.345}$$

where the values of A, B, and C are tabulated for many substances.

A relation more rigorous than Eq. (2.344) can be obtained by assuming a polynomial dependence of the evaporation-enthalpy change on temperature:

$$\widetilde{\Delta H}_{\text{ev}} = a + bT + cT^2 \tag{2.346}$$

Substituting in Eq. (2.343) and integrating it results in:

$$\ln p_i^\circ = \alpha - \frac{\beta}{T} + \gamma \ln T + \delta T \tag{2.347}$$

Exercise 2.10. Vapor-Pressure Estimation of Coefficients for Antoine's Equation

Experimental data for methanol vapor pressure were collected as a function of temperature and are listed in Table 2.16.

Estimate the parameters for Antoine's equation by correlating the experimental data of Table 2.16.

- **Solution**

Antoine's equation is frequently used for describing the vapor pressure of a pure compound as a function of temperature.

$$\log_{10} P = A - \frac{B}{T + C} \tag{2.348}$$

where A, B, and C are adjustable parameters that must be determined by nonlinear fitting on experimental vapor-pressure data as a function of temperature. The problem is solved by using a nonlinear least square fitting routine obtaining the following results:

Table 2.16 Methanol vapor pressures at different temperatures

T (K)	273	283	293	303	313	323	333	343
P (bar)	0.024	0.051	0.102	0.194	0.351	0.607	1.004	1.596

$$A = 7.30;\, B = 2.5113 \times 10^3;\, \text{and} \;\; C = 10.9174.$$

The described results were obtained using a MATLAB program reported as Electronic Supplementary Material, Data 2.1.

2.4.3 Vapor–Liquid Equilibrium (VLE) for a Multicomponent System at Moderate Equilibrium Pressure

The equilibrium between the phases in a multicomponent system is established when the temperature, the pressure, and the chemical potential of each component is equal in all phases, that is:

$$T^{\alpha} = T^{\beta} = T^{\gamma} = \ldots \tag{2.349}$$

$$P^{\alpha} = P^{\beta} = P^{\gamma} = \ldots \tag{2.350}$$

$$\mu_i^{\alpha} = \mu_i^{\beta} = \mu_i^{\gamma} = \ldots \tag{2.351}$$

where α, β, and γ are the phases in equilibrium.

As already seen, the last equation also can be expressed in terms of fugacities:

$$f_i^{\alpha} = f_i^{\beta} = f_i^{\gamma} = \text{etc.} \tag{2.352}$$

The fugacity can be written as:

$$f_i^{\alpha} = f_{i,r}^{\alpha} x_i^{\alpha} \gamma_i^{\alpha} \tag{2.353}$$

where $f_{i,r}$ is the fugacity of i in a reference state, that could be, for example, assumed as the fugacity of the pure component at the same temperature of the mixture; x_i is the molar fraction of i in the mixture and γ_i is the activity coefficient of i in the mixture; and $\gamma_i^{\alpha} = 1$ corresponds to an ideal behavior. If we want to describe the vapor–liquid equilibrium, we must first write:

$$f_i^{v} = f_i^{1}, \text{and hence}: \quad f_{i,R}^{V} y_i \gamma_i^{V} = f_{i,R}^{1} x_i \gamma_i^{1} \tag{2.354}$$

The fugacity of the vapor can be expressed more conveniently as:

$$f_i^{V} = \phi_i P y_i \tag{2.355}$$

where ϕ_i = fugacity coefficient

$$\phi_i = f_{i,R}^V \gamma_i^V / P \tag{2.356}$$

If $\phi_i = 1$ and $\gamma_i^V = 1, f_{i,R}^V$ exactly corresponds to the pressure, P, and the system is ideal. This occurs for real gases at low pressure. As a matter of fact, we can write:

$$\lim \phi_i = 1 \quad \text{for} \quad P \to 0 \tag{2.357}$$

The dependence of ϕ_i on the pressure is described by the following relation:

$$\phi_1 = \exp \frac{1}{RT} \int_0^P \left(v_i^G - \frac{RT}{P} \right) \mathrm{d}P \tag{2.358}$$

The molar volume, V_i^G, can be determined with an EOS. However, for moderate pressures, it is convenient to use the virial equation for its simplicity.

The determination of the fugacity of the liquid first requires the definition of the reference state of each component i, $f_i^1 = x_i \gamma_i^1 f_{i,R}^1$. Considering the temperature, the most convenient reference state is that of the pure component at the same temperature of the mixture. Considering the pressure, the most opportune reference state is suggested by the Gibbs–Duhem equation:

$$\sum_i x_i \mathrm{d} \ln \gamma_i^1 = 0 \tag{2.359}$$

which is valid at P and T constant. This equation can be integrated if a relation $\gamma_i = \gamma_i (x_i)$ is known. As by varying the composition the pressure changes, too, the calculated values of γ must be re-conduced to a given value of pressure that could be the pressure of reference. This can be performed with the relationship:

$$\gamma_i^{(P^R)} = \gamma_i^{(P)} \exp \int_P^{P^R} \frac{\overline{V}_i^1}{RT} \mathrm{d}P \tag{2.360}$$

where V_i^1 is the partial molar volume of i in the liquid phase. By introducing this expression in that of the liquid fugacity, we obtain:

$$f_i^1 = x_i \gamma_i^{(P^R)} \quad f_{i,R}^1 \quad \exp \int_{P^R}^{P} \frac{\overline{V}_i^1}{RT} \mathrm{d}P \tag{2.361}$$

$f_{i,R}^1$ is the fugacity of the pure liquid i at T and P^{R} (reference pressure) equal to:

$$f_{i,R}^{1} = P_i^S \phi_i^S \exp \int_{P_i^S}^{P^R} \frac{V_i^1}{RT} \mathrm{d}P \tag{2.362}$$

where P_i^S is the vapor pressure of the pure component. This derives as a consequence of the symmetric assumption conventionally imposed $\gamma_i^{PR} \rightarrow 1$ for $x_i \rightarrow 1$. In general, it is assumed that $P^R = 0$ because it is at low pressure, γ_I, independent of this variable. It is then possible to write:

$$P^R = P_i^S \quad \text{and} \quad \overline{V}_i^1 = V_i^1; \quad \text{(therefore,)}$$

$$f_i^1 = \gamma_i^1 x_i P_i^S \varphi_i^S \exp \int_{P_i^s}^{P} \frac{V_i^1}{RT} \mathrm{d}P \tag{2.363}$$

where φ_i^S can be determined in the already described manner. The expression under the exponential function is defined as the Poynting correction and at low pressure is near unity. In conclusion, vapor–liquid equilibrium can be expressed in a complete way with the following relation:

$$y_i \varphi_i P = x_i \gamma_i P_i^S \varphi_i^S \exp \int_{P_i^S}^{P} \frac{V_i^1}{RT} \mathrm{d}P \tag{2.364}$$

where, instead of γ_i^1, it has been written γ_i. Therefore, we can also write:

$$K_i = \frac{y_i}{x_i} = \frac{\gamma_i P_i^S \phi_i^S \exp \int_{P_i^S}^{P} \frac{V_i^1}{RT} \mathrm{d}P}{\phi_i P} \tag{2.365}$$

At low pressure, the fugacity coefficients are approximately equal and can be eliminated; the Poynting correction can then be assumed to be equal to 1, and consequently, it results in:

$$y_i P = x_i \gamma_i P_i^S \quad K_i = \frac{y_i}{x_i} = \frac{\gamma_i P_i^S}{P} \tag{2.366}$$

At low pressure, the nonideality is normally restricted to the liquid phase and needs only the calculation of the activity coefficients γ_i. When $\gamma_i = 1$, the behavior is ideal and is characterized by the combination of the Dalton law ($P = \sum_1^N P_i$) and the Raoult law ($P_i = x_i P_i^s$). If we want to be more rigorous, the problem of vapor–liquid equilibrium, at moderate pressure and with the addition of the determination of γ_i for the liquid phase, also requires the determination of both the coefficient of fugacity, ϕ_I, for vapor phase and $\phi_{\mathrm{i}}^{\mathrm{S}}$ for the liquid one. This can be performed by using the virial EOS. For this purpose, we can write:

$$\ln \phi_i = \int_0^P \frac{\overline{z}_i - 1}{P} \mathrm{d}P \tag{2.367}$$

$$\overline{z}_i = \frac{P\overline{V}_i}{RT} = 1 + \frac{BP}{RT} \tag{2.368}$$

The values of B for a multicomponent mixture can be calculated using the following mixing rule:

$$B_{\mathrm{M}} = \sum_{i=1}^{m} \sum_{j=1}^{m} y_i y_j B_{ij} \tag{2.369}$$

and the binary coefficients B_{ij} can be calculated by using the correlation given by Hayden and O'Connel (1975). Finally, we have:

$$\ln \phi_i = \left(2 \sum_{j=1}^{m} y_j B_{ij} - B_{\mathrm{M}} \right) \frac{P}{RT} \tag{2.370}$$

The virial equation truncated at the second term can be employed for pressure $<$ 10–20 bars.

Exercise 2.11A. Example of Vapor–Liquid Equilibrium Calculation for a Binary Mixture at Low Pressure: System at Constant Temperature Considering the binary system of benzene and acetonitrile, the experimental vapor–liquid data at a constant temperature of 318.15 K are listed in Table 2.17, while Table 2.18 lists the Antoine equation parameters for the pure-component vapor pressures. Using the Margules equation for determining liquid-phase activity coefficients, estimate the interaction coefficients, A_{12} and A_{21}.

Build two standard plots for binary vapor–liquid equilibrium: X–Y and P–X–Y. These plots should report, for comparison, the experimental data of Table 2.17 and the continuous curves calculated by solving the equilibrium relations. Build an additional surface plot with the objective function behavior in dependence of the two adjustable parameters.

- **Solution**

Activity coefficients can be calculated for a binary system using the Margules equations as follows:

Table 2.17 Vapor–liquid experimental data of the binary system benzene and acetonitrile[a]

N	P (kPa)	x_1	y_1
1	28.851	0.0167	0.0484
2	32.651	0.1088	0.2251
3	34.784	0.2047	0.3241
4	36.157	0.3142	0.4013
5	37.010	0.4320	0.4854
6	37.157	0.4734	0.5038
7	37.117	0.5855	0.5639
8	36.664	0.6909	0.6330
9	35.810	0.7779	0.6982
10	33.131	0.9256	0.8469
11	31.251	0.9753	0.9318

[a]Data from Smith (1972)

Table 2.18 Antoine equation parameters for pure benzene and acetonitrile

Component	A	B	C
1 Benzene ([a])	6.89272	1203.531	219.888
2 Acetonitrile ([b])	4.27873	1355.374	–37.853

[a]P in mmHg and T in °C: $\log_{10}(P^{\circ}) = A - B/(T + C)$
[b]P in bar and T in K: $\log_{10}(P^{\circ}) = A - B/(T + C)$

$$\begin{aligned} \ln \gamma_1 &= x_2^2[A_{12} + 2(A_{21} - A_{12})x_1] \\ \ln \gamma_2 &= x_1^2[A_{21} + 2(A_{12} - A_{21})x_2] \end{aligned} \tag{2.371}$$

$$\text{Equilibrium relations are}: \quad y_i P = x_i \gamma_i P_i^{\circ} \tag{2.372}$$

$$\text{Total pressure is}: \quad P = x_1 \gamma_1 P_1^{\circ} + x_2 \gamma_2 P_2^{\circ} \tag{2.373}$$

A mathematical routine can be used to search for the minimum of the following objective function with respect to parameters A_{12} and A_{21}:

$$f_{\text{obj}}(A_{12}, A_{21}) = w_y \sum_{j=1}^{\text{Ns}} \left(y_{1i}^{\text{exp}} - y_{1i}^{\text{calc}}\right) + w_P \sum_{j=1}^{\text{Ns}} \left(P_i^{\text{exp}} - P_i^{\text{calc}}\right) \tag{2.374}$$

In the previous expression, w_y and w_p are, respectively, the weights for the contributions of mole fraction and pressure in the overall objective function. In our calculations, we adopted $w_y = 1$ and $w_p = 10$. The optimization result is given by $A_{12} = 0.9951$ and $A_{21} = 1.0613$ (value of the objective function 5.054e-07). Then, by solving equilibrium relation, it is possible to construct the plots of Figs. 2.20, 2.21, and 2.22.

These results can be obtained using a MATLAB calculation program using the "lsqnonlin" command to search for the minimum of the objective function as Eq. (2.374) with respect to parameters A_{12} and A_{21}, as can be seen in the plot of

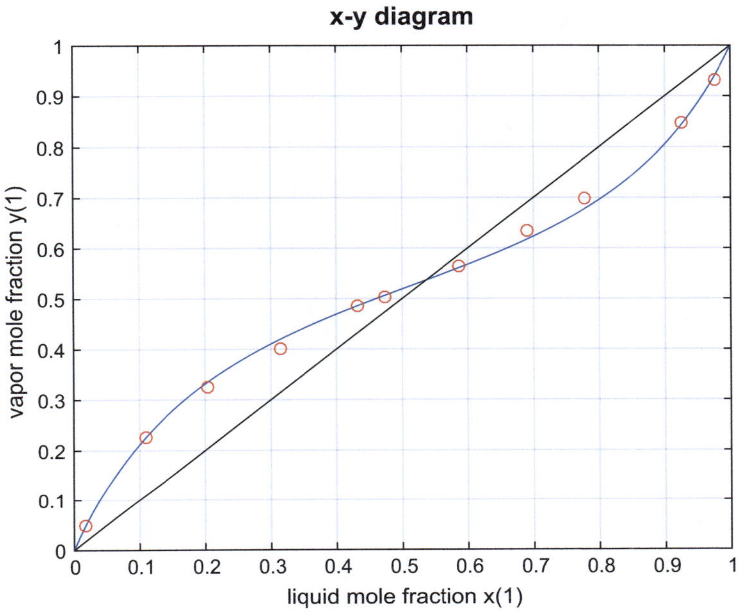

Fig. 2.20 *x–y* diagram for the binary mixture benzene and acetonitrile

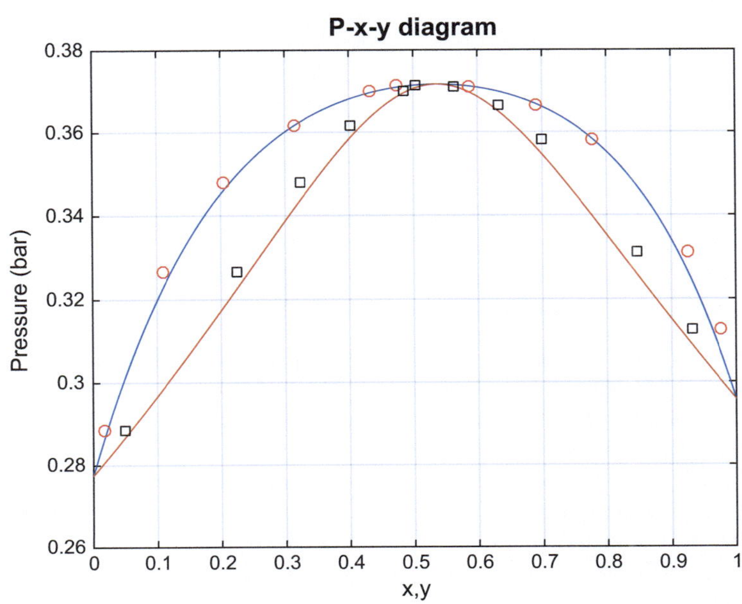

Fig. 2.21 *P–x–y* diagram for the binary mixture benzene and acetonitrile

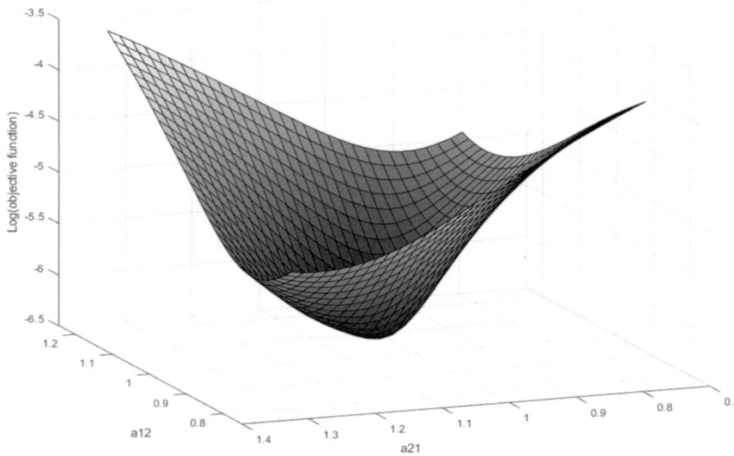

Fig. 2.22 Behavior of the objective function 2.369 as a function of the adjustable parameters A_{12} and A_{21}

Table 2.19 Vapor–liquid experimental data of the binary system benzene and acetonitrile[a]

N	T (K)	x_1	y_1
1	377.53	0.001	0.001
2	372.82	0.088	0.226
3	368.80	0.175	0.399
4	365.61	0.255	0.519
5	359.03	0.458	0.722
6	354.59	0.617	0.838
7	350.72	0.771	0.917
8	347.99	0.893	0.963
9	345.73	0.999	0.999

[a]Data from Dahal (2023)

Fig. 2.22 where a global minimum was found. The codes are present in the online version reported as Electronic Supplementary Material, Data 2.1.

Exercise 2.11B. Example of Vapor–Liquid Equilibrium Calculation for a Binary Mixture at Low Pressure: System at Constant Pressure
Let us consider the binary system made of toluene (1) and styrene (2) for which are available the experimental vapor–liquid data at a constant pressure of 225 mmHg (Dahal 2023). The experimental measurements are reported in Table 2.19, and in Table 2.20, the Antoine equation parameters for the pure-component vapor pressures are reported. Using the Margules equation for determining liquid-phase activity coefficients, estimate the interaction coefficients A_{12} and A_{21} in Eq. (2.371).

Build two standard plots for binary vapor–liquid equilibrium: X–Y and T–X–Y. These plots should report, for comparison, the experimental data of Table 2.18 and the continuous curves calculated by solving the equilibrium relations. Build an

Table 2.20 Antoine equation parameters for pure toluene and styrene

Component	A	B	C
1 Toluene ([a])	6.95805	1346.773	219.693
2 Styrene ([a])	7.06623	157.434	214.985

[a]P in mmHg and T in °C: $\log_{10}(P^{\circ}) = A - B/(T + C)$

additional contour plot with the objective function behavior in dependence of the two adjustable parameters.

- **Solution**

Activity coefficients can be calculated for a binary system using the Margules equations as already reported in relation 2.366 while the objective function to be minimized with respect to parameters A_{12} and A_{21} is now:

$$f_{\mathrm{obj}}(A_{12}, A_{21}) = w_y \sum_{j=1}^{\mathrm{Ns}} \left(y_{1i}^{\mathrm{exp}} - y_{1i}^{\mathrm{calc}}\right) + w_T \sum_{j=1}^{\mathrm{Ns}} \left(T_i^{\mathrm{exp}} - T_i^{\mathrm{calc}}\right) \tag{2.375}$$

In the previous expression, w_{y} and w_{T} are, respectively, the weights for the contributions of mole fraction and temperature in the overall objective function (2.375). In our calculations, we adopted $w_{\mathrm{y}} = 1$ and $w_{\mathrm{T}} = 1\mathrm{e}6$. The optimization result is given by $A_{12} = 0.0697$ and $A_{21} = 0.0507$ (value of the objective function 6.2109e+2). Then, by solving equilibrium relation, it is possible to construct the plots of Figs. 2.23, 2.24, and 2.25.

These results can be obtained using a MATLAB calculation program using the "lsqnonlin" command to search for the minimum of the objective function (Eq. 2.375) with respect to parameters A_{12} and A_{21}, as can be seen in the plot of Fig. 2.25 where a global minimum was found. The codes are present in the online version reported as Electronic Supplementary Material, Data 2.1.

Exercise 2.12A. Vapor–Liquid Equilibrium in a Multicomponent System: Dew Point Calculation

A vapor mixture, in which three components are present, is gradually cooled until a first liquid drop is formed at a total pressure of 300 mmHg. The molar composition of the vapor mixture is the following: 5% benzene (1), 40% toluene (2), and 55% di styrene (3). Antoine's constants for the considered components are listed in Table 2.21 (P in mmHg; T in °C).

- **Part 1.** By assuming an ideal behavior of both vapor and liquid phases, calculate the liquid composition and the temperature at which the condensation occurs (dew point temperature).
- **Part 2.** Repeat the calculation as in part 1 by fixing the vapor-phase mole fraction of component 1 (benzene) at 0.1 and changing the mole fraction of component 2 (toluene) from 0.1 and 0.8. The mole fraction of component 3 (styrene) is obtained by difference to unity.

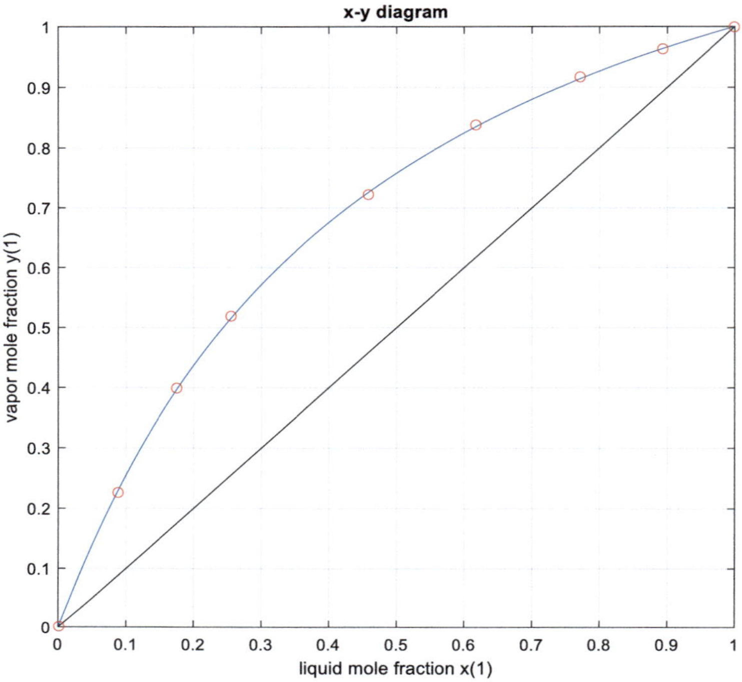

Fig. 2.23 *x–y* diagram for the binary mixture toluene and styrene

Build a plot in which the dew point temperature of the vapor mixture is reported as a function of mole fraction of component 2 (toluene) (note: $y_1 = 0.1$; $y_2 = 0.1{:}0.8$; $y_3 = 1 - y1 - y2$). Repeat the same sequence of calculations at the total pressure of 400 mmHg and draw the corresponding curve on the same plot.

- **Solution**

The problem of dew point temperature consists of solving a nonlinear function of the unknown temperature of the following type:

$$f(T) = \left(\sum_{i=1}^{3} x_i\right) - 1 = \left(\sum_{i=1}^{3} \frac{y_i P_{\text{tot}}}{P_i^o}\right) - 1 = 0 \tag{2.376}$$

Results Part 1 (Table 2.22)

The plot resulting by solving part 2 is reported in Fig. 2.26.

The described results were obtained using a MATLAB program reported as Electronic Supplementary Material, Data 2.1.

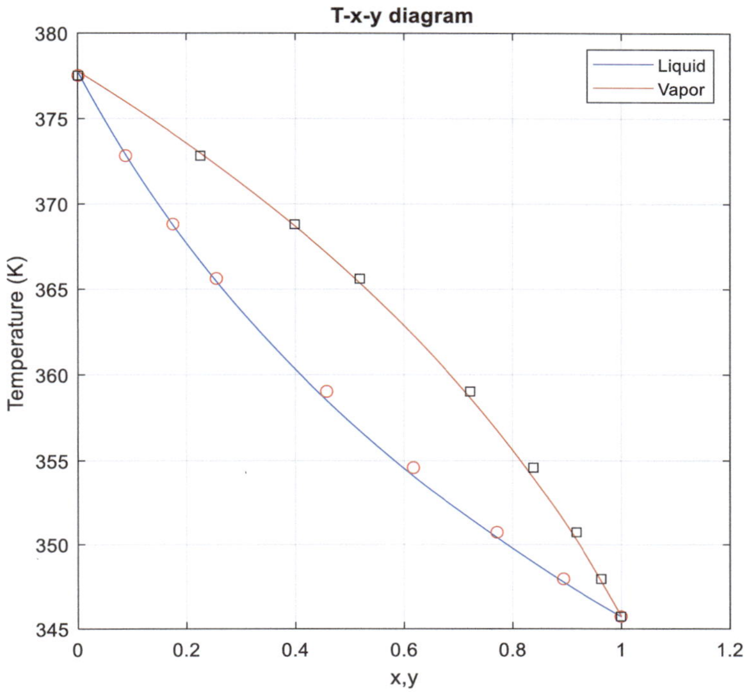

Fig. 2.24 T–x–y diagram for the binary mixture toluene and styrene

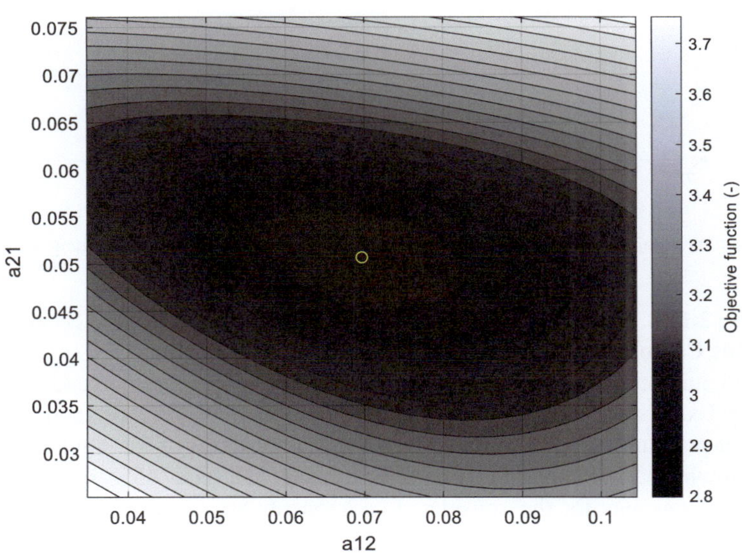

Fig. 2.25 Behavior of the objective function 2.370 as a function of the adjustable parameters A_{12} and A_{21}. The optimum value is indicated by the yellow circle

Table 2.21 Antoine's parameters[a]

Substance	*A*	*B*	*C*
Benzene	6.89272	1203.531	219.888
Toluene	6.95334	1343.943	219.377
Styrene	7.06623	1507.434	214.985

[a]P in mmHg and T in °C: $\log_{10}(P^o) = A - B/(T + C)$

Table 2.22 Results for the part 1 of Exercise 2.12A

Dew point – $P = 300$ mmHg	Mole fractions		
	Benzene	Toluene	Styrene
	0.0104	0.1994	0.7902
Dew point temperature = 102.61 (°C)			

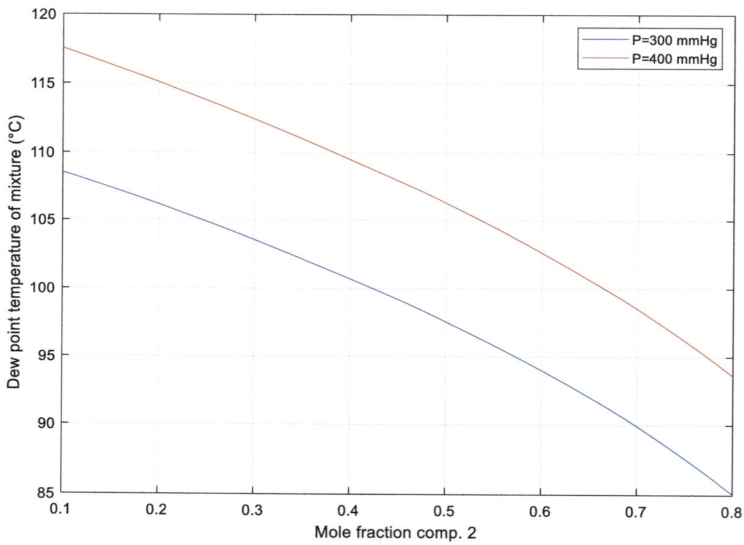

Fig. 2.26 Dew point of the mixture as a function of the mole fraction of component 2

Exercise 2.12B. Vapor–Liquid Equilibrium in a Multicomponent System: Bubble Point Calculation

A liquid mixture, in which are present the same components considered in the previous Exercise 2.12A, is gradually heated until a first amount of vapor is formed at a total pressure of 300 mmHg. The molar composition of the liquid mixture is the following: 20% benzene (1), 35% toluene (2), and 45% di styrene (3). Antoine's constants for the considered components are listed in Table 2.21 (P in mmHg; T in °C).

Table 2.23 Results for the part 1 of Exercise 2.12B

Bubble point – $P = 300$ mmHg	Mole fractions		
	Benzene	Toluene	Styrene
	0.5144	0.3468	0.1388
Bubble point temperature = 80.60 (°C)			

- **Part 1.** By assuming an ideal behavior of both vapor and liquid phases, calculate the vapor composition and the temperature at which the boiling occurs (bubble point temperature).
- **Part 2.** Repeat the calculation as in part 1 by fixing the vapor-phase mole fraction of component 1 (benzene) at 0.1 and changing the mole fraction of component 2 (toluene) from 0.1 and 0.8. The mole fraction of component 3 (styrene) is obtained by difference to unity.

Build a plot in which the bubble point temperature of the liquid mixture is reported as a function of mole fraction of component 2 (toluene) (note: $y_1 = 0.1$; $y_2 = 0.1{:}0.8$; $y_3 = 1 - y1 - y2$). Repeat the same sequence of calculations at the total pressure of 400 mmHg and draw the corresponding curve on the same plot.

Solution

The problem of bubble point temperature consists of solving a nonlinear function of the unknown temperature of the following type:

$$f(T) = \left(\sum_{i=1}^{3} y_i\right) - 1 = \left(\sum_{i=1}^{3} \frac{P_i^\circ x_i}{P_{\text{tot}}}\right) - 1 = 0 \tag{2.377}$$

Results Part 1 (Table 2.23)

The plot resulting by solving part 2 is reported in Fig. 2.27.

2.4.4 *The Equilibrium of Solubility at Moderate Pressures*

When in a mixture there are one or more incondensable gases if we maintain as before the symmetric norm $\gamma_i^{PR} \to 1$ for $x_i \to 1$, it is necessary to imagine the existence of a fictitious liquid in equilibrium with the incondensable gas because a pure liquid component, under the same conditions of pressure and temperature of the mixture, does not exist being that component under supercritical conditions. This procedure is sometime used when the considered component is close to the critical conditions. Vapor pressure and molar volume in the solution for this component will be determined by extrapolation, and it will be treated as a subcritical component.

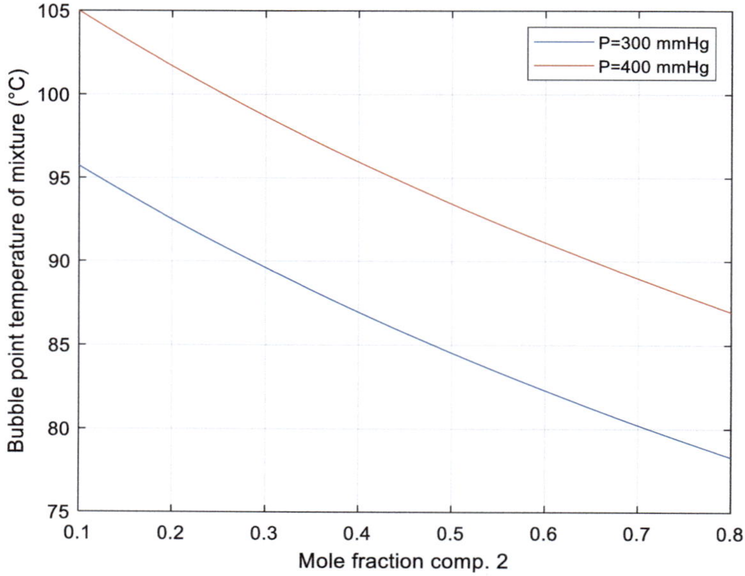

Fig. 2.27 Bubble point of the mixture as a function of the mole fraction of component 2

The alternative is to assume the asymmetrical norm, $\gamma_i^{PR} \rightarrow 1$ for $x_i \rightarrow 0$. In this case, $f_{i,R}^1$ corresponds to the solubility expressed as the Henry constant $H_{i,m}$, and we will have:

$$f_i^1 = \gamma^{*(P^R)} x_i H_{i,m}^{(P^R)} \exp \int_{P^R}^{P} \frac{V_i^1}{RT} \mathrm{d}P \cong \gamma_i^* H_{i,m} x_i \tag{2.378}$$

then we can write:

$$H_{i,m}^{(P^R)} = H_{i,m}^{\left(P_m^S\right)} \exp \int_{P_m^S}^{P^R} \frac{\overline{V}_i}{RT} \mathrm{d}P \tag{2.379}$$

where $\overline{V}_i$ is the molar partial volume of i at infinite dilution. The asterisk in γ_i^* indicates the choice of the asymmetric rule. The drawback of the asymmetric rule is that the Henry constant depends on both the solute and the solvent; consequently, problems arise in the description of multicomponent mixtures. Moreover, there is a lack of information about the values to be attributed to γ_i^*, and experimental data are therefore necessary.

2.4.5 Vapor–Liquid Equilibria and Gas Solubility in Liquids at Elevated Pressure

The vapor–liquid equilibria (VLE) at high pressures, for example, $P > 20$–30 atm, can be treated by following two different approaches: the gamma-phi method and a method using directly an EOS for describing both phases.

(a) **Gamma-phi method**

In this method, the two phases at equilibrium are considered separately by calculating the coefficient of fugacity with an EOS and γ_i with one of the previously described methods. However, it must be pointed out that at elevated pressure, γ_i is affected by the pressure, and all the data must be re-conduced to the pressure of reference with the relation:

$$\gamma_i^{(P^R)} = \gamma_i^{(P)} \exp \int_P^{P^R} \frac{\overline{V}_i^L}{RT} \mathrm{d}P \tag{2.380}$$

where $\gamma_i^{(P^R)}$ is independent of P, P^R is normally assumed equal to 0, and $\overline{V}_i^L$, the partial molar volume, can be written as:

$$\overline{V}_i^L = \left(\frac{\partial V}{\partial n_i}\right)_{P,T,nj} \neq V_i^L \tag{2.381}$$

and needs a particular procedure for the calculation.

(b) **Method based on the direct use of an EOS**

Starting from the equality $f_i^1 = f_i^V$, we can write:

$$y_i \varphi_i^V P = x_i \varphi_i^1; \text{therefore, } K_i = \frac{y_i}{x_i} = \frac{\varphi_i^1}{\varphi_i^V} \tag{2.382}$$

where with an EOS such as, for example, the RKS-EOS, it is possible to describe both the liquid and the vapor phase.

$$P = \underset{\text{(repulsive term)}}{\frac{RT}{V-b}} - \underset{\text{(actractive term)}}{\frac{a(T)}{V(V+b)}} \tag{2.383}$$

$$a_i = 0.42747 \frac{R^2 T_{\mathrm{Ci}}^2}{P_{\mathrm{Ci}}} \alpha(T_{\mathrm{Ri}}) \tag{2.384}$$

$$\alpha(T_{Ri}) = \left[1 + E_1\left(1 - T_{Ri}^{1/2}\right) + E_2\left(1 - T_{Ri}^{1/2}\right)^2 + E_3\left(1 - T_{Ri}^{1/2}\right)^3\right]^2 \tag{2.385}$$

where E_1, E_2, and E_3 are parameters that can be obtained from vapor pressures of the pure components. T_{ci} and P_{ci} are the critical values of T and P referred to the component i, while T_{Ri} is the reduced temperature.

The following mixing rules can be applied:

$$a = \sum_i \sum_j a_{ij} x_i x_j \tag{2.386}$$

$$a_{ij} = \left(a_i a_j\right)^{0.5} \left(1 - k_{a_{ij}}\right) \tag{2.387}$$

$$b = \sum_i \sum_j b_{ij} x_i x_j \tag{2.388}$$

$$b_{ij} = \frac{b_i + b_j}{2} \left(1 - k_{b_{ij}}\right) \tag{2.389}$$

Different attempts have been made for improving the mixing rules also considering the local composition, and this topic is still a matter of investigation.

2.4.6 *The Flash Unit*

The flash unit corresponds to a single-stage distillation unit, that is, it is the simplest example of separation by distillation. Imagine an apparatus such as the one shown in Fig. 2.28. A preheated mixture, F, is fed, under moderate pressure, into a vessel kept at a fixed temperature. Part of the mixture vaporizes, giving place to a stream of vapor of flow rate, V, and the liquid composition changes as a consequence of evaporation and is collected at the outlet of the vessel with a flow rate, L. Under ideal conditions, the vapor and liquid are in thermodynamic equilibrium.

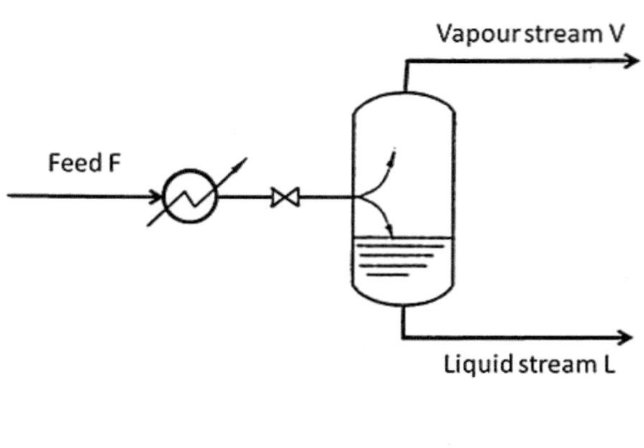

Fig. 2.28 Scheme of a flash unit

An overall mass balance can be written as:

$$F = V + L \tag{2.390}$$

If we consider any single component of the mixture, we can write for any generic component "i":

$$Fz_i = Vy_i + Lx_i \tag{2.391}$$

where z_i is the molar fraction of i in the entering mixture; y_i the molar fraction of i in the vapor phase; and x_i the molar fraction of i in the liquid phase.

By substituting F in this expression, we obtain:

$$Lz_i + Vz_i = Vy_i + Lx_i \tag{2.392}$$

Hence,

$$\frac{L}{V} = \frac{z_i - y_i}{x_i - z_i} \tag{2.393}$$

Then we can write that $K_i = y_i/x_i$, introducing y_i in the balance of i:

$$Fz_i = x_i(VK_i + L) \tag{2.394}$$

Therefore, it holds that:

$$x_i = \frac{F}{V}\frac{z_i}{(K_i + L/V)} = \left(1 + \frac{L}{V}\right)\frac{z_i}{(K_i + L/V)} \tag{2.395}$$

Remembering that $K_i = y_i/x_i$, y_i can be determined as:

$$y_i = \left(\frac{F}{V}\right)\frac{K_i z_i}{(K_i + L/V)} = \left(1 + \frac{L}{V}\right)\frac{K_i z_i}{(K_i + L/V)} \tag{2.396}$$

By operating at a prefixed pressure, the temperature inside the device will be determined by remembering that the following conditions must be respected:

$$\sum_i x_i = 1 \quad \sum_i y_i = 1 \tag{2.397}$$

Hence,

$$\left(\frac{F}{V}\right)\sum_i \frac{z_i}{K_i + L/V} = 1 \quad \left(\frac{F}{V}\right)\sum_i \frac{K_i z_i}{K_i + L/V} = 1 \tag{2.398}$$

In conclusion, if temperature and pressure are imposed and the composition of the inlet stream is known, the compositions of the vapor and liquid stream at the outlet of the vessel can be determined together with the overall amount of liquid and vapor. This calculation is relatively simple when the behavior of the mixture is ideal. In this case, we must introduce the laws of Raoult and Dalton:

Raoult law

$$p_i = x_i P_i^{\circ} \tag{2.399}$$

Dalton law

$$p_i = y_i P \tag{2.400}$$

and write: $x_i P_i^{\circ} = y_i P$; then it also holds that:

$$K_i = \frac{y_i}{x_i} = \frac{P_i^{\circ}}{P} \tag{2.401}$$

For solving the problem, it is sufficient to know the vapor pressure and its dependence on the temperature. More complicated is the case of a nonideal mixture. In this case, we need to know the activity coefficients of all the components for both the liquid and vapor phases, and—in the case of a multicomponent mixture—we need data for all the binaries. The previous equation, $x_i\, P^{\circ}{}_i = y_i\, P$, becomes:

$x_i \gamma_i P_i^{\circ} = y_i \Phi_i P$ and, consequently, the value of K_i can be determined as:

$$K_i = \frac{y_i}{x_i} = \frac{\gamma_i P_i^{\circ}}{\Phi_i P} \tag{2.402}$$

This suggests the opportunity to use the flash as laboratory device for determination of the activity and fugacity coefficients.

The energy balance for the flash unit can be written as follows:

$$Q_t = V\widetilde{H} + L\widetilde{h} - F\widetilde{h}_f \tag{2.403}$$

where $\widetilde{H}$ is the molar enthalpy, $\widetilde{h}$ is the molar enthalpy of the liquid, $\widetilde{h}_f$ is the molar enthalpy of the stream at the inlet, and $\widetilde{Q}_t$ is the heat to supply for unit of time and for unit of mass of fluid. The molar enthalpies of the liquids are normally negligible with respect to the ones of the vapor; hence, we can simply write $\widetilde{Q}_t \cong V\widetilde{H}$.

To choose the most appropriate conditions to adopt for the flash, it is opportune to know the bubble point temperature (T_{BP}) and the dew point temperature (T_{DP}) of the mixture and choose an intermediate value. The bubble point temperature is the temperature at which a heated mixture shows the formation of the first bubble of vapor, while the dew point temperature is the temperature at which a cooled vapor shows the formation of the first liquid drop. These two temperatures are correlated with the composition of the mixture. In fact, to calculate the bubble point temperature when the overall pressure and the composition are known, it is necessary to solve the equation:

$$\sum_{i=1}^{n} K_i x_i = 1 \tag{2.404}$$

If the vapor of a given composition is cooled at a constant pressure, the dew point temperature is determined by solving the following equation:

$$\sum_{i=1}^{n} \frac{y_i}{K_i} = 1 \tag{2.405}$$

In both cases, we must find the temperatures at which the observed equalities are respected.

In some cases, a reaction occurs inside the vessel, and the composition changes not only for the vapor–liquid equilibrium but also for the effect of the reaction. Because the reaction occurs with a rate that can be described by the kinetics, this aspect will be deepened in Chap. 4, which is devoted to the kinetics in homogeneous phase.

Exercise 2.13 A and B. Flash at Low and High Pressure
Exercise 2.13 A. Flash at Low Pressure

A mixture of four components (pentane, hexane, cyclohexane, and methanol) is submitted to flash evaporation (see scheme in Fig. 2.29) at atmospheric pressure and at temperature of 60 °C. The molar composition (z_i) of such a mixture is 25% of each component.

Antoine's constants for the considered components are listed in Table 2.24.

$$\log_{10} P = A - \frac{B}{T + C} \tag{2.406}$$

- **Part 1**

By assuming an ideal behavior of both vapor and liquid phases, calculate the liquid and vapor composition in equilibrium and the vapor fraction (V/F) of the system. Assume 1 mol for F as a calculation basis.

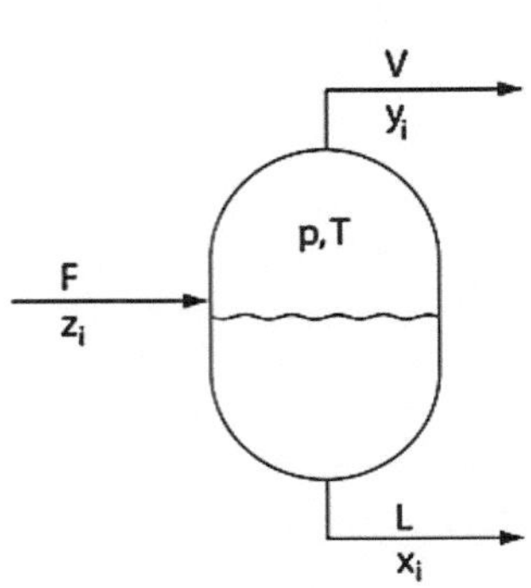

Fig. 2.29 Scheme of the flash of exercise

Table 2.24 Antoine's constants for the considered components[a]

Substances	A	B	C
Pentane	3.97786	1064.840	−41.136
Hexane	4.00139	1170.875	−48.833
Cyclohexane	3.93002	1182.774	−52.532
Methanol	5.20277	1580.080	−33.650

[a] P shown in bar and T in K

Table 2.25 Results for the part 1 of Exercise 2.13A

	Mole fractions		
$T = 60°C$, $P = 1$ atm	Feed	Liquid	Vapor
Pentane	0.2500	0.2009	0.0847
Hexane	0.2500	0.2571	0.3043
Cyclohexane	0.2500	0.2706	0.4717
Methanol	0.2500	0.2530	0.2710

- **Part 2**

Repeat the calculation as in part 1 by varying the temperature in the range 58.5–64.0 °C and fix the pressure at, respectively, 0.9, 1.0, and 1.1 atm. In correspondence of each temperature and pressure, calculate the *V*/*F* ratio. Build a plot in which *V*/*F* is reported as a function of temperature for each pressure.

- **Solution**

The flash problem can be solved by finding the root of the Rachford–Rice equation with the ratio *V*/*F* as unknown:

$$f\left(\frac{V}{F}\right) = \sum_{i=1}^{N_c} \frac{z_i(K_i - 1)}{\frac{V}{F}(K_i - 1)} = 0 \quad K_i = \frac{P_i^\circ}{P_{\text{tot}}} \tag{2.407}$$

- **Part 1**

The solution of the Rachford–Rice equation is $V/F = 0.1783$, which means that, in these T and P conditions, 17.83% of the feed is vaporized by flash operation. The composition of feed, liquid, and vapor streams is reported in the Table 2.25.

Results

- **Part 2**

The requested plot is shown in Fig. 2.30.

As expected, the *V*/*F* of the system increases as the temperature is increased at each chosen pressure.

Exercise 2.13 B. Flash at High Pressure

A mixture of two components (ethane and n-heptane) is submitted to flash evaporation (see scheme shown in Fig. 2.29) at a pressure of 10 bar and a temperature of 430 K. The molar composition (z_i) of such a mixture is 26.54% ethane and

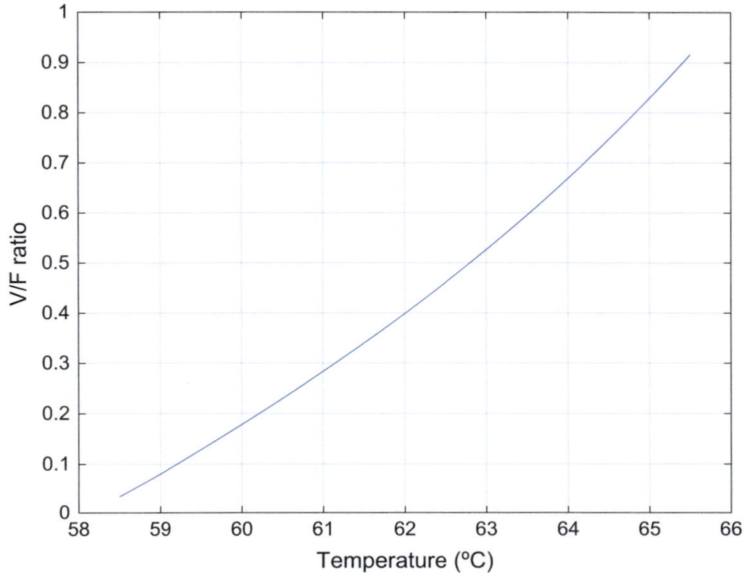

Fig. 2.30 *V/F* ratio calculated as a function of temperature for different pressures

Table 2.26 Critical data and acentric factor for the two mixture's components to be used in the application of RKS-EOS

Substances	T_c	P_c	ω
Ethane	305.4	48.8	0.099
n-Heptane	540.3	27.4	0.349

73.46% n-heptane. Critical data and acentric factor for the two mixture's components, to be used in the application of RKS-EOS, are listed in Table 2.26.

- Part 1

Calculate the *V/F* and composition of the two phases in equilibrium using the RKS-EOS. Report in a plot the convergence criterion as a function of the iteration number. The convergence criterion is defined as:

$$\left|\sum_{i=1}^{N_c}(y_i - x_i)\right| < \text{tolerance} \tag{2.408}$$

- Part 2

Repeat the calculation as in part 1 but varying the temperature in the range of 375–450 K and, in correspondence of each temperature, calculate the *V/F* ratio. Build a plot in which *V/F* is reported as a function of temperature.

- **Solution**

The solution procedure, for both parts 1 and 2, can be summarized in the following steps:

(1) Guess $L = 0.5$ and $x_i = y_i = z_i$.
(2) Solve EOS for z_L and z_V using the following relations:

$$P = \frac{RT}{V - b_m} - \frac{a_m}{V(V + b_m)} \tag{2.409}$$

$$a_i = 0.42748 \frac{\alpha_i (RT_{Ci})^2}{P_{Ci}} \tag{2.410}$$

$$b_i = 0.8664 \frac{(RT_{Ci})}{P_{Ci}} \tag{2.411}$$

$$m_i = 0.480 + 1.574\omega_i - 0.176\omega_i^2 \tag{2.412}$$

$$\alpha_i = \left[1 + m_i\left(1 - T_{Ri}^{0.5}\right)\right]^2 \tag{2.413}$$

$$a_m = \sum_{i=1}^{N_c} \sum_{i=1}^{N_c} y_i y_j \sqrt{a_i a_j} \tag{2.414}$$

$$b_m = \sum_{i=1}^{N_c} y_i b_i \tag{2.415}$$

The EOS can be rearranged in form of compressibility factor as follows:

$$z_m^3 - z_m^2 + \left(A_m - B_m - B_m^2\right) z_m - A_m B_m = 0 \tag{2.416}$$

with A_{m} and B_{m} defined as:

$$A_m = \frac{a_m P}{(RT)^2} \quad B_m = \frac{b_m P}{RT} \tag{2.417}$$

- **Part 3**

Calculate the fugacity coefficient for each component in the liquid and vapor phases:

$$\ln \Phi_i = \frac{b_i}{b_m}(z_m - 1) - \ln(z_m - B_m) - \frac{A_m}{B_m}\left(2\frac{a_i^{0.5}}{a_m^{0.5}} - \frac{b_i}{b_m}\right) ln\left(1 + \frac{B_m}{z_m}\right) \tag{2.418}$$

Calculate K_i as:

$$K_i = \frac{\Phi_i^L}{\Phi_i^V} \tag{2.419}$$

Check the convergence criterion with:

$$\left| \sum_{i=1}^{N_c} (y_i - x_i) \right| < \text{tolerance with } x_i = \frac{z_i}{K_i + L(1 - K_i)} \text{ and } y_i = K_i x_i \tag{2.420}$$

If the convergence tolerance is not satisfied, adjust L with the following relation:

$$L_{\text{new}} = L_{\text{old}} - \frac{\sum_{i=1}^{N_c} \frac{z_i(K_i - 1)}{K_i + (1 - K_i)L}}{\sum_{i=1}^{N_c} \frac{z_i(K_i - 1)^2}{[K_i + (1 - K_i)L]^2}} \tag{2.421}$$

By applying the described solution procedure, with a total of 15 iterations, part 1 of the exercise yields the following results (see also Fig. 2.31):

Results Part 1 (Table 2.27)

From the plot, it is evident that the correct solution has been achieved after only five iterations. The convergence criterion is satisfied with the function zeroed at 1e−16. By automatically repeating the procedure described previously (inserting the calculation in a for-loop) with different temperatures in the assigned range, the plot shown in Fig. 2.32 can be obtained.

By increasing the temperature, the amount of liquid produced by flash operation decreases gradually and, correspondingly, the amount of vapor increases. Above a temperature of approximately 450 K, only vapor is produced. All the described results were obtained using MATLAB programs reported as Electronic Supplementary Material, Data 2.1.

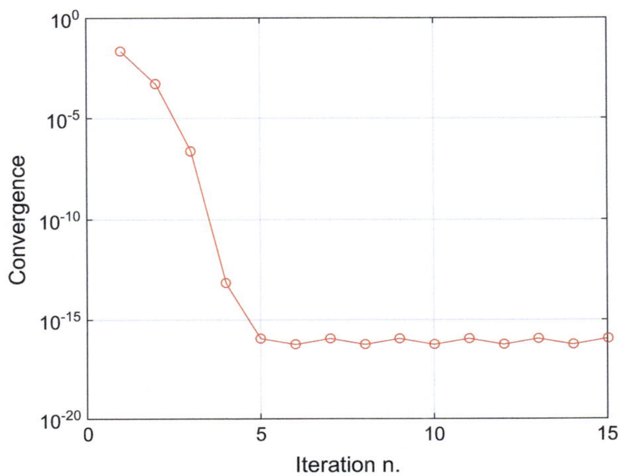

Fig. 2.31 Number of iterations to achieve convergence

Table 2.27 Results for the part 1 of Exercise 2.13B

Substances	Feed	Liquid	Vapor
Ethane	0.2654	0.0626	0.4876
n-Heptane	0.7346	0.9374	0.5124

Fun = 1.1102e-6; $V/F = 0.4771$, $L = 0.5229$ mol, $V = 0.4771$ mol

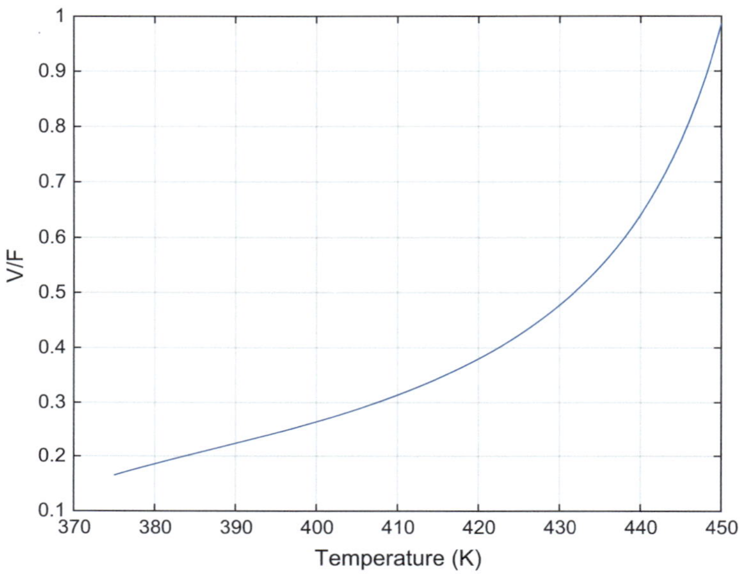

Fig. 2.32 *V/F* ratio calculated as a function of temperature

2.4.7 Vapor–Liquid Equilibrium and Distillation

In the previous section, we saw that in a flash apparatus, the vapor and liquid streams at the outlet of the vessel are, respectively, rich in volatile components and less volatile ones. Therefore, if these two streams are subject to further repeated flash operation, as in the scheme shown in Fig. 2.33, the emerging vapor stream is progressively enriched in more volatile components and the liquid stream is enriched in less volatile ones.

The described operation is not advantageous in practice because the number of pure components separated at the end is very small. If we follow the same principle but recycle the less purified stream, as in the scheme shown in Fig. 2.34, a good separation is obtained in the two collected streams of distillate and residual liquid.

Separation by distillation is based on this simple concept, that is, a series of different equilibrium stages in which the vapor phase is progressively enriched in more volatile components and liquid phase in the less volatile ones. The scheme operation shown in Fig. 2.34 is more simply realized in a rectifying column (tray

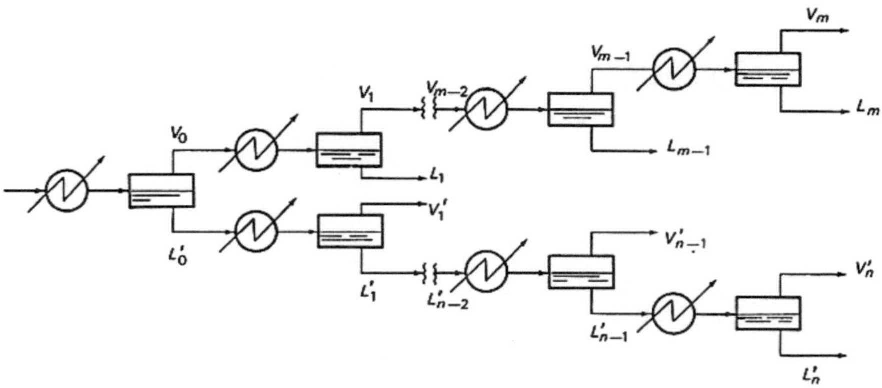

Fig. 2.33 Multistage flash operation (see Carrà 1977)

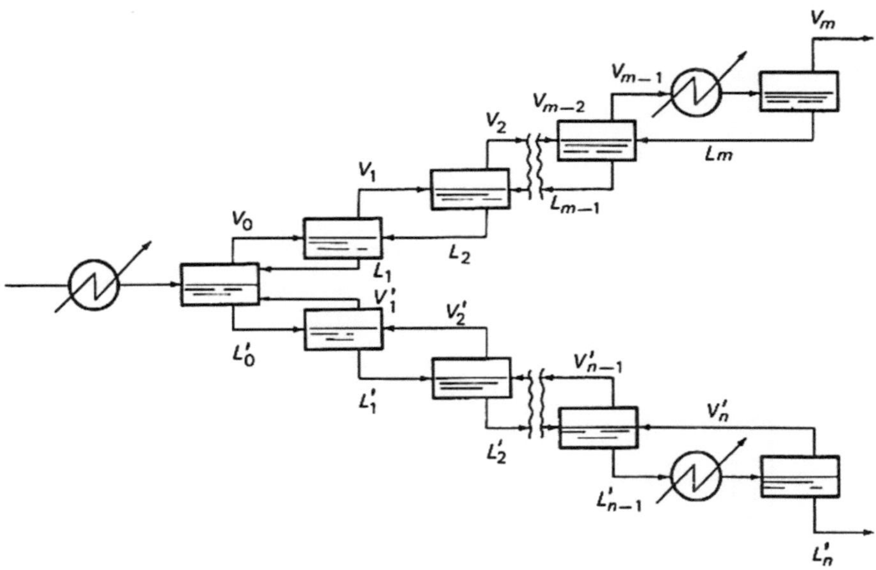

Fig. 2.34 Multistage flash operation with recycling (see Carrà 1977)

column), such as the scheme shown in Fig. 2.35a, b. As can be seen, we can identify a rectifying section over the feed inlet in which the most volatile components are progressively enriched and an exhaustion section in which the volatile components progressively disappear and less volatile ones are enriched. Two streams are collected, respectively, from the top and the bottom of the column: These are the distillate and the residual. Part of the distillate is recycled to improve the separation. Any single tray can be considered as a vapor–liquid equilibrium stage. Therefore,

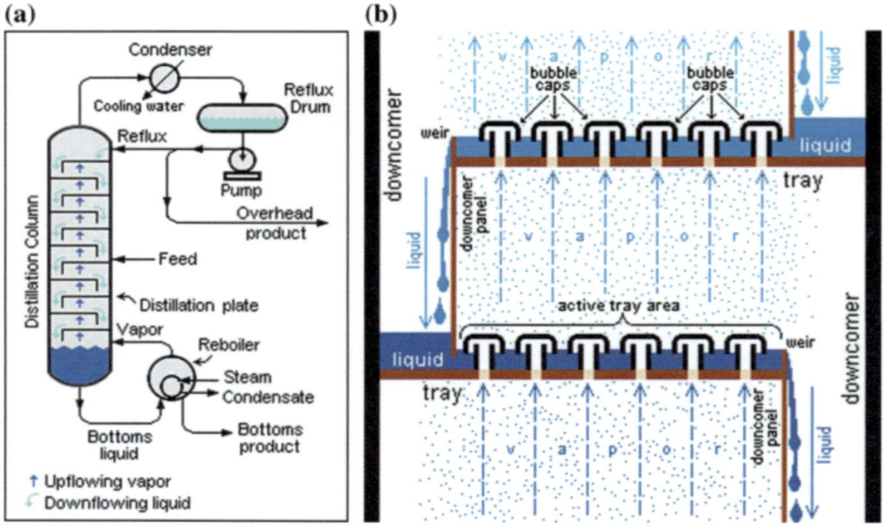

Fig. 2.35 (**a**) Scheme of a tray-distillation column. (**b**) Magnification of the tray structure

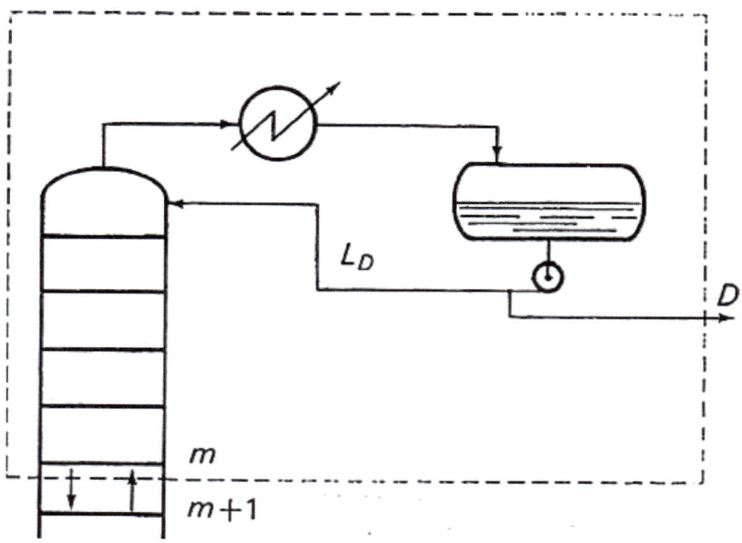

Fig. 2.36 The rectifying section of the tray column

mass and heat balance equations, similar to the ones seen for flash operations, can be applied to any single plate to describe what happens. Trays are realized in a way to favor the interface contact (see Fig. 2.35b).

Consider, for example, the rectifying section (see Fig. 2.36).

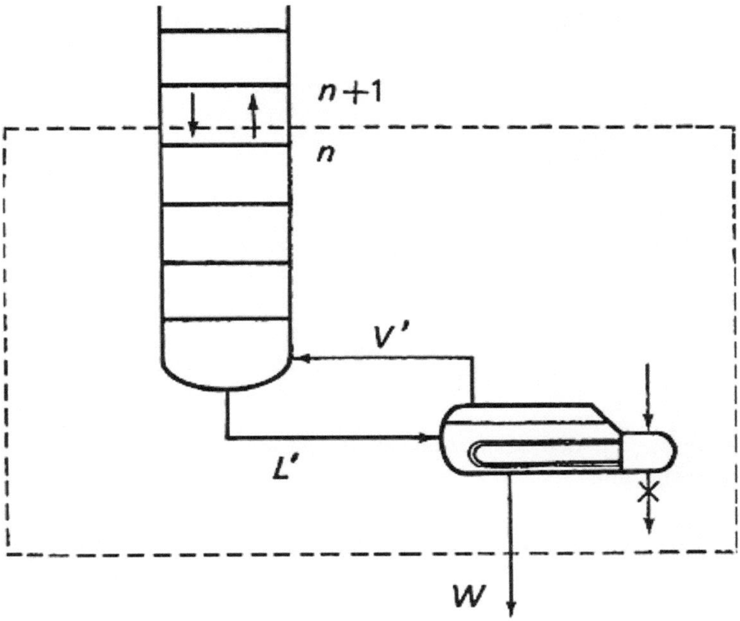

Fig. 2.37 The exhaustion section of the tray column

The mass balance for a generic tray "*m*" will be as follows:

(1) Overall mass balance on the tray:

$$V_{m+1} = L_m + D \tag{2.422}$$

(2) Balance on a single component "*i*":

$$L_m + Dy_{m+1,i} = L_m x_{m,i} + Dx_{D,i} \tag{2.423}$$

By combining these two balances, it is possible to obtain:

$$\frac{L}{V_{m+1}} = \frac{x_{D,i} - y_{m+1,i}}{x_{D,i} - x_{m,i}} \tag{2.424}$$

In the same way, considering the exhaustion section of Fig. 2.37, we obtain:

(1) Overall mass balance on the tray:

$$V_n + W = L_{n+1} \tag{2.425}$$

(2) Balance on a single component "*i*":

$$V_n y_{n,i} + W x_{W,i} = L_{n+1} x_{n+1,i} \tag{2.426}$$

Hence,

$$\frac{L_{n+1}}{V_n} = \frac{x_{W,i} - y_{n,i}}{x_{W,i} - x_{n+1,i}} \tag{2.427}$$

To solve the reported balance equations, we must introduce the vapor–liquid equilibrium equations:

$$\frac{y_i}{x_i} = K_i(T, P, \text{composition}) \tag{2.428}$$

and the heat balance for each tray. Considering the m tray, we have:

$$\widetilde{h}_{m-1} L_{m-1} + \widetilde{H}_{m+1} V_{m+1} = \widetilde{h}_m L_m + \widetilde{H}_m V_m \tag{2.429}$$

where $\widetilde{H}$ is the molar vapor enthalpy and $\widetilde{h}$ is the liquid molar enthalpy. It is known that $\widetilde{H} \gg \widetilde{h}$; therefore, we can neglect the heat flow coupled with the liquid flow. As a consequence, we can write:

$$\widetilde{H}_{m+1} V_{m+1} \simeq \widetilde{H}_m V_m \simeq \widetilde{H}_{m-1} V_{m-1} \tag{2.430}$$

A further simplification can be made when the vapor molar flow rate is constant. In this case, it holds that:

$$\widetilde{H}_{m+1} \simeq \widetilde{H}_m \simeq \widetilde{H}_{m-1} \tag{2.431}$$

However, the heat furnished to the reboiler determines the amount of vapor produced in the column, which must be the maximum possible but avoiding flooding in the column.

There are also other types of distillation columns as packed-bed columns. In this case, calculations are organized in such a way as to consider the column constituted by a series of theoretical plates characterized by the length of the column necessary to achieve the vapor–liquid equilibrium condition.

As we have seen, the tray column is a continuous unit operation, and therefore, it works under steady-state conditions. Consequently, all the profiles in the column (temperature, concentrations) are constant along time. Obviously, this is not true during the transient conditions before reaching the steady state. In this last case, and in the case of batch distillation, the equations to be solved are more complicated because temperature and composition change in both space and time.

Additional Solved Exercises

Exercise 2.14 Equilibrium Composition by Gibbs Energy Minimization

In a steam cracking unit, ethane is used as feed together with water vapor in a molar ratio ethane:water equal to 1:4. The unit operates at atmospheric pressure and at a temperature of 1000 K. The goal of this exercise is to calculate the product distribution at equilibrium. The adopted strategy is based on the minimization of the Gibbs energy of the mixture. The minimization is constrained by the mass balance on atoms that resulted distributed among the possible molecular species formed.

An additional task of the exercise is to estimate the effect of temperature on the equilibrium composition.

A list of the species that can be likely formed upon equilibrium attainment are the following:

$$CH_4\ C_2H_4\ C_2H_2\ CO_2\ CO\ O_2\ H_2\ H_2O\ C_2H_6$$

Gibbs energy of the gaseous mixture can be calculated as:

$$G=\sum_{j=1}^{n_S} n_j\mu_j \tag{2.432}$$

where n_S is the number of chemical compound present in the mixture at equilibrium conditions, n_j is the number of moles of jth specie, and μ_j is the chemical potential of jth specie defined as:

$$\mu_j=G_j^0+RT\ln x_j \tag{2.433}$$

And then:

$$\mu_j=G_j^0+RT\ln\frac{y_jP}{P^0} \tag{2.434}$$

As $P = 1$ atm, the function expressing Gibbs energy for the mixture can be written as:

$$\frac{G}{RT} = \sum_{j=1}^{ns} n_j \left(\frac{G_j^0}{RT} + \ln y_j \right) \tag{2.435}$$

Or by definition of mole fraction:

$$\frac{G}{RT} = \sum_{j=1}^{ns} n_j \left(\frac{G_j^0}{RT} + \ln \frac{n_j}{\sum_{k=1}^{ns} n_k} \right) \tag{2.436}$$

The Gibbs energy function expressed by the relation (2.436) must be minimized with respect to the number of moles at equilibrium, n_j, that should be regarded as unknown variables. These nine variables are subject to atomic balance constrains as in the feed (1:4 ethane:water), the following number of atoms are present, that must be conserved at equilibrium:

$$\begin{aligned} n_H^0 &= 14 \\ n_O^0 &= 4 \\ n_C^0 &= 2 \end{aligned} \tag{2.437}$$

The problem consists in a constrained minimization and can be solved with the "fmincon" function available in MATLAB. The code for the solution is present in the Electronic Supplementary Material, Data 2.1. In the Table 2.28, the solution is reported together with the data related to the standard Gibbs of pure substances.

The atomic balance constraints are fully respected, as indicated by the solver. Moreover, by observing the results in the previous table, it is evident that the amount of some components at equilibrium is negligible while some other are predominant in the exit stream of the steam cracker unit. It is interesting to observe that, even if ethane conversion is practically complete, the stoichiometry of the theoretical ethane steams reforming (2.438) is not fully respected.

Table 2.28 Results for Exercise 2.14

Component	G_j^0 (Kcal/mol)	n_j (mol)
CH_4	4.610	0.0664
C_2H_4	28.249	8.888e−8
C_2H_2	40.604	2.760e−21
CO_2	−94.61	0.5450
CO	−47.942	1.3900
O_2	0.000	4.560e−21
H_2	0.000	5.3500
H_2O	−46.030	1.5200
C_2H_6	23.130	1.600e−7

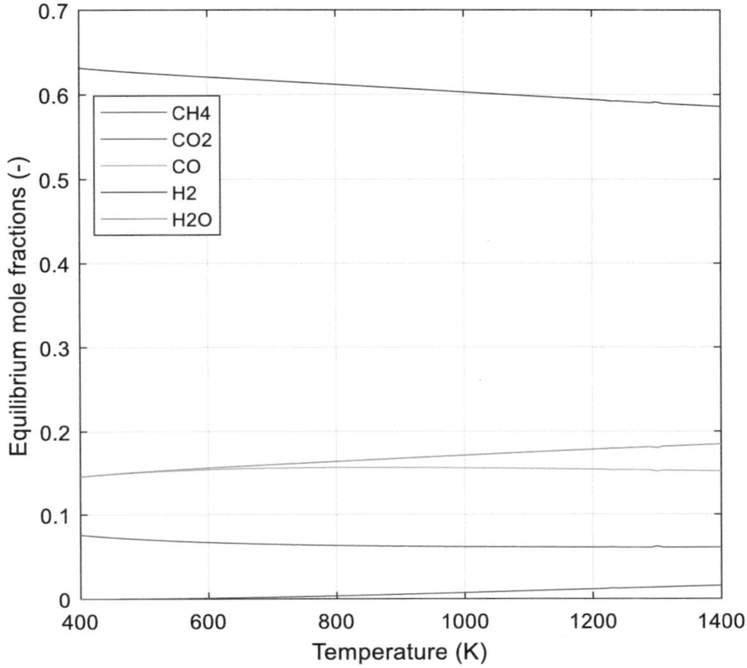

Fig. 2.38 Effect of temperature on equilibrium composition

$$C_2H_6 + 4H_2O \rightleftharpoons 2CO_2 + 7H_2 \tag{2.438}$$

The yield of hydrogen is lower that the theoretical one and a small amount of methane are also formed.

As an additional result, by observing the plot reported in Fig. 2.38, we can see that the effect of temperature, in the range 400–1400 K, is rather low.

Exercise 2.15 Reactive Flash: Epichlorohydrins

The process for production of epichlorohydrin from chlorohydrins has been studied (Carrà et al. 1979) and a particular attention has been devoted to the separation of the product. Even if the complete separation can be obtained in a multistage unit, such as a distillation column, a model for a single-stage reactive flash separation is useful to gain information about the influence of operating condition on epichlorohydrin yield.

The complete reaction scheme is the following:

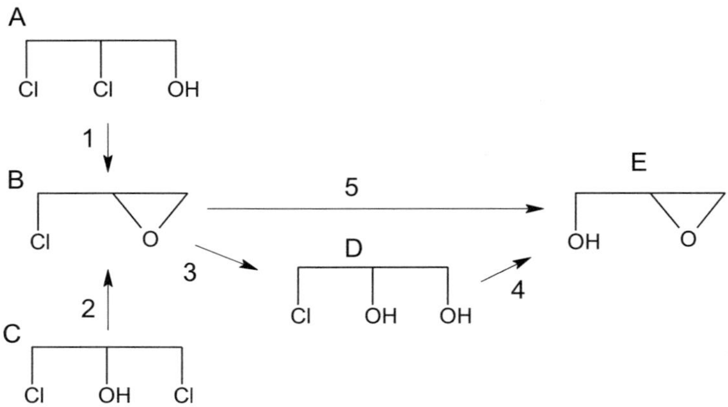

(2.439)

In the scheme above, A is αβ-dichlorohydrin, B is epichlorohydrin, C is α-γ-dichlorohydrin, D is α-monochlorohydrin, and E is glycidol. As the reaction 3 is fast while reaction 4 is slow, only one kinetic constant can be used for the production of glycidol so the presence of D can be neglected. The reactive mixture is then composed by five substances (A, B, C, and E plus water) and a complete model requires the simultaneous solution of mass balances and vapor–liquid equilibria relations.

The complete model for the flash unit can be written by using the following auxiliary variables:

$$\begin{aligned} \alpha &= V/F \\ \tau &= V_R/F \\ q &= Q/F \end{aligned} \tag{2.440}$$

where α is the vaporization ratio, τ is the contact time, and q is the heating ratio. The mass balance on a component is expressed as:

$$z_i = (1 - \alpha)x_i + \alpha y_i + \tau R_i \tag{2.441}$$

We must then couple with the set of individual mass balances, a group of vapor–liquid equilibria relationships, and a congruence equation as:

$$y_i = K_i x_i \tag{2.442}$$

$$\sum_i (x_i - y_i) = 0 \tag{2.443}$$

Moreover, the kinetics of the reactive system must be defined:

$$R_i = \sum_{k=1}^{3} \nu_{i,k} r_k \tag{2.444}$$

where R_i is the overall generation term for component i, while $\nu_{i,k}$ is the stoichiometric coefficient of the component i in the reaction k.

The calculation of vapor–liquid equilibria is fundamental as relations (2.442) must be solved in conjunction with the mass balance. Such calculation requires some specific equations and parameters as reported by Carrà et al. (1979). The equilibrium relation can also be written as follows by introducing the nonideality in liquid phase:

$$\frac{y_i}{x_i} = K_i = \frac{p_i^0 \gamma_i}{P} \tag{2.445}$$

Vapor pressure as a function of temperature is expressed by:

$$\ln p_i^0 = \frac{A_i}{T} + B_i \tag{2.446}$$

The nonideality of the liquid mixture can be accounted for by using an activity coefficient that can be calculated by using an infinite dilution activity coefficient (function of temperature, 2.449) and a correction due to a salt effect (2.447) that is also dependent from temperature (2.448).

$$\log \frac{\gamma_i}{\gamma_i^0} = K_s C_s \tag{2.447}$$

$$C_s = 0.064 - 0.0001475T \tag{2.448}$$

$$\log \gamma_i^0 = y_i^0 + a_i T + b_i T^2 \tag{2.449}$$

In the presented model, several parameters are necessary, and they have been collected and summarized in the Table 2.29.

Table 2.29 Parameters for the calculation of Exercise 2.15

Comp.	z_i	A_i	B_i	y_i^0	a_i	b_i	Ks
A	0.00108	6775	22.15	−17.752	0.148	−0.00025	0.4583
B	0.00252	6181	20.58	−219.484	1.2668	−0.0018	0.4552
C	0.00000	4847	19.06	−3.7742	0.05726	−0.0001	0.3791
H_2O	0.99640	5118	20.38	0.000	0	0	0
E	0.00000	5738	19.81	−3.7742	0.05726	−0.0001	0.3824

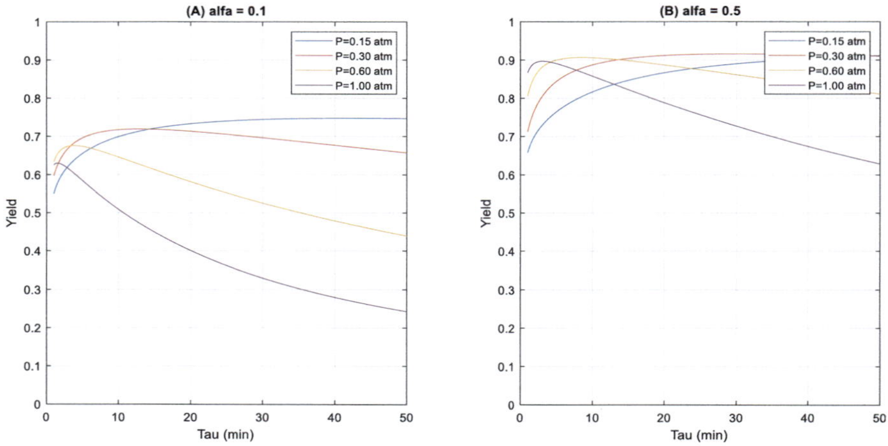

Fig. 2.39 Yield of epichlorohydrin vs. contact time if flash calculation at various pressure for (**a**) $\alpha = 0.1$ and (**b**) $\alpha = 0.5$

The solution strategy of the model can be as follows:

(1) Assuming fixed values for variable: pressure, degree of vaporization, and contact time.
(2) Perform numerical iteration on temperature until the convergence was attained.

This strategy has been implemented in a MATLAB program reported in the Electronic Supplementary Material, Data 2.1 and the following plot was produced in correspondence of two fixed values of degree of vaporization, with parametric curves for total pressure (Fig. 2.39).

Appendices

Appendix 1: Lydersen's Method: Increments for the Calculation of the Critical Variables

Functional groups	ΔT	ΔP	ΔV
Increments for groups not in a ring			
$-CH_3$	0.020	0.227	55
$-CH_2-$ (with one bond down)	0.020	0.227	55
$-CH<$ (with bonds up and down)	0.012	0.210	51

(continued)

Functional groups	ΔT	ΔP	ΔV
$-\overset{\mid}{\underset{\mid}{C}}-$	0.000	0.210	41
$=CH_2$	0.018	0.198	45
$=\underset{\mid}{C}H$	0.018	0.198	45
$=\underset{\mid}{C}-$	0.000	0.198	36
$=C=$	0.000	0.198	36
$\equiv CH$	0.005	0.153	(36)
$\equiv C-$	0.005	0.153	(36)
Increments for groups inside a ring			
$-CH_2-$	0.013	0.184	44.5
$-\overset{\mid}{\underset{\mid}{C}}H$	0.012	0.192	46
$-\overset{\mid}{\underset{\mid}{C}}-$	(−0.007)	(0.154)	(31)
$=\underset{\mid}{C}H$	0.011	0.154	37
$=\underset{\mid}{C}-$	0.011	0.154	36
$=C=$	0.011	0.154	36
Increments for alogens			
−F	0.018	0.224	18
−Cl	0.017	0.320	49
−Br	0.010	(0.50)	(70)
−I	0.012	(0.83)	(95)
Increments for groups containing oxygen			
−OH (alcohol)	0.082	0.06	(18)
−OH (phenol)	0.031	(−0.02)	(3)
−O− (not in a ring)	0.021	0.16	20
−O− (in a ring)	(0.014)	(0.12)	(8)
$-\overset{\mid}{C}=O$ (not in a ring)	0.040	0.29	60
$-\overset{\mid}{C}=O$ (in a ring)	(0.033)	(0.2)	(50)
$H\overset{\mid}{C}=O$ (aldehyde)	0.048	0.33	73

(continued)

Functional groups	ΔT	ΔP	ΔV
$-COOH$ (acid)	0.085	(0.4)	80
$-COO-$ (ester)	0.047	0.47	80
$=O$ (different from previous cases)	(0.02)	(0.12)	(11)
Increments for groups containing nitrogen			
$-NH_2$	0.031	0.095	28
$-\overset{}{\underset{\vert}{N}}H$ (not in a ring)	0.031	0.135	(37)
$-\underset{\vert}{N}H$ (in a ring)	(0.024)	(0.09)	(27)
$-\underset{\vert}{N}-$ (not in a ring)	0.014	0.17	(42)
$-\underset{\vert}{N}-$ (in a ring)	(0.007)	(0.13)	(32)
$-CN$	(0.060)	(0.36)	(80)
$-NO_2$	(0.055)	(0.42)	(78)
Increments for groups containing sulfur			
$-SH$	0.015	0.27	55
$-S-$ (not in a ring)	0.015	0.27	55
$-S-$ (in a ring)	(0.008)	(0.24)	(45)
$=S$	(0.003)	(0.24)	(47)
Other functional groups			
$-\overset{\vert}{\underset{\vert}{Si}}-$	0.03	(0.54)	–
$-\underset{\vert}{B}-$	(0.03)	–	–

(1) No increments are foreseen for hydrogen; (2) all the free bonds shown must be connected with atoms different from hydrogen; and (3) the values in brackets are not precise because they were based on few experimental data

Appendix 2: Group Contributions for Estimating C_p^0, $\Delta H_{f,298K}^0$, and ΔG_f^0

Groups	Heat capacity constants				ΔH_f^0 (298 K)	Constants for determining Gibbs free energy			
						300–600 K		600–1500 K	
	a	$b \times 10^2$	$c \times 10^4$	$d \times 10^6$		A	$B \times 10^2$	A	$B \times 10^2$
Part 1									
$-CH_3$	0.6087	2.1433	−0.0852	0.1135	10.25	−10.943	2.215	−12.310	2.436
$-CH_2-$	0.3945	2.1363	−0.1197	0.2596	−4.94	−5.193	2.430	−5.830	2.544
$-\overset{\vert}{\underset{\vert}{C}}H$	−3.5232	3.4158	−0.2816	0.8015	−1.29	−0.705	2.910	−0.705	2.910
$-\overset{\vert}{\underset{\vert}{C}}-$	−5.8307	4.4541	−0.4208	1.263	0.62	1.958	3.735	4.385	3.350
$H\backslash C=CH_2$ (H, /)	0.2773	3.4580	−0.1918	0.4130	15.02	13.737	1.655	12.465	1.762
$\backslash C=CH_2$ (/)	−0.4173	3.8857	−0.2783	0.7364	20.50	16.467	1.915	16.255	1.966
H\C=C/H (cis)	−3.1210	3.0860	−0.2359	0.5504	17.96	17.663	1.965	16.180	2.116
H\C=C\H (trans)	0.9377	2.9904	−0.1749	0.3918	17.83	17.187	1.915	15.815	2.062

(continued)

Groups	Heat capacity constants					Constants for determining Gibbs free energy				
						300–600 K		600–1500 K		
	a	$b \times 10^2$	$c \times 10^4$	$d \times 10^6$	ΔH_f^0 (298 K)	A	$B \times 10^2$	A	$B \times 10^2$	
H \ / C=C / \	−1.4714	3.3842	−0.2371	0.6063	−20.10	20.217	2.295	19.584	2.354	
\ / C=C / \	0.4736	3.5183	−0.3150	0.9205	30.46	25.135	2.573	25.135	2.573	
H \ C=C=CH$_2$ /	2.2400	4.2896	−0.2566	0.5908	49.47	49.377	1.035	48.170	1.208	
\ C=C=CH$_2$ /	2.6308	4.1658	−0.2845	0.7277	51.30	51.084	1.474	51.084	1.474	
H H \ / C=C=C / \	−3.1249	6.6843	−0.5766	1.743	55.04	52.460	1.483	52.460	1.483	
≡CH	–	–	–	–	27.10	27.048	−0.765	26.700	−0.704	
≡C–	–	–	–	–	27.38	26.938	−0.525	26.555	−0.550	
Part 2										
Groups for conjugated alkenes										
↔CH$_2$		–	–	–	–	(10.1)	5.437	0.675	4.500	0832
H / ↔C \		–	–	–	–	(12)	7.407	1.035	6.980	1.088
/ ↔C \		–	–	–	–	–	9.152	1.505	10.370	1.308

Groups	Heat capacity constants					Constants for determining Gibbs free energy				
						300–600 K		600–1500 K		
	a	$b \times 10^2$	$c \times 10^4$	$d \times 10^6$	ΔH_f^0 (298 K)	A	$B \times 10^2$	A	$B \times 10^2$	
⤡CH⤢		−1.4572	1.9147	−0.1233	0.2985	3.27	3.047	0.615	2.505	0.706
⤡C–⤢		−1.3883	1.5159	−0.0690	0.2659	5.55	4.675	1.150	5.010	0.988
⤡C↔⤢		0.1219	1.2170	−0.0855	0.2122	4.48	3.513	0.568	3.998	0.485
Corrections for cycloparaffin rings										
Three-atom rings		−3.5320	−0.0300	0.0747	−0.5514	24.13	23.458	−3.045	22.915	−2.966
Four-atom rings		−8.6550	1.0780	0.0425	0.0250	18.45	10.73	−2.65	10.60	−2.50
Five-atom rings (pentane)		−12.285	1.8609	−0.1037	0.2145	5.44	4.275	−2.350	2.665	−2.182
Five-atom rings (pentene)		−6.8813	0.7818	−0.0345	0.0591	–	−3.657	−2.395	−3.915	−2.150
Six-atom rings (hexane)		−13.3923	2.1392	−0.0429	−0.1865	−0.76	−1.128	−1.635	−1.930	−1.504
Six-atom rings (hexene)		−8.0238	2.2239	−0.1915	0.5473	–	−9.102	−2.045	−8.810	−2.071
Branched paraffins										
Side chain with ≥ 2 atoms		–	–	–	–	0.80	1.31	0	1.31	0
Three –CH– adjacent		–	–	–	–	−1.2	−2.13	0	2.12	0
–C– and –CH– adjacent		–	–	–	–	0.6	1.80	0	1.80	0

(continued)

Groups	Heat capacity constants				ΔH_f^0 (298 K)	Constants for determining Gibbs free energy				
						300–600 K		600–1500 K		
	a	$b \times 10^2$	$c \times 10^4$	$d \times 10^6$		A	$B \times 10^2$	A	$B \times 10^2$	
Two –C– adjacent		–	–	–	–	(5.4)	2.58	0	2.58	0
Part 3										
Branching in cycles with 5 atoms										
Single branching	–	–	–	–	0	−1.04	0	−1.69	0	
Double branching										
Position 1,1	–	–	–	–	0.30	−1.85	0	−1.19	−0.16	
Position cis-1,2	–	–	–	–	0.70	−0.38	0	−0.38	0	
Position trans-1,2	–	–	–	–	−1.10	−2.55	0	−0.945	−0.266	
Position cis-1,3	–	–	–	–	−0.30	−1.20	0	−0.370	−0.166	
Position trans-1,3	–	–	–	–	−0.90	−2.35	0	−0.800	−0.264	
Branching in cycles with 6 atoms										
Single branching	–	–	–	–	0	−0.93	0	0.230	−0.192	
Double branching										
Position 1,1	–	–	–	–	2.44	0.835	−0.367	1.745	−0.556	
Position cis-1,2	–	–	–	–	−0.20	−0.19	0	1.470	−0.276	
Position trans-1,2	–	–	–	–	−2.69	−2.41	0	0.045	−0.398	
Position cis-1,3	–	–	–	–	−2.98	−2.70	0	−1.647	−0.185	
Position trans-1,3	–	–	–	–	−0.48	−1.60	0	0.260	−0.290	
Position cis-1,4	–	–	–	–	−0.48	−1.11	0	−1.11	0	
Position trans-1,4	–	–	–	–	−2.98	−2.80	0	−0.995	−0.245	

Groups	Heat capacity constants				ΔH_f^0 (298 K)	Constants for determining Gibbs free energy			
						300–600 K		600–1500 K	
	a	$b \times 10^2$	$c \times 10^4$	$d \times 10^6$		A	$B \times 10^2$	A	$B \times 10^2$
Branching in aromatic rings									
Double branching									
Position 1,2	–	–	–	–	0.94	1.02	0	1.02	0
Position 1,3	–	–	–	–	0.38	−0.31	0	−0.31	0
Position 1,4	–	–	–	–	0.58	0.93	0	0.93	0
Triple branching									
Position 1,2,3	–	–	–	–	1.80	1.91	0	2.10	0
Position 1,2,4	–	–	–	–	0.44	1.10	0	1.10	0
Position 1,3,5	–	–	–	–	0.44	0	0	0	0
Groups containing oxygen									
−OH(primary)	6.5128	−0.1347	0.0414	−0.1623	−41.2	−41.56	1.28	−41.56	1.28
−OH(secondary)	6.5128	−0.1347	0.0414	−0.1623	−43.8	−41.56	1.28	−41.56	1.28
−OH(tertiary)	6.5128	−0.1347	0.0414	−0.1623	−47.6	−41.56	1.28	−41.56	1.28
−OH(quaternary)	6.5128	−0.1347	0.0414	−0.1623	−45.1	−41.56	1.28	−41.56	1.28
−O−	2.8461	−0.0100	0.0454	−0.2728	−24.2	−15.79	−0.85	–	–
−CHO−	3.5184	0.9437	0.0614	−0.6978	−29.71	−29.28	0.77	−30.15	0.83
\\ C=O /	1.0016	2.0763	−0.1636	0.4494	−31.48	−28.08	0.91	−28.08	0.91
−COOH	1.4055	3.4632	−0.2557	0.6886	−94.68	−98.39	2.86	−98.83	2.93
−COO−	2.7350	1.0751	0.0667	−0.9230	(−79.8)	−92.62	2.61	−92.62	2.61
	−3.7344	1.3727	−0.1265	0.3789	−21.62	−18.37	0.80	−16.07	0.40

(continued)

Groups	Heat capacity constants				ΔH_f^0 (298 K)	Constants for determining Gibbs free energy			
						300–600 K		600–1500 K	
	a	$b \times 10^2$	$c \times 10^4$	$d \times 10^6$		A	$B \times 10^2$	A	$B \times 10^2$
Part 4									
Groups containing nitrogen									
$-C \equiv N$	4.5104	0.5461	0.0269	−0.3790	36.82	30.75	−0.72	30.75	−0.72
$-N = C$	5.0860	0.3492	0.0259	−0.2436	(44.4)	46.32	−0.89	46.32	−0.89
$-NO_2$	1.0898	2.6401	−0.1871	0.4750	−7.94	−9.0	3.70	−14.19	4.38
$-NH_2$(aliphatic)	4.1783	0.7378	0.0679	−0.7310	3.21	2.82	2.71	−6.78	3.98
$-NH_2$(aromatic)	4.1783	0.7378	0.0679	−0.7310	−1.27	2.82	2.71	−6.78	3.98
\ NH / (aliphatic)	−1.2530	2.1932	−0.1604	0.4237	13.47	12.93	3.16	12.93	3.16
\ NH / (aromatic)	−1.2530	2.1932	−0.1604	0.4237	8.50	12.93	3.16	12.93	3.16
\ N − / (aliphatic)	−3.4677	2.9433	−0.2673	0.7828	18.94	19.46	3.82	19.46	3.82
\ N − / (aromatic)	−3.4677	2.9433	−0.2673	0.7828	8.50	19.46	3.82	19.46	3.82
$\nwarrow\!\searrow$ N $\leftrightarrow$ $\swarrow\!\nearrow$	2.4458	0.3436	0.0171	−0.2719	–	11.32	1.11	12.26	0.96

Groups	Heat capacity constants				ΔH_f^0 (298 K)	Constants for determining Gibbs free energy			
						300–600 K		600–1500 K	
	a	$b \times 10^2$	$c \times 10^4$	$d \times 10^6$		A	$B \times 10^2$	A	$B \times 10^2$
Groups containing sulfur									
–SH	2.5597	1.3347	–0.1189	0.3820	4.60	–10.68	1.07	–10.68	1.07
–S–	4.2256	0.1127	–0.0026	–0.0072	11.17	–3.32	1.42	–3.32	1.44
	4.0824	–0.0001	0.0731	–0.6081	(7.8)	–0.97	0.51	–0.65	0.44
Groups containing alogens									
–F	1.4382	0.3452	–0.0106	–0.0034	–	–45.10	0.20	–	–
–Cl	3.0660	0.2122	–0.0128	0.0276	–	–8.25	0	–8.25	0
–Br	2.7605	0.4731	–0.0455	0.1420	–	–1.62	–0.26	–1.62	–0.26
–I	3.2651	0.4901	–0.0539	0.1782	–	7.80	0	7.80	0

Units C_p° in kcal/(mol K); $\Delta H_{f,298K}^\circ$ in kcal/mol; and ΔG_f° in kcal/mol

References

Abrams, D., Prausnitz, J.M.: Statistical thermodynamics of liquid mixtures. A new expression for the excess Gibbs energy of partly and completely miscible systems. AIChE J. **21**, 116–128 (1975)

Ambrose, D.: Correlation and Estimation of Vapor-Liquid Critical Properties I. Critical Temperatures of Organic Compounds NPL Report Chem. 92. National Physical Laboratory, Teddington, UK (1978)

Ambrose, D.: Vapor-Liquid Critical Properties II. Critical Pressure and Critical Volume NPL Report Chem. 107. National Physical Laboratory, Teddington, UK (1980)

Anderson, J.W., Beyer, G.H., Watson, K.M.: Natl. Petrol News. **36**, R476–R483 (1944)

Bertucco, A., Barolo, M., Soave, G.: Estimation of chemical equilibria in high-pressure gaseous systems by a modified Redlich-Kwong-Soave equation of state. Ind. Eng. Chem. Res. **34**(9), 3159–3165 (1995)

Bodenstein, M.: Zersetzung und Bildung von Jodwasserstoff. Z. Physik. Chem. **22** (1897)

Carrà, S.: La Produzione Chimica: Processi e Operazioni Unitarie. In: Enciclopedia della Chimica. ISEDI (1977)

Carrà, S., Santacesaria, E., Morbidelli, M., Schwarz, P., Divo, C.: Synthesis of epichlorohydrin by elimination of hydrogen chloride from chlorohydrins. 2. Simulation of the reaction unit. I&EC Process Des. Dev. **18**(3) (1979)

Constantinou, L., Gani, R.: New group contribution method for estimating properties of pure compounds. AIChE J. **40**(10), 1697–1710 (1994)

Dahal, R., Uusi-Kyyny, P., Pokki, J., Alopaeus, V.: Isobaric vapor−liquid equilibrium of the binary mixtures toluene + styrene and styrene + α-methylstyrene. J. Chem. Eng. Data. **68**, 654–663 (2023)

Debye, P., Hückel, E.: Zur Theorie der Elektrolyte. Phys. Z. **24**, 185–206 (1923)

Dyson, D.C., Simon, J.M.: A kinetic expression with diffusion correction for ammonia synthesis on industrial catalyst. Ind. Eng. Chem. Fundam. **7**(4), 605 (1986)

Flory, P.J.: Thermodynamics of high polymer solutions. J. Chem. Phys. **10**, 51–61 (1942)

Francis, A.W.: The free energies of some hydrocarbons. Ind. Eng. Chem. **20**(3), 277–282 (1928)

Franklin, J.L.: Prediction of heat and free energies of organic compounds. Ind. Eng. Chem. **41**, 1070–1076 (1949)

Fredenslund, A.A., Russell, L.J., Prausnitz, J.M.: Group-contribution estimation of activity coefficients in nonideal liquid mixtures. AIChE J. **21**(6), 1086–1099 (1975)

Fredenslund, A.A., Gmehling, J., Rasmussen, P.: Vapor-Liquid Equilibria Using UNIFAC, A Group Contribution Method. Elsevier, Amsterdam (1977)

Gerasimov, Y.A., Dreving, V., Eremin, E., Kiselev, A., Lebedev, V., Panchenkov, G., Shlygin, A.: Physical Chemistry, vol. I. MIR Publisher, Moscow (1974)

Gillespie, L.J., Beattie, J.A.: The thermodynamic treatment of chemical equilibria in systems composed of real gases. I. An approximate equation for the mass action function applied to the existing data on the Haber equilibrium. Phys. Rev. **36**, 743 (1930)

Glasstone, S.: Thermodynamics for Chemistry. Van Nostrand, New York (1947)

Guéret, C., Daroux, M., Billaud, F.: Methane pyrolysis: thermodynamics. Chem. Eng. Sci. **52**(5), 815–827 (1997)

Hayden, J.G., O'Connel, J.P.: A generalized method for predicting second virial coefficients. Ind. Eng. Chem. Proc. Des. Dev. **14**(3), 209–216 (1975)

Hildebrand, J.H., Prausnitz, J.M., Scott, R.L.: Regular and Related Solutions. Van Nostrand Reinhold, New York (1970)

Hougen, O.A., Watson, K.M.: Chemical Process Principles, Part Two: Thermodynamics. Wiley, New York (1947)

Huggins, M.L.: Solutions of long-chain compounds. J. Chem. Phys. **9**, 440 (1941)

Joback, K.G., Reid, R.C.: Estimation of pure-component properties from group contributions. Chem. Eng. Commun. **57**, 233–243 (1987)

Kammerlingh Onnes, H.: Expression of the equation of state of gases and liquids by means of series. Commun. Phys. Lab. Univ. Leiden. **71**, 3–25 (1901)

Kay, W.B.: Density of hydrocarbon. Ind. Eng. Chem. **28**, 1014 (1936)

Kojima, K., Tochigi, K.: Prediction of Vapor-Liquid Equilibria by the ASOG Method. Kodanska, Elsevier (1979)

Larson, A.T.: The ammonia equilibrium at high pressure. J. Am. Chem. Soc. **46**, 367–372 (1924)

Lee, B.I., Kesler, M.G.: A generalized thermodynamic correlation based on three-parameter corresponding states. AIChE J. **21**(3), 510–527 (1975)

Lewis, G.N.: The law of physico-chemical change. Proc. Am. Acad. Arts Sci. **37**(4), 49–69 (1901)

Lewis, G.N., Randall, M.: Thermodynamics, 2nd edn (Revised by Pitzer, K.S., Brewer, L.). McGraw-Hill, New York (1961)

Lowry, T.M.: The uniqueness of hydrogen. J. Soc. Chem. Ind. **42**(3), 43–47 (1923)

Lydersen, A.L.: Estimation of Critical Properties of Organic Compounds. College of Engineering, Engineering Experiment Station, University of Wisconsin, Madison, WI (1955)

Margules, M.: Über die Zusammensetzung der gesättigten Dämpfe von ischungen. Sitzungsberichte der Kaiserliche Akadamie der Wissenschaften Wien Mathematisch-Naturwissenschaftliche Klasse II. **104**, 1243–1278 (1895)

McCann, D.W., Danner, R.P.: Prediction of second virial coefficients of organic compounds by a group contribution method. Ind. Eng. Chem. Process Des. Dev. **23**(3), 529–533 (1984)

Neumann, B., Kohler, G.: Die Gleichgewitchtsverhaltnisse bei der Wassergasreaktion im Temperaturbereich von 300 bis 1000°. Z. Elektrochem. **34**, 218–237 (1928)

Peneloux, A., Rauzy, E., Fréze, R.: A consistent correction for Redlich-Kwong-Soave volumes. Fluid Phase Equilib. **8**(1), 7–23 (1982)

Peng, D.Y., Robinson, D.B.: A new two-constant equation of state. Ind. Eng. Chem. Fundam. **15**(1), 59–64 (1976)

Pitzer, K.S.: The volumetric and thermodynamic properties of fluids. I. Theoretical basis and virial coefficients. J. Am. Chem. Soc. **77**, 3427–3433 (1955)

Redlich, O., Kwong, J.N.S.: On the thermodynamics of solutions. V. An equation of state. Fugacities of gaseous solutions. Rev. Chem. **44**(1), 233–244 (1949)

Reid, R.C., Prausnitz, J.M., Poling, B.E.: The Properties of Gases & Liquids. McGraw Hill, New York (1987)

Renon, H., Prausnitz, J.M.: Local compositions in thermodynamic excess functions for liquid mixtures. AIChE J. **14**(1), 135–144 (1968)

Rihani, D.N., Doraiswamy, L.K.: Estimation of heat capacity of organic compounds from group contributions. Ind. Eng. Chem. Fundam. **4**(1), 17–21 (1965)

Rossini, F.D., Pitzer, K.S., Arnett, R.L., Braun, R.M., Pimentel, G.C.: API Project 44, Selected Values of Physical and Thermodynamic Properties of Hydrocarbons and Related Compounds. Carnegie Press, Pittsburg (1953)

Scatchard, G.: Equilibrium in non-electrolyte mixtures. Chem. Rev. **44**(1), 7–35 (1949)

Scott, R.L.: Corresponding states treatment of nonelectrolyte solutions. J. Chem. Phys. **25**, 193 (1956)

Smith, B.D.: Thermodynamic excess property measurements for acetonitrile-benzene-n-heptane system at 45°C. J. Chem. Eng. Data. **17**, 71–76 (1972)

Soave, G.: Equilibrium constants from a modified Redlich-Kwong equation of state. Chem. Eng. Sci. **27**(6), 1197–1203 (1972)

Soave, G., Barolo, M., Bertucco, A.: Estimation of high-pressure fugacity coefficients of pure gaseous fluids by a modified SRK equation of state. Fluid Phase Equilib. **91**, 87–100 (1993)

Standing, M.B., Katz, D.L.: Density of natural gases. Trans. AIME. **146**, 140–149 (1942)

Stull, D.R., Westrum, E.F., Sinke, G.C.: The Chemical Thermodynamics of Organic Compounds. Wiley, New York (1969)

Su, G.J.: Modified law of corresponding states for real gases. Ind. Eng. Chem. **38**, 803–806 (1946)

Tsonopoulos, C.: An empirical correlation of second virial coefficients. A.I.Ch.E. J. **20**, 263 (1974) van der Waals, J.D.: On the continuity of the gaseous and liquid states. Ph.D. Thesis, University of Leiden, Leiden (1873)

Van Krevelen, D.W., Chermin, H.A.: Estimation of the free enthalpy (Gibbs free energy) of formation of organic compounds from group contributions. Chem. Eng. Sci. **1**, 66–80 (1951)

Van Laar, J.J.: The vapor pressure of binary mixtures. Z. Phys. Chem. **72**, 723 (1910)

Van Ness, H.C., Abbott, M.M.: Classical Thermodynamics of Nonelectrolyte Solutions. McGraw-Hill, New York (1982)

Vancini, C.A.: Synthesis of Ammonia. The Macmillan Press, London (1971)

Verma, K.K., Doraiswamy, L.K.: Estimation of heats of formation of organic compounds. Ind. Eng. Chem. Fundamen. **4**(4), 389–396 (1965)

Wallace, F.J., Linning, W.A.: Basic Engineering Thermodynamics. Pitman Paperbacks (1970)

Wilson, G.M.: Vapor-liquid equilibrium. XI. A new expression for the excess free energy of mixing. J. Am. Chem. Soc. **86**(2), 127–130 (1964)

Wisniak, Y.: Émile-Hilaire Amagat and the laws of fluids. Educacion Quimica. **17**(1), 86–96 (2005)

Chapter 3
The Role of Catalysis in Promoting Chemical Reactions

Abstract Some fundamental aspects of the catalytic phenomenon are described in the introductory part of this chapter. Next, some different catalyst classifications are described based on catalytic action (acid–base, redox, metallorganic, and enzymatic) or catalytic environment (homogeneous and heterogeneous). Different aspects of homogeneous catalysis are examined in detail considering acid–base catalysis, catalysis promoted by metal transition complexes, and enzymatic catalysis. Heterogeneous catalysis is then examined describing gas–solid interactions in particular and adsorption isotherms useful for determining gas–solid interface area. Different heterogeneous catalysts are described, such as acid–base solid catalysts based on metal oxides and their mixtures, metal oxide catalysts acting as semiconductors, zeolites, and unsupported and supported metal catalysts. The most commonly employed supports are described together with techniques for their preparation and impregnation. Finally, catalyst-forming procedures are briefly described. Some exercises and associated MATLAB code are provided online.

Keywords Mechanisms of catalysis · Homogeneous catalysis · Heterogeneous catalysis · Metallorganic catalysis · Enzymatic catalysis

3.1 Introduction to Catalytic Phenomena

We have seen that a reaction can occur spontaneously when it is thermodynamically favoured, but we have also clarified that a reaction is thermodynamically favoured simply when it passes from a higher energetic level to a lower one. However, thermodynamics gives no information about the path of this passage, and even if a reaction is highly favoured, often an energetic barrier prevents the occurrence of that reaction. We must consider also that the occurrence of a reaction is characterized by the breakage of one or more chemical bonds and the formation of new bonds. Whilst

---

Supplementary Information The online version contains supplementary material available at https://doi.org/10.1007/978-3-031-82502-6_3.

© The Author(s), under exclusive license to Springer Nature Switzerland AG 2025
R. Tesser, E. Santacesaria, *The Chemical Reactor from Laboratory to Industrial Plant*, https://doi.org/10.1007/978-3-031-82502-6_3

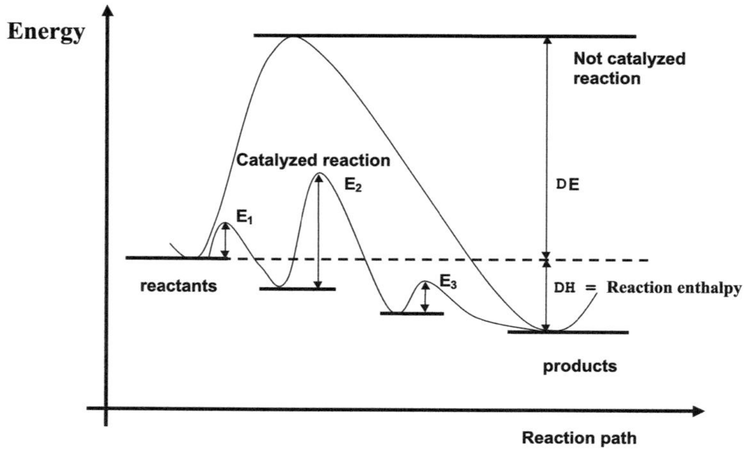

Fig. 3.1 Difference in the reaction path between a catalyzed and uncatalyzed reaction

the new bonds that can be formed are more stable, some energy is necessary for determining the breakage of the original bonds. In other words, any reaction, for starting, must be activated, and the energy required to do this is called "activation energy." The molecules have an average energetic level that is strictly related to temperature. The energy is due to translational, rotational, and vibrational motions of the molecules. Then molecules give place to collisions and are subject to reciprocal interaction. If the energy accumulated by a molecule, due to collisions or interactions, becomes greater than the activation energy, a chemical reaction can occur. Any molecule, for example, decomposes when we increase the temperature greatly, but if we desire to promote a specific reaction, we must activate selectively that reaction and cannot make this just by increasing the temperature. We must reduce the activation energy of the desired reaction, thus creating the conditions for a privileged reaction path. This can be obtained by choosing a suitable catalyst that is able to reduce the activation energy for obtaining the desired products by altering the reaction path. In other words, by using an appropriate catalyst, a shortcut for the reaction path is found, and the desired reaction occurs more quickly. A good catalyst must be selective, promoting only one reaction among all of the ones thermodynamically favoured. The catalyst changes the reaction mechanism, that is, it changes the sequence of elementary steps that occurs for a gradual transformation of the reactants in products. One or more intermediate species are formed during the reaction (Fig. 3.1). In the presence of a catalyst, the energetic barriers are lower and the intermediate-formed species corresponding to the maxima (Fig. 3.1) are more stable, that is, they are formed at a lower energy level than the overall maximum. For example, the thermal decomposition of ammonia shows an activation energy of 70.8 kcal/mol, whilst the same reaction catalyzed by a *W* catalyst occurs with an activation energy of 39.0 kcal/mol. The thermal decomposition of nitric

Table 3.1 Rates observed for ethylene hydrogenation promoted by different catalysts[a]

Catalyst	Relative reaction rates
W	1
Fe	10
Ni	60
Pt	200
Pd	1600
Rh	10,000

[a]The activation energy is, in all cases, approximately 10 kcal/mol

Table 3.2 Relative reaction rates determined for the decomposition of formic acid[a] performed at 300 °C in the presence of different solid catalysts[b]

Catalyst	Activation energy (Kcal/mol)	Relative reaction rates
Glass	24.5	1
Gold	23.5	40
Silver	31.0	40
Platinum	22.0	2000
Rhodium	25.0	10,000

[a]$HCOOH \rightarrow H_2O + CO$
[b] Gerasimov et al. (1974)

oxide shows an activation energy of 58.5 kcal/mol, whilst the same decomposition catalyzed by a Pt catalyst occurs with an activation energy of 32.5 kcal/mol.

However, activation energy is not the only factor characterizing catalytic action. In some cases, reaction rates are considerably different for different catalysts whilst the activation energy is quite similar. In Table 3.1, for example, the activities obtained by using different catalysts promoting the hydrogenation of ethylene are shown for comparison. The activation energy is approximately the same in all cases (approximately 10 kcal/mol), but—as can be seen—the relative rates are quite different.

This is due to the fact that the reactant molecules strongly interact with some atoms located on the surface with a particular geometrical configuration ("active centres"), thus giving place to chemical adsorption. Adsorbed molecules are responsible for the reaction. Therefore, the density of active sites on the solid surface can be very different, and this is the reason for the observed differences in the catalytic activities listed in Table 3.1. Another example of this type is shown in Table 3.2, in which the relative activities observed for the decomposition of formic acid, performed at 300 °C in the presence of different solid catalysts, are listed.

However, the effect of the catalyst is that of both orienting selectively the reaction and increasing the rate of the desired reaction; however, in any case, it cannot alter the thermodynamic equilibrium. Therefore, the characteristics required for a good catalyst include:

(a) Increased activity: The activity is more usually expressed as reaction rate referred to as the unit of mass of catalyst or as the conversion of the reagent for a given contact time.

Table 3.3 Examples of reactions occurring between carbon monoxide and hydrogen in the presence of different catalysts[a]

Catalyst	Occurring reaction
Ni	$CO + 3H_2 \rightarrow CH_4 + H_2O$
Cu	$CO + 2\ H_2 \rightarrow CH_3OH$
Co	$CO + H_2 \rightarrow$ olefins + paraffins
Ru	$CO + H_2 \rightarrow$ paraffins of high molecular weight

[a]Obviously, the operative conditions are different for each case

Table 3.4 Reactions of ethanol in the presence of different catalysts

Catalyst	Occurring reaction
Al_2O_3 (350 °C)	$C_2H_5OH \rightarrow C_2H_4 + H_2O$
Al_2O_3 (250 °C)	$2C_2H_5OH \rightarrow (C_2H_5)_2O + H_2O$
Cu (350 °C)	$C_2H_5OH \rightarrow CH_3CHO + H_2$
Copper chromite	$2C_2H_5OH \rightarrow CH_3COOC_2H_5 + 2H_2$
	Or
	$2C_2H_5OH \rightarrow CH_3CO\ CH_3 + 3H_2 + CO$
Zinc chromate	$2C_2H_5OH \rightarrow CH_2{=}CH\text{–}CH{=}CH_2 + H_2 + 2H_2O$
Sodium	$2C_2H_5OH \rightarrow C_4H_9OH + H_2O$

(b) Increased selectivity: The catalyst must favour the exclusive formation of the desired product.
(c) Increased stability: Regarding mechanical, thermal, and chemical resistance.

It is interesting to observe that the same reactants, in the presence of different catalysts, can give place to different products (Tables 3.3 and 3.4).

3.2 Catalyst Classification and Generalities

Different types of catalysts classification are possible. For example, we can distinguish, first of all, between homogeneous and heterogeneous catalysts. Homogeneous catalysts promote reaction occurring in a unique phase because the catalyst is dissolved in the same phase of the reactants. Heterogeneous catalysts are in a different phase with respect to the reactants, and the reaction occurs at the interface.

Another classification considers the catalytic action or mechanism independently of the previous classification (homogeneous or heterogeneous). We can recognize five different types of catalytic actions:

(a) **Acid–base catalysts (of Brønsted or Lewis type)**: Acid–base catalysts include many oxides and their mixtures, zeolites, some salts, and ionic-exchange resins. Many reactions are promoted by the acid–base action (e.g., hydrocarbon isomerization, cracking, alkylation, oligomerization, polymerization, hydration, dehydration, esterification, saponification, hydrochlorination/dehydrochlorination, polycondensation, etc.).

(b) **Redox catalysts**: These catalysts promote oxidation–reduction reactions. Redox catalysts can be metals, semiconductors oxides, metallic alloys, sulphides, metal ions, etc. The corresponding catalyzed reactions can be hydro-dehydrogenations, hydrogenolysis, oxychlorinations, oxidations, ammonoxidation, etc.
(c) **Polyfunctional catalysts**: These are catalysts that have both acid–base and redox properties for promoting a sequence of reactions requiring both of these properties.
(d) **Metallorganic catalysts**: These catalysts are normally constituted by metallorganic complexes and are used for promoting polymerization reactions (e.g., the Ziegler–Natta polymerization of propylene). Metallorganic complexes are also used in carbonylation, hydroformylation, and some hydrogenation reactions.
(e) **Enzymatic catalysts**: The catalyst in this case is a protein with a particular spatial configuration. Enzymes catalyze thousands of chemical reactions in the cells of our organism under very mild conditions of temperature. Some enzymes can be separated and used as a catalyst in some biochemical reactions, in particular in the pharmaceutical industry.

The described classification cannot be considered as a rigid separation of a class with respect to another. A catalyst can be classified in more than one of the mentioned classes. However, let us consider now, in more detail, the fundamental aspects of the first mentioned classification, that is, homogeneous catalysis.

3.3 Homogeneous Catalysis

As previously mentioned, we have homogeneous catalysis when a reaction occurs in a single phase containing both reagents and catalyst. Often, one of the reactants is gaseous and the reaction occurs in a liquid phase, where the gaseous reagent is dissolved with its own solubility. The solubility can be increased by increasing the pressure or by choosing a solvent in which the gas is more soluble. To increase the solubility of a gaseous reagent, it is important not only to increase the reaction rate by increasing its concentration but, in some cases, also to increase the stability of the catalytic complex responsible for the catalytic action. For example, carbonyl complexes of transition metals are stabilized by a high pressure of carbon monoxide (CO). The presence of a gaseous reagent can slow down the reaction as a consequence of a slow gas–liquid mass-transfer rate. This problem can be solved by increasing the gas–liquid interface through, for example, intense mixing.

Homogeneous catalysis is potentially more active and selective than the heterogeneous one because in this case, the catalyst is extremely dispersed (at the molecular level), and the active sites are uniform, therefore all exerting the same catalytic action. Despite these positive requisites, heterogeneous catalysis is largely preferred in industry because in homogeneous catalysis, it is difficult and expensive to separate and recover the catalyst from the reaction products. In some cases, the homogeneous catalyst can be heterogenized. For this purpose, intense research work has been devoted to this subject; however, heterogenized catalysts often lose part of their activity and selectivity. The main types of homogeneous catalysts are as follows: acid–base catalysts, transition-metal complexes catalysts, and enzyme catalysts.

3.3.1 Acid–Base Homogeneous Catalysis

As is well known, two types of acidity exist, one consisting of the exchange of a proton (Brønsted–Lowry acidity) and another in which an electron pair is exchanged (Lewis acidity). The first concept of acidity was proposed independently by Johannes Brønsted and Thomas Lowry in 1923. An acid, according to Brønsted–Lowry, is a species that is able to donate a proton, whilst a base is a proton acceptor as, for example, occurs in the following reaction (Brønsted 1923):

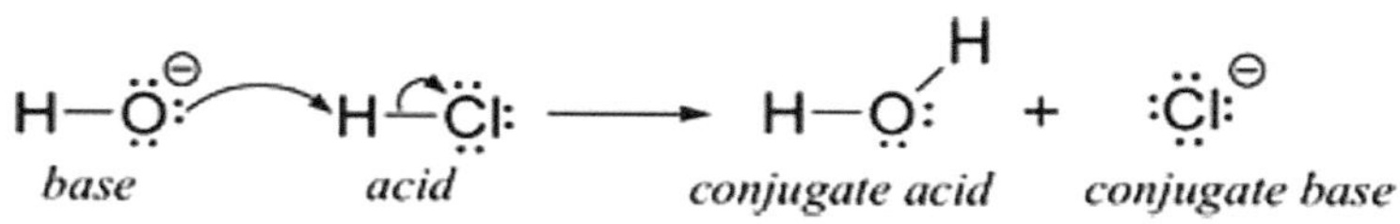

(3.1)

Lewis (1923), in the same year, wrote a monograph and extended this concept defining "acid" as an electron-pair acceptor and "base" as an electron-pair donor. Clearly, the Brønsted–Lowry theory is a particular case of the Lewis theory, as can be argued from the exchange of an electron pair occurring in the reaction (Eq. 3.1). However, a reaction between a Lewis acid and a Lewis base occurs in some cases without any proton exchange with the formation of a stable adduct, as can be seen in the following example:

$$RCl + AlCl_3 \rightarrow R^+ + AlCl_4^- \quad (3.2)$$

We can distinguish between "specific acid–base catalysis" and "general acid–base catalysis."

Specific Acid–Base Catalysis

In specific acid–base catalysis, the reaction rate depends on a specific acid present in the solution, that is, for example, the protonated form of the solvent. In specific acid–base catalysis, the first reaction step is therefore the protonation of the solvent B:

$$HA + B \leftrightarrow A^- + BH^+ \quad (3.3)$$

Then the reactant S is also protonated:

$$\mathrm{BH^+ + S \overset{k_1}{\underset{k_{-1}}{\leftrightarrow}} SH^+ + B} \tag{3.4}$$

Last, the reactant exchanges the proton with another base, B_1, or with the solvent B giving the product P.

$$\mathrm{SH^+ + B_1 \overset{k_2}{\rightarrow} P + B_1H^+} \tag{3.5}$$

General Acid–Base Catalysis

In general acid–base catalysis, all species able to donate protons can contribute in determining the overall reaction rate, that is, also un-dissociated acids and bases can contribute in determining the reaction rate.

3.3.2 Catalysis Promoted by Metal-Transition Complexes

Transition-metal complexes have found wide employment as homogeneous catalysts in many industrial processes. We remember, for example:

(a) Processes of heterolytic oxidation of ethylene to acetaldehyde (Wacker process); oxidation of propene to propene oxide (oxirane process); homolytic oxidation of hydrocarbons to acetic acid; oxidation of cyclo-hexane to cyclo-hexanol and cyclo-hexanone intermediates of adipic acid production; and oxidation of toluene to benzoic acid and of xylenes to phthalic acids
(b) Carbonylation of methanol to acetic acid; hydroformylation of alkenes to aldehydes or alcohols
(c) Oligomerization and polymerization of alkenes
(d) Process of polycondensation, such as polyethylene-terephthalate

In this type of catalysis, the reactions occur inside the coordination sphere of the metal (coordination chemistry). Therefore, the following are important: the number of ligands that a metal can coordinate; the geometry according to which the ligands are positioned around the metal; and the intensity of the metal–ligand bonds. Metals that give place to this type of catalysis are those of the *d*-block of the periodic system having the electronic configuration listed in Table 3.5.

Table 3.5 Electronic structure of some metals of the *d*-block of the periodic system

Some transition elements and their electronic configurations		
26 Fe(Ar)$3d^6 4s^2$	27 Co(Ar)$3d^7 4\ s^2$	28 Ni(Ar)$3d^8 4\ s^2$
44 Ru(Kr)$4d^7 5s^1$	45 Rh(Kr)$4d^8 5\ s^1$	46 Pd(Kr)$4d^{10} 5\ s^0$
76 Os(Xe)$4f^{14} 5d^6 s^2$	77 Ir(Xe)$4f^{14} 5d^7 5\ s^2$	78 Pt(Xe)$4f^{14} 5d^9 5\ s^1$
Noble gases		
(Ar) = $1s^2 2s^2 2p^6 3s^2 3p^6$	(Kr) = (Ar)$3d^{10} 4s^2 4p^6$	Xe = (Kr)$4d^{10} 5s^2 5p^6$

Transition metals have the possibility to give place to "dative" bonds involving their "*d*"-orbitals in the formation of complexes. In a complex, we can have σ or π bonds between the metal and an organic group. Moreover, whilst in the non-transition elements of the periodic system, the coordination number coincides with the oxidation state of the element, amongst the transition elements, the coordination number is always higher than the oxidation number, as will be seen later in the chapter. The transition metals, in the formation of complexes, tend to reach a stable electronic configuration, such as the one of the noble gases that follows in the periodic table, that is, an external electronic configuration of 18 electrons. This coordinatively saturated structure is the most stable.

Let us consider now some properties characterizing the complexes of the transition metals.

Oxidation Number (ON)
The oxidation number is the charge that remains on the metal when all of the ligands are removed.

For example:	$Fe\ (CO)_5$	ON = 0
	$Rh(PPh_3)_3Cl$	ON = +1
	$Ru(PPh_3)_4Cl_2$	ON = +2
	$Pt(PPh_3)_2Cl_2$	ON = +2

Considering the last complex, for example,

$$Pt(PPh_3)_2Cl_2 \rightarrow Pt^{++} + 2Cl^- + 2P(Ph)_3 \tag{3.6}$$

Electronic Configuration (*n*)
The electronic configuration, "*n*," is the difference between the electronic configuration in the oxidation state = 0 (n^o) and the NO.

For example:

		n^o − NO
$Fe(CO)_5$	$3d^6 4s^2$	8 − 0 = 8
$Rh(PPh_3)_3Cl$	$4d^8 5s^1$	9 − 1 = 8
$Ru(PPh_3)_4Cl_2$	$4d^7 5s^1$	8 − 2 = 6
$Pt(PPh_3)_2Cl_2$	$4f^{14} 5d^9 5s^1$	10 − 2 = 8

Maximum Coordination Number (CN_{max})
The maximum number of electronic pairs shared by the metal and the ligands can be obtained with the relation reported below:

		CN_{max}
	$Fe(CO)_5$	5
	$Rh(PPh_3)_3Cl$	5
$CN_{max} = (18 - n)/2$	$Ru(PPh_3)_4Cl_2$	6
With n = electronic configuration	$Pt(PPh_3)_2Cl_2$	5

Coordination Number and Molecular Geometry

According to the coordination number (CN), we can have different molecular geometry as follows:

CN = 2

In this case, we can have:

(a) Linear molecular geometry

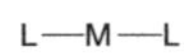

(b) Bent molecular geometry

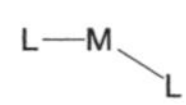

CN = 3

The possible geometries in this case are:

(a) Trigonal planar molecular geometry

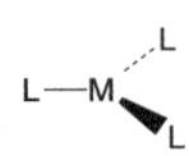

(b) Pyramidal geometry

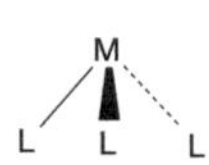

CN = 4

The possible geometries in this case are:

(a) Tetrahedral molecular geometry

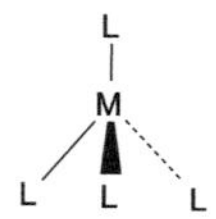

(b) Square planar molecular geometry

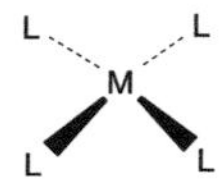

CN = 5

The possible geometries in this case are:

(a) Trigonal bipyramidal geometry

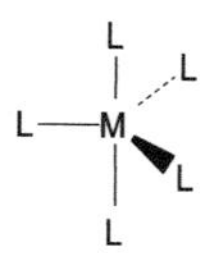

(b) Square-based pyramid molecular geometry

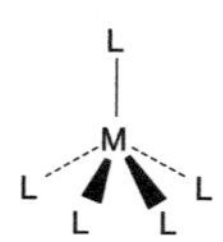

CN = 6

The structure in this case is the octahedral:

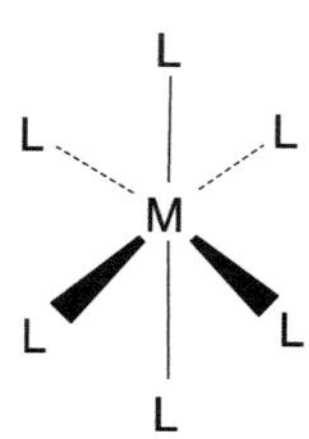

The catalytic mechanisms of the reactions promoted by the transition metal complexes are characterized by a series of consecutive steps such as:

(a) Reactions that contribute to the formation of coordinatively unsaturated species
(b) Reactions that contribute to the formation of coordinatively saturated species
(c) Reaction of insertion in which a ligand inserts between the metal and another ligand, thus leaving a free position in the coordination sphere of the metal
(d) Oxidative addition
(e) Reductive elimination

Reactions of type (a) are generally followed by reactions of type (b) with the entrance of a new ligand bonded to the metal. Reactions of type (c) are necessary for obtaining the product.

The hydroformylation of ethylene, for example, is catalyzed by a cobalt complex. The following consecutive equilibria probably occur:

Lack of ligand

$$\mathrm{Co\,H(CO)_4} \underset{+\mathrm{CO}}{\overset{-\mathrm{CO}}{\leftrightarrow}} \mathrm{Co\,H(CO)_3}\,\square \underset{-\mathrm{C_2H_4}}{\overset{+\mathrm{C_2H_4}}{\leftrightarrow}} \mathrm{CoH(CO_3)(C_2H_4)} \quad (3.7)$$

Then the following insertion reaction follows:

$$\begin{array}{ccccc} \mathrm{CH_3} & & \mathrm{CH_3} & & \\ | & & | & & \\ \mathrm{CH_2} & & \mathrm{CH_2} & & \\ | & & \vdots\ \ddots & & \\ (\mathrm{CO})_3\mathrm{Co}-\mathrm{CO} & \rightarrow & (\mathrm{CO})_3\mathrm{Co}-\mathrm{CO} & \rightarrow & (\mathrm{CO})_3\mathrm{Co}-\mathrm{CO}-\mathrm{CH_2}-\mathrm{CH_3} \end{array} \quad (3.8)$$

It is possible to identify two other possible reactions: oxidative addition and reductive elimination. When a ligand has just one electron available for a chemical bond, the other needed electron can be furnished by the metal by increasing its oxidation number. In the meantime, the coordination number and the geometrical configuration of the complex also change. An example of oxidative addition is as follows:

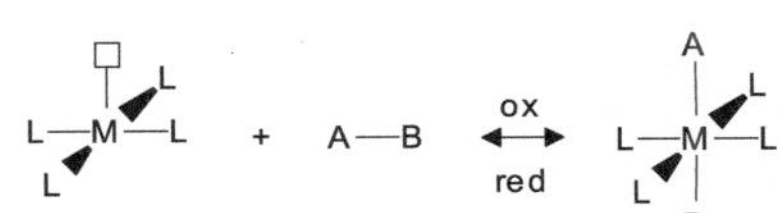

(3.9)

(A--B can be H_2, H_2O, CH_3 J etc.)

Reductive elimination is the reverse of the oxidative addition reaction.

Table 3.6 Metal complexes used as catalysts in some industrial processes

Hydrogenation	$Fe(CO)_5$, Co $H(CO)_4$, $RhClL_3$, $Ir(CO)L_2$, $[Ru\ Cl_6]^4-$, $[Co(CN)_5]^3-$
Hydroformylation	Co $H(CO)_4$, Rh $Cl(CO)L_2$
Carbonylation	Co $H(CO)_4$, Rh $Cl(CO)L_2$, $Ni(CO)_4$
Heterolytic oxidation	$[Pd\ Cl_4=]$
Homolytic oxidation	$Co(CH_3COO)_2$, Mn $(CH_3COO)_2$, or naphthenates, or acetylacetonate
Double-bond shifting	$[Fe\ H(CO)_4]-$, Co $H(CO)_4$, $[RhCl_3L]^{2-}$
Dimerization	$[Rh\ Cl_2\ L_2]^-$

L is a ligand, normally triphenylphosphine with the exception of reactions involving the double bond, the ligand being the double bond

In the carbonylation of methanol, for obtaining acetic acid in the presence of a cobalt complex, methyl iodide is added as activator. An oxidative addition of this type occurs:

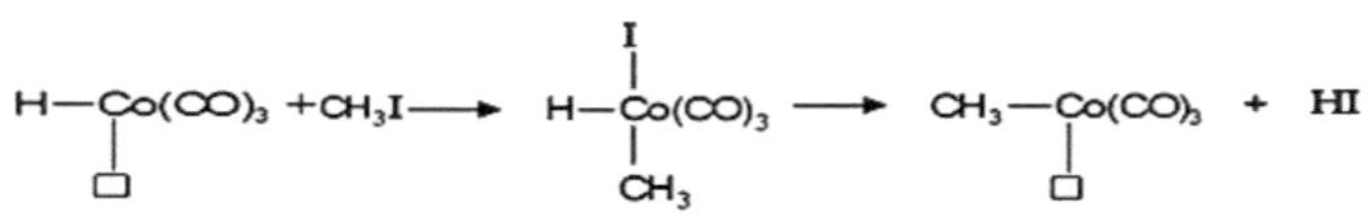

(3.10)

having the scope of bonding directly the methyl group to the metal. Then, we have the following steps:

$$CH_3\,Co\,(CO)_4 + H_2O \quad \rightarrow \quad CH_3\,COOH + \underset{\substack{|\\ *}}{HCo(CO)_3} \tag{3.11}$$

$$CH_3Co\,(CO)_4 + H_2O \rightarrow CH_3COOH + HCo(CO)_4 \tag{3.12}$$

In Table 3.6, some metal complexes used in industry as homogeneous catalysts are listed together with the corresponding catalyzed reactions.

3.3.3 *Enzymatic Catalysis*

Many biochemical reactions occurring in living organisms are catalyzed by enzymes. The ability of enzymes to catalyze biochemical reactions, under very mild conditions with very high selectivity, encouraged industry to use them to obtain different organic products that are difficult to produce using traditional catalysts. The reactions are conducted in industrial bioreactors to obtain either simple molecules (e.g., alcohols, carboxylic acids, acetone, etc.) or molecules with a complicated structure (e.g., antibiotics, proteins, etc.). The enzymes are proteins, which are polymeric chains constituted by a particular sequence of amino acids that assume a spatial configuration as, for example, the one shown in Fig. 3.2. The spatial

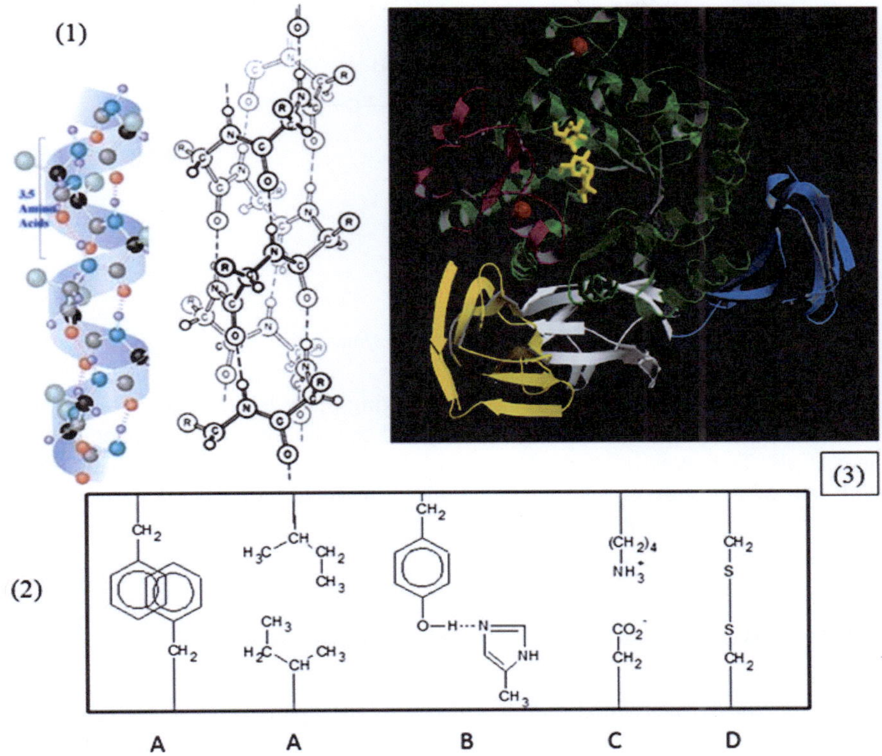

Fig. 3.2 (1) Structure of α-helix of a peptide chain and crossing hydrogen bonds (see Structural Biochemistry/Vol. 5. Wikibooks); (2) peptide chain interactions (**A** = hydrophobic interactions; **B** = hydrogen bonds; **C** = polar interactions; **D** = disulphide bonds); (3) example of the spatial configuration of the enzyme α-amylase. Reproduced with the permission of Nielsen and Borchert (2000) edited by Elsevier

configuration is due to both the sequence of amino acids characterizing that protein and the interactions between the amino acid functional groups in a folded chain. The main interactions are due to sulphur bridges, polar groups of different charge, and hydrogen bonds. Inside the enzyme molecule, together with the chain of amino acids (protein), cofactors assist the enzymes in their catalytic activity. Cofactors can be inorganic ions or particular organic molecules (non-protein). The organic cofactor is called a "coenzyme" and is located in a strategic point of the enzyme molecule where it binds the substrate favouring its transformation in products (see Fig. 3.3). In other words, the coenzyme can be recognized as a sort of "active site" of the enzyme, that is, the point at which the desired reaction effectively occurs with a mechanism described in Fig. 3.4. As can be seen, only one substrate can accommodate properly inside the enzyme structure to react. This explains the very high selectivity of enzymes.

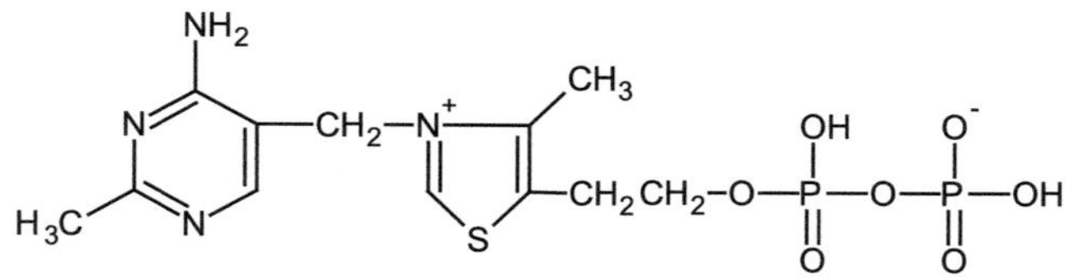

Thiamine pyrophosphate

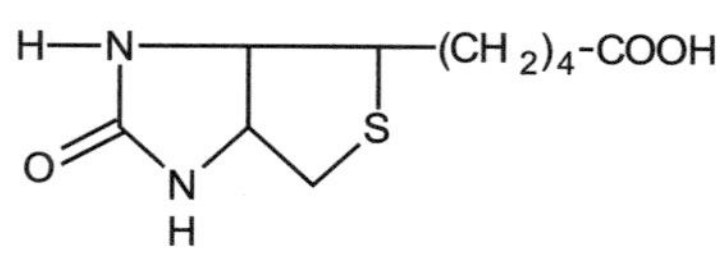

Fig. 3.3 Example of organic molecules acting as coenzyme

Enzymes are macromolecular structures giving place in water to colloidal solutions. The reactions normally occur quickly and selectively at room temperature and neutral pH. Temperature cannot be increased >50 °C to avoid protein denaturation with loss of the catalytic activity. In addition, pH must be kept in a narrow range, approximately 7, to avoid protein denaturation.

3.3.4 Heterogenization of Homogeneous Catalysts

It was previously mentioned that it is difficult and expensive to separate homogeneous catalysts from the products at the end of a reaction. This difficulty is well evidenced by examining the sequence of operations to be made for recovering, for example, a cobalt catalyst used for promoting alkene hydroformylation (see Fig. 3.5).

This difficulty pushed the research of new methods for anchoring homogeneous catalysts on the surface of a solid matrix. The simplest example of heterogenized catalysts are the acid and basic ionic exchange resins having H^+ and OH^- groups of uniform strengths distributed on the solid surface of a polymer. It is more difficult to anchor a metal transition complex on a solid surface. This can be achieved by introducing some particular functional groups on the solid surface that can provide strong interaction with the metal complex.

Phosphinic groups, for example, are good ligands and can be used at this purpose. If the solid matrix is a polymer, the phosphinic group can be introduced in the polymeric chain or through the polymerization of an already functionalized monomer:

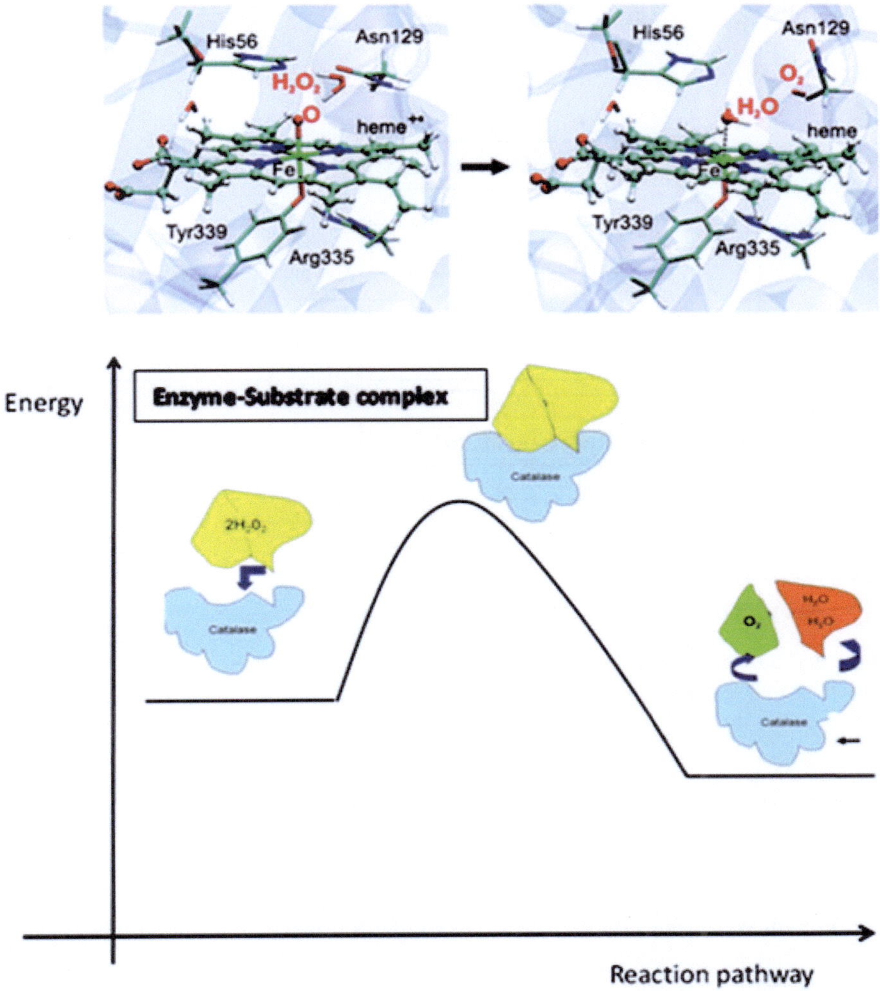

Fig. 3.4 Example of enzyme mechanism of reaction. The decomposition of hydrogen peroxide catalyzed by the enzyme catalase. Re-elaborated with permission from Prieto et al. (2009). Copyright ACS

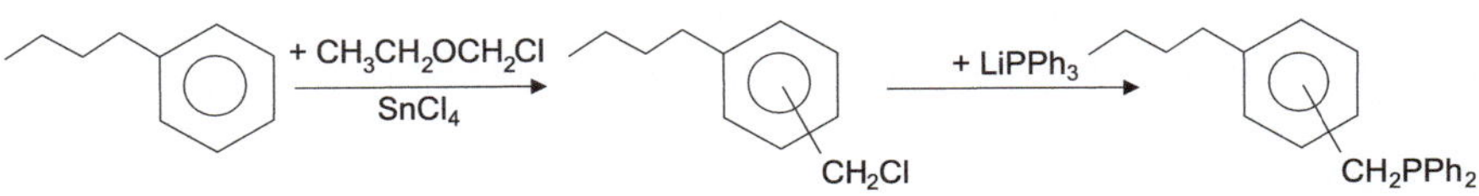

(3.13)

The functionalized polymer is then put in contact with the complex anchoring it to the surface.

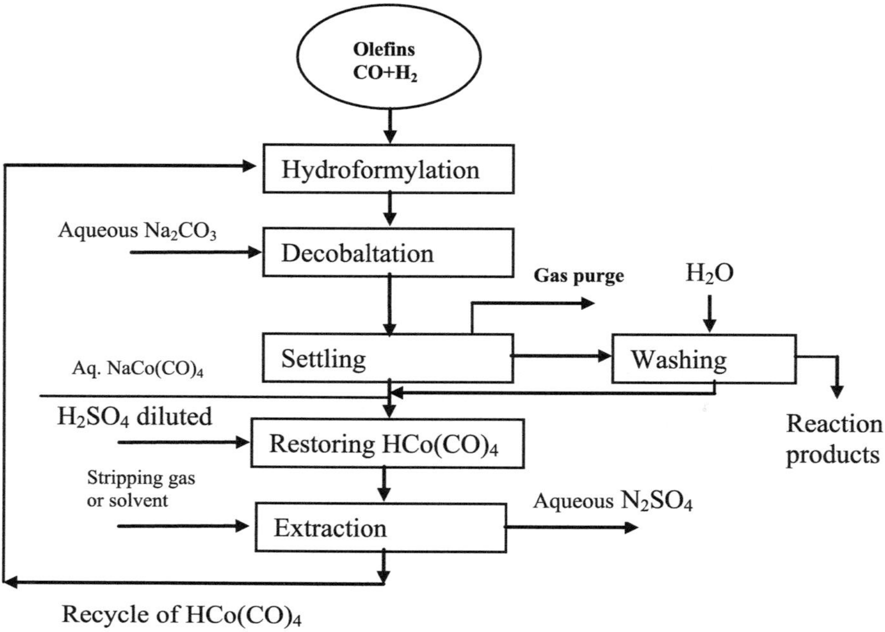

Fig. 3.5 Sequence of operations for recovering homogeneous cobalt catalyst after hydroformylation reaction

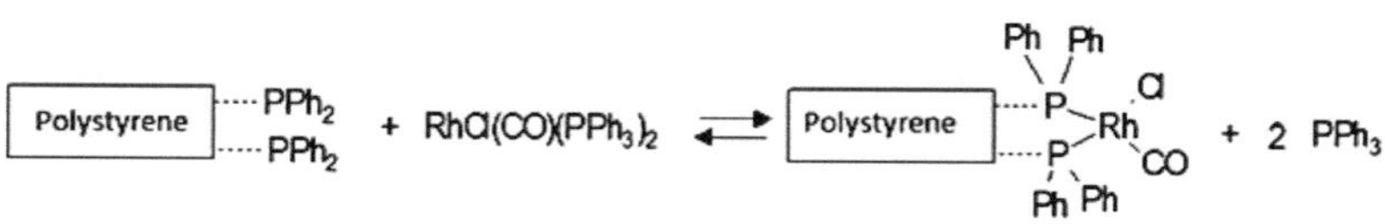

(3.14)

The functionalization also can be performed using inorganic materials that are rich in hydroxyl groups on the surface using the following sequence of reactions:

$$Ph_2PH + CH_2 = CH\ Si\ (OEt)_3 \rightarrow Ph_2P - CH_2 - CH_2 - Si(OEt)_3 \tag{3.15}$$

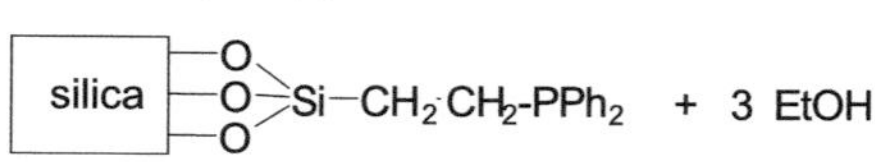

(3.16)

Apart from the advantage of easy separation, another advantage of heterogenized catalysts is the possibility to obtain a high concentration of active sites on the solid surface. Moreover, it is also possible to introduce polyfunctionality into these catalysts. A drawback is the low resistance to increased temperatures. In addition, the enzymes can be anchored on the surface of a solid matrix, such as polystyrene, porous glass, cellulose, etc. The anchoring, in this case, is obtained using functional groups that give bonds with the free amine groups, NH_2, of the enzyme molecule:

$$\boxed{\text{Cellulose--}}\text{--O CH}_2\text{-}\Phi\text{-N}_2{}^+\text{Cl}^- \quad + \quad \text{H}_2\text{ N--Protein}$$
$$\downarrow \tag{3.17}$$
$$\boxed{\text{Cellulose--}}\text{--O CH}_2\text{-}\Phi\text{-N}=\text{N-NH--Protein} + \text{HCl}$$

where Φ is an aromatic ring.

3.4 Heterogeneous Catalysis

3.4.1 Introduction

Heterogeneous catalysis occurs at the fluid–solid interface. Therefore, the activity will depend not only on the chemical composition of the surface but also on its extension. Clearly, it is convenient to prepare catalysts with a high specific surface area to reduce both the amount of catalyst to be used as well as the volume of the reactors. However, a high specific surface area indicates a porous structure characterized by narrow pores. The greater the surface, the narrower the pores; however, if the pores are too narrow, the internal diffusion of the molecules becomes slow, thus negatively affecting the reaction rate and altering the selectivity. The negative effect of the slow diffusion can be decreased by reducing the size of the catalyst pellets, but this increases the pressure decrease in a tubular packed reactor, thus increasing the energy consumption. All of this suggests that the catalyst must be prepared by optimizing both the size of the particles and the pore-size distribution. For gas–solid reactions, for example, particles range normally in size from 0.2 to 0.8 cm and have a void fraction of 40–50% determined by pores having average diameters in the range of 10^{-5} to 10^{-7} cm. The shape of the particles for gas–solid reactions can be cylindrical, spherical, or chips of irregular shape except for a fluidized bed reactor using powdered catalysts. The catalysts for liquid–solid reactions are normally in the form of powder. A solid can be used directly as catalyst, but more frequently, the true catalyst is dispersed as much as possible on the surface of a solid matrix called "support." In some cases, the support is a monolith. It is possible to demonstrate, by feeding poisoning molecules, that only a very small portion of the supported material is catalytically active in a certain reaction. This suggests, according to Taylor (1925), the existence of "active sites" that are responsible for the reaction. The heterogeneous catalysis occurs through a sequence of steps as follows:

(1) Chemical adsorption: Reactants are chemically adsorbed on the active sites.
(2) Surface reaction: Reaction occurs between the species adsorbed on the active sites or between an adsorbed specie and another not adsorbed coming from the fluid phase.
(3) Product desorption: Product molecules are released, thus liberating the active site for another reaction cycle.

The active sites of heterogeneous catalysts are not uniform. If we consider, for example, a perfect cubic crystal of a metal, we can have atoms located on the vertexes that are different from the atoms positioned on the corners that are different from the atoms located on the square faces. The chemical adsorption of a molecule on these different atoms is different in intensity and geometrical characteristics. If the crystals are irregular, the atoms on the surface are much less uniform. High selectivity requires that active sites are uniform. Normally, active sites are constituted by one or more atoms interacting with the reactant molecules, and a small difference in the geometrical location and in the interatomic distance can have dramatic effects in orienting one reaction with respect to another one as shown in the examples of reactions 3.18–3.19:

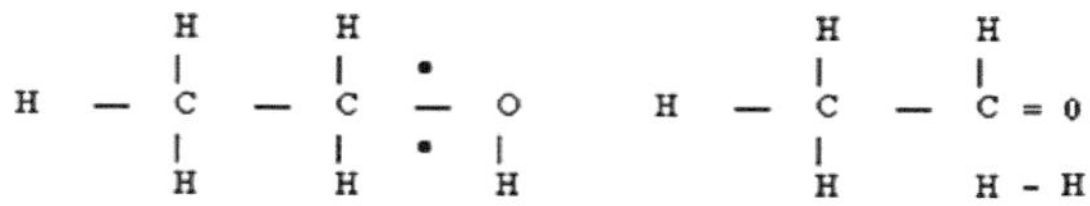
(3.18)

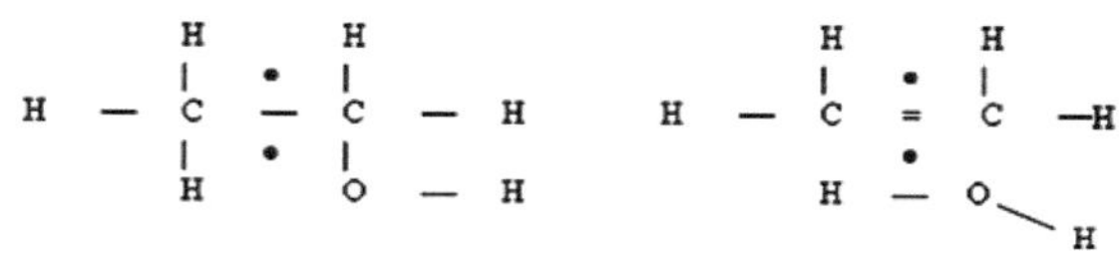
(3.19)

In conclusion, the surface of the solid catalysts does not have uniform composition. This can be demonstrated, for example, by measuring the differential heat of hydrogen adsorption as a function of the degree of the surface covering. A plot giving the trends of this type of measurement for different metals is shown in Fig. 3.6. In the case of a uniform surface, the adsorption heat would be constant, whilst—as can be seen in Fig. 3.6—the heat decreases with the surface coverage.

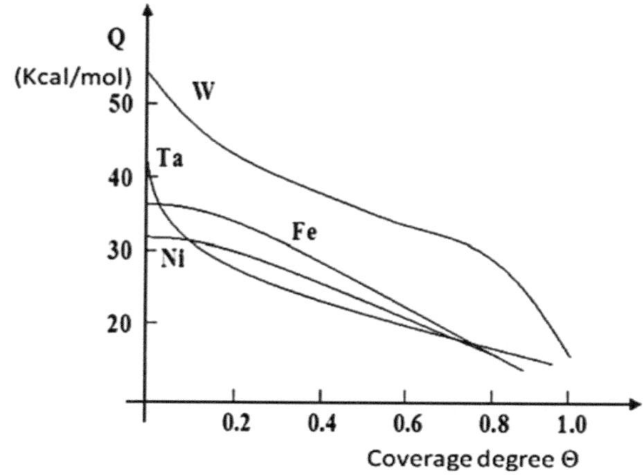

Fig. 3.6 Differential hydrogen adsorption heat on different metal films as a function of the degree of surface coverage. Re-elaborated the permission from Beek (1950). Copyright Royal Society of Chemistry (1950) [see also Smith (1981)]

Last, it is important to collect some other physical properties of the solid catalysts, such as the specific surface area, the porosity, and chemical adsorption of different substances.

3.4.2 Physical Adsorption, Specific Surface Area Measurement, and Porosity

The physical adsorption of a gas (adsorbate) on a solid surface (adsorbent) is characterized by weak interactions between the atoms of the solid surface and the gas molecules. These interactions are of the van der Waals type. The adsorption equilibrium is reached when the chemical potential of the adsorbed molecules and the gas is equal, that is:

$$\mu_s = \mu_g \tag{3.20}$$

being:

$$\mu_g = \mu_g^{\circ} + RT \ln f/f^{\circ} \tag{3.21}$$

where f is the fugacity of the non-adsorbed gas, and f^o is the fugacity of the gas in the reference state. At low pressure ($\lim P \to 0$), the fugacity can be identified with the pressure; therefore:

$$\mu_g = \mu_g^{\circ} + RT \ln P/P_0 \tag{3.22}$$

If we assume $P_0 = 1$ atm, it results in:

$$\mu_g = \mu_g^{\circ} + RT \ln P \tag{3.23}$$

For the adsorbed gas, we can write:

$$\mu_s = \mu_s^{\circ} + RT \ln a_s/a_s^{\circ} \tag{3.24}$$

where a_s is the activity of the adsorbed gas and a_s° is the activity in the standard state.

At a very low coverage degree of the surface, near zero, the activity can be identified with the concentration of the adsorbate on the surface, that is, $a_s = C_s$, whilst $a_s^{\circ} = 1$. Therefore, it is possible to write:

$$\mu_s = \mu_s^{\circ} + RT \ln C_s \tag{3.25}$$

Remembering then the equality (Eq. 3.20), it is possible to write:

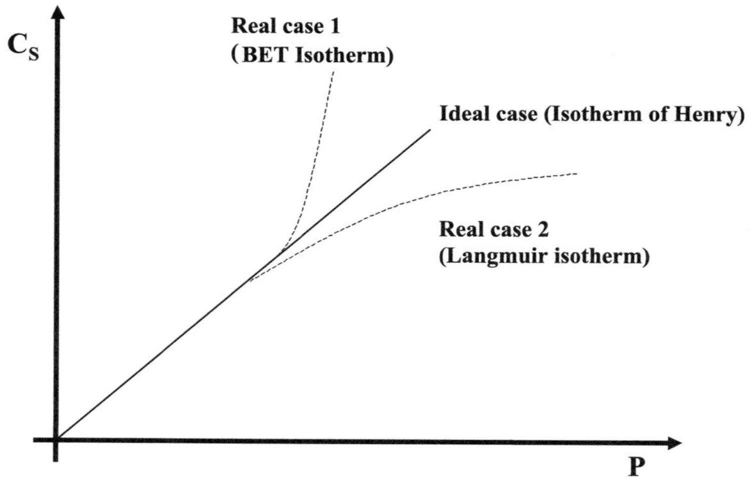

Fig. 3.7 Ideal and real behaviours observed in physical adsorption

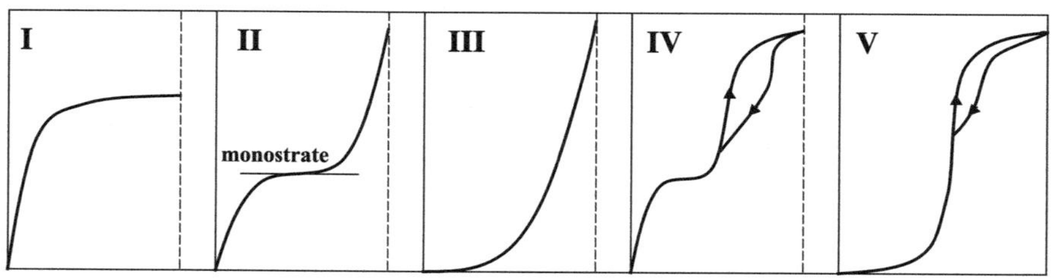

Fig. 3.8 Different types of real adsorption isotherms

$$\mu^{\circ} = \mu_g^{\circ} - \mu_s^{\circ} = -RT \ln C_s/P = -RT \ln K \tag{3.26}$$

$$K = C_s/P = \text{thermodynamic constant of adsorption (Henry isotherm).} \tag{3.27}$$

It is important to point out that K implicitly contains the reference states, and therefore, it does not have dimensional units. For a high degree of coverage, clearly a_s cannot be considered equal to C_S, and to describe the adsorption, it is necessary to introduce a coefficient of activity $\gamma_s(\Theta)$. Then it is possible to write:

$$K = \frac{C_s}{P} \gamma_s(\Theta) \tag{3.28}$$

where $\gamma_s(\Theta)$, the activity coefficient, depends on the coverage degree Θ. The introduction of the activity coefficient brings two different real experimentally observed behaviours, which are illustrated in Fig. 3.7.

The description of the real isotherms requires the definition of the function $\gamma_s = \gamma_s(\Theta)$. In gas–solid physical adsorption, we have five different real isotherms, which are represented in Fig. 3.8.

- **Isotherm I (Langmuir isotherm)**: The first of these isotherms, the Langmuir isotherm (1916), is particularly important because it is a reference situation for the other cases. As a matter of fact, the adsorption according to the Langmuir isotherm is limited to a monolayer of molecules. This isotherm can be observed in physical adsorption only in the case of relatively strong interactions between the molecules of the adsorbate and specific points of the adsorbent (specific localized adsorption). Despite the fact that Langmuir has suggested this isotherm to describe physical adsorption phenomena, this isotherm is more suitable for describing chemical adsorption and is important, in particular, to describe the competition between different molecules in the adsorption on a solid surface.
- **Isotherm II (BET isotherm)**: The BET isotherm (Brunauer et al. 1938) is characterized by the adsorption of an indefinite numbers of layers of molecules after the formation of a monolayer, which can be identified as a point of flex.
- **Isotherm III**: Isotherm III does not show the flex of the monolayer, which is due to very low adsorption heat. Also in this case, we have an indefinite number of adsorbed layers.
- **Isotherm IV**: Isotherm IV is a variation of the BET isotherm II with the adsorption of a definite number of layers as a consequence of a complete pore filling. Adsorption and desorption follow different pathways.
- **Isotherm V**: Isotherm V is a variation of isotherm III with the adsorption of a definite number of layers as a consequence of a complete pore filling. Adsorption and desorption follow different pathways.

In conclusion, we have seen that the Henry isotherm, $K = C_s/P$, is an ideal isotherm representing the thermodynamic reference for the adsorption phenomenon. The Langmuir isotherm represents an important reference for the real isotherm because it shows the progressive saturation of the monolayer on a solid characterized by uniform sites of adsorption. The adsorption in this case must be localized because it is characterized by a strong interaction between the molecules of the adsorbate and the adsorption sites on the solid surface. No interaction occurs between the adsorbed molecules.

Isotherm I: The Langmuir Isotherm

According to Langmuir (1916), it is possible to consider the described adsorption as a chemical reaction:

$$A + \sigma_f \underset{k_{-A}}{\overset{k_A}{\leftrightarrow}} \sigma_o \tag{3.29}$$

where σ_f and σ_o are, respectively, the number of sites free and occupied. Considering a finite number of sites, we can write: $\sigma_f + \sigma_o = \sigma_{tot}$ and assuming:

$$\Theta = \sigma_o/\sigma_{tot} \text{ and consequently } \sigma_f/\sigma_{tot} = (1 - \Theta) \text{ where } \Theta = \text{coverage degree of the sites} \tag{3.30}$$

On the basis of Eq. (3.29), we can write the adsorption rate as:

$$r = r_A - r_{-A} = k_A p_A \sigma_f - k_{-A}\sigma_o = k_A p_A \sigma_{tot}(1-\Theta) - k_{-A}\Theta\sigma_{tot} \tag{3.31}$$

At the equilibrium $r=0$; hence,

$$b = \frac{k_A}{k_{-A}} = \frac{1}{p_A}\frac{\Theta}{(1-\Theta)} \text{ with } b = b^o e^{-\Delta H/RT} \text{ and } b^o = e^{\Delta S/R} \tag{3.32}$$

Considering the premises, the Langmuir isotherm is valid for localized physical adsorption with strong specific interactions and for chemical adsorption on sites of uniform characteristics. As will be seen in a next chapter, which is devoted to kinetics, the main advantage of the Langmuir isotherm is the possibility to describe the competition in adsorption for a multi-component gaseous mixture. In this case, we can write:

$$b_i = \frac{1}{p_i}\frac{\Gamma_i}{\left(\Gamma^\infty - \sum_{j=1}^{N}\Gamma_j\right)} \tag{3.33}$$

where Γ_i is the surface concentration of the adsorbate "*i*" on the solid (mol/g); $\Gamma\infty$ is the maximum level of surface concentration corresponding to the monolayer; Γ_j is the surface concentrations of the adsorbates j (mol/g); and p_i is the partial pressure of i in equilibrium conditions.

As will be seen later, expressions of this type will be largely used in the kinetic expressions describing the reaction rates in heterogeneous catalysis.

Exercise 3.1. Description of Adsorption Using the Langmuir Model
A study published by Piccin et al. (2011) related to a red dye adsorption on chitosan reported several adsorption experiments under isothermal conditions, respectively, performed at 298 and 318 K. Two datasets taken from this work are listed in Table 3.7.

Describe the datasets of the two isotherms using the Langmuir model and calculate by fitting the adjustable parameters. Produce a plot in which both experimental data and calculated curves are reported.

Solution
The Langmuir equilibrium isotherm is represented by the following relation between the amount of a solute in the liquid phase and that of the same compound adsorbed on the solid:

$$q_e = \frac{q_{max} K C_e}{1 + K C_e} \tag{3.34}$$

where

Table 3.7 Experimental data of red dye adsorption on chitosan[a]

Ce (298 K)	1.879	5.888	21.139	33.389	88.218	117.926	166.129	214.849	252.781
	311.245	372.273	413.296	–	–	–	–	–	–
q_e (298 K)	2.371	3.654	5.474	6.257	6.848	7.323	7.363	7.326	7.557
	7.520	7.496	7.523	–	–	–	–	–	–
Ce (318 K)	51.676	106.942	142.794	171.475	244.746	287.780	317.503	403.621	486.159
q_e (318 K)	1.642	3.259	3.875	4.389	5.237	5.827	6.097	6.587	6.948

[a]Experimental data collected by Piccin et al. (2011); Ce in (mol/L) $\times 10^6$; q_e in (mol/g) $\times 10^4$

Table 3.8 Langmuir parameters determined by fitting

	K (L/mol)	q_{max} (mol/g) × 10^4
$T = 298$ K	0.1649 (0.1961)	7.5631 (7.512)
$T = 318$ K	0.0040 (0.0041)	10.6785 (10.614)

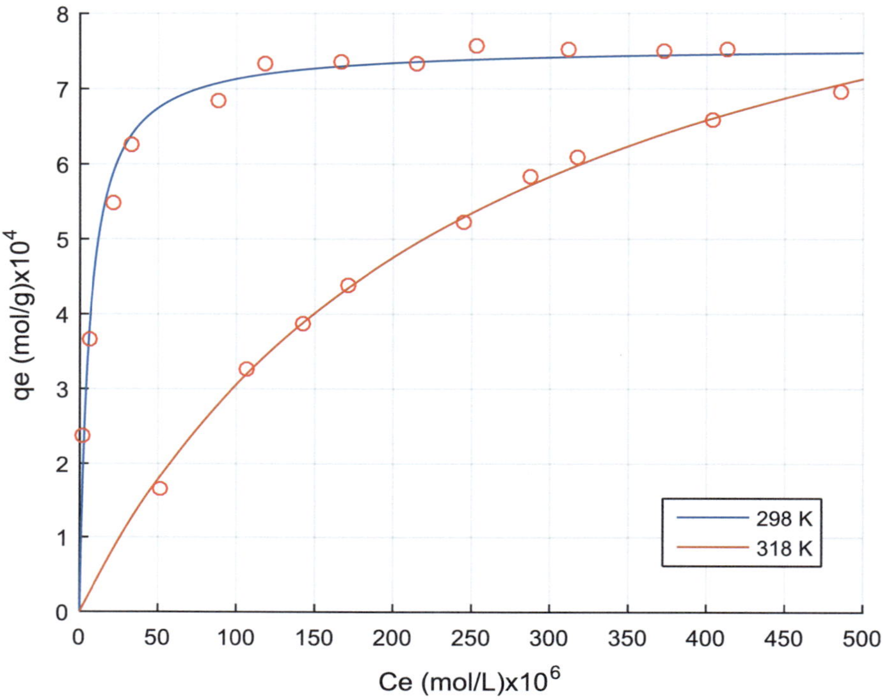

Fig. 3.9 Adsorption isotherms of red dye on chitosan at two different temperatures. Experimental points determined by Piccin et al. (2011)

q_e	amount of solute adsorbed on the solid (mol/g)
q_{max}	maximum adsorption capacity (mol/g)
K	Langmuir equilibrium constant (L/mol)
C_e	liquid-phase concentration (mol/L).

By fitting the expression of Langmuir isotherm to both datasets, the following parameters can be calculated (see Table 3.8).

With these parameters, the continuous curves of the simulations, shown below in Fig. 3.9, can be drawn. In Table 3.8, in parentheses, the values of parameters found in the cited paper are also listed. We can observe that there is good agreement between them. Another interesting observation is the fact that chitosan shows an adsorption behaviour in which adsorption capacity increases by decreasing the temperature. This was explained by Piccin et al. (2011) assuming a combination of

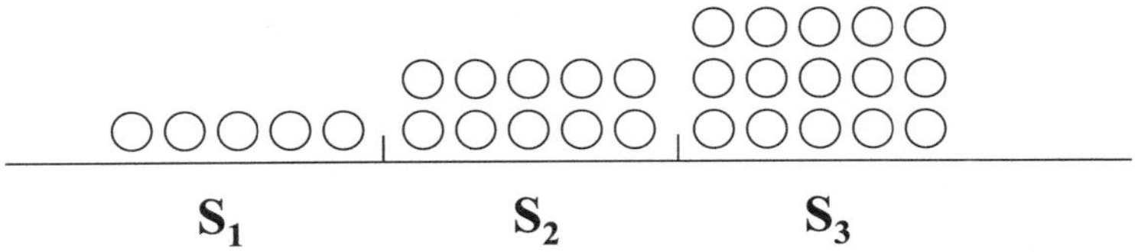

Fig. 3.10 Complexes formed on a solid surface as a consequence of the sequence of reactions

concurrent phenomena of physisorption and chemical interaction between the adsorbate and the solid adsorbent.

The described results were obtained using a MATLAB program available as Electronic Supplementary Material, Data 3.1.

Isotherm II: The BET Isotherm

The BET isotherm is quantitatively described with an expression derived from a quasi-chemical model proposed by Brunauer et al. (1938). First, three conditions, similar to the ones of the Langmuir model, are imposed:

(a) The surface is considered uniform.
(b) There are no lateral interactions between the adsorbed molecules.
(c) Adsorption is localized, that is, we admit the existence of adsorption sites.

Then we consider the following sequence of reactions:

$$\begin{aligned}
&\text{Vapour} + \text{free surface} \quad \leftrightarrow \quad \text{single complexes } S_1 \\
&\text{Vapour} + \text{single complexes } S_1 \quad \leftrightarrow \quad \text{double complexes } S_2 \\
&\text{Vapour} + \text{double complexes } S_2 \quad \leftrightarrow \quad \text{triple complexes } S_3 \\
&\text{etc.}
\end{aligned} \tag{3.35}$$

The effect of the reactions (Eq. 3.35) is shown in Fig. 3.10.

We can write the following sequence of equilibria:

$$K' = \frac{\Theta'}{p\Theta_o}; K'' = \frac{\Theta''}{p\Theta'}; K''' = \frac{\Theta'''}{p\Theta''} \text{ ecc.} \tag{3.36}$$

where Θ_o is the fraction of free surface, that is, not covered by adsorption.

The amount of gas adsorbed will be:

$$x = x_m\left(\Theta' + 2\Theta'' + 3\Theta''' + \cdots\right) \tag{3.37}$$

where x_m = capacity of the monolayer (mol/g).

Then considering that the interactions bringing to multiple complexes are weak and comparable with each other, we can write:

$$K'' \simeq K''' \cdots \simeq K_{\mathrm{L}} = 1/P_{\mathrm{s}} \tag{3.38}$$

K_{L}	condensation equilibrium constant.

Only the direct interaction between adsorbate–adsorbent correspondent to the first equilibrium is consistently different for the intensity of the interaction. By substituting in the equilibrium expression, the results of the described approximation we have:

$$\Theta'' = K'' P\Theta' = K_{\mathrm{L}} P\Theta' = \Theta' P/P_{\mathrm{s}} \tag{3.39}$$

$$\Theta''' = K''' P\Theta'' = (K_{\mathrm{L}}P)^2\Theta' = (P/P_{\mathrm{s}})^2\Theta' \ldots \text{etc;} \tag{3.40}$$

hence,

$$x = x_{\mathrm{m}}\ K'\ P\Theta_{\mathrm{o}}\ \left[1 + 2\ P/P_{\mathrm{s}} + 3(P/P_{\mathrm{s}})^2 + \cdots\right] \tag{3.41}$$

By series expansion, we obtain:

$$x = x_{\mathrm{m}} K'\ P\Theta_{\mathrm{o}}\ \frac{1}{(1 - P/P_{\mathrm{s}})^2} \tag{3.42}$$

Moreover, it is valid that:

$$\Theta_{\mathrm{o}} + \Theta' + \Theta'' + \Theta''' + \cdots = 1 \tag{3.43}$$

$$\Theta_{\mathrm{o}}\left\{1 + K'P\left[1 + P/P_{\mathrm{s}} + (P/P_{\mathrm{s}})^2 + \cdots\right]\right\} = 1 \tag{3.44}$$

and also:

$$\Theta_{\mathrm{o}}\left\{1 + K'P\frac{1}{(1 - P/P_{\mathrm{s}})}\right\} = 1 \tag{3.45}$$

By substituting:

$$X = X_{\mathrm{m}}\frac{K'P}{(1 - P/P_{\mathrm{s}})(1 + K'P - P/P_{\mathrm{s}})} \tag{3.46}$$

Writing then:

$$P = P_{\mathrm{s}} P/P_{\mathrm{s}} = (1/K_{\mathrm{L}})\,(P/P_{\mathrm{s}})\,C = K'/K_{\mathrm{L}} \tag{3.47}$$

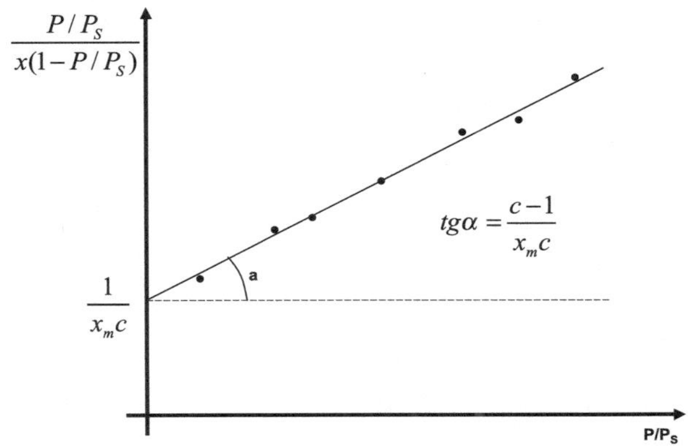

Fig. 3.11 BET isotherm expressed in linear form

$$x = \frac{x_m C(P/P_s)}{(1 - P/P_s)[1 + (C-1)P/P_s]} \text{BET Isoterma} \tag{3.48}$$

Rearranging this expression:

$$\frac{P/P_s}{x(1 - P/P_s)} = \frac{1}{X_m C} + \frac{C-1}{x_m C}\frac{P}{P_s} \tag{3.49}$$

We can put into a plot the quantity $\frac{P/P_s}{x(1-P/P_s)}$ as a function of P/P_S for obtaining a linear trend as shown in Fig. 3.11, in which the slope is $[(C - 1)/x_m C]$, and the intercept is $(1/x_m C)$.

The BET isotherm is useful for determining the specific surface area of solids using gases of known molecular encumbrance for adsorption. Nitrogen molecules, for example, at 77 K cover a surface area of 16.2 Å^2. The specific surface area of a solid, according to the BET isotherm, can be determined as:

$$S_{\text{BET}} = x_m \omega_{N_2} N_A \left(\text{m}^2/\text{g}\right) \tag{3.50}$$

As mentioned previously, $\omega_{N_2} = 16.2\,\text{A}^2$, and N_A is the Avogadro number $= 6.0232 \times 10^{23}$ (molecules/mole), where x_m is the amount of adsorbed nitrogen in the monolayer that can be determined from a plot such as the one shown in Fig. 3.11. C is the ratio K'/K_L, and it is possible to write:

$$C = \frac{K'}{K_L} = \frac{g' e^{\overline{Q}_i/RT}}{g_L e^{L/RT}} = g\ e^{(\overline{Q}-L)/RT} \tag{3.51}$$

where g' is related to the adsorption entropy; Q_1 is the heat of adsorption in the first layer; L is the heat of adsorbate condensation; $(Q_1 - L)$ is the net heat of adsorption; and g is related to the net entropy of adsorption.

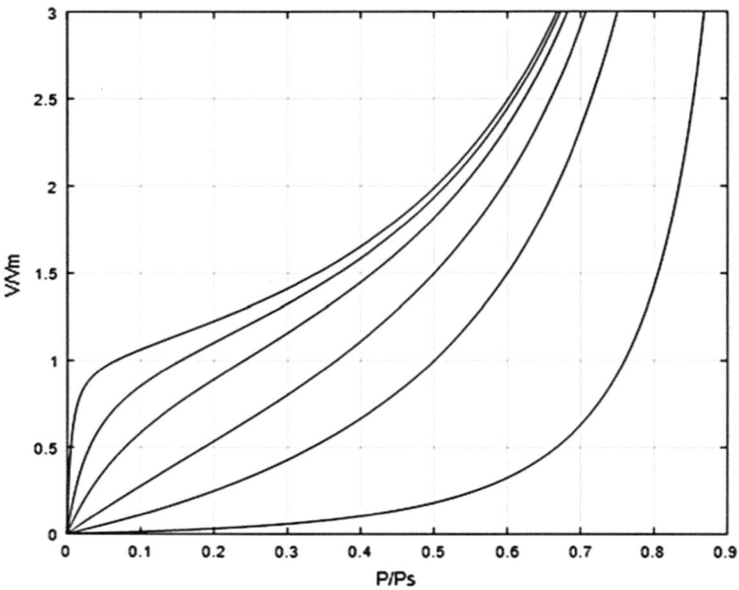

Fig. 3.12 The form of BET isotherms for different C values

- If $C \gg 1$ *e* P far from P_s we obtain → isotherm I
- $C < 1$ we obtain → isotherm III
- In the intermediate cases we obtain → isotherm II
- These behaviours are well represented in Fig. 3.12.

BET isotherms are collected with gas-volumetric apparatus, which were in the past such as the one shown in Fig. 3.13. Currently, completely automatic equipment is available on the market for the evaluation of the specific surface area based on the BET theoretical approach. An example of this type of apparatus (Sorptomatic, Thermo Fisher Scientific) is shown in Fig. 3.14. A scheme of the internal part of this apparatus is shown in Fig. 3.15. Software is supplied together with the apparatus for the determination of both the specific surface area with the BET method and the pore-size distribution using the method by Dollimore and Heal (1964).

Exercise 3.2. Determination of the Specific Surface Area of a Catalytic Support from Experimental Data of Nitrogen Adsorption

A sample of powdered silica was submitted to nitrogen-adsorption measurements for determining the specific surface area (m^2/g) using the BET method. The data collected are represented by a set of values of $P/P°$ ratio (total pressure over saturation pressure) as a function of adsorbed volume V. The experimental data are listed in Table 3.9.

As seen before, the mathematical expression of BET isotherm, useful for data interpretation, is of the type:

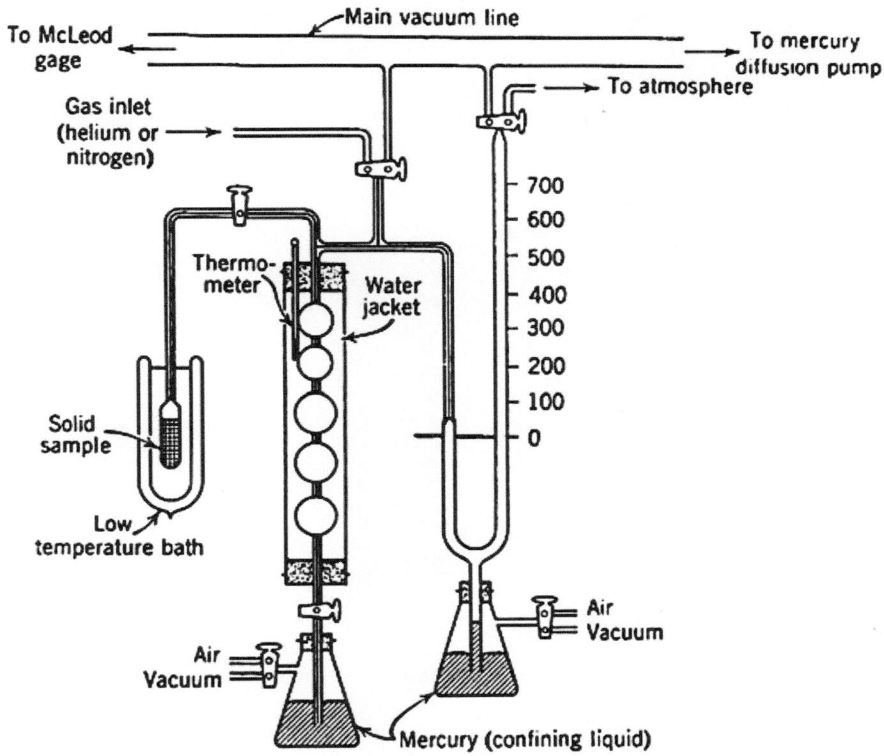

Fig. 3.13 Simplified scheme of a gas-volumetric apparatus for measuring nitrogen adsorption at 77 K. Published with permission from Anderson (1968). Copyright Elsevier

$$\frac{1}{V\left(\frac{P^{\circ}}{P}-1\right)}=\frac{C-1}{V_{\mathrm{m}}C}\left(\frac{P}{P^{\circ}}\right)+\frac{1}{V_{\mathrm{m}}C} \tag{3.52}$$

where C is a constant and V_{m} is the adsorbed volume corresponding to the solid surface covering with a monolayer of adsorbate molecules. This equation is expressed in linearized form ($y = Ax + B$) and can be directly used for the evaluation of the surface area of the solid sample under measurement.

Report, in a plot, the quantity on the left side of the previous equation as a function of the ratio P/P°, and interpolate the obtained data with a straight line. The slope and intercept of this line, A and B, represent, respectively, the quantities $(C-1)/(V_{\mathrm{m}}C)$ and $1/(V_{\mathrm{m}}C)$. From the numerical values of the slope and intercept, C and V_{m} can be calculated through the following formulae:

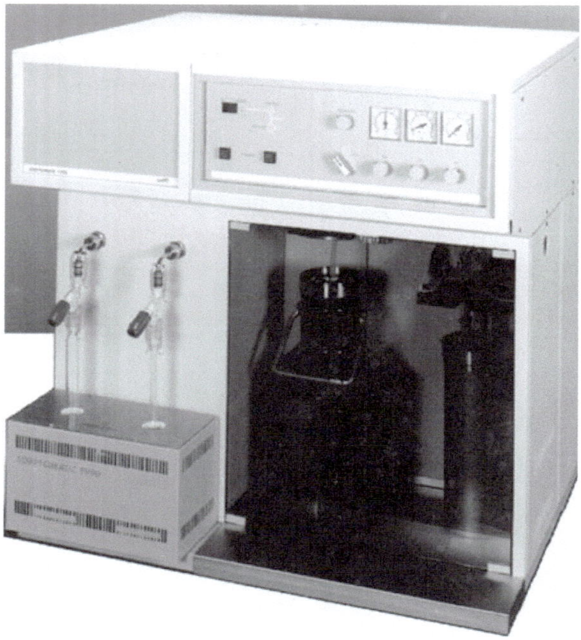

Gas reservoir

Fig. 3.14 An automatic gas-volumetric commercial apparatus for measuring nitrogen adsorption at 77 K (Instruction manual of Sorptomatic)

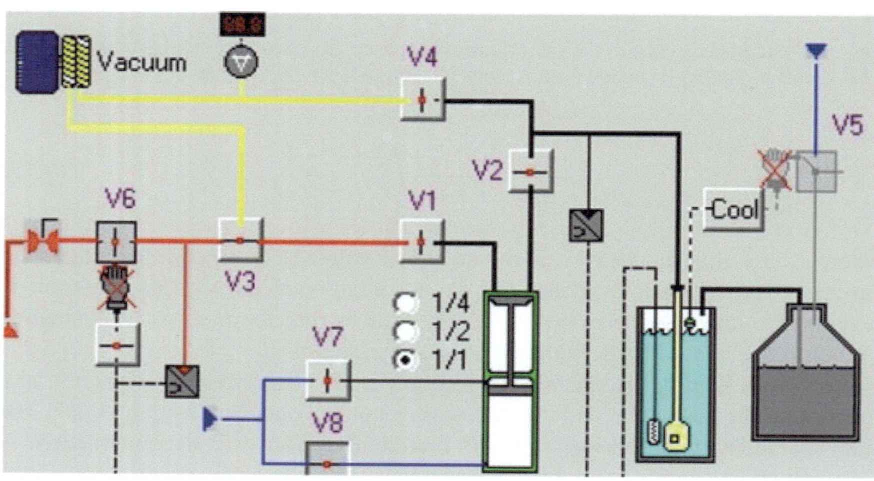

Fig. 3.15 A simplified scheme of the internal part of the apparatus shown in Fig. 3.14

Table 3.9 Experimental data of nitrogen adsorption on a silica support

$P/P°$	0.0002	0.0083	0.1601	0.1897	0.2114	0.2343	0.2582	0.2798	0.3026
	0.3263	0.3536	0.4657	–	–	–	–	–	–
V	10.42	26.55	42.49	44.18	45.41	46.80	48.25	49.65	51.17
	52.82	54.79	64.69	–	–	–	–	–	–

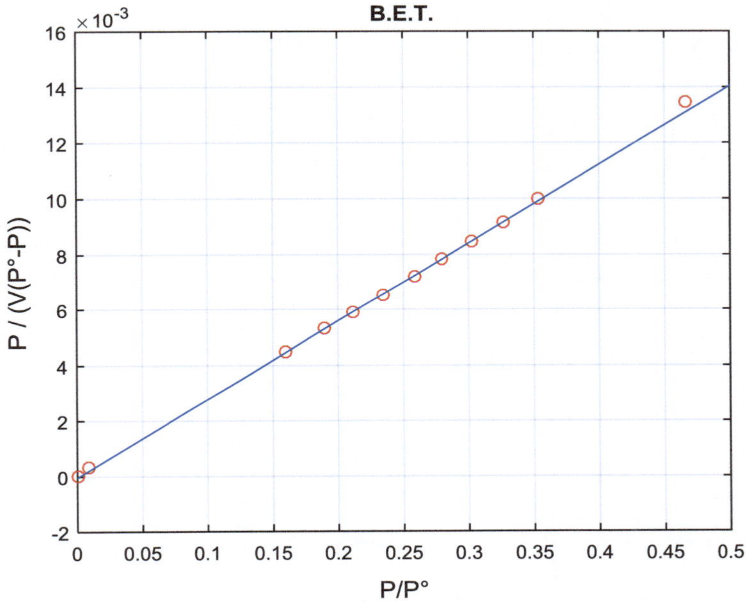

Fig. 3.16 BET plot obtained for the adsorption of nitrogen on a sample of silica

$$C = \frac{A}{B} + 1 \quad V_m = \frac{1}{BC} \tag{3.53}$$

From the value of V_m, it is possible to calculate the specific surface area of the solid A_s using the relation:

$$A_S = \frac{V_m}{22,415} N_A A_{N_2} \tag{3.54}$$

$N_A = 6.02 \times 10^{23}$ Avogadro number, and A_{N_2} is the surface area of a single nitrogen molecule.

Results:

$C = -635.8965$, $V_m = 35.5625$, and $A_s = 154.7333$.

The value of the specific area is, therefore, 155 m^2/g. The obtained plot is shown in Fig. 3.16.

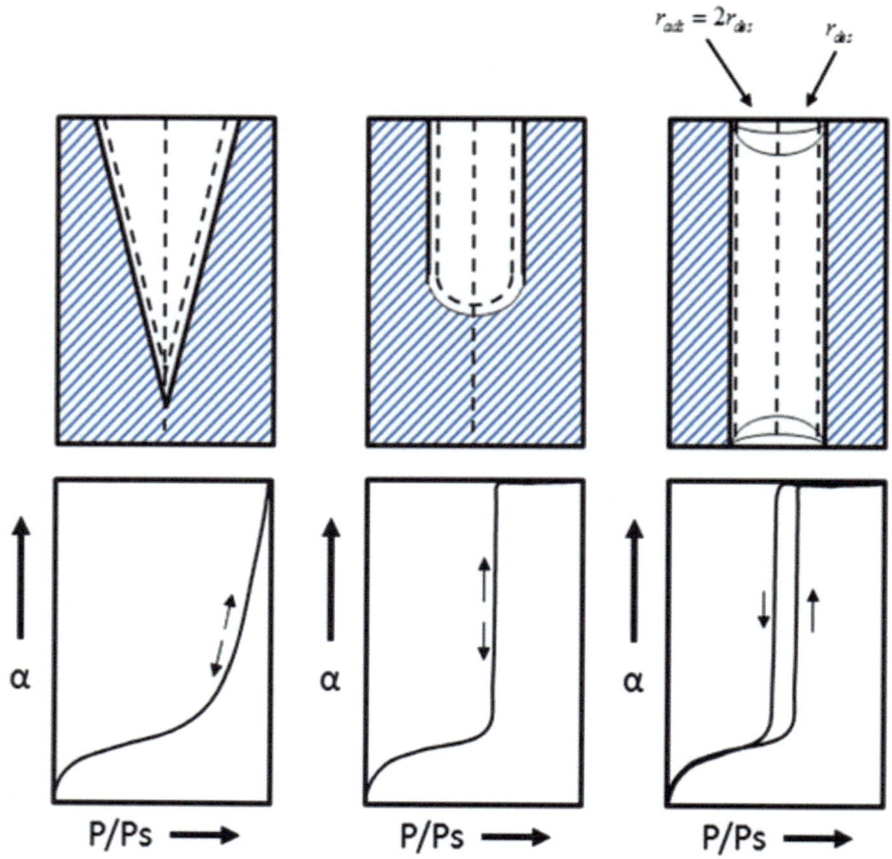

Fig. 3.17 Effect of pore shape on the isotherm form

The described results were obtained using a MATLAB program available as Electronic Supplementary Material, Data 3.1.

We have seen that isotherms IV and V give place to hysteresis phenomena, that is, the curve of the isotherm is different in adsorption with respect to desorption due to condensation of the adsorbate inside the pores. Different behaviours can be obtained according to the shape of the pores, as can be seen in Fig. 3.17.

The phenomenon can be quantitatively interpreted with the Kelvin equation, which can be derived from the equality of the chemical potentials in, respectively, vapour and liquid phase (Thomson and Kelvin 1871).

$$\ln \frac{P}{P_s} = -\frac{2\gamma V_M}{RT\, r_K} \tag{3.55}$$

where

r_K	Mean radius of the meniscus
γ	Surface tension of the liquid
V_M	Molar volume
P/P_s	Relative pressure at which condensation occurs.

Remembering then that $r_p = r_k \cos \Theta$ where $\Theta =$ contact angle and $r_p =$ pore radius:

$$r_p = -\frac{2\gamma V_M \cos \Theta}{RT \ln P/P_s} \tag{3.56}$$

The study of the shape of the isotherm, together with the use of the Kelvin equation, allows determination of the pore-size distribution. For cylindrical pores, the expression of Wheeler (1955) is valid:

$$V_P - V_A = \int_{r_p}^{\infty} (r_p - t)^2 L(r)\, dr \tag{3.57}$$

where V_P is the total volume of pores; V_A is the volume of pores with radius $< r_p$, $L(r) =$ distribution function; and t is the thickness of the adsorbed layer.

A popular, simple, and precise method for determining pore-size distribution from an adsorption–desorption experiment was reported by Dollimore and Heal (1964, 1970). The method of Dollimore and Heal consists of the application of a set of relations starting from measurements of the P/P° ratio and the adsorbed volume in a desorption experiment.

First, for each desorption step, a radius t (Å) from the Halsey equation (1948) is calculated:

$$t = 4.3 \left[\frac{5}{\ln(p^o/p)} \right]^{1/3} \tag{3.58}$$

The Kelvin radius r_k (Å) and pore radius r (Å) can be evaluated from the following relations:

$$r_k = -9.53/\ln(P/P^o) \qquad r = t + r_k \tag{3.59}$$

The average pore radius r_p (Å) is r averaged between steps:

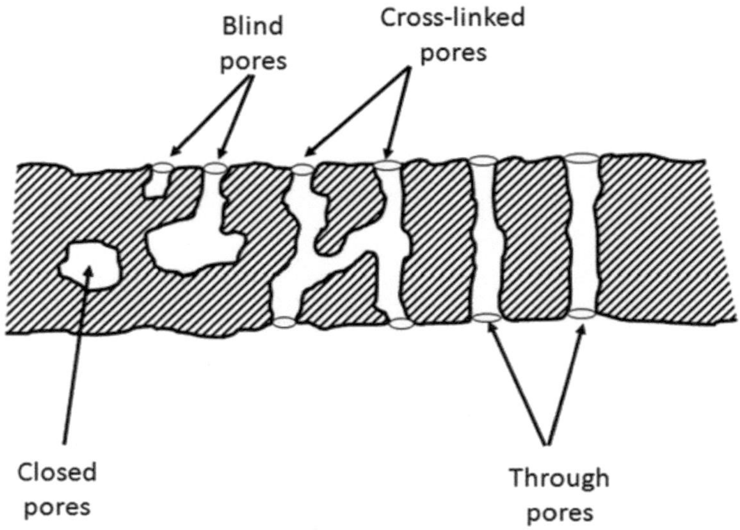

Fig. 3.18 Classification of pores based on their shape (Giesche 2006)

$$r_{\mathrm{p}} = \overline{r} \tag{3.60}$$

The amount of gas adsorbed or desorbed in each step can then be converted in an equivalent liquid volume V_{liq} (mL) by the relation that is valid for nitrogen:

$$V_{\mathrm{liq}} = 0.001555 V_{\mathrm{ads}} \quad \Delta V = \Delta V_{\mathrm{liq}} \tag{3.61}$$

The actual pore volume involved in a desorption step is:

$$\Delta V_{\mathrm{p}} = (\Delta V - A + B) R_n \quad \text{where } U = \Delta t \ \overline{t} \quad A = \Delta t \sum S_{\mathrm{p}} \quad B = U \sum 2\pi L_p \tag{3.62}$$

In previous equations, the symbol of summation is referred for the cumulative sum of terms until the present step. R_{n} is a factor that accounts for a multilayer adsorption, which is defined as:

$$R_{\mathrm{n}} = \frac{r_{\mathrm{p}}^2}{(\overline{r_{\mathrm{k}}} + \Delta t)^2} \tag{3.63}$$

Auxiliary relations for the calculation are related to pore surface area S_{p} and pore length L_{p} as follows:

$$S_{\mathrm{p}} = \frac{2\Delta V_{\mathrm{p}}}{r_{\mathrm{p}}} 2\pi L_{\mathrm{p}} = \frac{S_{\mathrm{p}}}{r_{\mathrm{p}}} \tag{3.64}$$

Pores in a solid can be classified according to their mean size. We can have micropores (radius $r < 20$ Å), mesopores (radius $20 < r < 1000$ Å), and macropores ($r > 1000$ Å). In addition, the shape of the pores is important. As can be seen in Fig. 3.18, we can have different types of pores.

Exercise 3.3. Determination of Porosity from Adsorption–Desorption Measurements

A popular, simple, and precise method for determining pore-size distribution from an adsorption–desorption experiment was reported by Dollimore and Heal in their papers of (1964) and (1970). In their main work, published in 1964, these authors collected the experimental data, which are listed in Table 3.10, related to the desorption of nitrogen previously adsorbed on silica gel.

From the data in Table 3.10, evaluate pore-size distribution by using the method suggested by Dollimore and Heal. Produce three plots: the first reporting the collected data as V_{ads} as a function of the $P/P°$ ratio, the second reporting the pore radius according to Eq. (3.58), and the third reporting $\Delta V_{\mathrm{p}}/\Delta r_{\mathrm{p}}$ as a function of r_{p}.

Solution

By applying the method of Dollimore and Heal described in the text, the two required plots can be constructed by using a MATLAB program available as Electronic Supplementary Material, Data 3.1. The obtained results are shown in Figs. 3.19a, b and 3.20.

From Fig. 3.20, it is possible to see that, for the examined sample of silica gel, the average size of the pores is approximately 20 °A. Very few pores have dimensions >50°A.

Porosity also can be determined using a mercury porosimeter. Mercury porosimetry allows evaluating the total pore volume or porosity, the skeletal and apparent density, and the pore-size distribution. Porosimeter cannot be used to analyze closed pores because mercury does not enter those pores. Mercury is pushed inside the pores by gradually increasing the pressure. Initially, only macropores are filled and then mesopores are filled; at the highest pressures, micropores are filled. A key assumption is that the pore shape is assumed in all cases to be of cylindrical geometry. For interpreting the results, the modified Young–Laplace equation, normally named the Washburn (1921) equation, is used:

$$\Delta P = \frac{2\gamma \cos\theta}{r_{\mathrm{pore}}} \tag{3.65}$$

where γ is the surface tension of mercury and θ is the contact angle between the solid and mercury. We must know these parameters, the measured pressure and the corresponding intruded volume, to evaluate the pore-size/pore-volume relation.

Table 3.10 Experimental data of nitrogen desorption previously adsorbed on silica gel[a]

P/P°	0.894	0.881	0.866	0.844	0.818	0.780	0.754	0.722	0.682
	0.628	0.556	0.538	0.519	0.499	0.477	0.453	0.428	0.401
	0.371	0.340	0.306	0.270	0.232	0.192	0.152	0.111	0.074
V_{ads} (mL STP)	337	337	337	337	336	335	334.5	334	332
	328	314	310	304	298	290	278	264	242
	231	225	218	208	198	189	179	166	154

[a]Dollimore and Heal (1964)

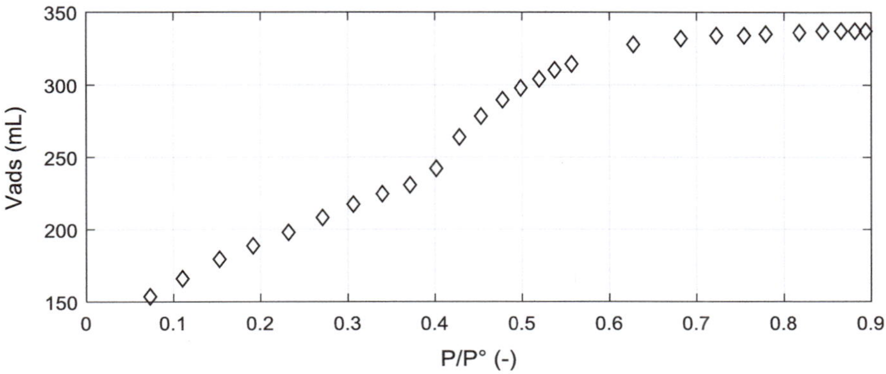

Fig. 3.19 Plot of (**a**) V_{ads} as a function of $P/P°$, (**b**) pore radius as a function of $P/P°$

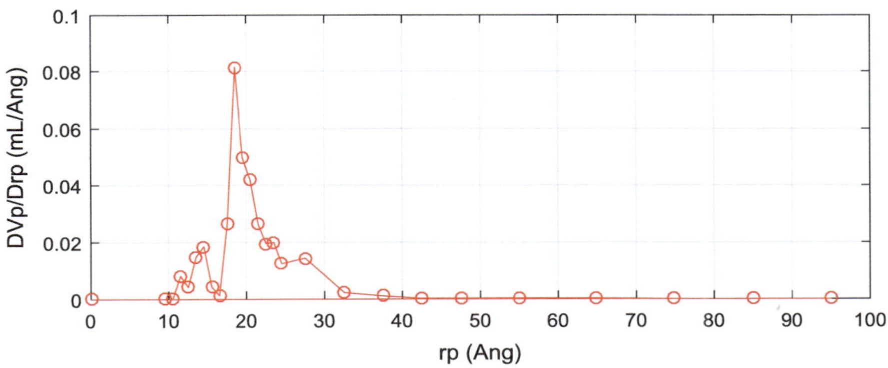

Fig. 3.20 Calculated pore distribution using the method of Dollimore and Heal

Exercise 3.4. Determination of Pore Volume, Density, and Pore-Size Distribution Using a Mercury Porosimeter

A sample of uranium dioxide in pellet form was submitted to mercury porosimetry after a sintering treatment of particles at 1000 °C for 2 h. At the beginning of the experiment, the pressure was 1.77 psi, and the amount of mercury displaced by the sample was 0.190 cm^3. The amount of sample was 0.624 g, and the data collected are listed in Table 3.11.

Using the data in Table 3.11, calculate the pore-volume distribution of the solid.

Solution

The solution of the exercise can be developed through the following steps:

(1) Refer the data of mercury volume to the mass of sample: $m_g = m/m_s$.

- m_g = volume of mercury referred to the mass of sample (g Hg/g of sample)
- m = volume of mercury collected by the instrument (g Hg)
- m_s = mass of sample submitted to the measurement (g of sample) (Table 3.12)

Table 3.11 Experimental data collected with a porosimeter for a sample of 0.624 g of uranium dioxide

P (psi)	116	310	344	364	410	456	484	540	620
	710	800	830	900	1050	1300	1540	1900	2320
	3500								
Hg vol. (cm^3)	0.002	0.006	0.010	0.014	0.020	0.026	0.030	0.038	0.050
	0.064	0.076	0.080	0.088	0.110	0.112	0.118	0.122	0.124
	0.125								

Table 3.12 Data of mercury volume intruded per gram of specimen

m_g (cm^3)/g)	0.0032	0.0096	0.0160	0.0224	0.0321	0.0417	0.0481	0.0609	0.0801
	0.1026	0.1218	0.1282	0.1410	0.1763	0.1795	0.1891	0.1955	0.1987
	0.2003	–	–	–	–	–	–	–	–

Table 3.13 Corrected values of mercury volume intruded per gram of specimen

m_{gc} (cm^3)/g)	0.1971	0.1907	0.1843	0.1779	0.1683	0.1587	0.1522	0.1394	0.1202
	0.0978	0.0785	0.0721	0.0593	0.0240	0.0208	0.0112	0.0048	0.0016
	0.0000	–	–	–	–	–	–	–	–

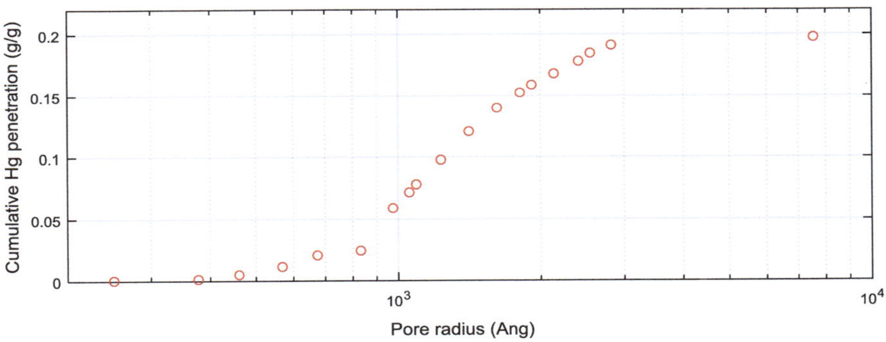

Fig. 3.21 Cumulative amount of Hg intruded as a function of pore radius

(2) Calculate the pore radius corresponding to each pressure with the relation $r = 8.75 \times 10^5/\mathrm{P}$

- r radius of pore (Å)
- P pressure (psi)

(3) Correct the m_g data of the previous table using the formula $m_{gc} = 0.2003$-mg corresponding to the volume m_g at a maximum pressure of 3500 psi. This allows calculation of the cumulative volume penetrated into the pores of the sample (Table 3.13).

With these data and the data of the radii, the first plot can be drawn (logarithmic scale on the x-axis).

(4) For each couple of data, an approximation of the derivative can be evaluated as:

$$\frac{\mathrm{d}V}{\mathrm{d}r} = \frac{m_{gc}^{i+1} - m_{gc}^{i}}{r^{i+1} - r^{i}} \tag{3.66}$$

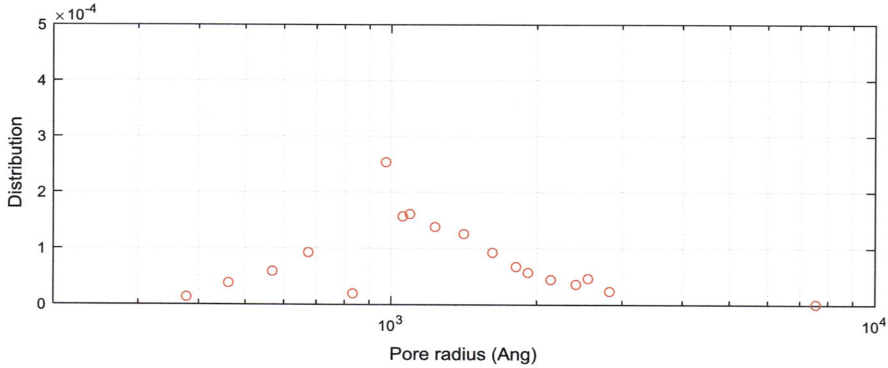

Fig. 3.22 Calculated pore-size distribution

The derivative calculated in step 4 can be plotted against pore radius (logarithmic scale on the *x*-axis) (see Fig. 3.21), thus obtaining the pore distribution shown in Fig. 3.22. Calculations were performed using a MATLAB program available as Electronic Supplementary Material, Data 3.1.

3.4.3 Chemical Adsorption

Chemical adsorption can be considered a reaction of a molecule coming from a fluid phase with an active site of a solid surface. A strong interaction is established, and new chemical bonds are formed. The molecule can be dissociated on the surface or bonded to a transition metal due, for example, to a dative bond. It is therefore possible to distinguish two types of chemical adsorption: **associative** or **dissociative**. The new bonds formed can be individuated by studying the chemisorptions with spectroscopic techniques. Because chemisorption is a chemical reaction, it is an activated process, normally favoured by temperature. In contrast, physical adsorption is favoured at low temperature and vanishes at a higher temperature. Figure 3.23 shows the behaviour of the chemical potential for, respectively, physical and chemical adsorption as a function of the distance of the molecules from the solid surface. Chemical adsorption, in this case, is assumed to be dissociative, that is, $X_2 + 2M \leftrightarrow 2\ XM$.

As can be seen, for passing from physical to chemical adsorption, a barrier of potential must be exceeded. This barrier corresponds to the activation energy of the adsorption reaction.

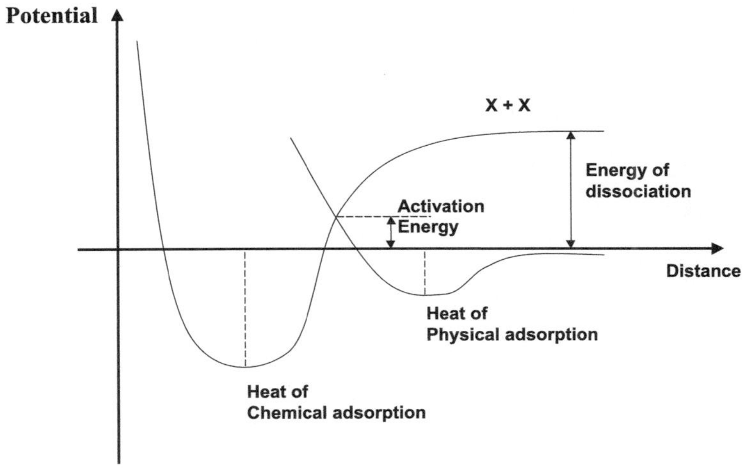

Fig. 3.23 Comparison between physical and chemical adsorption

3.4.4 Factors Determining Catalyst Deactivation: Poisoning, Aging, and Sintering

Catalyst deactivation can have a severe economic impact on chemical processes. It was seen that active sites are normally a very small portion of a catalyst surface; therefore, traces of impurities in the reactant flowing stream also can deactivate the catalyst through poisoning if the impurities yield a strong interaction with the active sites. It is possible to recognize two different types of poisoning: temporary poisoning and permanent poisoning. In the case of temporary poisoning, the catalytic activity can be restored after removing the poison from the surface. In the case of permanent poisoning, the activity cannot be restored because the poisoning molecules are strongly bonded to the active sites and cannot be easily removed. Iron catalysts, promoting ammonia synthesis, for example, are poisoned by the presence of water in a stream contacting the catalyst. Because the catalyst is normally produced in situ by reducing magnetite (Fe_3O_4) with hydrogen to iron, water is formed during the catalyst preparation and poisons the catalyst; however, when the water is removed, the activity is restored as can be seen in Fig. 3.24.

Other examples of temporary poisoning are those occurring on the surface of acid catalysts promoting the reactions of hydrocarbons cracking and isomerization. During these reactions, pitch and coke, formed as by-products, cover with a thick layer of the active sites deactivating the catalyst; however, by burning the coke deposit, the catalyst activity is completely restored.

Permanent poisoning is due to the irreversible reactions of some molecules, coming from the gaseous phase, with the active sites yielding stable chemisorbed species. Platinum catalysts, for example, are poisoned by traces of CO, CS_2, H_2S, PH_3, and arsenic compounds. In some cases, the poisoning is prompt and

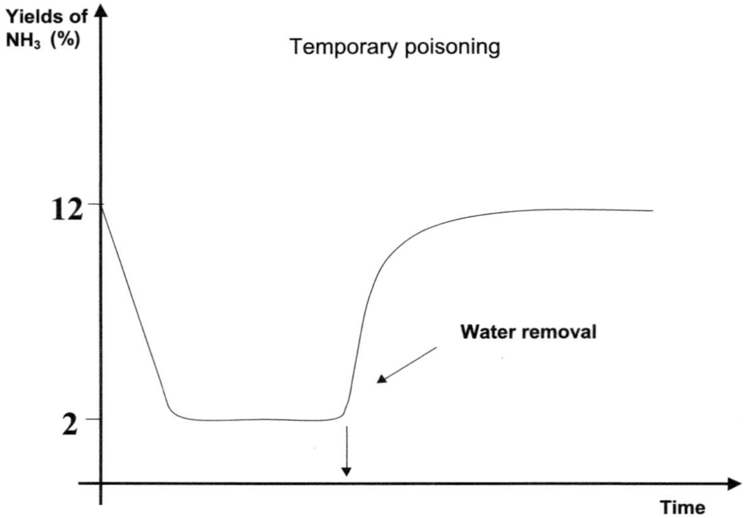

Fig. 3.24 Temporary poisoning effect of water on iron, a catalyst for the ammonia synthesis

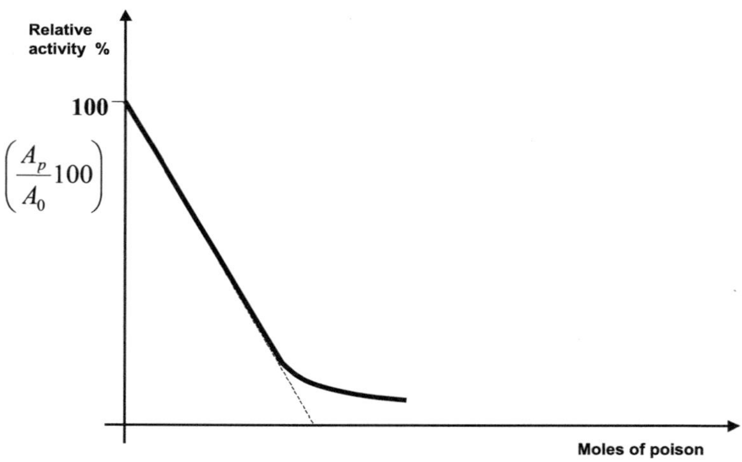

Fig. 3.25 An example of permanent catalyst poisoning

quantitative, thus allowing titration of the number of active sites on the surface. Permanent poisoning occurs as shown in Fig. 3.25. This behaviour can be interpreted with a relation of the type:

$$\frac{A_\mathrm{p}}{A_\mathrm{o}} = e^{-\alpha c} \tag{3.67}$$

where A_p is the activity of the poisoned catalyst; A_o is the activity of the fresh catalyst; α is the coefficient of poisoning; and c is the amount of chemisorbed poison.

For a low poison amount, a different relation can be used:

$$\frac{A_p}{A_o} = 1 - \alpha c \tag{3.68}$$

Catalysts also can deactivate as a consequence of sintering or aging. Small crystallites, for example, can be incorporated in larger crystals, thus reducing the overall specific surface area and consequently reducing the activity. To avoid sintering, some promoters are introduced in the catalyst composition to reduce the rate of diffusion on the solid state. Promoters normally are compounds having a very high melting point because, according to an empirical observation, diffusion in the solid state starts at an absolute temperature of approximately one-third of the melting point. However, the use of both a support of high specific surface area as well as high catalyst dispersion can prevent sintering. Finally, aging can be the consequence of (1) thermal shock during catalyst preparation or during the conduction of exothermic reactions; or (2) mechanical stress due to attrition.

3.4.5 A Brief Survey on Catalyst- and Support-Preparation Methods

Solid catalysts, respectively, can be categorized as unsupported and supported. Unsupported catalysts can be metals, oxides, or salts directly used as catalysts. In general, unsupported catalysts can be prepared by the following methods:

(a) Precipitation from a solution
(b) Thermal decomposition of opportune precursor compounds
(c) Chemical attack of a compound or of an alloy

In all cases, the scope is to obtain a specific surface area as large as possible.

The preparation of supported catalysts can be made using different techniques. Again, the scope is to disperse, as much as possible, the active catalyst on the surface of a suitable support. Therefore, the supports must have opportune properties, such as high specific surface area, optimal porosimetric structure, adequate acid–base characteristics, good mechanical resistance, and satisfactory thermal conductivity. Moreover, the support must be compatible with the component to be dispersed, thus favouring a stable dispersion due to a strong anchorage to the surface but without reducing the number of active sites and consequently depressing the activity. Examples of largely employed supports are alumina, silica, silica–alumina, diatomaceous earth, carbon black, titanium dioxide, zirconium dioxide, etc. Often, the supports have an active role in the catalytic processes as promoters or by introducing their own catalytic functionality. A support of alumina, for example, has acid and basic sites of Lewis type on the surface, whilst silica–alumina has strong acid sites of Brønsted type. The preparation of supported catalysts, normally, can be made by the following methods.

(a) **Impregnation by adsorption or wet impregnation**
An opportune amount of an active metal precursor is dissolved in a solvent. A powdered support is immersed in this solution. The precursor is partitioned: Part is chemisorbed on the support surface in equilibrium with the amount that remains dissolved in the solution.

(b) **Incipient wetness impregnation or dry impregnation**
An active metal precursor is dissolved in a solvent. The obtained solution is added, drop by drop, to a well-mixed powdered support. A limited amount of solution is added such that the volume of the solution is just equal to the overall volume of the pores. Solution enters the pores by capillary action, and at the end of the addition, the powder appears dry. Dry impregnation can be made also using the support in pellet form.

(c) **Precipitation in situ**
The support, in a thin powdered form, is immersed in a solution containing a precursor of the active part. By changing the conditions of the solution, the precursor gives place to precipitation. The precipitate is well mixed together with the support powder. The solid obtained is then filtered, dried, and pelletized.

(d) **Co-precipitation**
The support and catalyst are precipitated together from a unique solution.

3.4.6 Acid–Base Heterogeneous Catalysts

The definition of acid and base, valid for the homogeneous catalysts, is valid also for the solid surfaces if we consider the acid or basic character a peculiarity of the active sites. Active sites could be, therefore, acid or base of Lewis or Brønsted type if are able to exchange an electron pair or a proton. However, a difficulty arises because the sites on the surface are not uniform (surface heterogeneity), and thus the acid or basic strength is not the same for all of the sites on the surface: We have a distribution of sites of different strengths. Another complication derives from the fact that we cannot define a function for the acidity strength as the pH for the aqueous solution. It is possible to use an indirect way for expressing the acidity of a solid according to the approach suggested by Hammett and Deyrup (1932). It is known that some coloured organic substances, normally used as "pH indicators," are chemical detectors of the pH level of a solution because when protonated, they change the colour of the solution at a characteristic value of the pH. pH indicators are weak acids or weak bases. For a weak acid, in an aqueous solution, the following equilibrium is valid:

$$HInd + H_2O \rightleftarrows H_3O^+ + Ind^- \tag{3.69}$$

where HInd is the un-dissociated form of the indicator and Ind^- is the conjugated base of the indicator having a colour different from HInd. From this equilibrium, it is possible to derive, according to Henderson (1908) and Hasselbalch (1916), the relation:

$$pH = pK_a + \log[Ind^-]/[HInd] \tag{3.70}$$

when pH is equal to pK_a, the ratio $[Ind^-]/[HInd] = 1$. This corresponds to the point of colour change. Hammett defined a function of acidity H_o that for aqueous dilute solutions coincides with pH:

$$H_o = pK_a + \log[B]/[BH^+] \tag{3.71}$$

where B is a weak base subjected to the equilibrium of protonation:

$$B + H^+ \rightleftarrows BH^+ \tag{3.72}$$

Hammett used some anilines as bases B to evaluate the H_o function of strong acid, such as, for example, sulphuric acid determining a value of $H_o = -12$ for pure sulphuric acid. It should be pointed out that the more negative the $H_{0,}$ the stronger the acid. In particular, a value of -12 found for pure sulphuric acid is a reference for the definition of superacidity, that is, substances having a Hammett function H_o more negative than -12. Superacids include nafion (a fluorinated polymer $H_o = -12$); $H_3PW_{12}O_{40}$, $Cs_{2.5}H_{0.5}PW_{12}O_{40}$ (polyoxometalates-POM $H_o = -13.2$); SO_4^{2-}/ZrO_2 (sulphated zirconia $H_o = -16$); and magic acid $HSbF_6$ (able to protonate CH_4 $H_o = -20$). For determining the acid strength and the amount of the active sites on a solid, it is possible to make a titration of these sites with a base, for example, butylamine, dissolved in an appropriate solvent, not water or other protic solvents, in the presence of an indicator of known pKa until the occurrence of the colour changes. When colour changes, $H_o = pK_a$. In this way, all of the sites having acidity stronger than the determined value of H_o are titrated. The titration can be repeated using indicators with a different pK_a, thus selecting acid sites with different ranges of strength. After this procedure, a distribution curve of the acid strength can be constructed obtaining plots of the type shown in Fig. 3.26.

Titration occurs in the absence of water because water has a levelling effect on the acid strength giving always H_3O^+ as the most acid species. Other protic solvents have the same behaviour. Therefore, aprotic solvents—such as benzene, iso-octane, cyclo-hexane, or decalin—are used.

Another method for evaluating the acid strength of acid sites on a solid surface is to put the solid in contact with a flowing stream of gaseous ammonia. Salts are formed on the surface and can be decomposed by gradually increasing the temperature. The greater the acid strength of the sites, the higher the temperature needed for

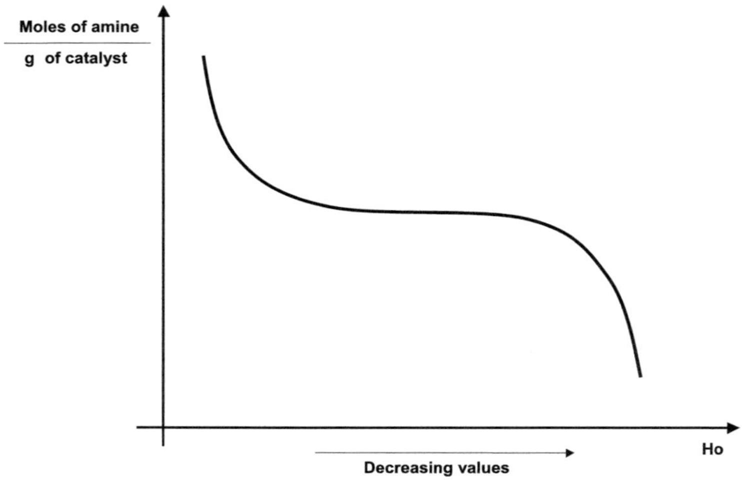

Fig. 3.26 Distribution curve of acid sites on a solid catalyst

decomposing the salts formed on the surface. By measuring the amount of desorbed ammonia, a quantitative evaluation of the number of sites in correspondence to a decomposition temperature is possible. A property strictly correlated with the acid strength can be also measured by measuring the decomposition heat with a differential calorimeter. Because some reactions are promoted by the acidity of solid catalysts and the kinetics is known, sometimes such reactions are used for evaluating the acidity of a catalyst simply by measuring the reaction rate at a given temperature and determining by poisoning the overall number of sites.

Last, IR spectroscopy is used to identify the properties of the different types of hydroxyls and different sites (Brønsted or Lewis) present on the surface by using probe molecules, such as ammonia or pyridine, for this characterization.

The most common acid catalysts are natural clays, acid zeolites, supported mineral acids (H_2SO_4, H_3PO_4, etc.), acid-exchange resins, some oxides and sulphides (in particular Al_2O_3, TiO_2, ZrO_2), and some mixed oxides (in particular, silica alumina). The most common basic catalysts are supported bases, anionic-exchange resins, some oxides (e.g., MgO, ZnO, CaO, and Al_2O_3), and some mixed oxides (e.g., MgO–SiO_2, CaO–SiO_2, MgO–Al_2O_3, etc.).

3.4.7 *Surface Acidity of Binary Mixed Oxides*

Metal oxides are largely employed in industry as catalysts. Their catalytic actions can be of different types. Metal oxides can act, for example, as transition metals or as semiconductors (as will be seen in a next section), but they mainly act as acids or bases. When two oxides are intimately mixed (e.g., by co-precipitating them from

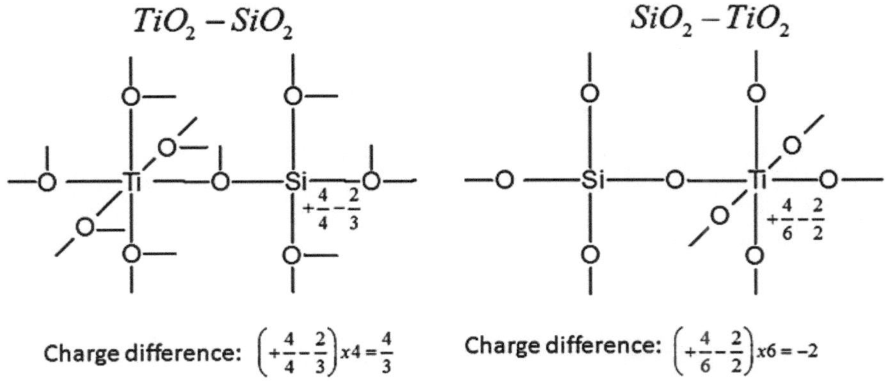

Fig. 3.27 Model structures for, respectively, TiO_2–SiO_2 with TiO_2 as the prevailing component and SiO_2–TiO_2 with SiO_2 as the major component

their solution in water, filtering, and calcinating), the binary mixed oxide formed can have properties completely different from the originating pure oxides. Tanabe et al. (1974) studied this phenomenon by examining the properties of 31 different binary mixed oxides and formulated a hypothesis that correctly predicts the acid behaviour of those mixed oxides in 90% of cases. The hypothesis explains the mechanism of acidity generation in binary oxides and predicts whether the acid sites will be of Brønsted or Lewis type. According to Tanabe et al., the acidity is generated by the excess of a negative or positive charge in a defined model structure of the binary oxide. The model structure is constructed by adopting the following two postulates:

(1) The coordination number of the positive elements of the two metal oxides is maintained after mixing.
(2) The coordination number of the negative element (the oxygen) becomes, in all cases, the one prevailing component.

For example, we can compare the behaviour of a mixture of $TiO_2 \cdot SiO_2$, in which TiO_2 is the major component, with that of a mixture of $SiO_2 \cdot TiO_2$, in which SiO_2 is the prevalent component. We have the following behaviour shown in Fig. 3.27.

In the first case, we have an excess of positive charge with an increase in the Lewis acidity; in the second case, we have an excess of negative charge with an increase in the Brønsted acidity. Considering the first case, the four positive charges of the silicon atom are distributed between four bonds, whilst the two negative charges of oxygen atom are distributed between three bonds ($-2/3$ of the valence unit is distributed to each bond). The difference in charge for one bond is $+1 - 2/3 = +1/3$, and considering all bonds, $1/3 \times 4 = +4/3$ is the excess of charge. To an excess of positive charge corresponds an increase of the Lewis acidity.

3.4.8 Zeolites, Structures, Properties, and Synthesis

Zeolites are crystalline silico-aluminates having the peculiarity of a porosity that is internal to the crystalline framework. These materials have crystals that correspond to a combination of a few polyhedral blocks as shown in Fig. 3.28.

A combination of these blocks can yield many different structures, such as the ones shown in Figs. 3.29, 3.30, and 3.31, which correspond to the structures of different existing zeolites.

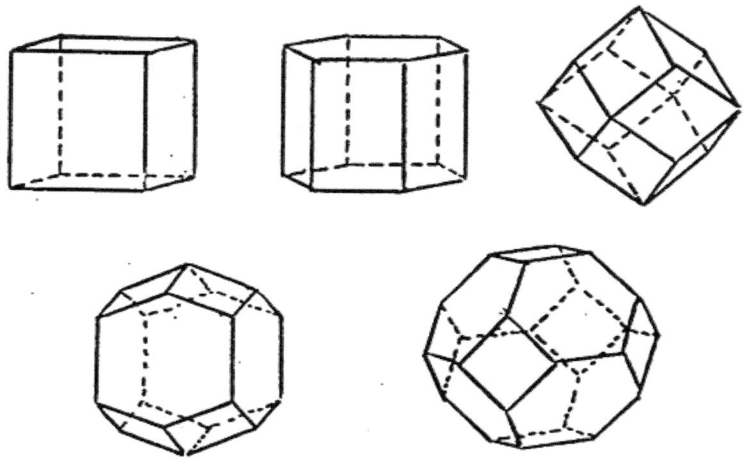

Fig. 3.28 Elemental polyhedral structure of zeolites

Fig. 3.29 Crystal structure of zeolite *A*

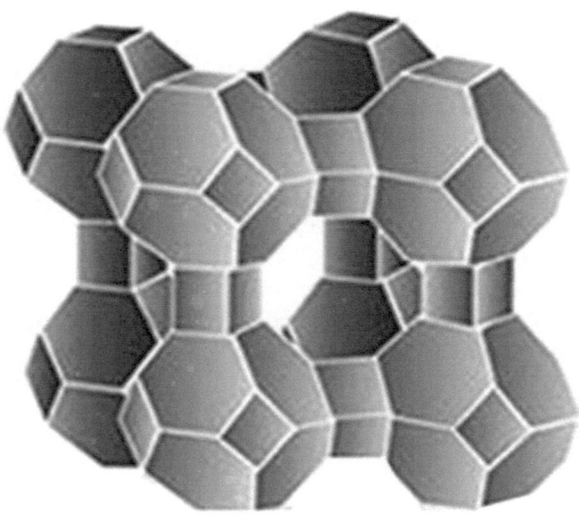

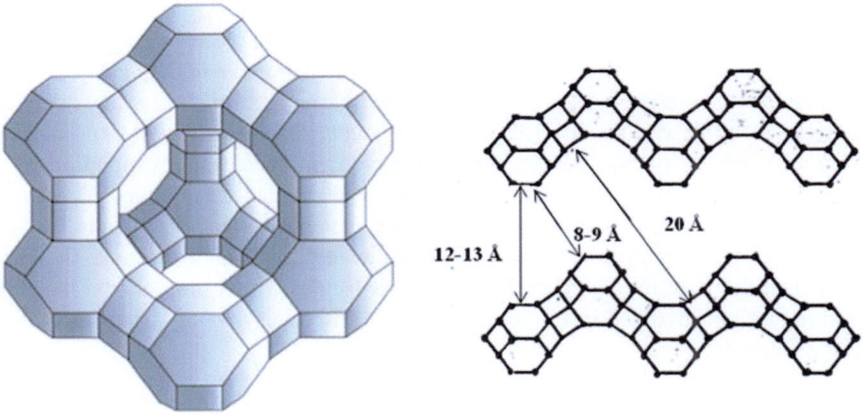

Fig. 3.30 Crystal structure of synthetic zeolites *X* and *Y* existing in nature as the mineral faujasite

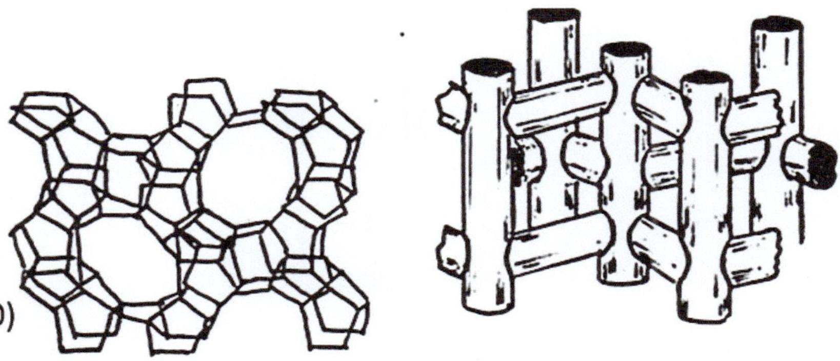

Fig. 3.31 Crystal structure of ZSM-5 zeolite

Table 3.14 Characteristics of the most used zeolites

Type of zeolite	$R =$ Silicon/aluminium	Pore diameter (Å)	Pore volume (cm^3/g)
A	1.0	3.5–5	0.3–0.4
X	1.0–1.5	7.4	0.3–0.4
Y	1.5–3.0	7.4	0.3–0.4
Mordenite	4.5–5.0	6.7–7.0	0.20
ZSM-5	2.5–35	5.4–5.6	0.1
ZSM-11	2.5–35	5.1–5.5	0.18

Zeolites most commonly employed as catalysts are the synthetic *X* and *Y* zeolites having the same crystal framework of the natural mineral faujasite, mordenite, ZSM-5, and ZSM-11. *X* and *Y* zeolites differs only for the ratio SiO_2/Al_2O_3, whilst the other zeolites are different for, respectively, the crystal structure, the average pore diameter, and the ratio SiO_2/Al_2O_3 as can be seen from the data listed in Table 3.14.

The internal crystalline porosity of the zeolites has relevant effects on the adsorption of the molecules, their diffusion, and their catalysis. Adsorption of the molecules is favoured by the strong electric fields existing inside the narrow zeolite channels due to the presence of metal ions and the oxygen of the framework. For example, a zeolite can absorb >20% b.w. of water. Zeolites often show a marked selectivity in the specific adsorption of particular molecules corresponding to a marked selectivity in the reactions in which those molecules are involved. However, a new type of selectivity was observed in the use of zeolite as catalyst, that is, "shape selectivity." Some molecules cannot enter in the zeolite channels because they are larger than the pore mouth. Therefore, these molecules, when present in a reacting mixture, are not involved in the reaction. Three different types of shape selectivity were observed:

(1) Shape selectivity can be related to the reagents, such as the one already described.

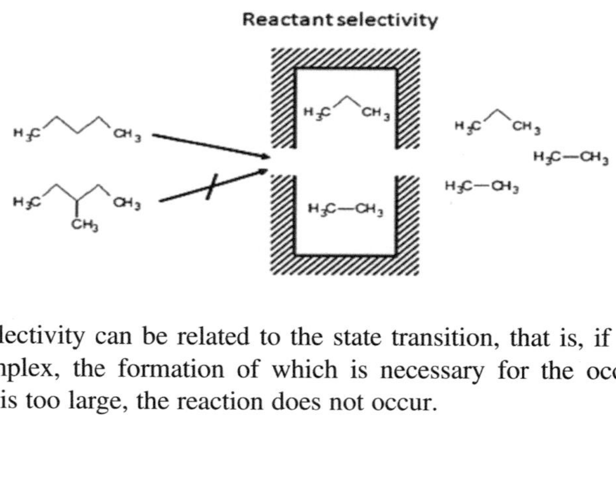

(2) Shape selectivity can be related to the state transition, that is, if the transition state complex, the formation of which is necessary for the occurrence of a reaction, is too large, the reaction does not occur.

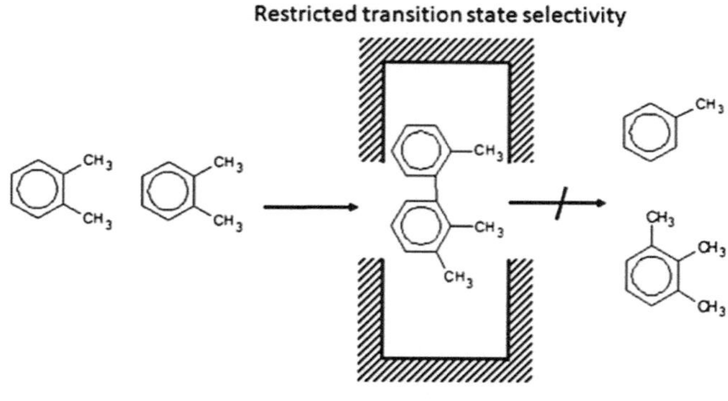

(3) Shape selectivity can be related to a product that is more stable inside the zeolite channel.

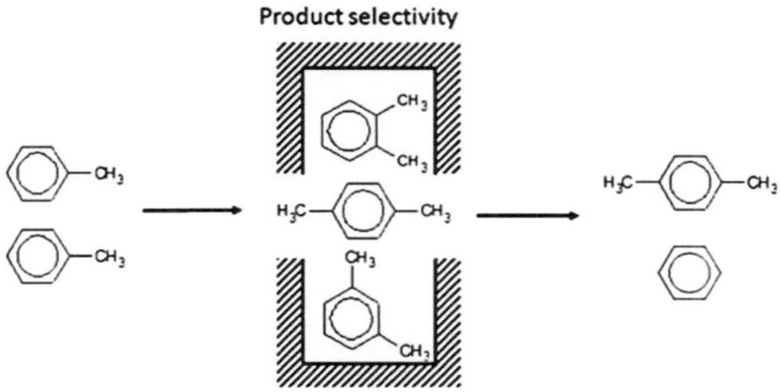

An example in which different types of shape selectivity are operative is the reaction of xylene isomerization promoted by ZSM-5 catalyst. In this process, ortho- and meta-xylenes are isomerized to para-xylene, which is more stable inside the zeolite channel due to its linear structure. However, at the exit of the zeolite pores, the para-structure is not so favoured, and if acid sites are present on the external surface, the reaction is reversed to reach thermodynamic equilibrium. To avoid this drawback, zeolite is externally poisoned with MgO, thus stopping the undesired reaction. Moreover, MgO avoids also the formation of coke on the pore mouth. Inside the pores, coke is not formed due to the shape selectivity related to the transition state complex.

Natural zeolites have been studied for many years, in particular, by Barrer (1978), who individuated their crystalline structure; however, zeolites remained just a curiosity until the discovery of methods for their synthesis [pioneering works of Breck (1974)]. Many different zeolites have been synthesized and have found many catalytic applications in industrial processes. Zeolites are used in particular for their Brønsted acid properties. Zeolite acid sites are similar to the silica–alumina sites of higher strength, but the sites of zeolites are much more uniform and numerous. Acid sites in *X* and *Y* zeolites can be obtained starting from synthesized zeolite, which contains sodium ions to neutralize the negative charge of the framework. The zeolite is put in contact, at boiling temperature, with an ammonia salt solution, and this treatment is repeated at least three times. Sodium is exchanged with ammonium ion for approximately 80%. Then zeolite is filtered, dried, and calcined. The ammonium salts on the surface are decomposed by heating releasing NH_3, whilst the proton remains anchored on the surface (decationated zeolite). Acid zeolites also can be obtained by exchanging sodium with bivalent cations (Ca, Sr, Ba) or trivalent cations (lantanides). Polyvalent cations cannot neutralize completely the framework charges for the rigidity of the crystal lattice and exerts with their positive charge an acid activity. Zeolites having a high content of silica, such as mordenite and ZSM-5, can be exchanged directly with mineral acids because they are more stable to acid chemical attack.

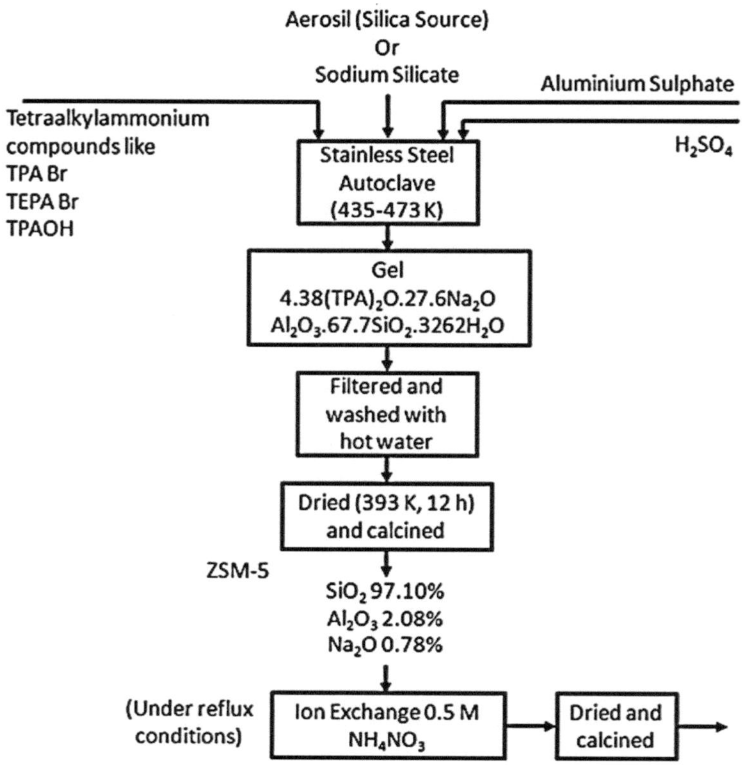

Fig. 3.32 Sequence of operations in the preparation of ZSM-5 zeolite. Published with permission from Jacobs et al. (1981). Copyright Elsevier (1981)

Zeolites *A*, *X*, and *Y* are obtained by crystallization under alkaline hydrothermal conditions of alumino-silicate gel of appropriate composition. Zeolites *A* are obtained at 80–100 °C, and zeolites *X* and *Y* are obtained at 100–120 °C. The use of organic cations instead of sodium for creating the alkaline environment allows the synthesis of other zeolites, such as mordenite, ZSM-5, silicalite, etc. The organic cation has the function of a templating agent remaining trapped in the zeolite cavity. In the case of mordenite, diethanolamine (DEA) is used as a templating agent, whilst for producing ZSM-5, tetra-propylammonium hydroxide is used with a sequence of operations as shown in Fig. 3.32. More recently, isolated atoms of titanium were introduced in a crystal lattice of silicalite (the same framework of ZSM-5). This type of titanium has shown a high selectivity in some oxidation reactions using hydrogen peroxide as reactant, and different new more convenient processes have been developed (propene oxide production, synthesis of quinone, production of caprolactam); see Millini et al. (1992), Notari (1988), Clerici (1991), Romano et al. (1990), and Clerici and Ingallina (1993).

3.4.9 Templating Mesoporous Zeolites

Currently, >170 different zeolite structure types are known, and new types appear every year. Many reviews have been published on the subject. However, <10% of discovered zeolites have not found practical use in industry because their mass transport to and from the active sites is hindered because the mouths of the pores are too narrow. The synthesis of a new class of zeolites, the mesoporous zeolites, is an attempt to overcome this problem. Such mesoporous zeolites contain, together with the crystalline micropore system, which is typical of other zeolites, a mesopore system that is not part of the crystal structure. Such pores can have diameters in the range of 2–50 nm. Mesoporous zeolites can be considered hierarchical porous material. Such materials normally can be prepared by templated methods that have the advantage of tailoring the pore size using a mesopore template of a characteristic size. Examples of such mesoporous structures are shown in Fig. 3.33.

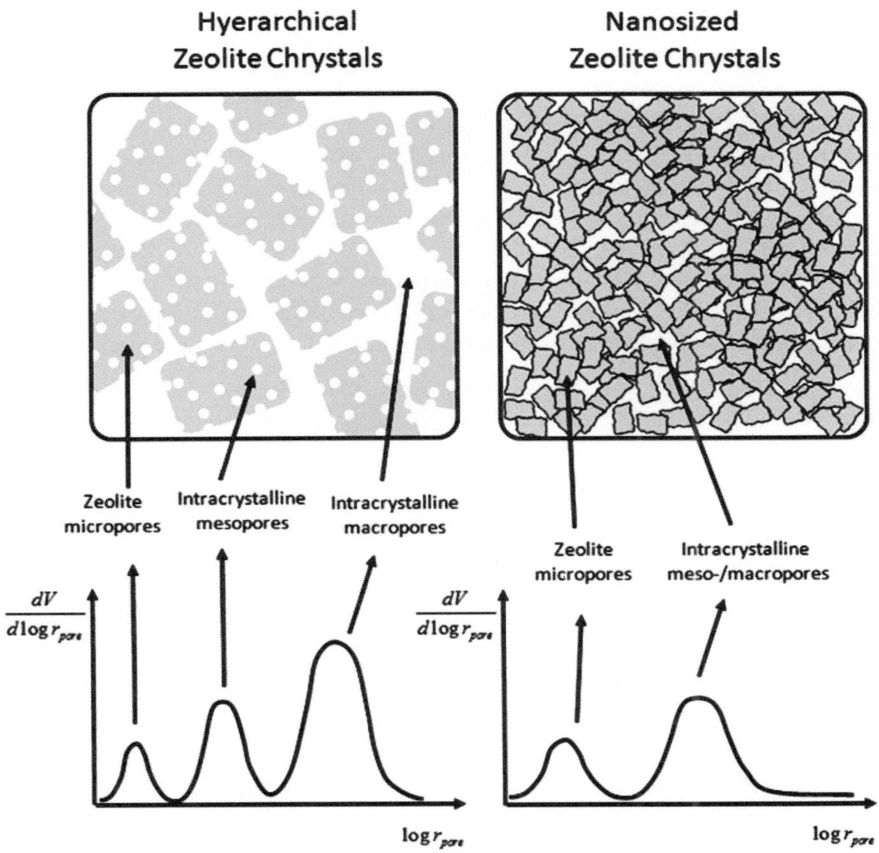

Fig. 3.33 Example of zeolite mesoporous structures

3.4.10 Catalytic Properties of Metal Oxides as Semiconductor

We have already discussed the acid–base properties of oxides and mixed oxides. Oxides can show also other catalytic actions, in particular the oxides of transition metals. In this case, oxides can give place to catalytic actions involving d-orbitals and changes in the oxidation number of the metal. A less common catalytic action is one related to their semiconductor properties. Normally, oxides are insulating materials, but some of them show semiconductor properties. In a semiconductor, electrons can be found in two different bands characterized by a difference of the energetic level: the **band of valence**, which is completely filled when the electrons are in their fundamental state, and the **band of conduction**, which exists at a higher energetic level and is populated by electrons coming from the band of valence as a consequence of an energetic input. The transition of an electron from the valence band to the band of conduction produces a "hole" of charge in the valence band ("cationic vacancy"). Electrical conductivity in semiconductors occurs through two different mechanisms of charge transfer. One is the migration of electrons occurring in the band of conduction, and the other is the migration of the charge holes in the valence band. An increase in temperature determines an increase in the number of electrons passing from the band of valence to the band of conduction, thus increasing the electrical conductivity of the oxide. Figure 3.34 compares the distribution of electrons between the two mentioned bands in, respectively, the metals, semiconductors, and insulating oxides.

As can be seen, the two bands are overlapped with metals. If E_g is > 3 eV, electrons do not have the possibility to migrate in the conduction band, and thus the electrical conductivity is very low (insulating materials); in the other cases, we have a more or less conductive semiconductor. Some pure substances without defects in their crystal structure have behaviour similar to that of a semiconductor

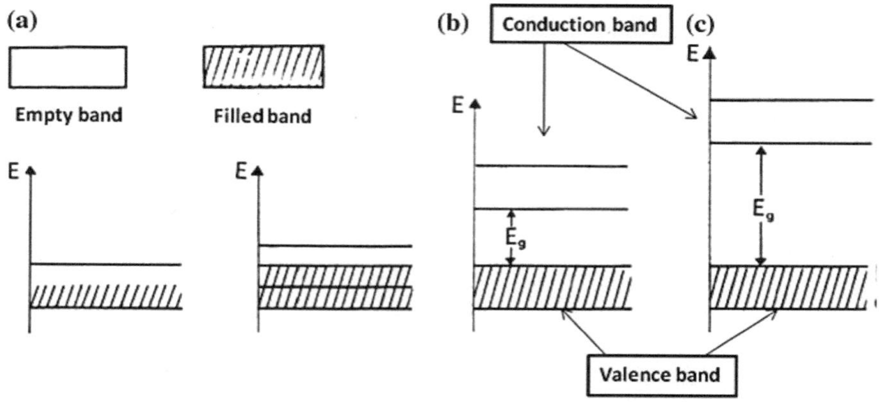

Fig. 3.34 Distribution of electrons in the conduction and valence bands: (**a**) metals, (**b**) semiconductors, and **c** insulating oxides. E_g at room temperature is <3 eV for the semiconductors, which is greater than this value for insulating material

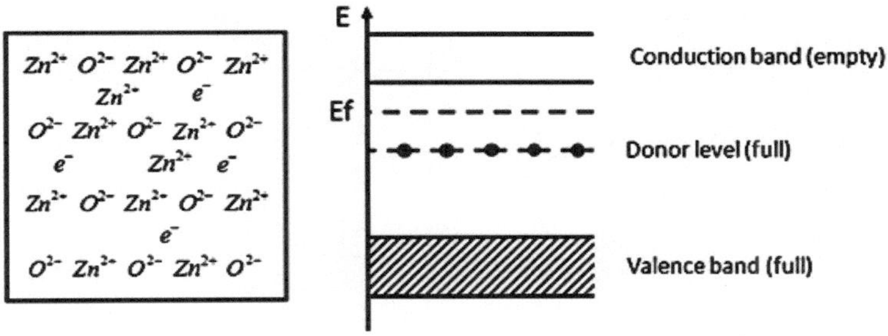

Fig. 3.35 Structure and bands for a type-n semiconductor. Example of ZnO having excess metal in the lattice

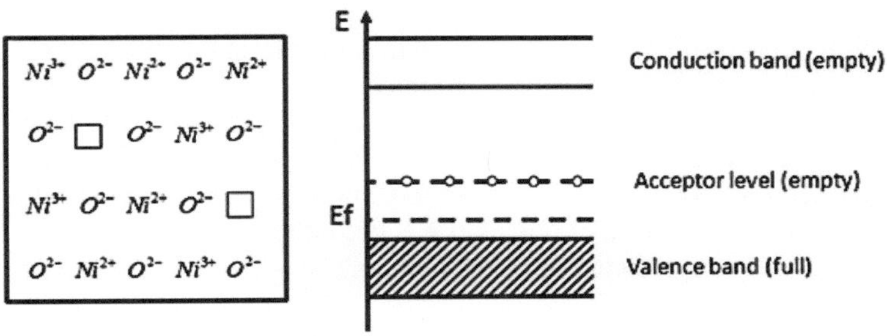

Fig. 3.36 Structure and bands for a type-p semiconductor. An example of Ni_2O_3 having an excess of oxygen in the lattice

(Ge, Si, etc.); these substances are called "intrinsic semiconductors." In other cases, the semiconductor properties are determined by the presence of defects in the structure or of impurities that create energetic levels intermediate between the two bands, thus reducing the energy gap for the transition of the electrons from the valence to the conduction band. An example of this type is represented by zinc oxide (ZnO), which has normally an excess of metal in the lattice (non-stoichiometric compound). The atoms of excess metal occupy interstitial positions, as shown in Fig. 3.35; the charges obviously will be neutralized, but the metal is easily ionized. The electrons released by the interstitial zinc will have a high mobility because they are not involved in any bond. These electrons will be in a band at an energetic level that is inferior to that of conduction (donor level), and therefore, they are more accessible to the electrons coming from the valence band. This type of semiconductor is defined as "type n" (negative) because the conductivity is imputable to the excess zinc in the lattice releasing electrons that fill the donor energetic level.

In the case of non-stoichiometric oxides (NiO, CuO, etc.) having an excess of anions in the lattice, because the anions have an atomic radius much greater than that of the cations, they cannot occupy the interstitial positions of the lattice. In this case, some cationic vacancies are created as shown in Fig. 3.36.

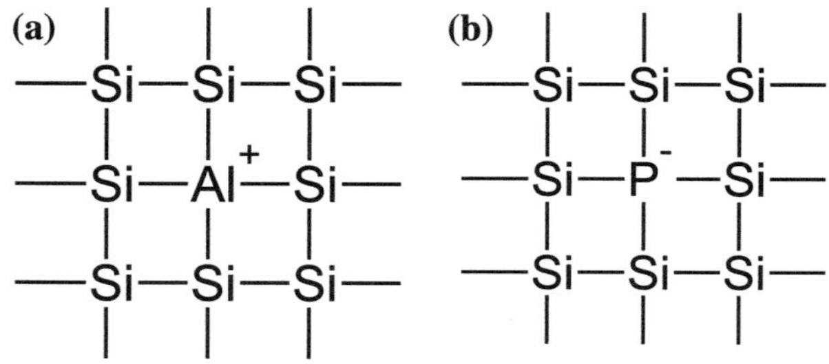

Fig. 3.37 An example of doping an intrinsic semiconductor: (**a**) Silicon doped with aluminium; (**b**) silicon doped with phosphorus

Because the excess oxygen in the lattice is in the form of O^{2-}, for satisfying the electroneutrality, we will have $2Ni^{3+}$ for any positive hole with the corresponding energetic level. This level is called the "acceptor level" and is intermediate between the band of valence and the conduction band. Some electrons will easily pass from the band of valence to the acceptor level, thus leaving a positive hole in the valence band. Electric conduction in this case occurs through the positive holes. These semiconductors are of "p-type" (positive). It is important to point out that the donor and acceptor levels, considering the small number of excess ions in the lattice, do not constitute conducting bands, and electric conduction always occurs in the band of conduction (by electrons) or the band of valence (by migration of positive holes).

The acceptor or donor energetic levels can be originated not only by the imperfect stoichiometry of the solid compounds but also by the presence of impurities. In this case, the semiconductors are called "doped semiconductors," respectively, of n-type or p-type. For example, the addition of a small percentage of foreign atoms in the regular crystal lattice of silicon or germanium, which are intrinsic semiconductors, produces dramatic changes in their electrical properties producing n-type and p-type semiconductors. Figure 3.37 shows what happens to silicon doped, respectively, with aluminium or phosphorus. Aluminium, because it is trivalent, creates deficiencies of valence electrons (holes), thus determining the behaviour of the p-type semiconductor, whilst the addition of phosphorus, a pentavalent impurity, increases the contribution of free electrons, thus greatly increasing the conductivity of silicon and giving place to an n-type semiconductor.

In addition, non-stochiometric semiconductors can be doped. For example, NiO can be doped with Li_2O or with Cr_2O_3. In the first case, the p-type character is enhanced, and in the second, it is weakened. The catalytic action of these materials is due to the availability of electrons or positive charges both characterized by high mobility.

Chemisorption is favoured by the transfer of electrons from a molecule to a solid and vice versa. Moreover, in the oxides of transition metals, the passage of electrons from one band to another and from a solid to a molecule occurs with a change in the valence of the metal. This behaviour is responsible for a particular catalytic cycle, called "redox," in which an organic compound is oxidized by the lattice oxygen with a reduction of the valence of the metal. Then the catalyst is re-oxidized in the presence of molecular oxygen. This catalytic cycle is sustained by the mobility of the anionic vacancies.

Chemisorptions on the surface of semiconductors can occur through two different mechanisms: One mechanism involves the existence of isolated active sites interacting with the reactant molecules through the coordinative unsaturation of the metal; the other one, on the contrary, involves the electronic properties of the solid. In this last case, the molecule is adsorbed as an ion, which could be a cation or an anion adsorbing on an n-type or p-type semiconductor. Therefore, we can have different types of chemisorption. For example, the chemisorption of molecular oxygen on a type-n semiconductor occurs as follows:

$$1/2\ O_2 + e^- \leftrightarrow O^-(\mathrm{ads}) \tag{3.73}$$

Species O^{2-} and O^- are reactive, which can be relieved using ESR techniques. The adsorption of CO on a p-type semiconductor will occur as:

$$\mathrm{CO(g)} \leftrightarrow \mathrm{CO}^+(\mathrm{ads}) + e^- \tag{3.74}$$

In both cases, the adsorption is contrasted by the charges created on the surface, and the monolayer is never reached. In other cases, the chemisorption is favoured, and the monolayer is completely filled. The doping of semiconductors can have dramatic effects on catalytic activity as observed, for example, by Parravano (1953). He showed that the addition of only 100 ppm of WO_3 to NiO reduced the activation energy for CO oxidation by half.

3.5 Preparation and Characterization of the Most Common Catalytic Supports

The most common catalytic supports are alumina, silica, silica–alumina, and carbon.

3.5.1 Alumina

There are different types of alumina because this compound shows different crystalline phases. When aluminium hydrate is precipitated from an aqueous solution of an aluminium salt, the nature of the precipitate depends on the precipitation conditions. However, even if an amorphous hydrogel is obtained, its dehydration by aging yields crystallization in one of the alumina structures, such as boehmite, bayerite, or gibbsite. It is not easy to control the preparation, and some empirical "recipes" are available to obtain the desired structure. However, the most commonly employed supports are γ-alumina or η-alumina (100–300 m^2/g). The most stable, obtained at

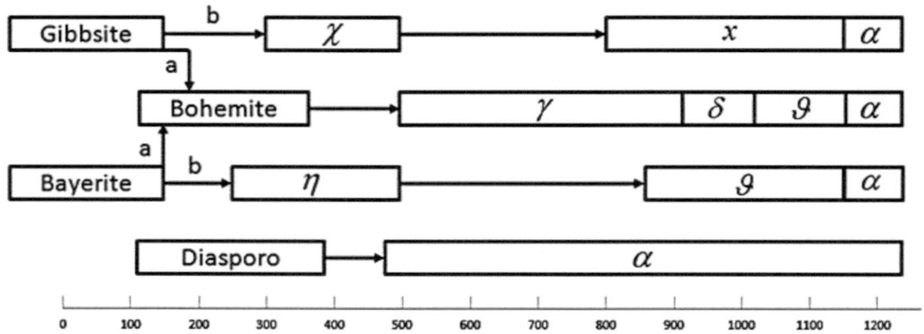

Fig. 3.38 The phase transformations of alumina

very high temperature, is α-alumina (corundum), but it is not useful as a support because it has a very low specific surface area (0.1–1.5 m^2/g). A scheme of all possible transformations is shown in Fig. 3.38.

Route (a) is characterized by pressure > 1 bar, wet air, heating rate > 1 °C/min, and particle size > 100 μm, whilst route (b) is characterized by pressure = 1 bar, flow of dry air, heating rate < 1 °C/min, and particle size < 10 μm. γ- and η-alumina are both stable in a range of 300–600 °C, maintain their specific surface area, and are increased in the stability by adding 1–4% SiO_2.

Alumina is an amphoteric oxide with an isoelectric point (IEP) = 8–9, that is, its surface contains both acid and basic sites. By calcination and re-hydration, we have the following equilibria:

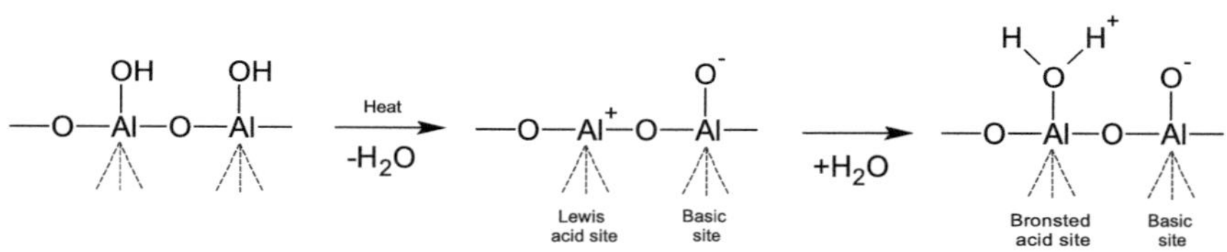

According to Peri (1965), five different types of hydroxyls—A through E—exist on the alumina surface as can be seen in Fig. 3.39. The difference is given by the number of O^{2-} groups surrounding the hydroxyls.

Tanabe (1970) titrated the acidity of alumina with amine titration method and obtained the results shown in Fig. 3.40.

An interpretation of one of the curves shown in Fig. 3.40 regarding the type of acidity on the surface is shown in Fig. 3.41. As can be seen, Brønsted acid sites are prevalent at lower temperatures, whilst at higher temperatures, these are transformed in Lewis acid sites.

The Lewis acid strength of the alumina sites can be enhanced by replacing surface hydroxyls with halogens, such as fluorine or chlorine.

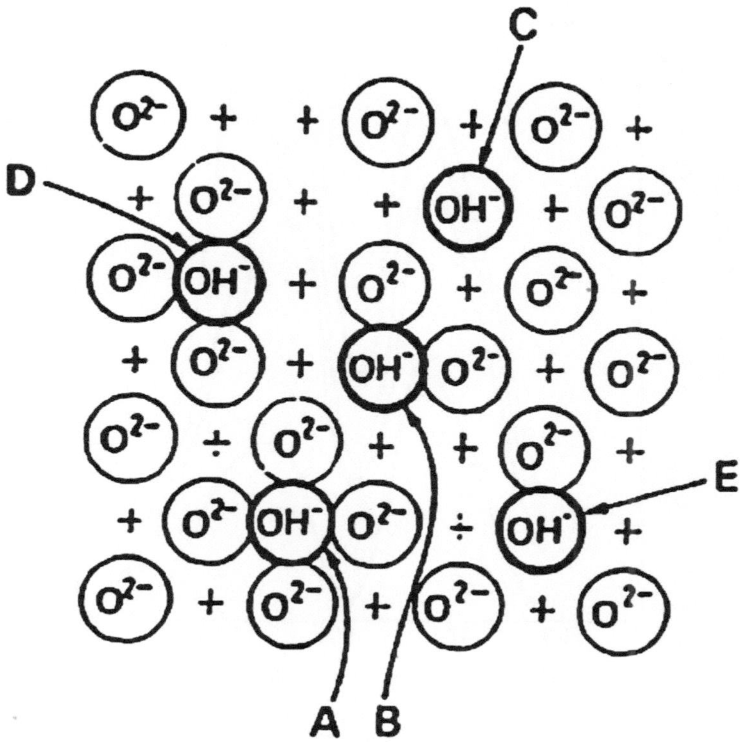

Fig. 3.39 Different types of hydroxyls on the surface of alumina. Reprinted with permission from Peri (1965). Copyright American Chemical Society (1965)

Exercise 3.5. Determination of Surface Acidity Distribution of a Heterogeneous Acid Catalyst

Wang et al. (2020) have studied the behaviour of Ni-exchanged MCM41 in the reaction of ethylene oligomerization. They registered TPD spectra on catalysts with different Ni content by using ammonia as probe molecule. In particular, for the catalyst with Si/Al ratio of 12.5, the spectra are reported in the following Fig. 3.42a.

The experimental data reported in Fig. 3.42a have been digitized directly from the paper of Wang et al. (2020) and are available in the Electronic Supplementary Material, Data 3.1 of the book website. The authors reported a detailed analysis of NH3-TPD spectra and concluded that five different types of acidic sites are present, characterized by different desorption activation energy and desorption temperature (Ea = 63, 90, 124, 150, and 175; Tdes = 423, 473, 515, 580, and 620). These values have been evaluated from all the NH_3-TPS experiments. The scope of this exercise is to evaluate, for the experiment on the catalyst with 12.5 Si/Al ratio, the distribution of sites with different strength and the desorption temperature of each of them.

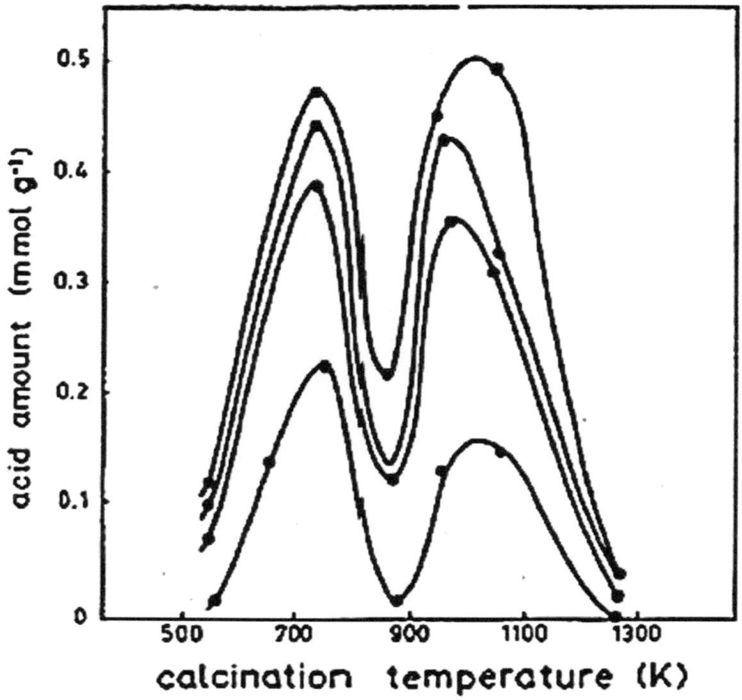

Fig. 3.40 Amounts of surface acidity of alumina as a function of calcination temperature at varying H_o values determined by amine titration using opportune indicators. The values of H_o explored are as follows: **(a)** <+3.3; **b)** <+1.5; **(c)** <−3.0; and **(d)** <−5.6. Published with permission from Tanabe (1970). Copyright Elsevier

Solution

The solution of the exercise can be obtained through deconvolution technique that consists in determination of the parameters of the five gaussian curves which sum correspond to the experimental spectrum signal. In other words, the overall signal is decomposed (deconvolved) into five separate signals related to each type of acidic site.

Mathematically, the calculate signal can be obtained from the following relation:

$$S_c(T) = \sum_{k=1}^{5} a_k \exp\left(-\frac{(T-b_k)^2}{c_k^2}\right) \tag{3.75}$$

This corresponds to the evaluation by fitting of 15 adjustable parameters. It is worth to note that parameters a represent the height of the individual peaks while b are related to the x-position of the peaks that is the desorption temperature. The exercise has been solved by using a MATLAB code reported in the Electronic Supplementary Material, Data 3.1 of the book and a graphical result is depicted in

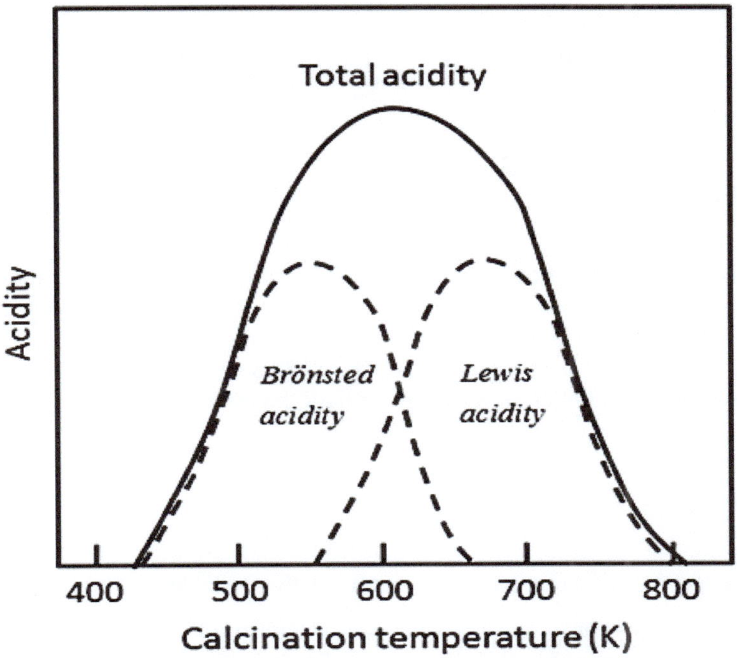

Fig. 3.41 Type of acidity existing on the alumina surface. Published with permission from Tanabe (1970). Copyright Elsevier

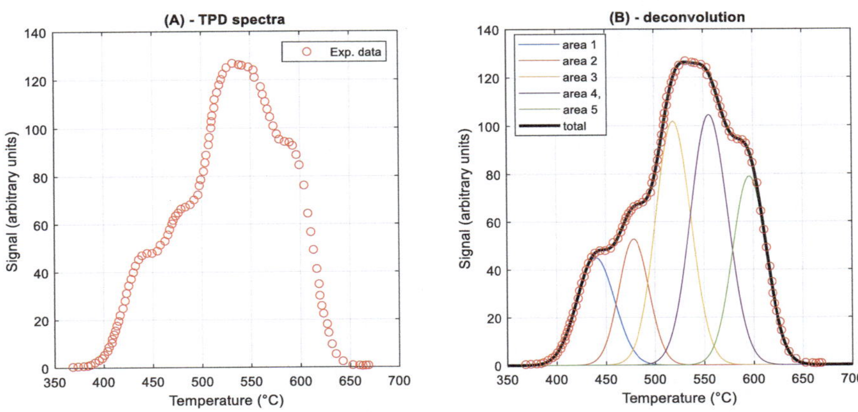

Fig. 3.42 NH_3-TPD spectra of (12.5)Ni-MM41. (**a**) spectra, (**b**) deconvolution of different sites of contribution

Fig. 3.42b. As it is possible to observe, the description of the signal is very good, and the five constituting peaks are clearly identified. In the following Table 3.15, the numerical results are reported.

Table 3.15 Numerical results of the calculations

Type of acidic site	*a*	*b*	*c*	Area	Area fraction
Ea = 63 kJ/mol	45.039	439.6	26.932	2149	0.126
Ea = 90 kJ/mol	52.755	478.3	21.659	2025	0.119
Ea = 124 kJ/mol	101.782	518.9	25.274	4559	0.267
Ea = 150 kJ/mol	104.463	55.4	26.645	4933	0.289
Ea = 175 kJ/mol	78.913	596.0	24.228	3389	0.199

By considering the values obtained for parameter *b* (calculated desorption temperature) they are in good agreement with those reported by the authors (reported above) also by considering that the present exercise is related to a single TPD spectra. Al last, the area fraction in the above table, is a quantitative measurement of the strength distribution of the acidic sites of the catalyst.

3.5.2 *Silica*

Silica supports of different types are commercially available:

(1) Diatomaceous earth, or kieselguhr, is a natural product constituted by microscopic skeletons of fossilized remains of diatoms. Kieselguhr contains 80–90% of silica and has a specific surface area in the range of 15–40 m^2/g.
(2) Silica gel is obtained by acidification of a solution of sodium silicate. A material with a very high specific surface area is obtained (>700 m^2/g) that has very narrow pores (2.5–5 nm). However, according to the preparation method, the specific surface area can change in the range of 200–800 m^2/g.
(3) Colloidal silica is a silica in a colloidal state that is stabilized with sodium or ammonium ions.
(4) Pyrogenic or fumed silica is a type of silica obtained in very thin non-porous particles (5–40 nm) obtained by performing $SiCl_4$ hydrolysis in a flame or by combustion of a silicon alkoxide. The surface area is in the range of 50–400 m^2/g.

The most commonly used support is silica gel, an amorphous material of high surface area obtained by precipitation from a solution of sodium silicate after acidification or by hydrolysis of alkoxides. The precipitate consists of particles of non-ordered arrangement of SiO_4 tetrahedra with silanol groups at the surface (hydrosol) such as shown in Fig. 3.43. The reticulation of these particles gives rise to a hydrogel, that is, a three-dimensional network retaining molecules of water. This process is called "gelation." An alternative path is sol flocculation, which can be obtained by neutralizing the micelle charges (see Fig. 3.44). The silica surface has a moderate acid character, and the IEP is approximately 1–2.

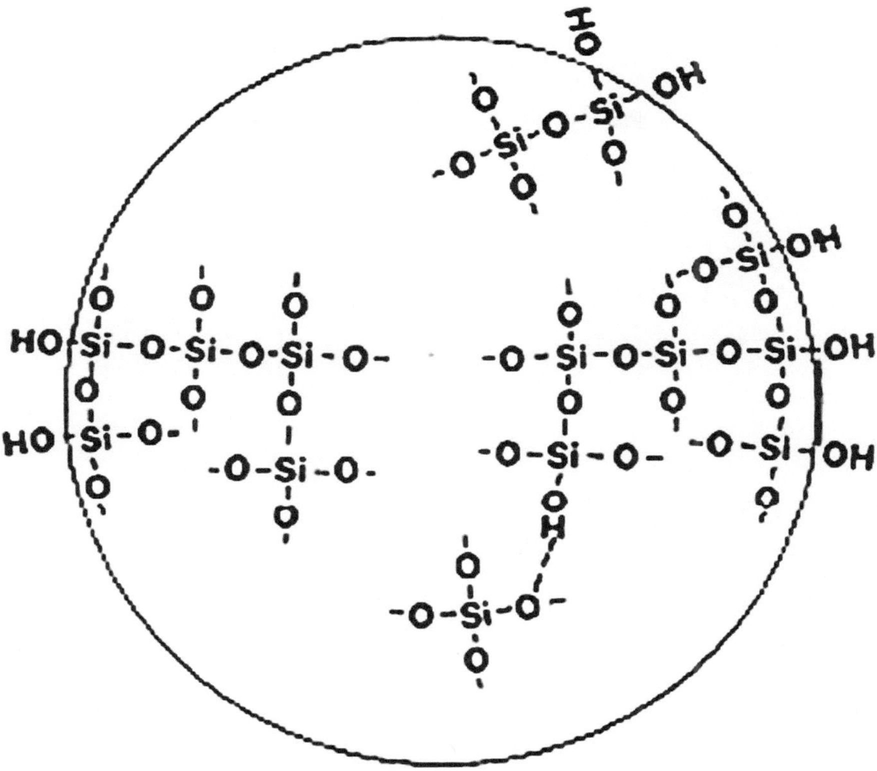

Fig. 3.43 A schematic representation of a precipitated colloidal silica particle; see Anderson (1975)

3.5.3 Silica–Alumina

Silica–alumina is the most important and known example of a binary mixed oxide in which acid–base properties are very different from the originating oxides silica and alumina. Silica, it has been seen, is an oxide of moderate acidity, whilst alumina normally shows a moderate basicity. Silica–alumina is characterized by a strong Brønsted acidity. In this case, the aluminium atoms are trapped inside the silica tetrahedral network, and a negative charge arises on aluminium as can be seen in Fig. 3.45.

The maximum acidity is shown for a composition containing approximately 30 wt% of alumina as shown in Fig. 3.46.

Silica–alumina is obtained by co-precipitation of both components or by precipitation of an alumina hydrogel in the presence of a freshly prepared silica hydrogel. The specific surface area can range from 350 to 600 m^2/g.

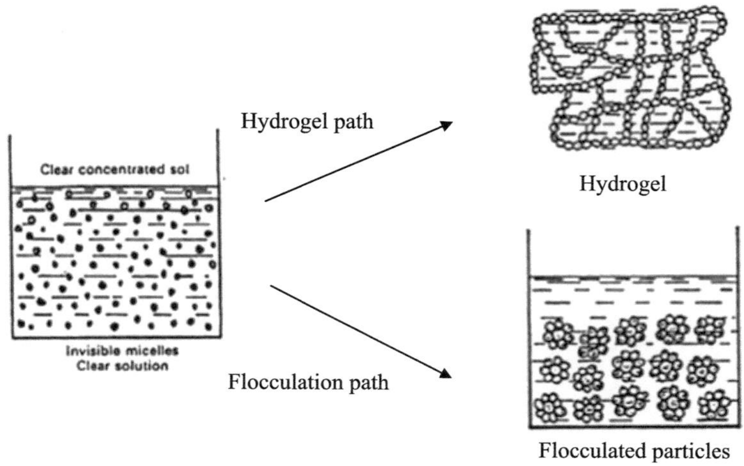

Fig. 3.44 Structures of silica hydrogel and of flocculated particles, which are both obtainable from a silica hydrosol. Reproduced with permission from Perego and Villa (1997). Copyright Elsevier (1997). See also Le Page et al. (1987)

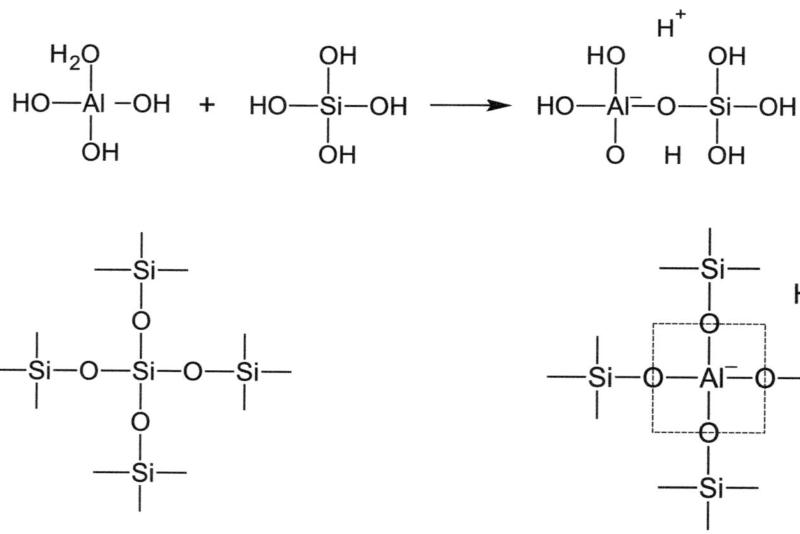

Fig. 3.45 The reason for the high Brønsted acidity shown by silico–alumina catalysts

3.5.4 Carbon

The types of carbon that can be used as catalytic supports include charcoal and carbon black. Charcoal is obtained by pyrolysis of natural or synthetic polymers at a temperature of approximately 700 °C. Activated charcoal is obtained by performing

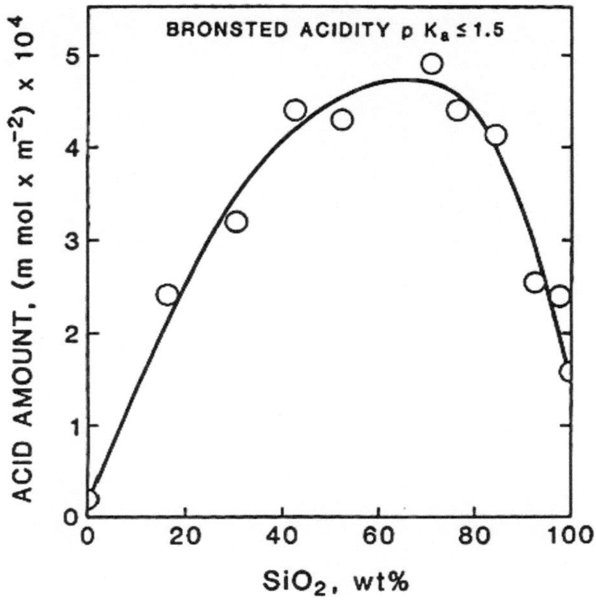

Fig. 3.46 Acidity of silica–alumina catalysts. Published with permission from Oblad et al. (1951). Copyright Elsevier (1951). See also Richardson (1992)

a controlled oxidation of charcoal at high temperature (approximately 900 °C). This last treatment eliminates residual products of pyrolysis from the surface and renders the internal pores of the carbon more accessible. Very high specific surface area can be obtained on the order of 1000 m^2/g. Similar products also can be obtained as by-products in many petrochemical productions for the incomplete combustion of heavy petroleum products (carbon black).

3.5.5 Monolithic Supports

These supports are usually made from ceramic materials—such as cordierite (magnesium silico-aluminate), mullite, or alumina—but metals have been used sometimes. Monolithic supports are constituted of a block of material containing large cavities (1–20 mm) of opportune shape favouring gas flow with a low pressure drop. The main use of these supports is for the automobile catalytic converters but other possible uses have been studied by many different authors despite the relatively low exposed surface area of these devices (200–2000 m^2/m^3). To favour the deposition of a metal on the surface, these materials are coated on the surface with a layer of approximately 30 μm of silica or alumina. Some examples of monolithic supports are shown in Figs. 3.47 and 3.48.

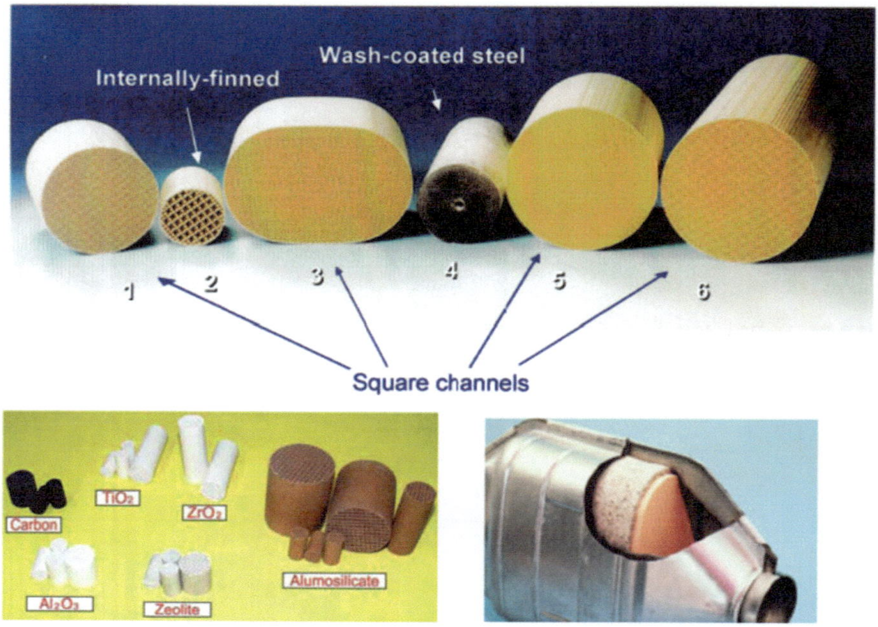

Fig. 3.47 Examples of monolithic supports and an example of monolithic support used as a catalytic converter in a car. Reprinted with permission from Kapteijn et al. (2001), copyright Elsevier (2001), and from the website of the Boreskov Institute of Catalysis

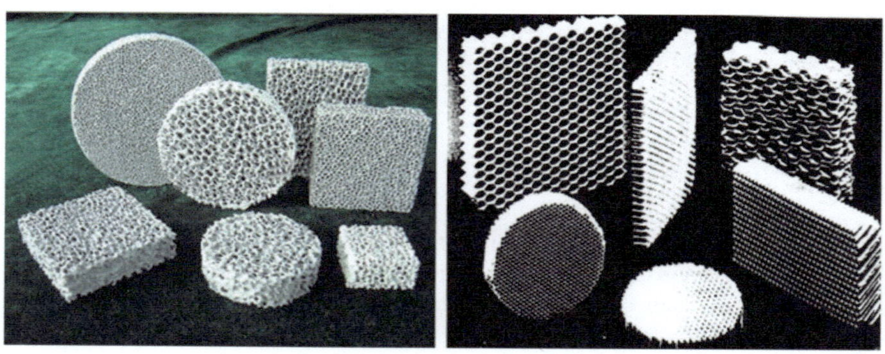

Fig. 3.48 Additional examples of commercial monolithic supports

3.5.6 Metal Catalysts

Metal catalysts can be unsupported or supported. Unsupported catalysts can be non-porous, that is, in the form of films or wire mesh, or porous, that is, in powdered form (platinum and palladium black, nickel Raney, etc.). Unsupported porous catalysts are extremely reactive with oxygen in the air and must be protected by covering them with a liquid solvent. For preparing platinum or palladium black with a high specific surface area, it is best to use a concentrated solution of the precursor (e.g., chloroplatinic acid) and a strong reducing agent (hydrazine, N_2H_4; sodium borohydride, $NaBH_4$, etc.). In this way, a large number of nuclei of precipitation are formed, and a highly dispersed precipitate is obtained. Unsupported metal catalysts are subjected to sintering, in particular when heated at high temperature. By sintering, the catalytic particles gradually increase in size, and consequently, both the specific surface area and the catalytic activity decrease. To make metal catalyst more stable to sintering, it is best to prepare them by reduction of the corresponding oxides mixed with substances acting as promoters and/or support. In this way, the iron catalyst used in ammonia synthesis from nitrogen and hydrogen is, for example, prepared. Iron is obtained by reducing magnetite (Fe_3O_4) containing uniformly dispersed amounts of Al_2O_3 and K_2O with hydrogen at high temperature. Both of these compounds are not reduced, and alumina prevents the sintering of the iron crystallite, whilst K_2O neutralizes the acid sites generated by alumina.

The production of Ni-Raney is made starting from a Ni-Al alloy by dissolving aluminium with a sodium hydrate solution. Ni obtained in this way is highly disperse and reactive. The preparation of supported metal is normally made by impregnation according to two different procedures:

(a) **Incipient Wetness Impregnation or Dry Impregnation**

An active metal precursor is dissolved in a solvent. The obtained solution is added, drop by drop, to a well-mixed powdered support. A limited amount of solution is added such that the volume of the solution is equal to the overall volume of the pores. The solution enters the pores by capillary action, and at the end of the addition, the powder appears dry. A large part of the metal precursor interacts with the surface of the support, thus giving place to chemical adsorption, whilst the residual remains dissolved. Then the solvent is removed by drying and calcining. In this way, all of the catalyst precursor is deposited on the surface, a large part is highly dispersed because it is chemisorbed, and the remaining part gives place to greater crystallites that precipitate when the solubility product is reached. This operation can be optimized by dosing the amount of precursor in such a way as to minimize the amount of precursor precipitated with respect to the one that is chemisorbed. Obviously, the choice of a support strongly interacting with the molecules of the precursor also is fundamental for obtaining a good result in dry impregnation. Dry impregnation also can be performed by using the support in pellet form.

(b) **Wet Impregnation**

An appropriate amount of an active metal precursor is dissolved in a solvent. A support in the form of powder or pellets is immersed in this solution. The precursor is partitioned, and part is chemisorbed on the support surface in equilibrium with the amount that remains dissolved in the solution. Adsorption normally occurs by ionic exchange. This adsorption depends on (1) the characteristics of the surface of the support, (2) the nature of the available precursor, (3) the employed solvent, and (4) the acid–base character of the precursor

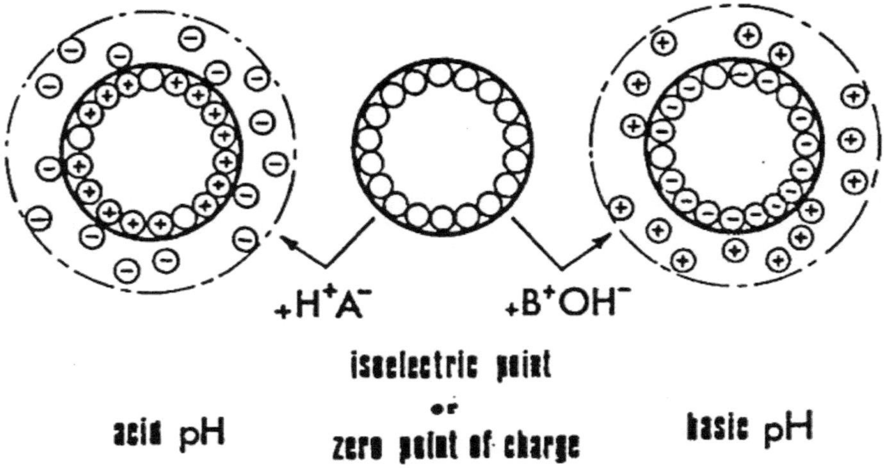

Fig. 3.49 Schematic view of the surface polarization of an oxide induced by the pH of a contacting solution. Reproduced with permission from Perego and Villa (1997). Copyright Elsevier (1997). See also Brunelle (1978)

solution. It is important to obtain highly dispersed catalysts not only for increasing the activity but also for reducing the amount of metal to be used, which is an important aspect considering that often the metals used are very expensive noble metals, such as platinum, palladium, rhodium, ruthenium, etc. To obtain a high dispersion of the metal on the support, some theoretical aspects must be considered. Normally employed supports have many hydroxyls on their surface with an amphoteric character. Therefore, in an acid environment, the hydroxyls of the support react as follows:

$$\mathrm{S-OH + H^{+}A^{-} \leftrightarrow S-OH_2^{+}A^{-}} \tag{3.76}$$

In contrast, in a basic environment, we have:

$$\mathrm{S-OH + B^{+}OH^{-} \leftrightarrow SO^{-}B^{+} + H_2O} \tag{3.77}$$

In the first case, the surface is charged positively, whilst in the second, it is charged negatively as shown in Fig. 3.49. However, there is a pH at which no charge is accumulated on the surface; this pH condition is defined as the "isoelectric point" (IP) or "point of zero charge" (PZC). The IP is a characteristic of the solid defining the acid or basic character of its surface. Solids characterized by a low IP preferably adsorb cations. If we must support platinum, for example, in this case, it is best to use a cationic precursor of the type: $Pt(NH_3)_4^{++}$. If the support has, on the contrary, a high IP, an anionic precursor, such as $PtCl_6^{=}$, is more suitable. Hence, on a silica support having a PZC = 2, the first precursor will be used, whilst on an alumina

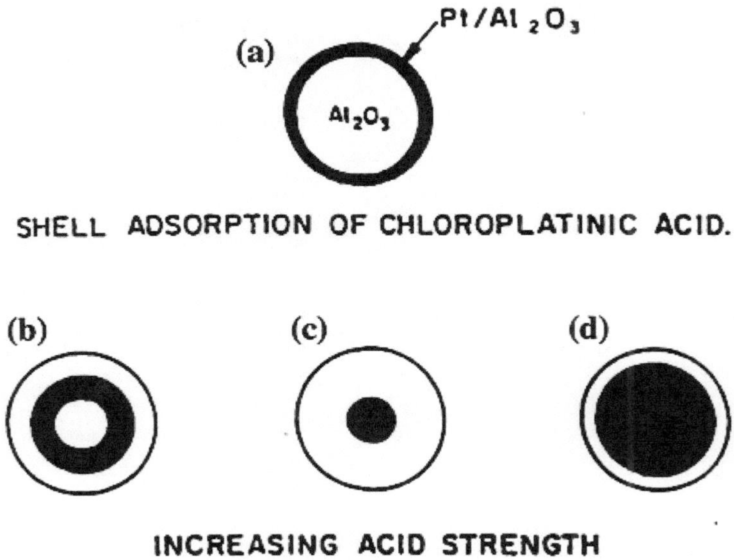

Fig. 3.50 Metal concentration profiles in catalytic particles obtained due to the presence of acids competing in adsorption with chloroplatinic acid

support of PZC = 8–9, the use of H_2PtCl_6 as precursor is preferable. Last, it is possible to use the competition in the ionic adsorption, for example, between $PtCl_6^{=}$ and Cl^- for favouring or depressing the adsorption of the precursor.

By choosing the competitive ion and dosing its relative concentration, it is possible to deposit the metal in a layer external to a catalytic pellet or, alternatively, in an annular layer internal to the particle or concentrated at the centre of the pellet as shown in Fig. 3.50. Because some reactions, affected by intra-particle mass-transfer limitation, occur mainly in a particular zone of the catalytic particle, in the described cases, the metal is deposited exactly where the chemical reaction occurs.

In some cases, two metals in alloy are used as catalyst. The method of preparation in this case has the scope to obtain exclusively the alloy of the desired composition without the presence of the other different phases.

The catalytic action of the metal can change with the dispersion. This can be understood by considering a perfect cube-octahedral crystallite as shown in Fig. 3.51. On this crystal, we can recognize atoms with different degrees of unsaturation. Some atoms are positioned on the faces of the crystal, some atoms are on the edges, and some atoms are on the vertices. As can be seen in Fig. 3.51, the abundance of atoms, respectively located on faces, edges, and vertices, is quite different. However, the ratios between the different types of atoms change considerably by changing the size of the crystallites (Fig. 3.52).

It is possible to foresee, in agreement with the findings of Boudart (1985), that the selectivity of some reactions is affected by the dispersion ("structure-sensitive reactions"), whilst on the contrary, some are not ("structure-insensitive reactions").

Fig. 3.51 Example of a perfect structure of metal cube-octahedral crystallite

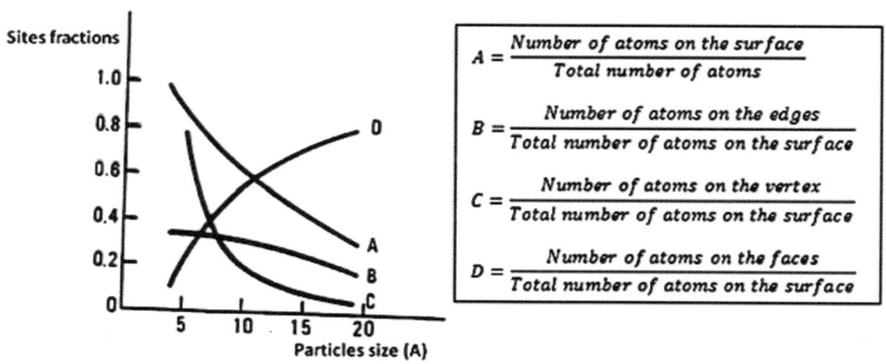

Fig. 3.52 Distribution of metal atoms in crystallite of different sizes

However, the surface of dispersed metals shows many irregularities, such as the ones shown in Fig. 3.53, that could be active sites of catalysis.

The dispersion of metal catalysts can be determined by measuring the specific surface area of the supported catalyst, that is, the overall surface of the metal referred to the mass of the metal. The greater the specific surface area, the greater the dispersion.

The specific surface area can be determined by adsorbing, under opportune conditions, H_2 or CO. The used methods can be static or dynamic (with a flowing stream or by feeding pulses of H_2 or CO). These methods are not rigorous for uncertainty in the stoichiometry of the reactions; in some cases, phenomenon known as "spillover" (i.e., migration of part of the hydrogen adsorbed on the metal toward the support) occurs.

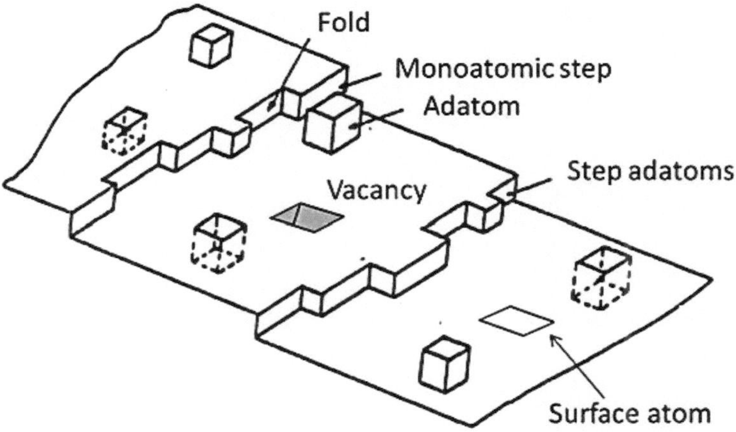

Fig. 3.53 Possible defects on a solid surface

The catalytic action of the metal is complicated by the possibility of the overlap of different mechanisms whose contribution is unknown. Metals can intervene with their electronic properties of conductive materials; they can yield the formation of metallorganic complexes for localized interaction between one or more atoms of the metal with the organic reactant; and, in the case of transition metals, adsorption and successive reaction can be favoured by the interaction with the electrons of *d*-orbitals. However, it is possible to classify the chemisorption on a metal surface as associative or dissociative. Hydrogen, for example, can give both these types of chemisorption according to the metal employed as catalyst:

(3.78)

Ethylene can yield associative adsorption by breaking the π bond and forming two new σ bonds with two atoms of the metal surface:

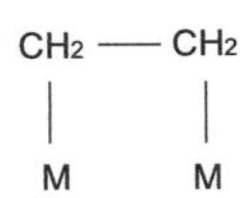

(3.79)

Alternatively, adsorption on a single metallic atom can occur by forming a coordination complex of the type shown in Fig. 3.54.

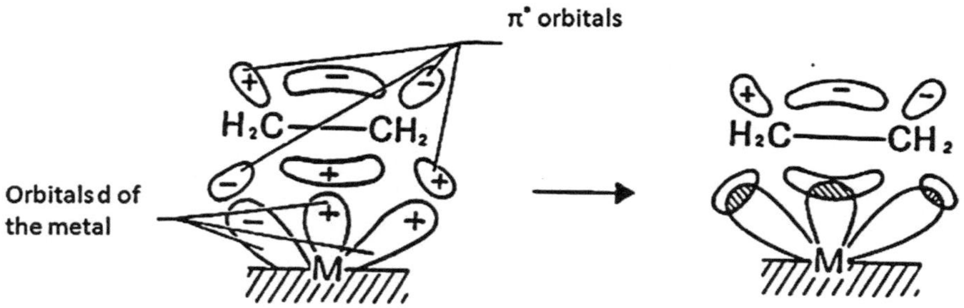

Fig. 3.54 Associative adsorption of ethylene on a single atom of the surface of a metal catalyst

Table 3.16 Different types of catalyst shapes[a]

Shape	Size	Type of reactor
Extrudate	$d = 1$–5 mm; $l = 5$–20 mm	Fixed-bed reactor
Pellet	$d = 3$–10 mm; $l = 3$–10 mm	Fixed-bed reactor
Granule, bead	$d = 1$–5 mm	Fixed-bed reactor
Sphere	$D = 1$–10 mm	Fixed-bed reactor, moving-bed reactor
Microspheroidal powder	$D = 20$–100 μm	Fluidized-bed reactor, slurry reactor

[a]Data from Doesburg and van Hoof (1993)

3.6 Catalyst Forming

Size and shape of catalyst particles are important for industrial purpose, in particular, for favouring catalytic activity avoiding mass- and heat-transfer limitation and limiting pressure drop in the reactor. The type of reactor chosen for performing a reaction clearly imposes the size and shape of the catalyst particles to be used as well as the modality of preparation. In a fluid-bed reactor, for example, a catalyst in the form of powder, which is resistant to the attrition, would be used. In a slurry reactor, again a powdered catalyst is preferred, but it must be resistant to contact with the liquid reagents as well as stirring stress. In a gas–solid tubular reactor, a pelletized catalyst is normally used that has a spherical or cylindrical geometry. In this last case, the particles size must be as small as possible to avoid mass-transfer limitation, thus maximizing the catalytic activity; however, too small a size gives place to high pressure drop and excess energy consumption for feeding the reagents. A compromise between these opposite requirements determines the optimal size, whilst the shape depends on both the method of preparation and a specific design devoted to optimal heat transfer. Table 3.16 lists different types of catalyst shapes.

The variety of possible shapes can be seen in Fig. 3.55.

Fig. 3.55 Miscellaneous commercial catalyst shapes suitable for fixed-bed reactors

3.6.1 Forming Microgranules

In the laboratory, crushing and grinding is often a common practice for obtaining powdered catalysts. This procedure is not used in industry because powder obtained in this way has a broad particle size distribution. Moreover, such powders have low mechanical resistance. Spray-drying is the technique normally employed to produce micro-spherical materials. A hydrogel or a sol is sprayed through nozzles in a heated chamber. Depending on the temperature of this chamber, either drying or both drying and calcination of the falling drops occur. Figure 3.56 shows a spray-dryer apparatus.

Figure 3.57 shows an example of microgranules of relatively uniform size falling in the range of 380–570 μm.

3.6.2 Forming Grains

Grains can be obtained by different methods such as (1) extrusion and wet pressing, (2) dry tableting, (3) granulation, and (4) drop coagulation (sol-gel method).

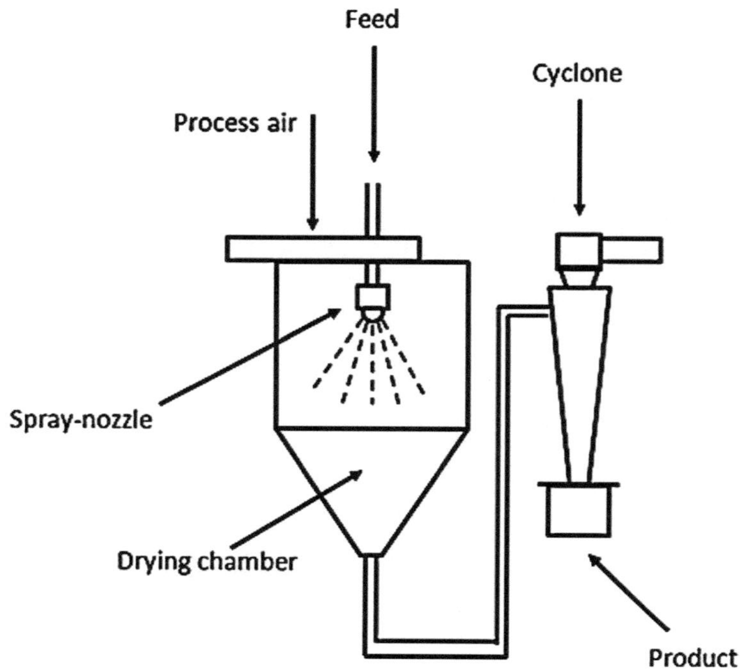

Fig. 3.56 Simplified scheme of a spray-dryer apparatus suitable to prepare microgranular catalysts. Reproduced with permission from Perego and Villa (1997). Copyright Elsevier (1997). See also Doesburg and van Hoof (1993)

1. **Extrusion and wet pressing**

   A wet paste is forced through a screw driver (see Fig. 3.58). The slurry is forced through holes in the end plate. Usually, the holes are circular, but they can also be made in the shape of lobes or stars. The extrudate at the hole exit is dried and consequently hardens, thus maintaining its shape. A rotating knife is used to cut the ribbon at a prefixed length. Extrudates are irregular in shape and length, are porous, and have low mechanical resistance, but they are easy to prepare and inexpensive compared with other preparation methods. Some additives are useful to obtain a paste suitable for extrusion. Monolith and honeycomb support are also manufactured by extrusion.

2. **Dry tableting**

   In this case, a dry powder is pressed in a pelleting press. Powder is poured into a cylindrical cavity and pressed at a prefixed pressure using a piston (Fig. 3.59). Very high pressures, from 100 to 4000 atm, are normally applied. Some additives—such as lubricants (mineral oil, talc, graphite, stearates) and binders (boehmite, clays)—favour pelletization by increasing adhesion between the grains of powder. Pellets are then ejected. These pellets have a uniform size

Fig. 3.57 Microgranules obtained by spray-drying technique. Diameters are in the range of 380–570 μm [Courty and Duhaut (1974, p. 861); see also Perego and Villa (1997)]

and shape, have high mechanical strength, and have a moderate porosity. This method of catalyst preparation is the most expensive, but it yields the best results.

3. **Granulation**
   Granules have normally spherical geometry. Granules can be obtained using an inclined rotating dish (Fig. 3.60). Small particles are fed into the dish. In the meantime, slurry is sprayed onto the particles. The particle surface becomes wet, and granules start to grow maintaining their spherical geometry.

4. **Drop coagulation**
   It also is possible to obtain spheres by "drop coagulation" from sols suspended in an immiscible liquid by inducing simultaneous gelation, ripening, and forming. The method is well illustrated in Fig. 3.61.

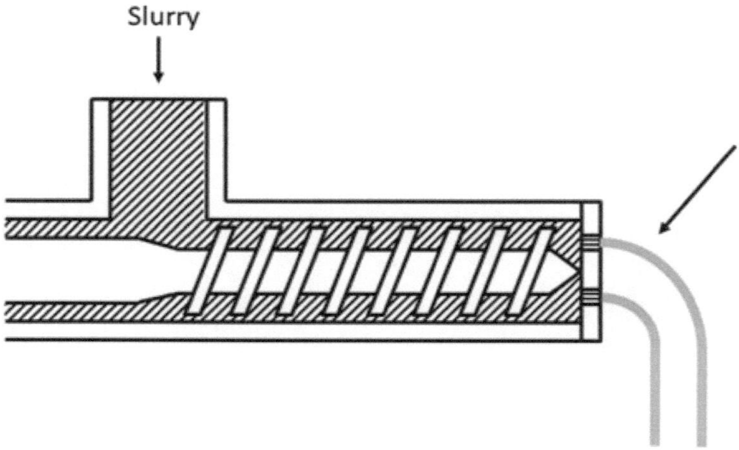

Fig. 3.58 Extrusion device

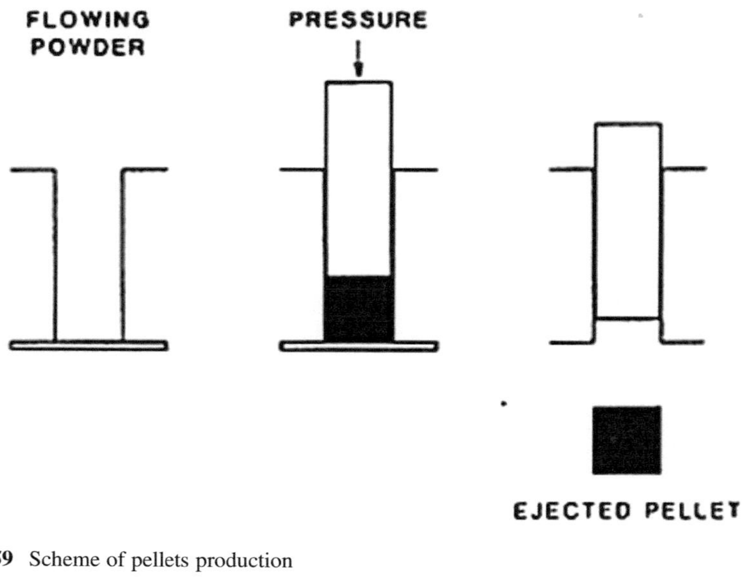

Fig. 3.59 Scheme of pellets production

An aqueous sol or a hydrogel is forced through the holes in a plate at the top of a column containing oil, in which the gel is immiscible. Oil is kept at a temperature >100 °C. Drops fall slowly through the oil and harden before reaching the bottom of the column. At an appropriate pH level, spherical drops give place to gelation, and pH is increased by a flow of ammonia following drying and ripening. In this way, beads of silica–alumina used in catalytic cracking can be produced.

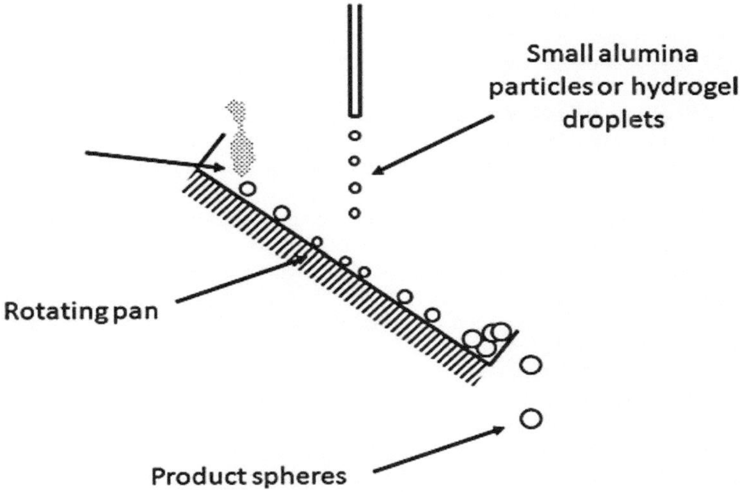

Fig. 3.60 Production of catalyst with a spherical shape using rotating-disk technique. Reproduced with permission from Doesburg and van Hoof (1993). Copyright Elsevier (1993)

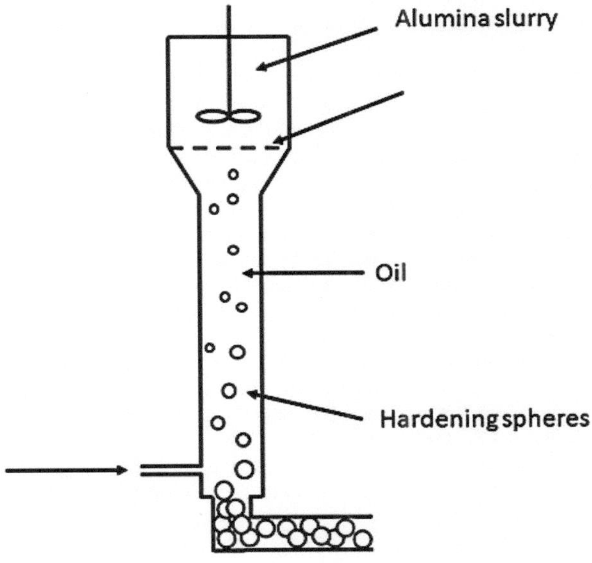

Fig. 3.61 Production of spherical particles by drop coagulation [from Richardson (1992); see also Perego and Villa (1997)]

References

Anderson, R.B.: Experimental Methods in Catalytic Research. Academic Press, New York (1968)
Anderson, J.R.: Structure of Metallic Catalysis. Academic Press (1975)
Barrer, R.M.: Zeolites and Clay Minerals as Sorbents and Molecular Sieves. Academic Press, London (1978)

Beek, O.: Hydrogenation catalysts. Discuss. Faraday Soc. **8**, 118–128 (1950)

Boudart, M.: Heterogeneous catalysis by metals. J. Mol. Catal. **30**, 27–38 (1985)

Breck, D.W.: Zeolite Molecular Sieves: Structure, Chemistry and Use. Wiley-Interscience, London (1974)

Brønsted, J.N.: Einige Bemerkungen über den Begriff der Säuren und Basen. Recl. Trav. Chim. Pays-Bas. **42**(8), 718–728 (1923)

Brunauer, S., Emmett, P.H., Teller, E.: Adsorption of gases in multimolecular layers. J. Am. Chem. Soc. **60**(2), 309–319 (1938)

Brunelle, J.P.: Preparation of catalysts by metallic complex adsorption on mineral oxides. Pure Appl. Chem. **50**, 1211 (1978)

Clerici, M.: Oxidation of saturated hydrocarbons with hydrogen peroxide, catalyzed by titanium silicalite. Appl. Catal. **68**(1–2), 249–261 (1991)

Clerici, M., Ingallina, P.: Epoxidation of lower olefins with hydrogen peroxide and titanium silicalite. J. Catal. **140**, 71 (1993)

Courty, P., Duhaut, P.: Rev. Inst. Franc. Petrole. **XXIX-6**, 861 (1974)

Doesburg, E.B.M., van Hoof, J.H.C.: Preparation of catalyst supports and zeolites in catalysis in an integrated approach to homogeneous. In: Moullijn, J.A., van Leeuven, P.W.N.M., van Santen, R.A. (eds.) Chapter 8 in Heterogeneous and Industrial Catalysis. Elsevier, Amsterdam (1993)

Dollimore, D., Heal, G.L.: An improved method for the calculation of pore-size distribution from adsorption data. J. Appl. Chem. **14**, 109–114 (1964)

Dollimore, D., Heal, G.L.: Pore-size distribution in typical adsorbent systems. J. Colloid Interf. Sci. **33**(4), 508–519 (1970)

Gerasimov, Y.A., Dreving, V., Eremin, E., Kiselev, A., Lebedev, V., Panchenkov, G., Shlygin, A.: Phys. Chem. **1** (1974) (MIR Pu. Moscow)

Giesche, H.: Mercury porosimetry: a general (practical) overview. Part. Syst. Charact. **23**, 1–11 (2006)

Halsey, G.: Physical adsorption in non-uniform surfaces. J. Chem. Phys. **16**, 931–937 (1948)

Hammett, L.P., Deyrup, A.J.: A series of simple basic indicators. I. The acidity functions of mixtures of sulfuric and perchloric acids with water. J. Am. Chem. **54**(7), 2721–2739 (1932)

Hasselbalch, K.A.: Die Berechnung der Wasserstoffzahl des Blutes aus derfreien und gebundenen Kohlensäure desselben, und die Sauerstoffbindung des Blutes als Funktion der Wasserstoffzahl. Biochemische Zeitschrift. **78**, 112–144 (1916)

Henderson, L.J.: Concerning the relationship between the strength of acids and their capacity to preserve neutrality. Am. J. Physiol. **21**, 173–179 (1908)

Jacobs, P.A., Beyer, H.K., Valyon, J.: Properties of the end members in the Pentasil-family of zeolites: characterization as adsorbents. Zeolites. **1**, 161–168 (1981)

Kapteijn, F., Nijhuis, T.A., Heiszwolf, J.J., Moulijn, J.A.: New non-traditional multiphase catalytic reactors based on monolithic structures. Catal. Today. **66**(2–4), 133–144 (2001)

Langmuir, I.: The constitution and fundamental properties of solids and liquids. Part I. Solids. J. Am. Chem. Soc. **38**(11), 2221–2295 (1916)

Le Page, J.F., Miller, R.L., Miller, E.B., Limido, J.: Applied Heterogeneous Catalysis, Design, Manufacture, Use of Solid Catalysts. Edition Technip, Paris (1987)

Lewis, G.N.: Valence and the Structure of Atoms and Molecules. Chemical Catalog, New York (1923)

Millini, R., Massaro, E., Perego, G., Bellussi, G.: Framework composition of titanium silicalite. J. Catal. **137**(2), 497–503 (1992)

Nielsen, J.E., Borchert, T.V.: Protein engineering of bacterial α-amylases. Biochem. Biophys. Acta. **1543**, 253–274 (2000)

Notari, B.: Innovation in zeolite material science. Synthesis and catalytic properties of titanium containing zeolites. Study Surf. Sci. Catal. **37**, 413 (1988)

Oblad, A.G., Milliken, T.H., Mills, G.D.: Chemical Characteristics and Structure of Cracking Catalysts; Advances in Catalysis, vol. 3. Academic, New York (1951)

Parravano, G.: The catalytic oxidation of carbon monoxide on nickel oxide. I. Pure nickel oxide. J. Am. Chem. Soc. **75**(6), 1448–1451 (1953)

Perego, C., Villa, P.: Catalysts preparation methods. Catal. Today. **34**, 281–305 (1997)

Peri, J.B.: A model for the surface of γ-alumina. J. Phys. Chem. **69**(1), 220 (1965)

Piccin, J.S., Dotto, G.L., Pinto, L.A.A.: Adsorption isotherms and thermochemical data of FD&C red n° 40 binding by Chitosan. Braz. J. Chem. Eng. **28**(2), 295–304 (2011)

Prieto, M.A., Biarnes, X., Vidossich, P.: The molecular mechanism of the catalase reaction. J. Am. Chem. Soc. **131**(33), 11751–11761 (2009)

Richardson, J.T.: Principles of Catalysts Development. Springer (1992)

Romano, U., Esposito, A., Maspero, F., Neri, F., Clerici, M.: New developments in selective oxidation. Study Surf. Sci. Catal. **55**, 33 (1990)

Smith, J.M.: Chemical Engineering Kinetics. McGraw-Hill Book, New York (1981)

Tanabe, K.: Solid Acids and Bases. Kodansha Tokio and Academic Press, New York (1970)

Tanabe, K., Sumiyoshi, T., Shibata, K., Kiyoura, T.: A new hypothesis regarding the surface acidity of binary metal oxides. Bull. Chem. Soc. Jpn. **47**(5), 1064 (1974)

Taylor, H.S.: A theory of the catalytic surface. Proc. R. Soc. A Math. Phys. Eng. Sci. **108**(745), 105–111 (1925)

Thomson, W., Kelvin, L.: On the equilibrium of vapour at a curved surface of liquid. Philos. Mag. **42**, 448 (1871)

Wang, C., Wang, L., Wu, G., et al.: Quantitative relationship between activity and acid site distribution in the oligomerization of ethylene over MCM-41 catalyst. Catal. Lett. **150**, 429–437 (2020). https://doi.org/10.1007/s10562-019-02938-w

Washburn, E.W.: The dynamics of capillary flow. Phys. Rev. **17**, 273 (1921)

Wheeler, A.: Reaction rates and selectivity in catalyst pores in catalysis. In: Emmet, P.H. (ed.) Rheinold, vol. II, p. 105. Academic Press, New York (1955)

Chapter 4
Kinetics of Homogeneous Reactions and Related Mechanisms

Abstract In this chapter, the study of kinetics is introduced by examining homogeneous reactions. The relationship between kinetic law and reaction mechanism is stressed introducing the concepts of elementary steps, slow-rate-determining steps, and steady-state conditions. The rate law is derived from a reasonable hypothesis of a reaction mechanism and is introduced in the mass balance equation that differs according to the employed reactor, that is, well-stirred batch, semi-batch, continuous reactor, or alternatively, tubular-packed bed reactor. The characterization methods for determining the ideal behavior of these different types of reactors are described. The schemes of multiple reactions (equilibrium, competitive, and consecutive reactions) are discussed and a general approach is given to the kinetic description of complex schemes. The kinetic approach to radical chain and enzymatic reactions is examined on a theoretical basis, enriched with different examples and exercises. Many other examples of kinetics are reported, with exercises and MATLAB code, associated with these exercises, available online.

Keywords Kinetics and reaction mechanisms · Rate-determining step · Batch · Semi-batch and continuous reactors · Scheme of multiple complex reactions · Kinetics of radical chain reactions · Kinetics of enzymatic reactions

4.1 Introduction

A chemical reaction occurs because a chemical bond of a molecule is broken, and a new bond is formed. Different methods are possible for determining the bond rupture in a molecule, for example, by heating the substance at high temperature or by submitting it to a high-energy electromagnetic field or an electrical shock. All of these methods transfer to the molecule an amount of energy greater than the bond energy, and this is the reason for the rupture occurring. By plotting the internal

---

Supplementary Information The online version contains supplementary material available at https://doi.org/10.1007/978-3-031-82502-6_4.

© The Author(s), under exclusive license to Springer Nature Switzerland AG 2025

R. Tesser, E. Santacesaria, *The Chemical Reactor from Laboratory to Industrial Plant*, https://doi.org/10.1007/978-3-031-82502-6_4

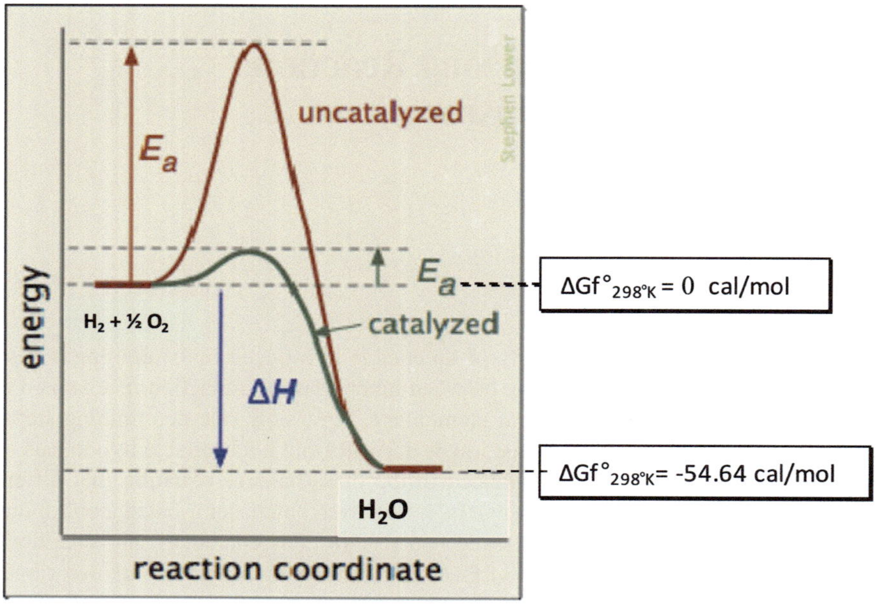

Fig. 4.1 Reaction coordinate for the → reaction H_2 + ½ O_2 → H_2O

energy of H_2- and O_2-reacting molecules as a function of the reaction path to form water (Fig. 4.1), it can be seen that great **activation energy** must be reached for the occurrence of the reaction. The transformation from reagents to products, in this case, is largely favored from the thermodynamic point of view (the product H_2O is much more stable than the reacting elements); however, the higher the activation energy, the lower the probability that the reaction will start. Therefore, although the thermodynamic suggests that this reaction is largely favored, it does not occur in practice at room temperature; therefore, it requires an external intervention, such as a flame, an electrical shock, or the presence of an active catalyst. The system activated in one of these ways could become very fast with the reaction occurring with an explosion. The conclusion is that the thermodynamic can foresee only the initial and final status of a reaction giving no information about the reaction rate, which is the subject of study of chemical kinetics. It is possible to increase the reaction rate, that is, to increase the probability of the reaction occurring, by strongly decreasing the energy barrier. This goal is normally reached by introducing a catalyst in the reacting system. As we already saw in the previous chapter, the catalyst, by interacting with the reacting molecules, allows us to obtain the product passing through a number of sequential steps, with each having an activation energy lower than the one of the un-catalyzed reaction (see Figs. 3.1 and 4.1).

In the presence of a catalyst, the reaction proceeds quickly at temperatures lower than in absence of the catalyst. Another peculiarity of a good catalyst is that one only reaction is selectively promoted. The sequence of the mentioned catalytic steps constitutes the "reaction mechanism," which is useful to know, as will be seen later, for determining a reliable expression of the reaction rate as a function of the composition and of the temperature, that is, for determining a "kinetic law" expression of the type:

$$r = f(\text{composition}, \ \text{temperature}) \tag{4.1}$$

Chemical reactions of industrial interest, with very few exceptions, occur in the presence of a catalyst. As previously seen, the catalyst can be a chemical species soluble in the reaction environment, and this case is normally classified as a "homogeneous catalytic reaction." The kinetic approach to homogeneous catalytic reactions will be the main subject of this chapter.

4.2 Relation Between the Kinetic Law and the Reaction Mechanism

We previously mentioned that a reaction occurs when a bond is broken and another is formed. There are two ways to break a bond: (1) **homolytic cleavage** or **homolysis**,

$$A{:}B \longrightarrow A\cdot + B\cdot \tag{4.2}$$

which occurs with the formation of unpaired electrons and unstable radical species (**radical mechanism**); and (2) **heterolytic cleavage** or **heterolysis**,

$$A{:}B \longrightarrow A^{+} + {:}B^{-} \tag{4.3}$$

in which ionic species are formed because one atom gets both of the shared electrons (ionic mechanism). Certain factors—such as the size and shape of the molecules as well as their polarizability—are important in kinetics. It first must be pointed out that the stoichiometry of a reaction does not reveal anything about the complexity of its reaction mechanism, that is, a reaction occurring with a very simple stoichiometry can be characterized by a complex mechanism. The stoichiometry is simply the result of a sum of different consecutive elementary reaction steps. Therefore, in very few cases, the mechanism and the stoichiometry coincide. An example of this type could be the unimolecular decomposition of some substances occurring at high temperature. A clear example showing that the stoichiometry cannot give useful information about the reaction mechanism is represented by the following reaction:

$$2\ MnO_4^- + 10\ Fe^{++} + 16H^+ \rightarrow 2\ Mn^{++} + 10\ Fe^{+++} + 8\ H_2O \tag{4.4}$$

This reaction requires, for occurring in a single step, the contemporary collision of 28 bodies, which clearly is an impossible event. Another significant example is the reaction of acetone with iodine catalyzed by the presence of acids:

$$CH_3COCH_3 + I_2 \xrightarrow{H^+} CH_3COCH_2I + H^+I^- \tag{4.5}$$

The reaction rate is independent of iodine concentration, although iodine is one of the reactants and is present in both of the products' molecules. This simply means that this reaction does not occur with a simple mechanism. The stoichiometry, in many cases, is the result of a complex mechanism, which is characterized by the sequence of many **elementary steps**. At this point, it is fundamental to point out that the elementary steps, contrary to the overall stoichiometry, obey simple kinetic laws, in which the reaction rate is proportional to the concentration (more rigorously to the activity) of the involved reactants raised to the stoichiometric coefficient appearing in the reaction. Therefore, for determining a reliable **kinetic law**, on a theoretical basis, it is important to define a reasonable reaction mechanism. Then two different approaches are possible by introducing two hypotheses corresponding to reasonable approximations: (1) the hypothesis of the "slow step" or the **rate-determining step** (RDS) and the hypothesis of the **steady-state condition**. The first hypothesis considers the overall reaction rate, which is completely determined just by one step of the mechanism being it much slower than all of the others. The kinetic law can easily be written for this slow step, but because the species involved are probably intermediate unstable species, the concentration of which cannot be experimentally determined, we will consider all of the steps preceding the slow one in the equilibrium and so evaluate the concentrations of the unstable intermediate as a function of measurable species involved in such equilibria. A simple example can better illustrate this method. Let us consider the following simple reaction:

$$A \rightarrow B \tag{4.6}$$

we can write for this reaction a simple mechanism of the type:

$$A \leftrightarrows C^* \rightarrow B \tag{4.7}$$

where C^* is an activated and unstable species generated by A.

If the second reaction is considered slow and, therefore, the RDS, we can write:

$$r = k_1[C^*] \tag{4.8}$$

As mentioned, normally we do not have the possibility to measure the concentration of the unstable species C^*, but we can consider the first reaction at equilibrium:

$$K_e = \frac{[C^*]}{[A]} \quad \text{therefore} \quad [C^*] = K_e[A] \tag{4.9}$$

We now have $[C^*]$ as a function of a measurable concentration A; hence, by substituting $[C^*]$ in Eq. (4.8), we have:

$$r = k_1 K_e[A] = k_{\text{app}}[A] \tag{4.10}$$

As you can see, we obtained a **kinetic law expression**, in which the reaction rate depends on the concentration of A (a measurable entity). The kinetic constant appearing in this expression is apparent, corresponding to the product of a true kinetic constant of the elementary step [Eq. (4.8)] times the equilibrium constant [Eq. (4.9)]. The dependence from the temperature of these two constants, k_c and K_e, are formally identical, but the physical means are quite different. The kinetic constant obeys the Arrhenius law, which defines the dependence of the kinetic constants on temperature:

$$k = k^o e^{-\Delta E/\text{RT}} \tag{4.11}$$

where ΔE is the activation energy, R is the constant of gas, T is the absolute temperature, and k^o is a pre-exponential factor. This relation indicates that the kinetic constant increases exponentially with temperature. In contrast, the equilibrium constant, K_e, obeys the law of van't Hoff for a limited range of the temperature:

$$K_e = K_e^o e^{-\Delta H/\text{RT}} \tag{4.12}$$

where ΔH is the enthalpy change of the reaction. When the reaction is exothermic, ΔH is negative, and the equilibrium constant decreases with the temperature, while the opposite occurs for endothermic reactions. If we consider now the dependence on temperature of the apparent kinetic constant k_{app}, we can write:

$$k_{\text{app}} = k_{\text{app}}^o e^{-(\Delta H + \Delta E)/\text{RT}} \tag{4.13}$$

Because ΔH and ΔE could have opposite signs, ΔE being always positive and ΔH very often negative, it must be considered that the observed activation energy normally will be lower than the one of the true kinetic constant.

The assumption of the steady-state condition for an intermediate specie means that the mentioned species disappears because of the reaction with the same rate of its formation, that is, the concentration of this species can be considered approximately constant during time, and its derivative with respect to the time will be null. By considering the same very simple monomolecular reaction of the previous example, $A \rightarrow B$, a possible mechanism could be the one suggested by Lindemann (1922):

$$A + A \underset{r_{-1}}{\overset{r_1}{\rightleftarrows}} A^* + A \tag{4.14}$$

$$A^* \xrightarrow{r_2} P \tag{4.15}$$

According to this mechanism, the collision between two A molecules gives place to an activated A^* molecule characterized by a high internal energy, and for this reason, it is unstable and ready to react. This molecule has two possibilities of reaction: It can return to the original state, A, or it can yield a new molecule, P. Considering all of the reactions of the mechanism as elementary steps, we can write:

$$\frac{dC_A^*}{dt} = r_1 - r_{-1} - r_2 = k_1 C_A^2 - k_{-1} C_A^* C_A - k_2 C_A^* \tag{4.16}$$

by applying to A^* the steady-state assumption, we can write: $dC_A^*/dt = 0$; therefore,

$$k_1 C_A^2 - k_{-1} C_A^* C_A - k_2 C_A^* = 0 \tag{4.17}$$

Hence,

$$C_A^* = \frac{k_1 C_A^2}{k_{-1} C_A + k_2} \tag{4.18}$$

Therefore, the rate of P formation, that is, the reaction rate, can be written as:

$$r = k_2 C_A^* = k_2 \frac{k_1 C_A^2}{k_{-1} C_A + k_2} \tag{4.19}$$

As can be seen, again we obtained a kinetic law expression, which is a function of the measurable quantity C_A, which is the scope of the introduction of the "**steady-state approximation**." It is then interesting to observe that at very low A concentration, we can assume $k_{-1} C_A \ll k_2$. In this case, the reaction rate becomes:

$$r \simeq k_1 C_A^2 \tag{4.20}$$

In the opposite case, $k_{-1} C_A \gg k_2$ and, consequently,

$$r \cong \frac{k_2 k_1}{k_{-1}} C_A = k_{app} C_A \tag{4.21}$$

This approach explains very well the experimental observation that for some unimolecular reactions, by progressively increasing the concentration of reagent, we pass from a linear to a quadratic dependence of the reaction rate on the reagent concentration. A rate law, such as that of Eq. (4.19), describes the experimental behavior in the entire field of reagent concentrations, thus justifying the validity of the suggested mechanism.

4.3 Elementary Background of Kinetics

4.3.1 Reaction-Rate Definition: Relation with Mass Balance Equations

The scope of kinetics is to study the reaction rate in relation to the composition and temperature, that is, to determine the **rate law**:

$$\begin{aligned} r &= f(\text{composition, temperature}) \\ &= (\text{amount of reagent consumed or product formed/time}) \end{aligned} \tag{4.22}$$

This relation can be determined with an experimental approach and verified as in the examples reported before, that is, through a theoretical approach based on a reasonable reaction mechanism. Moreover, the rate law must obey the thermodynamic constraints, and the kinetic parameters would have an understandable physical means.

The dimension of the reaction rate must be chosen according to your best convenience as (1) the amount of a reacted reagent (in moles, mass, or volume)/time or (2) the amount of formed product (moles, mass, or volume)/time. It also is possible to refer to the amount reacted or formed for time unit to mole, weight, or surface of the catalyst. The most convenient dimension is dictated by the type of reactor used for measuring the reaction rate and the related balance equation. The most general balance equation related to a component (i) can be written as:

$$\begin{aligned} &\text{Inlet flow rate of (i)} - \text{outlet flow rate of (i)} \\ &\pm \text{rate of (i) formation or disappearing} \\ &= \text{(i) accumulation} \end{aligned} \tag{4.23}$$

Consequently, referring to a molar flow rate of (i), we can write, in the most general case, for a reaction in which the overall number of moles does not change with time, and the volume flow rate F is constant:

$$Fc_i^{\text{in}} - Fc_i^{\text{out}} \pm Vr_i = V\frac{\text{d}c_i}{\text{d}t} \tag{4.24}$$

In the case of a well-stirred batch reactor, $F = 0$ and $r_i = \pm \text{d}c_i/\text{d}t$, while for a well-stirred continuous reactor, the accumulation is null, and the mass balance equation becomes the algebraic equation:

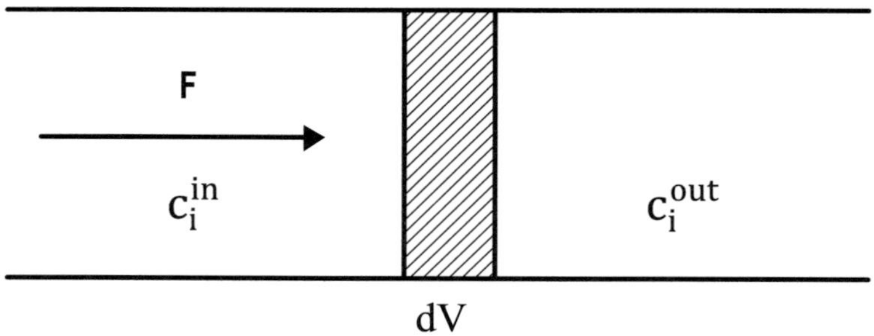

Fig. 4.2 Mass balance in a tubular reactor

$$r_i = \frac{Fc_i^{\mathrm{in}} - Fc_i^{\mathrm{out}}}{V} \tag{4.25}$$

Under semi-batch conditions, we can have c_i^{in} or, alternatively, c_i^{out} equal to zero. A more complex situation occurs in continuous tubular reactors in which a plug flow is operative. As will be discussed later in the text in more detail, plug flow corresponds to a flow in which all of the molecules stay in the reactor for the same length of time. In this case, the mass balance must be made on an infinitesimal portion of the reactor volume as shown in Fig. 4.2.

$$F\mathrm{d}c_i = r_i\mathrm{d}V \quad r_i = F\frac{\mathrm{d}c_i}{\mathrm{d}V} = f(\text{composition, temperature}) \tag{4.26}$$

where F is the volumetric flow rate (L/time), c_i is the concentration of i (mol/L), r_i is the reaction rate (mol/L time), and $\mathrm{d}V$ is the infinitesimal portion of reactor volume (L).

By integrating Eq. (4.26) from $V = 0$ to $V = V_R$, we obtain:

$$\frac{V_R}{F} = \int_{c_i^{\mathrm{in}}}^{c_i^{\mathrm{out}}} \frac{\mathrm{d}c_i}{r_i} \tag{4.27}$$

As you can see, for solving the integral, it is necessary to know the expression:

$$r_i = f\ (c_i,\ \text{temperature}) \tag{4.28}$$

The solution of the integral allows the evaluation of c_i^{out} together with a profile of c_i along the reactor.

However, in all cases, the rate law, that is, the relation:

$$r_i = f(\text{composition, temperature}) \tag{4.29}$$

remains always the same and is a univocal way to define the reaction rate in any situation. In addition, by considering the reaction stoichiometry, it is possible to give only one expression for the reaction rate for all of the reagents and products. Consider, for example, the reaction:

$$4\ NH_3 + 5\ O_2 \rightarrow 4\ NO + 6\ H_2O \tag{4.30}$$

and assume ammonia as a reference component. According to the stoichiometry, we can write:

$$\begin{aligned} r_i = -r_{NH_3} = -5/4\ r_{O_2} = r_{NO} = 3/2\ r_{H_2O} \\ = f(\text{composition, temperature}) \end{aligned} \tag{4.31}$$

that is, only one rate law can describe the evolution with time of all of the components involved in the reaction.

4.3.2 *Reaction Order and Formal Kinetics*

Consider a simple reaction, such as this one:

$$A + 2B \rightarrow C \tag{4.32}$$

Initially, we have no idea about what kinetic law the reaction will obey. Therefore, the simplest preliminary kinetic approach is to test experimentally a so-called **power-law expression**:

$$r = k^o \exp(-\Delta E/RT) C_A^\alpha C_B^\beta C_C^\gamma \tag{4.33}$$

where α is the reaction order with respect to A, β is the reaction order with respect to B, γ is the reaction order with respect to C, and $\alpha + \beta + \gamma$ is the overall reaction order. As previously mentioned, a reaction, such as that shown in Eq. (4.32), could be the result of a complex mechanism with many successive elementary steps. Usually, for simple mechanism, the reaction-rate expression is a function of the reactants A and B concentrations and not of product C; however, in some cases, a contribution of the product can be observed, for example, when a product catalyzes the reaction or when it interacts with the catalyst, thus causing a depressive effect. However, a general rule is that the reaction orders are small integer numbers, and very seldom does the overall reaction order reach a value ≥ 3. Fractional or negative orders of the different components suggest a complicated mechanism. The catalyst concentration must be considered in this analysis and can enter in the power law with its own order. After

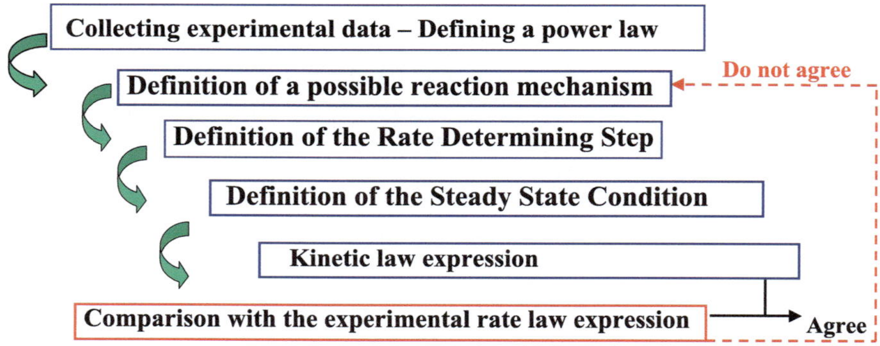

Fig. 4.3 Scheme of the method for determining the most reliable kinetic law

this experimental empirical approach, by observing the obtained reaction orders, it is opportune to write different possible reaction mechanisms, thus introducing the slow-step hypothesis or that of the steady-state condition and determining different expressions for the kinetic law until one of these expressions is in a good agreement with the experimentally founded **power law**. A scheme of the general suggested approach is shown in Fig. 4.3.

As previously mentioned, the reaction orders are often low-integer numbers; therefore, it is useful to consider what happens when we have reactions with order 0, 1, or 2. In particular, it is useful to consider what happens in a batch reactor which is the most common type used in the laboratory for conducting homogeneous reactions. By writing the mass balance for this particular case, we must integrate the corresponding differential equations:

(a) Single reaction		Rates expressions	Integrated relations	Time to reduce C_A to $C_A^\circ/2$
→	Order 0 $A \rightarrow B$	$r = -\frac{dC_A}{dt} = \frac{dC_B}{dt} = K_o$	$(C_A^\circ - C_A) = k_o t$	$C_A^\circ = \frac{k_o t_{1/2}}{2}$
→	Order 1 $A \rightarrow B$	$r = -\frac{dC_A}{dt} = k_1 C_A$	$\ln \frac{C^\circ{}_A}{C_A} = k_1 t$	$\ln 2 = k_1 t_{1/2}$
→	Order 2 $2A \rightarrow B$	$r = -\frac{dC_A}{dt} = k_1 C_A^2$	$\left(\frac{1}{C_A} - \frac{1}{C_A^\circ}\right) = k_2 t$	$\frac{1}{C_A^\circ} = k_2 t_{1/2}$

(4.34)

As you can see, for evaluating the reaction order of the considered reactions, it is sufficient to plot the concentration functions of the left members of the integrated relations as a function of time. A linear plot is obtained when the correct relation has been plotted. Another interesting aspect is to consider the time necessary to reduce

C_A° to $C_A^\circ/2$. This time depends on C_A° for the 0 order reaction, is independent of C_A° for reaction with order 1, and depends on $1/C_A^\circ$ for a reaction of order 2.

The last aspect that must be known in evaluating the reaction orders is the possibility of isolating the determination of the reaction order of one component by using an excess of the others. The order of the excess components can be considered with reasonable approximation equal to zero because of very low conversion due to the component that is present in a lower amount. Thus, by repeating the operation and isolating another component, all of the reaction orders can be experimentally determined.

4.3.3 Laboratory Reactors for Studying Single Fluid-Phase Reactions

The scope of the experiences performed in laboratory reactors is to determine the following for a certain reaction: (1) the operative conditions under which the reaction occurs (temperature, pressure, ratio between the reagents); (2) the type and amount of catalyst to be used; (3) the scheme of the occurring reactions, evaluating in particular the side reactions and the undesired by-products; and (4) the rate of both the main reaction and the reactions yielding by-products. The experimental runs for studying homogeneous reactions, occurring in one single fluid phase, can be performed in a discontinuous laboratory reactor (batch reactor) or in a continuous reactor. A scheme of the two mentioned reactors, for reactions performed in the liquid phase, is shown in Fig. 4.4.

In some cases, semi-batch conditions are adopted. This is useful, for example, when one of the reagents must be added slowly because the reaction is extremely exothermic or when the reaction product is more volatile than the reagents and is subtracted, in the vapor phase, from the reaction environment to shift the equilibrium to the right. We have, therefore, Fig. 4.5.

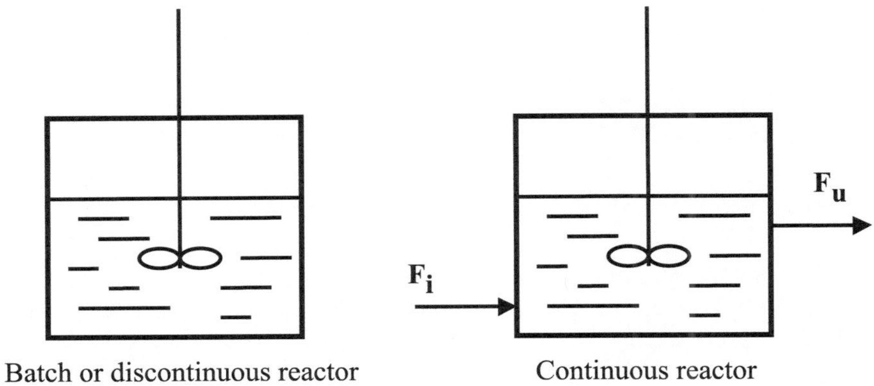

Fig. 4.4 Scheme of reactors used for studying homogeneous reactions

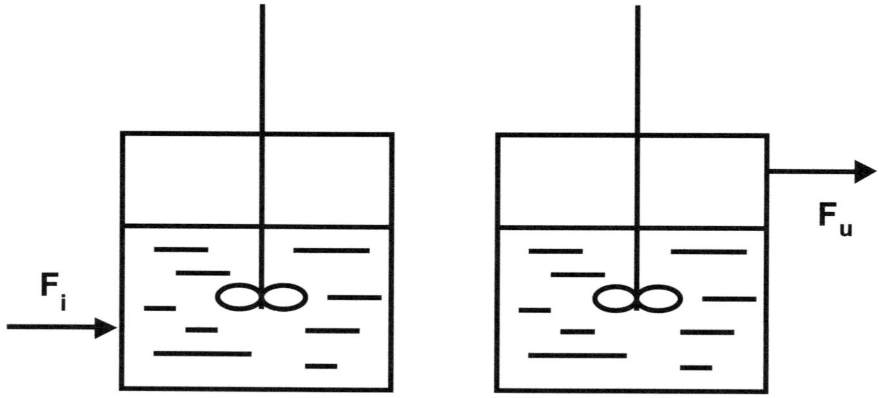

Fig. 4.5 Schemes of semi-batch reactors

Rewriting the mass balance referred to the moles of the component of "*i*" for the most general case:

Moles of *i* accumulated	=	moles of *i* at the inlet	−	moles of *i* at the outlet	+	moles of *i* generated	−	moles of *i* consumed
(1)		(2)		(3)		(4)		(5)

In the batch reactor	→	terms (2) and (3) = 0
In the continuous reactor	→	term (1) = 0
Under semi-batch conditions	→	term 2 or term 3 = 0

Terms (4) and (5) contain the reaction-rate expression:

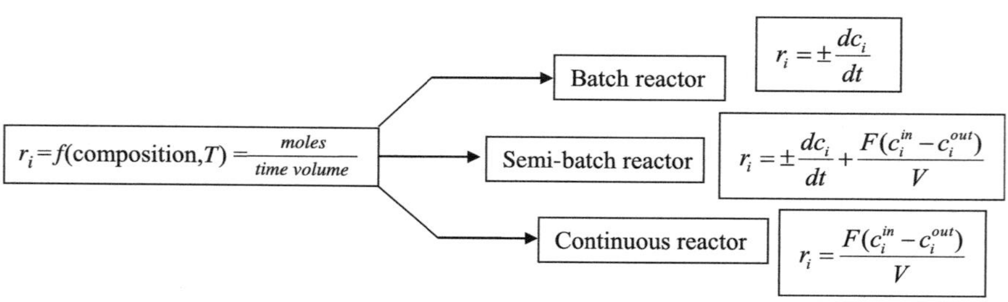

For a semi-batch reactor, the mass balance is clearly more complex, containing both a differential and an algebraic term for each reaction component "*i*" and, as seen, two options are possible one considering $C_i^{\text{in}} = 0$ and another one with $C_i^{\text{out}} = 0$.

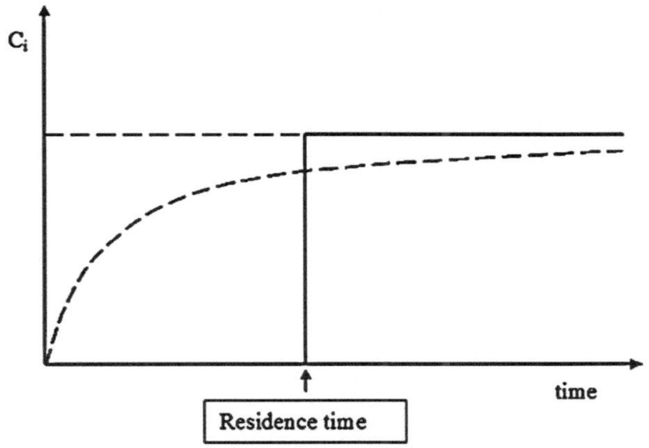

Fig. 4.6 CSTR ideality experimental step test (positive step)

Experimental data are collected in the batch reactors as evolution with time of the "*i*" concentrations. By plotting these concentration data as a function of time, the slopes of the obtained curves are the measured reaction rates of "*i*" formed if it is a product (sign "+" of the derivative) or "*i*" disappeared if it is a reagent (sign "−" of the derivative).

In the case of a continuous reactor, a steady-state condition must be reached, which means a constant composition must be attained inside the reactor. We impose the inlet concentration of the reagent "*i*", the volumetric flow rate F, and the volume of the liquid in the reactor V and determine the concentration at the reactor outlet as a consequence of the reaction rate. In this case, we obtain just one experimental point for each steady-state condition. For determining the unknown f (composition, T), clearly, we must have many experimental points; therefore, in the continuous reactor, we will change the ratio F/V and/or the inlet concentration of "*i*" to obtain many different steady-state conditions corresponding to different values of C_i^{out} and different values of reaction rates. In a continuous reactor, the inverse of F/V contains the dimension of time; therefore, the ratio V/F is called "residence time," which corresponds to the average amount of time that the molecules stay in the reactor. If a continuous reactor is well mixed, we define it as a continuous stirred-tank reactor (CTSR), that is, an ideal reactor, in which the residence times of the entering molecules are distributed according to a Gaussian distribution. An ideality test, for evaluating the mixing effect in a reactor, can be performed experimentally by feeding in a fluid stream a step of a new component "*i*" and measuring the concentration of this component at the exit. If the reactor is well mixed, the evolution of the "*i*" concentration over time has a trend, such as the one shown as a dotted line in Fig. 4.6.

Alternatively, in a flow stream containing "*i*," this flow is stopped, and we evaluate how the "*i*" component disappears over time (see Fig. 4.7).

This curve can be quantitatively interpreted by first defining $\theta = t\frac{F}{V}$, which is a dimensionless function of the residence time, V/F. It is then possible to write:

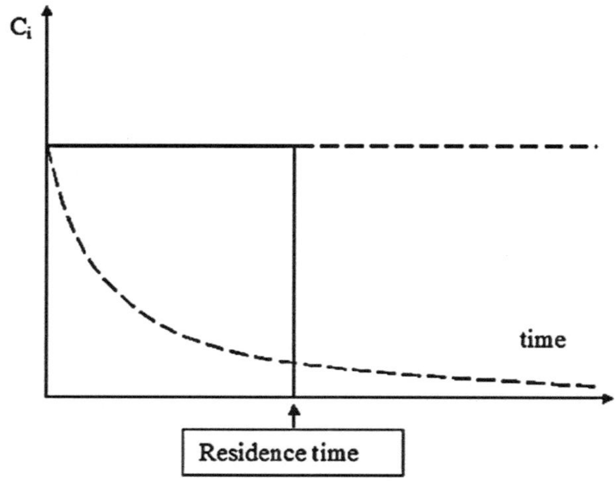

Fig. 4.7 CSTR ideality experimental step test (negative step, washout)

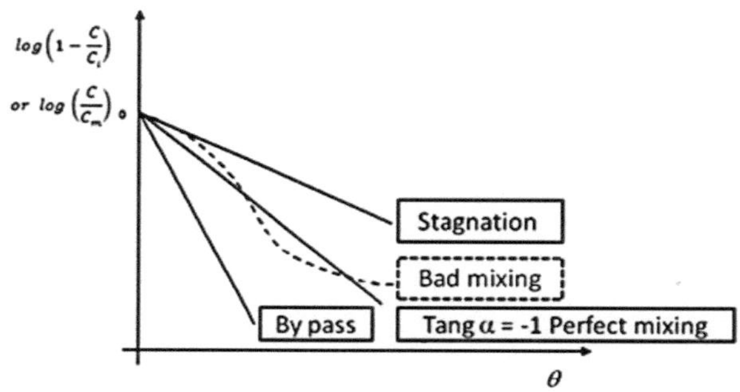

Fig. 4.8 Elaboration of CSTR ideality experimental data

$$1-\frac{C}{C}=e^{-\theta} \quad \text{and} \quad \log\left(1-\frac{C}{C_i}\right)=-\theta \tag{4.35}$$

By putting $\log\left(1-\frac{C}{C_i}\right)$ in a plot against θ, an ideal reactor yields a straight line with a slope of 45° as can be seen in Fig. 4.8.

Another possibility is to feed a pulse of the component "*i*" in a flow stream of another fluid and measure the evolution of the "*i*" concentration at the exit. In this case, the "*i*" concentration decreases according to the relation $\frac{C}{C_m}=e^{-\theta}$ as shown in Fig. 4.9.

In this case, a plot of $\log\frac{C}{C_m}$ against θ gives a straight line (see Fig. 4.8). As seen in Fig. 4.8, any deviation from the described behaviors means that the reactor is affected by the bypass of the flow (i.e., a portion of the flow is not involved in the stirring) or by stagnation (i.e., a part of the reactor volume is not involved in the stirring).

In conclusion, the batch reactor has the advantage of transient behavior that allows to evaluate, with only one kinetic run, all of the reaction rates related to

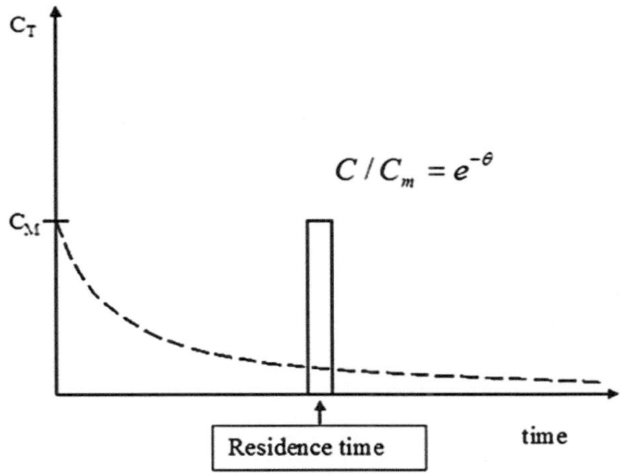

Fig. 4.9 Exemplification of CSTR ideality pulse test

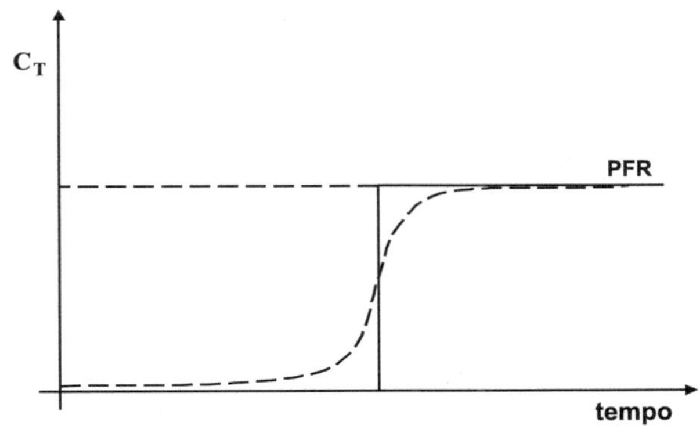

Fig. 4.10 Step test for a PFR: comparison between ideal and real behavior

different concentrations of the reagents because these change over time. However, the drawback is that we must determine these concentrations, and often this determination is difficult or impossible—requiring the withdrawal and analysis of the reaction mixture at different times—and the difficulty is related to the range of reaction time. Some reactions occur in a span of hours without any problems withdrawing and analyzing the reaction mixture at different reaction times; however, some reactions occur in a span of minutes or seconds. In this case, the use of a continuous reactor is preferred.

Another type of continuous reactor is the so-called plug flow reactor (PFR). This normally is a tubular reactor containing inert or catalytic particles for favoring local intense mixing. An ideal PFR is characterized by molecules all having the same residence time. In a step test, such as the one described previously for characterizing the well-mixed CSTR, the ideal behavior results in a continuous line (Fig. 4.10), while a real approximated behavior moderately deviating from that of the PFR results in a dotted line.

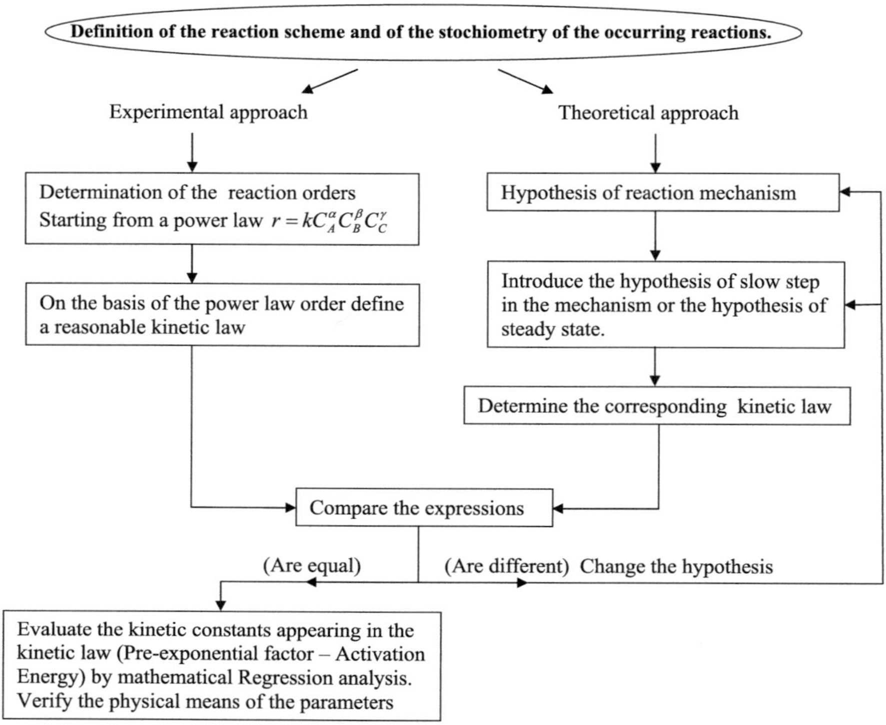

Fig. 4.11 Sequential approach in the determination of the kinetic law of a reaction and related parameters

Clearly, the step test can be experimentally performed by analyzing the composition at the outlet of the reactor.

In any case, as already mentioned, the kinetic objective is to find a relation $r = f$ (composition, temperature) able to correlate all of the experimental data. This can be performed by following the sequential approach shown in Fig. 4.11.

To start with the laboratory, runs require a series of choices as follows:

(A) The choice of the most suitable reactor regarding size and shape
(B) The choice of operative conditions (flow rates, pressure, temperature, inlet composition, stirring rate)
(C) Choice of systems for feeding the reagents
(D) Choice of systems of measure and control (pressure, temperature, flow rate)
(E) Choice of analytical methods and their tuning

In conclusion, the kinetic runs for modeling an industrial reactor are programmed in a three-level sequence.

First Level

The first level has the scope to evaluate the reaction scheme, the stoichiometry of all of the occurring reactions, and the kinetic laws of the first attempt for the main

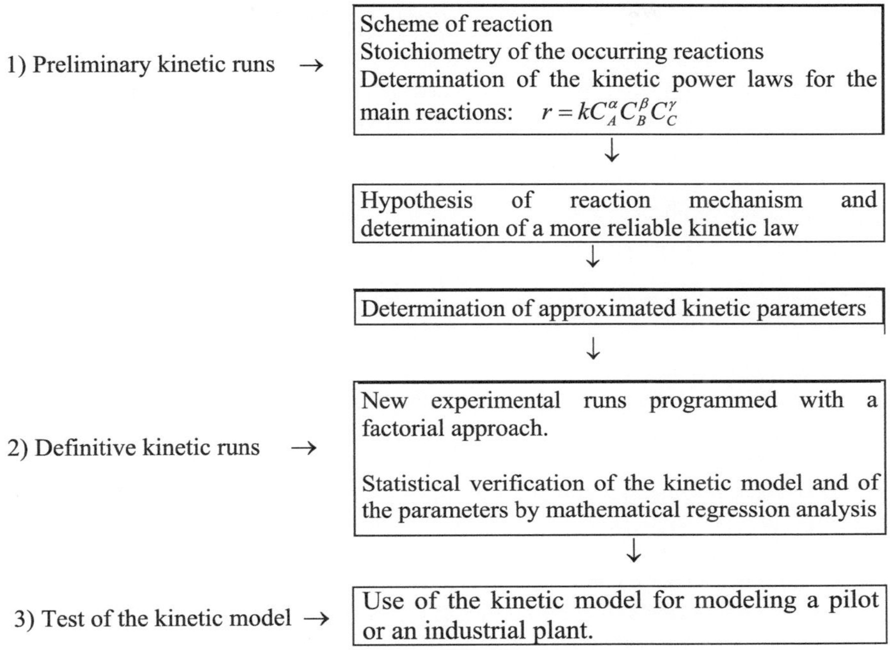

Fig. 4.12 A three-level approach for correctly modeling an industrial reactor

occurring reactions with determination of the related parameters with a satisfactory approximation.

Second Level

In the second level, kinetic runs are programmed according to a factorial approach, and the kinetic laws are tested on these runs with a mathematical regression analysis giving information about the reliability of the kinetic model on a statistical basis. If the model is not adequate, other kinetic equations are tested as explained in the previous scheme.

Third Level

The third level is verification of the developed model in the scale-up, that is, the kinetic equations found are used for modeling a pilot plant or, directly, an industrial reactor. The three-level sequence approach is shown in Fig. 4.12.

It is important to point out that reactor modeling uses both the kinetic expressions obtained from a kinetic study of a reaction and similar mass balance equations; however, in this case, the unknown is simply the volume of the reactor necessary for obtaining the desired conversion and yield. In contrast, in the kinetic investigation, which is normally performed in the laboratory, a mathematical expression (the kinetic law) able to describe the system behavior must be individuated.

4.3.4 Hints on the Factorial Programming of Kinetic Runs

When after a preliminary investigation a kinetic model for a given reaction is available to evaluate more reliable kinetic parameters, a factorial program of new kinetic runs is opportune. Consider as an example a reaction system in which there are three variables to keep under control (V_1, V_2, and V_3), for example, temperature, pressure, and overall flow rate. The simplest factorial approach is to consider two different levels of experimentation: one high (high temperature, high pressure, high flow rate) and one low. To cover completely the experimentation space (see Fig. 4.13) in this type of a system, we must perform at least 2^3 kinetic runs plus 2 reference runs in correspondence with the center of the chosen space of the variables. This must be done before and after the 8 runs to evaluate the experimental error, thus ensuring the possibility of a complete statistical analysis of the system.

Then the 8 + 2 kinetic runs must be opportunely programmed with a sequence of the type listed in Table 4.1.

All of the experimental runs will give data of reaction rate $r = f(V_1, V_2, V_3)$, which is foreseen by the kinetic model determined with the preliminary investigation. By assuming the parameters of that model to be a first approximation, more correct and reliable parameters can be obtained by submitting all of the experimental runs performed to mathematical nonlinear regression analysis minimizing the following objective function:

$$\phi(\beta) = \sum_{i=1}^{N} i\left[X_i^{\exp} - X_i^{\text{calc}}(\beta)\right]^2 \tag{4.36}$$

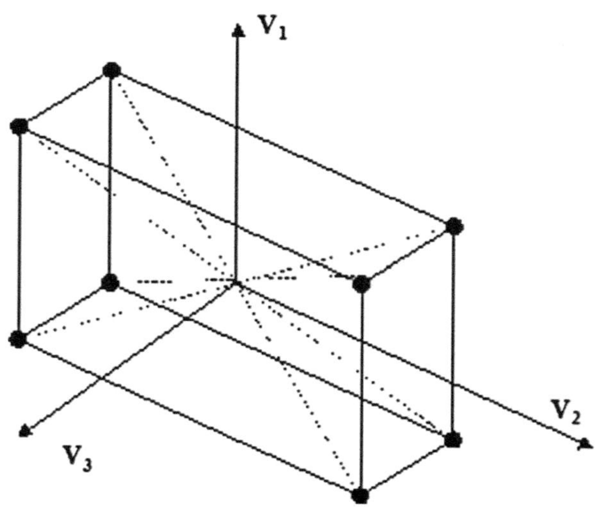

Fig. 4.13 Choice of the space of the variables in a factorial program

Table 4.1 An example of a factorial program of kinetic runs

Experiments/variables	V_1	V_2	V_3
1	0	0	0
2	+	+	+
3	+	+	−
4	+	−	+
5	+	−	−
6	−	+	+
7	−	+	−
8	−	−	+
9	−	−	−
10	0	0	0

where X is the conversion of one of the reagents, N is the number of experimental runs submitted to mathematical regression analysis, and β is the vector of the parameters to be determined.

If the variables are 2, a minimum of $2^2 = 4$ runs must be performed; if the variables are 4, a minimum of $2^4 = 16$ runs, alternating high and low levels with the same criterion, must be used for the case of three variables.

4.3.5 *Exercises on the Evaluation of Reaction Order and the Simulation of Kinetic Runs*

Exercise 4.1. Reaction of Bromine with Formic Acid

The reaction is:

$$Br_{2\,(aq)} + HCOOH\,(aq) \rightarrow 2\ Br^{-}(aq) + 2\ H^{+}(aq) + CO_2\ (gas) \tag{4.37}$$

It is easy to follow this reaction by observing the solution discoloring from brown to transparent as a consequence of bromine consumption. The reaction at 25 °C occurs at a rate that can be determined from the data listed in Table 4.2.

Estimate an approximated value of the kinetic constant from the initial reaction rate assuming a first-order kinetic law. Evaluate more precisely the kinetic constant and simulate the run.

Results

The initial reaction rate can be determined from the difference between the two first points of the Br_2 concentration divided by the corresponding time elapsed, that is:

$$(0.0120 - 0.0101)/50 = 0.0019/50 = 0.000038\ \text{mol/s}$$

Assuming a first-order law, it is easy to calculate the approximated kinetic constant as:

Table 4.2 Experimental data for bromine reaction

Time (s)	Br_2 (M)
0	0.0120
50	0.0101
100	0.00846
150	0.00710
200	0.00596
250	0.00500
300	0.00420
350	0.00353
400	0.00296

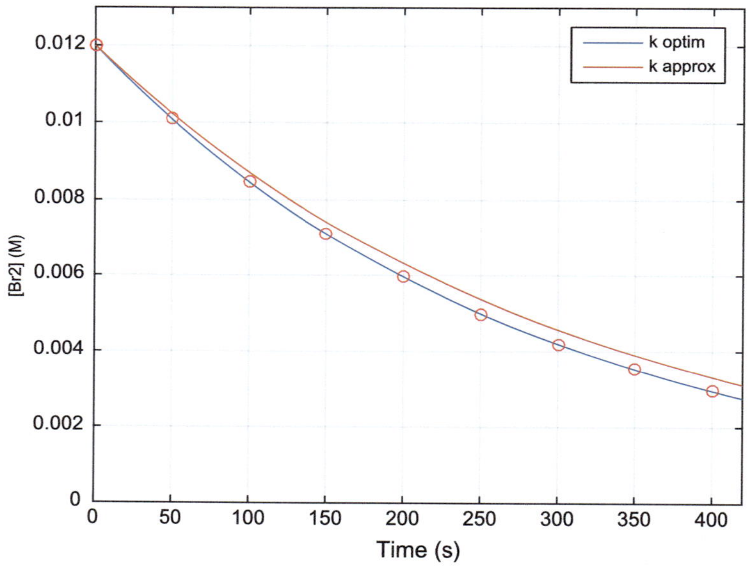

Fig. 4.14 Evolution of bromine concentration with time

$$k = r/[Br_2] \quad \text{which results in} \quad k_{\text{approximated}} : \; 0.0032103$$

The parameter giving the best fitting of the experimental points can be found using a MATLAB program available as electronic supplementary material (ESM), Data 4.1.

$$K_{\text{optimized}} : \; 0.0034993$$

In Fig. 4.14, it is possible to appreciate the difference in the run simulation with, respectively, the approximated and the optimized kinetic parameters.

Table 4.3 Experimental data of iodine titrated with sodium thiosulfate at different time

Time (min)	mL $Na_2S_2O_3$ (0.15 N)
0	49.3
30	42.0
60	35.6
90	30.2
120	25.7
150	21.8
180	18.5
240	13.8
360	7.3
480	4.8

Exercise 4.2. Conversion of N-chloroacetanilide into P-chloroacetanilide
The reaction is made using acetic acid as a solvent and occurs according to the following scheme:

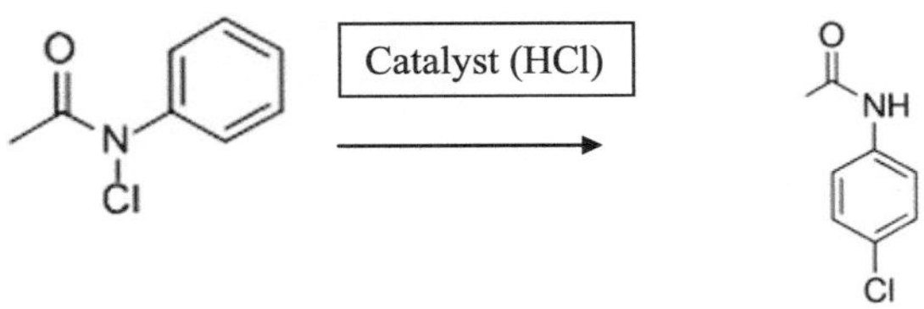

(4.38)

As for the N-chloroacetanilide, in contact with KI, it develops iodine, and it is easy to follow the evolution of the reaction simply by titrating developed iodine with sodium thiosulfate. In Table 4.3, the amount of titrant used at different times is listed. Sodium thiosulfate was 0.15 N. Evaluate the evolution with time of the moles of N-chloroacetanilide and determine the kinetic constant.

Results
The best-fitting kinetic constant resulted in: $k = 0.005369$.

The interpolation of the experimental data can be appreciated in Fig. 4.15.

These results were obtained using a MATLAB program available as electronic supplementary material, Data 4.1.

Exercise 4.3. Hydrolysis of Aspirin
Aspirin (acetylsalicylic acid) in the presence of water hydrolyses yields salicylic and acetic acids according to the reaction:

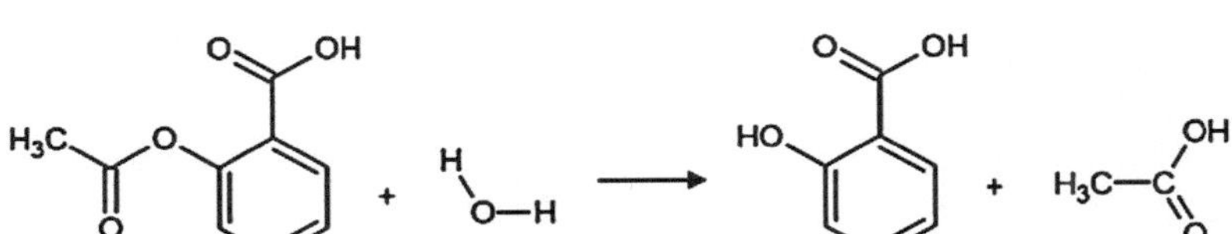

(4.39)

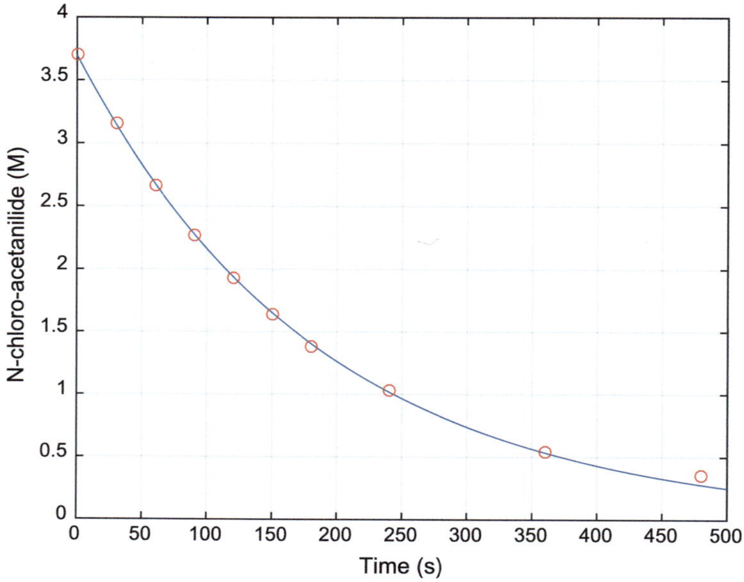

Fig. 4.15 Kinetics of N-chloroacetanilide reaction

Table 4.4 Kinetic data for aspirin hydrolysis

Time (h)	Aspirin (M)	Salicylic acid (M)
0	5.55×10^{-3}	0
2.0	5.51×10^{-3}	0.040×10^{-3}
5.0	5.45×10^{-3}	0.10×10^{-3}
10	5.35×10^{-3}	0.20×10^{-3}
20	5.15×10^{-3}	0.40×10^{-3}
30	4.96×10^{-3}	0.59×10^{-3}
40	4.78×10^{-3}	0.77×10^{-3}
50	4.61×10^{-3}	0.94×10^{-3}
100	3.83×10^{-3}	1.72×10^{-3}
200	2.64×10^{-3}	2.91×10^{-3}
300	1.82×10^{-3}	3.73×10^{-3}

The reaction is promoted by the acid environment and although slow at pH 7, it becomes fast in an acidic environment, such as the human stomach. At pH 7 and 37 °C, the hydrolysis rate can be derived from the data listed in Table 4.4.

Evaluate the reaction order and the kinetic parameter by linearization of $-\ln(C/C°)$ or $(1/C - 1/C°)$ versus time. With the estimated kinetic constant, simulate the run.

Results

Kinetic constant = 0.0037158 (h^{-1})

From the linearization, the resulting reaction is of the first order with respect to aspirin (see Fig. 4.16).

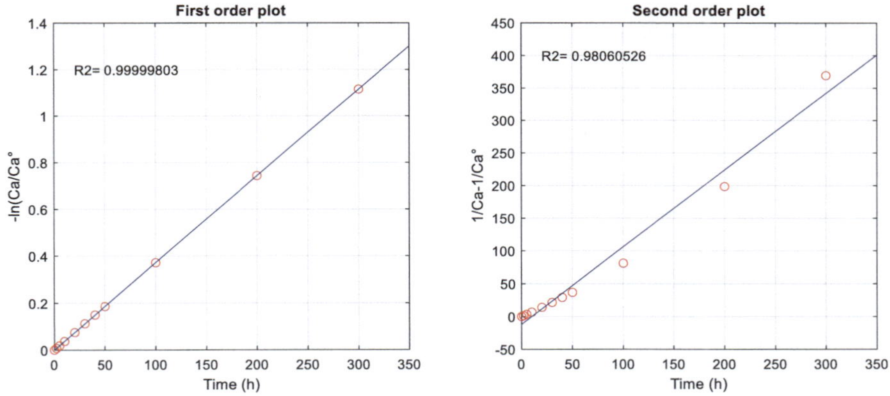

Fig. 4.16 Linearization of $-\ln(C/C°)$ and $(1/C - 1/C°)$ versus time

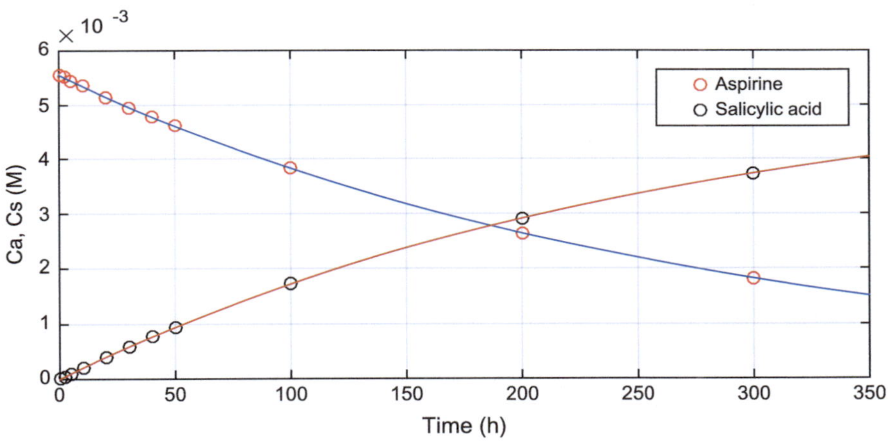

Fig. 4.17 Simulation of aspirin hydrolysis at pH 7 and temperature 37 °C

As can be seen, the trend expected for a first order is obtained. The kinetic parameter giving the best fit is $k_1 = 0.0037158$ (h^{-1}), and the simulation of the run is shown in Fig. 4.17.

The described results were obtained using a MATLAB program available as electronic supplementary material, Data 4.1.

Exercise 4.4. Hydration of Ethylene Oxide to Ethylene Glycol

Brönsted et al. (1929) studied the kinetics of the hydration of ethylene oxide to glycol ethylene by measuring the overall change of volume occurring as a consequence of the reaction. For this purpose, they evaluated this increase in volume by observing the increase of the height, *H*, of the liquid in a capillary tube situated over the reactor. The results, obtained at 20 °C in the presence of a small amount of an acid catalyst (0.00757 mol/L of $HClO_4$), are listed in Table 4.5.

Table 4.5 Some experimental data collected by Brönsted et al. (1929)

Time (min)	H (cm)	Time (min)	H (cm)
0	18.48	270	15.47
30	18.05	300	15.22
60	17.62	330	15.00
90	17.25	360	14.80
120	16.89	390	14.62
240	15.70	1830	12.29

The initial concentration of ethylene oxide was 0.12 mol/L. Because water is present in great excess, the reaction order with respect to this reagent can be assumed to be equal to zero. The reaction after 1830 min was completed. Evaluate the reaction order of ethylene oxide and the kinetic constant.

Solution

By assuming that the density of the system remains constant, because the overall variation of volume is very small, and neglecting the reaction order of water because present in excess, it is possible to write:

$$C_{\mathrm{EO}}^{o} = \alpha C_{\mathrm{EO}}^{o} = \alpha (H_o - H_{\mathrm{end}}) \quad \text{and} \quad C_{\mathrm{EO}} = \alpha (H - H_{\mathrm{end}}) \tag{4.40}$$

Therefore:

$$\ln \frac{C_{\mathrm{EO}}^{o}}{C_{\mathrm{EO}}} = \ln \frac{(H_o - H_{\mathrm{end}})}{(H - H_{\mathrm{end}})} = k_{293} t \tag{4.41}$$

Figure 4.18 shows the first order. The kinetic constant resulted in $k_{293} = 2.2506 \times 10^{-3}\,\mathrm{min}^{-1}$.

It is interesting to note that Brönsted et al. evaluated a change of the volume corresponding to only 0.16%. This observation gives an idea of the sensitivity and precision of the method adopted for this study.

The described results were obtained using a MATLAB program available as electronic supplementary material, Data 4.1.

Exercise 4.5. Kinetics of Sucrose Hydrolysis to Glucose and Fructose (Inversion of Sucrose) Catalyzed by a Mineral Acid

Sucrose, in the presence of a dilute mineral acid or dilute enzyme invertase, yields the following hydrolysis reaction:

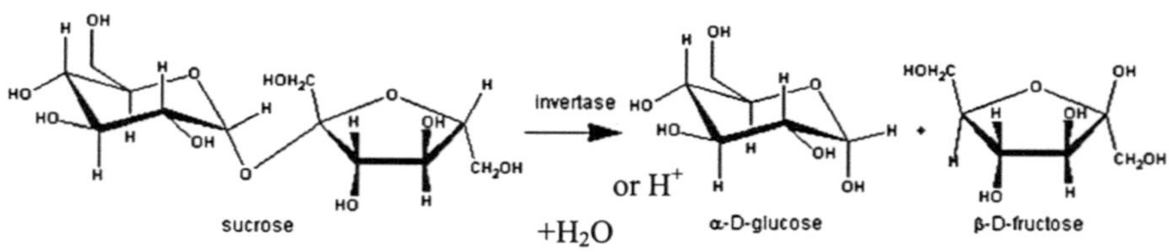

(4.42)

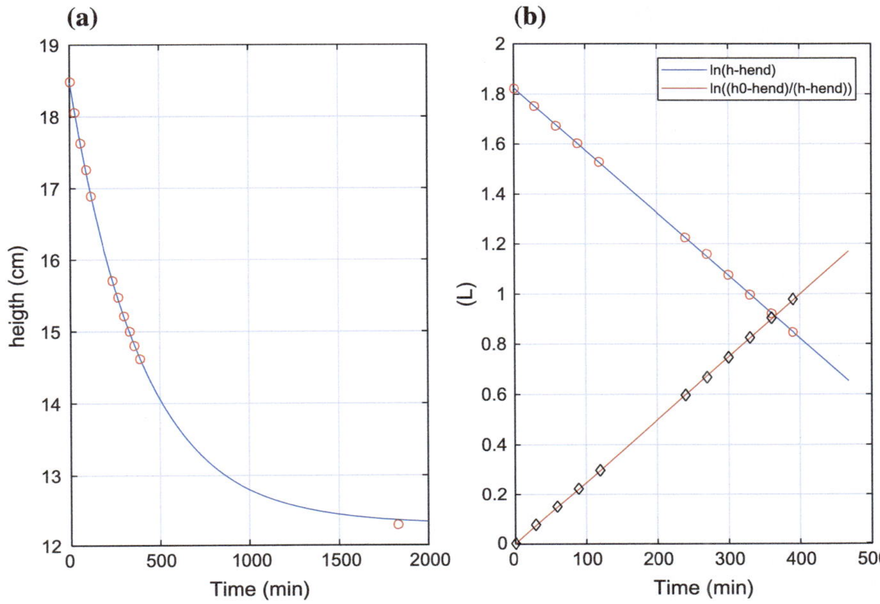

Fig. 4.18 (**a**) Simulation of the run and (**b**) linearization of the *H* function

The kinetics of the reaction can be followed easily because sucrose rotates polarized light clockwise $[\alpha]_D = +66.5°$ (polarimetric degree) while the contrary occurs for the mixture glucose $[\alpha]_D = +52°$ and fructose $[\alpha]_D = -92°$. In particular, fructose rotates the light anticlockwise. The average polarimetric degree of the mixture is approximately $[\alpha]_D = -20°$. Therefore, by observing the solution with a polarimeter, it is possible to evaluate the fraction of sucrose reacted over time.

Considering that water is present as a solvent in a great excess and that the concentration of the mineral acid does not change over time, we can write the following rate law:

$$-\frac{dC_{\text{sucrose}}}{dt} = k_{\text{eff}} C^r_{\text{sucrose}} \quad \text{being} \quad k_{\text{eff}} = C^s_{\text{H}_2\text{O}} C^t_{\text{H}^+} \tag{4.43}$$

Very probably, $r = s = t = 1$, and this is what we want to show. In this reaction, the rotation angle measured at the polarimeter decreases with the conversion for the sucrose disappearing as well as for the increase of both glucose and fructose. In an experiment, 20 g of sucrose were dissolved in 100 mL of water. Then 25 mL of this solution was mixed with 25 mL of HCl 4 M at 25 °C. The values of polarized light rotation were read at the polarimeter at different times. The obtained values are listed in Table 4.6.

Optical rotation α is a linear function of the concentration of each molecule in solution, and we can write:

Table 4.6 Experimental data for sodium thiosulfate

Time (min)	Rotation angle
5	23.2283
10	21.5802
16	19.6750
20	18.4947
30	15.7674
60	9.3221
120	1.8528
150	−0.2591
200	−2.4403

$$\alpha = \sum_i A_i C_i \tag{4.44}$$

At time 0 only sucrose is present; hence,

$$\alpha^{\circ} = A_{\text{sucrose}} C^{\circ}_{\text{sucrose}} \tag{4.45}$$

At the end of the reaction, we assume that no residual sucrose is present. Therefore:

$$\alpha_{\text{end}} = A_{\text{glucose}} C_{\text{glucose}-\text{end}} + A_{\text{fructose}} C_{\text{fructose}-\text{end}} \tag{4.46}$$

By subtracting

$$\alpha^{\circ} - \alpha_{\text{end}} = \left(A_{\text{sucrose}} - A_{\text{glucose}} - A_{\text{fructose}}\right) C^{\circ}_{\text{sucrose}} \tag{4.47}$$

We can also write that:

$$\alpha - \alpha_{\text{end}} = \left(A_{\text{sucrose}} - A_{\text{glucose}} - A_{\text{fructose}}\right) C_{\text{sucrose}} \tag{4.48}$$

By dividing these two expressions, we obtain:

$$\frac{\alpha - \alpha_{\text{end}}}{\alpha^{o} - \alpha_{\text{end}}} = \frac{C_{\text{sucrose}}}{C^{\circ}_{\text{sucrose}}} \tag{4.49}$$

Remembering that:

$$-\frac{\mathrm{d}C_{\text{sucrose}}}{\mathrm{d}t} = k_{\text{eff}} C_{\text{sucrose}} \tag{4.50}$$

It results

$$\ln\left(\frac{C_{\text{sucrose}}}{C^{\circ}_{\text{sucrose}}}\right) = \ln\left(\frac{\alpha - \alpha_{\text{end}}}{\alpha^{o} - \alpha_{\text{end}}}\right) = -k_{\text{eff}}t \tag{4.51}$$

As can be seen, the kinetic constant derived from the change of polarized light rotation is the same as the constant obtained from the actual concentration. In conclusion, optical-rotation data can directly be used in curve fitting for determining kinetic parameters. This is true in general for any property that is a linear function of the concentration.

To demonstrate that the reaction follows a first-order kinetic law, evaluate the kinetic constant. Continuing, we can now write:

$$\alpha = (\alpha^{o} - \alpha_{\text{end}}) \exp(-k_{\text{eff}}t) + \alpha_{\text{end}} \tag{4.52}$$

The fitting of the values of rotation light listed in Table 4.6 requires the evaluation of three parameters, which are:

$$a = (\alpha^{o} - \alpha_{\text{end}}), \quad b = k_{\text{eff}} \text{ and } c = \alpha_{\text{end}} \tag{4.53}$$

This can be made by performing mathematical regression analysis on the light-rotation data.

Results

The best-fitting curve is shown in Fig. 4.19.

The parameters obtained by fitting are:

$a = 30.036$	$b = 0.012335$	$c = -4.986$
$\alpha_0 = 25.05$	$k_{\text{eff}} = 0.012335$	$\alpha_{\text{end}} = -4.986$

The described results were obtained using a MATLAB program available as electronic supplementary material, Data 4.1.

Exercise 4.6. Decomposition of N_2O_5

N_2O_5 decomposes according to the following reaction:

$$N_2O_5 \rightarrow 2NO_2 + \tfrac{1}{2}O_2 \tag{4.54}$$

This reaction has an historical importance because it was the first monomolecular reaction studied for its kinetics. The reaction occurs in both homogeneous gas phase and in solution using different solvents. Data collected in the gas phase at 25 °C starting with an initial pressure of 268.7 mm Hg are listed in Table 4.7.

The reaction occurs in two steps:

$$N_2O_5 \rightarrow N_2O_4 + \tfrac{1}{2}O_2 \tag{4.55}$$

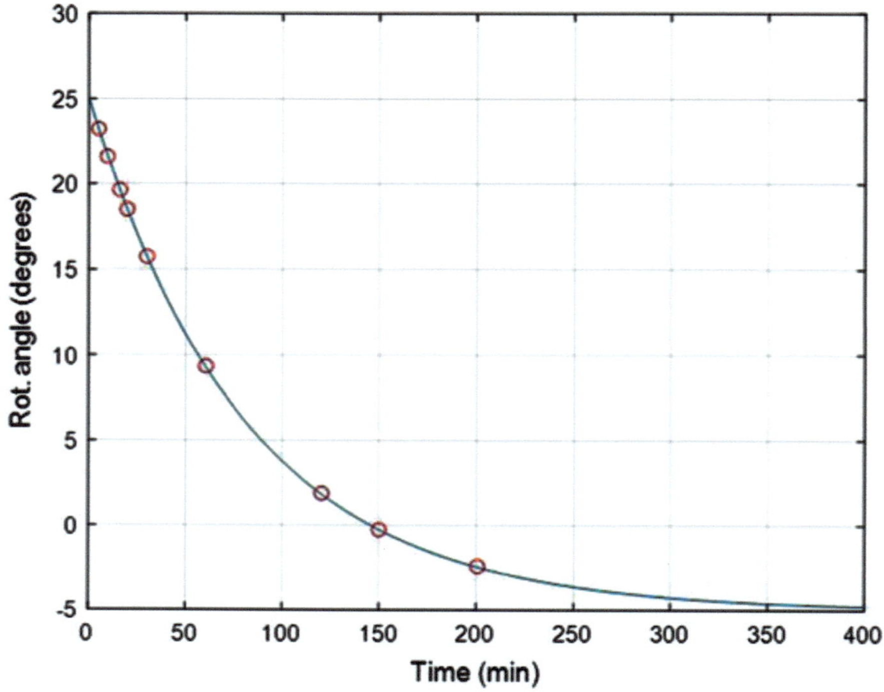

Fig. 4.19 Fitting obtained for the rotation angle as a function of time

Table 4.7 Experimental data collected in homogeneous gaseous phase

Time (min)	Pressure (mm Hg)
0	268.7
20	293.0
40	302.2
60	311.0
80	318.9
100	325.9
120	332.3
140	338.8
160	344.4
Long time	473.0

$$N_2O_4 \leftrightarrows 2\ NO_2 \tag{4.56}$$

The first reaction is an RDS. The second reaction is fast and can be considered always at equilibrium. Thus, it is correct to study the overall kinetics following the reaction $N_2O_5 \rightarrow N_2O_4 + \frac{1}{2}O_2$.

Considering a first-order kinetic law, evaluate the kinetic constant and simulate the curve of pressure along the time. As mentioned, the same reaction occurs also in

Table 4.8 Experimental data collected in CCl_4 solution

Time (min)	Concentration (mol/L)
0	2.33
3.06	2.08
5.31	1.91
8.76	1.67
14.45	1.36
19.96	1.11
31.28	0.72
38.58	0.55
52.40	0.34

solution, for example, in CCl_4. The results of the decomposition performed at 45 °C in CCl_4 are listed in Table 4.8.

Evaluate also in this case the kinetic constant and interpolate the experimental points. According to Panchenkov and Lebedev (1976), the kinetic constant for the reaction performed in CCl_4 is $K = 2.8 \times 10^{13} \exp(-24{,}075/RT)$ (s^{-1}) (ΔE = cal/mol). Compare the value calculated with this relation at the same temperature of the run with the kinetic constant obtained from the experimental points listed in Table 4.8.

Eyring and Daniels (1930) made a different experience by measuring the volume of oxygen released at different times always using CCl_4 as a solvent and 45 °C temperature. Data obtained are the following:

Time (s)	162	409	1721	3400	very long time
Volume of O_2	3.41	7.78	23.00	29.33	32.60

Calculate the kinetic constant from this run and compare it with the previous one.

At last, Alberty and Silbey (1992) gave the following kinetic constants at different temperatures for this reaction:

T (K)	273	298	308	318	328	338
$10^5\ k$	0.0787	3.46	13.5	49.8	150	487

Evaluate the activation energy and the pre-exponential factor, and compare with the activation energy furnished by Panchenkov and Lebedev.

Part 1. Gas-Phase Reaction

Results

The kinetic constant giving the best-fitting result was $k = 2.0914\text{e}{-05}$ 1/mm Hg, and the result of the simulation is shown in Fig. 4.20.

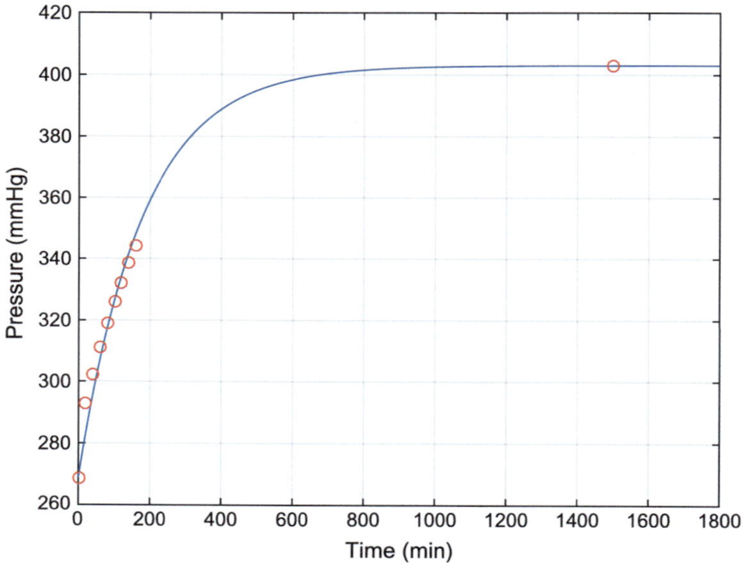

Fig. 4.20 Simulation of the curve of pressure as a function of time

Part 2. Reaction in CCl_4
Results

The kinetic constants determined, respectively, by calculation and by regression on the experimental points are as follows:

k optimized: 0.037331 (L/min); *k* calculated 0.04765 (L/min)

The plot of the simulation run is shown in Fig. 4.21.

Part 3. Arrhenius Plot for Activation Energy and Pre-exponential Factor
Results

The activation energy and pre-exponential factor resulted in $E_a = 24{,}590.65$ (cal/mol).

ln*A* = 31.273 (L/min). The Arrhenius plot is shown in Fig. 4.22.

All of the described results were obtained using a MATLAB program available as electronic supplementary material, Data 4.1.

Exercise 4.7. Evaluation of the Reaction Order for the Reaction Between Pyridine and Ethyl Iodide

Winkler and Hinshelwood (1935) studied the reaction between pyridine and ethyl iodide:

$$C_5H_5N + CH_3CH_2I \rightarrow (CH_3CH_2)C_5H_5N^+ + I^- \quad (4.57)$$

under isothermal conditions, at 100 °C, starting from 0.1 mol/L of both the reactants. As can be seen, iodine ions are formed during the reaction; therefore, it is easy to

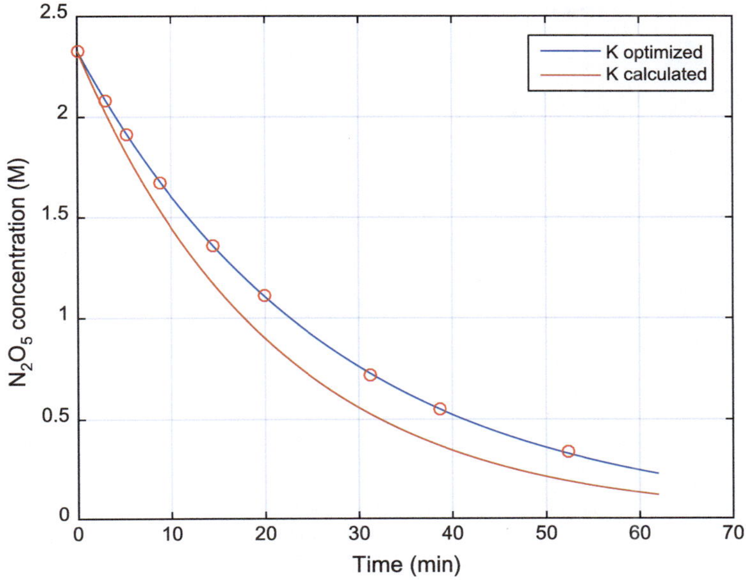

Fig. 4.21 Simulation curves with, respectively, the optimized kinetic constant and the one calculated from Panchenkov and Lebedev

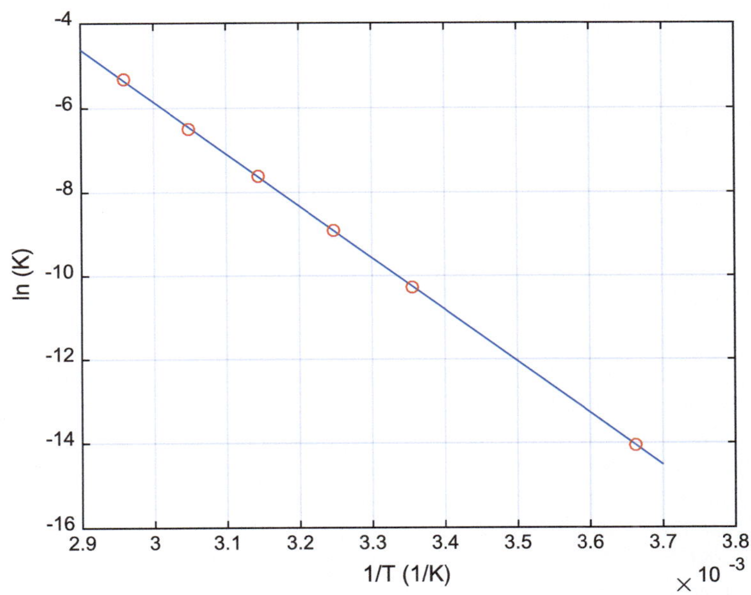

Fig. 4.22 Arrhenius plot of the data from Alberty and Silbey (1992)

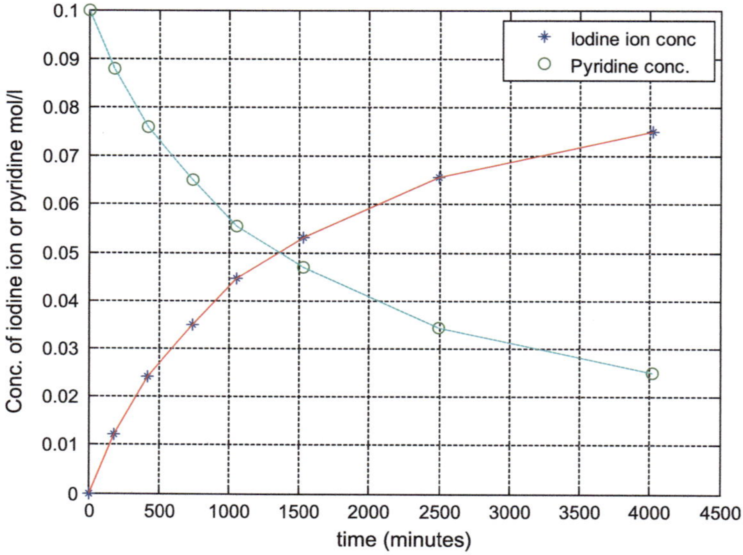

Fig. 4.23 Experimental data collected by Winkler and Hinshelwood (1935)

Table 4.9 Experimental data expressed in terms of pyridine conversion versus time[a]

Run	Time (min)	Conversion (%)	Reaction rate * 10^5 (mol/L min)
0	0	0	–
1	179	12.0	6.76
2	420	24.0	4.98
3	737	35.0	3.47
4	1054	44.5	3.00
5	1530	53.0	1.79
6	2497	65.5	1.29
7	4020	75.0	0.62

[a]An approximated value of the reaction rate is also listed

follow the reaction by measuring the concentration of I– at different times. The experimentally obtained results are shown in Fig. 4.23.

From the points shown in Fig. 4.23, it is possible to evaluate the evolution with time of the conversion of pyridine and estimate from the slopes an approximated value of the reaction rate. The obtained values are listed in Table 4.9.

Evaluate the reaction order, considering the possibility of first- or second-order kinetics and from the data, at different temperatures given below, evaluate the activation energy. Last, simulate by numerical integration the curves of both pyridine disappearance and iodine formation.

Table 4.10 Estimation of the kinetic constants from the data of reaction rates

Time (s)	k_1 (min^{-1}) * 10^3	k_2 (L/mol/min) * 10^2	C_P	C_{EJ} (mol/L)
0	–	–	0.1	0.1
179	0.7142	0.76	0.0880	0.0880
420	0.6534	0.75	0.0760	0.0760
737	0.5845	0.73	0.0650	0.0650
1054	0.5586	0.76	0.0555	0.0555
1530	0.4935	0.74	0.0470	0.0470
2497	0.4262	0.76	0.0345	0.0345
4020	0.3448	0.75	0.0250	0.0250

Solution

As required, we consider the two possibilities of the first and second order by writing:

$$r = -\frac{dC_P}{dt} = k_1 C_P \tag{4.58}$$

$$r = -\frac{dC_P}{dt} = k_2 C_P C_{EJ} = k_2 C_P^2 \tag{4.59}$$

The initial concentrations of pyridine and ethyl iodide are equal and remain equal with stoichiometry of 1:1. Different methods can be used to evaluate the reaction order.

(1) Determination of the kinetic constant from the experimental data

A first approach could be the evaluation, from the seven available experimental data, of the values of, respectively, k_1 and k_2. The calculated values are listed in Table 4.10.

As can be seen, k_2 for the runs performed at 100 °C is approximately constant (average value 0.0075); on the contrary k_1 is variable; therefore, the reaction is of the second order.

(2) Integral method

By putting in a plot $-\ln(C_T/C_T°)$ for the first order or, alternatively, $(1/C_T - 1/C_T°)$ for the second order, against time we obtain the plots shown in Fig. 4.24.

As can be seen, a linear trend is obtained only by putting $(1/C_T - 1/C_T°)$ against the time in agreement with a second-order kinetic law.

(3) Differential method

It also is possible to arrive at the same conclusion by following the "differential method." The slopes of iodine ion production with time correspond to the reaction rate, that is:

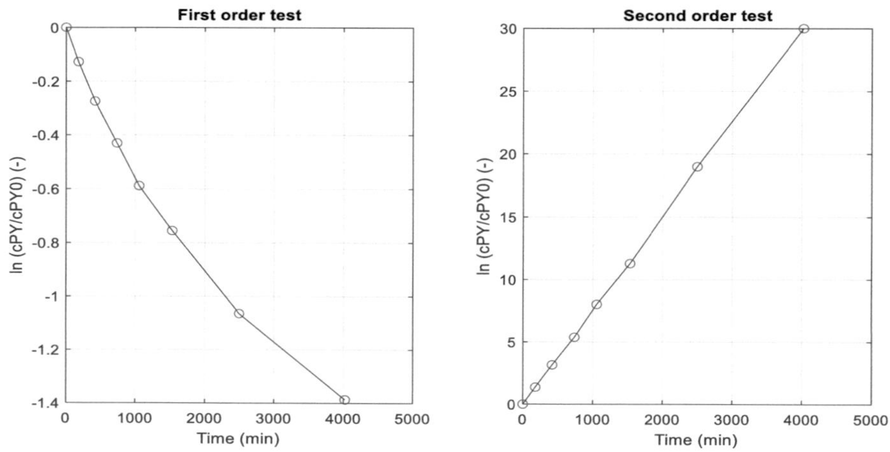

Fig. 4.24 Plots for individuating the reaction order

Table 4.11 Elaboration of the experimental data

Concentration (mol/L)		$r = \frac{\Delta C_{I^-}}{\Delta t}$ (mol/L min) * 10^5	$-\log r$
C_I^-	C_P		
0	0.1	6.76	9.6102
0.0120	0.0880	4.98	9.9076
0.0240	0.0760	3.47	10.2688
0.0350	0.0650	3.00	10.4154
0.0445	0.0555	1.79	10.9331
0.0530	0.0470	1.29	11.2562
0.0655	0.0345	0.62	11.9849
0.0750	0.0250	–	–

$$r = -\frac{dC_P}{dt} = \frac{dC_{I^-}}{dt} \tag{4.60}$$

For the first order, we have:

$$\log r = \log k_1 + \log C_P \tag{4.61}$$

In contrast, for a second order, we have:

$$\log r = \log k_2 + \log C_P^2 = \log k_2 + 2 \ \log C_P \tag{4.62}$$

Table 4.11 lists the experimental data and the values of reaction rates calculated from the slopes.

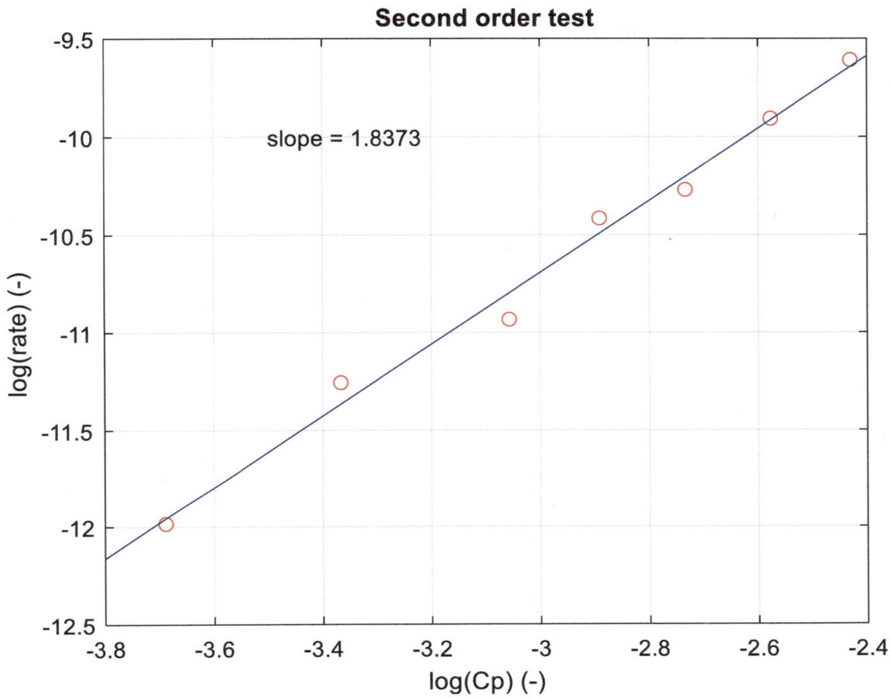

Fig. 4.25 Plot of log r against log C_p

By putting in a plot log r as a function of log C_P (see Fig. 4.25), we obtain—in agreement with the second-order law—a straight line with a slope = 2.

(4) By simulating the experimental data

Last, another possibility is to perform analytical or numerical integration of the alternative kinetic equations:

$$-\frac{\mathrm{d}C_P}{\mathrm{d}t} = r \tag{4.63}$$

$$\frac{\mathrm{d}C_{I^-}}{\mathrm{d}t} = r \tag{4.64}$$

Testing the best-fitting result by assuming $r = k_1 C_P$ or $r = k_1 C_P^2$. By mathematical regression analysis, it is possible to obtain in the meantime the parameter giving the best fit, and the simulation is shown in Fig. 4.26.

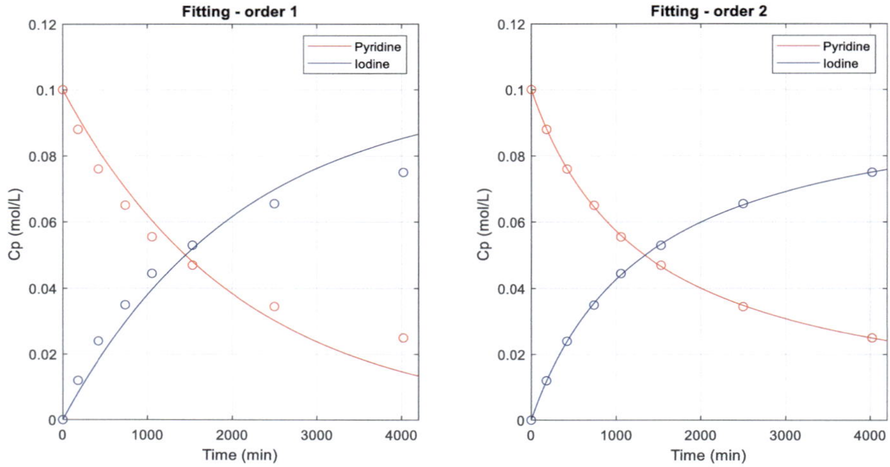

Fig. 4.26 Best-fitting result obtained by assuming first- and second-order kinetic law

Table 4.12 Effect of the temperature on the kinetic constant

T (°C)	k_2 (mol/L min)
80.5	0.0025
99.7	0.0076
117.6	0.0196
139.4	0.0642

Results

First order

$k_1 = 0.00047878\ (1/\min)$

Average % err = **10.6261**

Second order

$k_2 = 0.0074804\ (\mathrm{L}/(\mathrm{mol}\ \ \min)$Average % err = **0.49352**

Arrhenius parameters

$E_a = -15{,}897.99\ \ \mathrm{cal/mol}$

$A = 16{,}343{,}178.574$

At this point, the second-order kinetic constants calculated for the reaction, at different temperatures, are listed in Table 4.12.

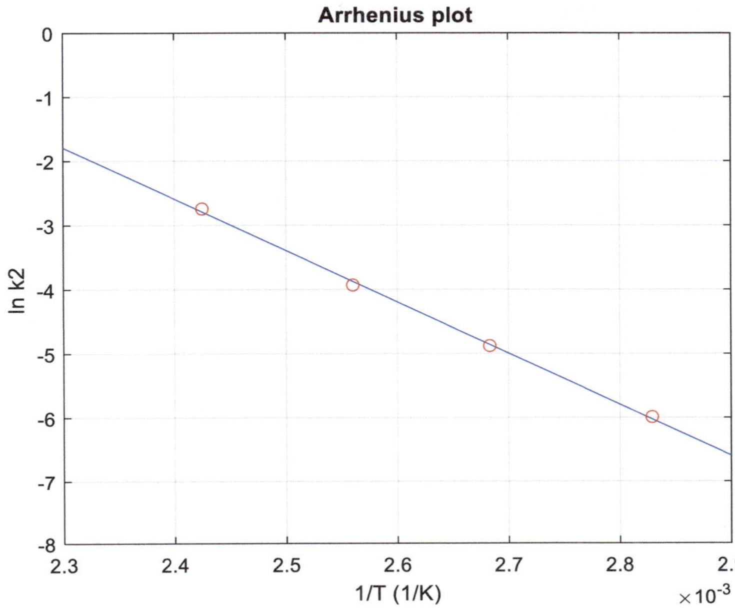

Fig. 4.27 Arrhenius plot for the reaction of pyridine with ethyl iodide

By arranging the data of Table 4.12 into an Arrhenius plot, it is possible to see that the trend is linear as expected and that the activation energy (the average slope of the straight line) is approximately 14,900 cal/mol (see Fig. 4.27).

All of the described results were obtained using a MATLAB program available as electronic supplementary material, Data 4.1.

Exercise 4.8. Reaction Order for the Reduction of NO with Hydrogen
Hinshelwood and Green (1926) studied the kinetics of the reaction:

$$2NO + 2H_2 \rightarrow N_2 + 2H_2O \tag{4.65}$$

As can be seen from the stoichiometry, the reaction occurs by a decrease in the number of moles that at constant volume indicates a decrease in pressure. For a correct evaluation, it is important to avoid water condensation. Hinshelwood and Green (1926) performed numerous runs at 826 °C. Some obtained data are collected in Table 4.13.

Evaluate the reaction order by considering the possibility of a second or third order. Estimate the kinetic constant. Hinshelwood et al. performed numerous other different runs changing the initial pressure of both NO and H_2. By using the kinetic law previously derived and the related kinetic constant, which simulates two runs with different initial pressures, for example, $p^{\circ}_{NO} = 300$ mm Hg and $p^{\circ}_{H_2} = 404$ and $p^{\circ}_{NO} = 232$ mm Hg and $p^{\circ}_{H_2} = 313$.

Table 4.13 Experimental data collected by Hinshelwood and Green (1926)[a]

Time (s)	8	13	19	26	33	43	54	69	87	110	140	204	310	Long time
ΔP (mm Hg)	10	20	30	40	50	60	70	80	90	100	110	120	127	144.5

[a]Data collected at $T = 826$ °C; initial NO pressure = 406 mm Hg; initial H_2 pressure = 289 mm Hg

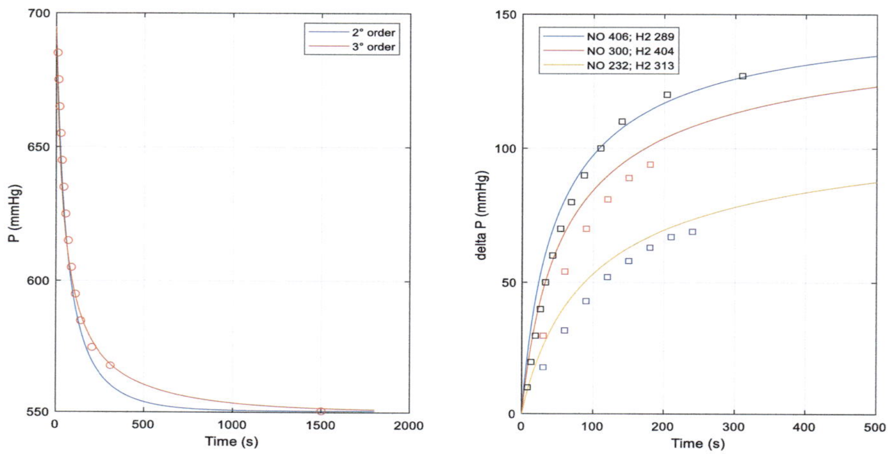

Fig. 4.28 Plots discriminating between the second and third order. Effect of the change in the initial ratio between the reactants

Results

The parameters giving the best fit are:

$$\text{second order } k = 4.4013\text{e} - 05$$
$$\text{third order } k = 1.486\text{e} - 07$$

The obtained plots shown in Fig. 4.28 are in agreement with third-order kinetics. Moreover, as can be seen by decreasing the initial NO pressure, the rates decrease very much. Moreover, it is interesting to observe that even if the fitting is performed only on the data at 406 °C and with an initial hydrogen pressure of 289 mmHg, also the other data at 300 and 232 °C are quite well described.

The described results were obtained using a MATLAB program available as electronic supplementary material, Data 4.1.

4.3.6 Complex Reactions

A reagent or a product can be involved in more than one single reaction. We can have three different possibilities of this type: equilibrium reactions, competitive reactions, and consecutive reactions.

$$\text{Equilibrium reactions} \quad A \leftrightarrows B \tag{4.66}$$

$$\text{Competitive reactions} \quad A \begin{matrix} \nearrow B \\ \searrow C \end{matrix} \tag{4.67}$$

$$\text{Consecutive reactions} \quad A \rightarrow B \rightarrow C \tag{4.68}$$

Equilibrium Reactions

Let us consider, first of all, the equilibrium reactions. The simplest cases are isomerization reactions of the first order from both sides. Equilibrium reactions occur with a trend, such as the one shown in Fig. 4.29.

As can be seen from Fig. 4.29, the equilibrium reactions are characterized by the fact that the reaction stops when the equilibrium composition is reached and the reactants, therefore, are not completely consumed. By assuming a first-order reaction on both sides, we can write:

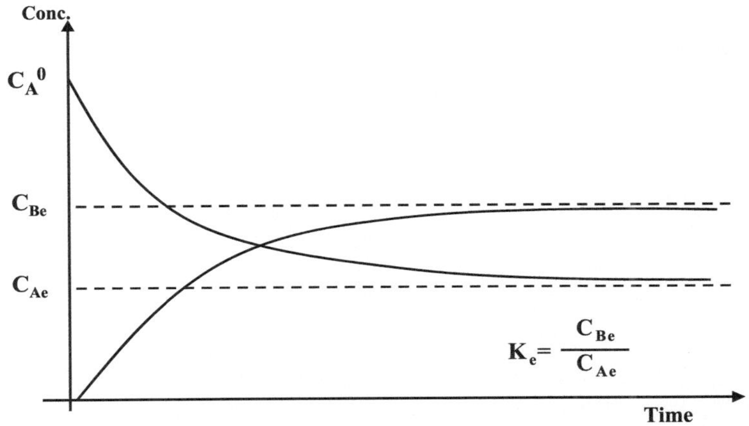

Fig. 4.29 Example of the evolution with time of the concentrations of, respectively, reactant A and product B for a simple equilibrium reaction of the type $A \rightleftarrows B$

$$r = k_1 C_A - k_{-1} C_B = k_1 \left(C_A - \frac{k_{-1}}{k_1} C_B \right) = k_1 C_A \left(1 - \frac{1}{K_e} \frac{C_B}{C_A} \right) \tag{4.69}$$

where $K_e = k_1/k_{-1}$ is the thermodynamic equilibrium constant. In this simplest case, $C_A^o = C_A + C_B$, then remembering that the conversion of A is defined as:

$$\lambda = \frac{A \text{ reacted molecules}}{A \text{ initial molecules}} = \frac{C_A^o - C_A}{C_A^o} = 1 - \frac{C_A}{C_A^o} \tag{4.70}$$

The rate equation becomes:

$$r = k_1 C_A^o (1 - \lambda) \left(1 - \frac{1}{K_e} \frac{\lambda}{(1 - \lambda)} \right) \tag{4.71}$$

When the kinetic law of the forward reaction is known, it is not necessary to study the kinetics of the backward reaction because this must be consistent with the thermodynamic equilibrium. Consider, for example, the reaction:

$$aA + bB \rightleftarrows mM + Nm \tag{4.72}$$

At equilibrium, we have:

$$K_e = \frac{k_r}{k_{-r}} = \frac{C_M^m C_N^n}{C_A^a C_B^b} \tag{4.73}$$

$$\begin{aligned} r = \vec{r} - \overleftarrow{r} &= k_r C_A^a C_B^b - k_{-r} C_M^m C_N^n = k_r \left(C_A^a C_B^b - \frac{1}{K_e} C_M^m C_N^n \right) \\ &= k_r C_A^a C_B^b \left[1 - \frac{1}{K_e} \frac{C_M^m C_N^n}{C_A^a C_B^b} \right] = \vec{r} \left[1 - \frac{1}{K_e} \frac{C_M^m C_N^n}{C_A^a C_B^b} \right] \end{aligned} \tag{4.74}$$

This expression is in agreement with the thermodynamic constraints. In fact, if the reaction is at equilibrium, $r = 0$, and we have,

$$1 = \frac{1}{K_e} \frac{C_M^m C_N^n}{C_A^a C_B^b} \quad \text{that is,} \quad K_e = \frac{C_M^m C_N^n}{C_A^a C_B^b} \tag{4.75}$$

The thermodynamic consistency must be verified also when the reaction orders are not integer numbers. It must be valid:

$$r = \vec{r} \left[1 - \left(\frac{\Pi_i C_i^{\alpha_i}}{K_e} \right)^n \right] \quad \text{with } n = \text{positive number} \tag{4.76}$$

An example of the necessity to respect the thermodynamic consistency when we write the equilibrium expression is the following reaction:

$$Cl_2 + CO \rightleftarrows COCl_2 \tag{4.77}$$

which was already mentioned in Chap. 1. As a matter of fact, the kinetic law of the forward reaction is an expression of the type:

$$\vec{r} = k_r C_{CO} C_{Cl_2}^{3/2} \tag{4.78}$$

Therefore, the kinetic law of the reverse reaction must be:

$$\vec{r} = k_{-r} C_{COCl_2} C_{Cl_2}^{1/2} \tag{4.79}$$

to respect the thermodynamic constraint:

$$K_e = \frac{C_{COCl_2} C_{Cl_2}^{1/2}}{C_{CO} C_{Cl_2}^{3/2}} = \frac{C_{COCl_2}}{C_{CO} C_{Cl_2}} \tag{4.80}$$

The fractional order suggests a complex mechanism for this reaction. In fact, in this case, a complex radical mechanism is operative.

Exercise 4.9. Kinetics and Equilibrium in the Isomerization of β-Hydroxy Crotonic Ester to Acetoacetic Ester
The reaction:

(4.81)

is one of the simplest examples of an equilibrium reaction obeying a first-order law in both directions. From the experimental data listed in Table 4.14, evaluate the kinetic constants of both the forward and backward reaction and the equilibrium constant. With the determined parameters, simulate the experimental points.

Table 4.14 Experimental data of β-hydroxy crotonic ester

Time (h)	Λ
0	0.634
71.8	0.723
145.5	0.785
215.8	0.826
264.3	0.848
333.3	0.870
383.5	0.879
478.3	0.894
506.0	0.900

[a]Temperature = 25 °C. $n_{HC}^{o} = 1$ = initial moles of reactant $\lambda_e = 0.922$ = conversion at equilibrium

Solution

We can write:

$$-\frac{\mathrm{d}n_{\mathrm{HC}}}{\mathrm{d}t}=\frac{\mathrm{d}n_{\mathrm{AA}}}{\mathrm{d}t}=k_1 n_{\mathrm{HC}}^{o}(1-\lambda)-k_{-1}n_{\mathrm{HC}}^{o}\lambda \tag{4.82}$$

However, at equilibrium $-\frac{\mathrm{d}n_{\mathrm{HC}}}{\mathrm{d}t}=\frac{\mathrm{d}n_{\mathrm{AA}}}{\mathrm{d}t}=0$, and hence:

$$k_1 n_{\mathrm{HC}}^{o}(1-\lambda_e)-k_{-1}n_{\mathrm{HC}}^{o}\lambda_e=k_1(1-\lambda_e)-k_{-1}\lambda_e=0 \tag{4.83}$$

In conclusion:

$$K_e=\frac{k_1}{k_{-1}}=\frac{\lambda_e}{1-\lambda_e} \tag{4.84}$$

Then it also is possible to write:

$$\frac{\mathrm{d}\lambda}{\mathrm{d}t}=k_1\left[(1-\lambda)-\frac{\lambda}{K_e}\right] \tag{4.85}$$

and remembering that $K_e=\frac{\lambda_e}{1-\lambda_e}$:

$$\frac{\mathrm{d}\lambda}{\mathrm{d}t}=k_1\left(1-\frac{\lambda}{\lambda_e}\right) \tag{4.86}$$

By integrating:

$$\int_{\lambda_o}^{\lambda_e}\frac{\mathrm{d}\lambda}{\left(1-\frac{\lambda}{\lambda_e}\right)}=k_1\int_0^t \mathrm{d}t \tag{4.87}$$

$$-\lambda_e\log(\lambda_e-\lambda)+\lambda_e\log(\lambda_e-\lambda_o)=k_1 t \tag{4.88}$$

$$\lambda_e\log\frac{(\lambda_e-\lambda_o)}{(\lambda_e-\lambda)}=k_1 t \tag{4.89}$$

$$k_1=\frac{\lambda_e}{t}\log\frac{(\lambda_e-\lambda_o)}{(\lambda_e-\lambda)} \tag{4.90}$$

The equilibrium conversion can be determined with a thermodynamic theoretical approach or, alternatively, by experimental determination as made here. Another possibility is to evaluate this value as a parameter of the kinetic model by mathematical regression analysis.

Results

The kinetic parameters giving the best-fitting result are reported below in Fig. 4.30 and in the same figure, the agreement in the simulation is shown.

The described results were obtained using a MATLAB program available as electronic supplementary material, Data 4.1.

Exercise 4.10. Kinetics and Equilibrium of HI Synthesis and Decomposition: An Example of a Double-Sided Second-Order Equilibrium Reaction

Part 1

Remember Fig. 2.5, where the evolution over time of both the reaction of HI synthesis and decomposition is shown with the aim to evaluate the equilibrium of the reaction. From that figure, it is possible to derive the kinetics of both synthesis and decomposition. Therefore, we have taken from Fig. 2.5 the experimental points listed in Table 4.15.

These runs were performed by Bodenstein (1894) at 717.66 K. The equilibrium conversion according to Bodenstein can be determined at different temperatures with the relation:

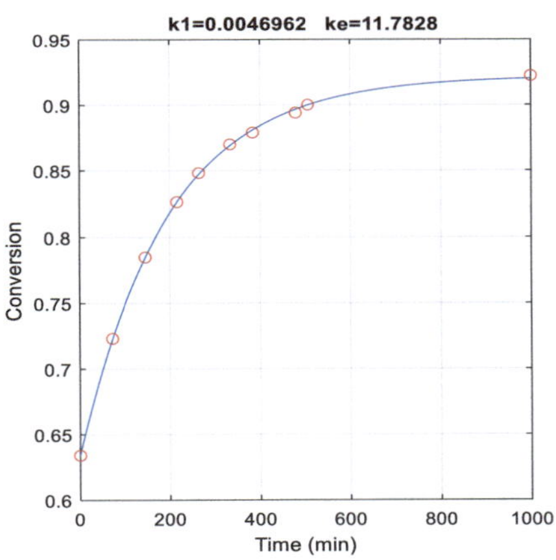

Fig. 4.30 Simulation of the experimental data

Table 4.15 Experimental data for synthesis and decomposition of HI[a]

Formation of HI									
Time (min)	0	6	13	20	22	28	38	48	60
HI moles fraction	0	0.29	0.45	0.54	0.60	0.66	0.71	0.73	0.76
HI decomposition									
Time (min)	0	13	22	33	42	52	62	74	81
HI moles fraction	1	0.96	0.91	0.865	0.84	0.825	0.81	0.79	0.78

[a]Temperature = 717.66 K

$$\lambda = 0.1376 + 7.22 \times 10^{-5} T + 2.576 \times 10^{-7} T^2 \quad T = \text{Temperature } (^\circ\text{C}) \tag{4.91}$$

At 717.66 K, the equilibrium constant in vapor phase Kp has a value of 46. The forward reaction started with 1 mol of hydrogen and 1 mol of iodine loaded in a reactor of 1 L volume, while the backward reaction started with 2 mol of HI at the same temperature and volume. Evaluate the forward kinetic constant and the equilibrium constant by fitting the data of the previous table. Compare your results with the kinetics given in the literature by Bamford and Tipper (1969); see also Missen et al. (1999).

$$\text{Forward reaction } 1.3 \times 10^{11} \ \exp(38{,}979/RT) = (\text{L/mol s}) \tag{4.92}$$

$$\text{Backward reaction } 7.9 \times 10^{10} \ \exp(43{,}971/RT) = (\text{L/mol s}) \tag{4.93}$$

In both cases, the activation energy is expressed in cal/mol. In the first case, we have:

$$r_{\text{forward}} = -\frac{\mathrm{d}p_{H_2}}{\mathrm{d}t} = -\frac{\mathrm{d}p_{I_2}}{\mathrm{d}t} = \frac{1}{2}\frac{\mathrm{d}p_{HI}}{\mathrm{d}t} = k_2\left(p_{H_2}p_{I_2} - \frac{p_{HI}^2}{K_e}\right) \tag{4.94}$$

$$\frac{\mathrm{d}\lambda}{\mathrm{d}t} = k_2\frac{n^o\mathrm{RT}}{V}\left[(1-\lambda)^2 - \frac{4\lambda^2}{K_e}\right] \tag{4.95}$$

where λ is the conversion of the reactant H_2.

While in the second case it results in:

$$r_{\text{backward}} = -\frac{\mathrm{d}p_{HI}}{\mathrm{d}t} = 2\frac{\mathrm{d}p_{H_2}}{\mathrm{d}t} = 2\frac{\mathrm{d}p_{I_2}}{\mathrm{d}t} = k_{-2}\left(p_{HI}^2 - K_e p_{H_2}p_{I_2}\right) \tag{4.96}$$

$$\frac{\mathrm{d}\lambda_1}{\mathrm{d}t} = k_{-2}\frac{n^o\mathrm{RT}}{V}\left[(1-\lambda_1)^2 - \frac{\lambda_1^2}{4K_e}\right] \tag{4.97}$$

where λ_1 is the conversion of the reactant HI.

Results

The experimental data at $T = 717.66$ K, taken from Gerasimov et al. (1974), were submitted to regression for determining both the kinetic and equilibrium constants for HI synthesis and decomposition. The values found for the best-fitting constants are $k_s = 0.06980$ L/(mol min), $K_e = 51.8$. The results are shown in the plot as blue curves. By using, on the contrary, the previously kinetic constant and a value of $K_e = 46$, the red curves are obtained, with a very good agreement with the nonlinear fitting method (see Fig. 4.31).

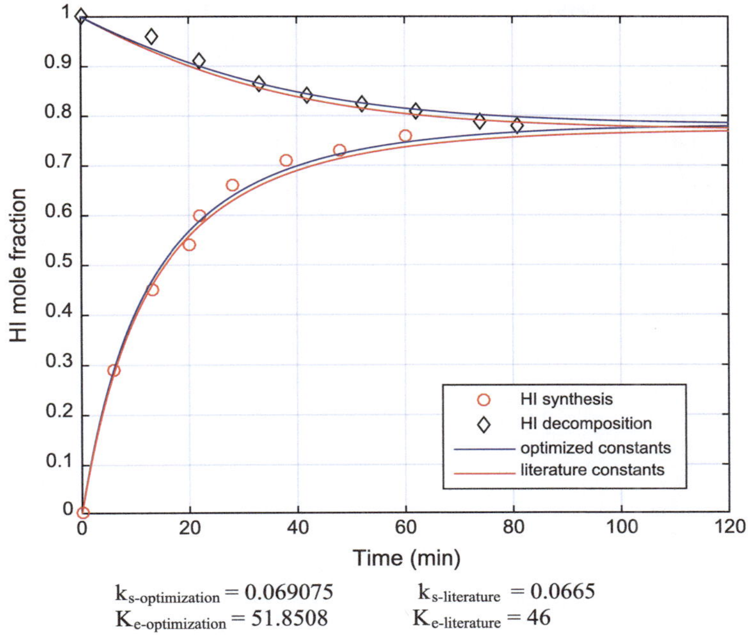

Fig. 4.31 Simulation of the kinetic runs performed by Bodenstein et al. at 717.66 K

Part 2

Different researchers investigated the kinetics of HI decomposition, and Smith (1981) has furnished an analytical solution to the problem. Here we repeat the exercise, but we apply a numerical integration method. The occurring reaction is:

$$2HJ \leftrightarrows H_2 + J_2$$

The kinetics of the reaction was studied, in particular, by Kistiakowski (1928) in a closed bulb of a known volume kept at the reaction temperature of 321.4 °C for different lengths of time. At the end of the reaction, the bulb was cooled, and the reaction mixture analyzed. The experimental data are listed in Table 4.16.

At 321.4 °C, the equilibrium conversion results in $\lambda_e = 0.187$. Evaluate by numerical integration the kinetic constant and the average scattering due to experimental error, thus producing a parity plot for all of the runs listed in table.

Solution

The equations to be solved are the same as already seen before related to the backward reaction, and we simply have to integrate Eq. (4.95). A numerical integration currently allows one to solve the problem more easily.

Table 4.16 Experimental data from Kistiakowski (1928) and Smith (1981)

Run	Time (s)	% HI decomposed	Volume bulb (cm^3)	$[HI]^\circ$ g mol/L
1	82.8	0.826	51.38	0.02339
2	172.8	2.567	59.80	0.03838
3	180.0	3.286	51.38	0.04333
4	173.1	3.208	7.899	0.04474
5	81.0	2.942	7.899	0.1027
6	57.56	2.670	7.899	0.1126
7	61.32	4.499	7.899	0.1912
8	19.2	2.308	7.899	0.3115
9	18.0	2.202	7.899	0.3199
10	16.8	2.071	7.899	0.3279
11	17.4	2.342	7.899	0.3464
12	17.7	2.636	7.899	0.4075
13	18.0	2.587	7.899	0.4228
14	23.4	4.343	7.899	0.4736
15	6.0	2.224	3.28	0.9344
16	5.4	1.903	0.778	0.9381
17	8.16	3.326	0.781	1.1380
18	5.4	2.741	0.713	1.2310

Results

$$k_2 = 0.0039967$$
$$K_e = 0.01322$$

The parity plot obtained is shown in Fig. 4.32.

All of the described results were obtained using a MATLAB program available as electronic supplementary material, Data 4.1.

Exercise 4.11. Synthesis of Acetone–Cyanohydrin

Svirbely and Roth (1953) studied the kinetics of the reaction:

$$CH_3COCH_3 + HCN \rightleftarrows \begin{array}{ccc} CH_3 & & CN \\ & C & \\ CH_3 & & OH \end{array} \tag{4.98}$$

at 25 °C. The reaction was performed in the aqueous phase starting with a solution containing initially 0.0758 mol/L of HCN and 0.1164 mol/L of acetone. The concentration of HCN varied as listed in Table 4.17.

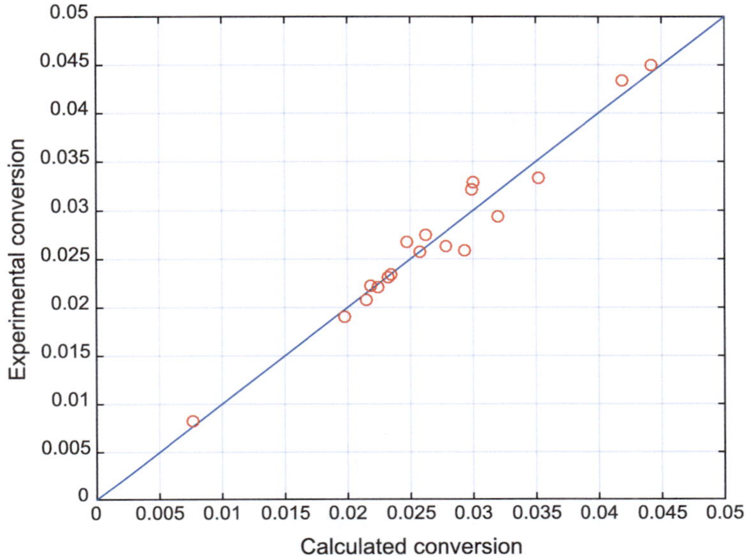

Fig. 4.32 Parity plot obtained in simulating the experimental conversion

Table 4.17 Kinetic experimental data for the synthesis of acetone–cyanohydrin

Time (min)	4.37	73.2	172.5	265.4	346.7	434.4
Concentration of HCN (mol/L)	0.0748	0.0710	0.0655	0.0610	0.0584	0.0557

The equilibrium concentration of HCN was 0.0366 mol/L, while the equilibrium concentration of acetone was 0.0772 mol/L. Determine the kinetic equation, the forward kinetic constant, and the equilibrium constant.

Results

The obtained parameters by finding the best-fitting result of the experimental data are:

$$k_1 = 0.0080642$$
$$K_e = 13.8531$$

The kinetic equation follows the reaction stoichiometry, that is, the second order for the forward reaction and the first order for the backward reaction.

In Fig. 4.33, the agreement obtained in reproducing the experimental points can be appreciated. The same plot also shows the evolution with time of the acetone concentration. As can be seen in this last case, the equilibrium point is perfectly found by the kinetic model.

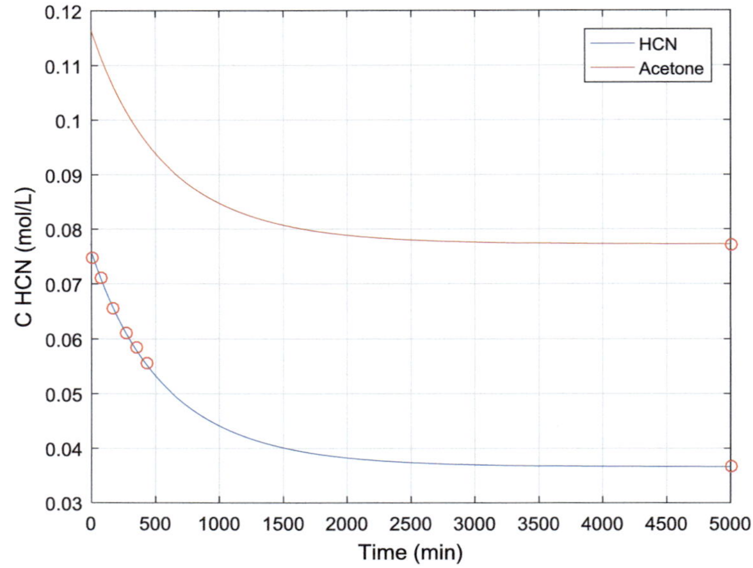

Fig. 4.33 Simulation of the available experimental data for the synthesis of acetone–cyanohydrin

Table 4.18 Experimental data for the hydrolysis of tert-butyl bromide

Time (h)	Concentration of $(CH_3)_3CBr$ (mol/L)
0	0.1039
3.15	0.0896
4.10	0.0859
6.20	0.0776
8.20	0.0701
10.0	0.0639
13.5	0.0529
18.3	0.0353
26.0	0.0270
30.8	0.0207

The described results were obtained using a MATLAB program available as electronic supplementary material, Data 4.1.

Exercise 4.12. Hydrolysis of Tert-Butyl Bromide to Tert-Butyl Alcohol
The reaction:

$$(CH_3)_3C - Br + H_2O \rightleftarrows (CH_3)_3C - OH + HBr \quad (4.99)$$

occurs at 25° by following the kinetics listed in Table 4.18.

Find the rate law and related parameters.

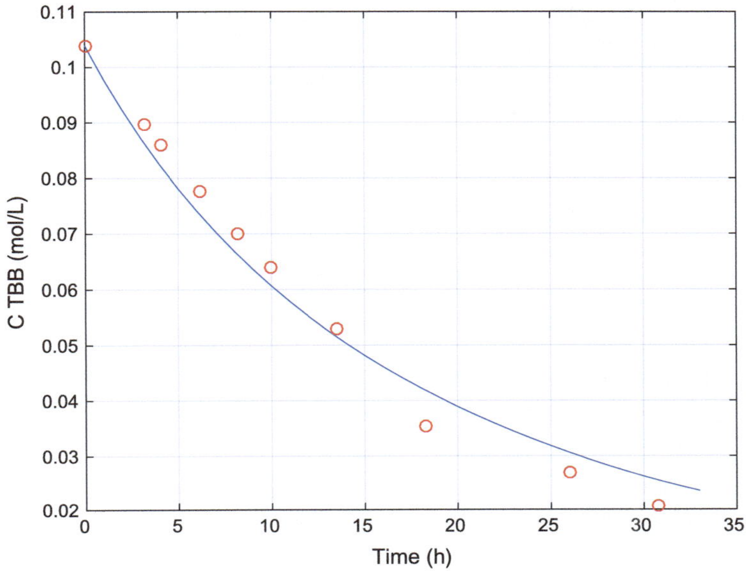

Fig. 4.34 Best-fitting option obtained for the experimental data for the hydrolysis of tert-butyl bromide

Results

The experimental data are well fitted with a kinetic equation of both sides in second order, and the determined best-fitting parameters are as follows:

$$k_1 = 0.30675$$
$$K_e = 1639.354$$

Figure 4.34 shows the obtained fitting of the experimental data.

The described results were obtained using a MATLAB program available as electronic supplementary material, Data 4.1.

Exercise 4.13. Equilibrium Reaction Between Methyl Iodide and Dimethyl-p-Toluidine

Laidler (1950) and Smith (1981) reported some kinetic data related to the reaction:

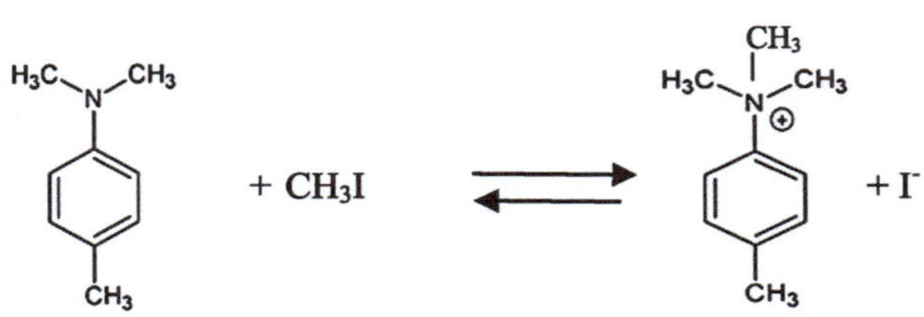

(4.100)

Starting from reactants having the same concentration of 0.05 mol/L, the following kinetic data were obtained (Table 4.19).

The equilibrium constant is 1.43, and the reaction order is 2. By employing the numerical integration, find the kinetic constant giving the best fit, and simulate the kinetic runs.

Results

The experimental data are well fitted with the following parameters:

$$k_1 = 0.42568\ (\mathrm{L/(mol\ s)})$$
$$K_e = 1.43\ (-)$$

The obtained fitting is shown in Fig. 4.35.

The described results were obtained using a MATLAB program available as electronic supplementary material, Data 4.1.

Competitive Reactions

In some cases, a reactant is involved simultaneously in two or more reactions. Let us consider the simplest case of just two competitive reactions of the first order occurring in a batch reactor:

Table 4.19 Experimental data for the reaction between methyl iodide and dimethyl-p-toluidine

Time (min)	Fraction reacted
10.2	0.175
26.5	0.343
36.0	0.402
78.0	0.523

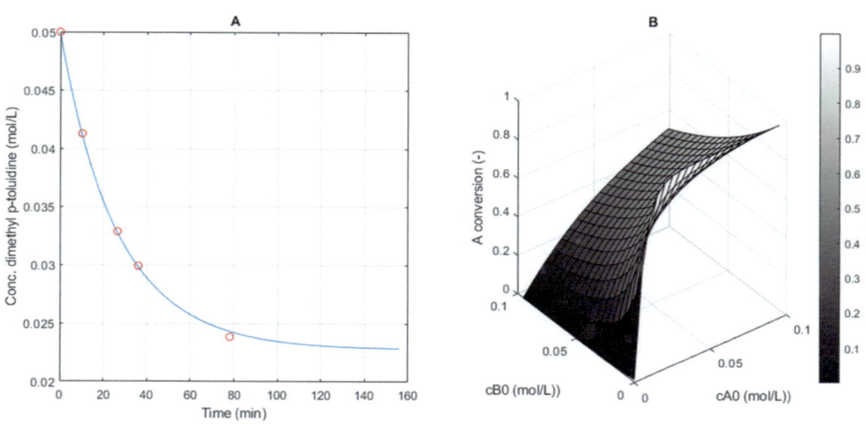

Fig. 4.35 Simulation of the experimental data for the reaction between methyl iodide and dimethyl-p-toluidine. (**a**) Time profile of dimethyl-p-toluidine concentration; (**b**) Surface plot of dimethyl-p-toluidine as a function of initial concentration of toluidine and methyl iodide

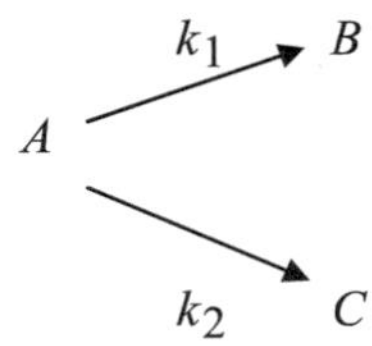

(4.101)

We can write:

$$r_A = -\frac{dC_A}{dt} = \frac{dC_B}{dt} + \frac{dC_C}{dt} = k_1 C_A + k_2 C_A = (k_1 + k_2) C_A \tag{4.102}$$

This relation integrated becomes:

$$\ln \frac{C_A^{\circ}}{C_A} = (k_1 + k_2)t \tag{4.103}$$

It is interesting to observe that the ratio r_1/r_2 remains constant and is equal to the ratio of the obtained products:

$$\frac{r_1}{r_2} = \frac{k_1 C_A}{k_2 C_A} = \frac{C_B}{C_C} \tag{4.104}$$

independent of time. This behavior allows us to recognize easily the competitive reactions having the same reaction order. It is important to point out that we also can have competitive reactions for elementary-step reactions inside a reaction mechanism, and this can influence the reaction kinetic law if these reactions are slow. A classic example of this reaction type is the introduction of a functional group on an aromatic ring containing an orienting group. Ortho-, meta-, and para-derivatives with constant ratios will be obtained simultaneously independently of time. Table 4.20 lists the results obtained for the nitration of different derivatives of benzene performed at 30 °C.

The experimental kinetic data of competitive reactions in a concentration/time plot show a trend, such as the one shown in Fig. 4.36.

Exercise 4.14. Production of Dinitrobenzenes Through Competitive Reactions Nitration of aromatic compounds is made using a mixture of concentrated H_2SO_4 and HNO_3 (sulfonitric mixture). The nitrating agent is the nitronium ion, NO_2^+, which is formed as a consequence of the reaction:

Table 4.20 Compositions resulting from the nitration of some derivatives of benzene performed at 30 °C

	Percentage of		
Compound	Ortho	Meta	Para
Nitrobenzene	8.1	90.9	1.0
Benzoic acid	22.3	76.5	1.2
Methyl benzoate	25.7	69.8	4.5
Ethyl benzoate	27.7	66.4	5.9

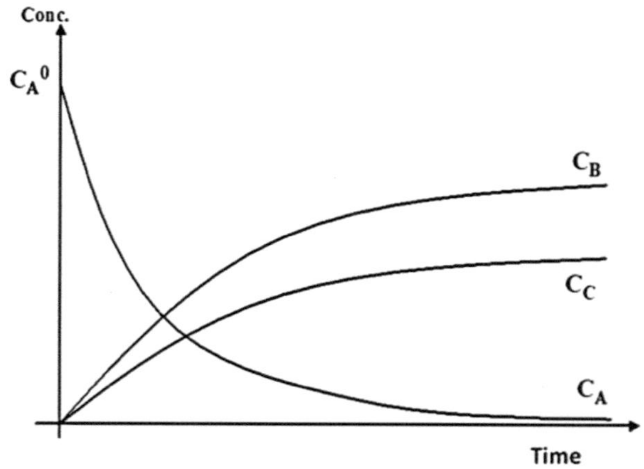

Fig. 4.36 Trends of reactants and products for competitive reactions

$$2H_2SO_4 + HNO_3 \rightleftarrows NO_2^+ + H_3O^+ + HSO_4^- \tag{4.105}$$

and the reaction mechanism for a generic nitration of an aromatic compound is:

$$AR + NO_2^+ \xrightarrow{\text{Slow}} \left[ARNO_2^+\right] \xrightarrow{\text{Fast}} ARNO_2 + H^+ \tag{4.106}$$

Starting from nitrobenzene as an organic reactant, we have:

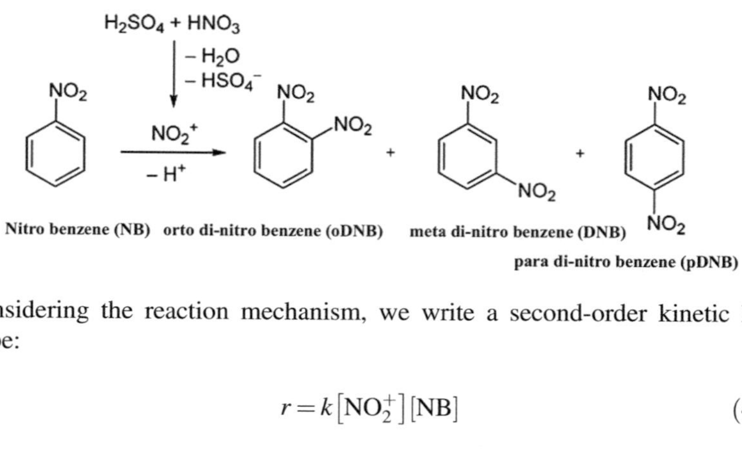

Considering the reaction mechanism, we write a second-order kinetic law of the type:

$$r = k\left[NO_2^+\right][NB] \tag{4.107}$$

Considering then the $\left[NO_2^+\right]$ in equilibrium, we write that:

$$\left[NO_2^+\right] = \frac{K_e[H_2SO_4]^2[HNO_3]}{[H_3O^+]\left[HSO_4^-\right]} \tag{4.108}$$

On the basis of this theoretical approach, the reaction rate is:

$$r = k[\mathrm{NB}]\frac{K_e[\mathrm{H_2SO_4}]^2[\mathrm{HNO_3}]}{[\mathrm{H_3O^+}][\mathrm{HSO_4^-}]} = k\frac{K_e}{[\mathrm{H_2O}]K_{\mathrm{H^+}}}[\mathrm{NB}][\mathrm{HNO_3}] = \frac{k_{\mathrm{app}}}{[\mathrm{H_2O}]}[\mathrm{NB}] \times [\mathrm{HNO_3}] \tag{4.109}$$

Westheimer and Kharasch (1946) studied the kinetics of this reaction and accepted the described mechanism, which also was justified by the great influence of the sulfuric acid concentration on the reaction rate. Nevertheless, they interpreted their results by assuming a simple bimolecular kinetics of the type:

$$r = k[\mathrm{NB}][\mathrm{HNO_3}] \tag{4.110}$$

and this kinetic law also is generally accepted by Hughes et al. (1950), who showed the existence and activity of the nitronium ion, and by many other researchers.

Westheimer et al. found, at 25 °C and using H_2SO_4 85%, an overall kinetic constant of 0.31 (L/mol min), and the value reached a maximum for H_2SO_4 90% of 4 (L/mol min). According to Vollhardt (1987), the formation of meta di-nitro benzene (mDNB) is largely favored, and 93% of this compound is obtained in the product distribution, while 5% of oDNB and 2% of the pDNB are, respectively, obtained from the three simultaneously occurring reactions.

Imagine that the production is made, at 25 °C, using H_2SO_4 98% as solvent (density 1.84) feeding it in a continuous well-stirred tank reactor with 10 L of volume having an overall flow rate changing from 0 to 700 L/h. The concentration of pure NB (density 1.2) and HNO_3 (fed as solution 60 wt% density 1.34) at the inlet, respectively, are 2 and 2.5 mol/L, while the stream of sulfuric acid is 11 mol/L.

Solve the mass balance equations and evaluate the concentration of any component "*i*" at the outlet of the reactor for different flow rates from 0 to 700 L/h, putting in a plot both the conversion of nitrobenzene and the concentration of all of the products as a function of the flow rate.

Solution

The mass balance for any component "*i*" is the following:

$$\frac{Q}{V}(C_i - C_{io}) - \sum_{j=1}^{NR} V_{ij} r_j = 0 \tag{4.111}$$

The volume of the reactants fed results from the calculation of approximately 200 cm^3 of HNO_3 60 wt%/L of the mixture, 200 cm^3 of DB/L of the mixture, and 600 cm^3 of H_2SO_4 98 wt% of/L of the mixture.

Considering an overall consumption of the reactant NB as the sum of the rates giving mDNB, oDNB, and pDNB, the overall kinetic constant 4 (L/mol min) can be partitioned in $\mathbf{K_{mDNB}} = 4 \times 0.93 = 3.72$, $\mathbf{K_{oDNB}} = 4 \times 0.05 = 0.2$, and $\mathbf{K_{pDNB}} = 4 \times 0.02 = 0.08$.

Results

The mass balance for the CSTR reactor (Eq. 4.111) can be solved iteratively by varying the feed flow rate Q in the range of 0–700 L/h. In correspondence with each value of Q, the nonlinear equation system can be solved for unknown concentrations.

The results of the calculations are shown in Fig. 4.37. The left plot shows the expected decrease of nitrobenzene conversion as the feed flow rate is increased then decreasing per residence time. The right plot shows the reactor-outlet concentration of all of the components involved in the reaction system. We can observe that the concentrations of oDNB and pDNB are practically independent of the resident time and are very low because their kinetics of formation is quite low. Moreover, as nitrobenzene conversion decreases, water is formed, and mDNB slightly decreases for lower residence times. The profile of nitric acid changes accordingly.

The described results were obtained using a MATLAB program available as electronic supplementary material, Data 4.1.

Consecutive Reactions

Consecutive reactions are reactions occurring in a sequence, such as in the following scheme:

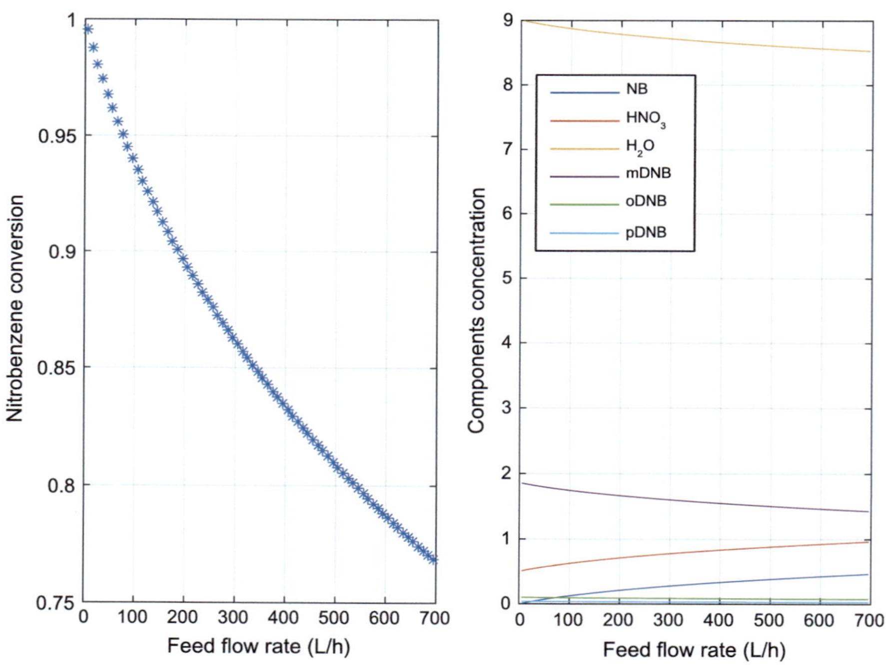

Fig. 4.37 Conversion of nitrobenzene as a function of volumetric feed flow rate (left panel) and the concentration of various components (right panel)

$$A \xrightarrow{k_1} B \xrightarrow{k_2} C \tag{4.112}$$

By assuming for simplicity the first order for all of the occurring reactions, we can write for a batch reactor:

$$r_A = -\frac{\mathrm{d}C_A}{\mathrm{d}t} = k_1 C_A \tag{4.113}$$

$$r_B = \frac{\mathrm{d}C_B}{\mathrm{d}t} = k_1 C_A - k_2 C_B \tag{4.114}$$

$$r_C = \frac{\mathrm{d}C_C}{\mathrm{d}t} = k_2 C_B \tag{4.115}$$

By integrating the first expression, we obtain:

$$C_A = C_A^{\circ} e^{-k_1 t} \tag{4.116}$$

By substituting C_A in the second relation and integrating with respect to C_B, we obtain:

$$C_B = C_{Ao} k_1 \left(\frac{e^{-k_1 t}}{k_2 - k_1} + \frac{e^{-k_2 t}}{k_1 - k_2} \right) \tag{4.117}$$

Moreover, if there is no change in the number of total moles and the initial concentration of B and C is null, we can write:

$$C_A^o = C_A + C_B + C_C \tag{4.118}$$

By plotting the trends of, respectively, A through C, we obtain behavior, such as the one shown in Fig. 4.38.

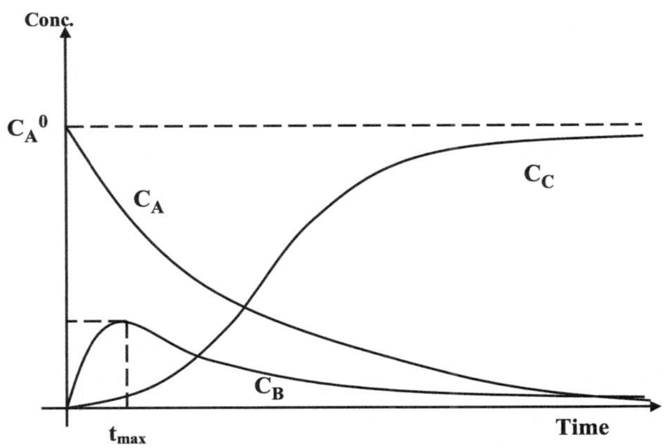

Fig. 4.38 Trends of reactant and products for consecutive reactions

The concentration of B shows a maximum that can be evaluated by equating to zero the derivative dC_B/dt. It can be obtained:

$$t_{\max} = \frac{\ln(k_2/k_1)}{k_2 - k_1} \tag{4.119}$$

In addition, it is possible to write:

$$C_{B_{\max}} = C_{Ao}\left(\frac{k_1}{k_2}\right)^{\frac{k_2}{k_2-k_1}} \tag{4.120}$$

From these two relations, it is possible to evaluate an estimation of the kinetic constants, k_1, and k_2, from the coordinates of the maximum characteristic point. This approach is valid if the consecutive reaction rates are comparable. If one of the reactions is slow, the slow-step hypothesis can be applied to the reaction scheme in a similar way as used for the reaction mechanisms. A plot, such as the one shown in Fig. 4.38, with an intermediate species passing through a maximum, is typical of a consecutive reaction and allows us to recognize this type of reaction. When an intermediate component is promptly transformed into another product, the steady-state approach is often reasonably applied to describe the kinetics.

Exercise 4.15. Schematic Approach to a Competitive–Consecutive Reaction
In a batch reactor, the following generic reactions occur:

$$A + B \rightleftarrows 2C + D \quad r_1 = k_1 C_a C_b - k_{-1} C_c^2 C_d \quad [\mathrm{L/(mol\ \ min)}] \tag{4.121}$$

$$A + E \rightarrow F \quad r_2 = k_2 C_a C_e \quad [\mathrm{L/(mol\ \ min)}] \tag{4.122}$$

The initial concentrations are:

$$\begin{aligned} C_a^{\circ} &= 30 \ \mathrm{mol/L} \\ C_b^{\circ} &= 22 \ \mathrm{mol/L} \\ C_c^{\circ} &= 1.5 \ \mathrm{mol/L} \\ C_d^{\circ} &= 0.2 \ \mathrm{mol/L} \\ C_e^{\circ} &= 12 \ \mathrm{mol/L} \\ C_f^{\circ} &= 0 \ \mathrm{mol/L} \end{aligned}$$

The kinetic constants are:

$$\begin{aligned} k_1 &= 0.020 \\ k_{-1} &= 0.041 \\ k_2 &= 0.011 \end{aligned}$$

The mass balance for each of the six components is:

$$\frac{dC_A}{dt} = -r_1 - r_2 \tag{4.123}$$

$$\frac{dC_B}{dt} = -r_1 \tag{4.124}$$

$$\frac{dC_C}{dt} = +2r_1 \tag{4.125}$$

$$\frac{dC_D}{dt} = +r_1 \tag{4.126}$$

$$\frac{dC_E}{dt} = -r_2 \tag{4.127}$$

$$\frac{dC_F}{dt} = +r_2 \tag{4.128}$$

Solve this ODE system and make a plot describing the evolution with time of the components from 0 to 5 h. In a separate plot, draw the evolution with time of the conversion of the reactant A.

Solution

The integration of the ODE system [Eqs. (4.123) through (4.128)] will yield the results shown in Figs. 4.39 and 4.40.

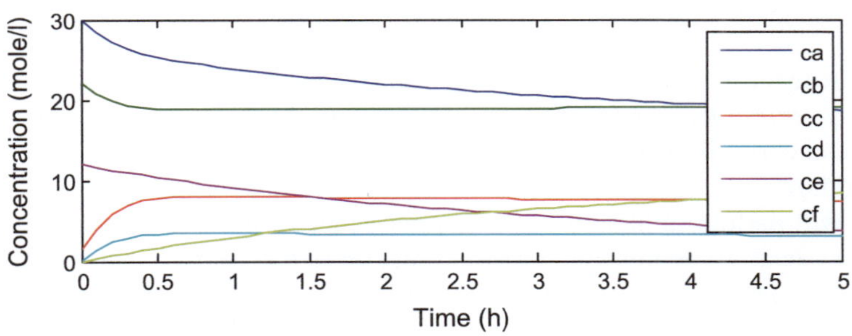

Fig. 4.39 Evolution with time of the concentration of reactants and products

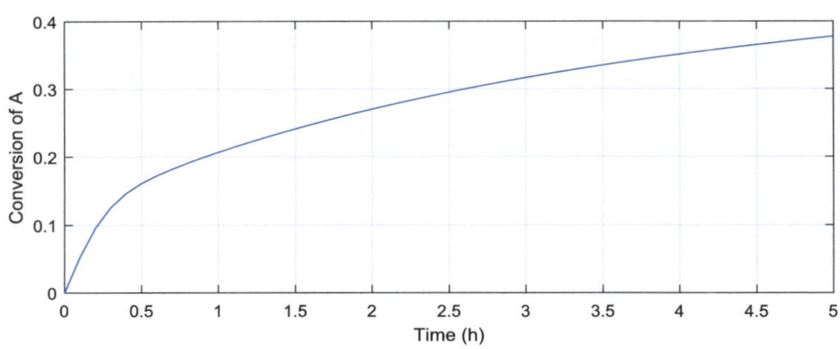

Fig. 4.40 Conversion of the reactant A with time

The described results were obtained using a MATLAB program available as electronic supplementary material, Data 4.1.

Exercise 4.16. Schematic Approach to Consecutive and Competitive–Consecutive Reactions

Part 1

Consider the following reaction scheme:

$$A \rightarrow B \rightarrow C \tag{4.129}$$

The two occurring reactions obey a first-order kinetic law; therefore, we can write:

$$\frac{dC_a}{dt} = -k_1 C_a \tag{4.130}$$

$$\frac{dC_b}{dt} = k_1 C_a - k_2 C_b \tag{4.131}$$

$$\frac{dC_c}{dt} = k_2 C_b \tag{4.132}$$

The kinetic constants and initial concentrations are, respectively:

$$k_1 = 1 \ \text{h}^{-1} \quad k_2 = 2 \ \text{h}^{-1} \quad t = 0 \quad C_a^\circ = 2 \ \text{mol} \quad C_b^\circ = C_c^\circ = 0 \ \text{mol}.$$

Evaluate the evolution with time of the all concentrations, and construct a plot of these concentrations as a function of time. Individuate in this plot the characteristic points and correlate these points with the kinetic constant.

Solution

We will perform a numerical integration of the ordinary differential equations (ODE) system: To do this, a MATLAB program was developed that is available as electronic supplementary material, Data 4.1.

Results

Integration of the ODE system [Eqs. (4.130) through (4.132)] yields the results shown in Fig. 4.41.

Part 2

Consider now another schematic example of competitive–consecutive reactions:

$$A \rightarrow B$$
$$A + A \rightarrow C$$

Assume that the first reaction obeys a first-order kinetic law, while the second reaction is of the second order. Therefore, we can write:

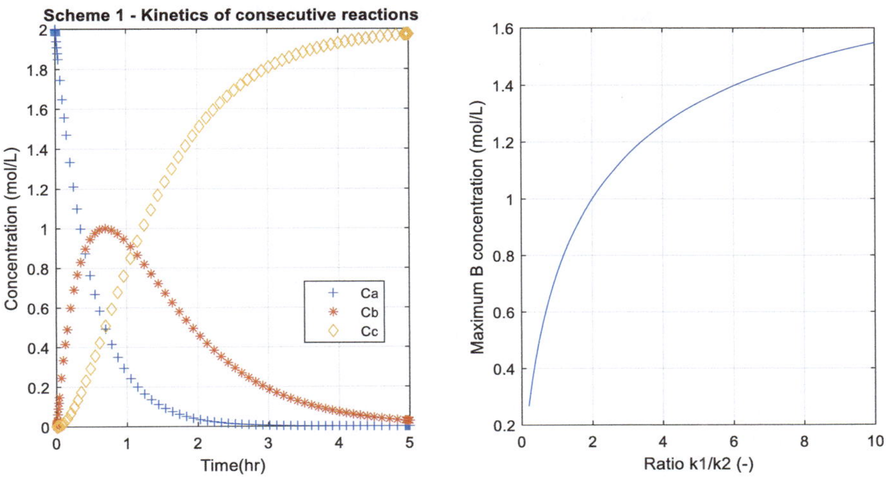

Fig. 4.41 **Left**: Evolution with time of the concentration of reactants and products for consecutive–competitive reactions. **Right**: Maximum concentration of B as a function of kinetic constants ratio

$$\frac{\mathrm{d}C_a}{\mathrm{d}t} = -k_1 C_a - k_2 C_a^2 \tag{4.133}$$

$$\frac{\mathrm{d}C_B}{\mathrm{d}t} = k_1 C_a \tag{4.134}$$

$$\frac{\mathrm{d}C_C}{\mathrm{d}t} = k_2 C_a^2 \tag{4.135}$$

Assume the same values for the following kinetic constants:

$$k_1 = 1\ \mathrm{h}^{-1} \quad k_2 = 2\ \mathrm{h}^{-1} t = 0 \quad C_a^{\circ} = 2\ \mathrm{mol} \quad C_b^{\circ} = C_c^{\circ} = 0\ \mathrm{mol}.$$

Evaluate the evolution with time of all components of the reaction system and construct the related plot.

Solution

The main program remains unchanged with the exception of the MATLAB "called" function. Integration of the ODE system [Eqs. (4.133) through (4.135)] yields the results shown in Fig. 4.42.

The described results were obtained using a MATLAB program available as electronic supplementary material, Data 4.1.

Exercise 4.17. A Practical Example of Competitive–Consecutive Equilibrium Reactions: Hydrocarbon Alkylation and Dimerization

The following alkylation and dimerization reactions occur simultaneously in a CSTR of 100 L kept at 500 K and 2 atm:

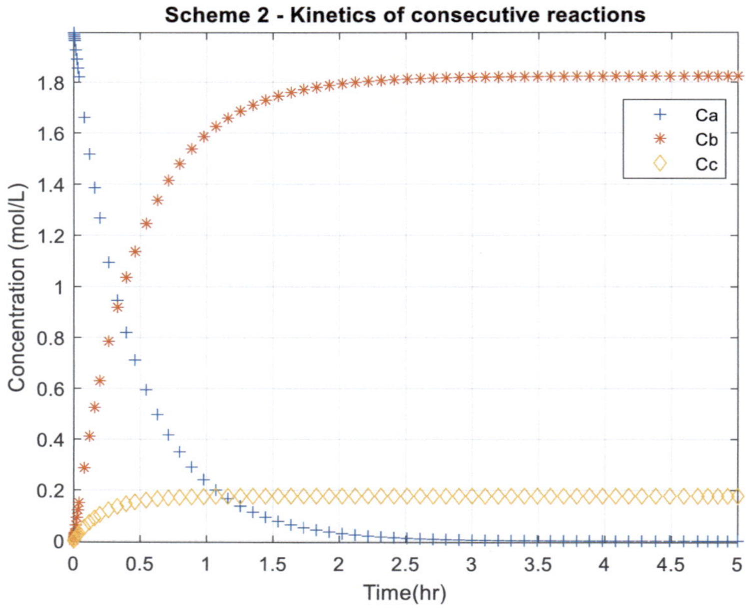

Fig. 4.42 Left: Evolution with time of the concentration of reactants and products for consecutive–competitive reactions

$$(1)\quad C_3H_6 + C_4H_{10} \rightleftarrows C_7H_{16}\quad \text{(dimethylpentane)} \tag{4.136}$$

$$(2)\quad 2C_3H_6 \rightleftarrows C_6H_{12} \tag{4.137}$$

The equilibrium constants are, respectively, $Kp_1 = 34.5$ and $Kp_2 = 7.19$.

Consider in both cases a second-order kinetic law with forward kinetic constants that are, respectively, $k_1 = 2$ and $k_2 = 1$ (L/mol min). Evaluate the following:

(a) The equilibrium composition by assuming an initial composition of 3 mol of C_3H_6 and 1 of C_4H_{10}.
(b) Considering the same initial composition, evaluate the evolution with residence time of both the reactants and products. What is the more convenient flow rate for producing dimethylpentane?

Solution

Assuming that $C_3H_6 = A$, $C_4H_{10} = B$, $C_7H_{16} = C$, and $C_6H_{12} = D$, the rates of the two reactions can be written as:

$$r_1 = k_1 p_A p_B \left(1 - \frac{p_C}{K_{p_1} p_A p_B}\right) \tag{4.138}$$

$$r_2 = k_2 p_A^2 \left(1 - \frac{p_D}{K_{p_2} p_A^2}\right) \tag{4.139}$$

The mass balance related to the four different components is:

$$\frac{QP}{RTV_R}\left(x_i - x_i^o\right) - \sum_{j=1}^{NR} \nu_{ij} r_j \tag{4.140}$$

where x_i is the molar fraction of "i."

The equilibrium composition can be determined by solving the system:

$$f_1(y,z) = K_{p_1} p_A p_B - p_C = 0 \tag{4.141}$$

$$f_2(y,z) = K_{p_2} p_A^2 - p_D = 0 \tag{4.142}$$

$$p_i = P_{tot} x_i \quad x_i = \frac{n_i}{n_{tot}} \tag{4.143}$$

Considering:

y	reacted moles for reaching the equilibrium in reaction 1
z	reacted moles for reaching the equilibrium in reaction 2

we have:

$$n_A = n_A^\circ - y - 2z \quad n_B = n_B^\circ - y \quad n_C = y \quad n_D = z \tag{4.144}$$

$$n_{tot} = n_A^\circ + n_B^\circ - y - z \tag{4.145}$$

Therefore,

$$f_1(y,z) = K_{p_1} p_A p_B - p_C = K_{p_1} P_{tot}^2 \frac{\left(n_A^o - y - 2z\right)\left(n_B^o - y\right)}{\left(n_A^o + n_B^o - y - z\right)^2} - P_{tot} \frac{y}{\left(n_A^o + n_B^o - y - z\right)} = 0 \tag{4.146}$$

$$f_2(y,z) = K_{p_2} p_A^2 - p_D = K_{p_2} P_{tot}^2 \frac{\left(n_A^o - y - 2z\right)^2}{\left(n_A^o + n_B^o - y - z\right)^2} - P_{tot} \frac{z}{\left(n_A^o + n_B^o - y - z\right)} = 0 \tag{4.147}$$

Results

By solving the system of two equations corresponding to different times, the plot of Fig. 4.43 can be obtained.

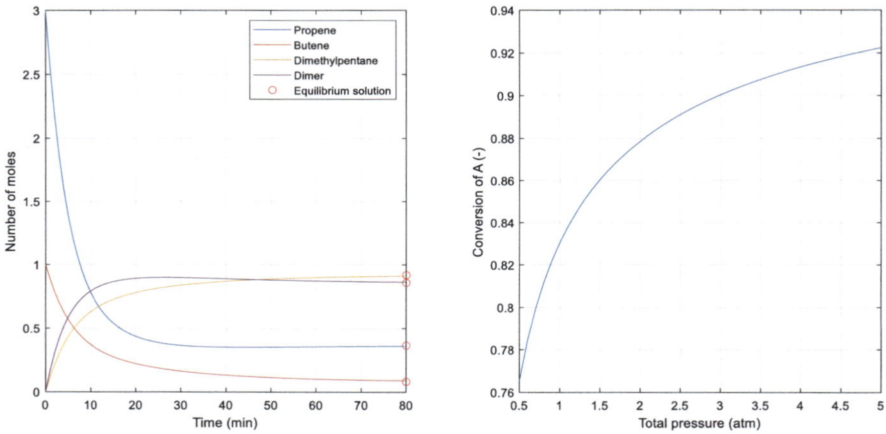

Fig. 4.43 **Left:** Evolution of the composition with residence time. **Right**: Conversion of A as a function of the total pressure

The described results were obtained using a MATLAB program available as supplementary information electronic supplementary material, Data 4.1.

4.3.7 *Complex Reaction Scheme: A Unified Approach*

In many cases, one of the problems with studying the kinetics of a reaction is to identify a correct reaction scheme, in particular, when this scheme is complex. Normally, we know what we enter in a reactor and can evaluate by analysis what comes out. However, often it is not easy to understand how many reactions occur, that is, to know the actual scheme of the reaction. However, it is useful to remember that, independently of the complexity of the reaction scheme, this can be considered as a combination of single reactions or a combination of less complex schemes, such as the ones previously described (equilibrium, competitive, and consecutive reactions). Moreover, any single reaction appearing in the scheme occurs through a mechanism characterized by a sequence of elementary steps, and the mechanism is the determinant for giving a reliable expression of the rate law. However, to face the kinetic description of a complex reaction scheme, a general unified approach is possible. Let us consider, for example, the following scheme:

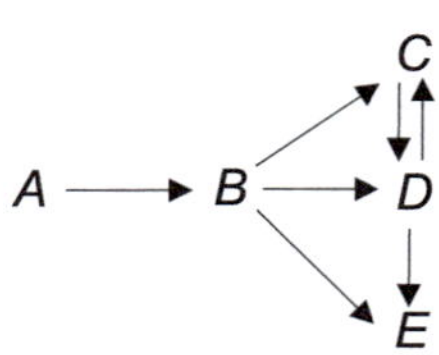

(4.148)

As can be seen, this scheme can be considered, as previously mentioned, a combination of equilibrium, competitive, and consecutive reactions or a combination of seven different single reactions. It is first important to define the stoichiometry of all of the seven reactions by writing in a more general way:

$$\begin{aligned} &\alpha_{11}A_1 + \alpha_{21}A_2 \ldots\ldots\ldots + \alpha_{n1}A_n = 0 \\ &a_{12}A_1 + \alpha_{22}A_2 \ldots\ldots\ldots + \alpha_{n2}A_n = 0 \\ &\ldots\ldots\ldots\ldots\ldots\ldots\ldots\ldots\ldots\ldots\ldots\ldots \\ &\alpha_{1m}A_1 + \alpha_{2m}A_2 \ldots\ldots\ldots + \alpha_{nm}A_n = 0 \end{aligned} \tag{4.149}$$

where α_{ij} are the stoichiometric coefficients assumed to be negative for the reagents and positive for the products. It is then possible to write the stoichiometric matrix:

$$\alpha_{ij} = \begin{vmatrix} \alpha_{11} & \alpha_{21} & \ldots & \alpha_{n1} \\ \alpha_{12} & \alpha_{22} & \ldots & \alpha_{n2} \\ \ldots & \ldots & \ldots & \ldots \\ \alpha_{1m} & \alpha_{2m} & \ldots & \alpha_{nm} \end{vmatrix} \tag{4.150}$$

Defining r_{ij} as the rate of i formation in the reaction, j, we can write for each reaction:

$$r_j = \frac{r_{ij}}{\alpha_{ij}} = \frac{r_{2j}}{\alpha_{2j}} = \ldots\ldots \frac{r_{ij}}{\alpha_{ij}} \quad \text{where } j = 1, 2, \ldots m \tag{4.151}$$

The overall rate of A_i formation can be obtained as the algebraic sum of the rates of A_i formation in each reaction in which A_i is involved, obviously also considering the stoichiometric coefficient; therefore,

$$r_i = \sum_{j=1}^{m} \alpha_{ij} r_j \tag{4.152}$$

After this analysis, it remains just a problem to be solved, that is, define the rate laws corresponding to each r_j.

This approach can more easily be understood with an example. According to Santacesaria et al. (1990a, b), methanol homologation occurs through the following reactions:

$$(1) \quad CH_3OH + CO + H_2 \xrightarrow{r_1} CH_3CHO + H_2O \tag{4.153}$$

$$(2) \quad CH_3CHO + H_2 \xrightarrow{r_2} CH_3CH_2OH \tag{4.154}$$

$$(3)\quad CH_3CHO + 2CH_3OH \overset{r_3}{\rightleftarrows} CH_3CH(OCH_3)_2 + H_2O \tag{4.155}$$

$$(4)\quad 2CH_3OH + CO \xrightarrow{r_4} CH_3COOCH_3 + H_2O \tag{4.156}$$

Assuming:

$$\mathbf{1-CH_3OH,\ 2-CH_3CHO,\ 3-CH_3CH_2OH,\ 4-CH_3CH(OCH_3)_2,}$$
$$\mathbf{5-CH_3COOCH_3,\ 6-H_2O,\ 7-CO,\ 8-H_2}$$

It is easy to write the stoichiometric matrix (see Table 4.21).

Negative terms correspond to reacting species while positive terms are related to products formed. Therefore, in the case of a batch reactor, we can write the following kinetic model, which is able to describe the evolution with time of any single component:

$$\begin{aligned}
&\frac{dn_1}{dt} = (-r_1 - 2r_3 - 2r_4)V_L && \frac{dn_5}{dt} = r_4 V_L\\
&\frac{dn_2}{dt} = (r_1 - r_2 - r_3)V_L && \frac{dn_6}{dt} = (r_1 + r_3 + r_4)V_L\\
&\frac{dn_3}{dt} = r_2 V_L && \frac{dn_7}{dt} = (-r_1 - r_4)V_L\\
&\frac{dn_4}{dt} = r_3 V_L && \frac{dn_8}{dt} = (-r_1 - r_2)V_L
\end{aligned} \tag{4.157}$$

where n_i are the number of moles of the i components at time t and V_L is the liquid volume in which the reactions occur. Clearly, the balance equations will be different for a continuous reactor, but the right side (containing the reaction rates from r_1 to r_4) will appear in any case in the balance. For solving the system of differential equations (Eq. 4.157), we need the four mathematical expressions related to the four rate laws—$r_j = f$ (composition, temperature)—that were determined by interpreting the experimental data.

Finally, it must be pointed out that when the results are difficult to use to individuate the effective reaction scheme, a valuable approach is to identify the thermodynamically independent reactions in the manner described in Chap. 2 (see paragraph 2.2.8) and to study the kinetics of all of these reactions.

Table 4.21 Stochiometric matrix for Eqs. (4.153) through (4.156)

	Components $i \rightarrow$							
Reactions $j \downarrow$	1	2	3	4	5	6	7	8
1	−1	1	0	0	0	1	−1	−1
2	0	−1	1	0	0	0	0	−1
3	−2	−1	0	1	0	1	0	0
4	−2	0	0	0	1	1	−1	0

4.4 Description of the Reaction Mechanisms and Their Relation with Kinetics

4.4.1 Heterolytic Mechanisms and Kinetics

Four main different mechanisms can be classified in organic chemistry:

(a) Substitution { Nucleophilic (S_N1, S_N2); Electrophilic }

(b) Addition { Nucleophilic; Electrophilic }

(c) Elimination { Nucleophilic; Electrophilic }

Nucleophilic or electrophilic indicates the type of chemical attack initiating the mechanism.

(d) Transposition

Let us consider briefly for each case the effect of the mechanism on the kinetic law expression.

4.4.2 Nucleophilic Substitutions of Type S_N1

These are reactions of the type:

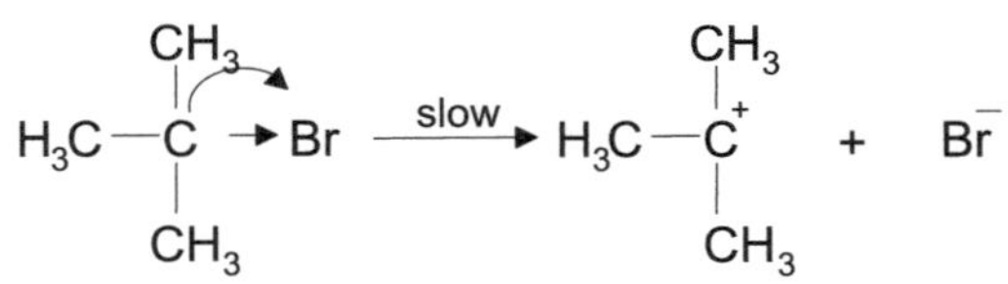

(4.158)

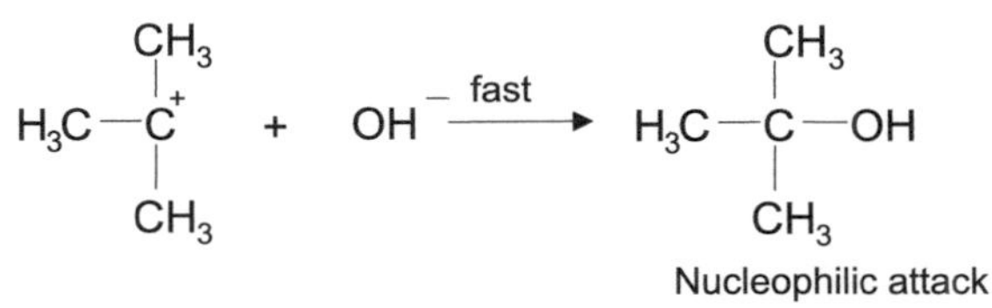

(4.159)

where Br is substituted by OH as a consequence of a nucleophilic attack occurring with first-order kinetics. Considering the slow elementary step (4.158), we can write:

$$r = k_1 [\text{R Br}] \tag{4.160}$$

4.4.3 Nucleophilic Substitutions of Type S_N2

The same reaction can occur with a different mechanism as follows:

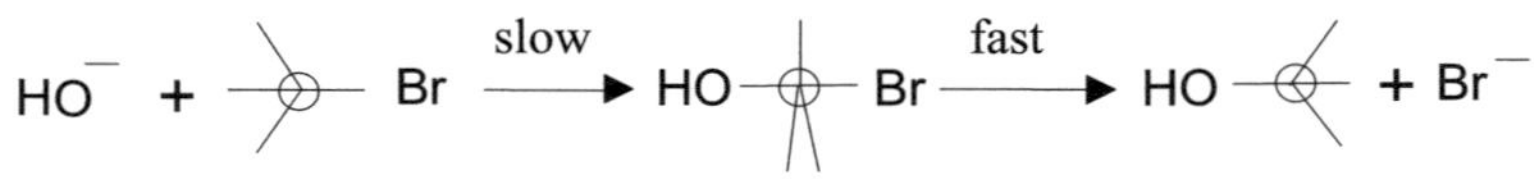

(4.161)

The kinetics, in this case, is of the second order:

$$r = k_2[\mathrm{R\,Br}][\mathrm{OH}^-] \tag{4.162}$$

It is interesting to observe that in this case, we have an inversion in the carbon-bond configuration (see reaction 4.161). By using, for the reaction, an asymmetric carbon bond, it is possible to show the occurrence of the inversion by stereochemistry.

4.4.4 Substitution with Electrophilic Attack

This mechanism occurs in reactions related to the substitution of hydrogens of the aromatic rings, such as nitration, sulfonation, and alkylation. In the case of nitration, for example, we have:

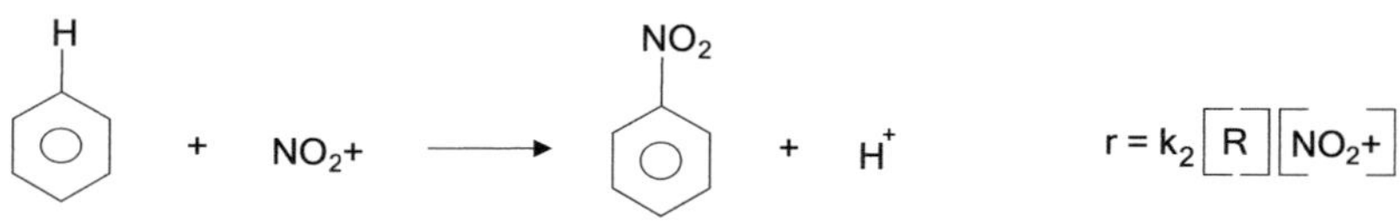

(4.163)

An accredited mechanism is the following:

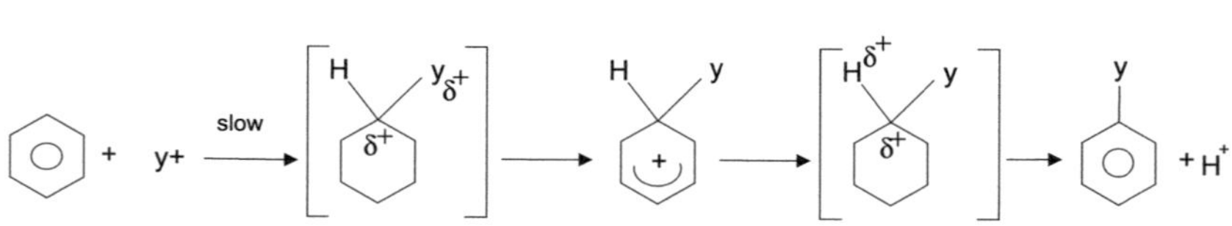

I1 I2 I3

(4.164)

It is obvious that the experimentally ascertained kinetic law is not suitable to define so complex a mechanism and other sophisticated techniques are thus necessary; nevertheless, the kinetics must be in agreement with the mechanism.

4.4.5 *Nucleophilic Additions*

An example of nucleophilic addition is the synthesis of cyanohydrins from carbonyl compounds:

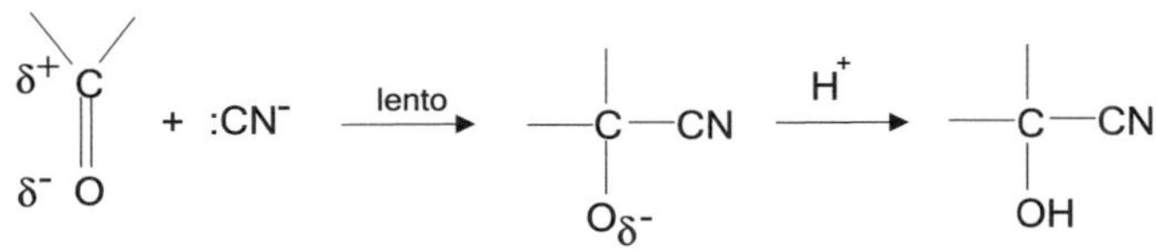

(4.165)

The reaction rate is:

$$r = k_2[\mathrm{R}][\mathrm{CN}^-] \tag{4.166}$$

4.4.6 *Electrophilic Additions*

These are a typical reaction of alkenes with acids:

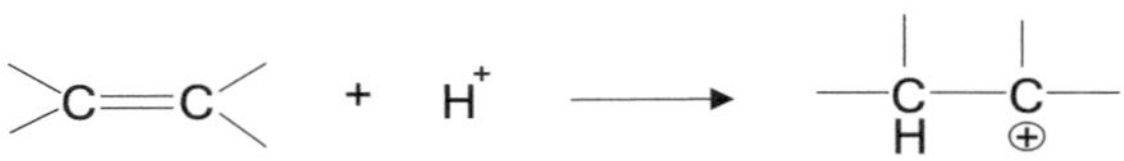

(4.167)

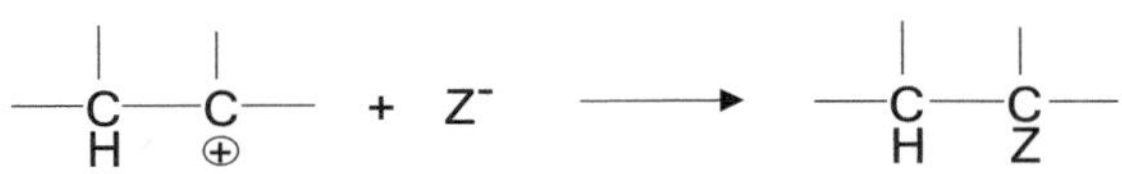

(4.168)

The reaction rate depends mainly on the carbocation stability.

4.4.7 *Nucleophilic Eliminations—E$_1$ (Monomolecular)*

These are reactions inverse to the elimination reactions and often are in competition with the S_N1 mechanism. For example:

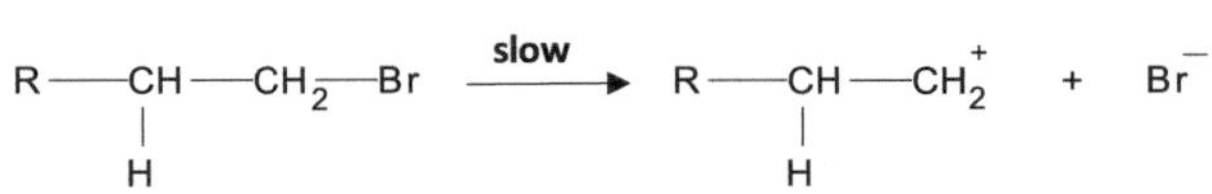

(4.169)

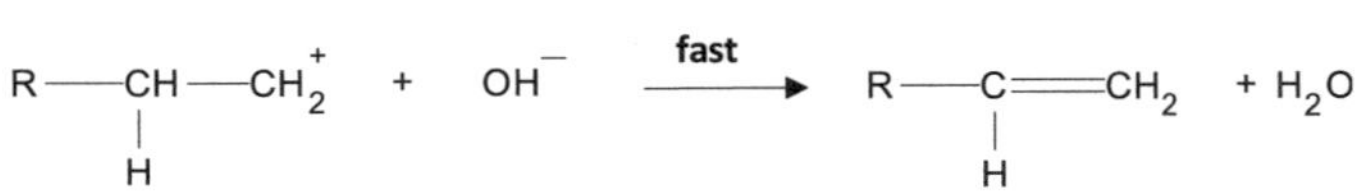

(4.170)

$$r = k_1[\mathrm{R}] \tag{4.171}$$

4.4.8 Nucleophilic Eliminations—E_2 (Bimolecular)

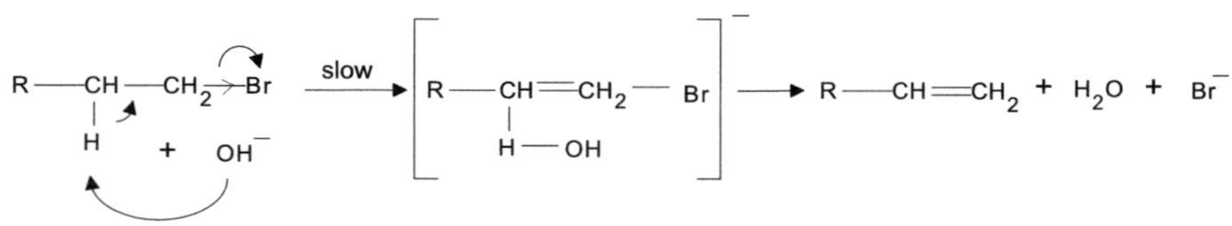

(4.172)

The kinetics are:

$$r = k_2[\mathrm{R}][\mathrm{OH}^-] \tag{4.173}$$

4.4.9 Electrophilic Eliminations

$$\mathrm{CH_3\text{-}CH_2OH} + \mathrm{H^+} \leftrightarrow \mathrm{CH_3CH_2\text{-----}}\underset{\mathrm{H}}{\underset{|}{\overset{\oplus}{\mathrm{OH}}}} \leftrightarrow \mathrm{CH_3CH_2}^{\oplus} + \mathrm{H_2O} \tag{4.174}$$

$$\mathrm{CH_3\text{-}CH_2}^{\oplus} \leftrightarrow \mathrm{CH_2} = \mathrm{CH_2} + \mathrm{H^+} \tag{4.175}$$

$$\mathrm{CH_3\text{-}CH_2}^{\oplus} + \mathrm{C_2H_5OH} \leftrightarrow \mathrm{C_2H_5\text{-}}\underset{\mathrm{H}}{\underset{|}{\overset{\oplus}{\mathrm{O}}}}\mathrm{\text{-}C_2H_5} \tag{4.176}$$

$$\mathrm{C_2H_5} - \underset{\mathrm{H}}{\underset{|}{\overset{\oplus}{\mathrm{O}}}} - \mathrm{C_2H_5} \leftrightarrow \mathrm{C_2H_5OC_2H_5} + \mathrm{H^+} \tag{4.177}$$

The rates are as follows:

$$r_{\text{olef}} = k_{\text{olef}}[C_2H_5OH][H^+] \tag{4.178}$$

$$r_{\text{alcohol}} = k_{\text{alcohol}}[C_2H_5OH][H^+] \tag{4.179}$$

4.4.10 *Molecular Rearrangement*

We can have in a molecule the shift of an electric charge, a functional group, or a bond. An example of a chemical bond shift is the following:

$$\underset{\displaystyle \overset{|}{X}}{CH_3\text{-}CH}\text{-}CH{=}CH_2 \rightarrow CH_3\text{-}CH{=}CH\text{-}CH_2X \tag{4.180}$$

The kinetic rate law will depend, as always, on the reaction mechanism.

Last, one example of a complex mechanism for a homogeneous reaction occurring in the presence of a transition metal complex is propene hydroformylation:

$$CH_3\text{-}CH{=}CH_2 + H_2 + CO \rightarrow CH_3CH_2CH_2CHO \tag{4.181}$$

catalyzed by a rhodium complex.

4.4.11 *Description of Catalytic Cycles*

An impressive way to represent the catalytic cycle is the following one:

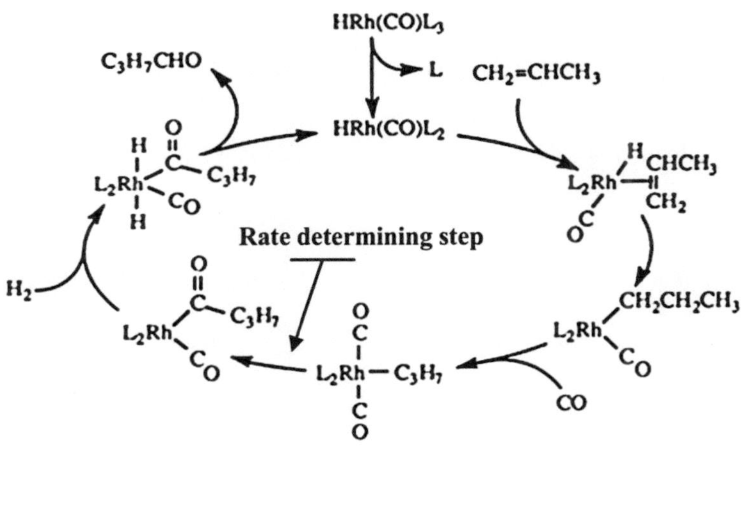

As can be seen, all of the elementary reaction steps are reported in a synthetic and clear way, and the RDS affecting the reaction rate is individuated.

The mechanism of methanol carbonylation, always in the presence of a rhodium complex, is reported below:

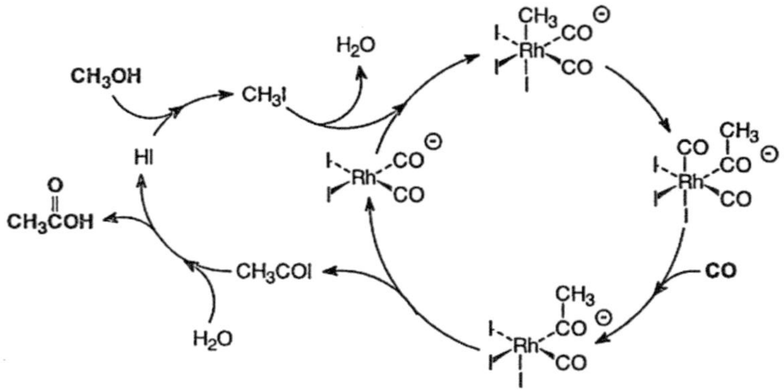

As mentioned in the previous chapter, iodine favors the oxidative addition of CH_3I, thus allowing the formation of a direct bond between the methyl group and the metal.

4.4.12 *The Wacker Process: An Example of Heterolytic Redox Reaction*

An interesting reaction is that of the heterolytic oxidation of ethylene to acetaldehyde (Wacker process) occurring in the presence of palladium:

$$Pd^{++} + C_2H_4 + H_2O \rightarrow Pd^{o} + CH_3CHO + 2H^{+} \quad (4.182)$$

Then palladium is oxidized with copper chloride (Cu^{++}):

$$Pd^{o} + 2Cu^{++} \rightarrow Pd^{++} + 2Cu^{+} \quad (4.183)$$

and copper (Cu^{+}) is, at last, reoxidized with oxygen, thus closing the catalytic cycle:

$$4Cu^{+} + O_2 + 4H^{+} \rightarrow 4Cu^{++} + 2H_2O \quad (4.184)$$

The experimental kinetic data fit quite well with the following kinetic expression:

$$r = k_1 \frac{\left[PdCl_4^{=}\right][C_2H_4]}{[Cl^-][H_3O^+]} + k_2 \frac{\left[PdCl_4^{=}\right][C_2H_4]}{[Cl^-]^3[H_3O^+]} \tag{4.185}$$

When $\left[PdCl_4^{=}\right] < 0.04$ M, the second term becomes negligible.
The most accredited mechanism for this reaction is the following:

$$\left[PdCl_4^{=}\right] + C_2H_4 \overset{\text{fast}}{\rightleftharpoons} [PdCl_3C_2H_4]^- + Cl^- \tag{4.186}$$

$$[PdCl_3C_2H_4]^- + H_2O \overset{\text{fast}}{\rightleftharpoons} [PdCl_2(H_2O)C_2H_4] + Cl^- \tag{4.187}$$

$$[PdCl_2(H_2O)C_2H_4] + H_4O \overset{\text{fast}}{\rightleftharpoons} [PdCl_2(OH)C_2H_4]^- + H_3O^+ \tag{4.188}$$

$$[PdCl_2(OH)C_2H_4]^- \xrightarrow{\text{slow}} [HOCH_2CH_2PdCl] + Cl^- \tag{4.189}$$

This last reaction corresponds to the insertion of ethylene between the metal and the hydroxyl changing the bond with the metal from π to σ bond. At last:

$$HOCH_2CH_2PdCl \xrightarrow{\text{fast}} CH_3CHO + Pd + HCl \tag{4.190}$$

If we apply the slow-stage hypothesis and assume reaction (4.189) as an RDS and all of the other steps at equilibrium, a kinetic law equal to the first term of Eq. (4.185) can be derived showing very good agreement between the experimentally determined kinetic law and that obtained from the described mechanism. If the concentration of $[PdCl_4^{=}] > 0.04$ M, the mechanism is complicated by the intervention of a dimer of the palladium complex. The described mechanism can be represented in a synthetic and impressive way through the following catalytic cycle:

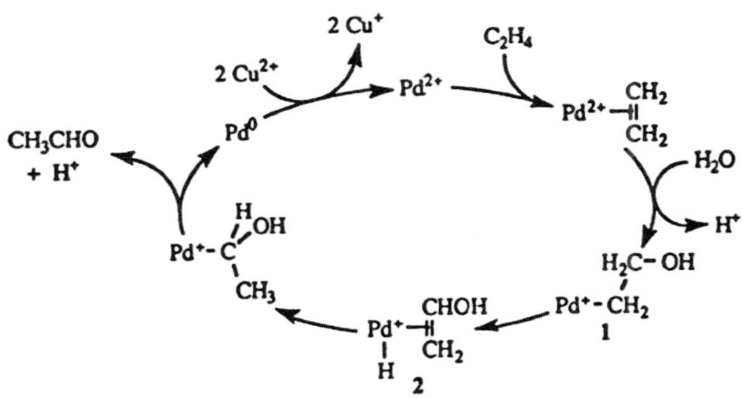

4.4.13 Kinetics of Reaction Catalyzed by Acid–Base

When the catalytic action is due to H+ or OH−, we can call this “specific acid–base catalysis.” When, on the contrary, the catalytic action is promoted either by Brönsted acid–base or by Lewis acid–base, the catalytic action is called “general acid–base catalysis.” Usually, the catalytic mechanism starts with an attack by an acid or a base on the organic substrate; however, before this occurs, the solvent gives place to protonation equilibrium:

$$HA + B \rightleftarrows A^- + BH^+ \tag{4.191}$$

The protonation of the substrate then occurs as follows:

$$BH^+ + S \underset{k_{-1}}{\overset{k_1}{\rightleftarrows}} SH^+ + B \tag{4.192}$$

At last, the protonated substrate reacts with another basis, B_1, or also with the solvent yielding the product P:

$$SH^+ + B_1 \xrightarrow{k_2} P + B_1H^+ \tag{4.193}$$

The substrate can also be attacked by the undissociated acid or base:

$$S + HA \underset{k_{-1}}{\overset{k_1}{\rightleftarrows}} SH^+ + A^- \tag{4.194}$$

The product formation can occur by de-protonation due to both the solvent or by the conjugated base A−. Considering water as the solvent, we have:

$$SH^+ + A^- \xrightarrow{k_2} P + HA \quad \text{Prototrophic mechanism} \tag{4.195}$$

$$SH^+ + H_2O \xrightarrow{k_2} P + H_3O^+ \quad \text{Protollithic mechanism} \tag{4.196}$$

As we have seen, many different species can promote acid–base catalysis, and we must consider different contributions to the reaction rate when these are significant. Because normally the acid and base catalytic species intervene with a first-order law in the kinetic expression, in the most general case, we can write:

$$r = \left\{k_{SH}[SH^+] + k_{HA}[HA] + k_{A^-}[A^-] + k_{H_3O^+}[H_3O^+] + k_{OH^-}[OH^-]\right\} \times [S] \tag{4.197}$$

Obviously, different terms reported in this relation can normally be neglected in a relatively narrow range of acidity, and by passing from acid to basic conditions, the terms that must be considered change.

4.4.14 *Radical Chain Reactions: Homolytic Mechanisms and Related Kinetics*

The radical chain reactions, more so than other reactions, show how important is to consider the kinetic aspects at the light of the reaction mechanism. Take, for example, the thermal decomposition of ethane to ethylene and hydrogen:

$$C_2H_6 \rightarrow C_2H_4 + H_2 \tag{4.198}$$

From experimental observation, we have a very simple first-order kinetic law:

$$r = k[C_2H_6] \tag{4.199}$$

A wrong conclusion could be that this reaction is simply a unimolecular decomposition. In fact, it is possible to show the formation of radicals during the reaction, and thus the mechanism is much more complicated than that of unimolecular decomposition occurring in different steps as follows:

$$C_2H_6 \xrightarrow{1} 2CH_3\cdot \quad \text{Initiation} \tag{4.200}$$

$$CH_3\cdot + C_2H_6 \xrightarrow{2} CH_4 + C_2H_5\cdot \quad \text{Propagation} \tag{4.201}$$

$$C_2H_5\cdot \xrightarrow{3} C_2H_4 + H\cdot \tag{4.202}$$

$$H\cdot + C_2H_6 \xrightarrow{4} C_2H_5\cdot + H_2 \tag{4.203}$$

$$H\cdot + C_2H_5\cdot \xrightarrow{5} C_2H_6 \quad \text{Termination} \tag{4.204}$$

By assuming the steady-state condition for all of the involved unstable radicals—$H\cdot$, $CH_3\cdot$ and $C_2H_5\cdot$—we obtain:

$$k_3[C_2H_5\cdot] - k_4[H\cdot] - k_5[H\cdot][C_2H_5\cdot] = 0 \tag{4.205}$$

$$2k_1[C_2H_6] - k_2[CH_3\cdot][C_2H_6] = 0 \tag{4.206}$$

$$k_2[CH_3\cdot][C_2H_6] - k_3[C_2H_5\cdot] + k_4[H\cdot][C_2H_6] - k_5[H\cdot][C_2H_5\cdot] = 0 \tag{4.207}$$

Remembering that the rate of C_2H_4 formation is:

$$r = r_3 = k_3[C_2H_5\cdot] \tag{4.208}$$

it is sufficient to evaluate the concentration of $[C_2H_5\cdot]$ from the previous three equations [Eqs. (4.205) through (4.207)]:

$$[C_2H_5\cdot] = \left\{ \frac{k_1}{2k_3} \pm \left[\left(\frac{k_1}{2k_3} \right)^2 + \frac{k_1 k_4}{k_3 k_5} \right]^{1/2} \right\} [C_2H_6] \tag{4.209}$$

Then, if $k_1 \ll k_3$, we obtain:

$$r = \left[\frac{k_1 k_3 k_4}{k_5} \right]^{1/2} [C_2H_6] \tag{4.210}$$

As can be seen, a very simple first-order kinetic law corresponds to a complex mechanism.

As previously mentioned, radical reactions can be distinguished from other reactions by a bond rupture occurring according to the following scheme:

Heterolytic breakage	Homolytic breakage	
$A—B \rightarrow A^{(-)} + B^{(+)}$	$A—B \rightarrow A\cdot + B\cdot$	(4.211)

After the homolytic breakage, we can have simple chain radical reactions, such as the one occurring in the synthesis of HBr:

$$Br_2 \rightleftarrows 2Br\cdot \tag{4.212}$$

$$Br\cdot + H_2 \rightleftarrows HBr + H\cdot \tag{4.213}$$

$$H\cdot + Br_2 \rightleftarrows HBr + H\cdot \tag{4.214}$$

where for each radical reacted, another is formed. Alternatively, we can have branched chain radical reactions when, for each reacted radical, two or more radicals are formed, while degenerate branched chain reactions are reactions in which intermediate compounds decompose, thus giving radicals. An example of a branched chain reaction is the one occurring between hydrogen and oxygen because of a spark:

$$H_2 + O_2 \rightarrow HO_2\cdot + H\cdot \quad \text{Initiation} \tag{4.215}$$

$$H\cdot + O_2 \rightarrow OH\cdot + O\cdot \quad \text{Branching} \tag{4.216}$$

$$O\cdot + H_2 \rightarrow OH\cdot + H\cdot \tag{4.217}$$

$$OH\cdot + H_2 \rightarrow H_2O + H\cdot \quad \text{etc.} \tag{4.218}$$

This scheme can be simplified by imagining one only radical supporting the chain:

$$\frac{1}{2}\mathrm{R}_2 + \cdots \rightarrow \mathrm{R.} + \cdots \quad \text{Initiation} \quad r_\mathrm{I} = f_\mathrm{I} \tag{4.219}$$

$$\mathrm{R}\cdot + \cdots \rightarrow \mathrm{R}\cdot \quad \text{Propagation} \quad r_\mathrm{II} = \alpha f_\mathrm{II}[\mathrm{R}\cdot] \tag{4.220}$$

$$\mathrm{R}\cdot + \cdots \rightarrow \quad \text{Products} \quad \text{Propagation} \quad r_\mathrm{III} = f_\mathrm{III}[\mathrm{R}\cdot] \tag{4.221}$$

$$\mathrm{R}\cdot + \text{wall} \rightarrow \quad 1/2\ \mathrm{R}_2 \quad \text{Termination} \quad r_\mathrm{IV} = P_p[\mathrm{R}\cdot] \tag{4.222}$$

$$\mathrm{R}\cdot + \text{gas} \rightarrow \quad 1/2\ \mathrm{R}_2 \quad \text{Termination} \quad r_\mathrm{V} = P_g[\mathrm{R}\cdot] \tag{4.223}$$

If $\alpha = 1$, the chain is linear; if $\alpha > 1$, the chain is branched.

By applying to [R·] the steady-state approximation, we can write:

$$\frac{\mathrm{d}[\mathrm{R}\cdot]}{\mathrm{d}t} = f_\mathrm{I} + \alpha f_\mathrm{III}[\mathrm{R}\cdot] - f_\mathrm{II}[\mathrm{R}\cdot] - P_p[\mathrm{R}\cdot] - P_g[\mathrm{R}\cdot] = 0 \tag{4.224}$$

By assuming $f_\mathrm{II} \simeq f_\mathrm{III}$, we have:

$$[\mathrm{R}\cdot] = \frac{f_\mathrm{I}}{P_p + P_g - f_\mathrm{II}(\alpha - 1)} \tag{4.225}$$

The overall reaction rate is:

$$\frac{\mathrm{d}X}{\mathrm{d}t} = \frac{f_\mathrm{I} f_\mathrm{III}}{P_p + P_g - f_\mathrm{II}(\alpha - 1)} \tag{4.226}$$

When $P_g + P_p \simeq f_\mathrm{II}(\alpha - 1)$, it results in $\frac{\mathrm{d}X}{\mathrm{d}t} \rightarrow \infty$, that is, the reaction becomes explosive.

The example shown in Fig. 4.44 shows how important is to identify a reliable reaction mechanism as a base of the kinetic expression. A power low correctly interpreting *A–B* and *C–D* segments of the curve is not able to represent the explosive behavior occurring between *B* and *C*.

The degenerate branching chain reaction is a particular radical branched chain reaction in which chain branching is a consequence of radical formation for the decomposition of an intermediate product, for example, a peroxide. An example of such type of reaction is hydrocarbons autoxidation, a reaction used in industry to produce acetic acid from hydrocarbons and many other products always by autoxidation. The reaction scheme, in this case, is the following one:

$$\text{Initiation} \quad \mathrm{In}_2 \xrightarrow{\mathrm{R}_i} \mathrm{In}\cdot \tag{4.227}$$

$$\mathrm{In}\cdot + \mathrm{RH} \rightarrow \mathrm{R}\cdot + \mathrm{InH} \tag{4.228}$$

Fig. 4.44 Kinetics of a branched chain reaction

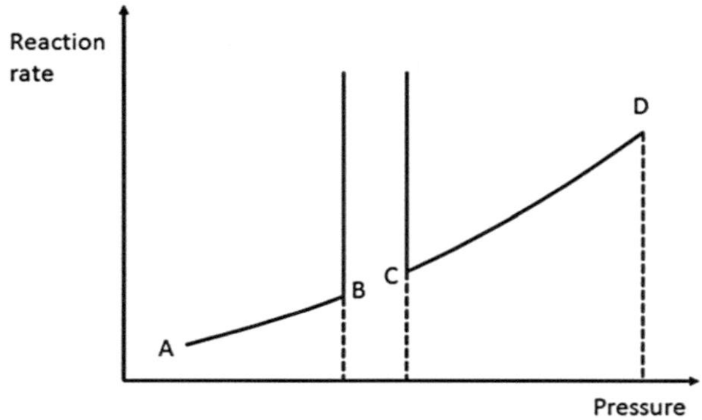

$$\text{Propagation} \quad R\cdot + O_2 \xrightarrow{\text{very fast}} RO_2\cdot \tag{4.229}$$

$$RO_2\cdot + RH \xrightarrow{k_p} ROOH + R\cdot \tag{4.230}$$

$$\text{Branching} \quad 2ROOH \xrightarrow{k_{d_2}} RO\cdot + RO_2\cdot + H_2O \tag{4.231}$$

$$ROOH \xrightarrow{k_{d_1}} RO\cdot + HO\cdot \tag{4.232}$$

$$\text{Termination} \quad 2R\cdot \rightarrow \text{inert products} \tag{4.233}$$

$$R\cdot + RO_2\cdot \rightarrow \text{inert products} \tag{4.234}$$

$$2RO_2\cdot \xrightarrow{k_t} \text{inert products} \tag{4.235}$$

The kinetic behavior of the system can be followed by considering the evolution with time of the organic substrate, the oxygen consumption, or the product formation. This behavior is normally characterized by an induction period or acceleratory phase, a steady-state phase characterized by the maximum reaction rate, and a deceleration phase. The kinetic behavior is similar to that of the autocatalytic systems as can be seen in Fig. 4.45. The induction period corresponds to initial radical formation; under steady-state conditions, the rate of radical formation is equal to the rate of their disappearance by termination. At last, termination becomes predominant, and the reaction rate decreases. To study the kinetics, it is opportune, in this case, to refer only to the steady-state conditions, that is, to refer to the maximum rate corresponding to the slope of the flex point, such as that shown in Fig. 4.45.

The radical $RO_2\cdot$ is predominant with respect to the other radical species because the reaction $R\cdot + O_2$ is very fast. At this point, two limit conditions can be considered. The first condition corresponds to the experimental conditions in which the formed hydroperoxide ROOH is stable and becomes the main reaction product. We have in this case a simple chain radical reaction. Under some other

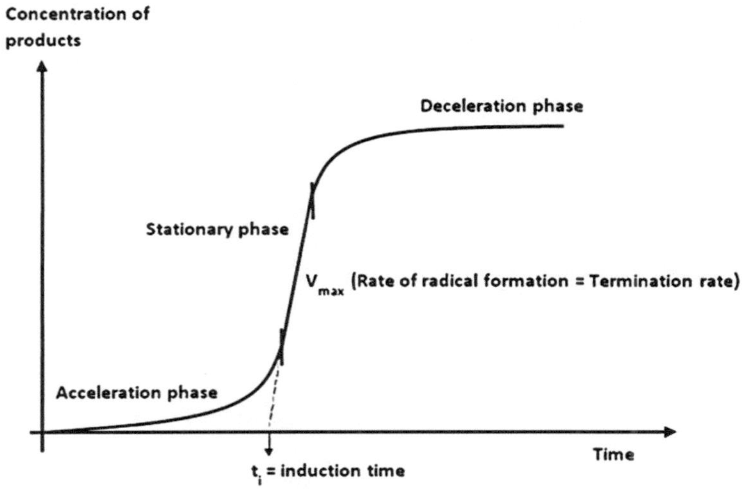

Fig. 4.45 Kinetic behavior of degenerate branching chain radical reaction in a batch reactor

conditions, for example, at high temperatures or in the presence of catalysts, hydroperoxide decomposes with mono or bimolecular stoichiometry, thus yielding radicals. The maximum reaction rate, in the first case, can easily be derived from the reaction mechanism and in batch conditions is:

$$-\frac{\mathrm{d}[\mathrm{O}_2]}{\mathrm{d}t} = k_o[\mathrm{R}\cdot][\mathrm{O}_2] - k_t[\mathrm{RO}_2\cdot]^2 \tag{4.236}$$

If the pressure of oxygen is high enough, we can write:

$$k_o[\mathrm{R}\cdot][\mathrm{O}_2] = k_p[\mathrm{RO}_2\cdot][\mathrm{RH}] + \mathrm{R}_i \tag{4.237}$$

and also:

$$\mathrm{R}_i = 2\ k_t[\mathrm{RO}_2\cdot]^2 \tag{4.238}$$

By combining the last relations, we obtain for the maximum rate of the simple radical chain reaction the following:

$$-\frac{\mathrm{d}[\mathrm{O}_2]}{\mathrm{d}t} = k_p\left(\frac{\mathrm{R}_i}{2k_t}\right)^{1/2}[\mathrm{RH}] + \frac{\mathrm{R}_i}{2} \tag{4.239}$$

Normally, R_i is very low, and the term $\mathrm{R}_i/2$ can be neglected; therefore, it results in:

$$-\frac{d[RH]}{dt}=-\frac{d[O_2]}{dt}=k_p\left(\frac{R_i}{2k_t}\right)^{1/2}[RH] \tag{4.240}$$

the validity of which has been tested in many cases. The ratio $k_p/\sqrt{2k_t}$ is called the "autoxidability of the substrate" and is a characteristic of the organic reagent used.

If the hydroperoxide is decomposed, yielding more radicals, the reaction becomes a degenerated branched chain reaction. In this case, the radicals come mainly from the hydroperoxide decomposition; however, in any case, the radical $RO_2\cdot$ is still predominant. If we assume a steady-state condition not only for the radicals but also for the hydroperoxide, we can write for a unimolecular decomposition of the hydroperoxide:

$$\frac{d[ROOH]}{dt}=k_p[RO_2\cdot][RH]-k_{d_1}[ROOH]=0 \tag{4.241}$$

By also applying the pseudo-steady-state condition to the radical, we have:

$$2\ k_{d1}[ROOH]=2\ k_t[RO_2\cdot]^2 \tag{4.242}$$

Therefore, it results in:

$$[RO_2\cdot]=\frac{k_p}{k_t}[RH] \quad [ROOH]=\frac{K_p^2}{k_t k_{d_1}}[RH]^2 \tag{4.243}$$

The rate of substrate consumption is:

$$-\frac{d[RH]}{dt}=3k_{d_1}[ROOH]=\frac{3k_p^2}{k_t}[RH]^2 \tag{4.244}$$

By considering the possibility of a different contribution of the two possible stoichiometries (unimolecular or bimolecular decomposition), we have:

$$-\frac{d[RH]}{dt}=\frac{n}{f}\left(\frac{k_p}{\sqrt{2k_t}}[RH]\right)^2 \tag{4.245}$$

where n is the number of radicals formed for each ROOH molecule decomposed and f is the fraction of RH consumed as a consequence of the $RO_2\cdot$ attack. For a bimolecular decomposition, for example, $n = 1$ and $f = 2/3$. This expression is a limiting rate. If this limit is exceeded, then a different mechanism is operative. This can occur, for example, in the presence of some particular catalysts. The term $k_p/\sqrt{2k_t}$ is the same of Eq. (4.240), that is, the substrate autoxidability.

The induction time can be eliminated by introducing an opportune initiator, such as a peroxide, that is unstable at the reaction temperature. The presence of a redox couple can alter considerably the behavior of a radical chain reaction by interacting with formed radicals or by promoting peroxide decomposition. In particular, the selectivity of the reaction in some cases is strongly enhanced. For this reason, in many industrial processes, cobalt or manganese salts, respectively, corresponding to the redox couples Co^{2+}/Co^{3+} and Mn^{2+}/Mn^{3+} are used. The kinetic expression in these cases can be more complicated and also often contains the catalyst concentration either in the oxidized or in the reduced form. Examples of industrial processes based on radical chain reactions are listed in Table 4.22.

The catalytic action in the autoxidation processes has not been completely clarified. However, it is bonded to the oxidoreductive properties of the ions of the metal used as catalyst. These ions with two different oxidation numbers intervene in some cases as an oxidant and in others as reducing agents interacting with the substrate, the reaction intermediates, and all of the radicals; that is, the ions interact with all of the steps of the reaction mechanism. However, the simplest action of the metallic catalysts is that of promoting hydroperoxide decomposition in a redox cycle of the following type:

$$\mathrm{ROOH} + \mathrm{M(II)} \rightarrow \mathrm{RO\cdot} + \mathrm{OH^-} + \mathrm{M(III)} \tag{4.246}$$

$$\mathrm{ROOH} + \mathrm{M(III)} \rightarrow \mathrm{RO_2\cdot} + \mathrm{H^+} + \mathrm{M(II)} \tag{4.247}$$

The sum of the two reactions corresponds to the bimolecular decomposition of the hydroperoxide seen in reaction mechanism (4.231). In this case, the maximum reaction rate is given by the previous equation (Eq. 4.245) with $n = 1$ and $f = 2/3$. In some other cases, we have a direct interaction of the metal with the substrate with the formation of a radical:

$$\mathrm{M(III)} + \mathrm{RH} \rightarrow \mathrm{M(II)} + \mathrm{H^+} + \mathrm{R\cdot} \tag{4.248}$$

After this reaction, the followings steps occur:

$$\mathrm{R\cdot} + \mathrm{O_2} \rightarrow \mathrm{RO_2\cdot} \tag{4.249}$$

$$\mathrm{RO_2\cdot} + \mathrm{RH} \rightarrow \mathrm{ROOH} + \mathrm{R\cdot} \tag{4.250}$$

$$\mathrm{M(II)} + \mathrm{ROOH} \rightarrow \mathrm{M(III)} + \mathrm{RO\cdot} + \mathrm{OH^-} \tag{4.251}$$

This behavior corresponds to the unimolecular peroxide decomposition obeying the limiting reaction rate to Eq. (4.245) with $n = 2$ and $f = 1/3$.

It is clear that other possible effects of the metals on the reaction mechanism cannot be interpreted with the described kinetic approach; also, no general approach has been developed describing the metal catalytic effects in the autoxidation processes because activities and selectivities are affected by a great number of factors, such as type of substrate, type of metal complex used, association of the catalyst molecules, temperature promoting the un-catalyzed mechanisms, nature of the solvent (polar, apolar), etc.

Table 4.22 Industrial processes based on homolytic oxidation

Raw material	Product	Catalyst	Temperature (°C)	Pressure (atm)	Yield (%)
Butane	Acetic acid	Co acetate	100–225	10–55	57
Fuel oil	Acetic acid	Co acetate	200	50	40
Waxes	Fatty acids	Mn salts in acetic acid	105–120	10–50	50–60
Propylene	Acetic acid	Co acetate	160–250	5–200	90–95
Pentenes	Methyl isobutyl ketone	Co naphthenate	45–75	10	90
Cyclohexane	Cyclohexanol + cyclohexanone	Co naphthenate	150–160	8–10	60–90
Toluene	Benzaldehyde	Co salts in acetic acid	140–160	3–5	60–90
Toluene	Benzoic acid	Co bromide	130–160	2–6	90–91
Benzoic acid	Phenol	Cu benzoate	230–240	1	89–93
p-xylene	Terephthalic acid	Co or Mn bromide	170–230	10–20	97–98
Cumene	Phenol + acetone		100–115	1	90–93
Acetaldehyde	Acetic acid	Mn acetate	50–80	8–10	95
Acetaldehyde	Acetic anhydride	Cu and Co acetate	45–60	1–3	95
Butyraldehyde	Butyric acid	Mn butyrate	65–75	5–6	95
Cyclohexanol + cyclohexanone	Adipic acid	Mn acetate	75–85	1	90

In conclusion, the study of the kinetics of the reactions occurring with a radical mechanism must be made under steady-state conditions, that is, by observing the behavior of the maximum rate in batch reactors or by employing a continuous reactor.

Exercise 4.18. Kinetics of HBr Synthesis
The kinetics of HBr synthesis from the elements was studied by Bodenstein and Lind (1906). The reaction stoichiometry is:

$$H_2 + Br_2 \rightarrow 2\ HBr \tag{4.252}$$

Bodenstein and Lind found a very complex kinetic behavior for this reaction denoting that the reaction mechanism also would be very complicated. In the early stages of the reaction, we have, in a batch reactor, the following kinetic behavior:

$$\frac{d[HBr]}{dt} = k[H_2][Br_2]^{1/2} \tag{4.253}$$

while, after the early stage, the following kinetic law was determined:

$$\frac{d[HBr]}{dt} = \frac{k[H_2][Br_2]^{1/2}}{1 + k'[HBr]/[Br_2]} \tag{4.254}$$

Any reaction mechanism would agree with these experimental observations.

Part 1

Considering that the reaction follows a radical mechanism, define a reasonable mechanism in agreement with the observed kinetic laws.

Part 2

Starting from the experimental points collected by Bodenstein and Lind and also reported in the text of Hougen and Watson (1961) [reproduced in Table 4.23], evaluate the two kinetic parameters of the overall kinetic law and their dependence on the temperature.

Solution of Part 1

The currently accepted mechanism was suggested by Polanyi (1920), who introduced for the first time the fundamental concept of the steady state for radicals, that is, all of the radicals are present in the system in a low amount because the rate of formation and disappearance can be considered approximately equal. We can write a sequence of reactions as:

$$Br_2 + M \xrightarrow{k_1} 2Br\cdot + M \quad \text{Initiation} \tag{4.255}$$

$$Br\cdot + H_2 \underset{k_{-2}}{\overset{k_2}{\rightleftarrows}} H\cdot + HBr \quad \text{Propagation/inihibition} \tag{4.256}$$

Table 4.23 Experimental data collected by Bodenstein and Lind (1906) on HBr synthesis

Run	1		2		3		4	
T (K)	498.8		524.5		550.6		574.4	
Initial conc. H_2 $C^o_{H_2} = 0.5637$ Initial conc. Br_2 $C^o_{Br_2} = 0.2947$ (g mol/ 22.4 L) x = Reacted moles of H_2 and Br_2 (g mol/ 22.4 L)	X	Time (min)	x	Time (min)	x	Time (min)	x	Time (min)
	0	0	0	0	0	0	0	0
	0.0160	840	0.023	150	0.0953	120	0.069	14.5
	0.0273	1500	0.042	300	0.1538	240	0.0985	24.5
	0.0408	2280	0.0831	600	0.1936	360	0.1262	34.5
	0.0516	3000	0.1076	900	0.2210	480	0.1644	49.5
	0.0635	3720	0.1353	1200	0.2460	600	0.2093	79.5
	0.0732	4500	0.1605	1500	0.2679	900	0.2306	99.5
	0.1025	6600	0.1773	1800	0.2797	1200	0.2502	124.5
	0.1404	9000	0.2181	2700	0.2852	1440	0.2619	149.5
			0.2473	3600				

Conc. – concentration

$$\text{H}\cdot + \text{Br}_2 \xrightarrow{k_3} \text{Br}\cdot + \text{HBr} \quad \text{Propagation} \tag{4.257}$$

$$2\text{Br}\cdot + \text{M} \xrightarrow{k_4} \text{Br}_2 + \text{M} \quad \text{Termination} \tag{4.258}$$

As can be seen, we have two chain carriers, which are, respectively, H· and Br·. The H–H bond is stronger than the H–Br bond; therefore, the backward of reaction (4.256) becomes important when a significant amount of HBr is accumulated. By assuming the steady-state condition for the radicals, it is possible to write:

$$\frac{\mathrm{d}[\mathrm{H}]}{\mathrm{d}t} = k_2[\mathrm{Br}][\mathrm{H}_2] - k_{-2}[\mathrm{H}][\mathrm{Br}] - k_3[\mathrm{H}][\mathrm{Br}_2] = 0 \tag{4.259}$$

$$\frac{\mathrm{d}[\mathrm{Br}]}{\mathrm{d}t} = 2k_1[\mathrm{M}][\mathrm{Br}_2] - k_2[\mathrm{Br}][\mathrm{H}_2] + k_{-2}[\mathrm{H}][\mathrm{HBr}] + k_3[\mathrm{H}][\mathrm{Br}_2] - 2k_4[\mathrm{M}][\mathrm{Br}]^2 \tag{4.260}$$

Adding Eqs. (4.259) and (4.260) gives:

$$2k_1[\mathrm{M}][\mathrm{Br}_2] - 2k_4[\mathrm{M}][\mathrm{Br}]^2 = \text{Rate of initiation} - \text{Rate of termination} = 0 \tag{4.261}$$

Therefore:

$$[\mathrm{Br}] = \left(\frac{k_1[\mathrm{Br}_2]}{k_4}\right)^{1/2} \tag{4.262}$$

Then it is possible to write:

$$[\mathrm{H}] = \frac{k_2[\mathrm{Br}][\mathrm{H}_2]}{k_2[\mathrm{HBr}] + k_3[\mathrm{Br}_2]} = \frac{k_2[\mathrm{H}_2]}{k_2[\mathrm{HBr}] + k_3[\mathrm{Br}_2]}\left(\frac{k_1[\mathrm{Br}_2]}{k_4}\right)^{1/2} \tag{4.263}$$

The reaction rate is:

$$\begin{aligned}\frac{\mathrm{d}[\mathrm{HBr}]}{\mathrm{d}t} &= k_2[\mathrm{Br}][\mathrm{H}_2] - k_{-2}[\mathrm{H}][\mathrm{HBr}] + k_3[\mathrm{H}][\mathrm{Br}_2] \\ &= \frac{2k_2(k_1/k_4)^{1/2}[\mathrm{Br}_2]^{1/2}[\mathrm{H}_2]}{1 + (k_2/k_3)[\mathrm{HBr}]/[\mathrm{Br}_2]}\end{aligned} \tag{4.264}$$

This is in agreement with the experimental observation of Bodenstein and Lind [see Eq. (4.252)].

In the initial stage of the reaction, [HBr] is much less than [Br_2] and Eq. (4.264) becomes:

$$\frac{\mathrm{d}[\mathrm{HBr}]}{\mathrm{d}t} = 2k_2(k_1/k_4)^{1/2}[\mathrm{Br}_2]^{1/2}[\mathrm{H}_2] \tag{4.265}$$

which is in agreement with the experimental observation [see Eq. (4.254)].

Solution of Part 2

To solve the problem requires expressing the species involved in the kinetic equation as a function of the conversion:

$$[\mathrm{H}_2] = C^{o}_{\mathrm{H}_2} - x \tag{4.266}$$

$$[\mathrm{Br}_2] = C^{o}_{\mathrm{Br}_2} - x \tag{4.267}$$

$$[\mathrm{HBr}] = 2x \tag{4.268}$$

and integrating the differential equation

$$-\frac{\mathrm{d}[\mathrm{H}_2]}{\mathrm{d}t} = -\frac{\mathrm{d}[\mathrm{Br}_2]}{\mathrm{d}t} = 2\frac{\mathrm{d}[\mathrm{HBr}]}{\mathrm{d}t} = r \tag{4.269}$$

The result is simulation of the experimental curves obtained by Bodenstein and Lind, and this was obtained using a MATLAB program available as electronic supplementary material, Data 4.1.

The obtained kinetic parameters are:

Kinetic parameters

– – – – – – – – – – – – – – – –

$$k \rightarrow \Delta E = 46,876.6026 \text{ cal/mol}$$

$$\ln A = 38.6363 \text{ – – – – – – – – – – – – – – –}$$

$$k_1 \rightarrow \Delta E_1 = 45,461.1871 \text{ cal/mol}$$

$$\ln A_1 = 42.104$$

In Fig. 4.46, the simulation of the experimental points for runs performed at different temperatures is shown in the first plot, while in a second plot, the kinetic constants are arranged in an Arrhenius plot.

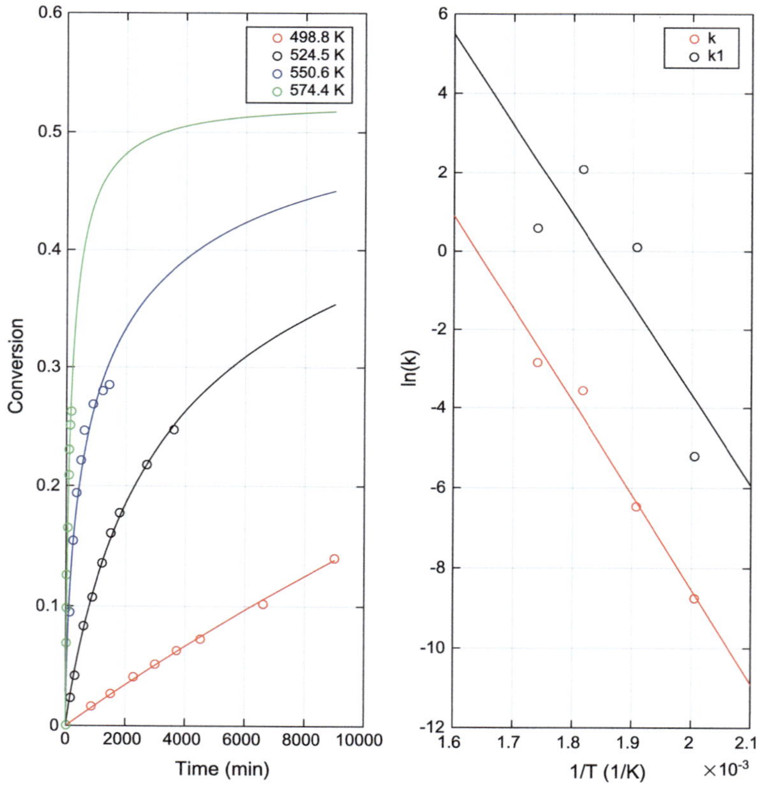

Fig. 4.46 Fitting of experimental data for HBr synthesis at different temperatures

4.4.15 *Kinetics of Enzymatic Reactions*

The simplest mechanism of enzyme catalytic action was described by Michaelis and Menten (1913). They first observed that the reaction rate is proportional to the enzyme concentration while the reaction rate depends on the substrate with the first-order law at the lower substrate concentrations and zero order at the higher substrate concentrations. This behavior was interpreted by Michaelis and Menten with the following mechanism:

$$S + E \underset{k_{-1}}{\overset{k_1}{\rightleftarrows}} ES \tag{4.270}$$

$$ES \xrightarrow{k_2} P + E \tag{4.271}$$

where S is the substrate, E is the enzyme, ES is the complex enzyme–substrate, and P is the product.

The reaction rate is the rate of product formation; therefore, assuming the described reaction step as an elementary step, we can write:

$$r = k[ES] \tag{4.272}$$

Assuming then the first reaction at equilibrium $[ES] = K_e\,[S]\,[E]$, and hence:

$$r = k\ K_e[S][E] \tag{4.273}$$

However, the initial enzyme concentration $[E]_o$ is partitioned between $[E]$ and $[ES]$; therefore, $[E]_o = [E] + [ES]$, and hence:

$$[E] = \frac{[E]_o}{1 + K_e[S]} \tag{4.274}$$

and

$$r = \frac{k K_e[S][E]_o}{(1 + K_e[S])} = \frac{k[S][E]_o}{\left(\frac{1}{K_e} + [S]\right)} \tag{4.275}$$

This mechanism explains well the behavior experimentally observed and shown in Fig. 4.47, according to which the reaction order of the substrate changes from 1 to 0 by increasing the substrate concentration.

Lineweaver and Burk (1934) plotted the reciprocal of the reaction rate as a function of the reciprocal of the substrate concentration, such as seen in Fig. 4.48. A linear plot is normally obtained, from which we can more easily derive the kinetic parameters.

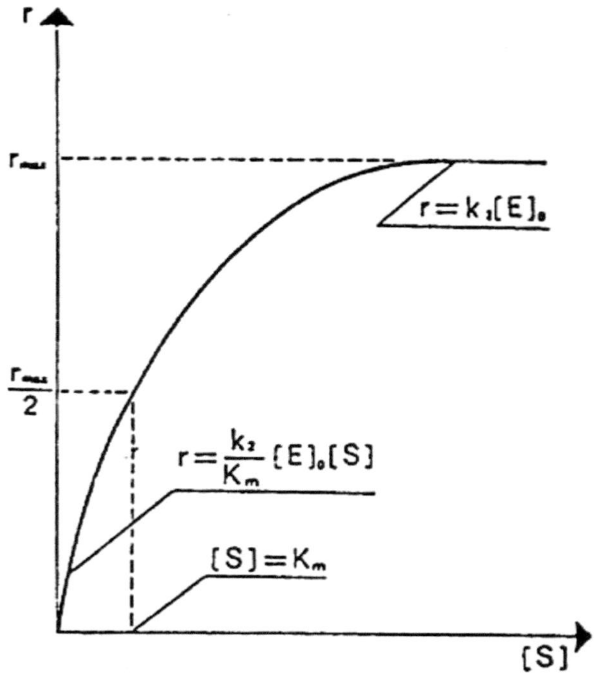

Fig. 4.47 Reaction rate of an enzymatic reaction as a function of the substrate concentration

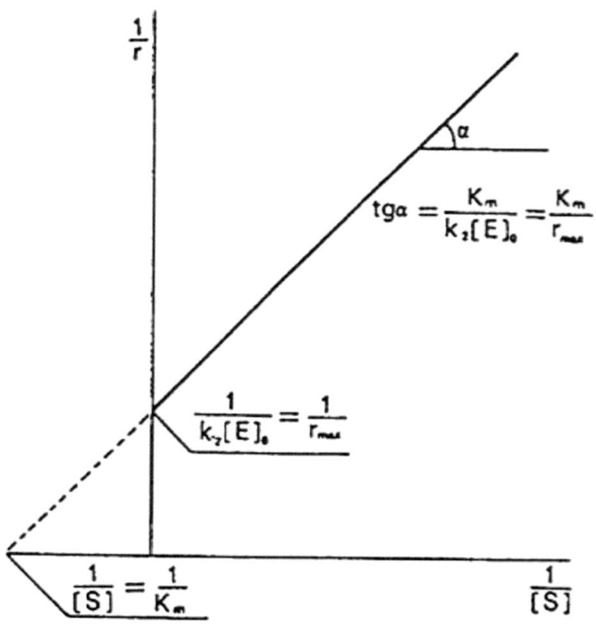

Fig. 4.48 Trend of $1/r$ against $1/[S]$. The Lineweaver–Burk plot better interprets the Michaelis–Menten kinetic model

Table 4.24 Kinetic experimental data

Urea (mM)	Initial reaction rate (μM CO_2/min)
5	30
10	50
20	80
50	100
100	115

The kinetic constant obeys the Arrhenius law, insofar as it depends on temperature, but at 50–60 °C enzyme loss occurs from their conformational structure and, therefore, there is loss of activity. In addition, pH is important for enzyme activity. Considering the amphoteric character of the protein that exists for any enzyme, there is an optimal value; however, to avoid denaturation, the pH cannot be either too acidic or too basic.

Exercise 4.19 The Michaelis–Menten Classical Approach Applied to the Hydrolysis of Urea by the Enzyme Urease

The enzyme urease promotes the hydrolysis of urea according to the following reaction:

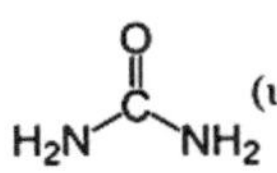

$$\text{(urea)} + H_2O \xrightarrow{\text{urease}\ (0.1\ \mu M)} CO_2 + 2\,NH_3 \quad (4.276)$$

According to the Michaelis and Menten (1913) classical approach, the initial reaction rate is determined for different concentrations of the "urea" substrate collecting data similar to those listed in Table 4.24.

The scheme of the occurring reactions is the following one:

$$E + S \underset{k_{-1}}{\overset{k_1}{\rightleftarrows}} ES \xrightarrow{k_2} E + P \quad (4.277)$$

Clearly, the initial reaction rate is characterized by:

$$\text{Initial reaction rate} = \frac{\mathrm{d}[P]}{\mathrm{d}t} = \nu_o = \frac{k_2 [E]_t [S]}{K_m + [S]} \quad \text{Michaelis – Menten equation} \quad (4.278)$$

$[E]_t$	Concentration of total enzyme
$[S]$	Concentration of free substrate

We can have three different possible situations:

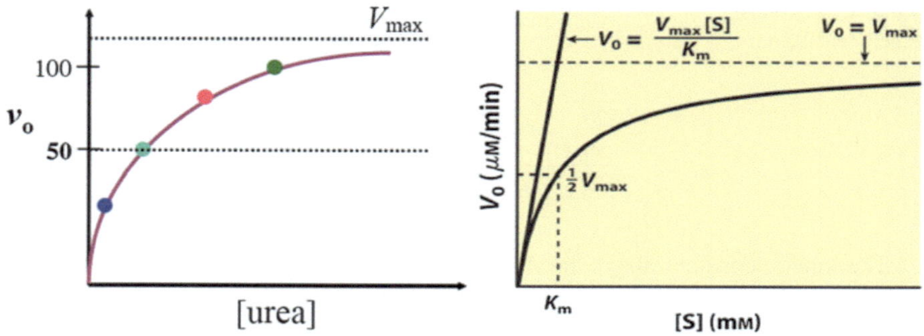

Fig. 4.49 Plot of urea hydrolysis experimental data and characteristic points of the curve

Fig. 4.50 Schematic example of Lineweaver–Burk plot

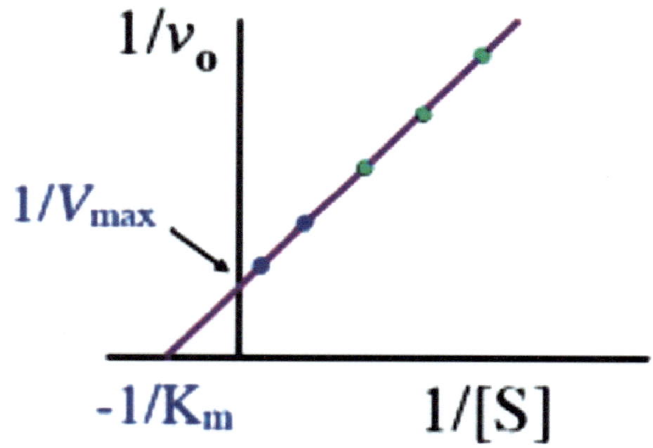

$$\begin{array}{lll} 1) & [S] = 0 & v_o = 0 \\ 2) & [S] = \text{a large value} & v_o = V_{max} \simeq k_2[E]_t \\ 3) & [S] = K_m & v_o = \frac{1}{2} V_{max} \end{array} \tag{4.279}$$

Consequently, it also is possible to write:

$$\nu_o = \frac{V_{max}[S]}{K_m + [S]} \tag{4.280}$$

From the experimental data plot shown in Fig. 4.49, it is possible, therefore, to evaluate the parameters of the model; however, following the approach suggested by Lineweaver and Burk (1934), it is easier to evaluate the parameters by considering a derived linearized plot:

By putting $1/\nu_o$ as a function of $1/[S]$, a linear trend is obtained, and the parameters are obtained from the plot shown in Fig. 4.50. Starting from Eq. (4.280), it is possible to write:

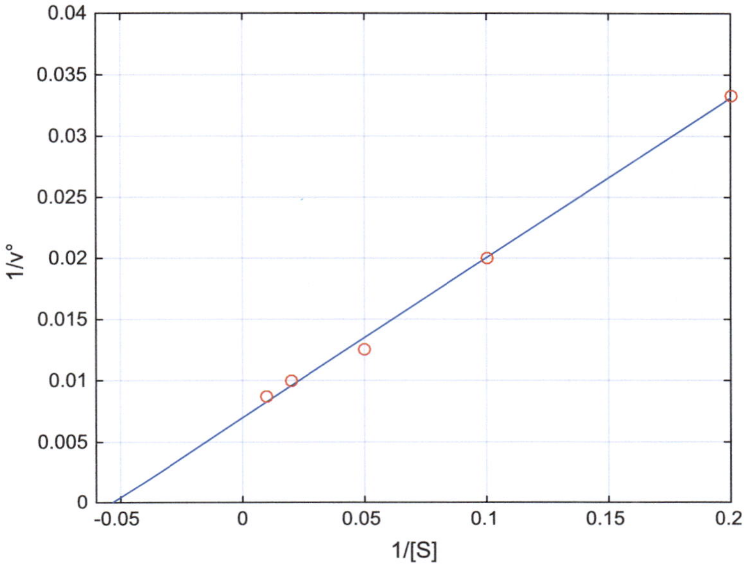

Fig. 4.51 Lineweaver–Burk plot for enzymatic urea hydrolysis

$$\frac{1}{\nu_o} = \frac{K_m}{V_{\max}}\left(\frac{1}{[S]}\right) + \frac{1}{V_{\max}} \tag{4.281}$$

Starting from the experimental data, construct the Lineweaver–Burk plot and evaluate the kinetic parameters. The results were obtained by using a MATLAB program available as electronic supplementary material, Data 4.1.

Kinetic parameters fitting the experimental points are as follows:

$$V_{\max} = 143.5442$$
$$K_m = 18.7727$$

The Lineweaver–Burk plot for this reaction is shown in Fig. 4.51.

Exercise 4.20 The Michaelis–Menten Work on Sucrose Hydrolysis Promoted by Invertase

Michaelis and Menten (1913) studied sucrose hydrolysis catalyzed by the invertase enzyme. The reaction is:

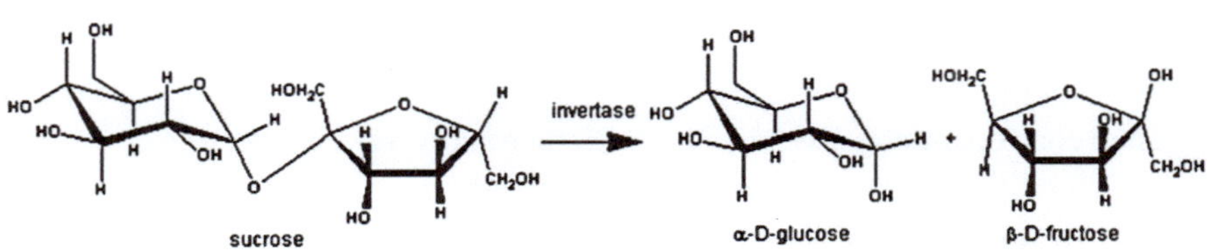

(4.282)

Table 4.25 Kinetic data collected by Michaelis and Menten for the hydrolysis of sucrose promoted by the enzyme invertase

Sucrose 20.8 mM		Sucrose 41.6 mM		Sucrose 83.0 mM		Sucrose 166.7 mM		Sucrose 333.0 mM	
Time (min)	$[P]/[S_o]$	Time (min)	$[P]/[S_o]$	Time (min)	$[P]/[S_o]$	Time (min)	$[P]/[S_o]$	Time (min)	$[P]/[S_o]$
17	0.331	10.25	0.115	49.5	0.352	8	0.035	7	0.016
27	0.452	30.75	0.372	90.0	0.575	16	0.064	14	0.032
38	0.611	61.75	0.615	125.0	0.690	28	0.108	26	0.053
62	0.736	90.75	0.747	151.0	0.766	52	0.198	49	0.092
95	0.860	112.70	0.850	208.0	0.900	82	0.300	75	0.140
1372	0.990	132.70	0.925	–	–	103	0.378	117	0.214
–	–	154.70	0.940	–	–	–	–	1052	0.983

As seen in a previous exercise, the reaction kinetics can be followed considering that sucrose rotates polarized light clockwise $[\alpha]_D = +66.5°$ (polarimetric degree) while the contrary occurs for the mixture of glucose $[\alpha]_D = +52°$ and fructose $[\alpha]_D = -92°$. Therefore, by observing the solution with a polarimeter, it is possible to evaluate the fraction of sucrose reacted along time. Michaelis and Menten collected the following experimental data (Table 4.25), which was reported in a translation of the work with comments and detailed description of the work by Johnson and Goody (2011).

$[P]/[S_o]$ is the ratio between the moles of product formed and the starting sucrose concentration (in practice, the sucrose conversion).

The most rigorous kinetic model for interpreting those experimental data is based on the following scheme:

$$E + S \overset{K_S}{\rightleftarrows} ES \xrightarrow{r} E + F + G \tag{4.283}$$

where E is the enzyme, ES is the complex enzyme–sucrose, and F and G are, respectively, fructose and glucose. According to Michaelis and Menten, glucose and fructose compete with sucrose in the formation of complexes with E. Therefore, we can also write:

$$E + F \overset{K_F}{\rightleftarrows} EF \tag{4.284}$$

$$E + G \overset{K_G}{\rightleftarrows} EG \tag{4.285}$$

According to the mass balance on the enzyme, it holds that $[E] = [E]_o - [ES] - [EF] - [EG]$.

Then considering the dissociation equilibria of the three complexes:

$$K_S = \frac{[E][S]}{[ES]} \quad K_F = \frac{[E][F]}{[EF]} \quad K_G = \frac{[E][G]}{[EG]} \tag{4.286}$$

By opportunely manipulating the previous equations, we can write:

$$[ES] = \frac{[S][E]_o}{[S] + K_S\left(1 + \frac{[F]}{K_F} + \frac{[G]}{K_G}\right)} \tag{4.287}$$

The reaction rate is, therefore,

$$\begin{aligned} r &= -\frac{d[S]}{dt} = \frac{d[F]}{dt} = \frac{d[G]}{dt} = k[ES] = \frac{k[E]_o[S]}{[S] + K_S\left(1 + \frac{[F]}{K_F} + \frac{[G]}{K_G}\right)} \\ &= \frac{k_{\text{eff}}[S]}{[S] + K_S\left(1 + \frac{[F]}{K_F} + \frac{[G]}{K_G}\right)} \end{aligned} \tag{4.288}$$

Michaelis and Menten did not know the initial enzyme concentration, but they knew that it was a constant value, so they grouped the two terms, $k_{\text{eff}} = k[E]_o$. They solved the problem with a long, tedious, and approximated approach starting from the evaluation of the initial reaction rates under different operative conditions and succeeded in determining $K_S = 16.7$ mM, $K_F = 58.8$ mM, $K_G = 91$ mM, and the ratio $k_{\text{eff}}/K_S = 0.0454 \pm 0.0032\ \text{min}^{-1}$, from which it is possible to calculate $r_{\max} = k[E]_o = 0.76 \pm 0.05$ mM/min.

Today by mathematical regression analysis, the constants of the kinetic model can be determined by numerical integration of the previously noted differential equation. Remembering that:

$$[S] = [S]_o - [F] \quad \text{and} \quad [F] = [G] \tag{4.289}$$

Evaluate the kinetic constants of the model and simulate the experimental data constructing a plot $[P]/[S]_o$ against time. Evaluate what happens if the competition in forming the complexes by F and G is neglected. Show that, in this case, the initial reaction rate is the same ($[F] = [G] = 0$) but that deviation occurs at high conversion.

The results were obtained using a MATLAB program available as electronic supplementary material, Data 4.1. The best-fitting kinetic parameters are the following ones:

$$k_{\text{eff}} = 0.76263 \quad K_s = 16.5351 \quad K_f = 62.8777 \quad K_g = 87.7299$$

A plot reporting the agreement between experimental and calculated data is shown in Fig. 4.52. Another plot shown in the same figure, the parity plot, allows us to appreciate the validity of the kinetic model.

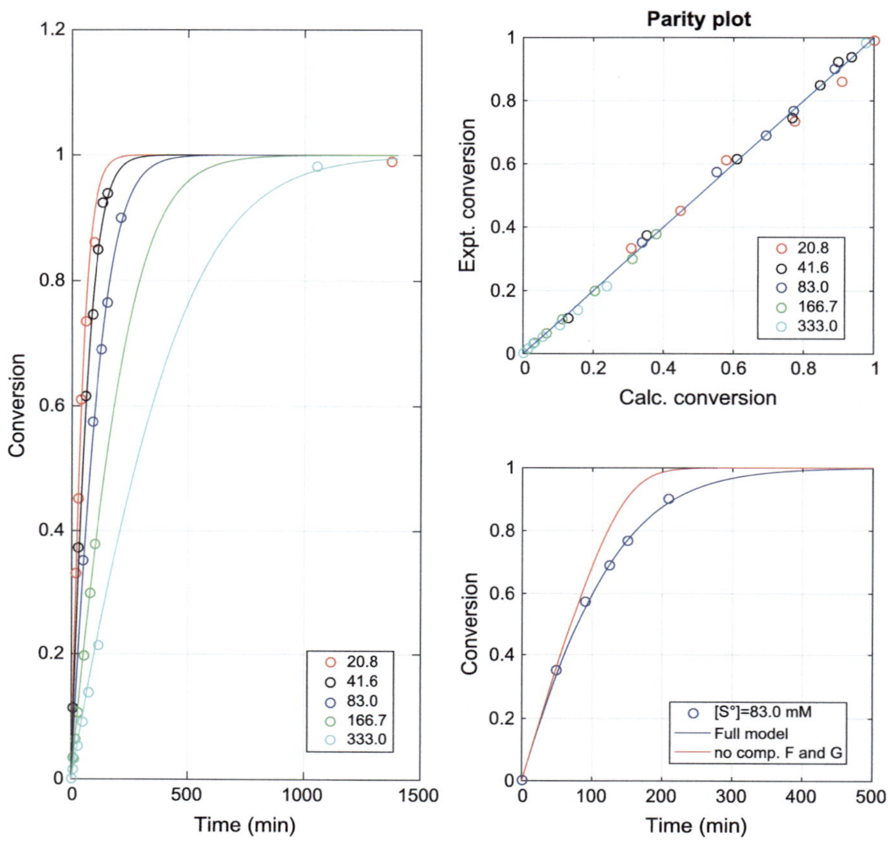

Fig. 4.52 Simulation of the kinetic runs performed starting with different sucrose concentrations. Parity plot related to all of the available data

Exercise 4.21 Fitting the Original Michaelis–Menten Data

In their classic paper, Michaelis and Menten (1913) developed a complete analysis for the interpretation of enzyme kinetics based on the production of fructose and glucose starting from sucrose and using invertase as the enzyme. At that time, the available computational facilities were very scarce and authors made extensive use of linearization techniques. Nowadays, it is instead possible to perform the direct integration of differential equations system resulting from mass balances. The complete reaction scheme is reported as follows:

$$
\begin{aligned}
&1) E + S \underset{k_{m1}}{\overset{k_1}{\rightleftarrows}} ES \\
&2) E + F \underset{k_{m2}}{\overset{k_2}{\rightleftarrows}} EF \\
&3) E + G \underset{k_{m3}}{\overset{k_3}{\rightleftarrows}} EG \\
&4) ES \xrightarrow{k_4} EFG \\
&5) EFG \xrightarrow{k_5} EF + G \\
&6) EFG \xrightarrow{k_6} EG + F
\end{aligned} \tag{4.290}
$$

In the multiple equilibria represented by relation (4.290), some of the parameters are considered adjustable (km_1, km_2, km_3, and k_4) while others are taken from the paper. In particular, the first three equilibria are modeled by assuming an arbitrary value of the forward reaction (100), and the reverse constant is determined by fitting. Reactions 4, 5, and 6 are assumed as irreversible. The remaining parameters are fixed at the following values: $k_5 = 16{,}200$, $k_6 = 10{,}500$.

The code developed for the solution of this exercise, reported in the ESM (electronic supplementary material), Data 4.1, allows for the determination of the adjustable parameters (scaled by a factor of 100 used as arbitrary forward constant): $km_1 = 15.63$, $km_2 = 59.71$, $km_3 = 89.56$, $k_4 = 76.73$. The obtained values are in good agreement with the literature values of $km_1 = 16.7$, $km_2 = 58.8$, and $km_3 = 91.0$. The quality of fit can be appreciated in Fig. 4.53.

4.5 Comparison of the Performances of CSTRs and PFRs

In Sect. 4.3.1, the mass balances, respectively, related to a CSTR and a PFR are shown [see Eqs. (4.25) and (4.26)]. By assuming the simplest situation of a first-order reaction and no change of moles and volume in the reactor, we can write for the two mentioned cases:

$$
F\left(c_i^{\text{in}} - c_i^{\text{out}}\right) = Vkc_i^{\text{out}} \quad \frac{V}{F} = \frac{\left(c_i^{\text{in}} - c_i^{\text{out}}\right)}{kc_i^{\text{out}}} \tag{4.291}
$$

$$
\frac{V}{F} = \int_{c_i^{\text{in}}}^{c_i^{\text{out}}} \frac{dc}{r_i} = \int_{c_i^{\text{in}}}^{c_i^{\text{out}}} \frac{dc}{kc} = \frac{1}{k} \int_{c_i^{\text{in}}}^{c_i^{\text{out}}} \frac{dc}{c} = \frac{1}{k} \ln\left(\frac{c_i^{\text{in}}}{c_i^{\text{out}}}\right) \tag{4.292}
$$

If we introduce the conversion, $\lambda = \frac{c_i^{\text{in}} - c_i^{\text{out}}}{c_i^{\text{in}}}$, it also is possible to write:

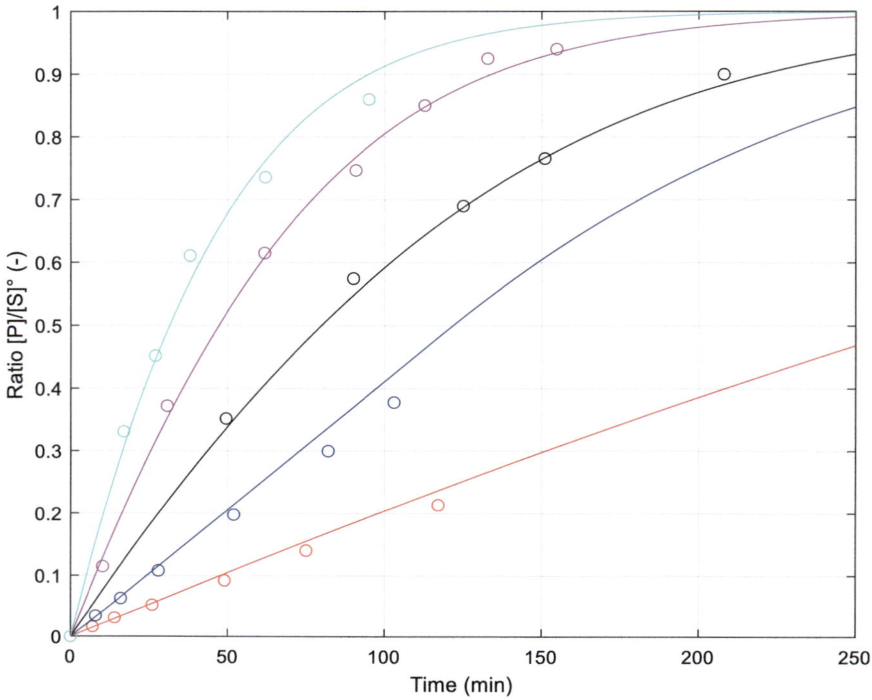

Fig. 4.53 Fitting of the data related to sucrose hydrolysis to fructose and glucose. In the plot, the ratio between the product concentration (fructose = glucose) and the initial substrate (sucrose)

$$\frac{V}{F} = \frac{\lambda_{\mathrm{CSTR}}}{k(1-\lambda_{\mathrm{CSTR}})} \quad \frac{V}{F} = \frac{1}{k}\ln\left(\frac{1}{(1-\lambda_{\mathrm{PFR}})}\right) \tag{4.293}$$

By working in the two cases with the same volumetric flow rate, F, and the same reactor volume, V, we can write the equality:

$$\frac{\lambda_{\mathrm{CSTR}}}{(1-\lambda_{\mathrm{CSTR}})} = \ln\left(\frac{1}{(1-\lambda_{\mathrm{PFR}})}\right) \tag{4.294}$$

By plotting λ_{PFR} as a function of λ_{CSTR}, it results that the conversion in a PFR is higher than in a CSTR and the difference increases for the higher values of λ, as can be seen in Fig. 4.54.

This also means that a smaller reactor volume is necessary to obtain the same conversion in a PFR with respect to a CSTR. In the case of consecutive reactions, selectivity also is affected, and the yields in PFR reactors are normally higher.

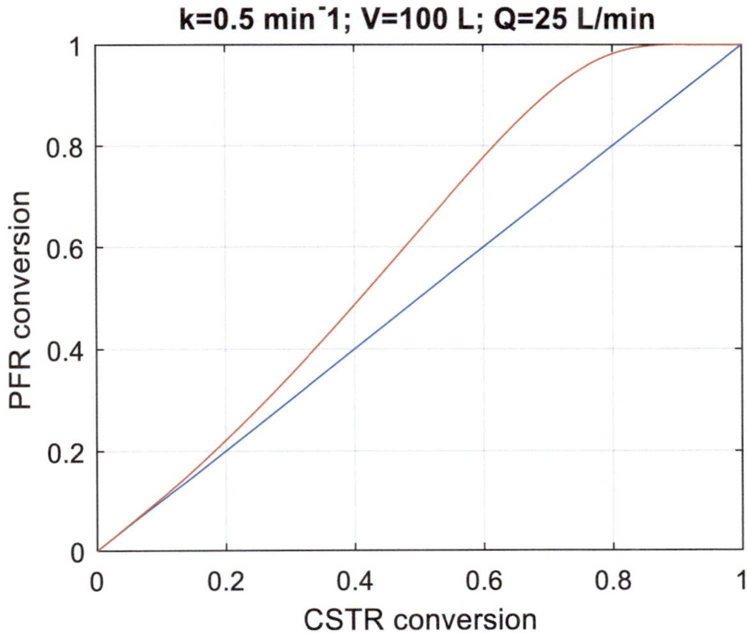

Fig. 4.54 Comparison between the performances—under the same operative conditions of, respectively, a PFR and a CSTR having the same volume—for a first-order reaction kinetics

It is interesting to mention that a battery of CSTR operating in series with progressive conversions approaches PFR performances as can be seen in the example shown in Fig. 4.55.

Exercise 4.22 Comparison Between the Performances of CSTRs and PFRs Construct the plots shown in Figs. 4.54 and 4.55. In practice, you must solve the two equations:

$$\lambda_{\mathrm{PFR}} = 1 - \exp[-k(V/Q)] \tag{4.295}$$

$$\lambda_{\mathrm{CSTR}} = 1 - \left[\frac{1}{1 + k\frac{(V/Q)}{n}}\right]^n \tag{4.296}$$

In Fig. 4.56, the conversion of the two types of reactors is reported as a function of the residence time. By increasing the value of the kinetic constant, the differences in performance between the two reactor configurations become more pronounced.

This can be performed using the MATLAB program available as electronic supplementary material, Data 4.1.

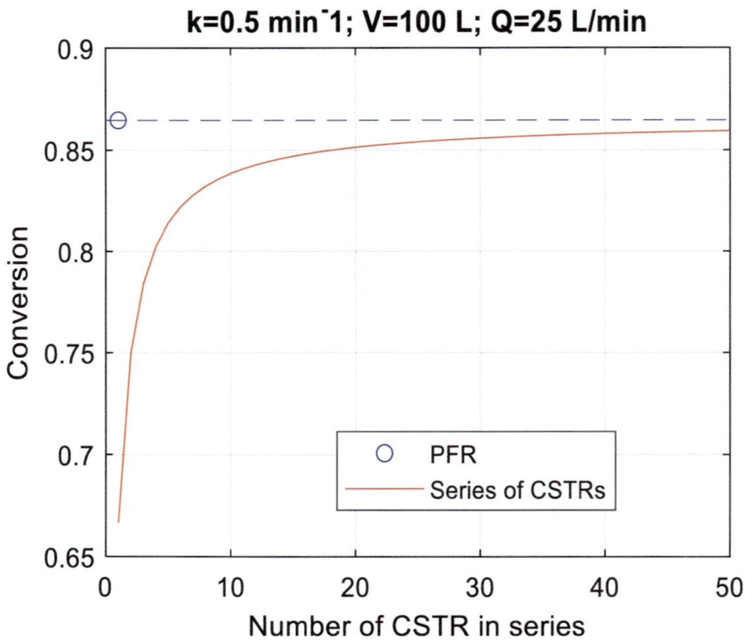

Fig. 4.55 Conversion obtained in series of CSTR versus a PFR

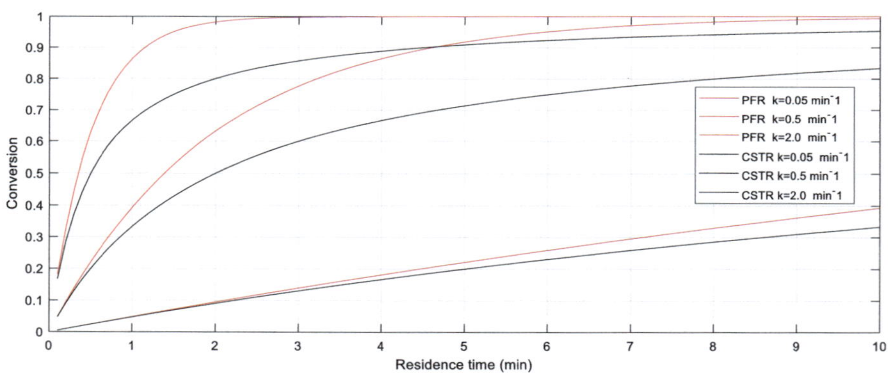

Fig. 4.56 Effect of kinetic constant on conversion, comparison between CSTR and PFR.

4.6 Gas-Phase Reactions and Kinetic Theory

As we have seen, the Arrhenius law suggests that the occurrence of a given reaction requires that the reagents overcome a threshold energy barrier called "activation energy." Two different complementary theories have tried to explain what occurs to molecules during a reaction. The first theory, the "collision theory," sustains that the activation energy is reached due to the collision between two molecules. The number of collisions between two molecules, A and B, per unit of time and volume, can be calculated by the "kinetic molecular theory." For the reaction:

$$A + B \rightarrow C \tag{4.297}$$

The number of collisions occurring between A and B is given by:

$$Z_{AB} = \left(\frac{\sigma_A + \sigma_B}{2}\right)^2 \frac{N^2}{10^6} C_A C_B \sqrt{8\pi kT \left(\frac{1}{M_A} + \frac{1}{M_B}\right)} \tag{4.298}$$

where σ is the molecular diameter, N is the Avogadro number, and k is the Boltzmann constant.

However, normally, the reaction rate is much lower than the number of collisions among the molecules of the reagents. This simply means that not all of the collisions yield a reaction, that is, not all of the collisions are active. Only the collision reaching and overcoming the activation energy for the involved molecules can be followed by the reaction. According to the Maxwell law, the fraction of bimolecular collisions overcoming a given energy value E_a corresponds to $e^{-E_a/RT}$. Therefore, we can write:

$$R = k C_A C_B = (\text{overall number of collisions}) \times (\text{fraction of active collisions}) C_A C_B \tag{4.299}$$

and hence,

$$r = Z_{AB} \frac{10^3}{N} e^{-E_a/RT} C_A C_B \tag{4.300}$$

It is interesting to observe that the dependence of k on temperature in this expression is the following one: $k = \text{cost} \times T^{1/2} \times e^{-E_a/RT}$. This dependence is slightly different from that postulated by the Arrhenius law; however, because the first term changes little compared with the second one, the approximation introduced by the Arrhenius law can be considered acceptable. The experimentally observed rate expressions are seldom in agreement with this theoretical approach, in particular, for the pre-exponential factor. To have a satisfactory agreement, it is opportune to introduce another **probability factor** or **steric factor**, which is a value that can change between 1 and 10^{-8}.

We will have then:

$$k = PZe^{\Delta E_a/RT} \tag{4.301}$$

The steric factor means that even if the involved molecules reached the required activation energy, the reaction does not necessarily occur because the collision did not occur according to a favorable orientation. In conclusion, the value of this theory is just that it defines a limit to the reaction rate, that is, the reaction rate cannot be greater than the number of potentially active molecular collisions (overall collisions × fraction of active collisions).

The second theory is known as the "theory of the transition state" or the "activated complex." This theory is not antithetic or alternative to the collision theory but complementary. In fact, in this theory, the collision is the first step of a reaction path. After the collision, the reagents yield an unstable compound in equilibrium that can react to give a new product or decompose to give the reagents. The two reaction steps can be considered successive elementary steps as shown in the following scheme:

$$A + B \underset{k_2}{\overset{k_1}{\rightleftharpoons}} AB^* \xrightarrow{k_3} P \tag{4.302}$$

By assuming the second step to be slow (RDS), the formation of the product, P, will occur with the rate: $r = k_3\, C_{AB^*}$.

Then we can write: $K_c^* = \frac{k_1}{k_2} = \frac{C_{AB^*}}{C_A C_B}$, and hence,

$$r = k_3 K_c^* C_A C_B \tag{4.303}$$

It is possible to show, based on the methods of statistical mechanics, that for all of the reactions, $k_3 = \frac{kT}{h}$, where k is the Boltzmann constant and h is the Plank constant, while

$$K_c^* = e^{\left(-\Delta G^*/RT\right)} = e^{\left(-\Delta H^*/RT\right)} e^{\left(\Delta S^*/R\right)} \tag{4.304}$$

$$r = \frac{kT}{h} e^{\left(\frac{\Delta S^*}{R}\right)} e^{\left(-\frac{\Delta H^*}{RT}\right)} C_A C_B \tag{4.305}$$

$$\text{Therefore}: \quad k_{\text{cin}} = \frac{kT}{h} e^{\left(\frac{\Delta S^*}{R}\right)} e^{\left(-\frac{\Delta H^*}{RT}\right)} \tag{4.306}$$

$$\text{and hence}: \quad \frac{\mathrm{d}(\ln k_{\text{cin}})}{\mathrm{d}T} = \frac{1}{T} + \frac{\Delta H^*}{RT^2} = \frac{\Delta H^* + RT}{RT^2} \tag{4.307}$$

Again, by comparing the last expression with the Arrhenius, we have $E_a = \Delta H^* + RT$, that is, the activation energy depends on the temperature, while for the Arrhenius law, it is a constant. However, the difference between E_a and ΔH^* is very low, and also in this case, the approximation introduced by the Arrhenius law is acceptable. The evaluation of ΔS^* and ΔH^* requires knowledge of the structure of the activated complex and normally is possible only for very simple reactions. However, also in this case, the theory is useful for evaluating whether or not the experimentally evaluated kinetic parameters have a physical mean.

4.7 Flash with Reaction

Exercise 4.23 Flash with Reaction: Esterification of Acetic Acid with Ethanol
The flash with reaction is a convenient operation where there is an equilibrium reaction in which the desired product is more volatile than the other component and is vaporized and continuously subtracted from the reaction environment, thus shifting the equilibrium to the right. Another advantageous condition is that of consecutive reactions in which the desired product is produced as intermediate and is volatile. In this last case, the product is subtracted before further reactions occur so increasing the yields.

The reaction we consider now is of the first type mentioned, that is, the esterification of acetic acid with methanol according to the following scheme:

$$\underset{1}{CH_3CH_2OH} + \underset{2}{CH_3COOH} \overset{K_e}{\rightleftarrows} \underset{3}{CH_3COOCH_2CH_3} + \underset{4}{H_2O} \tag{4.308}$$

The reaction is performed in a flash apparatus as represented in Fig. 4.57.

We calculate the equilibrium conversion, the flow rates, and the composition at the outlet of the flash apparatus.

Table 4.26 lists the Antoine parameters for all of the components, which are useful for evaluating the vapor pressures of the involved substances. The Antoine equation is as follows:

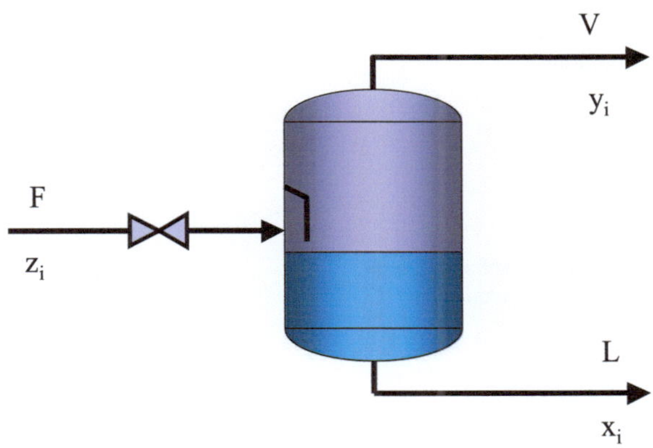

Fig. 4.57 Scheme of flash apparatus

Table 4.26 Data related to the pure components

i	Compound	A	B	C	z_i
1	Ethanol	8.11220	1592.864	226.184	0.50
2	Acetic acid	8.02100	1936.010	258.451	0.50
3	Ethyl acetate	7.10179	1244.951	217.881	0.00
4	Water	8.07131	1730.630	233.426	0.00

$$\log P = A - \frac{B}{T + C} \tag{4.309}$$

The A, B, and C constants listed in table are expressed in Torr, and the temperature T is in °C. The operative conditions for the flash are listed in Table 4.27.

Solution

We choose as reference the calculation of ethanol conversion. Therefore, we can arrange the following in Table 4.28.

If we put:

$$\lambda = \lambda_e F z_1 \tag{4.310}$$

Table 4.28 can be opportunely modified (see Table 4.29).

Generally, the equilibrium term can be written as:

$$F z_i + \nu_i \lambda \tag{4.311}$$

A mass balance for each component "i" can be written as:

$$F z_i + \nu_i \lambda = L x_i + V y_i \tag{4.312}$$

Considering the vaporized fraction as:

Table 4.27 Operative conditions

Feed rate (mol/h)	100
Temperature (°C)	92
Pressure (atm)	1
K equilibrium	35.4

Table 4.28 Reference relations for the calculation

i	Compound	Initial	At equilibrium
1	Ethanol	Fz_1	$Fz - \lambda_e F z_1$
2	Acetic acid	Fz_2	$Fz_2 - \lambda_e F z_1$
3	Ethyl acetate	0	$\lambda_e F z_1$
4	Water	0	$\lambda_e F z_1$

Table 4.29 Reference relations for the calculation

i	Compound	Initial	At equilibrium
1	Ethanol	Fz_1	$Fz_1 - \lambda$
2	Acetic acid	Fz_2	$Fz_2 - \lambda$
3	Ethyl acetate	0	λ
4	Water	0	λ

$$\alpha = V/F \tag{4.313}$$

by introducing this expression in the mass balance, we obtain:

$$z_i + \nu_i \frac{\lambda}{F} = (1-\alpha)x_i + \alpha y_i \tag{4.314}$$

Then it is possible to write:

$$\frac{y_i}{x_i} = K_i = \frac{P_i^0}{P} \quad \text{and then} \quad y_i = K_i x_i \tag{4.315}$$

and hence,

$$z_i + \nu_i \frac{\lambda}{F} = (1-\alpha)x_i + \alpha K_i x_i \tag{4.316}$$

Now it is possible to obtain an expression for x_i

$$x_i = \frac{z_i + \nu_i \frac{\lambda}{F}}{(1-\alpha) + \alpha K_i} \quad \text{or} \quad x_i = \frac{z_i + \nu_i \frac{\lambda}{F}}{T_i} \quad \text{with} \quad \mathrm{T}_i = (1-\alpha) + \alpha K_i \tag{4.317}$$

Considering then the expression of congruence:

$$\sum_i x_i = \sum_{\mathrm{i}} \frac{z_i + \nu_i \frac{\lambda}{F}}{T_i} = \sum_{\mathrm{i}} \frac{z_i}{T_i} + \sum_{\mathrm{i}} \frac{\nu_i \frac{\lambda}{F}}{T_i} = \sum_{\mathrm{i}} \frac{z_i}{T_i} + \frac{\lambda}{F} \sum_{\mathrm{i}} \frac{\nu_i}{T_i} = 1 \tag{4.318}$$

from this equation, it is possible to obtain a relation of λ as a function of the other variables;

$$\lambda = F\left(\frac{1 - \sum_{\mathrm{i}} \frac{z_i}{T_i}}{\sum_{\mathrm{i}} \frac{\nu_i}{T_i}}\right) \tag{4.319}$$

Using this equation for a fixed value of the vaporization ratio, α, it is possible to calculate the conversion. The chemical equilibrium is another constraint of the system that must be respected, that is:

$$\prod_i (x_i)^{\nu_i} = \prod_i \left(\frac{z_i + \nu_i \frac{\lambda}{F}}{T_i}\right)^{\nu_i} = \prod_i E_i = \mathrm{K_e} \quad \text{with} \quad E_i = \left(\frac{z_i + \nu_i \frac{\lambda}{F}}{T_i}\right)^{\nu_i} \tag{4.320}$$

The numerical solution of the last two nonlinear equations allows to evaluate λ and α and hence the amount of the streams at the outlet and the composition. The following procedure must be adopted:

1. Give a reasonable value to α.
2. Calculate the corresponding value of λ.
3. Control that the last equation is verified.
4. Assume another value of α and repeat the calculation.

The results of the calculations are reported in Table 4.30.

Vapor fraction: 0.77969; conversion 43.8%.

As we have seen, the equilibrium constant plays a key role in the performance of the flash operation, for reversible reaction. In Fig. 4.58, the effect of the equilibrium constant on the composition of the vapor and liquid phase is reported. The proposed exercise can be solved using the MATLAB program available as electronic supplementary material, Data 4.1.

Table 4.30 Results of flash calculations: liquid- and vapor-phase composition

i	Compound	Liquid mole fractions	Vapor mole fractions
1	Ethanol	0.0404	0.0679
2	Acetic acid	0.1141	0.0471
3	Ethyl acetate	0.2989	0.4775
4	Water	0.5466	0.4075

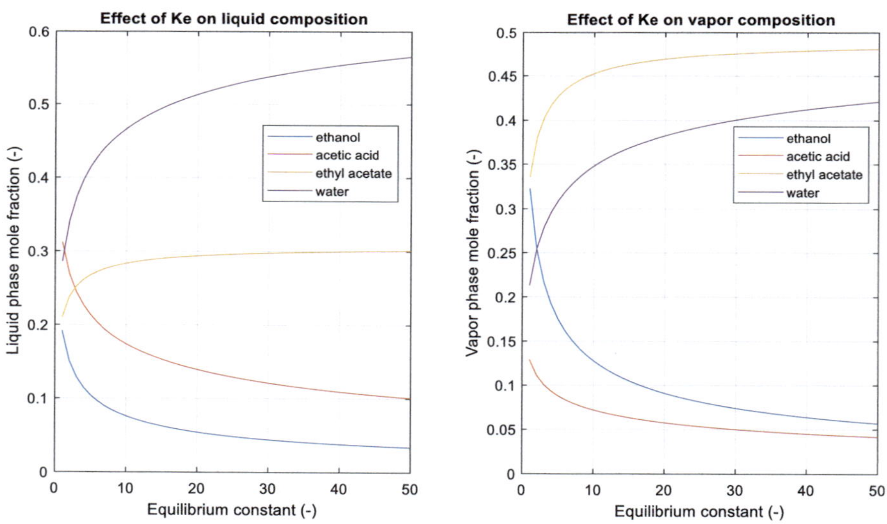

Fig. 4.58 Effect of the equilibrium constant on liquid- and vapor-phase composition

References

Alberty, R.A., Silbey, R.J.: Physical Chemistry. Wiley, New York (1992)

Bamford, C.H., Tipper, C.F.H.: Comprehensive Chemical Kinetics, vol. 2, p. 197. Elsevier, Amsterdam (1969)

Bodenstein, M.: Über die Zersetzung des Jodwasserstoffgases in der Hitze. Z. Physik. Chem. **13**, 56 (1894)

Bodenstein, M., Lind, S.C.: Geschwindigkeit der Bildung des Bromwasserstoffs aus seinen Elementen. Z. Physik. Chem. **57**, 168 (1906)

Brönsted, J.N., Kilpatrick, M., Kilpatrick, M.: Kinetic studies on ethylene oxides. J. Am. Chem. Soc. **51**(2), 428–461 (1929)

Eyring, H., Daniels, F.: The decomposition of nitrogen pentoxide in inert solvent. J. Am. Chem. Soc. **52**(4), 1472–1486 (1930)

Gerasimov, Y.A., Dreving, V., Eremin, E., KIselev, A., Lebedev, V., Panchenkov, G., Shlygin, A.: Physical Chemistry, vol. 1. MIR Pu, Moscow (1974)

Hinshelwood, C.N., Green, T.E.: The interaction of nitric oxide and hydrogen and the molecular statistics of termolecular gaseous reactions. J. Chem. Soc. **128**, 730 (1926)

Hougen, O.A., Watson, K.M.: Chemical Process Principles, Part Three; Kinetics and Catalysis. Wiley, New York (1947 and 1961)

Hughes, E.D., Ingold, C.K., Reed, R.I.: Kinetics and mechanism of aromatic nitration. Part II. Nitration by the nitronium ion, NO_2^+, derived from nitric acid. J. Chem. Soc. 2400–2440 (1950)

Johnson, K.A., Goody, R.S.: The original Michaelis constant: translation of the 1913 Michaelis–Menten paper. Biochemistry. **50**, 8264–8269 (2011)

Kistiakowski, G.B.: Homogeneous gas reaction at high concentration. I. Decomposition of hydrogen iodide. J. Am. Chem. Soc. **50**(9), 2315–2332 (1928)

Laidler, K.J.: Chemical Kinetics. Mc Graw Hill Book Co., New York (1950)

Lindemann, F.A.: Discussion on the radiation theory of chemical action. Trans. Faraday Soc. **17**, 598 (1922)

Lineweaver, H., Burk, D.: The determination of enzyme dissociation constants. J. Am. Chem. Soc. **56**, 658–666 (1934)

Michaelis, L., Menten, M.L.: Die kinetik der invertinwirkung. Biochem. Z. **49**, 333–369 (1913)

Missen, R.W., Mims, C.A., Saville, B.A.: Introduction to Chemical Reaction Engineering and Kinetics. Wiley (1999)

Panchenkov, G.M., Lebedev, V.P.: Chemical Kinetics and Catalysis. MIR, Moscow (1976)

Polanyi, M.: On the problem of the reaction velocity. Z. Elektrochem. **26**, 50 (1920)

Santacesaria, E., Di Serio, M., Ciambelli, P., Gelosa, D., Carrà, S.: Catalytic alkylation of phenol with methanol: factors influencing activities and selectivities. Appl. Catal. **64**, 101–117 (1990a)

Santacesaria, E., Di Serio, M., Gelosa, G., Carrà, S.: Kinetics of methanol homologation: part I. Behaviour of cobalt-phosphine-iodine catalysts. J. Mol. Catal. **58**(1), 27–42 (1990b)

Smith, J.M.: Chemical Engineering Kinetics. Mc Graw-Hill Book Co, New York (1981)

Svirbely, W.J., Roth, J.F.: Carbonyl reactions. I. The kinetics of cyanohydrin formation in aqueous solution. J. Am. Chem. Soc. **75**(13), 3106–3111 (1953)

Vollhardt, K.P.C.: Organic Chemistry. W.H. Freeman and Co, New York (1987)

Westheimer, F.H., Kharasch, M.S.: The kinetics of nitration of aromatic nitro compounds in sulfuric acid. J. Am. Chem. Soc. **68**(10), 1871–1876 (1946)

Winkler, C.A., Hinshelwood, C.N.: The factors determining the velocity of reactions in solution. The formation of quaternary ammonium salts in benzene solution. J. Chem. Soc., 1147–1151 (1935)

Chapter 5
Kinetics of Heterogeneous Reactions and Related Mechanisms

Abstract In this chapter, the influence of mass-transfer and adsorption rates on heterogeneous catalysis is described. As heterogeneous reactions normally occur between molecules adsorbed on the catalytic surface, that is, mainly inside catalyst particles, the concentration of the reactants inside the particles is always lower than that of the fluid bulk. In this way, concentration gradients are originated by the reaction and produce a mass-transfer flux of the reagents and products between the fluid bulk and the particles. The overall reaction rate is, therefore, represented by a sequence of steps: (1) external diffusion, (2) internal diffusion, (3) adsorption, (4) surface chemical reaction, (5) desorption, (6) internal diffusion, and (7) external diffusion. The main heterogeneous mechanisms, like Langmuir–Hinshelwood, are presented and different solved examples and exercises are developed, such as steam reforming of methanol, hydrogenation of iso-octenes, etc. The MATLAB code associated with these examples is available online.

Keywords Active sites · Residence time · Correlation matrix · Statistical tests

5.1 Introduction

Reaction rates in heterogeneous catalysis can be affected by mass-transfer or adsorption rates. The reaction normally occurs between molecules adsorbed on the catalytic surface, that is, mainly inside the catalyst particles. Therefore, due to the reaction, the concentration of the reactants inside the particles is always lower than that of the fluid bulk. In contrast, the concentration of the products is always greater. These concentration gradients, originated by the reaction, produce a continuous mass transfer of the reagents from the fluid bulk to the particles and of the products from the heat of the particles to the fluid bulk. Therefore, before and after the reaction, we can imagine the sequence of steps shown in Fig. 5.1.

Supplementary Information The online version contains supplementary material available at https://doi.org/10.1007/978-3-031-82502-6_5.

© The Author(s), under exclusive license to Springer Nature Switzerland AG 2025
R. Tesser, E. Santacesaria, *The Chemical Reactor from Laboratory to Industrial Plant*, https://doi.org/10.1007/978-3-031-82502-6_5

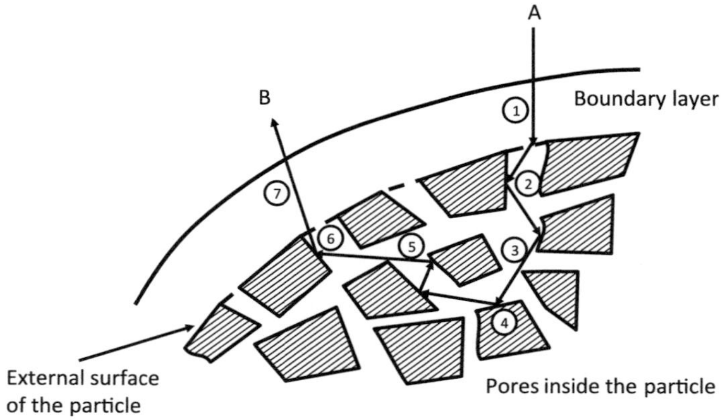

Fig. 5.1 Steps occurring in a fluid–solid catalytic reaction

Step 1: External diffusion of reactants. External diffusion of reactants is the diffusion of the reactants from the bulk fluid to the external surface of the catalytic pellet. Around the particle, there is a thin layer of static fluid (the **boundary layer**). The concentration gradient is normally located in this layer because, in a static fluid, mass transfer occurs with a slow molecular diffusion mechanism. As it will be seen in more detail in the next chapter, the thickness of the layer and the mass-transfer rate will depend on both the fluid dynamic characteristic of the system (stirred or in a flowing stream) and the reaction rate, with the latter affecting the gradient. However, it is important to point out that this step occurs before and separately from the other steps.

Step 2: Internal diffusion of the reactants. Internal diffusion of the reactants is the diffusion of molecules from the pore mouth, through the pores, until they reach an active site of the catalyst not occupied by other chemisorbed molecules. The fluid inside the pores can be considered static, and—again—the mass transfer occurs with the slow molecular diffusion mechanism. The mass-transfer rate in this case will depend on both the catalyst porosity and the size and shape of the catalytic particles. This step is not independent as the previous one; it occurs together with the adsorption and the surface chemical reaction.

Step 3: Adsorption of reactants. Adsorption of reactants involves the active sites of the internal solid surface. It depends on the possible interaction between the molecules and the catalyst surface.

Step 4: Chemical reaction. The chemical reaction occurs between molecules adsorbed on the surface.

Step 5: Desorption of products. Desorption of products involves products being desorbed from the surface.

Step 6: Internal diffusion of products. Internal diffusion of products occurs from the interior of the pellets to the pore mouth.

Step 7: External diffusion of products. External diffusion of products occurs through the boundary layer from the external surface of the particles into the fluid bulk.

Each of the abovementioned steps could affect the reaction rate, but usually kinetic runs are performed in the laboratory under conditions selected in such a way as to avoid any influence of the diffusion steps (steps no. 1, 2, 6, and 7) on the **intrinsic reaction rate**. However, also in the case of heterogeneous catalysis, it is useful to consider a reaction mechanism and introduce as a useful hypothesis the slow step becoming the "rate determining step" (RDS) or the steady-state approximation (SSA).

5.2 Definition and Evaluation of Reaction Rate, Mass Balance, and Kinetic Equations in Heterogeneous Fluid–Solid Systems

As already seen in the previous section, the performance of a chemical reactor working under isothermal conditions can be described by two kinds of equations: the mass balance and the kinetic equations. The mass balance equations refer to the time unit for the component "i," which can be written again in the most general case, referred to as a "tank reactor," as follows:

$$\underset{(1)}{\text{Input of i mass}} = \underset{(2)}{\text{Output of i mass}} + \underset{(3)}{\text{Mass of i reacted}} + \underset{(4)}{\text{Accumulation}} \tag{5.1}$$

Thus, considering a reaction occurring without a change in the number of moles, we can write the mole balance as follows:

$$\underset{(1)}{Fc_i^e} = \underset{(2)}{Fc_i^u} + \underset{(3)}{wr_i} + \underset{(4)}{\frac{\mathrm{d}c_i}{\mathrm{d}t}} \tag{5.2}$$

where F is the volumetric feed rate; c_i^e and c_i^u are, respectively, the molar concentrations of i at the entrance and at the exit of the reactor; w is the catalyst weight; t is the time; r_i is the reaction rate at which i appears (positive) or disappears (negative) expressed as moles of i per unit time per unit of catalyst weight; and c_i is the i concentration in the reactor.

In the case of a continuous reactor operating under steady-state conditions, term (4) of the mass balance can be neglected. In contrast, in the case of a batch reactor, terms (1) and (2) must be ignored, while in the case of a semi-batch reactor term (2) is normally null. It is easy to recognize that the reaction rate contained in term (3) can be experimentally determined in different ways according to the type of reactor in which it is measured. For a well-mixed batch reactor, for example, we have:

$$r_i = \pm \frac{1}{W}\frac{dc_i}{dt} = \frac{\text{moles of } i \text{ appeared or disappeared}}{\text{unit of catalyst weight} \times \text{unit of time}} \tag{5.3}$$

while, for a well-mixed continuous stirred-tank reactor (CSTR), we have:

$$r_i = \frac{F\left(c_i^e - c_i^u\right)}{W} \tag{5.4}$$

In the case of a continuous tubular reactor, the reaction rate changes along the catalytic bed yielding, under steady-state conditions, a constant concentration profile. Therefore, the mass balance must be applied, in this case, to an infinitesimal part of the reactor, for example, dW. Then, we will have:

$$r_i = F\frac{dc_i}{dW} \tag{5.5}$$

At each point of the reactor, we can have a different reaction rate depending on the composition of the reaction mixture and on the temperature at that point, that is:

$$r_i = f(\text{composition, temperature}) \tag{5.6}$$

This function is the kinetic equation or rate law, which is indispensable for describing the behavior of all previously mentioned types of reactors and for modeling them. The kinetic equation for a given reaction can be determined only by collecting and processing experimental data under different operative conditions. However, before determining the kinetic equations of the reactions involved in a process, it is opportune to individuate the reaction scheme and the stoichiometry of each single reaction concurring in the scheme.

5.3 Reaction Scheme, Stoichiometry, Thermodynamic Constraints, and Analysis of Reaction Networks

Two or more reagents put in a reactor could yield more products than expected. We can identify the products by analysis and determine them quantitatively; however, often we do not know their origin, that is, we ignore how many and what kind of reactions have occurred. Therefore, the first step of a kinetic study must be the identification of the reaction scheme and, consequently, of the stoichiometry of each reaction. A reaction scheme could be very complex; however, it will be always recognizable as a combination of simpler schemes of this type:

(a) Single reactions

$$A \rightarrow B \tag{5.7}$$

$$A + B \rightarrow C + D \tag{5.8}$$

(b) Equilibrium reactions

$$A \leftrightarrow B \tag{5.9}$$

(c) Parallel competing reactions

$$A \begin{array}{l} \nearrow B \\ \searrow C \end{array} \tag{5.10}$$

(d) Consecutive reactions

$$A \rightarrow B \rightarrow C \tag{5.11}$$

The stoichiometry of a single reaction allows to correlate the rate of appearing or disappearing of any reagent or product with the kinetic equation. If we have, for example, the reaction:

$$aA + bB \rightarrow mM + nN \tag{5.12}$$

The kinetic equation will be $r = f$ (composition temperature).

The rate of appearing products or disappearing reagents will be:

$$r_i = \alpha_i r \tag{5.13}$$

where α_i is the stoichiometry coefficient, which is positive for the products and negative for the reagents. If products and reagents are involved in more than one reaction, each occurring with its own rate and stoichiometry, r_i becomes:

$$r_i = \sum_{i=1}^{j} \alpha_{ij} \, r_j \tag{5.14}$$

Table 5.1 Matrix of the stoichiometric coefficients α_{ij}, for the reactions of Scheme 5.15

j \ i	A_1	A_2	A_3	A_4	A_5	A_6	
1	−1	−1/2	1	0	0	1	
2	0	−5/2	−1	1	0	2	
3	−1	−6	0	0	4	4	(5.16)
4	0	−11/2	−1	0	4	3	
5	0	−3	0	−1	4	1	

where, in this case, α_{ij} is the matrix of the stoichiometric coefficients. An example could be the oxidation of butene to maleic anhydride occurring with the following simplified scheme:

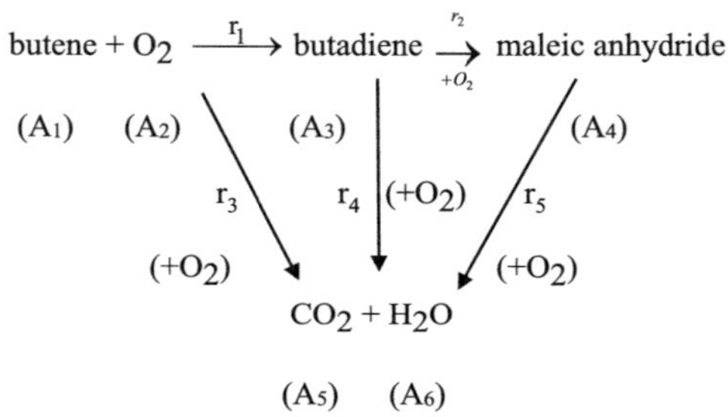

(5.15)

The matrix of the stoichiometric coefficients α_{ij} for the reactions of Scheme 5 is reported in Table 5.1.

As a consequence, we can write the following for each component:

$$\frac{dp_{A_1}}{d\tau} = -r_1 - r_3 \qquad \frac{dp_{A_2}}{d\tau} = -\frac{1}{2}r_1 - \frac{5}{2}r_2 - 6r_3 - \frac{11}{2}r_4 - 3r_5$$
$$\frac{dp_{A_3}}{d\tau} = r_1 - r_2 - r_4 \qquad \frac{dp_{A_4}}{d\tau} = r_2 - r_5 \quad \frac{dp_{A_5}}{d\tau} = 4(r_3 + r_4 + r_5)$$
$$\frac{dp_{A_6}}{d\tau} = r_1 + 2r_2 + 4r_3 + 3r_4 + r_5 \tag{5.17}$$

where τ is a function of the residence time equal to W/F for a tubular reactor and p_i are the partial pressures of the reaction components. As can be seen, the problem comes back to the determination of the five kinetic equations $r_j = f$ (composition, temperature). If the kinetic equations $r_1\_r_5$ are known, it is possible to reproduce the evolution of any single specie along the catalyst bed by simultaneously integrating the system of differential Eq. (5.17).

When equilibrium reactions occur in a reaction scheme, the yields of products involved in these reactions are subject to thermodynamic constraints. It is possible to define theoretically those yields as in the case of a complex reaction scheme. In this last case, we identify, with a simple procedure, the reactions that could be considered thermodynamically independent and that justify the presence of the observed products. Then, the equilibrium constants of the reactions are determined from data on the free energies of reagents and products. The thermodynamic constraints must be introduced in the kinetic equation. For a second-order reaction of the type $A + B \leftrightarrow M + N$, for example, we will write:

$$r = kp_A p_B \left(1 - \frac{1}{K_e} \frac{p_M p_N}{p_A p_B}\right) \tag{5.18}$$

where k is the kinetic constant of direct reaction and K_e is the equilibrium constant of the reaction.

When many different products are obtained in a reaction, it is not simple to correctly define a reaction scheme, and interpretation of the experimental data can be difficult and controversial. The o-xylene oxidation to phthalic anhydride is a reaction of industrial interest that has been intensively studied by many different researchers. A reaction scheme containing all of the possible reaction steps and side products is shown in the following scheme (5.19):

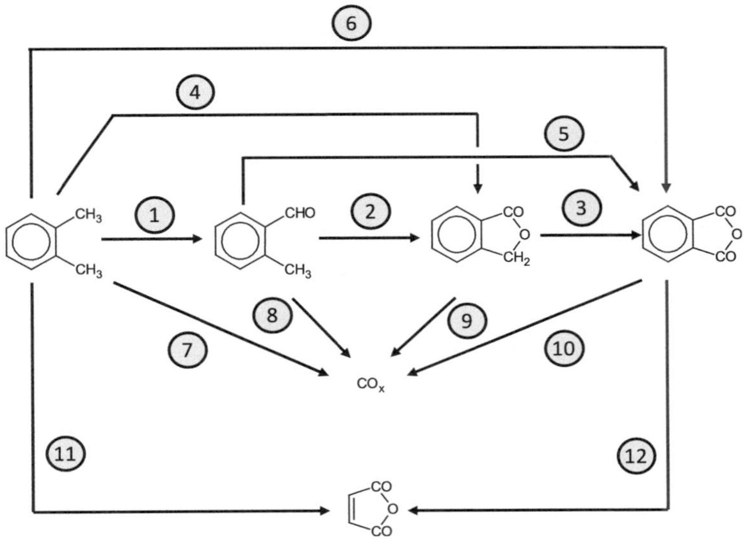

(5.19)

It is interesting to observe that in the kinetic approach, some of the reaction paths of this scheme were neglected by the researchers who studied this reaction, and the only agreement between the different researchers is that the reaction occurs through a "redox" mechanism (see Sect. 5.4.3). In Table 5.2, we can observe the disagreement of different investigators about the steps of the scheme that must be considered in the kinetic analysis.

Table 5.2 Reaction steps of Scheme (5.19) that effectively occur[a]

	Reaction steps considered											
References	1	2	3	4	5	6	7	8	9	10	11	12
Herten and Froment (1968)	x		x	x	x	x	x	x	x			
Vanhove and Blanchard (1975)	x	x	x		x			x		x		x
Calderbank et al. (1977)	x	x	x			x	x			x		
Yabrov and Ivanov (1980)	x	x	x	x	x	x	x	x	x	x		
Chandrasekharan and Calderbank (1980)	x				x	x	x	x				
Skrzypek et al. (1985)	x	x	x		x	x	x			x	x	
Saleh and Wachs (1987)	x	x	x		x		x				x	
Papageorgius et al. (1994)	x	x	x		x		x					
Dias et al. (1996)												
Anastasov (2003)	x	x	x			x	x			x		
Gimeno et al. (2008)	x	x	x			x	x			x		

[a]According to different researchers

5.4 Kinetic Equations Based on the Mechanisms of Chemical Adsorption and Chemical Surface Reaction: The Langmuir–Hinshelwood Kinetic Model

In the case of heterogeneous reactions, we do not know what steps are occurring on the surface between adsorbed molecules; therefore, the heterogeneous mechanisms are constructed considering chemical adsorption of reactants and chemical desorption of products as elementary steps occurring before and after the surface reaction, with the latter approximated to a single step. Therefore, considering the simplest reaction, $A \rightarrow B$, we can write a mechanism characterized by three consecutive reactions of the type:

$$\text{Reactant(A)} + \text{active site(S)} \rightleftarrows \text{Adsorbed reactant(AS)} \times \text{(adsorption reaction)} \tag{5.20}$$

$$\text{Adsorbed reactant(AS)} \rightarrow \text{adsorbed product(BS)(surface reaction)} \tag{5.21}$$

$$\text{Adsorbed product(BS)} \rightleftarrows \text{product(B)} + \text{active(S)(desorption reaction)} \tag{5.22}$$

Then we can assume, more frequently, one of these three reactions as a "slow step" (rate-determining step) considering the other steps at the equilibrium or, less frequently, when two reactions or all have comparable rates, we must assume the "steady-state approximation" for determining the rate-law expression. In any case, all the steps will have the same rate, that is, the reaction rate.

Considering a species, A, for example, reacting with a free catalytic site, σ_f, we can write the adsorption reaction as follows:

$$A + \sigma_f \underset{k_{-A}}{\overset{k_A}{\rightleftharpoons}} \sigma_A \tag{5.23}$$

where σ_A is a site occupied by A as a consequence of the chemical adsorption.

The adsorption rate, considered as an elementary step, as referred to the total number of sites, will be, therefore:

$$r_A = k_A p_A \frac{\sigma_f}{\sigma_t} - k_{-A} \frac{\sigma_A}{\sigma_t} \tag{5.24}$$

By introducing:

$$\Theta_A = \sigma_A / \sigma_t = \text{the adsorption coverage degree} \tag{5.25}$$

and remembering that $\sigma_t = \sigma_A + \sigma_B + \sigma_f$, we have $\sigma_f/\sigma_t = (1 - \Theta_A - \Theta_B)$; hence:

$$r_A = k_A p_A (1 - \Theta_A - \Theta_B) - k_{-A} \Theta_A \tag{5.26}$$

At the equilibrium, this rate is null; therefore:

$$b_A = \frac{k_A}{k_{-A}} = \frac{\Theta_A}{p_A (1 - \Theta_A - \Theta_B)} \tag{5.27}$$

where b_A is the adsorption equilibrium constant. This constant will depend on the temperature according to a vant' Hoff–type law, that is:

$$b_A = b_A^o \exp(-\Delta H_{\text{ads}} / RT) \tag{5.28}$$

Equation (5.27) is called the "Langmuir adsorption isotherm" and is rigorously valid for a uniform surface and localized adsorption without lateral interactions between the adsorbed molecules. As it will be seen, when these conditions are not observed other isotherms—such as those of Temkin, Freundlich, Fowler, etc.—must be considered; however, the Langmuir isotherm is particularly useful in kinetics because it well describes the competition in the adsorption of different molecules on the catalytic sites.

For example, product B can be adsorbed in competition with A on the free catalytic sites, and at the equilibrium, it holds:

$$b_B = \frac{\Theta_B}{p_B (1 - \Theta_A - \Theta_B)} \tag{5.29}$$

By considering together both the adsorption equilibria b_A and b_B, it is possible to demonstrate that:

$$\Theta_A = \frac{b_A p_A}{1 + b_A p_A + b_B p_B} \qquad \Theta_B = \frac{b_B p_B}{1 + b_A p_A + b_B p_B} \tag{5.30}$$

In conclusion, by assuming the surface reaction as a rate-determining step (RDS) and considering the reaction far from the equilibrium, the reaction rate will be:

$$r = k_S \Theta_A = k_s \frac{b_A p_A}{1 + b_A p_A + b_B p_B} \tag{5.31}$$

Then, if the adsorption terms ($b_A p_A$ + $b_A p_A$) < < 1, the reaction becomes of pseudo first order. Extending the example to more components and considering a generic '*i*' component, we can write:

$$\Theta_i = \frac{b_i p_i}{1 + \sum_{j=1_j}^{NC} b_j p_j} \tag{5.32}$$

The kinetic equations obtained assuming the surface reaction as rate-determining step, and the adsorption–desorption steps at equilibrium, is called the Langmuir–Hinshelwood (LH) kinetic model. The same mechanism also can be applied to a bimolecular reaction of the type:

$$A + B \rightarrow C \tag{5.33}$$

The corresponding kinetic law can easily be derived and is:

$$r = k_S \Theta_A \Theta_B = \frac{k_2 b_A b_B p_A p_B}{\left(1 + \sum_{j=1_j}^{NC} b_j p_j\right)^2} \tag{5.34}$$

If the adsorption on the surface occurs with dissociation, we can write:

$$A_2 + 2\sigma_f \rightleftarrows 2\sigma_A \tag{5.35}$$

Consequently:

$$r_A = k_A p_A (1 - \Theta A)^2 - k_{-A} \Theta_A^2 \tag{5.36}$$

at the equilibrium $r_A = 0$; therefore:

$$\Theta_A = \frac{\sqrt{b_A p_A}}{1 + \sqrt{b_A p_A}} \tag{5.37}$$

The surface reaction rate will change accordingly:

$$r = k_S \frac{\sqrt{b_A p_A}}{1 + \sqrt{b_A p_A}} \tag{5.38}$$

5.4.1 Dual-Site Mechanism

In some cases, chemical species require more than one catalytic site to yield a reaction, for example, two sites could be required and the adsorption on the second site is slow (dual-site mechanism):

$$A + \sigma_f \rightleftarrows A\sigma \tag{5.39}$$

$$A\sigma + \sigma_f \rightarrow A\sigma_2 \tag{5.40}$$

In this case, we can write:

$$r = k_s \Theta_A \Theta_f = k_s \Theta_A (1 - \Theta_A) = \frac{k_s b_A p_A}{(1 + b_A p_A)^2} \tag{5.41}$$

5.4.2 Eley–Rideal Mechanism

When an adsorbed molecule reacts with another one coming from the fluid phase, we have the Eley–Rideal mechanism, that is:

$$A\sigma + B \rightarrow \text{products} \tag{5.42}$$

$$r = k_s \Theta_A p_B = \frac{k_s b_A p_A p_B}{1 + b_A p_A} \tag{5.43}$$

5.4.3 Redox Mechanism According to Mars and Van Krevelen

Different catalytic oxidations occur through a redox mechanism, in which the active sites on the surface can be reduced or oxidized by the reactants according to the following reactions (Mars and Van Krevelen mechanism):

$$A+\Theta_{\mathrm{ox}}\overset{k_1}{\rightarrow}\mathrm{P}+\Theta_{\mathrm{red}} \tag{5.44}$$

$$\Theta_{\mathrm{red}}+\frac{1}{2}O_2\overset{k_2}{\rightarrow}\Theta_{\mathrm{ox}} \tag{5.45}$$

where Θ_{ox} and Θ_{red} are, respectively, the fraction of oxidized and reduced sites on the surface and *A* and *P* are, respectively, the reagent and the product. We can write, then, the following:

$$\Theta_{\mathrm{ox}}+\Theta_{\mathrm{red}}=1 \tag{5.46}$$

by introducing the "steady-state assumption"—that is, $r_1 = r_2$—we will have the following:

$$k_1p_A\Theta_{\mathrm{ox}}=k_2p_{O_2}^n\Theta_{\mathrm{red}} \tag{5.47}$$

Then, remembering that $\Theta_{red} = (1 - \Theta_{ox})$, it is possible to evaluate Θ_{ox}, and the reaction rate r_1 giving the product, *P*, will become:

$$r_1=\frac{k_1k_2p_Ap_{O_2}^{1/2}}{k_1p_A+k_2p_{O_2}^{1/2}}=\frac{k_1p_A}{1+\left(\frac{k_1}{k_2}\frac{p_A}{p_{O_2}^{1/2}}\right)} \tag{5.48}$$

The value of *n* is normally equal to one half. We can also have a mixed mechanism if the active sites are occupied by inactive molecules formed during the reaction or present in the fed composition. It is sufficient in this case to write $\Theta_{red} + \Theta_{ox} + \Theta_{occ} = 1$. The reaction-rate equation will become:

$$r=\frac{k_1p_A}{\left[1+\left(\frac{k_1}{k_2}\frac{p_A}{p_{O_2}^{1/2}}\right)\right]\left(1+\sum_j b_jp_j\right)} \tag{5.49}$$

5.4.4 Adsorption on Nonuniform Surfaces

As mentioned previously, in many cases, the surface of the catalysts is not uniform, and the active sites give place to more or less strong interactions with the reactants and products. The characteristics of the surface can be determined by measuring the heat developed during the adsorption of molecules for different degrees of surface coverage as shown in Fig. 5.2.

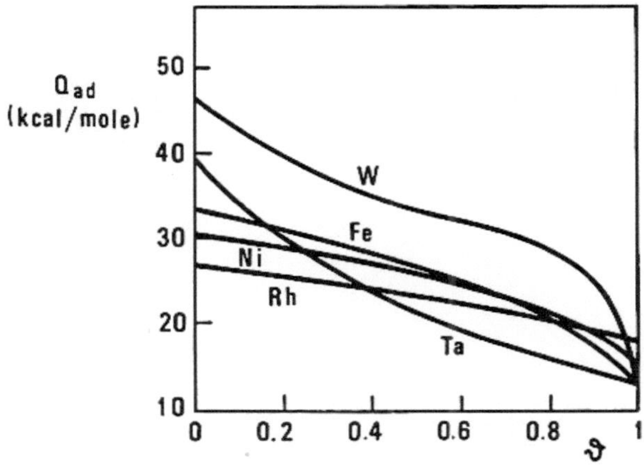

Fig. 5.2 Heat of adsorption of hydrogen on the surface of different metals. Published with permission from Beek O.; Hydrogenation Catalysts; Discuss. Faraday Soc. 8118–128 (1950). Copyright Royal Society of Chemistry (1950) (see also Smith [1981])

For describing the curves of Fig. 5.2, a distribution function of the type ρ (Q_{ads}) must be defined. ρ (Q_{ads}) dQ_{ads} gives the fraction of active sites with adsorption heat in the range Q_{ads} and (Q_{ads} + dQ_{ads}). Then, it is possible to write:

$$\int_{Q_{ads_1}}^{Q_{ads_2}} \rho(q_{ads})\,\mathrm{d}Q_{ads} = 1 \tag{5.50}$$

where Q_{ads1} and Q_{ads2} are, respectively, the minimum and maximum values of the adsorption heat. With a reasonable approximation, we can apply the Langmuir isotherm to each type of active center. By writing $\theta(Q_{ads})$—the fraction of active sites occupied having adsorption heat equal to Q_{ads}—the overall fraction of occupied active sites will be:

$$\begin{aligned} \varphi(P) &= \int_{Q_{ads_1}}^{Q_{ads_2}} \theta(Q_{ads})\;\rho(Q_{ads})dQ_{ads} \\ &= \int_{Q_{ads_1}}^{Q_{ads_2}} \frac{bP}{1+bP}\rho(Q_{ads})dQ_{ads} \\ &= \int_{Q_{ads_1}}^{Q_{ads_2}} \frac{P}{\frac{1}{b^{\circ}}\cdot e^{(-Q_{ads}/RT)}+P}\rho(Q_{ads})dQ_{ads} \end{aligned} \tag{5.51}$$

For solving the integral (5.51), it is necessary to give an expression to the distribution function $\rho(Q_{ads})$. It is possible to write, for example, $\rho(Q_{ads})$ = cost,

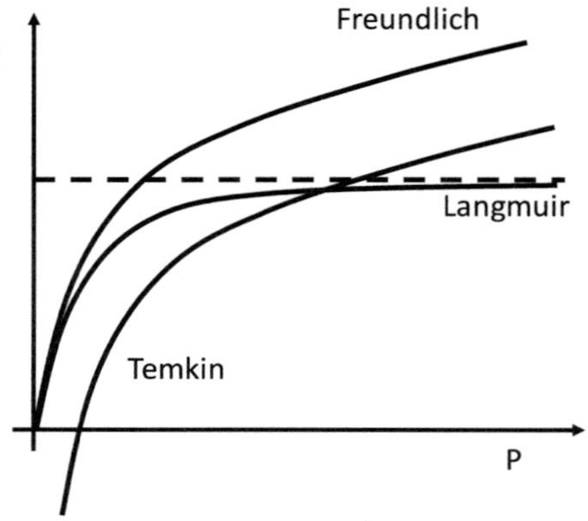

Fig. 5.3 A qualitative description of the adsorption according to the three different isotherms: *L* Langmuir, *F* Freundlich, *T* Temkin

that is, active sites with different adsorption heat have the same probability to exist. In this case, we have:

$$\varphi(P) = C_1 + C_2 \ln P \quad \text{(isotherm of Temkin)} \tag{5.52}$$

Another possibility is to assume: $\rho(Q_{\text{ads}}) = \alpha e^{-Q_{\text{ads}}}$. This means assuming a random distribution of the adsorption heats. Active sites with low adsorption heat are much more numerous than active sites with high adsorption heat. In this case, we will have:

$$\varphi(P) = C P^{\beta RT} = C P^n \text{(isotherm of Freundlich)} \tag{5.53}$$

In conclusion, we have seen three different types of isotherm: (1) the Langmuir isotherm characterized by a uniform surface; (2) the Freundlich isotherm, which does not go to saturation and can be used only for low-coverage degree; and (3) the Temkin isotherm, which has no physical mean at low-coverage degree and must be used only for a coverage degree near $\varphi(P) \simeq 0.5$. In Fig. 5.3, the behavior of the three described isotherms is qualitatively compared considering the coverage degree, q, as a function of the pressure, P.

Often, by studying the chemisorption rate, it has been experimentally observed (Elovich equations) that:

$$r_A = k_A P e^{-g\theta} = \text{adsorption rate} \tag{5.54}$$

$$r_{-a} = k_{-a} e^{h\theta} = \text{desorption rate} \tag{5.55}$$

At equilibrium:

$$r_a = r_{-a} \quad k_a P e^{-g\Theta} = k_{-a} e^{h\Theta} \tag{5.56}$$

it follows that:

$$\frac{k_a P}{k_{-a}} = e^{h\theta + g\theta} = e^{\theta(h+g)}$$
$$\theta = \frac{1}{h+g} \ln \frac{k_a}{k_{-a}} P = \frac{1}{f} \ln bP \tag{5.57}$$

This equation corresponds to the Temkin isotherm.

5.4.5 *The Kinetics for Heterogeneous Complex Reaction Systems*

Equilibrium Reactions

As was seen in Chap. 4, devoted to homogeneous reactions, in a complex reaction system, we can find the following: (a) equilibrium reactions, (b) simultaneous reactions, and (c) consecutive reactions. The same occurs for heterogeneous catalytic reactions. Let us consider the simplest equilibrium reaction:

$$A \rightleftarrows B \tag{5.58}$$

As previously described, we must consider a mechanism as occurring in three steps:

$$A + \text{active site(S)} \rightarrow \text{AS(adsorption reaction)} \tag{5.59}$$

$$\text{AS} \rightleftarrows \text{BS} \qquad \text{(surface reaction)} \tag{5.60}$$

$$\text{BS} \rightleftarrows B + \text{active site(S)} \quad \text{(desorption reaction)} \tag{5.61}$$

We have four possibilities: One of the three reactions is a rate-determining step; the other two must be considered to be in equilibrium (apply the hypothesis of the slow step); and two or all of the steps have a comparable rate (apply the steady-state approximation).

The rates of the three different steps are as follows:

Adsorption rate:

$$r_A = k_A \left[p_A (1 - \Theta_A - \Theta_B) - \frac{1}{b_A} \Theta_A \right] \tag{5.62}$$

Surface reaction rate:

$$r_S = k_S \left(\Theta_A - \frac{\Theta_B}{K_S} \right) \tag{5.63}$$

Desorption rate:

$$r_B = k_B\left[\Theta_B - \frac{p_B(1-\Theta_A-\Theta_B)}{b_B}\right] \tag{5.64}$$

If the surface reaction is the slow step, we know that:

$$\Theta_A = \frac{b_A p_A}{1+b_A p_A + b_B p_B} \quad \Theta_B = \frac{b_B p_B}{1+b_A p_A + b_B p_B} \tag{5.65}$$

Therefore:

$$r = r_S = k_S\left(\Theta_A - \frac{\Theta_B}{K_S}\right) = \frac{k_S b_A\left(p_A - \frac{b_B p_B}{b_A K_S}\right)}{(1+b_A p_A + b_B p_B)} = \frac{k_S b_A\left(p_A - \frac{p_B}{K_P}\right)}{(1+b_A p_A + b_B p_B)} \tag{5.66}$$

where K_S is the surface reaction equilibrium constant, that is, the equilibrium between the adsorbed species A and B, and $K_P = K_S b_A/b_B$ is the reaction gas phase equilibrium constant. In the case in which adsorption is the slow step, the surface reaction and desorption are considered at equilibrium, and we will have:

$$r_A = k_A\left(p_A\Theta_f - \frac{\Theta_A}{b_A}\right) \quad \text{Adsorption rate(RDS)} \tag{5.67}$$

$$r_S = k_S\left(\Theta_A - \frac{\Theta_B}{K_S}\right) \text{Surface reaction(at equilibrium)} \frac{r_S}{k_S} = 0 \qquad \Theta_A = \frac{\Theta_B}{K_S} \tag{5.68}$$

$$r_B = k_B\left(\Theta_B - \frac{p_B\Theta_f}{b_B}\right) \text{Desorption(at equilibrium)} \frac{r_B}{k_B} = 0 \qquad \Theta_B = \frac{p_B\Theta_f}{b_B} \tag{5.69}$$

Substituting Θ_A in the adsorption rate expression, we obtain:

$$r_A = k_A\Theta_f\left(p_A - \frac{p_B}{b_A b_B K_S}\right) \tag{5.70}$$

$$\Theta_f = (1-\Theta_A-\Theta_B) = \left(1 - \frac{p_B\Theta_f}{b_B K_S} - \frac{p_B\Theta_f}{b_B}\right) \tag{5.71}$$

Hence:

$$\Theta_f = \frac{1}{1+\frac{p_B}{b_B K_S}+\frac{p_B}{b_B}} \qquad r = r_A = \frac{k_A\left(p_A - \frac{p_B}{b_A b_B K_S}\right)}{\left(1+\frac{p_B}{b_B K_S}+\frac{p_B}{b_B}\right)} \tag{5.72}$$

Again, consider that at equilibrium $K_s = \Theta_B/\Theta_A$ and $K_P = p_B/p_A$ and that a relation between K_S and K_P exists and is $K_S = K_P\, b_B/b_A$, where K_P is the thermodynamic gas-phase equilibrium constant.

A similar approach can be developed when the rate-determining step is the desorption rate. In this case, adsorption and the surface reaction must be considered at equilibrium, and the result is:

$$r = r_B = \frac{k_B b_A K_S \left(p_A - \frac{p_B}{b_A b_B K_S}\right)}{1 + b_A p_A (1 + K_S)} \tag{5.73}$$

We have considered different cases in which the rate-determining step is known (adsorption, surface rate, or desorption), but in the case in which the rates are all comparable, we must assume a "steady-state condition" considering the rates of any single step and equating $r_A = r_S = r_B$. By eliminating Θ_A and Θ_B, we obtain in this case:

$$r = \frac{K_S b_A p_A - b_B p_B}{\left(\frac{K_S}{k_S} + \frac{K_S b_A}{k_A} + \frac{b_B}{k_B}\right) + \left[\frac{K_S}{k_S} + \frac{b_B(1+K_S)}{k_B}\right] b_A p_A + \left[\frac{K_S}{k_S} + \frac{b_A(1+K_S)}{k_A}\right] b_B p_B} \tag{5.74}$$

By observing the complexity of this expression for describing the kinetics of a very simple reaction, we can appreciate the great advantage of applying the "rate-determining step hypothesis."

Let us consider now the following bimolecular equilibrium reaction:

$$A + B \rightleftarrows C + D \tag{5.75}$$

By assuming the surface as the rate-determining step, it is possible to write:

$$r = k_s \Theta_A \Theta_B - k_{-s} \Theta_C \Theta_D = k_s \left(\Theta_A \Theta_B - \frac{\Theta_C \Theta_D}{K_S}\right) \tag{5.76}$$

$$\Theta_i = \frac{b_i p_i}{1 + \sum_j b_j p_j} \tag{5.77}$$

while:

Equilibrium constant in the gas phase

$$K_S = \left(\frac{\Theta_C \Theta_D}{\Theta_A \Theta_B}\right) = \left(\frac{b_C p_C b_D p_D}{b_A p_A b_B p_B}\right) = K_P \frac{b_C b_D}{b_A b_B} \tag{5.78}$$

$$r = k_S \frac{b_A b_B \left(p_A p_B - \frac{p_C p_D}{K_P}\right)}{\left(1 + b_A p_A + b_B p_B + b_C p_C + b_D p_D\right)^2} \tag{5.79}$$

$$r = k_S \frac{b_A b_B p_A p_B \left(1 - \frac{1}{K_P} \frac{p_C p_D}{p_A p_B}\right)}{\left(1 + \sum_{j=1}^{NC} b_j p_j\right)^2} \tag{5.80}$$

This is the most general equation, but often some terms at the denominator are negligible, and thus the expression can be simplified in practice. In addition, consider that adsorption equilibrium parameters generally vanish by increasing the temperature.

Simultaneous or Parallel Reactions
In the case of simultaneous parallel reactions, we have:

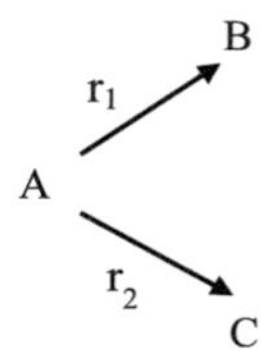

$$r_A = -r_1 - r_2; \quad r_B = r_1 = k_1 \Theta_A; \quad r_C = r_2 = k_2 \Theta_A$$

and

$$\Theta_A = \frac{b_A p_A}{1 + \sum_j b_j p_j} \tag{5.81}$$

A reaction scheme containing simultaneous or competitive reactions is easily recognizable because $r_1/r_2 = k_1/k_2 = p_B/p_C$.

Consecutive Reactions
In the case of consecutive reactions, we can have, for example:

$$A \xrightarrow{r_1} B \xrightarrow{r_2} C \tag{5.82}$$

Hence:

$$r_A = -r_1 = -k_1 \Theta_A \tag{5.83}$$

$$r_B = r_1 - r_2 = k_1 \Theta_A - k_2 \Theta_B \tag{5.84}$$

$$r_C = r_2 = k_2 \Theta_B \tag{5.85}$$

where $\Theta_A + \Theta_B + \Theta_C = 1$ and Θ_i can be expressed as in Eq. (5.32).

Consecutive reactions are recognizable because the concentration of the intermediate species—such as B—at different residence times passes through a maximum.

5.4.6 The Collection and Processing of Kinetic Data with the Scope of Determining the Kinetic Equation

The main scope of the collection and processing of kinetic data normally is the evaluation of the kinetic equation, that is, the relation according to which the reaction rate depends on temperature and composition. Only when this equation is known can the parameters appearing in the equation be exactly determined. The two abovementioned problems are different; it is opportune, therefore, to solve them with different approaches. Sequential experiences can be planned to determine the functional form of the kinetic equation by considering the influence of the different variables individually. In particular, experiences are useful in differential tubular reactors, CSTRs, or tubular reactors with external recirculation because they allow a simple elaboration of the reaction-rate data. However, also the integral tubular reactor can be usefully used with some difficulty. In this case, the conversion data must be related to the residence time, *W/F* (*W* is the mass of the catalyst and *F* is the reactant flow rate), in polynomial equations of the following type:

$$\lambda = a\left(\frac{W}{F}\right) + b\left(\frac{W}{F}\right)^2 + c\left(\frac{W}{F}\right)^3 + d\left(\frac{W}{F}\right)^4 \ldots \tag{5.86}$$

The parameters—a, b, c, d, etc.—can be evaluated by mathematical regression analysis of the experimental data. The derivative $d\lambda/d(W/F)$, corresponding to the reaction rates, can easily be calculated from Eq. (5.86) and applied to determine the exponent of a kinetic power law. For a reaction $A + B \rightarrow C$, the power law could be:

$$r = kp_A^{\alpha} p_B^{\beta} p_c^{\gamma} \tag{5.87}$$

where α, β, and γ can be small integers, fractions, positive, negative, or zero and give information on the form of the rate law. The problem of the integral reactor is to achieve isothermal conditions at each point of the reactor. When the functional form of the kinetic equation is known, kinetic parameters can be determined with accuracy by planning a factorial sequence of kinetic runs and submitting the results to statistical analysis.

5.4.7 Determination of Kinetic Parameters

The kinetic equation must be derived from experimental data of reaction rates, measured under different operative conditions, after the introduction of a reliable hypothesis of a reaction mechanism and the assumption of a rate-determining step. In this way, the kinetic equation has a physical mean, and the parameters appearing in the equation would obey all the thermodynamic constraints. Elementary steps, for example, cannot be trimolecular or more, and activation energy must be a positive number in the range 10 to approximately 50 kcal/mol, that is, in the range of the chemical bonds' energies. Values <6–7 kcal/mol suggest external diffusion limitation, while very high values >50 kcal/mol are suspected of the existence of thermal gradients. The pre-exponential factor would be inferior to the number of collisions of the molecules on the catalytic surface per time unit and per surface-area unit. This number can easily be calculated with the aid of the kinetic theory of gases and is (Boudart 1968):

$$A < \frac{1}{4}\left(\frac{8kT}{\pi m}\right)^{1/2} \tag{5.88}$$

Adsorption equilibrium parameters depend on the temperature as follows:

$$b_i = b_i^{\circ} \exp(-\Delta H/RT) = \exp(\Delta S/R)\exp(-\Delta H/RT) \tag{5.89}$$

The adsorption enthalpy changes ΔH are normally negative, falling in the range of -10 to -50 kcal, typical of the chemical bonds involved in the chemical adsorption. The adsorption entropy changes ΔS would be greater than the translational entropy and less than the overall entropy of the molecule in the gas phase. Both values can easily be calculated. The translational entropy can be calculated as:

$$S_t^{\circ} = \frac{3}{2}R\ \ln M + \frac{5}{2}R\ \ln T - Rnl P - 2,2298(u.e.) \tag{5.90}$$

while the total entropy of the molecules is normally tabulated values.

If the kinetic parameters obtained by elaborating the kinetic data have a physical mean, the related mechanism is supported and can give suggestions for the improvement of the catalytic activities and selectivities by opportunely modifying the catalyst surface.

5.4.8 Effects of Catalyst Dispersion, Sintering, and Poisoning on Kinetics

The specific surface area of a catalyst increases by increasing dispersion. The dispersion of a metal on a support, for example, can be defined as the ratio between the atoms on the surface and the total atoms of the metal. However, the surface atoms have different positions and properties. We can have atoms on a vertex, atoms on a corner, atoms on a plane, adatoms, and so on. The distribution of the different types of atoms can change strongly with the dispersion. As suggested by Boudart (1985), if a reaction is catalyzed on sites whose concentration strongly changes with the dispersion, specific activities for this reaction will strongly be affected by catalyst dispersion (reaction structure sensitive or demanding), while, on the contrary, when the reaction occurs on all the surface atoms, specific activities do not change with dispersion (reaction insensitive or facile). If more than one reaction occurs, dispersion also could affect selectivities. This fact explains the bad reproducibility in the behavior of some catalysts and for the change in activity and selectivity as a consequence of sintering.

Catalyst poisoning is another phenomenon responsible for changing catalyst activities and selectivities. Permanent poisons give strong chemical bonds with the catalytic sites and gradually decrease the activity to zero. Different kinetic equations have been proposed for considering the poisoning effect. Two different approaches have been considered. One approach considers poisoning as an independent phenomenon occurring with its own kinetics in reducing the initial activity:

$$\frac{A_{\text{residual}}}{A_o} = 1 - \alpha c \quad c = \text{conc.of the poison } \alpha = \text{poisoning coefficient} \tag{5.91}$$

The other approach considers the poisoning effect dependent on the kinetics of the occurring reactions. Permanent poisoning of the first type is useful for determining the density of active sites on a catalyst surface. Temporary poisoning, on the contrary, is the consequence of the competitive adsorption of molecules on the catalytic sites. These molecules reduce more or less the activity according to the value of their adsorption equilibrium constant, and the Langmuir—Hinshelwood model considers quantitatively this type of poisoning.

5.4.9 Theta Rule: Compensating Effect

As has been seen, the surface of solid catalysts is not usually uniform, and active sites are different for the interaction energy exchanged with the chemisorbed molecules. This difference affects the reaction rate mainly by altering the value of the activation energy. For this purpose, it is possible to introduce a distribution function of the type:

$$\varphi(E_S) = \alpha e^{hE_S} \tag{5.92}$$

This expression gives the fraction of active sites having activation energy between E_S and E_S + dE_S. If we consider a simple reaction, $A \rightarrow B$, occurring on a uniform catalytic surface, we can write:

$$r = k_r b_A p_A = \mathrm{A}\mathrm{e}^{-E_S/\mathrm{RT}} b_A p_A \tag{5.93}$$

On the contrary, for the simple reaction occurring on a nonuniform surface, we can write:

$$r = A b_A p_A \int_{E_{\min}}^{E_{\max}} \mathrm{e}^{-E_S/RT} \varphi(E_S) \mathrm{d}E_S \tag{5.94}$$

$$r = A b_A p_A \int_{E_{\min}}^{E_{\max}} \mathrm{e}^{-E_S/RT} \alpha \mathrm{e}^{hE_S} dE_S \tag{5.95}$$

$$r = \alpha A b_A p_A \int_{E_{\min}}^{E_{\max}} \mathrm{e}^{\left(h - \frac{1}{RT} E_S\right)} \mathrm{d}E_S \tag{5.96}$$

$$r = \frac{A \alpha b_A p_A}{h - \frac{1}{RT}} \left[\mathrm{e}^{\left(h - \frac{1}{RT}\right) E_{\max}} - \mathrm{e}^{\left(h - \frac{1}{RT}\right) E_{\min}} \right] \tag{5.97}$$

If $h < 1/RT$, the first term in the brackets can be neglected, while if $h > 1/RT$, the second term can be ignored. Therefore, it results:

$$r = k_d \exp\left[\left(h - \frac{1}{RT}\right) E\right] b_A p_A \tag{5.98}$$

where:

$$k_d = \pm \frac{A\alpha}{h - 1/RT} \tag{5.99}$$

E is the activation energy, in other words, the maximum or minimum value of Eq. (5.97). Equation (5.98) can also be written as:

$$r = k_d \mathrm{e}^{hE} \mathrm{e}^{-E/RT} b_A p_A = A_d \mathrm{e}^{-E/RT} b_A p_A \tag{5.99}$$

As can be seen from this equation, the pre-exponential factor A_d now depends on the temperature, and it has been observed to be correlated with the activation energy. In particular, it has been observed—by studying a reaction in the presence of different catalysts of the same type or reactions similar to the same catalyst—that Ad and E are correlated as shown qualitatively in Fig. 5.4.

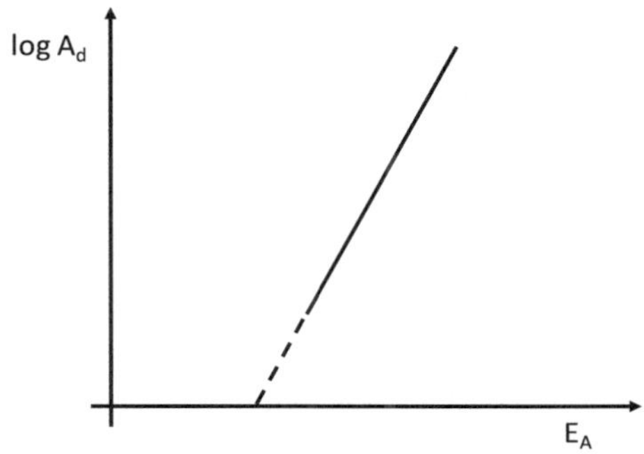

Fig. 5.4 Compensation effect, in other words, the correlation between the pre-exponential factor and activation energy because of the nonuniformity of the catalytic surface

5.5 Continuous Gas–Solid Laboratory Reactors

5.5.1 Tubular Reactors, the Ideal Conditions for the Laboratory Reactors: Plug Flow and Isothermal Conditions

Laboratory tubular reactors are constructed in such a way as to satisfy as much as possible two ideal conditions: a plug-flow fluid dynamic model and isothermal conditions. As seen in Chap. 4, a plug-flow reactor (PFR) is characterized by a flow stream, in which the fluid is well mixed in the radial but not the axial direction. All the molecules will have the same residence time inside the reactor (Fig. 5.5). The approach to the PFR condition can be experimentally determined by feeding a stream of a probe molecule inside a carrier gas and measuring its concentration at the reactor exit (step test). Clearly, the presence of catalyst particles inside the cylindrical tube favors the radial mixing, especially at a high flow rate.

In Fig. 5.5, it is possible to observe that the deviation from the "plug-flow ideal condition" gives place to a distribution of the concentration along the time, thus rendering more difficult the mathematical interpretation of the results. This aspect will be examined in more detail in Chap. 6, which is devoted to mass and heat transfer in reactors. We consider now only ideal systems (isothermal PFR or CSTR reactors) useful for individuating the "intrinsic kinetics" and determining the related parameters. For this purpose, plug-flow condition is surely obtained for a turbulent flow when:

$$L/\mathrm{d}_P \geq 30,$$

where being

L	reactor length
dp=	particle diameter

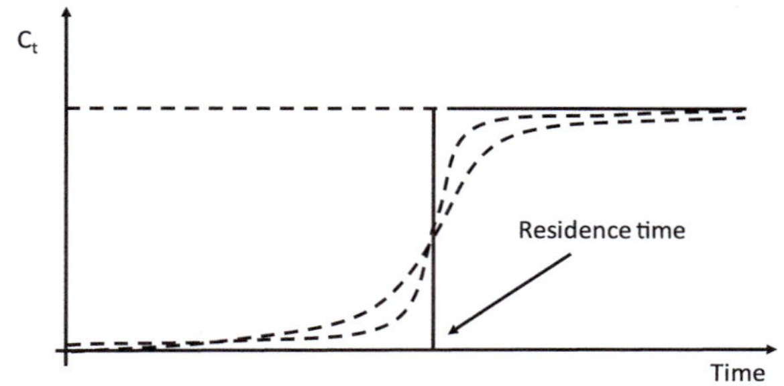

Fig. 5.5 Continuous line represents the ideal condition of plug flow, while the dotted lines show deviations from the ideal behavior

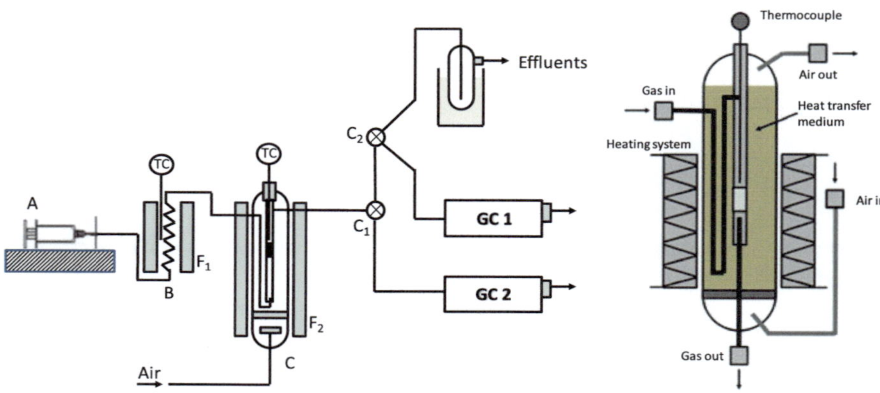

Fig. 5.6 Scheme of a laboratory plant with a plug-flow–isothermal reactor. *A* syringe pump for feeding a liquid or other system for continuous (not pulsed) flow, *B* preheater, F_1 heating oven, *C* reactor jacketed and immersed in a fluidized sand bath, *F*2 heating oven, *C*1 and *C*2 switching valves, and *TC* thermocouples

Moreover, under isothermal conditions, the reaction rate is exclusively a function of the composition; therefore, the second ideal condition has the scope to simplify the elaboration of the laboratory experimental data to determine the most reliable "intrinsic kinetic law" of the reaction. For obtaining isothermal conditions, there are different possibilities:

(a) Thermostat the reactor with $\begin{cases} \text{Molten metals} \\ \text{Molten salts} \\ \text{Fluidized bed of sand} \end{cases}$
(b) Dilution of the catalyst with inert particles
(c) Dilution of the reactants
(d) Reduction of the reactor size

An example of a laboratory plant is shown in Fig. 5.6. The reactor can be realized in glass, quartz, or stainless steel (when the reaction is performed under pressure). In the same figure, a scheme of the tubular reactor is also reported in more detail.

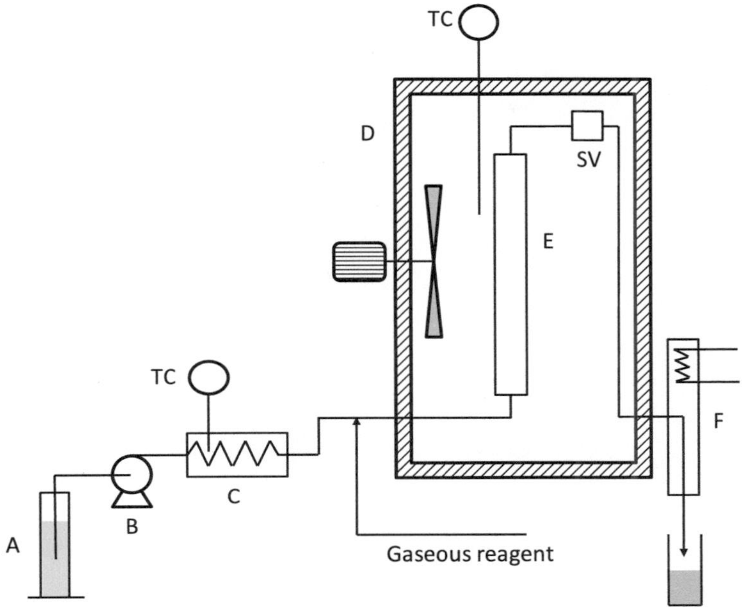

Fig. 5.7 Another example of a scheme of a laboratory with a plug-flow isothermal reactor. *A* liquid reagent reservoir, *B* micrometric pump, *C* preheater, *F* collecting vessel, *TC* thermocouples, and *SV* sampling valves for the analysis

Another example is shown in Fig. 5.7. In this case, a stainless-steel tube is put inside a vented oven. A device of this type can be used when the reaction is athermic. Otherwise, the tube must be immersed, as seen previously, in a fluidized bed of sand or in stirred or recirculated thermostating fluid. In the reported scheme, a liquid is fed by using a micrometric pump. Pellets of inert material are put inside the tube to mix the reactants well.

Another example of a stainless-steel tubular laboratory reactor is shown in Fig. 5.8. Also in this case, isothermal conditions are achieved with the employment of a fluidized bed of sand.

5.5.2 *Mass Balance for an Ideal Plug-Flow–Isothermal Reactor*

The mass balance in such type of reactors has already been described in Chap. 4. Considering the simplest reaction, $A \rightarrow B$, with A fed into the reactor with a molar flow rate, F. An example of the mass balance in this case is:

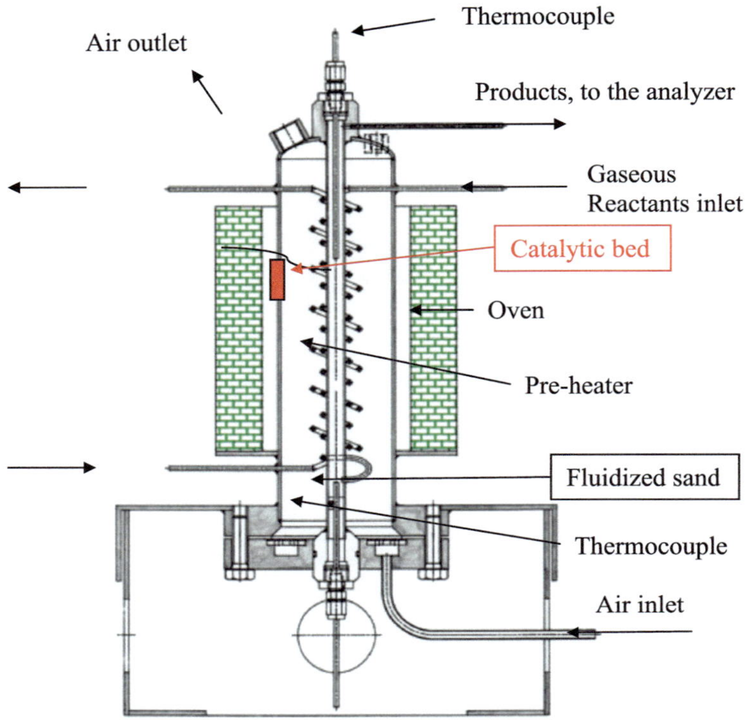

Fig. 5.8 A tubular laboratory reactor kept isotherm with a fluidized bed of sand

$$r_A \mathrm{d}W = \mathrm{F}\, \mathrm{d}\lambda_A \tag{5.100}$$

where:

λ_A	Conversion of the reagent A
F	Molar flow rate of A (mol/time)
W	Mass of catalyst
r_A	Reaction rate (mol/mass of catalyst*time)

The choice to express the reaction rate in "mol/mass of catalyst time" is arbitrary, and we can make other choices considering, for example, a flow rate in overall mass/ time or referring the rate to the unit volume of the reactor. The most opportune choice depends on the type of considered reaction. For example, if the number of moles changes because of the reaction, we know that the mass flow remains constant. In this case, a reaction rate expressed as "mass of reactant/mass of catalyst time" will be more convenient, and the mass balance will be arranged accordingly. By integrating between $W = 0$ and $W = W$, we obtain:

$$\frac{W}{F} = \int_{X_o}^{X_f} \frac{d\lambda_A}{r_A} \tag{5.101}$$

Our problem is now to evaluate the unknown equation, $r_A = r_A(\lambda_A)$, while remembering that

$$r_A = d\lambda_A / d(W/F) \tag{5.102}$$

5.5.3 Determination of Kinetics Using Integral Reactors

A reactor is considered "integral" when the reactant conversion is >5–10%. For this type of reactor, we must solve Eq. (5.101). To do this, experimental data of *A* conversion are collected for different residence time W/F and plotted as shown in Fig. 5.9.

The experimental points are opportunely interpolated, for example, with a polynomial equation, such as the following one:

$$X = a\left(\frac{W}{F}\right) + b\left(\frac{W}{F}\right)^2 + c\left(\frac{W}{F}\right)^3 + d\left(\frac{W}{F}\right)^4 \dots \tag{5.103}$$

Letter *a*, *b*, *c*, *d*—the coefficients of the polynomial equation—can be determined by mathematical regression analysis. In the plot shown in Fig. 5.9, r_A values are the slopes at any point of the continuous curve. Therefore, r_A can be easily determined as the derivative of Eq. (5.103). Then, the numerical values of r_A, found in this way, must be used for determining the reaction orders and a kinetic law corresponding to a reliable reaction mechanism. This procedure is long and complicated, and errors are possible, in particular for the difficulty of realizing isothermal conditions for the whole reactor length. In conclusion, the integral reactor has two advantages: (1) high precision in the analysis of the composition at the reactor outlet as a consequence of

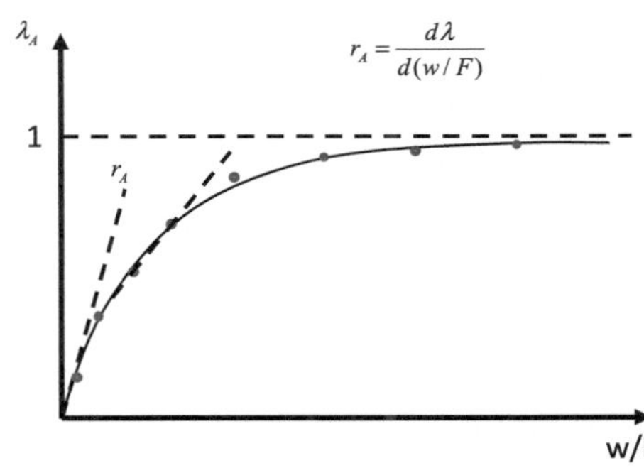

Fig. 5.9 Kinetic behavior of a tubular integral reactor

the high conversions and (2) the possibility of collecting many reaction rate data at a given temperature simply by changing the flow rate. The drawbacks are, on the contrary, as follows: (1) the difficulty of maintaining isothermal conditions in the whole reactor and (2) the complexity of data elaboration.

5.5.4 Determination of Kinetics with Differential Reactors

A tubular reactor is considered "differential" if the conversion of the reactant is <5%. For these low conversion values, it is possible to evaluate an average value of the reaction rate by using the approximate equation:

$$\overline{r}_A \cong \frac{F}{W} \Delta\lambda_A \tag{5.104}$$

The lower the value of $\Delta\lambda_A$, the more r_A approaches the correct value, as can be argued by the plot shown in Fig. 5.10. However, an error is always present in the evaluation of r_A using this method.

Considering the initial reaction rates, the eventual deactivating effect of the reaction products on the reaction rate can be determined by feeding a mixture containing the reactants and a product into the reactor. A comparison of the initial activity in the presence or the absence of that product allows us to individuate the effect of the product on the reaction rate. Repeating the operation for all the products allows us to evaluate the kinetic law and collect information about the reaction mechanism. In some cases, an integral reactor is used to produce the mixture to be fed to a differential reactor. Moreover, an integral reactor can be transformed in a sequence of differential reactors as shown in Fig. 5.11, where reagents and products are withdrawn and analyzed at different points of the reactor; the average temperature for any single portion of the reactor also is measured.

The main advantage of the differential reactor is direct observation of the factors influencing the reaction rate. For example, if we increase the concentration of the reactants by a factor of two and the reaction rate is doubled, the reaction is clearly of

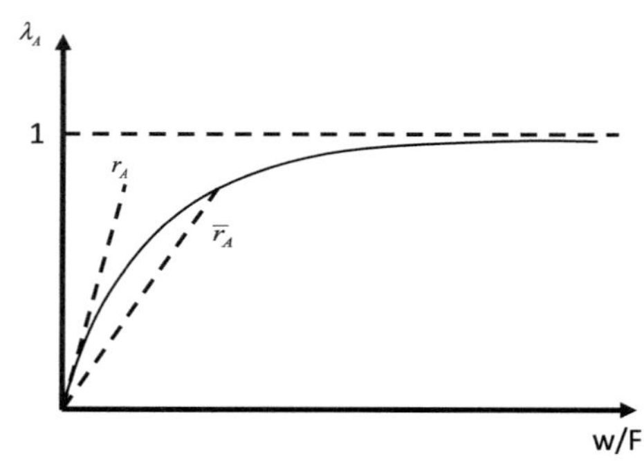

Fig. 5.10 The error made by evaluating r_A using the method of the differential reactor

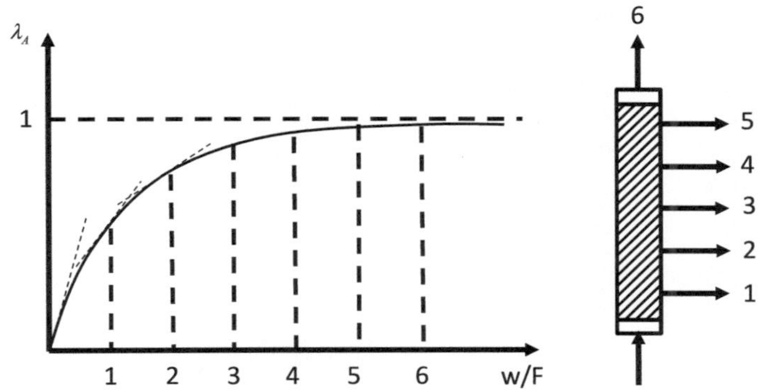

Fig. 5.11 An integral reactor constituting a sequence of differential reactors. The slopes are the averaged reaction rates

the first order. By changing the temperature, we can evaluate immediately the activation energy. If a product has a poisoning effect, it is sufficient to feed it together with the reagent and evaluate the effect on the reaction rate. In other words, by using the differential reactor, it is easier to individuate the reaction orders and the kinetic law of the reaction together with an approximated estimation of the kinetic parameters. Another advantage is the isothermal behavior, which is easily achieved because the reactor is very small. The disadvantage is the low precision of analysis of the reaction mixture at the reactor exit because of the obtained low conversions and the approximation introduced in the averaging the measurement of the reaction rate.

Exercise 5.1 Steam Reforming of Methanol

The steam reforming of methanol is an important reaction useful for producing hydrogen under mild conditions to be used, for example, in fuel cells. The reaction is:

$$CH_3OH + H_2O \rightarrow CO_2 + 3H_2 \tag{5.105}$$

As can be seen, methanol is a good hydrogen vector because it yields 3 mol for each mole of reacted methanol. This reaction, in the presence of CuO-ZnO-Al_2O_3 catalysts, occurs at relatively low temperatures (200–250 °C). Santacesaria and Carrà (1978) studied the kinetics of this reaction in the presence of three different commercial catalysts normally used for promoting the gas-shift reaction at low temperature, that is, the following reaction:

$$CO + H_2O \rightleftarrows CO_2 + H_2 \tag{5.106}$$

The kinetics of steam reforming was studied in a laboratory reactor such as the one shown in Fig. 5.6. Experimental data were collected in a temperature range from 150 to 200 °C. The adopted flow-rate conditions were high enough to avoid external

Table 5.3 Experimental data collected for the steam reforming of methanol on a CuO-ZnO-Al_2O_3 catalyst[a]

Run	T (°C)	Ratio H_2O/CH_3OH $R_{W/M}$	Conversion	Reaction rate (mol/h g) × 10^3
1	190	0	0.0636	14.00
2	190	1	0.0606	13.33
3	190	2	0.0516	11.36
4	190	4	0.0437	9.61
5	190	7	0.0355	7.81
6	190	9	0.0303	6.66
7	190	11	0.0267	5.88
8	172	1	0.0282	6.20
9	178	1	0.0359	7.90
10	184	1	0.0430	9.45
11	189	1	0.0514	11.30
12	193	1	0.0636	14.00
13	194	1	0.0673	14.80
14	202	1	0.1068	23.50

[a]For the runs made at RW/M = 1 methanol, the flow rate was 0.11 mol/h. The other runs were performed while keeping the overall flow rate constant

mass-transfer limitation allowing, in the meantime, to maintain the methanol conversion $>10\%$, that is, the PFR reactor operated under differential conditions, and initial reaction rates were so determined. The commercial catalysts were powdered and sieved to obtain granules of 40–60 mesh to avoid internal diffusion limitation. Catalyst, 0.5 g, diluted 1:2 with corundum having the same granulometry was put into the reactor, which was kept isothermal with an external fluidized bed of sand and using air for fluidization. The reactor was heated by an oven. Two chromatographs, connected to two different sampling valves heated at 130 °C, were used for the analysis of, respectively, H_2 and CO (nitrogen as carrier gas, first chromatograph) and CO_2, CH_3OH and H_2O (helium as carrier gas, second chromatograph). For one of the three catalysts, the collected experimental data are reported in Table 5.3.

Based on some experimental evidence, it has been proposed by the authors that the reaction occurs in two steps:

$$(1)\quad CH_3OH \rightarrow CO + H_2 (\text{rate determining step}) \tag{5.107}$$

$$(2)\quad CO + H_2O \rightleftarrows CO_2 + H_2 \tag{5.108}$$

By adding these two reactions, the steam-reforming reaction is obtained.

The first step would be slow, while the second is very fast, probably always at equilibrium. This hypothesis is confirmed by the following observations:

1. CO is present as a by-product but in a very small amount.
2. Methanol, fed in the absence of water on the same catalyst, at 190 °C, decomposes with kinetics of the following type:

$$r_M = k_{\mathrm{app}} p_M^{0.834} \tag{5.109}$$

This power law suggests a Langmuir–Hinshelwood kinetic law of the following type:

$$r_M = \frac{k_M b_M p_M}{1 + b_M p_M} \tag{5.110}$$

Considering the possibility that water can be adsorbed on the surface in competition with methanol, we can write for the steam reforming:

$$r = \frac{k_M b_M p_M}{1 + b_M p_M + b_W p_W} \tag{5.111}$$

Considering now the inverse of the reaction rate, it holds:

$$\frac{1}{r} = \frac{1}{r_M} + \frac{b_W}{k_M b_M} R_{W/M} = \overbrace{\frac{1}{k_M}\left(1 + \frac{1}{b_M p_M}\right)}^{\text{Intercept}} + \overbrace{\left(\frac{b_W}{k_M b_M}\right)}^{\text{Slope}} R_{W/M} \tag{5.112}$$

That is, the reciprocal of the reaction rate would be a linear function of the molar ratio of water to methanol.

Based on the isothermal data, collected for different $R_{W/M}$ initial values, we verify this hypothesis. Determine by extrapolation the value of r_M (consider that $R_{W/M} = 0$ it holds $1/r = 1/r_M$). Estimate then, by linear regression analysis, at 190 °C (7 runs), an approximate value of the parameters k_M, b_M, and b_W, considering in all cases a pressure of 1 atm. Last, evaluate the dependence on the temperature of the three parameters—k_M, b_M, and b_W—interpolating the experimental data collected at different temperatures.

Solution

Part 1

In the first part, linearization of the experimental data collected at 190 °C is performed (see Fig. 5.12). Extrapolation of the reciprocal of reaction rate for $R_{W/M} = 0$ brings us to a value of r_m of 0.0146

From this first approach, the values of slope and intercept have, respectively, been found: slope = 9.0226 and intercept = 68.5749

Part 2

Then, the isothermal data, at 190 °C, are submitted to mathematical regression analysis with the expression of reaction rate (see Fig. 5.13 [the obtained parity plot]):

$$r = \frac{k_M b_M p_M}{1 + b_M p_M + b_W p_W} \tag{5.113}$$

The following values of the kinetic parameters are estimated:

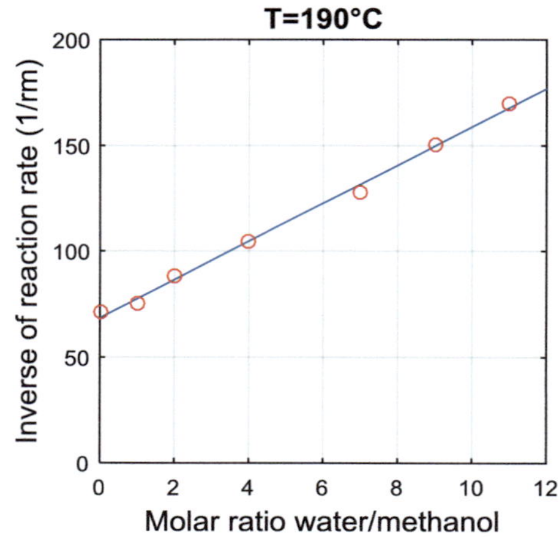

Fig. 5.12 Linear trend of the inverse of the reaction rate with the initial molar ratio of water to methanol for runs performed at 190 °C

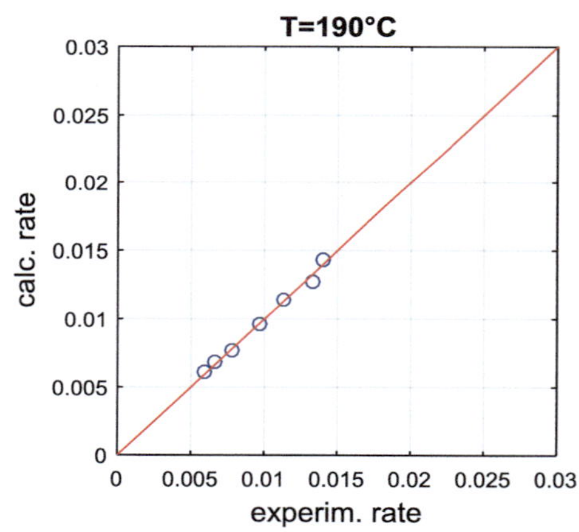

Fig. 5.13 Parity plot obtained for, respectively, the experimentally determined and calculated rates

$$k_M : 0.014266; b_M : 11626.2248; \text{and } b_W : 1381.5008$$

From these parameters, alternative values for slope and intercept can be calculated as follows: slope = 8.329 and intercept = 70.0997. These values are quite similar to those found in Part 1.

Part 3

Finally, all experimental data are simultaneously fitted by considering the temperature dependence of the parameters (see Fig. 5.14).

The corresponding obtained parameters are as follows:

$k_M°$: 6.1693e+09	ΔE_M: 24630
$b_M°$: 1.7486e-07	ΔHm: 18474
$b_W°$: 1.9619	ΔHw: 1571

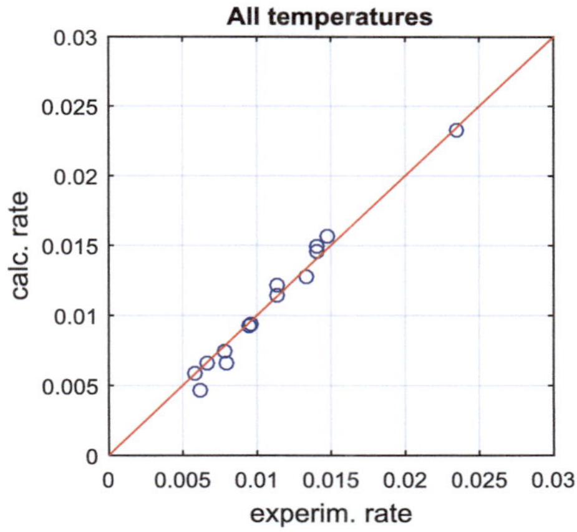

Fig. 5.14 Comparison between experimental reaction rates and rates calculated with the described model

Comments to the Obtained Results

The obtained parameters and related confidence intervals are as follows:

Parameters	C.I.	SD
km0 =	+6.2805e+09 +/− +1.9369e+11	8.3995e+10
bm0 =	+5.8870e-07 +/− +6.7307e-05	2.9188e-05
bw0 =	+1.9921e+00 +/− +2.0818e+03	9.0276e+02
ea =	+2.4640e+04 +/− +2.3328e+04	1.0116e+04
dhm =	+1.7360e+04 +/− +5.0431e+05	2.1870e+05
dhw =	+1.5980e+03 +/− +4.9841e+05	2.1614e+05

Correlation coefficients and test
R^2 = +0.972813
R^2 adj = +0.949511
f-test = 41.7467
e-test = 47.7105
FF-test = 67.2776

As we can see, most of the parameters are affected by an error, from a statistical point of view, that is greater than the parameter itself. It is interesting to observe that notwithstanding the parameters' wide confidence intervals, the overall data-fitting is quite good if we consider the parity plot and the high value of the R square. This high uncertainty on the parameters is mainly due to the strong correlation among them, as can be observed by inspecting the correlation matrix evaluated by the MATLAB code:

Parameter-correlation matrix					
1.0000	0.5412	−0.4925	0.9675	−0.5568	−0.5568
0.5412	1.0000	−0.9494	0.3192	−0.9675	−0.9675
−0.4925	−0.9494	1.0000	−0.2574	0.9955	0.9955
0.9675	0.3192	−0.2574	1.0000	−0.3287	−0.3287
−0.5568	−0.9675	0.9955	−0.3287	1.0000	1.0000
−0.5568	−0.9675	0.9955	−0.3287	1.0000	1.0000

In particular, adsorption enthalpy of methanol and water (parameters no. 5 and 6) are strongly correlated, and this means that many couples of these parameters are equally statistically correct and furnish practically the same quality of fit. From the correlation matrix, it is evident that different other parameters are correlated because the corresponding element of the matrix is close to 1 (as an absolute value). In real case, this type of situation can be faced by including more data in the analysis or, ultimately, by evaluating different candidate models.

All the described results were obtained by using a MATLAB program available as Electronic Supplementary Material, Data 5.1.

Exercise 5.2 Hydrogenation of Iso-octenes: Another Example of Kinetic Analysis Using a Differential Reactor

Hougen and Watson (1947), starting from a work published by Tschernitz et al. (1946), examined 17 different kinetic models based on all of the possible reaction mechanisms and rate-determining steps for the reaction of hydrogenation of iso-octenes to isooctanes in the presence of Ni-supported catalysts. The reaction was performed in a reactor that operated under differential conditions. The reaction studied is:

$$C_8H_{16}(\text{gas}) + H_2(\text{gas}) \rightleftarrows C_8H_{18}(\text{gas}) \tag{5.114}$$

Because the reaction was far from the equilibrium, the reverse reaction can be neglected. Iso-octenes were fed in liquid form and vaporized before arriving on the catalytic bed kept at a constant temperature. The mechanisms considered were:

1. Both molecules of the reactants are adsorbed on the catalytic sites, and the possible rate-determining steps (RDS) include:
   (a) Hydrogen molecule adsorption.
   (b) Iso-octene molecule adsorption.
   (c) Isooctane molecule desorption.
   (d) The reaction between molecules adsorbed on the surface.
2. Hydrogen gives place to dissociative adsorption and reacts with adsorbed molecules of iso-octenes. Possible RDS include:
   (e) Dissociative hydrogen molecule adsorption
   (f) Iso-octene molecule adsorption

(g) Isooctane molecule desorption
(h) Reaction between atoms of hydrogen adsorbed on the surface and adsorbed iso-octene molecules

3. The reaction between molecules of hydrogen adsorbed on the surface and molecules of iso-octenes coming from the gas phase. Possible RDS include:

(i) Hydrogen molecule adsorption
(j) Isooctane molecule desorption
(k) The reaction between molecules of hydrogen adsorbed on the surface and iso-octene molecules from the gas phase

4. Reaction from atomic hydrogen adsorbed on the surface and iso-octene molecules coming from the gas phase. Possible RDS include:

(l) Hydrogen molecule adsorption
(m) Isooctane molecule desorption
(n) Reaction between molecules of hydrogen adsorbed on the surface and iso-octene molecules from the gas phase

5. The reaction between iso-octene molecules adsorbed on the surface and hydrogen coming from as phase. Possible RDS include:

(o) Iso-octene molecule adsorption
(p) Isooctane molecule desorption
(q) Reaction between molecules of iso-octenes adsorbed on the surface and hydrogen molecules coming from the gas phase

From the examination of kinetic parameters, it was possible to reject 15 of them because the parameters were unreliable (for example negative adsorption constants), and only the mechanism + RDS (1d) and (2 h) resulted in being acceptable. Then, from the best fitting of the mechanism + RDS, (1d) resulted in being the best. Therefore, we can write for this mechanism the following reaction rate:

$$r=\frac{kb_u b_{H_2} p_u p_{H_2}}{\left(1+b_u p_u+b_s p_S+b_{H_2} p_{H_2}\right)^2} \tag{5.115}$$

where:

b_u	adsorption equilibrium constant of the unsaturated molecule of iso-octene
b_s	adsorption equilibrium constant of the saturated molecule of iso-octane
b_{H2}	adsorption equilibrium constant of the hydrogen molecules
k	Kinetic constant
p_i	partial pressures of iso-octene (u), isooctane (s), and hydrogen (H_2)

Table 5.4 Experimental runs performed at 200°C[a]

Run	Overall P	p_{H2}	p_U	p_S	Rate (g mol/g h)
1	1.000	0.482	0.100	0.508	0.00353
2	3.500	2.459	0.527	0.515	0.02500
3	3.490	2.450	0.530	0.515	0.03200
4	3.510	0.477	0.494	2.538	0.00553
5	1.510	0.514	0.540	0.455	0.00870
6	1.500	0.473	0.552	0.473	0.01392
7	1.500	0.470	0.558	0.473	0.00960
8	1.105	0.104	0.562	0.440	0.00514
9	3.520	0.450	2.840	0.230	0.01920
10	3.510	0.409	2.810	0.289	0.02060
11	2.500	0.484	1.075	0.942	0.01310
12	2.100	0.357	1.590	0.153	0.01860

[a]See Tschernitz et al. (1946)

Table 5.5 Experimental runs performed at 275 °C[a]

Run	Overall P	p_{H2}	p_U	p_S	Rate (mol/g h)
26	3.50	0.500	2.735	0.270	0.02010
27	1.10	0.482	0.560	0.062	0.01140
28	1.50	0.501	0.533	0.466	0.01340
29	1.50	0.475	0.553	0.471	0.00946
30	1.50	0.476	0.553	0.478	0.00871
31	1.10	0.101	0.543	0.451	0.00250
32	3.50	2.335	0.550	0.610	0.03380
33	3.50	2.540	0.452	0.510	0.02860
34	3.50	2.405	0.524	0.572	0.02970
35	1.10	0.489	0.098	0.513	0.00440
36	3.51	0.480	0.480	2.550	0.01106
37	2.50	0.555	0.299	1.645	0.00533
38	2.50	0.599	0.996	0.907	0.01885
39	2.10	0.438	1.523	0.136	0.01414
40	2.10	0.478	1.480	0.144	0.01590

[a]See Tschernitz et al. (1946)

The experimental data, collected at 200 and 275 °C, are reported, respectively, in Tables 5.4 and 5.5.

Evaluate the kinetic parameters and their dependence on the temperatures. Construct a parity plot.

Results

The best-fitting kinetic parameters, as evaluated by mathematical regression analysis, are reported below together with their confidence intervals and standard errors:

Parameters	CI	SD
kc0 =	+5.0276e+00 ± +1.8267e+01	+8.9678e+00
bu0 =	+6.9593e−03 ± +1.8282e−02	+8.9751e−03
bs0 =	+1.1342e−07 ± +5.0789e−07	+2.4934e−07
bh0 =	+1.5263e−03 ± +5.7542e−03	+2.8249e−03
ea =	+1.9161e+03 ± +3.8107e+03	+1.8708e+03
dhu =	−4.8838e+03 ± +2.7977e+03	+1.3735e+03
dhs =	−1.6039e+04 ± +3.8885e+03	+1.9090e+03
dhh =	−5.6060e+03 ± +4.0465e+03	+1.9865e+03

The corresponding parity plot is shown in Fig. 5.15.

It is interesting to observe, by examining the parameter-correlation matrix reported below, that some of the estimated parameters are correlated with each other. In the parameter-correlation table, couples of strongly correlated parameters are evidenced by underlined and bolded data. This can be a typical situation that occurs when all the experimental data, at different temperatures, are submitted simultaneously to a mathematical regression and when the overall number of parameters is high with respect to the number of experimental data.

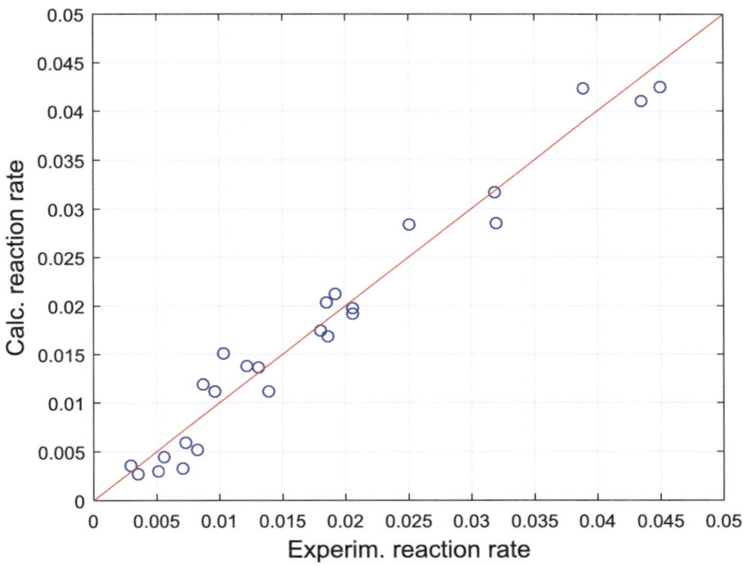

Fig. 5.15 Comparison between experimental reaction rates and rates calculated with the described model

Parameter-correlation matrix							
1.0000	−0.7753	−0.1299	−0.9197	**0.9955**	−0.7341	−0.0929	−0.8947
−0.7753	1.0000	0.2327	0.7498	−0.7831	0.9921	0.2569	0.7531
−0.1299	0.2327	1.0000	0.1012	−0.1470	0.2123	**0.9920**	0.1043
−0.9197	0.7498	0.1012	1.0000	−0.9312	0.7436	0.0986	**0.9946**
0.9955	−0.7831	−0.1470	−0.9312	1.0000	−0.7512	−0.1131	−0.9152
−0.7341	0.9921	0.2123	0.7436	−0.7512	1.0000	0.2468	0.7575
−0.0929	0.2569	**0.9920**	0.0986	−0.1131	0.2468	1.0000	0.1079
−0.8947	0.7531	0.1043	**0.9946**	−0.9152	0.7575	0.1079	1.0000

A possible alternative is to represent it by a regression made on each set of data at a fixed temperature and then construct then an Arrhenius plot in which the trend of lnK versus 1/T is shown. For the data of this exercise, such a plot is shown in Fig. 5.16:

The drawback of this approach is evident because the data are quite scattered, and an erroneous trend can be found—such as that of bu. (blue line and symbols)—that is physically not correct for an adsorption coefficient. All of the described results were obtained using a MATLAB program available as Electronic Supplementary Material, Data 5.1.

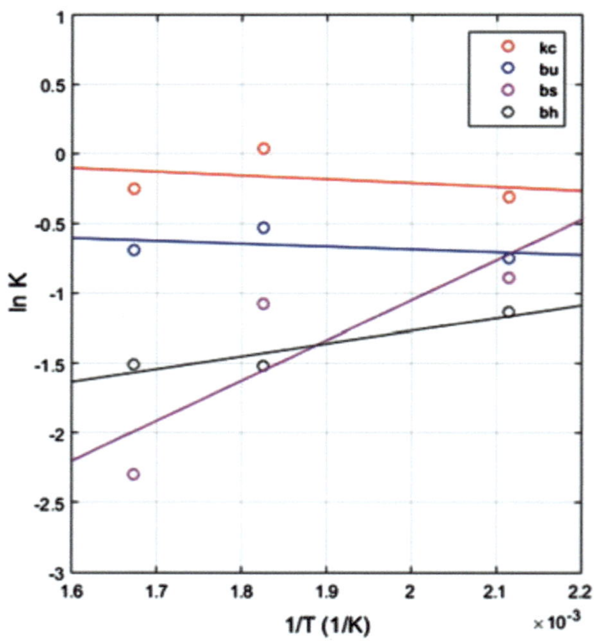

Fig. 5.16 Dependence of kinetic parameters on temperature

Exercise 5.3 Dehydrogenation of Ethanol to Acetaldehyde: Another Example of Kinetic Analysis by Differential Method

Franckaerts and Froment (1964) studied the reaction of ethanol dehydrogenation to acetaldehyde promoted by a $CuO/CoO/Cr_2O_3$ catalyst supported on asbestos and reduced in situ. The occurring reaction was:

$$CH_3CH_2OH(g) \rightarrow CH_3CHO(g) + H_2(g) \tag{5.116}$$

The reaction was studied in the gas–solid phase using an integral flow reactor with an internal diameter of 3.3 cm and a length of 110 cm. Different amounts of catalyst were used (3, 6, and 10 g) to operate with a wide range of space–time W/F_E (0.2, 0.4, 0.6, 0.88, and 1.6 g h/mol). Data of conversions were collected as a function of space–time for different temperatures by working at different pressures (1, 3, 4, 7, and 10 bars) by feeding ethanol normally containing 13.5 mol% of water. Some runs also were performed by feeding ethanol containing slightly different amounts of water and some different amounts of acetaldehyde.

A preliminary kinetic analysis was conducted according to the approach suggested by Hougen and Watson (1947) by elaborating, under differential conditions, the initial reaction rates, that is, the slopes of the curves conversion-space time for conversion = 0 with the scopes to individuate the reaction mechanism, the rate-determining step, and the approximate value of the kinetic parameters.

Franckaerts and Froment, in agreement with the approach suggested by Hougen and Watson (1947), demonstrated that a dual-site mechanism was operative and that the surface reaction was a rate-determining step due to the obtained plots of initial reaction rate as a function of total pressure with trends, such as the ones in the plot shown in Fig. 5.17.

The initial differential rates obtained, as a function of total pressure, by Franckaerts and Froment are reported in Table 5.6. Data were collected by feeding a mixture of ethanol and water containing 13.5 mol% of water.

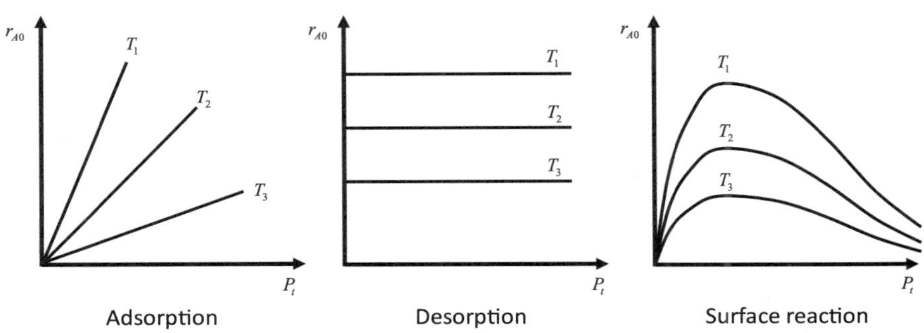

Fig. 5.17 Qualitative trends of the initial reaction rates as a function of total pressure when adsorption, desorption, or surface reaction are controlling the rate, according to Froment and Bischoff (1969)

Table 5.6 Initial reaction rates at different temperatures and total pressure

Run no.	T (°C)	r_o	p_E^o	Run	T (°C)	r_o	p_E^o
1	225	0.22	0.80	16	265	0.60	0.80
2	225	0.24	2.63	17	265	0.73	2.63
3	225	0.23	3.50	18	265	0.74	3.50
4	225	0.15	6.12	19	265	0.56	6.12
5	225	0.10	8.50	20	265	0.57	8.50
6	235	0.28	0.80	21	275	0.78	0.80
7	235	0.30	2.63	22	275	1.00	2.63
8	235	0.31	3.50	23	275	1.00	3.50
9	235	0.21	6.12	24	275	0.83	6.12
10	235	0.16	8.50	25	275	0.73	8.50
11	250	0.36	0.80	26	285	0.98	0.80
12	250	0.40	2.63	27	285	1.38	2.63
13	250	0.44	3.50	28	285	1.41	3.50
14	250	0.32	6.12	29	285	0.99	6.12
15	250	0.29	8.50	30	285	0.86	8.50

Part 1

Construct a plot r_o against p_E° for the different temperatures and verify the trend demonstrating the dual-site mechanism with the rate-determining step surface reaction.

Part 2

The dual-site mechanism with the surface reaction rate-determining step can be interpreted with a reaction rate law of the following type:

$$r = \frac{kb_E\left(p_E - \frac{p_A p_{H_2}}{K_E}\right)}{\left(1 + b_E p_E + b_A p_A + b_{H_2} p_{H_2} + b_w p_w\right)^2} \tag{5.117}$$

However, by considering only the initial rates r_o, we can neglect the reverse reaction and the negative contribution of the reaction products. The rate law expression is therefore simplified as follows:

$$r_o = \frac{kb_E p_E^o}{\left(1 + b_E p_E^o + b_w p_W^o\right)^2} \tag{5.118}$$

We can also write:

$$\sqrt{\frac{p_E^o}{r_o}} = \frac{1}{\sqrt{kb_E}} + \frac{b_E p_E^o}{\sqrt{kb_E}} + \frac{b_w p_W^o}{\sqrt{kb_E}} \tag{5.119}$$

The last term is a constant because water is not a reagent; therefore, we can write:

$$\sqrt{\frac{p_E^o}{r_o}} = a + bp_E^o + c \tag{5.120}$$

In conclusion, a straight line would be obtained by putting in a plot $\sqrt{\frac{p_E^o}{r_o}}$ as a function of p_E^o. The intercept corresponds to:

$$a + c = \frac{1}{\sqrt{kb_E}} + \frac{b_w p_W^o}{\sqrt{kb_E}} = \frac{1}{\sqrt{kb_E}}\left(1 + b_w p_W^o\right) \tag{5.121}$$

and the slope to:

$$b = \frac{b_E}{\sqrt{kb_E}} \tag{5.122}$$

The experimental data collected by Franckaerts and Froment (1964) using the mixture of ethanol–water containing 13.5 mol% of water are reported in the Table 5.7.

1. Construct the plot of $\sqrt{\frac{p_E^o}{r_o}}$ against p_E^o and verify that all the trends are linear. Evaluate the intercept and the slope at, respectively, 225 and 285 °C.
2. Evaluate by regression analysis the kinetic parameters: k^o, ΔE, $b_E{}^\circ$, ΔH_E, $b_W{}^\circ$, and ΔH_W, and verify the value of the intercept and the slope at, respectively, 225 and 285 °C. Verify if the contribution of water adsorption can be neglected.

Table 5.7 Experimental data collected by Franckaerts and Froment (1964) for ethanol dehydrogenation

Run no.	T (°C)	$\sqrt{\frac{p_E^o}{r_o}}$	p_E^o	Run	T (°C)	$\sqrt{\frac{p_E^o}{r_o}}$	p_E^o
1	225	1.90	0.80	16	265	1.20	0.80
2	225	3.20	2.63	17	265	1.80	2.63
3	225	3.80	3.50	18	265	2.10	3.50
4	225	5.90	6.12	19	265	3.20	6.12
5	225	8.66	8.50	20	265	3.80	8.50
6	235	1.70	0.80	21	275	1.00	0.80
7	235	2.90	2.63	22	275	1.55	2.63
8	235	3.31	3.50	23	275	1.80	3.50
9	235	5.20	6.12	24	275	2.66	6.12
10	235	7.12	8.50	25	275	3.42	8.50
11	250	1.50	0.80	26	285	0.87	0.80
12	250	2.50	2.63	27	285	1.37	2.63
13	250	2.83	3.50	28	285	1.46	3.50
14	250	4.30	6.12	29	285	2.33	6.12
15	250	5.50	8.50	30	285	3.10	8.50

3. Construct a parity plot comparing the experimental and calculated reaction rate.
4. Simulate the behavior of an isothermal fixed bed reactor packed with 400 g of catalyst and fed with, respectively, 50, 10, and 5 mol/h of ethanol, water, and hydrogen. Assume a total pressure of 5 atm.

Solution

Part 1

The plot r_o against p_E° for the different temperatures is shown in Fig. 5.18. As can be seen, the trend is the one of surface-reaction controlling (see plot on the right of Fig. 5.17).

Parts 2 and 3

Straight lines are obtained by putting in a plot $\sqrt{\frac{p_E^o}{r_o}}$ as a function of p_E^o as can be appreciated in Fig. 5.19; the obtained slopes and intercepts are reported in Table 5.8.

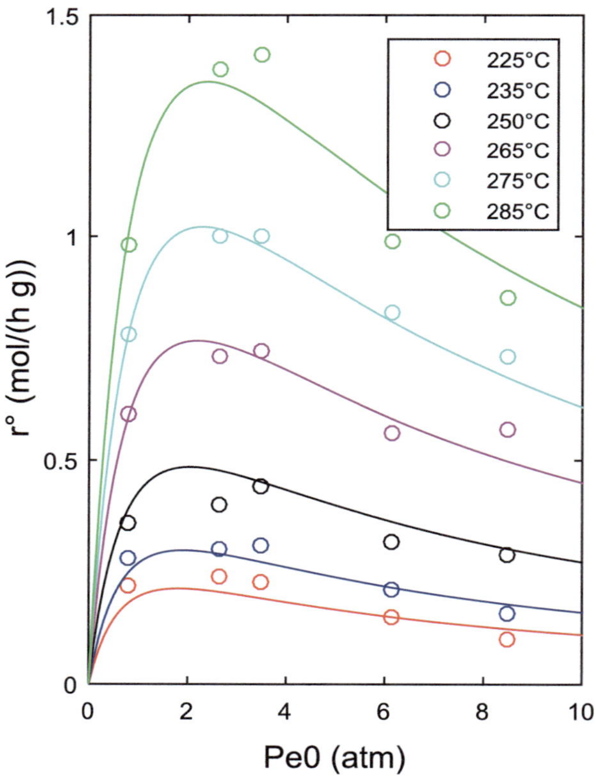

Fig. 5.18 Experimental r_o values against p_E^o for ethanol dehydrogenation evaluated by Franckaerts and Froment (1964) at different temperatures

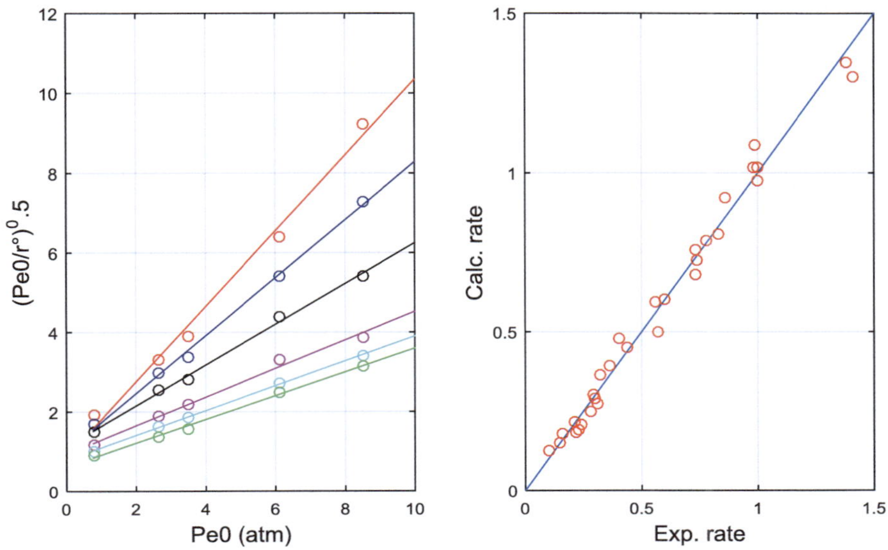

Fig. 5.19 Plot on the left shows the linear trend of $\sqrt{\frac{p_E^o}{r_o}}$ as a function of p_E^o. The plot on the right side is the obtained parity plot for the rates calculated with the described kinetic law and the parameters determined by regression

Table 5.8 Slopes and intercept obtained from the experimental data $\sqrt{\frac{p_E^o}{r_o}}$ as a function of p_E^o from linearization

Temperature (°C)	Slope	Intercept
225	0.949	0.853
235	0.728	1.000
250	0.512	1.123
265	0.359	0.930
275	0.312	0.783
285	0.298	0.615

The kinetic parameters determined by regression analysis on all the runs are:

Kinetic parameters
k0: 1972594657.4167
be0: 0.00052874
bw0: 0.27758
ΔEa: 18165.9086
ΔHe: −3711.8813
ΔHw: −2480.043

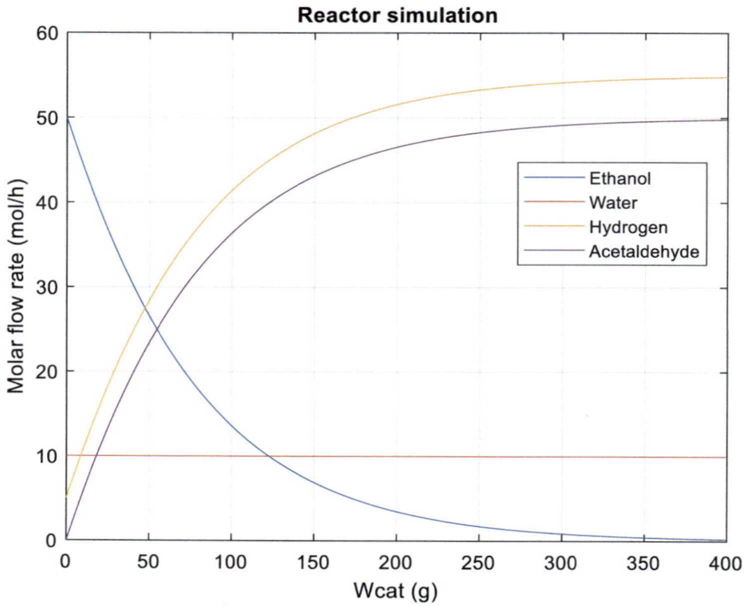

Fig. 5.20 Reactor profiles for molar flow rates for ethanol, water, hydrogen, and acetaldehyde

Part 4 (Fig. 5.20)

Exercise 5.4 Dehydrogenation of Ethanol to Acetaldehyde, A Complete Kinetic https://www.youtube.com/Analysis

Starting from the results of the previous exercise, consider now the complete kinetic law:

$$r = \frac{kb_E\left(p_E - \frac{p_A p_{H_2}}{K_E}\right)}{\left(1 + b_E p_E + b_A p_A + b_{H_2} p_{H_2} + b_w p_w\right)^2} \tag{5.123}$$

K_E can be estimated from the plot shown in Fig. 5.21.

1. Evaluate from the experimental data of Franckaerts and Froment (1964), reported in Table 5.9, the best kinetic parameters for runs performed at 1 atm and different temperatures. Assume, as a first attempt, the parameters already obtained in the previous exercise.

2. Simulate the experimental point of the table in a plot reporting the ethanol conversion as a function of the space velocities.

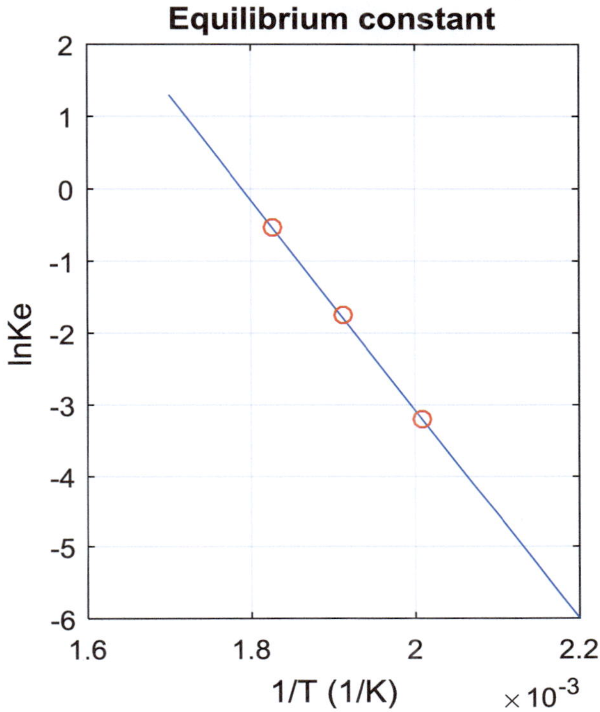

Fig. 5.21 Equilibrium constants at different temperatures for the reaction of ethanol dehydrogenation

Table 5.9 Experimental data on the dehydrogenation of ethanol to acetaldehyde[a]

Run	T (°C)	Conversion λ	W/F_E Kg cat h/kmol	Run	T (°C)	Conversion λ	W/F_E Kg cat h/kmol
1	225	0.033	0.19	19	265	0.091	0.19
2	225	0.060	0.39	20	265	0.156	0.39
3	225	0.072	0.59	21	265	0.211	0.59
4	225	0.109	0.88	22	265	0.280	0.88
5	225	0.154	1.50	23	265	0.363	1.50
6	225	0.158	1.61	24	265	0.370	1.61
7	235	0.044	0.19	25	275	0.118	0.19
8	235	0.076	0.39	26	275	0.196	0.39
9	235	0.109	0.59	27	275	0.263	0.59
10	235	0.142	0.88	28	275	0.338	0.88
11	235	0.193	1.50	29	275	0.362	1.50
12	235	0.195	1.61	30	275	0.368	1.61
13	250	0.063	0.19	31	285	0.148	0.19
14	250	0.111	0.39	32	285	0.245	0.39
15	250	0.150	0.59	33	285	0.325	0.59
16	250	0.202	0.88	34	285	0.417	0.88
17	250	0.268	1.50	35	285	–	–
18	250	0.275	1.61	36	285	–	–

[a]Data from Franckaerts and Froment (1964)

Solution

The best-fitting parameters are the following ones:
k0: 22301.7784
be0: 2.136e−05
ba0: 1.4266
bh0: 1.3265
ΔEa: 15770.8404
ΔHe: −7499.3995
ΔHa: −2418.713
ΔHh: −2418.716

The simulation of the kinetic runs is shown in Fig. 5.22. The same figure also shows the parity plot.

All the described results were obtained using a MATLAB program available as Electronic Supplementary Material, Data 5.1.

Exercise 5.5 Dehydrogenation of Methyl Cyclohexane to Toluene over Pt-Al_2O_3 Catalyst

Sinfelt et al. (1960) studied the kinetics of methylcyclohexane dehydrogenation to toluene using a catalyst of Pt (0.3 wt%) supported by impregnation on Al_2O_3 (surface area = 155 m^2/g). The adopted operative conditions were chosen in such a way as to obtain a low conversion degree; thus, the initial reaction rates (differential approach) were determined by calculating:

$$r = \frac{F\Delta\lambda}{W} \tag{5.124}$$

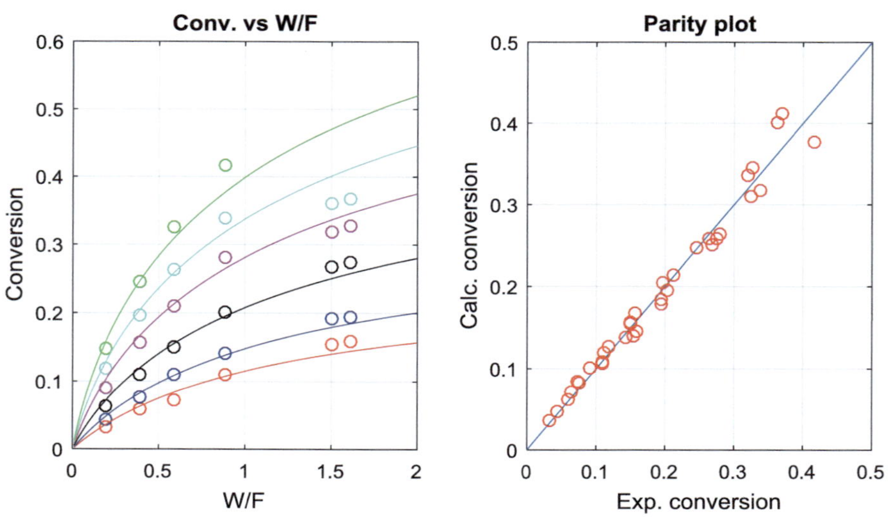

Fig. 5.22 Simulation of the kinetic runs performed by Franckaerts and Froment (1964) for ethanol dehydrogenation in a tubular reactor

Table 5.10 Experimental data for the dehydrogenation of cyclohexane to toluene

T (°C)	p_{MCH} (atm)	p_H (atm)	F/W (mol MCH/h g_{cat})	Conversion $\Delta\lambda$ %	R (mol T/h g_{cat})
315	0.36	1.1	0.21	5.6	0.012
315	0.36	3.0	0.20	5.8	0.012
315	0.07	1.4	0.21	4.1	0.0086
315	0.24	1.4	0.21	5.2	0.011
315	0.72	1.4	0.21	6.2	0.013
344	0.36	1.1	0.53	5.7	0.030
344	0.36	3.1	0.52	6.2	0.032
344	0.08	1.4	0.52	3.9	0.020
344	0.24	1.4	0.55	6.2	0.034
344	0.68	1.4	0.52	6.5	0.034
372	0.36	1.1	1.02	7.4	0.076
372	0.36	4.1	1.02	7.8	0.080
372	1.1	4.1	1.05	11.8	0.124
372	2.2	4.1	1.05	12.5	0.131

Toluene was the only obtained product. The experimental data collected by the authors are reported in Table 5.10. Rate data were collected at three different temperatures—315, 344, and 372 °C—by using the same amount of catalyst. The conversion level was kept in the range of 4−12%; therefore, the effect of the reverse reaction can be neglected. As can be observed, the reaction rate is poorly affected by the partial pressure of methylcyclohexane and not affected by the hydrogen partial pressure showing a nearly to zero order with respect to methylcyclohexane and a zero order with respect to hydrogen. Independent experiments have shown that by adding aromatics, such as benzene or xylene in a ratio 1:1 with methylcyclohexane, the reaction rates were moderately reduced by a factor of 0.8.

The authors interpreted initially the results considering a Langmuir–Hinshelwood kinetic model of the type:

$$r = \frac{kb_m p_m}{1 + b_m p_m} \tag{5.125}$$

with "m" referring to methylcyclohexane molecules.

1. Evaluate the kinetic parameters appearing in this equation and their dependence on the temperature.

As will be seen from the results of calculations, the physical mean of the adsorption enthalpy is not reliable; therefore, the authors reinterpreted the results by assuming that methylcyclohexane is promptly adsorbed onto the platinum surface, readily transforming in adsorbed toluene, but the desorption rate is slow. Thus, we can write a kinetic model of the type:

$$\mathrm{MCH}(g) \xrightarrow{k} \mathrm{Toluene_{ads}} \xrightarrow{k_2} \mathrm{Toluene\ (gas)} \tag{5.126}$$

By assuming steady-state conditions for the adsorbed toluene, it is possible to write:

$$\frac{\mathrm{dToluene_{ads}}}{\mathrm{d}t} = k_1 p_m (1 - \theta_T) - k_2 \theta_T = 0 \tag{5.127}$$

And for the reaction rate: $r = k_2\theta_T$.

Solving with respect to θ_T, we obtain:

$$r = \frac{k_2\left(\frac{k_1}{k_2}\right)p_m}{1 + \left(\frac{k_1}{k_2}\right)p_m} \tag{5.128}$$

which is mathematically identical to the previous rate law assuming $k = k_2$ and $b_m = k_1/k_2$. The equation is the same, but the physical meaning of the parameter changes with k_1 being the rate coefficient of methylcyclohexane adsorption and k_2 being the rate coefficient of toluene desorption rate.

2. Evaluate the values of k_1 and k_2 and their dependence on temperature.

Solution
Part 1

The parameters of Eq. (5.125) resulted in:
k0: 0.043428
bm0: 9.8845
ΔEa: 18103.476
ΔHm: −17552.789

Part 2

The parameters for Eq. (5.128) resulted in:
k01: 0.42926
k02: 0.043428
ΔEa1: 550.684
ΔEa2: 18103.477

The obtained parity plots are shown in Fig. 5.23.

The described results were obtained by using a MATLAB program available as Electronic Supplementary Material, Data 5.1.

Exercise 5.6 Oxidation of Butenes to Maleic Anhydride

Varma and Saraf (1978) studied the kinetics and mechanism of the oxidation of butenes (a mixture of 1-butene + cis and trans-2-butenes) to maleic anhydride in the

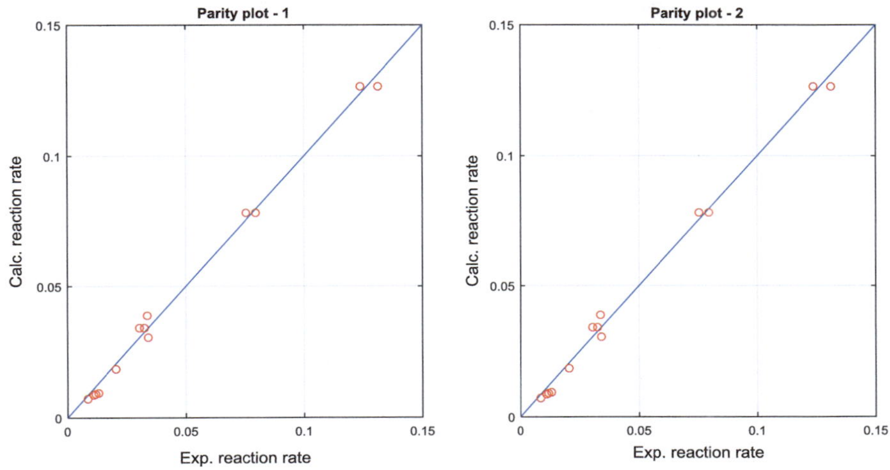

Fig. 5.23 Parity plots respectively obtained by adopting the two described models

presence of a catalyst, vanadyl phosphate, supported on silica gel (8−10 mesh). The catalyst was opportunely diluted with an inert material to avoid a hot spot and put in a stainless-steel tubular reactor of 2.5-cm diameter and 25-cm length. The reactor was immersed in a bath of melted salts to warrant the isothermicity of the reactor. No more than a 2 °C temperature increase was observed during the experimental runs. Butenes were fed to the reactor together with oxygen and nitrogen, with the latter used as diluent. The gases were preheated at the reaction temperature before their introduction into the reactor. The kinetics were studied in the temperature range of 350–390 °C; the flow rate was always taken as $>4 \times 10^{-3}$ mol/s; the residence time W/F was changed in the range of 1430 − 15,000 g s/mol; the partial pressure of butenes was changed in the range of 0.0078–0.018 atm; and the partial pressure of oxygen was kept constant and $>$ 0.208 atm.

Together with maleic anhydride, the desired product, different by-products were obtained as a consequence of side reactions as shown in Fig. 5.24. To justify all the obtained by-products, a reaction scheme of the following type would be considered:

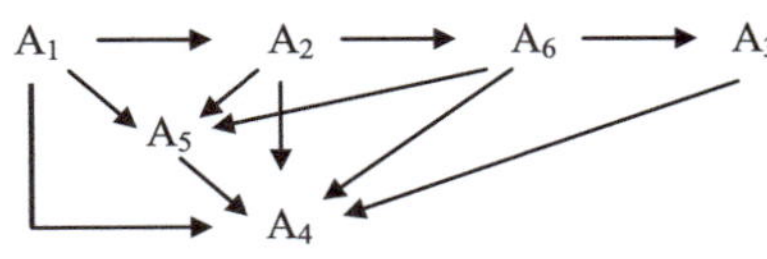

where acetic acid is neglected, and acetaldehyde and butyraldehyde are formed in very small amounts. This scheme constitutes 11 different reactions, too many for a significant kinetic analysis. However, it can be experimentally observed that butadiene is an intermediate reaction product yielding a maximum with a measurable concentration, while furan is always present in a negligible amount. Thus, we can neglect the reactions $A_2 \rightarrow A_6$, $A_6 \rightarrow A_3$, $A_6 \rightarrow A_5$, and $A_6 \rightarrow A_4$.

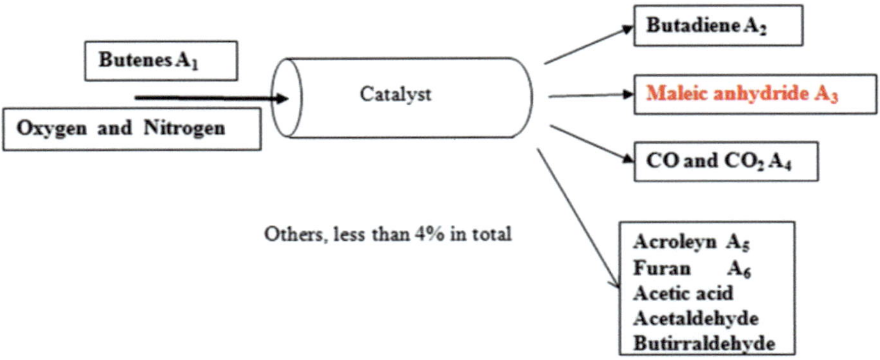

Fig. 5.24 Products obtained by oxidation of butenes

Then, we can neglect the formation of the secondary products that are obtained in a small amount. This allows to eliminate the reaction paths $A_1 \rightarrow A_5$, $A_2 \rightarrow A_5$, and $A_5 \rightarrow A_4$. The authors, through extensive experimentation, observed that the reaction $A_3 \rightarrow A_4$ also can be neglected. Thus, the scheme can be simplified as it follows:

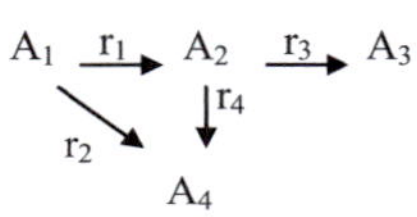

Now, we have only four reactions involving four chemical species: butenes, butadiene, CO–CO_2, and maleic anhydride. Many different kinetic runs were performed, at a given temperature (370 °C, 0.78 mols % of butenes in air), to evaluate how the product distribution changes with the residence time W/F_{tot}. Other runs also were performed, at different temperatures and keeping constant the concentration of butenes in the feeding flow (0.78 mols % in air), to evaluate the yields of maleic anhydride. Last, some runs were performed by changing the partial pressure of butenes at three different values of *W/F* keeping the temperature constant at 350 °C.

Let us consider the stoichiometric matrix:

	i			
j	A_1	A_2	A_3	A_4
1	−1	+1	0	0
2	−1	0	0	+4
3	0	−1	+1	0
4	0	−1	0	+4

Based on this stoichiometric matrix, we can write the mass balance as:

$$\frac{dp_{A_1}}{d\tau} = -r_1 - r_2 \frac{dp_{A_2}}{d\tau} = r_1 - r_3 - r_4 \frac{dp_{A_3}}{d\tau} = r_3 \frac{dp_{A_4}}{d\tau} = 4r_2 + 4r_4 \tag{5.129}$$

where $\tau = W/F_{tot}$. For solving this system of ordinary differential equation, we must define the expressions for r_1, r_2, r_3, and r_4. It has been assumed that all the considered reactions occur through a two-step Mars–Van Krevelen redox mechanism. The resulting reaction rates will be:

$$r_j = \frac{k_{jox} k_{2ox} p_{A_i} p_{O_2}}{\alpha k_{jox} p_{A_i} + k_{2ox} p_{O_2}} \tag{5.130}$$

in which α is the number of oxygen molecules required for mole of hydrocarbon oxidized. This expression can also be written as:

$$\frac{1}{r_j} = \frac{1}{k_{jox} p_{A_i}} + \frac{\alpha}{k_{2ox} p_{O_2}} \tag{5.131}$$

That is, the inverse of the reaction rate is a linear function of $1/p_{Ai}$ when the partial pressure of oxygen is kept constant.

1. Verify the mechanism by constructing a plot of $1/r_j$ against $1/p_{Ai}$ starting from the experimental data reported in Table 5.11.

In the conditions adopted by the authors, that is, low butane concentration (<1 mols %) and in the presence of excess oxygen, the oxidation of butenes resulted in a reaction of pseudo first order. This means that in the following equation:

Table 5.11 Rate of butane oxidation for different partial pressure at a constant temperature of 350 °C

Run	Rate × 10^5 (mol/g s)	Butene pressure (atm)
1	0.141	0.0178
2	0.128	0.0178
3	0.121	0.0178
4	0.130	0.0158
5	0.127	0.0158
6	0.121	0.0158
7	0.115	0.0128
8	0.112	0.0128
9	0.110	0.0128
10	0.107	0.0099
11	0.103	0.0099
12	0.100	0.0099
13	0.093	0.0077
14	0.082	0.0077
15	0.081	0.0077

Table 5.12 Kinetic parameters estimated by mathematical regression on the experimental data

Reaction j	Activation energy (Kcal/mol)	Pre-exponential factor (mol/g-cat atm h)
1	22.1	1.5867×10^7
2	17.2	1.8866×10^5
3	20.3	1.4432×10^7
4	18.8	1.4670×10^6

Table 5.13 Experimental data of product distribution collected at 370 °C

Y_{A1}	Y_{A2}	Y_{A3}	Y_{A4}	(W/Ft) $\times 10^{-3}$ gcat s/mol
1.000	0.000	0.000	0.000	0.00
0.713	0.122	0.045	0.159	1.36
0.665	0.120	0.068	0.168	2.27
0.554	0.127	0.100	0.200	2.82
0.379	0.127	0.218	0.270	4.72
0.261	0.095	0.322	0.356	6.54
0.247	0.068	0.318	0.368	7.09
0.150	0.022	0.398	0.439	9.18
0.136	0.018	0.384	0.436	10.09

$$r_j = \frac{k_{jox} k_{2ox} p_{A_i} p_{O_2}}{\alpha k_{jox} p_{A_i} + k_{2ox} p_{O_2}} \tag{5.132}$$

the term, $\alpha k_{jox} p_{A_i}$, is very small with respect to $k_{2ox} p_{O_2}$, and the rates become simply:

$$r_j \simeq k_{jox} p_{A_i} \tag{5.133}$$

The main reactions 1 and 2 are of pseudo first order. If the others also are assumed to obey a first-order law, then we have:

$$r_1 = k_1 p_{A_1} \quad r_2 = k_2 p_{A_1} \quad r_3 = k_3 p_{A_2} \quad r_4 = k_4 p_{A_2} \tag{5.134}$$

The kinetic parameters found by the authors for the four mentioned reactions are listed in Table 5.12.

With these parameters, we can evaluate the product distribution, at 370 °C, for different residence time W/F_{tot}. Compare your results with the experimental data reported in Table 5.13.

Solution

In the left plot shown in Fig. 5.25, it is possible to appreciate the linearity of the trend of $1/r_j$ against $1/p_{Ai}$ justifying the proposed mechanism. In the right plot shown in Fig. 5.23, the experimental data of Table 5.13 were simulated.

All the described results were obtained using a MATLAB program available as Electronic Supplementary Material, Data 5.1.

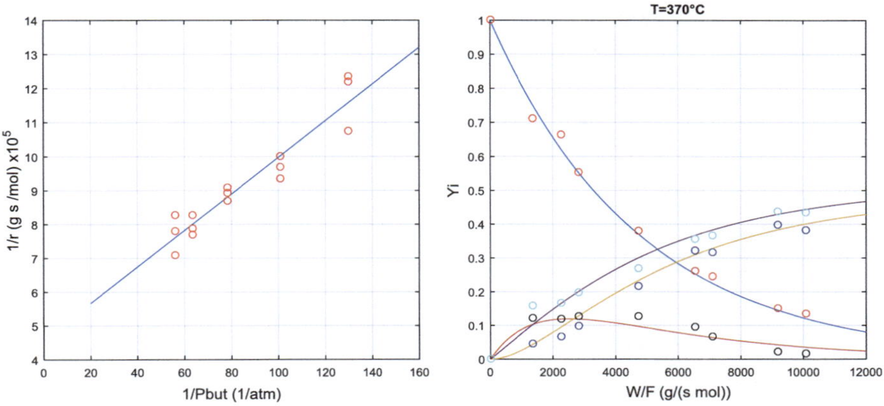

Fig. 5.25 Demonstration of $1/r_j$ against $1/p_{Ai}$ linearity trend (left). Simulation of the experimental data (right)

Table 5.14 Experimental initial rate data collected by Thaller and Thodos (1960)

T (°C)	P_{tot} (bar)	r_o (mol/g_{cat} h)	T (°C)	P_{tot} (bar)	r_o (mol/g_{cat} h)
371	1.0	0.195	315.5	3.0	0.0420
371	2.0	0.189	315.5	0.22	0.0295
371	4.0	0.188	315.5	1.0	0.0410
371	9.0	0.198	302	1.0	0.0227
371	12.0	0.190	302	3.0	0.0277
315.5	1.0	0.0392	302	5.0	0.0255
315.5	7.0	0.0416	302	7.0	0.0217
315.5	4.0	0.0416	302	9.6	0.0183
315.5	10.0	0.0326	288	1.0	0.0115
315.5	14.6	0.0247	288	3.0	0.0161
315.5	5.5	0.0415	288	2.0	0.0146
315.5	8.5	0.0376	–	–	–

Exercise 5.7 Dehydrogenation of Sec-Butyl Alcohol to Methyl Ethyl Ketone
Thaller and Thodos (1960) (see also Bischoff and Froment (1962)) studied the mechanism of the dehydrogenation of isobutanol to methyl ethyl ketone by measuring differential initial reaction rates. They observed a change of the reaction mechanism governing the rate law by increasing the temperature. At the lower temperature, the surface reaction was controlling, while at higher temperatures, the hydrogen desorption became the slow step. The obtained experimental data are listed in Table 5.14.

In this exercise, it will be pointed out that when two or more reaction steps in a mechanism are both controlling, the kinetic expressions become very complicated, and the integral approach consequently becomes unsuitable to individuate the reaction mechanism. It is imperative, in these cases, to elaborate the initial reaction rates (differential method) that allows to highly simplify, as will be seen, the reaction rate expression to be used to discriminate among the different possible mechanisms.

The reaction stoichiometry is:

$$CH_3CHOHCH_2CH_3 \rightarrow CH_3COCH_2CH_3 + H_2 \tag{5.135}$$

Put isobutanol = I, methyl ethyl ketone = K, and hydrogen = H_2

Single Rate-Controlling Step: Reaction Surface Controlling

Consider the following reaction mechanism:

Step 1	I + σ ⇄ Iσ Adsorption of I on a single site
Step 2	Iσ + σ ⇄ Kσ + H_2σ Surface reaction
Step 3	Kσ ⇄ K + σ Desorption of K
Step 4	H_2σ ⇄ H_2 + σ Desorption of H_2

If the surface reaction is a rate-determining step, we can write:

$$r = k\left(\Theta_I\Theta_F - \frac{\Theta_K\Theta_{H_2}}{K_{es}}\right) \tag{5.136}$$

Adsorption and desorption are in equilibrium; therefore:

$$b_I = \frac{\Theta_I}{p_I\Theta_F} \quad b_K = \frac{\Theta_K}{p_K\Theta_F} \quad b_{H_2} = \frac{\Theta_{H_2}}{p_{H_2}\Theta_F} \tag{5.137}$$

$$\Theta_I + \Theta_K + \Theta_{H_2} + \Theta_F = 1 \tag{5.138}$$

Substituting

$$b_I p_I \Theta_F + b_K p_K \Theta_F + b_{H_2} p_{H_2} \Theta_F + \Theta_F = 1 \tag{5.139}$$

$$\Theta_F = \frac{1}{1 + b_I p_I + b_K p_K + b_{H_2} p_{H_2}} \tag{5.140}$$

Substituting all the Θ_i terms in the surface rate expression, we obtain:

$$r = k\left(b_I p_I \Theta_F^2 - \frac{b_K p_K b_{H_2} p_{H_2} \Theta_F^2}{K_{es}}\right) = \frac{k\left(b_I p_I - \frac{b_K p_K b_{H_2} p_{H_2}}{K_{es}}\right)}{\left(1 + b_I p_I + b_K p_K + b_{H_2} p_{H_2}\right)^2} \tag{5.141}$$

It is possible then to demonstrate that $K_{es} = \frac{b_K b_{H_2}}{b_I} K_e$ with K_e being the thermodynamic equilibrium constant of the overall reaction. Finally, it is possible to write:

$$r = \frac{kb_I\left(p_I - \frac{p_K p_{H_2}}{K_e}\right)}{\left(1 + b_I p_I + b_K p_K + b_{H_2} p_{H_2}\right)^2} \tag{5.142}$$

For differential initial reaction rates, this expression can be simplified as:

$$r_{\text{init}} = \frac{kb_I p_I}{(1 + b_I p_I)^2} \tag{5.143}$$

This equation can be used for interpreting runs performed at lower temperature.

Single Rate-Controlling Step: Hydrogen Desorption Controlling

In this case, we can consider at equilibrium the first three steps and can write, therefore:

$$b_I = \frac{\Theta_I}{p_I \Theta_F} \quad b_K = \frac{\Theta_K}{p_K \Theta_F} \quad K_{es} = \frac{\Theta_K \Theta_{H_2}}{\Theta_I \Theta_F} \tag{5.144}$$

Moreover, it holds:

$$\Theta_I + \Theta_K + \Theta_{H_2} + \Theta_F = 1 \tag{5.145}$$

$$b_I p_I \Theta_F + b_K p_K \Theta_F + K_{es} \frac{\Theta_I \Theta_F}{\Theta_K} + \Theta_F = 1 \text{ but} \frac{\Theta_I}{\Theta_K} = \frac{b_I p_I}{b_K p_K} \tag{5.146}$$

and hence:

$$\Theta_F = \frac{1}{1 + b_I p_I + b_K p_K + \frac{K_{es} b_I p_I}{b_K p_K}} \tag{5.147}$$

Remembering that:

$$K_{es} = \frac{b_K b_{H_2}}{b_I} K_e \quad \Theta_F = \frac{1}{1 + b_I p_I + b_K p_K + \frac{K_e b_{H_2} p_I}{p_K}} \tag{5.148}$$

The hydrogen desorption rate is:

$$\begin{aligned} r_d &= k_d\left(\Theta_{H_2} - b_{H_2} p_{H_2} \Theta_F\right) = k_d\left(\frac{K_{es} b_I p_I \Theta_F}{b_K p_K} - b_{H_2} p_{H_2} \Theta_F\right) \\ &= k_d \Theta_F\left(\frac{K_{es} b_I p_I}{b_K p_K} - b_{H_2} p_{H_2}\right) \end{aligned} \tag{5.149}$$

$$r_d = k_d \Theta_F\left(\frac{K_e b_{H_2} p_I}{p_K} - b_{H_2} p_{H_2}\right) \tag{5.150}$$

Finally:

$$r_d = \frac{k_d K_e b_{H_2}\left(p_I - \frac{p_{H_2} p_K}{K_e}\right)}{p_K\left(1 + b_I p_I + b_K p_K + \frac{K_e b_{H_2} p_I}{p_K}\right)} \tag{5.151}$$

By also considering the initial reaction rates, this expression is highly simplified by imposing

$$p_{\mathrm{K}} = p_{\mathrm{H}\,2} = 0$$

$$r_{d\,-\mathrm{init}} = \frac{k_d}{b_{H_2}} \tag{5.152}$$

However, the control of hydrogen desorption, which occurs at a higher temperature, is not complete because the passage from a mechanism to another is gradual, and the two steps—surface reaction and hydrogen desorption—therefore have comparable rates. Hence, one only expression must be derived considering both these steps as controlling. To do this, it is sufficient to consider both the reactant adsorption and ketone desorption at equilibrium and then equate the two controlling reaction rates. It seems a simple procedure, but the mathematical elaboration for obtaining the rate expression is complicated and tedious. The complete elaboration can be found in a work published by Bischoff and Froment (1962). We report here just the expression simplified for describing differential initial reaction rates:

$$r_{\mathrm{init}} = \left[k_d + \frac{k_d^2}{2k}\frac{(1 + b_I p_I)^2}{b_I p_I}\right] - \sqrt{\left[k_d + \frac{k_d^2}{2k}\frac{(1 + b_I p_I)^2}{b_I p_I}\right]^2 - k_d^2} \tag{5.153}$$

This expression would be valid in all the temperature range. If only one of the steps is controlling, for example, if k_r is extremely high, then we have the control of hydrogen desorption, and the rate expression becomes simply:

$$r_{\mathrm{init}} = k_d \tag{5.154}$$

In contrast, if k_d is extremely high, we have surface reaction control, and the equation reduces to the already found equation:

$$r_{\mathrm{init}} = \frac{k b_I p_I}{(1 + b_I p_I)^2} \tag{5.155}$$

1. Apply the surface controlling kinetic law in the temperature range 288–315.5 °C; evaluate the kinetic parameters by regression; and then simulate with this kinetic law all the kinetic runs for observing the deviation at the higher temperature.
2. Apply the hydrogen desorption controlling kinetic law to interpretation of the runs performed from 315.5 to 371 °C for determining the kinetic parameters, and

then simulate all the runs to observe the deviation of the model at lower temperatures.

3. Consider the rate law with two steps as controlling (surface reaction and hydrogen desorption), and evaluate the best-fitting kinetic parameters of this new model.

Solution
Part 1

Determined parameters for reaction surface–controlling kinetic model
k0: 0.11246
bI0: 0.3909
ΔE_a: 14090.92
ΔH_I: 2245.51

Part 2

Determined parameters for hydrogen desorption–controlling kinetic model
k_d0: 0.023594
ΔE_{ad}: 11251.57

Part 3

Determined parameters for the full model
k0: 0.39907
bI0: 0.57562
kd0: 0.049853
ΔEa: 26004.60
ΔH_I: −3623.42
ΔE_{ad}: 8022.08

Figure 5.26 shows all the corresponding parity plots.

The described results were obtained using a MATLAB program available as Electronic Supplementary Material, Data 5.1.

5.5.5 *Tubular Reactor with External Recycle*

Tubular reactors with external recycle combine the advantages of both integral and differential tubular reactors. A simplified scheme of a tubular reactor with an external recycle is shown in Fig. 5.27. In this case, the flow stream at the exit is mainly recycled, while only a small portion of this flow stream is collected and analyzed. The ratio between the recycled flow stream F_R and that withdrawn at the exit F normally is $>$15–20. The mass balance in this case is reported below (Eqs. 5.156–5.157).

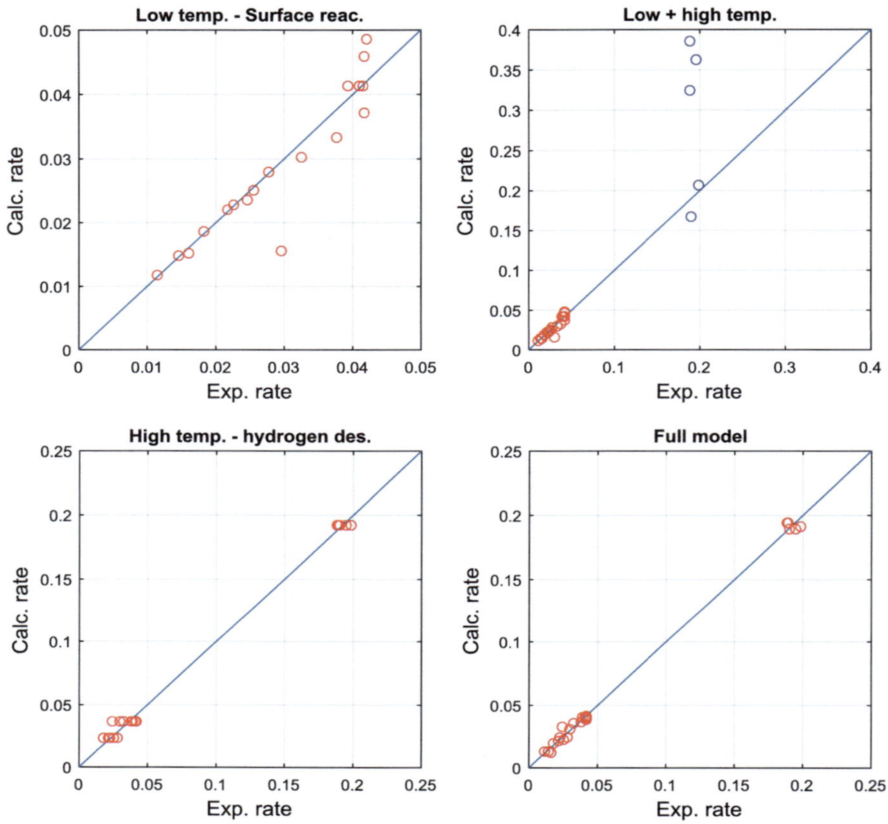

Fig. 5.26 Parity plots obtained for the different tested models

$$F y_{Ao} + F_R y_{\mathrm{Af}} = (F + F_R) y_{Ai} \tag{5.156}$$

$$y_{Ai} = \frac{1}{\left(\frac{F_R}{F} + 1\right)} y_{Ao} + \frac{F_R/F}{\left(\frac{F_R}{F} + 1\right)} y_{Af} \tag{5.157}$$

where F and F_R are the molar flow rates, respectively, at the reactor inlet and recycled; y_{Ai}, y_{Ao}, and y_{Af} are the molar fractions of the reactant A, respectively, initial at the reactor inlet and final. For high F_R/F ratios, $y_{Ai} \simeq y_{Af}$ and the behavior of the reactor approach that of a CSTR; therefore:

$$r = \frac{y_{Ao} - y_{Af}}{y_{Ao}(W/F)} = \frac{F}{W} \Delta\lambda_A \tag{5.158}$$

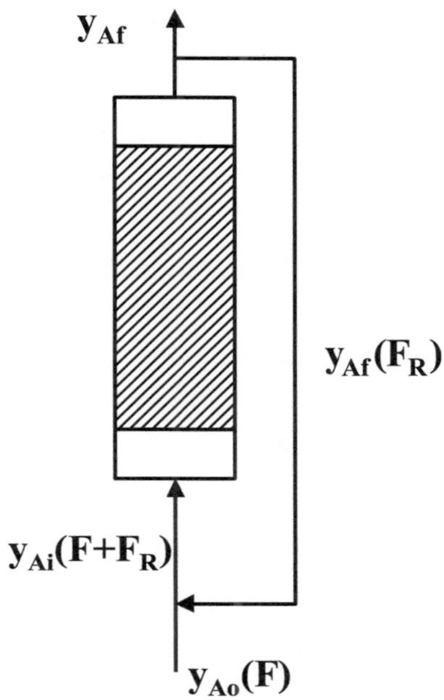

Fig. 5.27 Scheme of a tubular reactor with an external recycle

The interpretation of the kinetic data in this case is simple, as in the case of the differential reactor. However, we have high conversions as in the integral reactors. The only drawback of this reactor is the difficulty of maintaining the recycle pump and the recycled stream at the same temperature as the reactor. Another problem is to warrant the seal in all of the apparatus.

5.5.6 Adiabatic Tubular Reactors

The use of these reactors for kinetic studies is suggestive because in a single kinetic run we can change both concentrations and temperatures, thus obtaining the reaction-rate data for different conditions. In this case, together with the mass balance, we must consider also the energy balance in evaluating the change of temperature along the reactor. For this purpose, we can write:

$$\begin{cases} \dfrac{d\lambda}{dZ} = \dfrac{A}{F} r \\ \dfrac{dT}{dZ} = \dfrac{A}{F} \dfrac{(-\Delta H)}{\overline{C}_p} r \end{cases} \tag{5.1595.160}$$

With

λ	Conversion
Z	Reactor length
A	Reactor section
F	Molar flow rate
r	Reaction rate
$\overline{C_p} =$	Average specific heat of the fluid

Dividing both of the terms of the two equations, we obtain:

$$\frac{\mathrm{d}\lambda}{\mathrm{d}T} = \frac{\overline{C}_p}{\Delta H} \tag{5.161}$$

and integrating:

$$\lambda = \mathrm{cost} + \frac{\overline{C}_p}{\Delta H} T \tag{5.162}$$

As can be seen, the conversion is simply a linear function of the temperature. The reaction rates can be obtained as seen for the integral reactor by a polynomial interpolation of the longitudinal temperature profile and successive derivation. The main drawback of this reactor is the presence of interface gradients that affect kinetic data and cause errors in data elaboration.

Tubular adiabatic reactors are often used in industry, for example, in the case of a strongly exothermic reaction by performing the reaction in several steps and cooling the flowing fluid between two successive steps. The same approach can be followed for endothermic-reaction heating, in this case, the flowing fluid between two successive steps.

5.5.7 Non-isothermal, Non-adiabatic Tubular Reactors

When a reaction is strongly exothermic or endothermic it is difficult, in a tubular packed bed reactor, to keep a constant temperature along the catalytic bed, especially in industrial reactors using catalytic pellets with equivalent diameter > 0.5 cm. In this case, we also must solve—together with the mass balance related to any single component—the overall energy balance, that is:

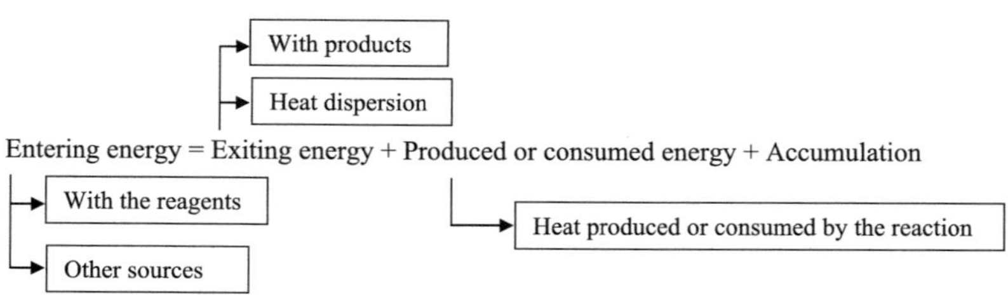

For a constant molar flow rate, F, we can write, for example:

Entering-exiting energy; Heat exchanged with the external environment

$$Q\,\delta\,\overline{C_p}\,dT = A\,dZ\,r_j\left(-\Delta H_j\right) - b\,dZ\,U\,(T - T_s) \tag{5.163}$$

Heat produced or consumed by the reaction

where Q is the volumetric flow rate, δ is the fluid density, A is the reactor section, Z is the reactor length, r_j is the reaction rate j, ΔH_j is the enthalpy change of reaction j, $\overline{C_p}$ is the average specific heat of the reaction fluid, b is the cylindrical tube perimeter, U is the overall heat-transfer coefficient, and T_S is the temperature of the refrigerating or heating fluid.

Remembering the mass balance equations and coupling mass and heat balance for n different occurring reactions we will have:

$$\frac{d\lambda_i}{dZ} = \frac{A}{Q\delta}\sum_{j=1}^{n}\alpha_{ij}r_j \tag{5.164}$$

$$\frac{dT}{dZ} = \frac{A}{Q\delta\overline{C_p}}\sum\nolimits_{j=1}^{n}\alpha_{ij}r_j\left(-\Delta H_j\right) - \frac{bU}{Q\delta\overline{C_p}}(T - T_s) \tag{5.165}$$

$$\frac{dT_S}{dZ} = \frac{bU}{Q_S\delta_S\overline{C_{ps}}}(T - T_s) \tag{5.166}$$

where Q_S is the volumetric flow rate of the external freezing or heating flowing fluid, δ_S is the density of the external flowing fluid, and $\overline{C_{ps}}$ is the specific heat of the thermostating fluid. The initial conditions normally will be $\lambda_i = 0$, $T = T°$, and $Ts = Ts°$.

By solving this equation system, it is possible to describe the profiles along the reactor of all the concentrations of the reactants and products as well as the temperature profile. However, in this case, we must also evaluate independently the thermal parameters U and $\overline{C_p}$ in $\overline{C_{ps}}$.

5.5.8 Pulse Reactors

This type of reactor normally is exclusively employed for the fast screening of heterogeneous catalysts. Pulse reactors are very small reactors, opportunely thermostated, and put immediately before a gas-chromatographic column. An inert gas, the same one used in the gas-chromatographic column, is fed to the reactor as a carrier fluid. A small pulse of the reactants is injected inside the reactor and when passing through the catalytic bed yields the reaction. The obtained products and the unreacted reagents are conveyed by the carrier gas directly into the gas-chromatographic column where they are separated and analyzed. From the analysis, we can evaluate the reactants' conversion. A kinetic elaboration is possible and easy only if the occurring reaction is of the first order. The order of reaction can be deduced by the symmetry of the gas-chromatographic peaks. For a first-order reaction, the peak symmetry after the reaction is conserved because, in the case of a first-order reaction, the fraction of the reacted substance does not depend on its initial concentration. Remember the following relationship:

$$\ln \frac{C_o}{C} = kt \tag{5.167}$$

By assuming residence time = reaction time, the kinetic constant can be estimated from the conversion as:

$$k = \frac{V}{LA\theta} \ln\left(\frac{1}{1-\lambda}\right) \tag{5.168}$$

where:

V	is the volumetric flow rate.
A	is the section of the reactor.
L	is the catalytic bed length.
θ	is the void degree of the catalytic bed.
λ	is the conversion.

Then, by putting $\ln\{\ln[1/(1 - \lambda)]\}$ against $10^3/T$, it is possible to estimate an approximated value of the activation energy from the following equation:

$$\text{loglog}\left(\frac{1}{1-\lambda}\right) = \text{costant} - \frac{\Delta E}{RT} \tag{5.169}$$

However, by comparing the shape of the peaks of unreacted reagents with the same obtained bypassing the reactor, the reaction order can be argued. As mentioned previously, the most important information that can be achieved with this type of reactor is the catalyst activity (as conversion data) and the selectivity for a defined catalyst. It is easy to change the catalyst inside the reactor and make a comparison of the data obtained from many proven catalysts. In conclusion, with this technique, fast catalytic screening can be performed.

5.5.9 *Gas–Solid CSTRs*

Gas–solid CSTRs are advantageous for collecting and interpreting laboratory kinetic data because, under ideal conditions, the composition inside the reactor is the same as that of the reactor outlet. Therefore, the reaction rate can be directly evaluated from the concentration difference of any component "i" from the inlet to the outlet of the reactor or, more simply, by considering the conversion of a generic component, A, the reaction rate can be determined with the following algebraic expression:

$$r_A = F\lambda_A / W \tag{5.170}$$

As can be seen, the reaction rate is proportional to the conversion. The ideal condition is the one of a perfectly mixed reactor, and the laboratory equipment must warrant this condition thanks to a vigorous stirring system. Under these conditions, the reacting gas inside the reactor has a uniform temperature. The drawbacks of these reactors include the presence of a wide void space inside the reactor where homogeneous uncatalyzed reactions, for example, radical reactions, can occur. Void space also can yield nonideal mixing behavior by volume or flow segregation giving an effect of stagnation or bypass. An accurate design of these reactors to avoid the abovementioned negative phenomena is therefore important. Preliminary mixing tests, as the ones described in Chap. 4, are opportune to evaluate the absence of bypass or stagnation effects. Another inconvenience is the thermal characteristic of these reactors, which—for their structure—can be considered approximately adiabatic. Therefore, in the presence of strongly exothermic or endothermic reactions, the temperature inside the reactor increases or decreases with respect to the initial condition depending on the catalytic activity. In conclusion, in preliminary kinetic runs, it is not possible to foresee the reaction temperature. Another problem of these reactors is the inability to realize a perfect seal.

Many different commercial gas–solid CSTR reactors are available that differ in terms of the reciprocal position of the impeller and of the catalytic bed (Fig. 5.28). The optimization of the design is oriented to limit the portion of void space.

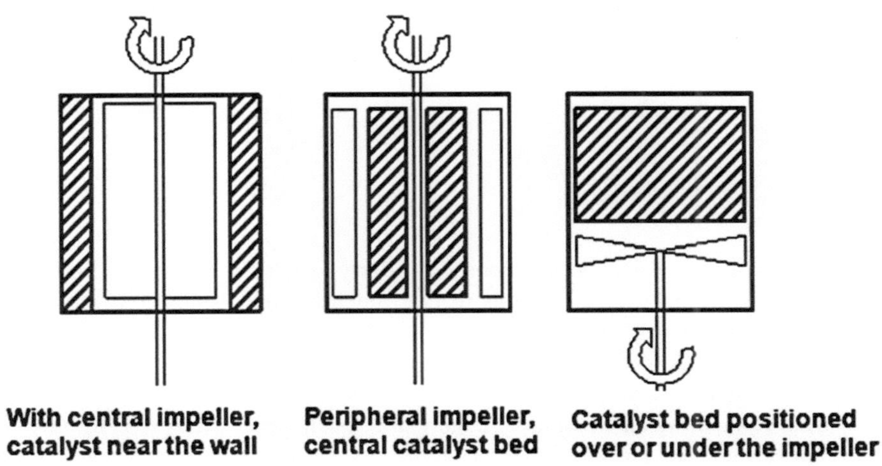

Fig. 5.28 CSTRs with catalytic fixed bed differently positioned inside the reactor. The catalytic bed corresponds to the dashed zone

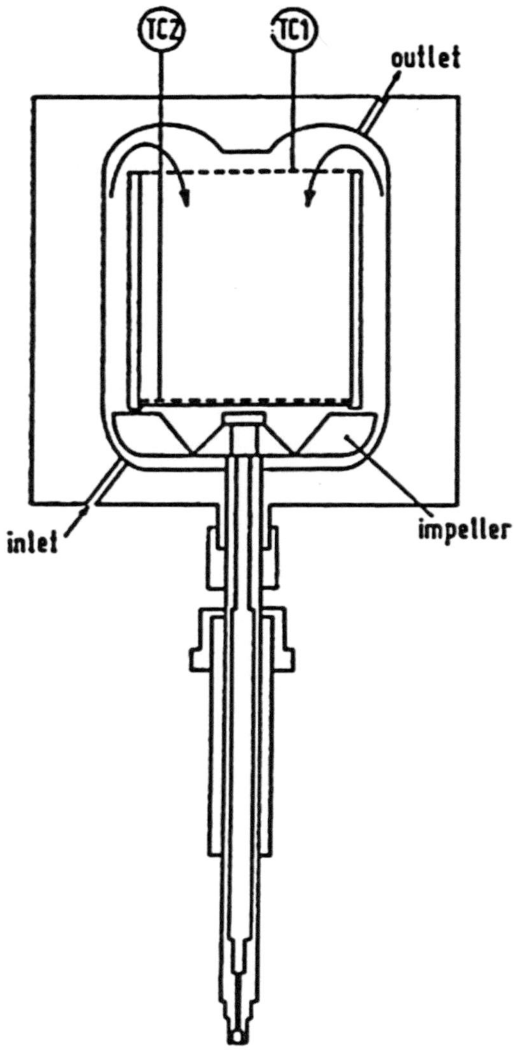

Fig. 5.29 Gas–solid CTSR with internal recycle (Berty reactor). The impeller rotates thanks to a MagneDrive system. Published with permission from Santacesaria E., Morbidelli M., Carrà S., Kinetics of the catalytic oxidation of methanol to formaldehyde; Chem. Eng. Sci.; 36, 909–918 (1981), Copyright Elsevier (1981)

A particular CSTR has been proposed by Berty (1979). A scheme of the Berty reactor is shown in Fig. 5.29. In this reactor, intense internal recycle occurs. The advantage of this reactor is the high measurable space velocity of the gas passing through the catalytic bed. In practice, the catalytic bed works under fluid dynamic conditions resembling those of a section of an industrial tubular reactor.

Figure 5.30 shows an example of a laboratory plant with a Berty CSTR.

Another interesting CST gas–solid laboratory reactor is the one proposed by Carberry (1964), having a rotating basket containing the catalyst, as shown in

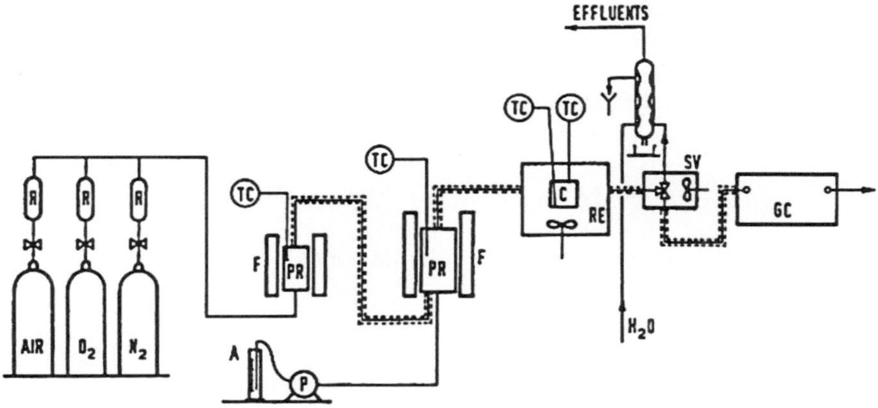

Fig. 5.30 Scheme of laboratory equipment for studying the oxidation of methanol to formaldehyde by using a Berty reactor. *R* rotameters, *F* ovens, *PR* preheaters, *TC* thermocouples, *RE* Berty reactor, *C* catalyst basket, *SV* thermostated sampling valve, *GC* online gas-chromatograph. Published with permission from Santacesaria E., Morbidelli M., Carrà S., Kinetics of the catalytic oxidation of methanol to formaldehyde; Chem. Eng. Sci.; 36, 909–918 (1981), Copyright Elsevier (1981)

Fig. 5.31 Scheme of the Carberry reactor with rotating catalyst basket

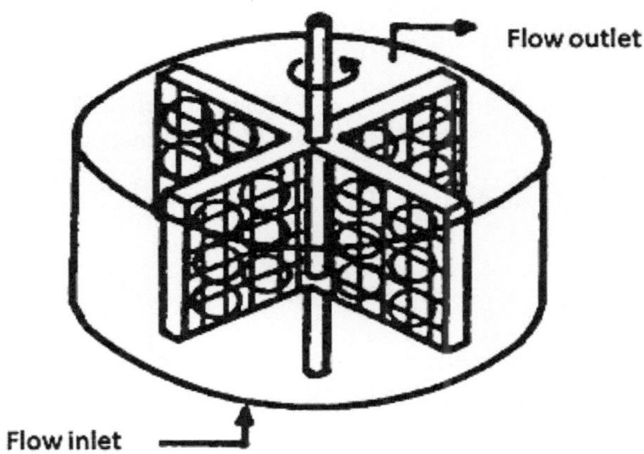

Fig. 5.31. This reactor solves the problem of eliminating any void space inside the reactor, thus reducing the possibility of the occurrence of homogeneous competing reactions.

The drawbacks of this reactor include the following: (1) the catalyst is subjected to strong mechanical stress and crumbles easily, and (2) it is not possible to insert a thermocouple inside the catalytic bed.

Exercise 5.8 Kinetics of the Oxidation of Methanol to Formaldehyde Determined Using a Berty CSTR

Santacesaria et al. (1981) studied the kinetics of methanol oxidation of formaldehyde promoted by iron molybdate as a catalyst. The kinetic data were collected using a Berty CSTR. The overall oxidation reaction is:

$$CH_3OH + 1/2\,O_2 \rightarrow HCHO + H_2O \tag{5.171}$$

A reasonable reaction scheme is the following one:

$$CH_3OH + \frac{1}{2}O_2 \xrightarrow{(1)} CH_2O + \frac{1}{2}O_2 \xrightarrow{(3)} CO + H_2O$$

$$CH_2O \xrightarrow[+2CH_3OH]{(2)} CH_2(OCH_3)_2 + H_2O$$

$$CH_2O \xrightarrow[+\frac{1}{2}O_2+CH_3OH]{(4)} HCOOCH_3 + H_2O$$

All the by-products are formed from formaldehyde. Thus, we can define an overall conversion as:

$$\lambda = \frac{\text{Moles of methanol reacted}}{\text{Moles of methanol fed}} \tag{5.172}$$

and evaluate the overall reaction rate as:

$$r = \frac{F}{W}\Delta\lambda \tag{5.173}$$

with F being methanol molar flow rate (mol/h).

Eventually, we also can define a conversion to any single reaction product as:

$$\lambda_i = \frac{\text{Moles of methanol reacted to give i}}{\text{Moles of methanol fed}} \tag{5.174}$$

However, in this exercise, we will consider only the overall reaction rate.

The catalyst was a commercial catalyst of iron molybdate with a ratio of Mo/Fe = 2.5 containing a small amount of CoO (1.8 wt%). The surface area was 6.0 m^2/g, and the porosity was 0.236. Catalyst, 15 g, in the form of pellets characterized by a hollowed cylindrical shape (0.15-cm internal diameter, 0.45-cm external diameter, and 0.45-cm height) were placed in a stainless-steel cylindrical basket (4.45-cm diameter and 7-cm height) mixed with Raschig rings of glass of approximately the same size of the catalyst pellets. The basket was put inside the reactor. Under the catalyst basket an impeller, connected with a MagneDrive assembly, allowed intense mixing of the reacting gases. The kinetics was studied in a temperature range of 200–250 °C. All the runs were performed at atmospheric pressure by changing in the feed the molar ratio O_2/CH_3OH (a), the ratio N_2/CH_3OH (b), and the ratio H_2O/CH_3OH (c). The operative conditions were chosen in such a way as to exclude mass-transfer limitations except for the runs performed at 250 °C showing a small effect of the internal diffusion (catalyst effectiveness = 0.94). The obtained results in terms of overall oxidation conversion as a function of F/W are listed in Table 5.15.

The authors, examining different rival kinetic models, found that a Mars–Van Krevelen redox mechanism, corrected for the deactivating effect of water, is suitable for interpreting all the experimental data, that is:

$$CH_3OH + \sigma - \text{ox} \xrightarrow{r_1} HCHO + \sigma - \text{red} \tag{5.175}$$

$$\sigma - \text{red} + 1/2O_2 \xrightarrow{r_2} \sigma - \text{ox} \tag{5.176}$$

by assuming the steady-state condition $r_1 = r_2$. Moreover, if some redox sites are occupied, by water, for example, we can write:

$$r = \frac{k_1 k_2 p_M p_{O_2}^{1/2}}{k_1 p_M + k_2 p_{O_2}^{1/2}} (1 - \nu_{\text{occ}}) \tag{5.177}$$

Remembering the Langmuir–Hinshelwood equation related to the adsorption equilibrium:

$$\nu_{\text{occ}} = \frac{b_w p_w}{1 + b_w p_w} \tag{5.178}$$

The rate equation becomes, therefore:

$$r = \frac{k_1 k_2 p_M p_{O_2}^{1/2}}{k_1 p_M + k_2 p_{O_2}^{1/2}} \cdot \frac{1}{(1 + b_w p_w)} \tag{5.179}$$

Starting from the experimental data, evaluate the kinetic parameters and their temperature dependence. Construct a parity plot of the calculated and experimental overall conversions. Put in an Arrhenius plot the kinetic constants k_1 and k_2 and the adsorption equilibrium constant b_w.

Table 5.15 Kinetic data collected by Santacesaria et al. (1981) for the steam reforming of methanol by using a Berty CSTR

Run no.	T (°C)	F/W	a	b	c	$\Delta\lambda$
1	211	0.03290	0.99400	3.97620	0.00000	0.0641
2	210	0.03290	0.99400	3.97620	0.00000	0.0597
3	208	0.03290	0.99400	3.97620	0.00000	0.0563
4	207	0.03290	0.99400	3.97620	0.00000	0.0514
5	206	0.03290	0.99400	3.97620	0.00000	0.0502
6	205	0.03290	0.99400	3.97620	0.00000	0.0465
7	204	0.03290	2.44368	3.97620	0.00000	0.0544
8	205	0.03290	2.44368	3.97620	0.00000	0.0471
9	204	0.03290	3.97610	3.97620	0.00000	0.0409
10	204	0.03290	5.96415	3.97620	0.00000	0.0415
11	205	0.11030	0.29850	1.18690	0.00000	0.0194
12	205	0.37530	0.08710	0.34870	0.00000	0.0078
13	204	0.37530	0.08710	0.34870	0.00000	0.0078
14	203	0.37530	0.08710	0.34870	0.00000	0.0076
15	202	0.03290	0.99400	6.95830	0.00000	0.0357
16	203	0.03290	0.99400	6.95830	0.00000	0.0371
17	203	0.03290	0.99400	8.94630	0.00000	0.0356
18	246	0.03290	0.99400	3.97620	0.00000	0.2415
19	248	0.03290	0.99400	3.97620	0.00000	0.266
20	246	0.03290	2.44368	3.97620	0.00000	0.255
21	245	0.03290	2.24436	3.97620	0.00000	0.244
22	244	0.03290	2.44368	3.97622	0.00000	0.240
23	243	0.03290	3.97610	3.97620	0.00000	0.2251
24	242	0.03290	3.97618	3.97622	0.00000	0.2272
25	241	0.03290	5.96410	3.97620	0.00000	0.2460
26	240	0.03290	5.96415	3.97622	0.00000	0.201
27	248	0.11030	0.29850	1.18690	0.00000	0.1438
28	252	0.11030	0.29850	1.18690	0.00000	0.1574
29	249	0.11030	0.29850	1.18690	0.00000	0.1537
30	246	0.03290	0.99400	5.42580	0.00000	0.248
31	244	0.03290	0.99400	5.42580	0.00000	0.229
32	242	0.03290	0.99400	5.42580	0.00000	0.227
33	240	0.37530	0.08710	0.34870	0.00000	0.0316
34	247	0.37530	0.08710	0.34870	0.00000	0.037
35	251	0.37530	0.08710	0.34870	0.00000	0.0348E
36	250	0.03290	0.99400	6.95330	0.00000	0.22905
37	248	0.03290	0.99400	6.95830	0.00000	0.2342
38	243	0.03290	0.99400	8.94635	0.00000	0.2025
39	241	0.03290	0.99400	8.94635	0.00000	0.1790
40	240	0.03290	0.99400	8.94630	0.00000	0.1773
41	223	0.03290	0.99400	3.97620	0.00000	0.1201
42	222	0.03290	0.99400	3.97620	0.00000	0.112

(continued)

Table 5.15 (continued)

Run no.	T (°C)	F/W	a	b	c	$\Delta\lambda$
43	220	0.03290	0.99400	5.42580	0.00000	0.093725
44	220	0.03290	0.99400	6.95830	0.00000	0.08790
45	219	0.03290	0.99400	6.95830	0.00000	0.0923
46	227	0.03290	0.99400	8.94635	0.00000	0.1165
47	226	0.03290	0.99400	8.94635	0.00000	0.1074
48	225	0.03290	0.99400	8.94635	0.00000	0.1020
49	225	0.03290	2.44368	3.97622	0.00000	0.1158
50	226	0.03290	2.44368	3.97622	0.00000	0.1188
51	225	0.03290	3.97618	3.97622	0.00000	0.114
52	226	0.03290	3.97618	3.97622	0.00000	0.1169
53	225	0.03290	3.97610	3.97620	0.00000	0.1149
54	225	0.03290	5.96410	3.97620	0.00000	0.1088
55	224	0.03290	5.96425	3.97622	0.00000	0.1102
56	228	0.11030	0.29850	1.18690	0.00000	0.0646
57	230	0.11030	0.29850	1.18690	0.00000	0.0754
58	202	0.02250	1.45640	5.82520	1.04690	0.0102
59	201	0.01000	3.26330	3.05360	5.13760	0.00585
60	200	0.01000	3.26330	3.05360	5.13760	0.00710
61	201	0.02990	1.09430	4.37740	0.22720	0.0161
62	202	0.02990	1.09430	4.37740	0.22720	0.0175
63	246	0.02250	1.45640	5.82520	1.04690	0.1289
64	235	0.01000	3.26330	3.05360	5.13760	0.1061
65	236	0.01000	3.26330	3.05360	5.13760	0.0904
66	251	0.02990	1.09430	4.37740	0.22720	0.2213
67	204	0.02250	1.45640	5.82570	1.04690	0.0090
68	239	0.02250	1.45640	5.82570	1.04690	0.0797
69	243	0.02250	1.45640	5.82570	1.04690	0.1556
70	245	0.02250	1.45640	5.82570	1.04690	0.1655
71	201	0.01000	3.26330	3.05360	5.13760	0.0052
72	240	0.01000	3.26330	3.05360	5.13760	0.1293
73	235	0.01000	3.26330	3.05360	5.13760	0.0969
74	248	0.02990	1.09430	4.37740	0.22720	0.1280
75	250	0.02990	1.09430	4.37740	0.22720	0.2801

Solution

Kinetic parameters determined by regression analysis
k10: 506568094.4459
k20: 2316475.6828
bw0: 8.0189e−06
ΔE_{a1}: 22520.9331
ΔE_{a2}: 17372.1353
ΔHw: −14932.9523

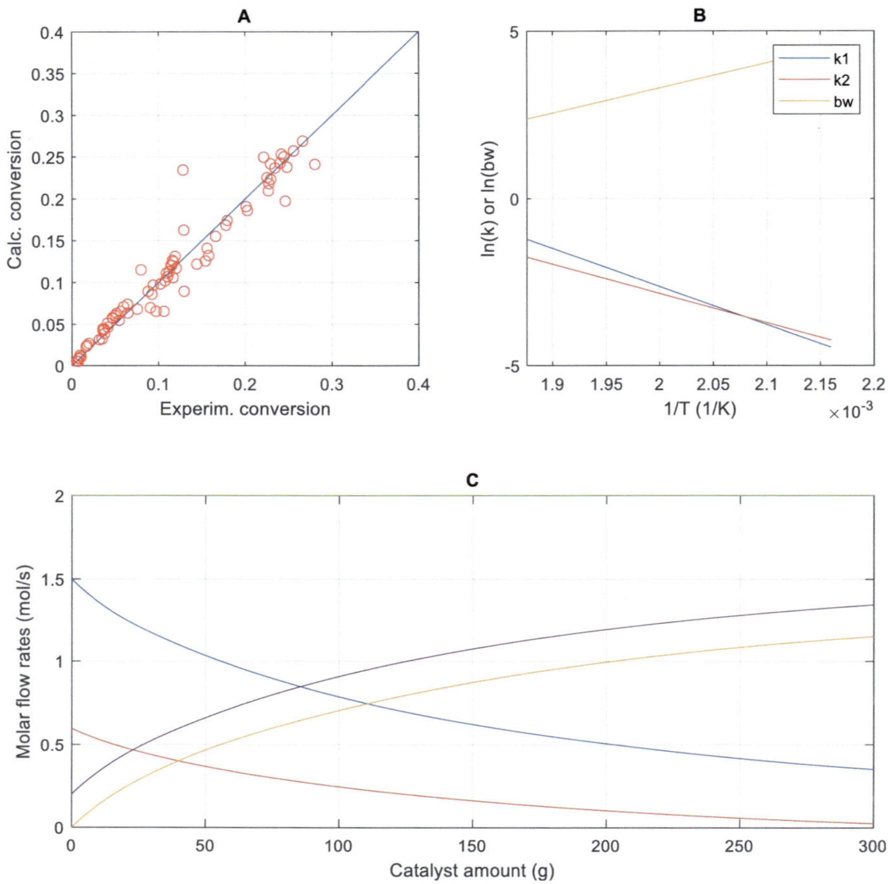

Fig. 5.32 (**a**) The parity plot comparing experimental and calculated activities; (**b**) Arrhenius-type plot of the kinetic and adsorption parameters; (**c**) packed bed reactor simulation: 250 g of catalyst, $T = 250$ °C, feed: methanol 1.5, oxygen 0.6, formaldehyde 0, water 0.2, nitrogen 2.0

Figure 5.32 shows the parity plot (left), which allows to appreciate the goodness of the model, and an Arrhenius-type plot (right) of the kinetic parameters.

The described results were obtained using a MATLAB program available as Electronic Supplementary Material, Data 5.1.

5.6 Thermal Behavior of Gas–Solid CSTRs

CSTR gas–solid reactors normally work under isothermal conditions because, as a consequence of the vigorous stirring, a uniform concentration and temperature in all points of the reactor are obtained. However, under transient conditions, we do not know how much the temperature increases or decreases inside the reactor in the case of strongly exothermic or endothermic reactions. Therefore, we do not know how much heat we must subtract from/introduce into the system to keep constant a desired temperature. This happens because the heat released or adsorbed by a reaction depends not only on the enthalpy change but also on the reaction rate, which we do not initially know. Therefore, to keep the thermal behavior of a CSTR under control, again we must solve the energy balance coupled with the mass balance. The mass balance in this case can be written as:

$$Q(C^o - C) = rV_R \tag{5.180}$$

$$Q\delta\overline{C_p}(T^o - T) = r(-\Delta H)V_R - UA(T - T_S) \tag{5.181}$$

where Q is the volumetric flow rate, δ is the fluid density, A is the reactor section, r is the reaction rate, ΔH is the enthalpy change of reaction, $\overline{C_p}$ is the average specific heat of the reaction fluid in the temperature range T^o–T, U is the overall heat transfer coefficient, and T_S is the temperature of the refrigerating or heating fluid, which is assumed to be constant.

If more than one reaction occurs, the two balances become:

$$Q(C^o - C) = V_R \sum_{j=1}^{n} \alpha_{ij} r_j \tag{5.182}$$

$$Q\delta\overline{C_p}(T^o - T) = V_R \sum_{j=1}^{n} \alpha_{ij} r_j(-\Delta H_j) - UA(T - T_S) \tag{5.183}$$

It is possible to demonstrate that a multiplicity of steady states is possible for a slight change of the parameters. In particular, for a single reaction three different steady-state conditions—two stable and one unstable—are sometimes possible, while the number of possible steady-state conditions increases when the reactions are more than one. This is due to the nonlinear dependence of reaction rate on the temperature, while heat removal follows a linear trend. Considering a first-order nonreversible reaction and constructing a plot of the heat generated or adsorbed, respectively, by (1) the reaction compared with that adsorbed by the reactant fluid plus (2) the heat exchanged with the external environment, we have a sigmoid curve for the reaction and a linear trend for corresponding to the other terms, as will be seen in Figs. 5.33 and 5.34. The intersection points are the solutions of Eq. (5.183), that is, they are the possible steady-state conditions.

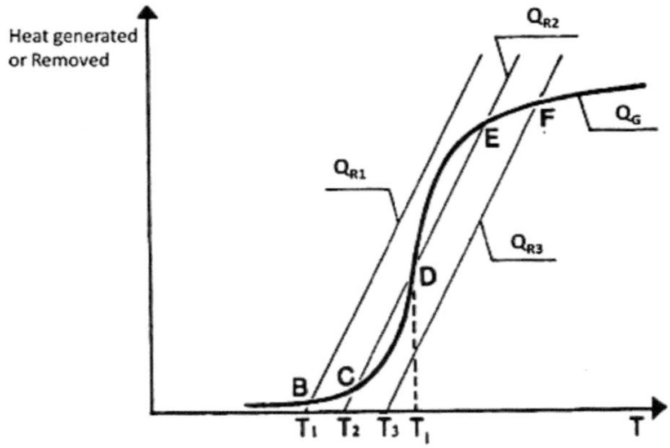

Fig. 5.33 Heat generated by the reaction, Q_G, the heat removed for different temperatures of fed gases, Q_R. The reaction is of the first order and is irreversible

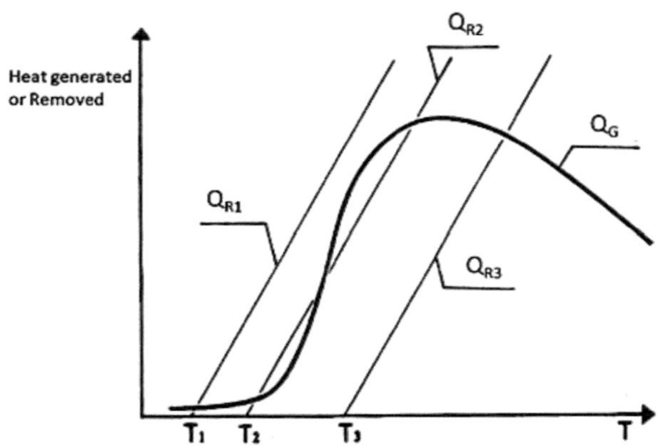

Fig. 5.34 Heat generated by the reaction, Q_G, the heat removed for different temperatures of fed gases, Q_T. The reaction is of the first order and is reversible

5.7 Fluidized Bed Reactors

When a gas–solid reaction is extremely exothermic, it is possible to operate under an isothermal condition only by using a fluidized bed reactor. In this case, the catalyst is powdered, and the gas containing the reactants is bubbled inside the powder. The catalyst particle must be small but resistant to attrition. To obtain the bubbling effect, the gas-flow rate must be kept relatively high, and the solid particles are consequently well-mixed e (see Fig. 5.35). Another advantage of these reactors is the possibility to operate under autothermal conditions, that is, the heat of the reaction keeps the temperature of the reactor at the desired level. Then, the use of relatively small particles avoids a mass-transfer limitation inside the particles and the consequent use of a small amount of catalyst. Disadvantages are as follows: (1) mechanical abrasion along with erosion of pipes and internal parts of the equipment, (2) attrition of the particles yielding smaller particles requiring a cyclone for recovering thinner

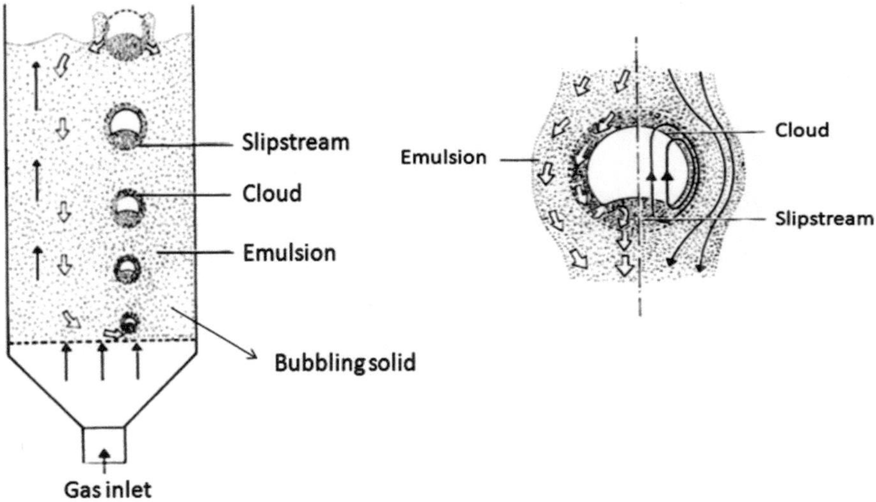

Fig. 5.35 Boiling fluidized bed and magnification of gas-bubble behavior. Model suggested by Davidson and Harrison (1963)

particles, (3) pressure decrease of gas that is high requiring high energy consumption (the fluid dynamic is complicated, and this renders the scale-up difficult); and (iv) large bubbles of gas yield bypassing effects.

This type of reactor is commonly employed in the industry but seldom used in the laboratory for kinetic studies because, as previously mentioned, the fluid dynamic conditions differ from small to large reactors and thus are not easily predictable.

To obtain fluidization, it is necessary to reach the value of the gas-flow rate to which a stable value of pressure decrease ΔP_f (fluidization pressure drop) corresponds. By increasing the gas-flow rate, ΔP_f increases linearly until reaching the limiting constant value, ΔP_f. ΔP_f can be calculated as:

$$\Delta P_f = \left(\delta_s - \delta_f\right)\left(1 - \varepsilon_f^o\right) g H_f \tag{5.184}$$

where δ_s and δ_f are the densities, respectively, of the solid and the fluid; ε_f° is the vacuum degree when fluidization starts; g is the gravity acceleration; and H_f is the height of the fluidized bed. The minimum flow rate necessary to obtain solid fluidization corresponds to a minimum fluidization rate of the gas, u_{mf}. The gas-surface velocity is normally taken to be much greater than u_{mf} to obtain the boiling effect. However, under these conditions, the fluid dynamic behavior of the system is extremely complicated because it depends on many factors such as the following: (1) aggregation of the catalyst particles due to the adhesive properties of the material with the formation of some stagnant zones in the reactor, (2) disruption of particles by attrition giving place to a broad size distribution of catalytic particles, and (3) formation of large bubbles in which the gas does not contact the surrounding solid (bypass). By changing the reactor size, the fluid dynamic behavior changes in unpredictable ways. For these reasons, the scale-up is often unreliable. As can be seen in Fig. 5.35, the bubbles of gas increase in size from bottom to top, and we can recognize three different zones: (1) a zone that is internal to the bubble, in which the gas is not in contact with the solid and therefore cannot react; (2) a zone (cloud and slipstream), in which the gas and solid are strictly in contact and well mixed; and (3) a zone (emulsion), where there is gas–solid contact, but the system is less mixed. Many different models have been suggested to describe how the reaction rate changes in a fluidized bed reactor. In particular, Davidson and Harrison (1963) described in detail the behavior of the gas bubbles (Fig. 5.35). They observed that all the bubble properties—such as the ascending rate, the thickness of the cloud, and the recirculation rate—are simple functions of the bubble sizes. Starting from the findings of Davidson and Harrison, Kunii and Levenspiel (1968) developed a model based on the following four assumptions:

(a) The bubbles are of uniform size and uniformly distributed in the fluidized bed.
(b) The bubble motion occurs according to the model of Davidson and Harrison.
(c) Each bubble brings up a small amount of solid, which then returns toward the bottom.
(d) The emulsion corresponds to the conditions corresponding to the minimum flow for obtaining fluidization, and the relative velocity of gas and solid does not change.

Then, a mass balance can be made, respectively, for the gas and the solid considering a two-zone model as shown in Fig. 5.36.

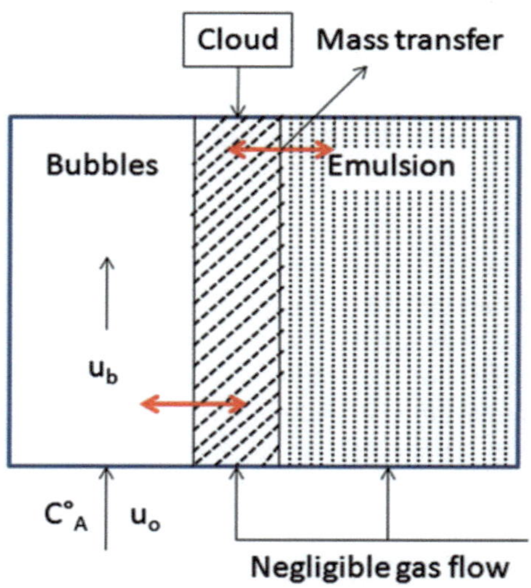

Fig. 5.36 Model of boiling fluidized bed

The mass balances related to the reactant A will be:

$$(\text{Disappearance of } A \text{ from the bubble zone})$$
$$= (\text{Reaction inside the bubble zone}) + (\text{Transfer of } A \text{ to the cloud solid})$$
$$(\text{Transfer of } A \text{ to the cloud solid}) = (\text{Reaction inside the cloud zone})$$
$$+ (\text{Transfer of } A \text{ to the emulsion})$$
$$(\text{Transfer of } A \text{ to the emulsion}) = (\text{Reaction inside the emulsion zone})$$

As can be seen, the mass-transfer coefficients from one zone to another are very important in this case. Therefore, this model will be developed in more detail in the next chapter, where the mass-transfer aspects will be extensively considered. The reaction terms will clearly contain the intrinsic kinetic eventually determined in the laboratory using different reactors.

5.8 Planning Experimental Runs and Elaborating Kinetic Data Using a Statistical Approach

Experimental runs can be planned with the same criteria described in Chap. 4 (see Figs. 4.11 and 4.12) related to the kinetics of homogeneous reactions. For a single reaction, we must first evaluate a power law that approximately individuates the reaction order with respect to the reactants, the catalyst, and—eventually—the products. The differential method applied to the initial reaction rate is useful to evaluate the reaction orders and the kinetic expression, and it gives some information about the mechanism. Reaction products normally have negative orders by competing with the reactants in adsorption on active sites. When the order of one of the products is positive, we have an autocatalytic effect. On the basis of the obtained power law, one or more reaction mechanisms (Langmuir–Hinshelwood, dual-site, Rideal, redox, etc.) are verified until the best-fitting of all experimental runs is obtained, and the kinetic parameters are determined. The kinetic parameters must have physical meaning.

When many different reactions occur, we must first determine the most probable reaction scheme. As we have already seen, this is not an easy task because, for schemes characterized by many occurring reactions, different alternative schemes could be compatible with the experimental observations. In any case, we can individuate the thermodynamic independent reactions, as described in Chap. 2, and study the kinetics of those reactions.

To discriminate between different kinetic models, statistical analysis could be useful.

When we face the problem of the development of a mathematical model, we first must solve the task of determining unknown parameters, especially for nonlinear models. The advantages of a mathematical model that correctly describes the physical reality are several, and among them, the most interesting are fast response and time-saving in problem-solving, data interpolation and extrapolation, collecting frequently dispersed information, the possibility to embed the developed model into more complex models, possibility to simulate a physical system without experimental activity, etc.

Many books and reviews have been published on the topic of parameter estimation; however, only short practical considerations regarding mathematical and statistical concepts and useful formulas are reported in this chapter. In our treatment, a procedure for discriminating amongst rival models on a statistical bases also will be given.

Parameter estimation is the process of using observations derived from an experimental plan to develop mathematical models that adequately represent the system characteristics. The developed model consists of a group of equations (algebraic or differential), in which a finite set of parameters appears, the values of which are estimated using estimation techniques. Fundamentally, the estimation approach is based on the least-squares minimization of error between the model response and the experimental system response; usually, the error function (objective function) is the sum of the squares of these differences (residuals).

In a general way, the model can be expressed as follows:

$$\widehat{y}=f(x,\beta) \tag{5.185}$$

where f is the function defining the model (algebraic or differential), x is the vector of the n experimental observation (independent variables), β is the vector of p unknown parameters that must be estimated, $\widehat{y}$ is the vector of calculated dependent variables (response of the model), and $\overline{y}$ is the average value of experimental y data.

The estimation of p, unknown β parameters, is made by searching for the minimum of the least square sum of the type:

$$\min \sum_{i=1}^{n}\left[y_i-f\left(x_i,\beta_j\right)\right]^2 \quad j=1,2\ldots p \tag{5.186}$$

where y_i are the experimental observations related to the dependent variables. Essentially, two main classes of algorithms are adopted in the literature for solving the problem of searching for the minimum of a least squares nonlinear regression function (Eq. 5.186): The first class is that of deterministic algorithms, such as gradient-based, simplex; the second class includes stochastic and evolutionary algorithms, such as particle swarm, genetic algorithm, pattern search, etc. In the MATLAB environment, practically all of these algorithms are present individually but also as a combination of two or more of them. Once an estimation of the parameters for a certain model has been found, a further question is how reliable are the estimated parameters? What is the error or confidence interval of them? Are these parameters correlated to one another (i.e., a value of one parameter influences the value of the other)?

The standard deviation (SD) of the estimated values of the parameters can be calculated with a mathematical procedure that starts from the Jacobian matrix calculated as:

$$J_{i,j} = \frac{\partial f(x_i, \beta)}{\partial \beta_j} \quad i = 1, \ldots, n \quad j = 1, \ldots, p \tag{5.187}$$

The resulting matrix is made of n rows and p columns. From this matrix, the covariance matrix can be evaluated, with standard matrix algebra operations, through the expression:

$$V = s^2 \left[J^T J\right]^{-1} \tag{5.188}$$

where s is the root mean square error, also called "standard deviation," of the residues, which is defined as:

$$s = \sqrt{\frac{\sum_{i=1}^{n} (y_i - \hat{y}_i)^2}{n - p}} \tag{5.189}$$

The importance of the covariance matrix V defined by Eq. (5.188) can be better appreciated by considering that it can be used directly for estimating the standard error (SE) on each parameter of the model as the square root of the diagonal elements of V:

$$s_j = \sqrt{V_{jj}} \quad j = 1, \ldots, p \tag{5.190}$$

In many cases, we are more interested in the confidence intervals of the parameters than the parameter SD, s_j. These are symmetric intervals centered on the estimated value of each parameter having an amplitude that is proportional to the chosen probability. The concept of confidence interval is indeed a statistical concept and represents the interval, around the estimated value of the parameter, inside which the true value of the parameter can be found with a certain probability. The higher the probability level, the lower the risk of missing the true value and the wider the confidence interval for that parameter. Usually, a probability level of 95% is chosen (corresponding to a risk of $\alpha = 5\%$), and this means that the confidence interval can be evaluated as a product of the SE of the parameter, s_j (Eq. 5.190) and a multiplier factor evaluated on statistical bases; usually the inverse Student t distribution is used for the calculation of this last multiplier factor as:

$$t_{\text{inv}}\left(1 - \frac{\alpha}{2}, n - p\right) \tag{5.191}$$

The quantity $n - p$ represents the degrees of freedom of the system evaluated as the difference between the number of experiments (n) and the number of parameters in the model (p). The complete equation for associating a confidence interval on each parameter is therefore:

$$\beta_j \pm s_j \cdot t_{\text{inv}}\left(1 - \frac{\alpha}{2}, n - p\right) \tag{5.192}$$

From this equation, it is evident that the lower the chosen risk (and the higher is the probability), the wider the confidence interval on the parameters of the model.

Another useful information that can be achieved from statistical considerations is which parameters are correlated and how much this correlation is important. The answer to this question is in the correlation matrix, which can be evaluated from the elements of the covariance matrix V, as follows:

$$\text{Corr}_{ij} = \frac{V_{ij}}{\sqrt{V_{ii}}\sqrt{V_{jj}}} \tag{5.193}$$

The correlation matrix is characterized by diagonal elements equal to 1, while each off-diagonal element (i, j) represents the correlation between the corresponding parameters i and j. The closer an off-diagonal element is to 1, the higher the correlation between the corresponding parameters. Obviously, the correlation matrix is symmetrical with respect to the diagonal.

5.8.1 Quality of Fit and Model Selection

Another critical issue involved in the elaboration of experimental data is related to the quality of fit. In other words, how accurately does the model describe the experimental data? Can we express quantitatively the goodness of the experimental data-fitting? The well-known parameter R^2 (correlation coefficient or determination coefficient) is a measure of data-description quality and is defined as:

$$R^2 = 1 - \frac{\sum_{i=1}^{n} (y_i - \widehat{y}_i)^2}{\sum\limits_{i=1}^{n} (y_i - \overline{y})^2} \tag{5.194}$$

In this expression, $\overline{y}$ represents the numerical average of the experimental data. A value of R^2 close to 1 indicates generally a good quality of fit. In any case, the only use of R^2 is not sufficient to judge if a model is adequate for the description of experimental data because this parameter does not contain information about the number of adjustable parameters present in the model. Moreover, the use of R^2 also is not recommended for the selection of a model between different candidates. More suitable for these purposes is the "adjusted" R^2 (R^2_{adj}), which is defined as:

$$R^2_{\mathrm{adj}} = 1 - \frac{(1 - R^2)(n-1)}{(n-p-1)} \tag{5.195}$$

This coefficient is sensitive to the number of adjustable parameters (p) and is always lower than R^2. With adjusted R^2_{adj}, it is possible to statistically compare different models with different numbers of parameters.

For the purpose of model comparison, selection, or discrimination, many other statistical tests have been proposed in the specialized books and literature, and even if a deep discussion of these tests is outside of the scope of this book, some practical formulas are reported below.

First, we must define three quantities that are functions of experimental and calculated y data:

$$\mathrm{RSS} = \sum_{i=1}^{n} (y_i - \hat{y}_i)^2 \tag{5.196}$$

$$\mathrm{TSS} = \sum_{i=1}^{n} (y_i - \bar{y})^2 \tag{5.197}$$

$$\mathrm{ESS} = \sum_{i=1}^{n} (\hat{y}_i - \bar{y})^2 \tag{5.198}$$

The most-used expressions for statistical tests are the following:

$$F = \frac{(R^2/p)}{(1-R^2)/(n-p-1)} \quad \text{Fisher test} \tag{5.199}$$

$$F_{\mathrm{E}} = \frac{(\mathrm{TSS}-\mathrm{RSS})/p}{(\mathrm{RSS})/(n-p)} \quad \text{Fisher E-test} \tag{5.200}$$

$$F_{\mathrm{m}} = \frac{\mathrm{ESS}/(p-1)}{(\mathrm{RSS})/(n-p)} \quad \text{Modified Fisher test} \tag{5.201}$$

In all cases, the values of the statistical tests from Eq. (5.199) to (5.201) should be compared with tabulated reference values to assess the statistical significance of a single mode; however, when used for comparing different models, the highest calculated value of the test should correspond to the best model. Obviously, this is only a statistical inference, and the choice among rival models also must be made by considering the physical meaning of the parameter.

5.9 Additional Exercises

Exercise 5.9 Oxidative Dehydrogenation of Ethanol to Acetaldehyde in the Presence of a V_2O_5/TiO_2-SiO_2 Catalyst

Tesser et al. (2004) studied the kinetics of the oxidative dehydrogenation of ethanol to acetaldehyde in the presence of a V_2O_5/TiO_2-SiO_2 catalyst. The exothermicity of the oxidation reactions compensates for the endothermicity of the dehydrogenation, even if the selectivity is lowered. The following reaction scheme has been individuated by the authors:

$$CH_3CH_2OH + 1/2\,O_2 \rightarrow CH_3CHO + H_2O \tag{5.202}$$

$$CH_3CHO + 1/2\,O_2 \rightarrow CH_3COOH \tag{5.203}$$

$$CH_3COOH + 2O_2 \rightarrow 2CO_2 + 2H_2O \tag{5.204}$$

$$2\,CH_3CH_2OH + CH_3CHO \rightleftarrows CH_3CH(OCH_2CH_3)_2 + H_2O \tag{5.205}$$

$$2CH_3CH_2OH \rightarrow CH_3CH_2OCH_2CH_3 + H_2O \tag{5.206}$$

The first three oxidation reactions occur through a redox mechanism that is similar for the three reactions. The most reliable kinetic equations resulted in the following:

$$r_1 = \frac{k_1 P_1}{1 + \frac{k_1 P_1}{k_{ox} P_{O_2}^{1/2}}} \tag{5.207}$$

$$r_2 = \frac{k_2 P_2}{1 + \frac{k_2 P_2}{k_{ox} P_{O_2}^{1/2}}} \tag{5.208}$$

$$r_3 = \frac{k_3 P_3}{1 + \frac{k_3 P_3}{k_{ox} P_{O_2}^{1/2}}} \tag{5.209}$$

where 1 is ethanol, 2 is acetaldehyde, 3 is acetic acid, 4 is acetal, 5 is carbon dioxide, 6 is oxygen, 7 is water, and 8 is ethyl ether. A simpler kinetic law was imposed on reactions 4 and 5 because their contribution is very low:

$$r_4 = k_4 P_2 \tag{5.210}$$

$$r_5 = k_5 P_1^2 \tag{5.211}$$

where k_{ox} is the rate constant for the catalytic-site reoxidation.

The best-fitting parameters found by the authors are listed in Table 5.16.

By feeding a stochiometric amount of ethanol and oxygen at 180 °C and 1 atm and considering an ethanol conversion of 0.75, calculate the flow rates necessary to

Table 5.16 Arrhenius parameters for the considered reactions

	$\ln k = \ln A\text{-}Ea/RT$	
Reaction	$\ln A$	E_a (kcal/mol)
Ethanol → acetaldheyde	12.61 ± 1.75	10.9 ± 1.4
Catalytic-site reoxidation	10.47 ± 1.46	11.7 ± 1.2
Acetaldheyde → acetic acid	27.04 ± 25.95	35.2 ± 21.3
Acetic acid → CO_2	44.56 ± 33.32	47.1 ± 27.4
Ethanol + acetaldheyde → acetals	5.46 ± 5.08	7.5 ± 4,2
Ethanol → ethyl ether	48.64 ± 22.01	47.0 ± 18.1

produce approximately 50 tons/h of acetaldehyde. Assume the absence of any mass-transfer limitation working with pellets of an apparent diameter of 0.5 cm and a void degree of 0.4.

Imagine using a multi-tube reactor to obtain this production. Any tube has a diameter of 2 inches. Considering the reactor to be isotherm, how many tubes, of what length, are necessary? Last, consider in detail the behavior of one of these tubes. The wall temperature is kept at 180 °C by the recirculating thermostating fluid. Evaluate how the change of temperature profile responds by changing the coefficient of heat exchange from 0 (adiabatic) to infinite (isotherm). Consider just two intermediate conditions, and for one of these, also describe the profile of ethanol conversion as well as that of the different product yields.

Results

Part 1. In this case, the reactor is operated isothermally at 180 °C, and the results of the simulation are presented in Figs. 5.37 plot 1 and 5.37 plot 2. As we can observe, the overall amount of catalyst necessary to produce approximately 50 tons/h of acetaldehyde (by imposing an ethanol conversion of 75%) is 26 tons. By considering the necessity of a good thermal exchange for operating isothermally, the catalyst should be arranged in a set of tubes. As an approximate evaluation of the number of tubes that can be used, the following calculations can be developed:

Tube diameter	5 cm
Tube cross-section	20 cm^2
Tube length	10 m
Tube volume	20 dm^3
Catalyst density	1.5 g/cm^3
Amount of catalyst in a tube	30 kg
Number of tubes needed	870

Part 2. In this second part of the exercise, five different global heat-exchange coefficients, U, were used (see Fig. 5.37 plots 3–7). Different behaviors of the system can be observed, passing from an isothermal system (U = 1e-4), similar to that seen in Part 1, to an almost adiabatic system (U = 1e-7), in which a hot spot ≤300 °C is observed (see Fig. 5.37 plot 5). By decreasing the heat-exchange

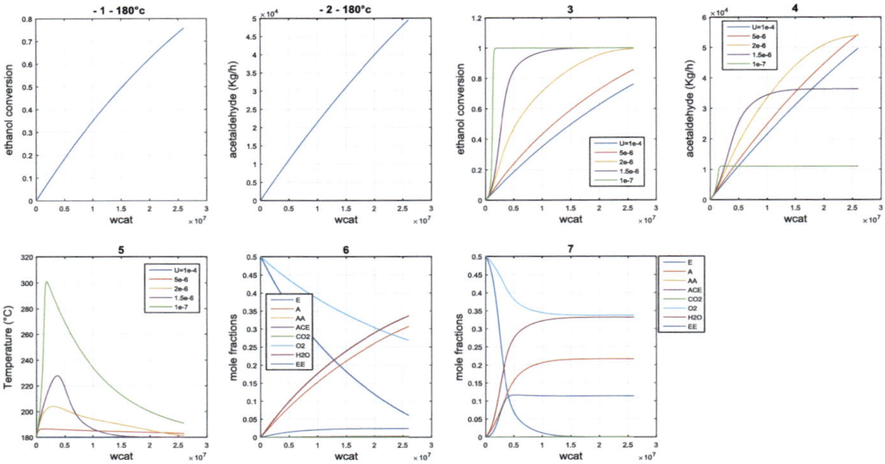

Fig. 5.37 Results obtained in the simulations as a function of catalyst loading; plot 1, ethanol conversion at $T = 180$ °C; plot 2, acetaldehyde productivity at 180 °C; plot 3: ethanol conversion for various efficiencies of thermal exchange; plot 4, acetaldehyde productivity for various efficiencies of thermal exchange; plot 5, temperature profile for various efficiencies of thermal exchange; plot 6, composition profiles at 180 °C; plot 7, composition profiles for adiabatic reactor ($U = 1$e-7)

coefficient, a general increase of the reactor temperature is obtained and, as expected, ethanol conversion also increases (Fig. 5.37 plot 3). By observing Fig. 5.37 plot 4 (acetaldehyde productivity) and the trend of the composition profiles in Fig. 5.37 plot 6 and plot 7, it is evident that, in correspondence with a higher ethanol conversion, a decrease of acetaldehyde is realized. The lower U, corresponding to a practically adiabatic behavior, yields production of approximately only 10 tons/h instead of the 50 tons/h assumed as a target. The greater temperatures achieved with low thermal exchange are detrimental to acetaldehyde production as ethanol dehydration becomes relevant when this compound reaches a mole fraction of approximately 0.12 (Fig. 5.37 plot 7) instead of 0.02 in the case of constant-temperature operation (Fig. 5.37 plot 6). Another consideration is that, for adiabatic operation, the system rapidly reaches the equilibrium composition along the bed, when the ethanol is quite completely consumed.

All the described results were obtained using the MATLAB programs available as Electronic Supplementary Material, Data 5.1.

Exercise 5.10 Alkylation of Phenol with Methanol on H-ZSM5 Zeolite
In this exercise, it will be explained how, in some cases, a complicated reaction scheme can be simplified with lumped kinetic models. Santacesaria et al. (1990) studied the alkylation of phenol to cresols using an acid H-ZSM5 zeolite. The reaction scheme is complicated because all the reaction products are alkylating agents competing with methanol. We can recognize 12 different possible reactions that can be grouped into two different reaction types: (1) reactions alkylating to the

Table 5.17 Alkylation reactions occurring starting from phenol and methanol

Number	Reaction	Alkylation to …
1	ϕOH + CH_3OH → ϕOCH_3 + H_2O	Oxygen
2	ϕOH + CH_3OH → cresols + H_2O	Ring
3	ϕOCH_3 + ϕOCH_3 → methylanisoles + ϕOH	Ring
4	ϕOCH_3 + ϕOH → cresols + ϕOH	Ring
5	ϕOCH_3 + CH_3OH → methylanisoles + H_2O	Ring
6	Cresols + CH_3OH → methylanisoles + H_2O	Oxygen
7	Cresols + CH_3OH → xylenols + H_2O	Ring
8	Cresols + ϕOCH_3 → methylanisoles + ϕOH	Oxygen
9	Cresols + ϕOCH_3 → xylenols + ϕOH	Ring
10	Methylanisoles + ϕOH → 2 cresols	Ring
11	2 Methylanisoles → cresols + xylenols	Ring
12	Methylanisoles + cresols → cresols + xylenols	Ring

oxygen and (2) reactions alkylating to the ring having the same kinetic expression. Another complication is that phenol is both a reactant and a product of some reactions. The overall reaction has been studied in the temperature range of 260–350 °C. The complete reaction scheme is listed in Table 5.17.

The composition at the outlet of the reactor can be determined by solving the following system of differential equations:

$$\frac{F}{W}\frac{\mathrm{d}y_1}{\mathrm{d}L} = -r_1 - r_2 - r_{10} + r_3 + r_8 + r_9 \tag{5.212}$$

$$\frac{F}{W}\frac{\mathrm{d}y_2}{\mathrm{d}L} = -r_1 - r_2 - r_5 - r_6 - r_7 \tag{5.213}$$

$$\frac{F}{W}\frac{\mathrm{d}y_3}{\mathrm{d}L} = +r_1 - 2r_3 - r_4 - r_5 - r_8 - r_9 \tag{5.214}$$

$$\frac{F}{W}\frac{\mathrm{d}y_4}{\mathrm{d}L} = +r_1 + r_2 + r_5 + r_6 + r_7 \tag{5.215}$$

$$\frac{F}{W}\frac{\mathrm{d}y_5}{\mathrm{d}L} = +r_2 + r_4 + 2r_{10} + r_{11} - r_6 - r_7 - r_8 - r_9 \tag{5.216}$$

$$\frac{F}{W}\frac{\mathrm{d}y_6}{\mathrm{d}L} = +r_3 + r_5 + r_6 + r_8 - r_{10} - 2r_{11} - r_{12} \tag{5.217}$$

$$\frac{F}{W}\frac{\mathrm{d}y_7}{\mathrm{d}L} = +r_7 + r_9 + r_{11} + r_{12} \tag{5.218}$$

where 1 is phenol, 2 is methanol, 3 is anisole, 4 is water, 5 is cresols, 6 is methyl anisole, and 7 is xylenols, L is the relative length of the reactor, F is the molar feed rate, and W the catalyst weight. A Rideal mechanism was found to be reliable for the following reactions:

$$A + \sigma \rightarrow A\sigma$$

$$A\sigma + \mathrm{ROH} \rightarrow \mathrm{ROCH_3} + \sigma$$

$$A\sigma + \mathrm{R} \rightarrow \mathrm{RCH_3} + \sigma$$

where R is an aromatic ring simple or already alkylated and ROH is an aromatic molecule containing a hydroxyl. Consequently, two kinetic expressions were formulated: one for describing the alkylation to the oxygen:

$$r_{\text{O-alk}} = \frac{\eta_o k_o Y_{Ai} Y_{\mathrm{POH}i} P^2}{1 + b_{\mathrm{AA}} P \sum_i Y_{\mathrm{AA}i} + b_{\mathrm{A}} P y_4} \tag{5.219}$$

which is valid for r_1, r_6, and r_8, and another to describe the ring alkylation:

$$r_{R-alk} = \frac{\eta_R k_R Y_{Ai} Y_{Ri} P^2}{1 + b_{\mathrm{AA}} P \sum_i Y_{\mathrm{AA}i} + b_A P y_4} \tag{5.220}$$

which is valid for all the other cases. The subscript A_i is related to each alkylating agent, while AA_i is related to the aromatic components having alkylating properties, such as anisoles and methylanisoles. The subscript $\mathrm{POH_i}$ is related to any component that can be alkylated at the oxygen as phenol and cresols. η_o and η_R are the catalyst-efficiency factors related to the slow intra-cristalline diffusion of reactants and products into the crystalline network of the zeolite. These two terms can be considered together with the kinetic constants.

Another complication of this reaction is the catalyst deactivation. Fortunately, the deactivation occurs independently according to a kinetic law of the type:

$$\frac{A_t}{A_o} = \frac{1}{1 + \varepsilon t} \tag{5.221}$$

By submitting all the kinetic runs performed, the authors found the best-fitting kinetic parameters (Table 5.18).

Table 5.18 Kinetic parameters for the proposed model

	ε (h^{-1})	$\eta_o k_{\mathrm{O-Alk}}$ $\left(\frac{\mathrm{mols}}{\mathrm{hatm}^2}\right)$	$\eta_R k_{\mathrm{R-Alk}}$ $\left(\frac{\mathrm{mols}}{\mathrm{hatm}^2}\right)$	b_{AA} (atm^{-1})	b_{A} (atm^{-1})
Kinetic parameters at 300 °C	0.066	4752	2544	210	1200
Activation energies or enthalpy change in (cal/mol)	–	11,000	22,000	−5000	−20,000
	Phenol conversion		Anisole	Cresols	Methylanisoles
Mean error (%)	13.5		4.9	6.3	12.3
Correlation index	0.97		–	–	–

Using the kinetic parameters reported in Table 5.18, calculate the conversions of the product distribution of the runs performed at 300 °C considering 1 g of catalyst while considering the catalyst deactivation. Compare the obtained results with the experimental ones reported in Table 5.19. Then, evaluate the optimal residence time, at this temperature, for obtaining cresols. Last, how do the yields in cresols change with the temperature?

Results

The described results were obtained using a MATLAB program available as Electronic Supplementary Material, Data 5.1 (Fig. 5.38).

Exercise 5.11 Esterification of Acetic Acid with Ethanol in Vapor Phase in the Presence of H-Y Zeolite

Santacesaria et al. (1983) studied the kinetics of acetic acid esterification with ethanol in vapor phase using H-Y zeolite as a catalyst. The acetic acid esterification reaction is normally performed in the liquid phase in the presence of a mineral acid as a catalyst. Under these conditions, at the ethanol boiling point, the equilibrium yield for a stochiometric mixture of the reactants is approximately 66% of conversion to ethyl acetate with problems in the successive step of separation due to the formation of an azeotrope. However, it is known that the same reaction is thermodynamically much favored when performed in the vapor phase, yielding a conversion equilibrium 95% (Othmer [1958]). In the vapor phase, it is convenient to use a heterogeneous catalyst, and Santacesaria et al. (1983) used as a catalyst a decationized H-Y zeolite in pellets of 0.13-cm diameter. A tubular PFR was used with an internal diameter of 1 cm and a length of 36 cm. Some spherical balls of 0.3-cm diameter were placed before the catalytic bed to obtain a plug flow before entering the reactor. The used apparatus is shown in Fig. 5.7. Hawes and Kabel (1968) suggested the following equation to calculate the equilibrium constant:

$$\log K_e = \frac{649}{T} + 0.042 \tag{5.222}$$

It is important to point out that acetic acid gives place to the formation of a dimer that, very probably, is not involved in the esterification reaction. An equation for calculating the dimerization equilibrium constant has been suggested by Potter et al. (1955):

$$\log K_d = \frac{3000}{T} + 10.149\left(\text{mmHg}^{-1}\right) \tag{5.223}$$

A long transient time is necessary for reaching steady-state conditions because zeolite absorbs a great number of reactants (approximately 10% b.w.). Kinetic runs were performed at different temperatures in the range of 150–200 °C, at atmospheric pressure, which significantly changed the ratio between the reactants and changing the space velocity, F/W, as can be noted in Table 5.19. All the runs reported in Table 5.20 reached the steady-state condition.

Table 5.19 Kinetic runs performed for >200 h at 300 °C using 1 g of catalyst

Time (h)	Feed rate (mL/min)	Ratio of PhOH to CH_3OH	Conversion (%)	Anisole	o–cresol	p–cresol	m–cresol	o–MA	p–MA	m–MA	2–6 Xy	Other Xy	F/W
0	0.26	2:1	17.7	53.5	25.4	12.3	2.1	0.7	2.2	0.5	0.9	2.3	142.0
1.5	0.26	2:1	15.9	57.6	23.6	11.7	1.7	0.7	2.0	0.5	0.5	1.8	142.0
22	0.26	2:1	11.9	62.6	21.8	11.0	1.5	0.6	1.6	–	–	0.9	142.0
26	0.26	2:1	11.1	62.8	21.6	11.0	1.5	0.6	1.7	–	–	0.9	142.0
46	0.26	2:1	10.6	63.6	21.2	10.7	1.5	0.6	1.6	–	–	0.9	142.0
68	0.26	2:1	7.3	43.7	21.3	10.7	1.5	–	1.9	–	–	0.9	142.0
69	0.40	2:1	5.9	64.9	21.4	10.9	1.5	–	1.4	–	–	–	218.4
72	0.13	2:1	10.5	60.8	22.1	11.2	1.5	0.9	2.2	–	–	1.3	71.0
80	0.26	2:1	7.9	63.8	21.2	10.8	1.6	–	1.7	–	–	0.9	142.0
98	0.26	2:1	7.1	65.0	21.0	10.6	1.5	–	1.6	–	–	–	142.0
130	0.26	2:1	5.6	66.9	19.6	9.6	1.6	0.7	1.6	–	–	–	142.0
150	0.26	2:1	5.1	65.6	20.2	10.3	1.5	0.7	1.6	–	–	0.8	142.0
153	0.26	1:1	5.8	61.3	22.0	11.1	1.6	0.8	1.9	–	–	1.3	119.8
177	0.26	1:1	5.7	64.2	21.3	10.8	1.6	0.6	1.5	–	–	–	120.0
181	0.26	4:1	3.7	69.9	18.1	9.0	1.4	–	1.6	–	–	–	156.4
183	0.16	4:1	5.5	68.0	19.0	9.6	1.5	0.6	1.3	–	–	–	78.2
206	0.26	2:1	3.8	65.5	20.3	10.3	1.8	–	1.4	–	–	0.8	142.0

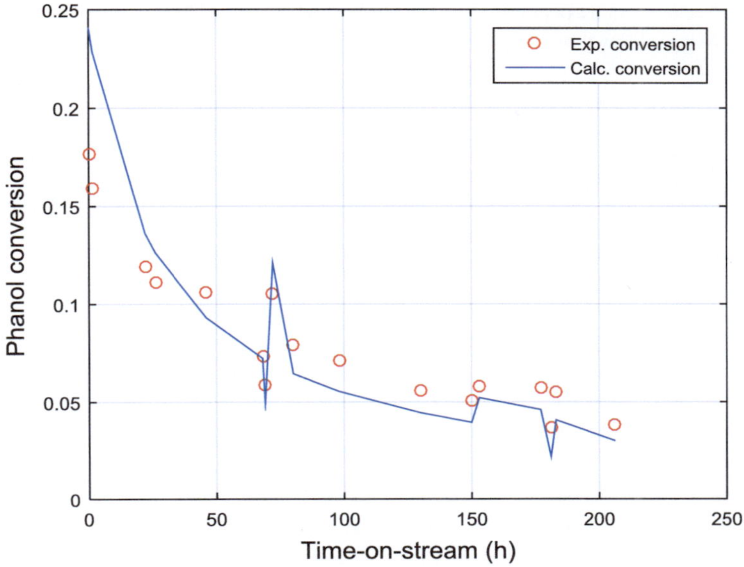

Fig. 5.38 Experimental and calculated phenol conversion along the time-on-stream (Exercise 5.10)

If we assume the stoichiometric number of moles of acetic acid in the liquid mixture of reagent to be:

$$W_1^0 = n_1^0 + 2n_2^0 = \frac{V_1 d_1}{M_1} \tag{5.224}$$

and considering the following relationship for the dimerization equilibrium:

$$K_d = \frac{n_2^0 \left(n_1^0 + n_2^0 + n_3^0\right)}{\left(n_1^0\right)^2} \tag{5.225}$$

the initial composition of the reagents in vapor phase will be:

$$n_1^0 = \frac{-n_3^0 + \sqrt{\left(n_3^0\right)^2 + (4K_d + 1)\left(2n_3^0 + W_1^0\right)W_1^0}}{(4K_d + 1)} \tag{5.226}$$

$$n_2^0 = \frac{W_1^0 - n_1^0}{2} \tag{5.227}$$

$$n_3^0 = \frac{V_3 d_3}{M_3} \tag{5.228}$$

Table 5.20 Conversions obtained for the esterification reaction by changing the temperature, residence time, and initial molar ratio of the reagents[a]

Run	T (°C)	n_{Ac}/n_{EtOH}	W = catalyst (g)	F_{tot} (cm^3/min)	F/W $\left(\frac{\text{EtOH mols}}{\text{h kg}}\right)$	Experimental conversion (%)	Calculated conversion (%)
1	150	2:1	5.04	0.71	47.94	15.18	16.42
2	150	2:1	5.08	0.41	27.68	22.75	26.20
3	150	2:1	5.08	1.50	101.28	8.50	8.28
4	151	2:1	14.00	1.00	24.52	28.10	29.92
5	151	2:1	14.00	2.00	4.90	79.50	77.45
6	151	3:1	14.00	1.00	18.39	32.00	32.74
7	151	1:1	14.00	1.00	36.78	24.83	24.20
8	152	2:1	5.04	0.67	44.22	17.33	19.06
9	152	10:1	5.04	0.67	12.43	22.80	23.99
10	152	29:1	5.04	0.67	4.56	31.43	25.39
11	152	1:3	5.04	0.71	108.70	9.87	8.58
12	152	1:5	5.04	0.71	120.77	6.54	6.25
13	153	1:10	5.04	0.67	124.33	6.51	4.04
14	153	1:5	5.04	0.67	128.21	4.09	2.88
15	153	1:29	5.04	0.67	132.20	2.49	1.60
16	182	10:1	5.04	0.67	12.43	58.95	59.14
17	182	1:3	5.04	0.71	108.70	22.46	20.13
18	182	1:5	5.04	0.67	113.97	14.94	13.92
19	184	2:1	5.04	0.41	27.68	63.59	65.34
20	201	10:1	5.04	0.67	12.43	82.69	81.75
21	201	1:5	5.04	0.67	113.97	17.37	17.90
22	201	1:3	5.04	0.67	102.57	26.72	27.60

[a]Conversions were calculated using the Rideal model

The composition obviously changes due to the effect of the reaction; accordingly, the equilibrium equation becomes:

$$K_d = \frac{n_2(n_1 + n_2 + n_3 + n_4 + n_5)}{(n_1)^2} \tag{5.229}$$

where:

$$n_1 = \frac{-n_3^0(1+\lambda) + \sqrt{\left[n_3^0(1+\lambda)\right]^2 + W_1\left[W_1 + 2n_3^0(1+\lambda)\right](4K_d+1)}}{(4K_d+1)} \tag{5.230}$$

$$n_2^0 = \frac{W_1 - n_1}{2} = \frac{W_1^0 - n_3^0\lambda - n_1}{2} \tag{5.231}$$

$$n_3 = n_3^0(1-\lambda) \tag{5.232}$$

$$n_4 = n_5 = n_3^0\lambda \tag{5.233}$$

where λ is the conversion of ethanol. For each value of λ, the composition of the reaction mixture is known. In interpreting the experimental data, it was assumed that only monomeric acetic acid reacts. To individuate the kinetic model to be adopted, a pseudo second kinetic law was adopted to interpret the runs performed at 150 °C by changing the initial ratio of the reactant ($n_{Ac}{}^{\circ}/n_{Al}{}^{\circ}$), that is, the following kinetic equation was tested:

$$r = k_2\left(x_1x_3 - \frac{x_4x_5}{K_e}\right)P^2 \tag{5.234}$$

Figure 5.40 shows the obtained results. As can be seen, the kinetic model works well for a large range of reactant ratios, but the apparent kinetic constant dramatically decreases for low ratio of acetic acid and ethanol Fig. 5.39.

The model clearly is unsuitable; therefore, two other kinetic models were tested: (1) a surface bimolecular model (BS); and a Rideal model (*R*):

$$r = \frac{k_{app}b_1}{(1+b_1p_1)^2}\left(p_1p_3 - \frac{p_4p_5}{K_e}\right) \quad \text{BS model} \tag{5.235}$$

$$r = \frac{kb_1}{(1+b_1p_1)}\left(p_1p_3 - \frac{p_4p_5}{K_e}\right) \quad \text{R model} \tag{5.236}$$

Both of the kinetic models were subjected to statistical analysis, to determine the kinetic parameters, by searching for the minimum of the objective function:

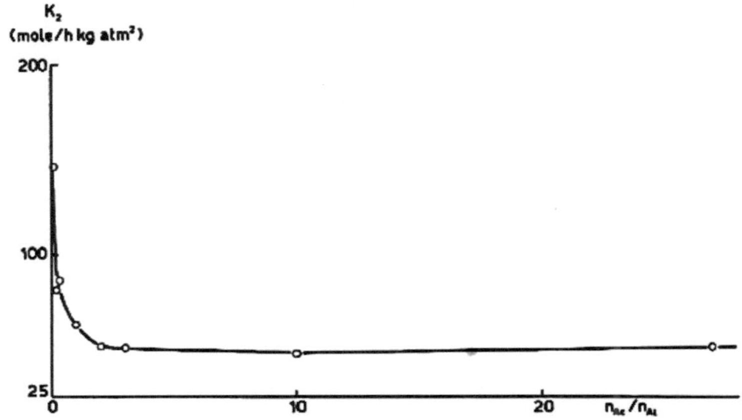

Fig. 5.39 Dependence of the second-order kinetic constant on the reactant ratio

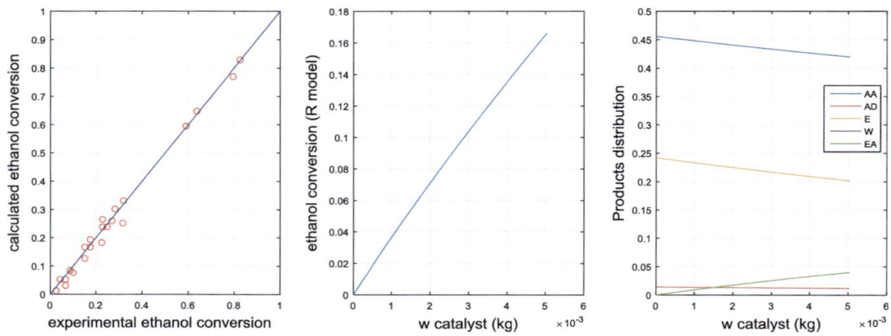

Fig. 5.40 Results obtained for the Rideal model (R model)

$$\Phi(\beta) = \sum_{1}^{N} \left(\lambda_{\text{exp}} - \lambda_{\text{calc}}\right)^2 \tag{5.237}$$

where β is the set of parameters. In Fig. 5.40, results are reported regarding the Rideal model. The plot on the left is the parity plot for ethanol conversion, where the overall agreement between experiments and model can be appreciated. The plot in the center of Fig. 5.40 is the calculated trend of ethanol conversion for the run no.1 of Table 5.19, while in the plot on the right, the corresponding product distribution is reported.

- Construct the plot at 150 °C for the Rideal model (run no. 1 in Table 5.19).
- Evaluate the kinetic parameters for the two alternative considered models.

Results

BS (Bimolecular Surface) Model

Parameters	Confidence intervals	Standard deviation
ka ref. = +1.4445e + 02	± +9.6549e + 00	+4.5956e+00
bi ref. = +4.8415e−01	± +2.7119e–02	+1.2908e−02
ea = +1.8934e + 04	± +1.1607e + 03	+5.5249e+02
dhb = −4.8582e + 03	± +1.5488e + 02	+7.3721e+01
k0 = +8.7691e+11 (see article)		
bi0 = +1.4953e−03 (see article)		

Correlation coefficients and test
R^2 = +0.988176
R^2 adj = +0.985394
f-test = 355.203
e-test = 376.0973
FF-test = 522.5966

Parameter-correlation matrix			
1.0000	−0.1144	−0.4253	−0.5724
−0.1144	1.0000	−0.2925	0.3207
−0.4253	−0.2925	1.0000	0.3337
−0.5724	0.3207	0.3337	1.0000

R (Rideal) Model

Parameters	Confidence intervals	Standard deviation
ka ref. = +1.2372e + 02	± +7.7110e + 00	+3.6703e+00
bi ref. = +4.3427e−01	± +3.5202e−02	+1.6756e−02
ea = +1.9302e + 04	± +1.5972e + 03	+7.6023e+02
dhb = −4.9319e + 03	± +2.5008e + 02	+1.1903e+02
k0 = +1.1640e + 12(see article)		
bi0 = +1.2286e−03(see article)		

Correlation coefficients
R^2 = +0.988950
R^2 adj = +0.986350
f-test = 380.3544
e-test = 402.7282
FF-test = 557.8542

Parameter-correlation matrix			
1.0000	−0.2738	−0.6902	−0.3055
−0.2738	1.0000	−0.1732	0.3815
−0.6902	−0.1732	1.0000	−0.0209
−0.3055	0.3815	−0.0209	1.0000

In the previous tables, a complete statistical analysis is reported for the comparison of the two considered models: BS and R. For each model, the statistical analysis includes, first of all, the fitting parameters, their confidence intervals, and standard deviation. A second group of information is related to correlation coefficients and statistical tests useful for discrimination among different models. The last table in the statistical analysis is the parameter correlation matrix, in which values near to unity (except the diagonal) indicate that the corresponding couple of parameters are correlated.

The described results were obtained using a MATLAB program available as Electronic Supplementary Material, Data 5.1.

Exercise 5.12 Esterification of Acetic Acid with Ethanol in Vapor Phase in the Presence of Decationized Zeolite H-Y: Modeling an Industrial Reactor Using the kinetic law of Rideal with the related parameters, calculate the overall volume of an adiabatic reactor in four different stages. The feeding rate is 2200 kg/h of ethanol containing 6% of water and 5400 kg of acetic acid containing 0.3% of water. The reaction enthalpy change is

$\Delta H_R = -3310$ cal/mole of reacted ethanol. The first three stages of the reactor must be sized to yield 28% of ethanol conversion. Feeding the reagents at 150 °C and restoring this temperature after each stage, calculate for the first three stages:

- Volume of each stage
- Amount of catalyst that must be loaded
- Amount of heat that must be removed after each stage
- Increase in temperature after each stage

The last stage will bring the overall conversion to 0.93, that is, near to equilibrium. Then, determine the following:

- Amount of ethyl acetate produced per hour
- Final composition
- Volume of the last stage
- Overall volume of catalyst that must be loaded in the reactor
- Overall weight of catalyst that must be used
- Amount of heat that must be removed between the third and fourth stages
- Temperature at the exit of the reactor

Compare the obtained results with those of an isothermal reactor working at 150°. Consider using a tubular reactor with a diameter of 2 m. The catalyst density in the reactor

$\delta_{cat} = 648$ kg/m^3. The average specific heat is $\overline{C_{sp}} = 380$ cal/kg °K.

Solution

To solve this problem requires solving the following system of differential equations, which is valid, stage-by-stage, for an adiabatic system:

$$F_{EtOH} d\lambda = r dW \quad \text{Material balance} \tag{5.238}$$

$$F_{TOT} \overline{C_{sp}} dT = (-\Delta H_R) r dW \quad \text{Thermal balance} \tag{5.239}$$

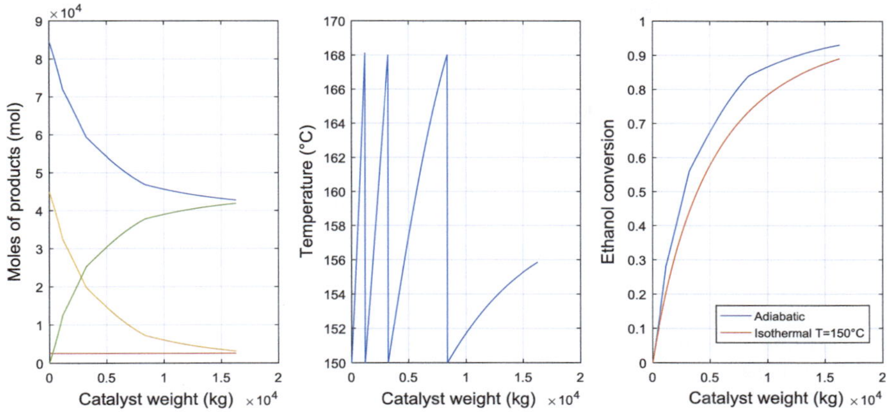

Fig. 5.41 Esterification of acetic acid with ethanol in vapor phase. Left, product distribution; middle, temperature profile along catalytic beds; right, ethanol conversion along the catalytic beds

By knowing the ethanol conversion, it is easy to evaluate the composition. Under isothermal conditions, the thermal balance can be neglected, and only the mass balance equation must be solved.

Results

The first three beds convert 28% of ethanol each (28 × 3 = 84%), while the fourth reaches a conversion of 93%. The catalyst weights in these beds are, respectively, as follows: 1180, 2035, 5155, and 7900 kg (total weight = 16,270 kg). Some of the obtained results are plotted in Fig. 5.41.

Exercise 5.13 Reactor for Ethanol Dehydrogenation to Ethyl Acetate in the Presence of Copper Chromite Catalyst

Carotenuto et al. (2013) studied the kinetics of ethanol dehydrogenation to ethyl acetate using copper chromite promoted by barium chromate and supported on alumina as a catalyst. The kinetic runs performed at 20–30 bars, on a tubular reactor packed-bed reactor containing, respectively, 2 and 50 g of catalyst, were successfully interpreted considering:

1. The following simplified reaction scheme:

$$CH_3CH_2OH \rightleftarrows CH_3CHO + H_2 \quad \Delta H_1 = 16.45\,\text{kcal/mol} \tag{5.240}$$

$$CH_3CHO + CH_3CH_2OH \rightleftarrows CH_3COOCH_2CH_3 + H_2 \quad \Delta H_2 = -10.37\,\text{kcal/mol} \tag{5.241}$$

$$2CH_3CHO \rightarrow \text{Other products} \tag{5.242}$$

2. The following main reaction mechanisms:

The acetaldehyde formation can be described through the following reaction mechanism:

$$C_2H_5OH + \sigma_0 \rightleftarrows \sigma_{EtOH} \tag{5.243}$$

$$\sigma_{EtOH} + \sigma_0 \rightarrow \sigma_{AcH} + \sigma_{H_2} \quad \text{rate} - \text{determining step} \tag{5.244}$$

$$\sigma_{AcH} \rightleftarrows \sigma_0 + CH_3CHO \tag{5.245}$$

$$\sigma_{H2} \rightarrow \sigma_0 + H_2 \tag{5.246}$$

The rate-determining step should be the reaction between the chemisorbed ethanol and a catalyst void site. The mechanism suggested by Carotenuto et al. (2013) for the formation of ethyl acetate is:

$$C_2H_5OH + \sigma_0 \rightleftarrows \sigma_{EtOH} \tag{5.247}$$

$$CH_3CHO + \sigma_0 \rightleftarrows \sigma_{AcH} \tag{5.248}$$

$$\sigma_{EtOH} + \sigma_{AcH} \rightleftarrows \sigma_{EA} + \sigma_{H2} \quad \text{rate} - \text{determining step} \tag{5.249}$$

$$\sigma_{EA} \rightleftarrows \sigma_0 + CH_3COOC_2H_5 \tag{5.250}$$

$$\sigma_{H2} \rightleftarrows \sigma_0 + H_2 \tag{5.251}$$

3. The following kinetic expressions:

$$r_1 = \frac{k_1 b_{EOH} P_{EtOH} \left(1 - \frac{1}{Ke_1} \frac{P_{AcH} P_{H_2}}{P_{EtOH}}\right)}{\left(1 + b_{EtOH} P_{EtOH} + b_{AcH} P_{AcH} + b_H P_H + b_{EA} P_{EA}\right)^2} \tag{5.252}$$

$$r_2 = \frac{k_2 b_{EOH} b_{AcH} P_{EtOH} P_{AcH} \left(1 - \frac{1}{Ke_2} \frac{P_{EA} P_{H2}}{P_{EtOH} P_{AcH}}\right)}{\left(1 + b_{EtOH} P_{EtOH} + b_{AcH} P_{AcH} + b_H P_H + b_{EA} P_{EA}\right)^2} \tag{5.253}$$

$$r_3 = k_3 P_{AcH}^2 \tag{5.254}$$

The reaction of acetaldehyde to other components (by-products) has been simplified to a pseudo second-order reaction because it is characterized by a very low conversion. The best-fitting parameters obtained by interpreting all the experimental runs performed are listed in Table 5.21.

1. Verify these parameters by constructing the plot of conversion versus space time, W/F (gh/mol), and comparing with the experimental points reported in Table 5.22.

2. This reaction system is singular because, looking at the reaction scheme, we can observe that ethanol transformation to ethyl acetate passes through the formation of acetaldehyde and that this is an endothermic reaction ($\Delta H \simeq 17$ kcal/mol),

Table 5.21 Kinetic parameters of the LHHW dual-site model as determined by regression analysis on the experimental runs performed with 2 g of catalyst[a]

Kinetic constants at 220 °C	Dimension	Activation energy (Kcal/mol)
$k_1 = 97.1 \pm 6.8$	mol/(g_{cat} h atm)	36.25 ± 4.35
$k_2 = 0.089 \pm 9.8 \times 10^{-3}$	mol/(g_{cat} h atm^2)	12.95 ± 0.65
$k_3 = 0.0011 \pm 7.8 \times 10^{-4}$	mol/(g_{cat} h atm^2)	$1.6 \times 10^{-4} \pm 1.8 \times 10^{-5}$
Adsorption parameters at 220 °C		Adsorption enthalpy (Kcal/mol)
$b_{EtOH} = 10.4 \pm 0.83$	(atm^{-1})	-25.53 ± 2.55
$b_{AcH} = 98.4 \pm 12.80$	(atm^{-1})	-7.02 ± 0.35
$b_{EA} = 41.2 \pm 4.94$	(atm^{-1})	-13.91 ± 0.14
$b_H = 2.5 \times 10^{-4} \pm 3.5 \times 10^{-5}$	(atm^{-1})	-13.34 ± 1.47

[a]See Carotenuto et al. (2013)

Table 5.22 Conversions and selectivities obtained for different space–times

Run no.	W/F (g h/mol)	Conversion to ethanol	Fraction of acetaldehyde	Fraction of ethyl acetate
1	0	0	0	0
2	1.15	0.14	0.06	0.54
3	4	0.25	0.04	0.80
4	20.85	0.46	0.05	0.91
5	32	0.47	0.015	0.96
6	97	0.54	0.01	0.98

while the successive reaction is moderately exothermic ($\Delta H \simeq -9.5$ kcal/mol). This means that if using a tubular reactor, it is opportune to separate it into different stages (at least two) that are differently heated to maintain the system approximately isotherm. In the first part, we must furnish heat, and in the second, we must cool the reactor. If we operate a unique adiabatic reactor feeding the reactant at 235 °C, we can observe initially a decrease in the temperature followed by a progressive moderate increase. Verify this aspect for a reactor 1 m of diameter programmed for producing 5 tons/h of ethyl acetate. What is the length of the reactor? Depict the temperature and concentration profiles along the reactor length.

Separate the adiabatic reactor into three successive reactors of equal length with heating or cooling between them in order to always start with the same initial temperature. How much heat must be exchanged for each stage? Depict the profiles for each step.

Solution

Part 1

The following system of differential equations must be solved:

$$\frac{dF_{\text{EtOH}}}{dZ} = -W(r_1 + r_2) \tag{5.255}$$

$$\frac{dF_{\text{AcH}}}{dZ} = W(r_1 - r_2 - 2r_3) \tag{5.256}$$

$$\frac{dF_{\text{AcOEt}}}{dZ} = Wr_2 \tag{5.257}$$

$$\frac{dF_{\text{H}_2}}{dZ} = W(r_1 + r_2) \tag{5.258}$$

$$\frac{dF_{\text{others}}}{dZ} = Wr_3 \tag{5.259}$$

The presence of nitrogen as an inert gas and a small amount of hydrogen must be considered.

Part 2

The mass balance equations related to reactions 1 and 2 must be coupled with the corresponding heat balance:

$$\frac{dT}{dZ} = \frac{W}{F}\left[\frac{(\Delta H_1)}{\overline{C_{p1}}}r_1 + \frac{(\Delta H_2)}{\overline{C_{p2}}}r_2\right] \tag{5.260}$$

Results

Case no. 1. Test for kinetics

In Fig. 5.42 (part no. 1), ethanol conversion and product selectivity are reported, and agreement with the experimental data is shown.

Case no. 2. Single adiabatic reactor

In this case, a production of 5000 kg/h of ethyl acetate can be achieved with a reactor packed with 35 tons of catalyst in a tube of 1-m diameter and approximately 40-m length (Fig. 5.42 [part nos. 2–4]).

Case 3. Three reactors in series with intermediate heating

By arranging three reactors in series, better performances can be achieved. The same production, as in Case no. 2, is obtained with three reactors having a total length of 9.3 m and with the amount of catalyst divided into three sections of, respectively, 3, 3, and 2 tons. Between the first and second beds and the second and third beds, the mixture was heated to the original temperature of 235 °C (see Fig. 5.42 [part no. 5]). In Fig. 5.42 (part nos. 6 and 7), the ethanol conversion and amount of ethyl acetate produced are reported.

Case 4. Single reactor with heating jacket

Using a single reactor equipped with a heating jacket, in which the temperature was set at 235 °C, the operation is also more convenient. A catalytic bed of 5 tons of catalyst with a length of approximately 5.8 m is sufficient to achieve the

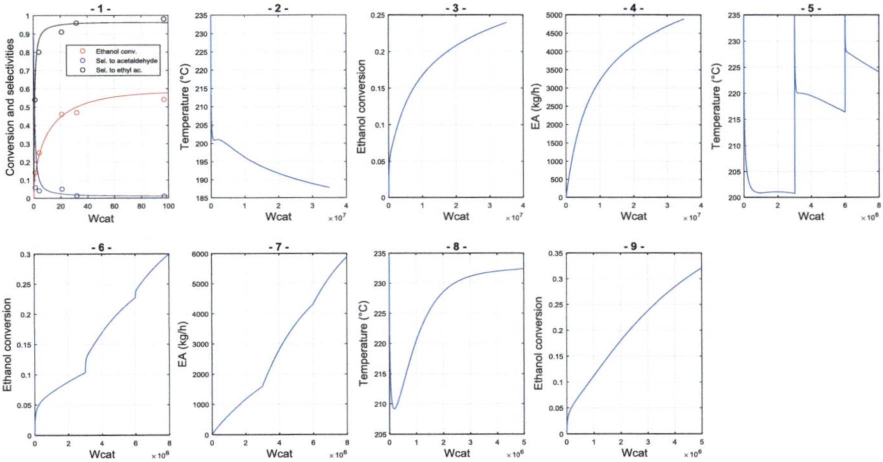

Fig. 5.42 Results of the calculations are explained below

Table 5.23 Experimental data of sulfur dioxide oxidation in fixed bed (Smith 1981)

Run no.	Rate (mol/(h gcat))	pSO_3 (atm)	pSO_2 (atm)	pO_2 (atm)
1	0.02	0.0428	0.0255	0.1860
2	0.04	0.0331	0.0352	0.1900
3	0.06	0.0272	0.0409	0.1930
4	0.08	0.0236	0.0443	0.1950
5	0.10	0.0214	0.0464	0.1960
6	0.12	0.0201	0.0476	0.1970

desired production of ethyl acetate. In Fig. 5.42 (part no. 8) and (part no. 9), temperature and conversion profiles are reported. In Fig. 5.42 (part no. 8), it is interesting to observe that with this reactor configuration (overall heat-transfer coefficient arbitrarily set at 20), an initial decrease in temperature is realized. As the gradient between the reactor and the jacket increases, more heat is furnished to the system, and the temperature rapidly is recovered up to the set value of 235 °C.

All the described results were obtained using a MATLAB program available as Electronic Supplementary Material, Data 5.1.

Exercise 5.14 Oxidation of Sulfur Dioxide to Sulfur Trioxide
Smith (1981) reported a set of experimental data regarding the oxidation reaction of SO_2 to SO_3 in a fixed bed reactor, in which a catalyst based on platinum supported on alumina was used. The experimental data are reproduced in Table 5.23.

A suitable expression for the description of the reaction rate, on a heterogeneous solid catalyst, was also reported by Smith as the following expression:

$$r = \frac{kK_o^{0.5}\left[p_{SO_2}p_{O_2}^{0.5} - \frac{1}{K_e}p_{SO_3}\right]}{A + Bp_{SO_3}} \tag{5.261}$$

The experiments were conducted at a fixed temperature of 480 °C, and the equilibrium constant has the following trend with temperature:

$$K_e = e^{\frac{22000}{RT} - 10.5} \tag{5.262}$$

This exercise aims to determine by fitting the four adjustable parameters k, K_o, A, and B. Produce a parity plot reporting experimental and calculated reaction rate in comparison.

Results
The fitting furnished the following results:

$k = 1.1096$	$K^\circ = 5.9033$	$A = 4.7374$	$B = -23.5893$

The parity plot is reported in the following Fig. 5.43. The agreement between the experimental and calculated values is very good.

All the described results were obtained using a MATLAB program available as Electronic Supplementary Material, Data 5.1.

Exercise 5.15 Pulse Test for the Evaluation of Peclet Number
As we have seen in the present chapter, packed bed reactors are extensively used in the laboratory practice of heterogeneous catalysis. For a reliable kinetic study, these

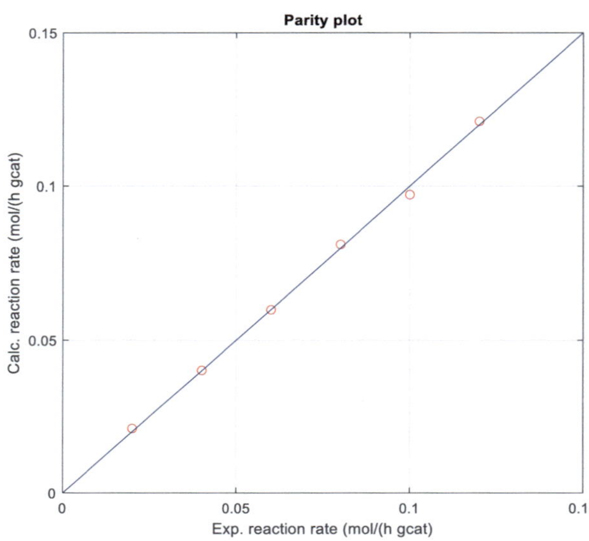

Fig. 5.43 Parity plot: calculated vs experimental reaction rate

reactors must be characterized from the point of view of the dispersion or, in other words, the deviation from plug-flow behavior. Leitao et al. (1994) have studied a packed bed reactor with the following characteristics:

Length	5.2 cm
Particle diameter	0.155 cm
Volumetric flow rate	11.1 mL/min
Void fraction	0.47 –
Tube diameter	1.6 cm

The experimental data (Leitao et al. 1994) of dimensionless outlet concentration are the following.

Dimensionless time θ:								
0.3786	0.5136	0.5745	0.6248	0.7757	0.7889	0.8180	0.8657	0.8974
0.9133	0.9371	1.1039	1.1383	1.2892	1.3369	1.4348	1.5169	1.6175
1.6466	1.6995	1.7525	1.7922	1.8954	1.9378	1.9590	2.0781	2.1231
2.2714	2.3428	2.4355	2.6473	2.8273	3.0205	3.2138	3.4017	
$E(\theta)$:								
0.0454	0.2808	0.4276	0.5557	0.9460	0.9740	1.0217	1.0544	1.0695
1.0765	1.0835	0.9577	0.9309	0.7328	0.6792	0.5313	0.4474	0.3250
0.3216	0.2575	0.1957	0.1946	0.1351	0.1118	0.1107	0.0676	0.0548
0.0245	0.0210	0.0175	0.0035	0.0070	0.0070	0.0117	0.0047	

The following function is used to represent the dispersion inside a packed bed reactor in terms of Peclet number:

$$E(\theta) = \frac{e^{-(1-\theta)^2/\left(\frac{4\theta}{Pe}\right)}}{2\left(\frac{\pi\theta}{P_e}\right)^{0.5}} \tag{5.263}$$

The scope of the exercise is the determination of the Peclet dimensionless number (Pe) that describes the behavior of the pulse experiment.

Results

A suitable MATLAB program code was developed to perform the fitting operation (see Electronic Supplementary Material, Data 5.1). A value of 14.59 for Pe number was calculated, and the results of the fitting are shown in Fig. 5.44. The relatively low value of Pe number (well below 100) is an index of the high deviation of the reactor from the plug-flow behavior.

All the described results were obtained using a MATLAB program available as electronic supplementary material, Data 5.1.

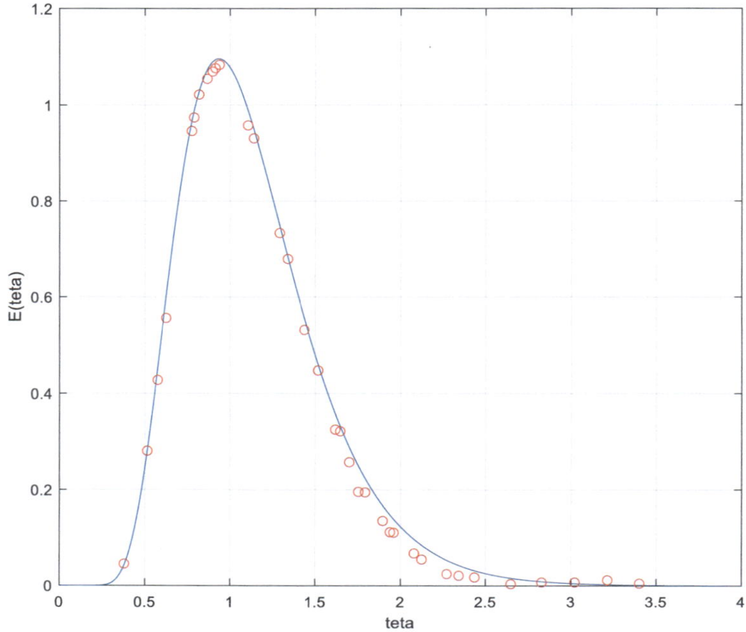

Fig. 5.44 Fitting of the pulse experiment data on tubular packed bed reactor data from (Leitao et al. 1994)

References

Anastasov, A.I.: Chem. Eng. Process. **42**, 151–165 (2003)

Berty, J.M.: Testing commercial catalysts in recycle reactors; catal. Rev. Sci. Eng. **20**, 75 (1979)

Bischoff, K.B., Froment, G.F.: Rate equations for consecutive heterogeneous processes. I&EC Fundamental. **1**(3), 195–200 (1962)

Boudart, M.: Kinetics of Chemical Processes. Prentice-Hall Inc (1968)

Boudart, M.: Heterogeneous catalysis by metals. J. Mol. Catal. **30**, 27–38 (1985)

Calderbank, P.H., Chandrasekharan, K., Fumagalli, C.: The prediction of the performance of packed-bed catalytic reactors in the air-oxidation of *o*-xylene. Chem. Eng. Sci. **32**, 1435–1443 (1977)

Carberry, J.J.: Designing laboratory catalytic reactors; Ind. Eng. Chem. **56**, 39 (1964)

Carotenuto, G., Tesser, R., Di Serio, M., Santacesaria, E.: Kinetic study of ethanol dehydrogenation to ethyl acetate promoted by a copper/copper-chromite based catalyst. Catal. Today, 202–210 (2013)

Chandrasekharan, K., Calderbank, P.H.: Kinetics of the catalytic air-oxidation of o-xylene measured in a tube-wall-catalytic reactor. Chem. Eng. Sci. **35**(1–2), 341–347 (1980)

Davidson, J.F., Harrison, D.: Fluidized Particles. Cambridge University Press, New York (1963)

Dias, C.R., Farinha, P.M., Bond, G.C.: Oxidation of *o*-Xylene to phthalic anhydride over V_2O_5/TiO_2 catalysts part 4; mathematical modelling study and analysis of the reaction network. J. Catal. **164**(2), 347–351 (1996)

Franckaerts, J., Froment, G.F.: Kinetic study of the dehydrogenation of ethanol. Chem. Eng. Sci. **19**, 807 (1964)

Froment, G.F., Bischoff, K.B.: Chemical reactor analysis and design. Wiley, New York (1969)

Gimeno, M.P., Gascon, J., Tellez, C., Herguido, J., Menedez, M.: Selective oxidation of *o*-xylene to phthalic anhydride over V_2O_5/TiO_2: kinetic study in a fluidized bed reactor. Chem. Eng. Process. **47**(9–10), 1844–1852 (2008)

Hawes, R.W., Kabel, R.L.: Thermodynamic equilibrium in the vapor phase esterification of Acetic acid with ethanol. AICHE J. **14**(4), 606–611 (1968)

Herten, J., Froment, G.F.: Kinetics and product distribution in oxidation of o-xylene on a vanadium pentoxide catalyst. Ind. Eng. Chem. Proc. Des. Dev. **7**(4), 516–526 (1968)

Hougen, O.A., Watson, K.M.: Chemical process principles, part two; thermodynamics. Wiley, New York (1947)

Kunii, D., Levenspiel, O.: Bubbling bed model for kinetic processes in fluidized beds. Gas-solid mass and heat transfer and catalytic reactions. Ind. Eng. Chem. Process. Des. Dev. **7**, 481–492 (1968)

Leitao, A., Carlos, P., Santos, S., Rodrigues, A.: Tracer experiments in fixed beds: effects of flow maldistribution on the estimation of transport kinetic parameters. Chem. Eng. J. **53**, 193–199 (1994)

Othmer, K.: Encyclopedia of Chemical Technology. Wiley-Interscience, New York (1958)

Papageorgius, J.N., Abello, M.C., Froment, G.F.: Kinetic modeling of the catalytic oxidation of *o*-xylene over an industrial V_2O_5-TiO_2 (anatase) catalyst. Appl. Catal. A Gen. **120**(1), 17–43 (1994)

Potter, A.E., Bender, P., Ritter, H.L.: The vapor phase association of acetic-d3 acid-d. J. Phys. Chem. **59**, 250–254 (1955)

Saleh, R.Y., Wachs, I.E.: Reaction network and kinetics of o-xylene oxidation to phthalic anhydride over V_2O_5/TiO_2(anatase) catalysts. Appl. Catal. **31**(1), 87–98 (1987)

Santacesaria, E., Carrà, S.: Cinetica dello steam reforming del metanolo. La Rivista dei Combustibili, vol XXXII. **7–8**, 227–232 (1978)

Santacesaria, E., Di Serio, M., Gelosa, G., Carrà, S.: Kinetics of methanol homologation: Part I. Behaviour of cobalt-phosphine-iodine catalysts. J. Molec. Catal. **58**(1), 27–42 (1990)

Santacesaria, E., Morbidelli, M., Carrà, S.: Kinetics of the catalytic oxidation of methanol to formaldehyde. Chem. Eng. Sci. **36**, 909–918 (1981)

Santacesaria, E., Gelosa, D., Danise, P., Carrà, S.: Vapor-phase esterification catalyzed by decationized zeolites. J. Catal. **80**, 427–436 (1983)

Sinfelt, J.H., Hurwitz, H., Shulman, R.A.: Kinetics of methylcyclohexane dehydrogenation over Pt—Al2O3. J. Phys. Chem. **64**(10), 1559–1562 (1960)

Skrzypek, J., Grzesik, M., Galantowicz, M., Solinski, J.: Kinetics of the catalytic air oxidation of o-xylene over a commercial V2O5-TiO2 catalyst. J. Chem. Eng. Sci. **40**(4), 611–620 (1985)

Smith, J.M.: Chemical engineering kinetics. Mc Graw-Hill Book Co, New York (1981)

Tesser, R., Maradei, V., Di Serio, M., Santacesaria, E.: Kinetics of the oxidative dehydrogenation of ethanol to acetaldehyde on V_2O_5/TiO_2-SiO_2 catalysts prepared by grafting. Ind. Eng. Chem. Res. **43**, 1623–1633 (2004)

Thaller, L.H., Thodos, G.: The dual nature of a catalytic reaction: The dehydrogenation of sec-butyl alcohol to methyl ethyl ketone at elevated pressures. A.I.Ch.E.J. **6**(3), 369–373 (1960)

Tschernitz, J., Bornstein, S., Beckmann, R.B., Hougen, O.A.: Trans. Am. Inst. Chem. Engrs. **42**, 883–903 (1946)

Vanhove, D., Blanchard, M.: Catalytic oxidation of o-xylene. J. Catal. **36**(1), 6–10 (1975)

Varma, R.L., Saraf, D.N.: Oxidation of butene to maleic anhydride: I. Kinetics and mechanism. J. Catal. **55**(3), 361–372 (1978)

Yabrov, A.A., Ivanov, A.: Response studies of the mechanism of o-xylene oxidation over a vanadium-titanium oxide catalyst. React. Kinet. Mech. Catal. **14**(3), 347–351 (1980)

Chapter 6
Kinetics of and Transport Phenomena in Gas–Solid Reactors

Abstract The fundamental laws of mass, heat, and momentum transport are briefly presented in this chapter and their relationships with chemical kinetics depicted. The approach presented starts with estimation of basic properties like viscosity, thermal conductivity, and mass diffusivity, all of which are of fundamental importance in describing transport phenomena. Transport phenomena originate gradients in temperature, pressure, and concentration that are the driving forces for transformations occurring in a system. Two scales of transformation can be considered, i.e., molecular and macroscopic. Molecular-transport phenomena are normally much slower than macroscopic ones, and this can result in a limitation on chemical reaction rates. In this chapter, detailed examples of basic property calculations are reported as are examples of chemical gas–solid reactions limited by diffusional resistance (like ammonia oxidation). Finally, a great part of the chapter is dedicated to the evaluation of catalyst effectiveness factor. MATLAB code associated with the examples in this chapter is available online.

Keywords Catalytic particles · Boundary layer · Thermal conductivity · Feed mass flow rate · Reynolds number

6.1 Fundamental Laws of Transport Phenomena

A system can be considered at equilibrium when the composition, pressure, and temperature are uniform at any point. In contrast, if differences exist, then transformations spontaneously occur with mass or energy transfer, giving place to a gradual evolution toward equilibrium conditions. Thermal, pressure, and concentration gradients are the driving forces for these transformations, which can be considered at two different levels: the molecular level and the macroscopic level.

Supplementary Information The online version contains supplementary material available at https://doi.org/10.1007/978-3-031-82502-6_6.

© The Author(s), under exclusive license to Springer Nature Switzerland AG 2025
R. Tesser, E. Santacesaria, *The Chemical Reactor from Laboratory to Industrial Plant*, https://doi.org/10.1007/978-3-031-82502-6_6

Molecular-transport phenomena are normally much slower than the macroscopic ones; therefore, those phenomena can limit chemical reaction rates. In this chapter, we will see how and when these limitations occur.

In a fluid, a pressure gradient originates a bulk motion of the fluid from the high-pressure to the low-pressure zone. The mass-transfer encounters the internal resistance given by the fluid viscosity, and the motion can be interpreted with the Newton law:

$$\tau = \pm\mu \frac{\partial u_x}{\partial z} = \text{Force of internal friction per unit surface area} \tag{6.1}$$

where τ is the force of internal friction per unit surface area, μ is the viscosity coefficient of the fluid, z is the coordinate normal to the direction of the motion x, and u_x is the speed component in the same direction.

A temperature gradient determines a heat flow from higher to the lower temperature, and the flux occurs according to the Fourier's law:

$$q = -k\frac{\partial T}{\partial z} = \text{Heat transmitted per unit of time per unit of surface area in the } z \text{ direction} \tag{6.2}$$

where k is the thermal conductivity of the fluid and T is the absolute temperature.

When a concentration gradient related to a component is operative, a mass flow of this component from higher to lower concentration occurs according to Fick's Law:

$$N_i = -D_i\frac{\partial C_i}{\partial z} = \text{Number of } i \text{ moles diffusing in unit of time per unit of surface area in } z \text{ direction.} \tag{6.3}$$

It is opportune to remark the similarity of the three fundamental laws of transport: This similarity is justified because these laws can be derived from a unique physical model based on the molecular properties. In fact, transport phenomena are influenced from both the motion of the molecules and their interactions. A unique general equation of the transport can be derived from the kinetic molecular theory of gases:

$$G(y) = n\bar{u}\lambda\frac{\partial y}{\partial z} \tag{6.4}$$

where λ is the free mean path of the molecules, $\bar{u}$ is the mean velocity of the molecules, n is the molecular concentration and y is the transported property.

On the basis of the kinetic molecular theory, Chapman (1916) and Enskog (1917), translation by Brush (1965) (see also Chapman and Cowling (1970)), independently proposed the following relations for evaluating μ, k, and D_i, respectively, as a function of molecular properties:

$$\mu = 2.6693 \times 10^{-5} \frac{\sqrt{MT}}{\sigma^2 \Omega_\mu} \left(\frac{\text{g}}{\text{cm s}}\right) \tag{6.5}$$

$$k = 1.989 \times 10^{-4} \frac{\sqrt{T/M}}{\sigma^2 \Omega_\mu} \left(\frac{\text{cal}}{\text{cm s K}}\right) \tag{6.6}$$

$$D_{12} = \frac{1.858 \times 10^{-3} \sqrt{\frac{T^3 (M_1 + M_2)}{M_1 M_2}}}{P \sigma_{12}^2 \Omega_D} \left(\frac{\text{cm}^2}{\text{s}}\right) \tag{6.7}$$

These relations are valid for very low pressure of gases, that is, approaching the properties of ideal gases. In these relations, M is the molecular weight, and σ is the kinetic diameter of the molecules in Angström. Ω_μ and Ω_D, named "collision integrals," are tabulated value functions of $k_B T/\varepsilon$ (adimensional temperature), where k_B is the Boltzmann constant $= 1.38 \times 10^{-16}$ erg/K and ε is a molecular interaction parameter. Both σ and ε can be evaluated from the intermolecular potential relation of Lennard–Jones:

$$\varphi(r) = 4\varepsilon_{ij} \left[\left(\frac{\sigma_{ij}}{r}\right)^{12} - \left(\frac{\sigma_{ij}}{r}\right)^6 \right] \tag{6.8}$$

$$\sigma_{ij} = \left(\sigma_i + \sigma_j\right)/2 \quad \text{(arithmetic mean)} \tag{6.9}$$

$$\varepsilon_{\text{ij}} = \sqrt{\varepsilon_i \varepsilon_j} \quad \text{(geometric mean)} \tag{6.10}$$

where r is the distance between two molecules.

The parameters ε and σ for a given molecule can be determined from the critical properties. For example, $\varepsilon/k_B = 0.77\ T_C$ and $\sigma = 0.841\ {V_C}^{1/3}$, where k_B is the Boltzmann constant; T_C is the critical temperature; and V_C is the critical volume. However, in Appendix 1, a table is reported with the values of ε/k_B and σ for some representative compounds [taken from Satterfield and Sherwood (1963)]. The values of the collision integrals, valid for apolar molecules, are reported in many textbooks, tabulated as Ω_μ and Ω_D as a function of the adimensional temperature $T^* = k_B T/\varepsilon$. All data reported in the literature were interpolated with the following equation:

$$\Omega_i = 10^{\left(ax^6 + bx^5 + cx^4 + dx^3 + ex^2 + fx + g\right)} \quad \text{where} \quad x = \log_{10}(T^*) \tag{6.11}$$

In Table 6.1, the parameters necessary for solving Eq. (6.11) are reported.

By applying this correlation to the determination of the collision integrals, average absolute percent errors of 0.1396 and 0.2343% are obtained for, respectively, Ω_D and Ω_μ. In Appendix 2, a MATLAB procedure for determining the

Table 6.1 Parameters useful for the determination of the collision integrals Ω_μ and Ω_D

Collision integrals	*a*	*b*	*c*	*d*	*e*	*f*	*g*
Ω_D	−0.0120	0.0877	−0.2146	0.1426	0.1948	−0.4848	0.1578
Ω_μ	−0.0165	0.1204	−0.3011	0.2360	0.1708	−0.4922	0.1997

parameters a through *g* by mathematical regression analysis is reported together with the numerical values of Ω_D and Ω_μ taken from the literature.

In the case of polar molecules, the molecular interactions are better described by the Stockmayer relation instead of the Lennard–Jones equation:

$$\varphi(r) = 4\varepsilon_o\left[\left(\frac{\sigma}{r}\right)^{12} - \left(\frac{\sigma}{r}\right)^{6} - \frac{m_d^2}{r^3}\phi(\vartheta)\right] \tag{6.12}$$

where ε_o and σ have the same mean of the corresponding parameters of Lennard–Jones but different numerical values, m_d is the dipole moment of the molecule, and $\phi(r)$ is a parameter describing the geometrical disposal of the interacting molecules. In Appendix 3, the parameters of the Stockmayer relation for some different substances are reported (taken from Forni 1979). For polar molecules, the collision integrals Ω_μ and Ω_D depend not only on $T^* = k_B T/\varepsilon$ but also on another parameter, δ:

$$\delta = \frac{\mu_d^2}{2\varepsilon_o\sigma^3} \tag{6.13}$$

Again, the values of Ω_μ and Ω_D are tabulated in different textbooks as a function of both T^* and δ. All of the data reported in the literature were interpolated with the following equations:

$$f_1 = d_1 + d_2\frac{\delta}{T^*} + d_3\frac{\delta^2}{T^{*2}} + d_4\frac{\delta^3}{T^*} + d_5\delta^4 \tag{6.14}$$

$$f_2 = \frac{f_1\delta^{1.5}}{K^{\log_{10}(T^*)}} \tag{6.15}$$

$$f_3 = a_1x^6 + a_2x^5 + a_3x^4 + a_4x^3 + a_5x^2 + a_6x + a_7 \tag{6.16}$$

$$\Omega_i = 10^{f_2+f_3} \quad \text{where} \quad x = \log_{10}(T^*) \quad i = D \text{ or } \mu \tag{6.17}$$

The coefficients for the functions related to Ω_D and Ω_μ are summarized in Table 6.2.

By applying these correlations to the collision integrals, average absolute percent errors of 1.62 and 1.85% are obtained for, respectively, Ω_D and Ω_μ. In Appendix 4, a MATLAB procedure for determining the parameters of the model by mathematical regression analysis is described and the numerical values of Ω_D and Ω_μ, taken from the literature (see Forni 1979), are reported.

Table 6.2 Parameters for determining the collision integrals Ω_D and Ω_μ by solving the equation system (Eqs. 6.14–6.17)

Parameters	Collision integrals	
	Ω_D	Ω_μ
d_1	0.066225	0.067498
d_2	−0.002888	−0.002375
d_3	0.0000707	0.0000618
d_4	−0.0000626	−0.0000865
d_5	−0.0000785	−0.0001022
K	4.507	3.934
a_1	0.010254	0.010948
a_2	−0.033249	−0.039147
a_3	−0.014026	−0.005066
a_4	0.096320	0.105559
a_5	0.068759	0.049660
a_6	−0.434055	−0.425989
a_7	0.163439	0.203885

In the presence of more than two components, the properties of the mixture must be defined through an opportune averaging procedure. In the case of the diffusion coefficient of an i component in the mixture of N components, for example, we must first evaluate all of the binary diffusion coefficients $D_{i,j}$ and then calculate $D_{i,m}$ with the following relation proposed by Fairbanks and Wilke (1950):

$$D_{i,m} = \frac{1 - y_i}{\sum_{j \neq i}^{N} \frac{y_j}{D_{i,j}}} \tag{6.18}$$

where y_i is the molar fraction. For a mixture, the other properties—such as μ_{mix} and k_{mix}—also must be opportunely averaged following, for example, the semiempirical approach suggested by Fairbanks and Wilke (1950):

$$\mu_{\text{mix}} = \sum_{i=1}^{N} \frac{y_i \mu_i}{\sum_{j=1}^{N} y_j \phi_{ij}} \qquad k_{\text{mix}} = \sum_{i=1}^{N} \frac{y_i k_i}{\sum_{j=1}^{N} y_j \phi_{ij}} \tag{6.19}$$

$$\phi_{ij} = \frac{1}{\sqrt{8}} \left(1 + \frac{M_i}{M_j}\right)^{-1/2} \left[1 + \left(\frac{\mu_i}{\mu_j}\right)^{1/2} \left(\frac{M_j}{M_i}\right)^{1/4}\right]^2 \tag{6.20}$$

where M_j and M_i are molecular weight.

Binary diffusion coefficients in liquid phase are difficult to predict because diffusivities depend on both concentration and non-ideality of the solution. However, it must be pointed out that liquid-phase diffusivity coefficients are three or four orders less than the corresponding values in gas phase (normally approximately 1×10^{-5} cm^2/s). Several correlations have been proposed in the literature for estimating liquid diffusivities, but all of them are valid only at very low solute concentration. The most reliable theoretical approach is the one of Stokes–Einstein (1905). Einstein proposed the following correlation:

$$D_{12} = \frac{k_B T}{6\pi r_1 \mu_2} \quad \text{or} \quad \frac{D_{12}}{k_B T} = \text{constant} \tag{6.21}$$

where k_B = Boltzmann's constant = 1.38×10^{-16}, T = temperature, (K) r_1 = solute molecule radius, and μ_2 = solvent viscosity. This relation is valid for dilute solution of large spherical molecules. More popular and employed is the empirical correlation of Wilke and Chang (1955) derived from the previous one:

$$D_{12} = 7.4 \times 10^{-10} \frac{T(X_2 M_2)^{1/2}}{\mu V_1^{0.6}} \tag{6.22}$$

where D_{12} is obtained in cm^2/s, V_1 is the molar volume of the diffusing solute at its normal boiling point in cm^3/g mol (see Appendix 5), M_2 is the molecular weight of the solvent, μ is the viscosity of the solution in poises, T is the temperature in K, and X_2 is an empirical association parameter of the solvent. X_2 corresponds to a value of 2.6 for water, 1.9 for methanol, 1.5 for ethanol, and 1 for nonassociated solvents, such as benzene, ether, heptane, etc. However, this correlation also is valid for dilute solutions, and the employment in other conditions can give place to errors. Moreover, it is important to point out that D_{12} and D_{21}, in contrast to the gas-phase binaries, normally have a different value. In conclusion, for a correct determination of the diffusion coefficient in liquid phase, the correct approach is experimental evaluation. Experimental values have been reported by different authors, in particular, by Johnson and Babb (1956). Extrapolation of the experimental data can be eventually made by using, with some caution, the Wilke–Chang relation.

Other semiempirical correlations have been employed for the estimation of viscosities and thermal conductivities. Viscosity can be evaluated, for example, with the relation:

$$\mu = \frac{h_P N_{Av}}{V_b} \exp\left(\frac{0.408 \ \lambda_{ev}}{RT}\right) \tag{6.23}$$

where N_{Av} is the Avogadro number, h_P is Plank's constant, V_b is the molar volume at the normal boiling point (see Appendix 5 and Exercise 6.1), and λ_{ev} is the vaporization heat. The viscosity of a mixture can be evaluated by making a logarithmic average of the type:

$$\log\left(\frac{1}{\mu_{mix}}\right) = \sum_{i=1}^{N} y_i \log\left(\frac{1}{\mu_i}\right) \tag{6.24}$$

Table 6.3 Some properties of the reaction components

Substance	M_W (g/mol)	T_C (K)	σ (Å)	ε/k_B (K)
1-Benzene	78.11	562.1	5.270	440
2-Hydrogen	2.016	33.3	2.915	38
3-Cyclohexane	84.16	553.2	6.093	324

In contrast, for calculating the thermal conductivity of a pure liquid, the Bridgman (1923) equation can be employed:

$$k = 3k_B a \left(\frac{N_{Av}}{V_b}\right) \tag{6.25}$$

where a is the speed of sound into the liquid (m/s) and the other symbols have already been described. Data of sound speed can be found at: http://webbook.inst.gov/. The thermal conductivity of nonassociated binary mixture can be calculated as:

$$k_{mix} = k_1 w_1 + k_2 w_2 - 0.72(k_1 - k_2) w_1 w_2 \tag{6.26}$$

Exercise 6.1
Estimation of the Diffusion Coefficients in a Reacting Mixture of Apolar Molecules

Consider the hydrogenation reaction of benzene to cyclohexane. The reactants are fed into the reactor with a molar ratio hydrogen/benzene = 4. The reaction occurs at 250 °C and 5 atm.

1. Evaluate the diffusion coefficient of benzene in hydrogen at the reactor inlet.
2. Evaluate the diffusion coefficient $D_{i,m}$ of benzene in the reacting mixture after a 50% benzene conversion.
3. Show in a plot how $D_{1,m}$ changes with the conversion.

Some properties of the reaction components are reported in Table 6.3; the other parameters necessary for the calculations are reported in the Appendixes reported at the end of this chapter.

Solution
Part 1

At the inlet of the reactor, we have only two components—benzene and hydrogen—and we must evaluate, therefore, only the D_{12} coefficient with the Chapman–Enskog formula (6.7) assuming $D_{12} = D_{21}$. From data reported in Appendix 5, calculate the molar volume V_b for, respectively, benzene (1) and hydrogen (2):

$$V_{b1} = 14.8 \times 6 + 3.7 \times 6 - 15 = 96$$
$$V_{b2} = 14.3$$

Then, with the relation $\sigma = 1.18\ V_{\mathrm{b}}^{1/3}$, evaluate σ_1 and σ_2:

$$\sigma_1 = 1.18 \times 96^{1/3} = 5.40 \quad \sigma_2 = 1.18 \times 14.3^{1/3} = 2.86$$

As we can see, the calculated values are in a satisfactory agreement with the values reported in Table 6.3 albeit determined in a different way:

$$\sigma_{12} = \frac{1}{2}(\sigma_1 + \sigma_2) = 4.125$$

Evaluate now $\frac{k_{\mathrm{B}}T}{\varepsilon_{12}}$

$$\frac{k_{\mathrm{B}}T}{\varepsilon_1} = 1.3\frac{T}{T_{\mathrm{c}}} = 1.3\frac{523}{562.1} = 1.209 \qquad \frac{k_{\mathrm{B}}T}{\varepsilon_2} = 1.3\frac{T}{T_{\mathrm{c}}} = 1.3\frac{523}{33.3} = 20.42$$

$$\varepsilon_{12} = \sqrt{\varepsilon_1\varepsilon_2} = \frac{k_{\mathrm{B}}T}{\sqrt{1.209 \times 20.42}} = \frac{k_{B}T}{4.97} \quad \frac{k_{\mathrm{B}}T}{\varepsilon_{12}} = 4.97$$

From Eq. (6.11) or Appendix 2, we can evaluate the value of the collision integral $\Omega_{D12} = 0.8422$.

$$D_{12} = \frac{1.858 \times 10^{-3}\sqrt{\frac{T^3(M_1 + M_2)}{M_1 M_2}}}{P\sigma_{12}^2\Omega_{D12}} = \frac{1.858 \times 11.96\sqrt{\frac{78.11 + 2.016}{78.11 \times 2.016}}}{5 \times 4.125^2 \times 0.8422}$$

$$= \frac{15.85}{71.65} = 0.22\left(\frac{\mathrm{cm}^2}{\mathrm{s}}\right)$$

Part 2

For evaluating the diffusion coefficient of benzene into the reacting mixture, we must evaluate all of the binary coefficients, that is, also D_{13} and D_{32} according to the same procedure adopted for calculating D_{12}. We must first evaluate all data related to cyclohexane:

$$V_{\mathrm{b}3} = 14.8 \times 6 + 3.7 \times 12 - 15 = 118.2$$

$$\sigma_3 = 1.18 \times 118.2^{1/3} = 5.79$$

$$\sigma_{13} = \frac{1}{2}(\sigma_1 + \sigma_3) = 5.59$$

$$\sigma_{32} = \frac{1}{2}(\sigma_3 + \sigma_2) = 4.32$$

$$\frac{k_{\mathrm{B}}T}{\varepsilon_3} = 1.3\frac{T}{T_{\mathrm{c}}} = 1.3\frac{523}{553.2} = 1.229$$

$$\varepsilon_{13} = \sqrt{\varepsilon_1 \varepsilon_3} = \frac{k_B T}{\sqrt{1.209 \times 1.229}} = \frac{k_B T}{1.219} \quad \frac{k_B T}{\varepsilon_{13}} = 1.219 \quad \Omega_{D13} = 1.32$$

$$\varepsilon_{32} = \sqrt{\varepsilon_2 \varepsilon_3} = \frac{k_B T}{\sqrt{20.42 \times 1.229}} = \frac{k_B T}{5} \quad \frac{k_B T}{\varepsilon_{32}} = 5 \quad \Omega_{D32} = 0.8422$$

$$D_{13} = \frac{1.858 \times 10^{-3} \sqrt{\frac{T^3 (M_1 + M_3)}{M_1 M_3}}}{P \sigma_{13}^2 \Omega_{D13}} = \frac{1.858 \times 11.96 \sqrt{\frac{78.11 + 84.16}{78.11 \times 84.16}}}{5 \times 5.59^2 \times 1.32}$$

$$= \frac{3.49}{206.23} = 0.017 \left(\frac{\text{cm}^2}{\text{s}} \right)$$

$$D_{32} = \frac{1.858 \times 10^{-3} \sqrt{\frac{T^3 (M_3 + M_2)}{M_3 M_2}}}{P \sigma_{32}^2 \Omega_{D12}} = \frac{1.858 \times 11.96 \sqrt{\frac{84.16 + 2.016}{84.16 \times 2.016}}}{5 \times 4.32^2 \times 0.8422}$$

$$= \frac{15.83}{78.58} = 0.20 \left(\frac{\text{cm}^2}{\text{s}} \right)$$

The reaction started with a molar ratio hydrogen/benzene = 4. When the conversion of benzene is 50%, we have a composition in molar fraction: $y_1 = 0.142$, $y_2 = 0.714$, and $y_3 = 0.142$. Therefore,

$$D_{1,m} = \frac{1 - y_1}{\frac{y_2}{D_{12}} + \frac{y_3}{D_{13}}} = \frac{1 - 0.142}{\frac{0.714}{0.22} + \frac{0.142}{0.017}} = \frac{0.858}{3.245 + 8.352} = 0.074 \ \left(\text{cm}^2/\text{s}\right)$$

$$D_{2,m} = \frac{1 - y_2}{\frac{y_1}{D_{21}} + \frac{y_3}{D_{23}}} = \frac{1 - 0.714}{\frac{0.142}{0.22} + \frac{0.142}{0.20}} = \frac{0.286}{0.645 + 0.71} = 0.210 \ \left(\text{cm}^2/\text{s}\right)$$

$$D_{3,m} = \frac{1 - y_3}{\frac{y_1}{D_{31}} + \frac{y_2}{D_{32}}} = \frac{1 - 0.142}{\frac{0.142}{0.017} + \frac{0.714}{0.20}} = \frac{0.858}{3.57 + 8.352} = 0.072 \ \left(\text{cm}^2/\text{s}\right)$$

Part 3

The initial number of moles of the reactants are n_B^o for benzene and $n_{H_2}^o$. Considering then

$$n_T^o = n_B^o + n_{H_2}^o$$

After the reaction, we will have:

$$n_T = n_B + n_{H_2} + n_{CH} = n_B^o(1 - \lambda) + \left(n_{H_2}^o - 3 n_B^o \lambda \right) + n_B^o \lambda = n_B^o (1 - \lambda) + \left(4 n_B^o - 3 n_B^o \lambda \right) + n_B^o \lambda$$

$$n_T = n_B^o (5 - 3\lambda)$$

Therefore, the molar fractions will be:

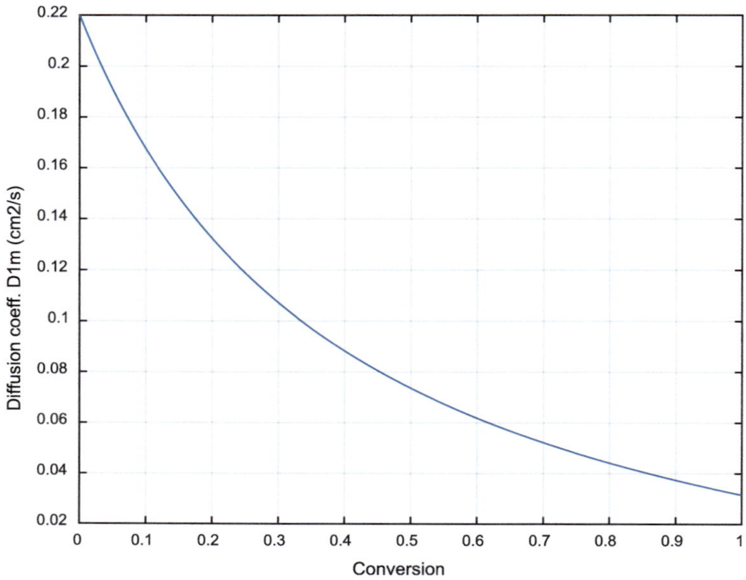

Fig. 6.1 Evolution of $D_{1,m}$, the molecular diffusion coefficient of benzene, in the hydrogenation reaction mixture as a function of the conversion

$$y_1 = \frac{n_B}{n_T} = \frac{(1-\lambda)}{(5-3\lambda)} \quad y_2 = \frac{n_{H_2}}{n_T} = \frac{(4-3\lambda)}{(5-3\lambda)} \quad y_3 = \frac{n_{CH}}{n_T} = \frac{\lambda}{(5-3\lambda)}$$

Hence,

$$D_{1,m} = \frac{1-\frac{(1-\lambda)}{(5-3\lambda)}}{\frac{\frac{(4-3\lambda)}{(5-3\lambda)}}{D_{12}} + \frac{\frac{\lambda}{(5-3\lambda)}}{D_{13}}} = \frac{2(2-\lambda)}{\frac{(4-3\lambda)}{D_{12}} + \frac{\lambda}{D_{13}}}$$

The calculation of $D_{1,m}$ at a different conversion was performed using a MATLAB program. The obtained results are reported in Fig. 6.1.

The described results were obtained using a MATLAB program available as Electronic Supplementary Material, Data 6.1.

Exercise 6.2
Estimation of the Viscosity and Thermal Diffusivity Coefficient for Gaseous Mixtures of Nonpolar Molecules

1. Evaluate the viscosity of the gaseous mixture of the previous exercise (6.1), that is, a mixture of benzene, hydrogen, and cyclohexane. Calculate the specific composition corresponding to benzene reacted at 50% starting from an initial mixture of hydrogen and benzene with a molar ratio of 4. The temperature is 250 °C.

Table 6.4 Some properties of the reaction components

Substance	M_W (g/mol)	T_C (K)	V_C (cm^3/mol)
1-Benzene	78.11	562.1	260
2-Hydrogen	2.016	33.3	65
3-Cyclohexane	84.16	553.2	308

2. Also evaluate, for the same mixture, the thermal conductivity coefficient.

Solution

Part 1

Some properties of the reaction components are reported in Table 6.4; some other parameters necessary for the calculations are reported in the Appendixes reported at the end of this chapter.

The viscosity can be estimated with the Chapman–Enskog formula (6.5). Because all of the components of the mixture are apolar, the necessary parameters can be derived from the Lennard–Jones approach. First, we must calculate for each component a value of σ_i and of the integral collision Ω_i. Index 1 is referred to benzene, index 2 to hydrogen, and index 3 to cyclohexane; therefore,

$$\sigma_1 = 0.841\, V_{C1}^{1/3} = 0.841 \times 260^{1/3} = 5.368 \quad \mathring{A}$$
$$\sigma_2 = 0.841\, V_{C2}^{1/3} = 0.841 \times 65^{1/3} = 3.381 \quad \mathring{A}$$
$$\sigma_3 = 0.841\, V_{C1}^{1/3} = 0.841 \times 308^{1/3} = 5.68 \quad \mathring{A}$$

Evaluate now the ε_i/k_B values using the relations $\frac{\varepsilon_i}{k_B} = 0.77 T_{Ci}$ and then $\frac{k_B T}{\varepsilon_i}$ and $\Omega_{\mu i}$ (from Appendix 2)

$$\frac{\varepsilon_1}{k_B} = 0.77 T_{C1} = 0.77 \times 562.1 = 432.82 \text{ K} \quad \frac{k_B T}{\varepsilon_1} = \frac{523}{432.82} = 1.208 \quad \Omega_{\mu 1} = 1.4470$$
$$\frac{\varepsilon_2}{k_B} = 0.77 T_{C2} = 0.77 \times 33.1 = 25.64 \text{ K} \quad \frac{k_B T}{\varepsilon_2} = \frac{523}{25.64} = 20.40 \quad \Omega_{\mu 2} = 0.7464$$
$$\frac{\varepsilon_3}{k_B} = 0.77 T_{C3} = 0.77 \times 553.2 = 425.96 \text{ K} \quad \frac{k_B T}{\varepsilon_3} = \frac{523}{425.96} = 1.228 \quad \Omega_{\mu 3} = 1.4364$$

Calculate now the viscosity of the pure compounds using the Chapman–Enskog formula:

$$\mu_1 = 2.6693 \times 10^{-5} \frac{\sqrt{\text{TM}_1}}{\sigma_1^2 \Omega_{\mu 1}} = 2.6693 \times 10^{-5} \frac{\sqrt{523 \times 78.11}}{5.368^2 0.9464} = 1.294 \times 10^{-4} \text{ (Poise)}$$
$$\mu_2 = 2.6693 \times 10^{-5} \frac{\sqrt{\text{TM}_2}}{\sigma_2^2 \Omega_{\mu 2}} = 2.6693 \times 10^{-5} \frac{\sqrt{523 \times 2.016}}{3.381^2 0.7432} = 1.016 \times 10^{-4} \text{ (Poise)}$$
$$\mu_3 = 2.6693 \times 10^{-5} \frac{\sqrt{\text{TM}_3}}{\sigma_3^2 \Omega_{\mu 3}} = 2.6693 \times 10^{-5} \frac{\sqrt{523 \times 84.16}}{5.68^2 0.9464} = 1.209 \times 10^{-4} \text{ (Poise)}$$

To evaluate the viscosity of the reaction mixture at 50% benzene conversion, it is necessary to know the composition in molar fraction y_1, y_2, and y_3. Considering that we know the total number of moles as

$$
\begin{aligned}
&n_{\text{tot}} = n_1 + n_2 + n_3, \qquad \text{with} \qquad n_1 = n_1^o(1-\lambda),\\
&n_2 = 4\ n_1^o - 3n_1^o\lambda \qquad \text{and} \qquad n_3 = n_1^o\lambda.\\
&n_{\text{T}} = n_1^o(1-\lambda) + 4n_1^o - 3n_1^o\lambda + n_1^o\lambda\\
&= n_1^o(1-\lambda+4-3\lambda+\lambda) = n_1^o(5-3\lambda)
\end{aligned}
$$

Therefore:

$$
\begin{aligned}
y_1 &= (1-\lambda)/(5-3\lambda) = 0.5/(5-1.5) = 0.1428\\
y_2 &= (4-3\lambda)/(5-3\lambda) = 2.5/3.5 = 0.7143\\
y_3 &= \lambda/(5-3\lambda) = 0.1428
\end{aligned}
$$

The viscosity of the mixture related to the calculated composition can be determined with Eqs. (6.19) and (6.20). Equation (6.19) also can be written as:

$$
\mu_{\text{mix}} = \frac{y_1\mu_1}{y_1\phi_{11} + y_2\phi_{12} + y_3\phi_{13}} + \frac{y_2\mu_2}{y_1\phi_{21} + y_2\phi_{22} + y_3\phi_{23}} + \frac{y_3\mu_3}{y_1\phi_{31} + y_2\phi_{32} + y_3\phi_{33}}
$$

Equation (6.20) must be used for calculating the ϕ_{ij} values. ϕ_{11}, ϕ_{22}, and ϕ_{33} are equal to 1, so we must calculate the other six values:

$$
\varphi_{12} = \frac{1}{\sqrt{8}}\left(1+\frac{78.11}{2.016}\right)^{-1/2}\left[1+\left(\frac{1.294\times10^{-4}}{1.016\times10^{-4}}\right)^{1/2}\left(\frac{2.016}{78.11}\right)^{1/4}\right]^2 = 0.12
$$

$$
\varphi_{13} = \frac{1}{\sqrt{8}}\left(1+\frac{78.11}{84.16}\right)^{-1/2}\left[1+\left(\frac{1294\times10^{-4}}{1.209\times10^{-4}}\right)^{1/2}\left(\frac{84.16}{78.11}\right)^{1/4}\right]^2 = 1.07
$$

$$
\varphi_{21} = \frac{1}{\sqrt{8}}\left(1+\frac{2.016}{78.11}\right)^{-1/2}\left[1+\left(\frac{1.016\times10^{-4}}{1.294\times10^{-4}}\right)^{1/2}\left(\frac{78.11}{2.016}\right)^{1/4}\right]^2 = 3.59
$$

$$
\varphi_{23} = \frac{1}{\sqrt{8}}\left(1+\frac{2.016}{84.16}\right)^{-1/2}\left[1+\left(\frac{1.016\times10^{-4}}{1.209\times10^{-4}}\right)^{1/2}\left(\frac{84.16}{2.016}\right)^{1/4}\right]^2 = 3.87
$$

$$
\varphi_{31} = \frac{1}{\sqrt{8}}\left(1+\frac{84.16}{78.11}\right)^{-1/2}\left[1+\left(\frac{1.209\times10^{-4}}{1.294\times10^{-4}}\right)^{1/2}\left(\frac{78.11}{84.16}\right)^{1/4}\right]^2 = 0.93
$$

$$\varphi_{32} = \frac{1}{\sqrt{8}}\left(1 + \frac{84.16}{2.016}\right)^{-1/2}\left[1 + \left(\frac{1.209 \times 10^{-4}}{1.016 \times 10^{-4}}\right)^{1/2}\left(\frac{2.016}{84.16}\right)^{1/4}\right]^2 = 0.11$$

In conclusion, we have:

	I_j	1	2	3
$\Phi_{ij}=$	1	1	0.12	1.07
	2	3.59	1	3.87
	3	0.93	0.11	1

By substituting the ϕ_{ij} in the previous relation, we obtain:

$$\mu_{\text{mix}} = \frac{0.1428 \times 1.294 \times 10^{-4}}{0.1428 + 0.7143 \times 0.12 + 0.1428 \times 1.11} + \frac{0.7143 \times 1.016 \times 10^{-4}}{0.1428 \times 3.59 + 0.7143 + 0.1428 \times 3.87} + \frac{0.1428 \times 1.209 \times 10^{-4}}{0.1428 \times 0.93 + 0.7143 \times 0.11 + 0.1428}$$

$$\mu_{\text{mix}} = 1.37 \times 10^{-4} \text{ Poise}$$

Clearly, this result largely depends on the values of σ_i and ε_i/k_B, which are often tabulated, and the result is more correct if these values are more precise.

Part 2

It is easy now to evaluate the thermal conductivity of the mixture by first determining k_1, k_2, and k_3 using Eq. (6.6). This results in:

$$k_1 = \frac{1.989 \times 10^{-4}\sqrt{\frac{523}{78.11}}}{1.4470 \times 5.368^2} = 1.235 \times 10^{-5}\left(\frac{\text{cal}}{\text{cm s K}}\right)$$

$$k_2 = \frac{1.989 \times 10^{-4}\sqrt{\frac{523}{2.016}}}{0.7464 \times 3.381^2} = 3.75 \times 10^{-4}\left(\frac{\text{cal}}{\text{cm s K}}\right)$$

$$k_3 = \frac{1.989 \times 10^{-4}\sqrt{\frac{523}{84.16}}}{1.4354 \times 5.68^2} = 1.07 \times 10^{-5}\left(\frac{\text{cal}}{\text{cm s K}}\right)$$

Last, the value of k_{mix} can be determined with relations similar to the ones employed for the evaluation of μ_{mix}, that is,

$$k_{\mathrm{mix}} = \frac{y_1 k_1}{y_1\phi_{11} + y_2\phi_{12} + y_3\phi_{13}} + \frac{y_2 k_2}{y_1\phi_{21} + y_2\phi_{22} + y_3\phi_{23}} + \frac{y_3 k_3}{y_1\phi_{31} + y_2\phi_{32} + y_3\phi_{33}}$$

where ϕ_{ij} are the same previously calculated. Hence, we have:

$$k_{\mathrm{mix}} = \frac{0.1428 \times 1.235 \times 10^{-5}}{0.1428 + 0.7143 \times 0.117 + 0.1428 \times 1.07}$$

$$+ \frac{0.7143 \times 3.75 \times 10^{-4}}{0.1428 \times 3.59 + 0.7143 + 0.1428 \times 3.86}$$

$$+ \frac{0.1428 \times 1.07 \times 10^{-5}}{0.1428 \times 0.93 + 0.7143 \times 0.11 + 0.1428}$$

$$k_{\mathrm{mix}} = 1.595 \times 10^{-4}\ (\mathrm{cal/cm\ s\ K})$$

Exercise 6.3
Estimation of the Viscosity, Conductivity, and Molecular Diffusion Coefficients for Gaseous Mixtures of Polar Compounds

Consider the reaction of ethanol dehydrogenation:

$$CH_3CH_2OH \rightarrow CH_3CHO + H_2$$

The reaction occurs on a catalyst at 300 °C and atmospheric pressure starting from pure ethanol.

Evaluate the molecular diffusion coefficients, viscosity, and thermal conductivity of ethanol in the reaction mixture for an ethanol conversion of 50%.

Solution

Part 1 Determining the Molecular Diffusion Coefficient

The approach is similar to the one adopted in the Exercise 6.1. The only difference is that the Stockmayer potential must be considered, instead of the Lennard–Jones potential, for the estimation of the parameters related to ethanol and acetaldehyde, while the Lennard–Jones potential will be employed for hydrogen. Some properties of the molecules are listed in Table 6.5.

Solution

Part 1 Determination of the Molecular Diffusion Coefficient of Ethanol

At the inlet of the reactor we have only one component, ethanol. After the reaction, we will have three components, and the related diffusion coefficients can be evaluated with the Chapman–Enskog formula (Eq. 6.7).

From the data reported in Appendix 3, calculate the molar volume V_b for, respectively, ethanol (1), hydrogen (2), and acetaldehyde (3) (see Treybal 1955).

Table 6.5 Some properties of the reaction components

Substance	M_W (g/mol)	T_C (K)	V_C (cm^3/mol)	σ (Å)	ε/k_B (K)
1-Ethanol	46.07	516.3	168.0	4.310	415
2-Hydrogen	2.016	33.3	64.9	2.915	38
3-Acetaldehyde	44.05	461.0	154.0	–	–

$$V_{b1} = 14.8 \times 2 + 3.7 \times 6 + 7.4 = 59.2$$
$$V_{b2} = 14.3$$
$$V_{b3} = 14.8 \times 2 + 3.7 \times 4 + 7.4 = 51.8$$

Then, with the relation $\sigma = 1.18 V_b^{1/3}$, evaluate all σ_i:

$$\sigma_1 = 1.18 \times 59.2^{1/3} = 4.60$$
$$\sigma_2 = 1.18 \times 14.3^{1/3} = 2.86$$
$$\sigma_3 = 1.18 \times 51.8^{1/3} = 4.40$$

By applying the Stockmayer correlations for polar compounds without and with hydrogen bonds, $\sigma = 0.785\ V_C{}^{1/3}$ (without hydrogen bonds) or $\sigma = 36.9\ V_C{}^{1/3}\ Z_C{}^{2.75}$ (with hydrogen bonds).

Considering the first relation suitable for acetaldehyde and the second for ethanol; because hydrogen is considered apolar, we obtain:

$$\sigma_1 = 36.9 \times 168^{1/3} \times 0.240^{2.75} = 4.02$$
$$\sigma_2 = 1.18 \times 14.3^{1/3} = 2.86$$
$$\sigma_3 = 0.785 \times 154^{1/3} = 4.20$$

As we can see, these last calculated values are in a satisfactory agreement with the values reported in Table 6.5 but determined in a different way.

$$\sigma_{12} = \frac{1}{2}(4.02 + 2.86) = 3.44$$
$$\sigma_{13} = \frac{1}{2}(4.02 + 4.20) = 4.11$$
$$\sigma_{23} = \frac{1}{2}(2.86 + 4.20) = 3.53$$

Evaluate now $\frac{k_B T}{\varepsilon_{ij}}$:

For this purpose, we follow again the Stockmayer potential for polar compounds, with the exception of hydrogen being considered apolar, by applying two relations for polar compounds containing or not containing a hydrogen bond:

$$\frac{\varepsilon}{k_B} = 0.897 T_C \quad \text{(without hydrogen bond)}$$
$$\frac{\varepsilon}{k_B} = 0.00331 \frac{T_C}{Z_C^4} \quad \text{(with hydrogen bond)}$$

$$\frac{\varepsilon_1}{k_B} = 0.00331 \frac{516.3}{0.2402^4} = 513.38 \quad \frac{\varepsilon_2}{k_B} = 29.87 \quad \frac{\varepsilon_3}{k_B} = 0.00331 \frac{461.0}{0.2236^4} = 608.22$$
$$\frac{k_B T}{\varepsilon_1} = \frac{573}{513.38} = 1.116 \qquad \frac{k_B T}{\varepsilon_2} = \frac{573}{29.87} = 19.18 \quad \frac{k_B T}{\varepsilon_3} = \frac{573}{608.22} = 0.942$$

$$\varepsilon_{12} = \sqrt{\varepsilon_1 \varepsilon_2} = \frac{k_B T}{\sqrt{1.116 \times 19.18}} = \frac{k_B T}{4.62}$$
$$\varepsilon_{13} = \frac{k_B T}{\sqrt{1.116 \times 0.942}} = \frac{k_B T}{1.025}$$
$$\varepsilon_{23} = \frac{k_B T}{\sqrt{19.18 \times 0.942}} = \frac{k_B T}{4.25}$$

Because the two polar components are mixed with hydrogen, we use the Lennard–Jones integral collision parameters of the mixtures, that is:

$$\frac{k_B T}{\varepsilon_{12}} = 4.62 \quad \Omega_{D12} = 0.855 \frac{k_B T}{\varepsilon_{13}} = 1.025 \quad \Omega_{D13} = 1.42 \frac{k_B T}{\varepsilon_{23}} = 4.25$$
$$\Omega_{D13} = 0.87$$

$$D_{12} = \frac{1.858 \times 10^{-3} \sqrt{\frac{T^3 (M_1 + M_2)}{M_1 M_2}}}{P \sigma_{12}^2 \Omega_{D12}} = \frac{1.858 \times 13.721 \sqrt{\frac{46.07 + 2.016}{46.07 \times 2.016}}}{1 \times 3.44^2 \times 0.855}$$
$$= \frac{18.34}{10.11} = 1.81 \ \left(\frac{\text{cm}^2}{\text{s}}\right)$$

$$D_{13} = \frac{1.858 \times 13.721 \sqrt{\frac{46.07 + 44.05}{46.07 \times 44.05}}}{1 \times 4.11^2 \times 1.42} = \frac{5.37}{23.98} = 0.223 \ \left(\frac{\text{cm}^2}{\text{s}}\right)$$

$$D_{23} = \frac{1.858 \times 13.721 \sqrt{\frac{44.05 + 2.016}{2.016 \times 44.05}}}{1 \times 3.53^2 \times 0.87} = \frac{18.39}{10.84} = 1.696 \ \left(\frac{\text{cm}^2}{\text{s}}\right)$$

The composition at 50% ethanol conversion is: $y_1 = y_2 = y_3 = 0.33$; therefore:

$$D_{1,m} = \frac{1 - y_1}{\frac{y_2}{D_{12}} + \frac{y_3}{D_{13}}} = \frac{1 - 0.333}{\frac{0.333}{1.81} + \frac{0.333}{0.223}} = \frac{0.666}{0.184 + 1.493} = 0.397 \ (\text{cm}^2/\text{s})$$

Part 2 Determination of the Mixture Viscosity

The viscosity can be estimated with the Chapman–Enskog formula (Eq. 6.5). Because ethanol and acetaldehyde are polar molecules, the necessary parameters can be derived from the Stockmayer potential parameters, while for hydrogen, which is apolar, the Lennard–Jones parameters are more suitable. First, we must evaluate for each component a value of σ_i and of the integral collision $\Omega_{\mu i}$. From the previous calculation, we have:

$$\sigma_1 = 4.02 \quad \sigma_2 = 2.86 \quad \sigma_3 = 4.20$$

Stockmayer further parameter:

$$\frac{k_B T}{\varepsilon_1} = 1.116$$

$$\delta = \frac{m_d^2}{2\varepsilon_1 \sigma^3} = \frac{\left(1.69 \times 10^{-18}\right)^2}{2 \times 513.38 \times 1.38 \times 10^{-16} \times \left(4.39 \times 10^{-8}\right)^3} = 0.25$$

In correspondence of the two values found, $\Omega_{\mu 1} = 0.9389$

$$\frac{k_B T}{\varepsilon_2} = 15.07 \quad \Omega_{\mu 2} = 1.5644$$

$$\frac{k_B T}{\varepsilon_3} = 0.942 \quad \delta = \frac{m_d^2}{2\varepsilon_3 \sigma^3} = \frac{\left(2.7 \times 10^{-18}\right)^2}{2 \times 608.22 \times 1.38 \times 10^{-16} \times \left(4.33 \times 10^{-8}\right)^3} = 0.535$$

In correspondence of the two values found, $\Omega_{\mu 3} = 0.9555$.

Calculate now the viscosity of the pure compounds using the Chapman–Enskog formula:

$$\mu_1 = 2.6693 \times 10^{-5} \frac{\sqrt{\mathrm{TM}_1}}{\sigma_1^2 \Omega_{\mu 1}} = 2.6693 \times 10^{-5} \frac{\sqrt{573 \times 46.07}}{4.02^2 \times 0.939} = 2.85 \times 10^{-4} \quad \text{Poise}$$

$$\mu_2 = 2.6693 \times 10^{-5} \frac{\sqrt{\mathrm{TM}_2}}{\sigma_2^2 \Omega_{\mu 2}} = 2.6693 \times 10^{-5} \frac{\sqrt{573 \times 2.016}}{2.86^2 \times 1.564} = 0.71 \times 10^{-4} \quad \text{Poise}$$

$$\mu_3 = 2.6693 \times 10^{-5} \frac{\sqrt{\mathrm{TM}_3}}{\sigma_3^2 \Omega_{\mu 3}} = 2.6693 \times 10^{-5} \frac{\sqrt{573 \times 44.05}}{4.20^2 \times 0.955} = 2.51 \times 10^{-4} \quad \text{Poise}$$

To evaluate the viscosity of the reaction mixture at 50% ethanol conversion, it is necessary to know the composition in molar fractions y_1, y_2, and y_3. As previously explained, $y_1 = y_2 = y_3 = 0.33$. The viscosity of the mixture related to the calculated composition can be determined with Eqs. (6.19) and (6.20). Equation (6.19) also can be written as:

$$\mu_{\mathrm{mix}} = \frac{y_1 \mu_1}{y_1 \phi_{11} + y_2 \phi_{12} + y_3 \phi_{13}} + \frac{y_2 \mu_2}{y_1 \phi_{21} + y_2 \phi_{22} + y_3 \phi_{23}} + \frac{y_3 \mu_3}{y_1 \phi_{31} + y_2 \phi_{32} + y_3 \phi_{33}}$$

Equation (6.20) must be used for calculating the ϕ_{ij} values. ϕ_{11}, ϕ_{22}, and ϕ_{33} are equal to 1, so we must calculate the other six values:

$$\phi_{12} = \frac{1}{\sqrt{8}}\left(1+\frac{46.07}{2.016}\right)^{-1/2}\left[1+\left(\frac{2.85\times10^{-4}}{0.71\times10^{-4}}\right)^{1/2}\left(\frac{2.016}{46.07}\right)^{1/4}\right]^2 = 0.266$$

$$\phi_{13} = \frac{1}{\sqrt{8}}\left(1+\frac{46.07}{44.05}\right)^{-1/2}\left[1+\left(\frac{2.85\times10^{-4}}{2.51\times10^{-4}}\right)^{1/2}\left(\frac{44.05}{46.07}\right)^{1/4}\right]^2 = 1.042$$

$$\phi_{21} = \frac{1}{\sqrt{8}}\left(1+\frac{2.016}{46.07}\right)^{-1/2}\left[1+\left(\frac{0.71\times10^{-4}}{2.85\times10^{-4}}\right)^{1/2}\left(\frac{46.07}{2.016}\right)^{1/4}\right]^2 = 1.512$$

$$\phi_{23} = \frac{1}{\sqrt{8}}\left(1+\frac{2.016}{44.07}\right)^{-1/2}\left[1+\left(\frac{0.71\times10^{-4}}{2.51\times10^{-4}}\right)^{1/2}\left(\frac{44.05}{2.016}\right)^{1/4}\right]^2 = 1.595$$

$$\phi_{31} = \frac{1}{\sqrt{8}}\left(1+\frac{44.05}{46.07}\right)^{-1/2}\left[1+\left(\frac{2.51\times10^{-4}}{2.85\times10^{-4}}\right)^{1/2}\left(\frac{46.07}{44.05}\right)^{1/4}\right]^2 = 0.961$$

$$\phi_{32} = \frac{1}{\sqrt{8}}\left(1+\frac{44.05}{2.016}\right)^{-1/2}\left[1+\left(\frac{2.51\times10^{-4}}{0.71\times10^{-4}}\right)^{1/2}\left(\frac{2.016}{44.05}\right)^{1/4}\right]^2 = 0.259$$

	Ij	1	2	3
Φ_{ij}=	1	1	0.266	1.0419
	2	1.512	1	1.5953
	3	0.961	0.259	1

By substituting ϕ_{ij} in the previous relation, we obtain:

$$\mu_{mix} = \frac{0.33\times2.85\times10^{-4}}{0.33+0.33\times0.266+0.33\times1.042} + \frac{0.33\times0.71\times10^{-4}}{0.33\times1.512+0.33+0.33\times1.595} + \frac{0.33\times2.51\times10^{-4}}{0.33\times0.961+0.33\times0.259+0.33}$$

$$\mu_{mix} = \frac{2.85\times10^{-4}}{1+0.266+1.042} + \frac{0.71\times10^{-4}}{1.512+1+1.595} + \frac{2.51\times10^{-4}}{0.961+0.259+1}$$

$$\mu_{mix} = 2.53\times10^{-4}\ \text{(Poise)}$$

Part 3 Determination of thermal conductivity

To evaluate thermal conductivity, we first determine k_1, k_2, and k_3 using Eq. (6.6). This results in:

$$k_1 = \frac{1.989 \times 10^{-4}\sqrt{\frac{573.16}{46.07}}}{4.02^2 \times 0.94} = 0.461 \times 10^{-4}\left(\frac{\text{cal}}{\text{cm s K}}\right)$$

$$k_2 = \frac{1.989 \times 10^{-4}\sqrt{\frac{573.16}{2.016}}}{2.86^2 \times 1.56} = 2.62 \times 10^{-4}\left(\frac{\text{cal}}{\text{cm s K}}\right)$$

$$k_3 = \frac{1.989 \times 10^{-4}\sqrt{\frac{573.16}{44.05}}}{4.20^2 \times 0.95} = 0.42 \times 10^{-4}\left(\frac{\text{cal}}{\text{cm s K}}\right)$$

Last, the value of k_{mix} can be determined with relations similar to the ones employed for the evaluation of μ_{mix}, that is:

$$k_{mix} = \frac{y_1 k_1}{y_1\phi_{11} + y_2\phi_{12} + y_3\phi_{13}} + \frac{y_2 k_2}{y_1\phi_{21} + y_2\phi_{22} + y_3\phi_{23}} + \frac{y_3 k_3}{y_1\phi_{31} + y_2\phi_{32} + y_3\phi_{33}}$$

where ϕ_{ij} are the same previously calculated. Hence, we have:

$$k_{mix} = \frac{0.33 \times 0.461 \times 10^{-4}}{0.33 + 0.33 \times 0.266 + 0.33 \times 1.042} + \frac{0.33 \times 2.62 \times 10^{-4}}{0.33 \times 1.512 + 0.33 + 0.33 \times 1.595}$$
$$+ \frac{0.33 \times 0.39 \times 10^{-4}}{0.33 \times 0.961 + 0.33 \times 0.259 + 0.33}$$
$$k_{mix} = \frac{0.461 \times 10^{-4}}{1 + 0.266 + 1.042} + \frac{2.62 \times 10^{-4}}{1.512 + 1 + 1.595} + \frac{0.39 \times 10^{-4}}{0.961 + 0.259 + 1}$$
$$k_{mix} = 1.01 \times 10^{-4}(\text{cal/cm s K})$$

Exercise 6.4
Estimation of Molecular Diffusion Coefficients in a Liquid Mixture

Part 1 It has been mentioned that the Wilke–Chang relation is valid for determining the diffusion coefficient D_{12} for dilute solution. It also has been mentioned that normally for liquid phase mixtures $D_{12} \neq D_{21}$. Evaluate with the Wilke–Chang relation the diffusion coefficient D_{12} of ethanol in water and compare it with the experimental values reported in Table 6.6. Consider then that the last concentration corresponds to a diluted solution of water in ethanol. How is D_{21} calculated by the Wilke–Chang relation agrees with the experimental value?

$$D_{12} = 7.4 \times 10^{-8}\frac{T(X_2 M_2)^{\frac{1}{2}}}{\mu_2 V_1^{0.6}} = 7.4 \times 10^{-8}\frac{298(2.6 \times 18)^{\frac{1}{2}}}{1 \times (14.8 \times 2 + 3.7 \times 6 + 7.4)^{0.6}}$$
$$= 1.3 \times 10^{-5}\ (\text{cm}^2/\text{s})$$

The viscosity of water is expressed in cpoise. The molar volumes were determined with the additive values in Appendix 5. As can be seen, the calculated

Table 6.6 Experimental diffusivity of ethanol in water at 25 °C

Binary system	x_1	$D_{12} \times 10^5$ (cm²/s)
Ethanol–water	Near 0	1.24
	0.05	1.13
	0.275	0.41
	0.50	0.90
	0.70	1.40
	0.95	2.20

[a]Data from Johnson and Babb (1956), see also Bird et al. (1960). The first value was taken from Satterfield and Sherwood

diffusion coefficient is comparable with the value determined at the lowest concentration ($x_1 \approx 0$). Let us consider now the D_{21} estimated value:

$$D_{21} = 7.4 \times 10^{-8} \frac{T(X_1 M_1)^{\frac{1}{2}}}{\mu_1 V_2^{0.6}} = 7.4 \times 10^{-8} \frac{298(1.5 \times 46)^{\frac{1}{2}}}{1.1 \times (3.7 \times 2 + 7.4)^{0.6}}$$
$$= 3.3 \times 10^{-5} \ (\mathrm{cm^2/s})$$

As it can be seen, the agreement in this case is not good, probably because the concentration is too high. However, it has been shown that $D_{12} \neq D_{21}$.

Part 2 Although the Wilke–Chang relation has drawbacks, it can be usefully employed for estimating the effect of the temperature on diffusivity. Starting, for example, from a known value of D_{12}, at a given temperature, and the viscosity of the solvent at the two temperatures, we can write:

$$\frac{D_{12}^{T1} \mu_{T1}}{T_1} = \frac{D_{12}^{T2} \mu_{T2}}{T_2}$$

For example, Treybal (1955) reported an example in which the diffusivity of mannitol $C_6H_{14}O_6$ in water was determined at 20 °C (viscosity of water 1.005 cp) using the Wilke–Chang expression obtaining $D_{12} = 0.644 \times 10^{-5}$. The experimental value was 0.56×10^{-5}. Then, starting from the observed value, he calculated the diffusivity at 70 °C (viscosity of water 0.4061 cp) as:

$$\frac{0.56 \times 10^{-5} \times 1.005}{293} = \frac{D_{12}^{343} \times 0.4061}{343} \quad D_{12}^{343} = 1.62 \times 10^{-5} \ \mathrm{cm^2/s}$$

The experimental value is 1.56×10^{-5}, which is in satisfactory agreement with the calculated value.

Estimate the liquid diffusivity of methanol in water at 15 °C.

$$D_{12} = 7.4 \times 10^{-8} \frac{T(X_2M_2)^{\frac{1}{2}}}{\mu_2 V_1^{0.6}} = 7.4 \times 10^{-8} \frac{288(2.6 \times 18)^{\frac{1}{2}}}{1.14x(14.8 + 3.7 \times 4 + 7.4)^{0.6}}$$
$$= 1.465 \times 10^{-5} \ (\mathrm{cm^2/s})$$

With D_{12} of methanol in water at 15 °C = 1.465 × 10^{-5} cm^2/s, evaluate the corresponding value at 100 °C.

$$\frac{1.465 \times 10^{-5} \times 1.14}{288} = \frac{D_{12}^{343} \times 0.2821}{373} \quad D_{12}^{343} = 7.67 \times 10^{-5} \ \mathrm{cm^2/s}$$

It is also possible to obtain an approximated estimation of the diffusivity of a component in a given solvent by comparing this diffusivity with that of another component in the same solvent at the same temperature. It is possible, for example, to calculate D_{EW}, the diffusivity coefficient of ethanol in water, from the corresponding values determined for methanol in water D_{MW} at the same temperature of 15 °C.

$$\frac{D_{\mathrm{EW}}}{D_{\mathrm{MW}}} = \left(\frac{V_{\mathrm{bM}}}{V_{\mathrm{bW}}}\right)^{0.6}$$
$$D_{\mathrm{EW}} = 1.465 \times 10^{-5} \left(\frac{14.8 + 3.7 \times 4 + 7.4}{14.8 \times 2 + 3.7 \times 6 + 7.4}\right)^{0.6} = 1.105 \times 10^{-5} \ \mathrm{cm^2/s}$$

Corresponding to a value of 1.22 × 10^{-5} at 25 °C is in satisfactory agreement with the observed value.

6.2 Kinetics and Transport Phenomena in Gas–Solid Reactors

6.2.1 Mass and Heat Transfer From a Fluid to the Surface of a Catalytic Particle

As described in the introduction of Chap. 5, when a solid porous catalyst is employed to promote a chemical reaction, the reaction mainly occurs inside the catalytic particles consuming reagents, yielding products, and absorbing or releasing heat. The chemical reaction is therefore responsible for increasing both temperature and concentration gradients. These gradients first originate inside the pores of the catalytic particles where the molecular diffusion occurs together with the chemical reaction. For higher reaction rates, we also can have significant gradients inside the boundary layer at the external fluid–solid interface. The following possibilities can be distinguished:

1. Reaction rates depend on the extension of the overall catalytic surface area. In this case, the gradients are negligible, and we can measure directly the true chemical reaction rate. This is the preferred condition for laboratory kinetic runs, as seen in the previous chapter.
2. Reaction rates are limited by resistance to the internal diffusion of the reagents or products. In this case, we have a concentration profile for each component inside the particles. In particular, the concentration of the reactants decreases with respect to the concentration on the external catalytic surface.
3. When the reaction is exothermic or endothermic, we can also have a temperature profile inside the catalyst particles.
4. A temperature gradient can arise also at the boundary layer. As a consequence, the temperature of the fluid flowing can be different from that of the external catalytic surface.
5. When the reaction rate is very high, it can be limited by resistance to the external diffusion at the boundary layer, that is, in this case, the reaction occurs mainly on the external surface of the particle, and the reactant concentration on the catalyst surface is lower than the one in the fluid bulk.

All of the abovementioned situations can be seen as particular kinetic regimes and can be adequately treated. In the case of a gas–liquid–solid reactor, for example, gradients 3 and 4 can be neglected for the high thermal conductivity of the liquid wetting the solid. By increasing the reaction rate, the regimes involved progressively pass from possibility 1 to possibility 5. However, it must be pointed out that external mass transfer occurs as a consecutive independent step preceding the chemical reaction, which occurs inside the catalyst particles. In contrast, internal mass transfer occurs together with the reaction, and the two phenomena—both physical and chemical—must be considered together.

The motion of a fluid inside a catalytic bed can be turbulent, but as already mentioned, at the interface, a stagnant thin film of the fluid exists, the so-called boundary layer, through which the mass flux occurs by a slow process characterized by molecular diffusion. This process also is normally accompanied by a slow thermal flux. Both the concentration and the temperature gradients are therefore located at the boundary layer. Figure 6.2, for example, shows that the boundary layer is located around a spherical particle, while Fig. 6.3 qualitatively shows the possible gradients occurring inside the boundary layer considering, respectively, an exothermic and an endothermic reaction.

As can be seen in Fig. 6.3, in both cases (a and b), the concentration decreases from the bulk to the catalyst surface, while the temperature decreases in the case of the occurrence of an endothermic reaction and increases when the reaction is exothermic. Starting from Fick's law and by assuming a linear profile for the concentration gradient (see Fig. 6.3), the molar flow rate can be written as:

$$N_i = k_c(c_{ib} - c_{is}) = \text{moles of } i \text{ diffused/time } b \text{ (external surface area)} \quad (6.27)$$

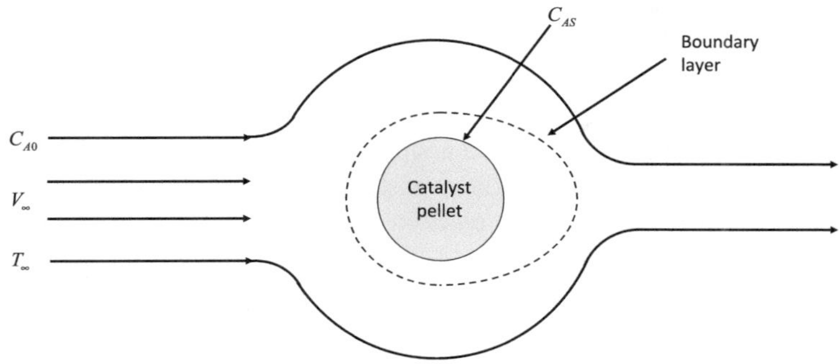

Fig. 6.2 Boundary layer around the surface of a spherical catalytic particle

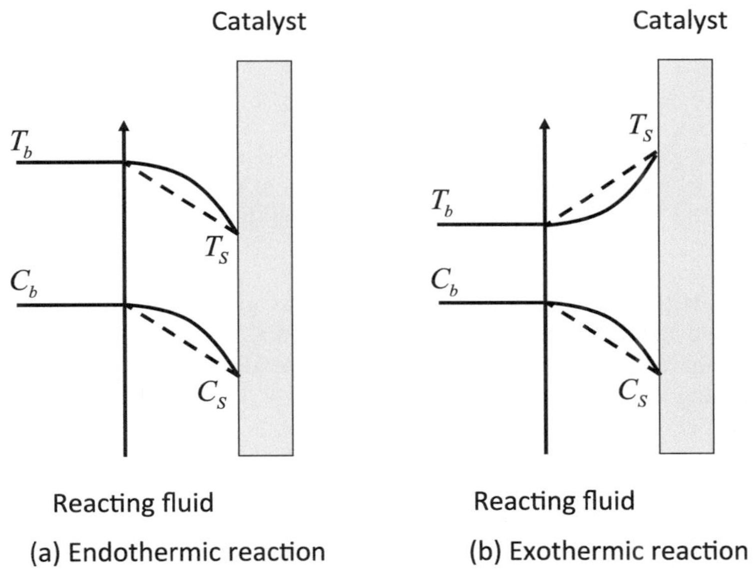

Fig. 6.3 Possible gradients of concentration and temperature occurring inside the boundary layer. A reasonable approximation is to assume linear profiles in any case (dotted lines)

where c_{ib} is the concentration of i in the fluid bulk, c_{is} the concentration of i on the solid surface, and k_c is the mass-transfer coefficient. For a gaseous mixture, the following relation is more convenient:

$$N_i = K_g(p_{ib} - p_{is}) \tag{6.28}$$

where p_{ib} and p_{is} are, respectively, the partial pressures of i in the bulk of the fluid and on the solid surface. k_c and k_g are related to the molecular diffusion coefficient D_{12} and to the thickness δ of the boundary layer, as in the following relations:

$$k_c = \frac{D_{12}}{\delta} \quad k_g = \frac{D_{12}}{\delta RT} \tag{6.29}$$

that is, $k_c = k_g$ RT.

Under steady-state conditions, for a nonporous catalyst or for a very fast reaction occurring just on the external surface of the catalytic particle, the flux of the component i must be equal to the reaction rate. Assuming for simplicity a first-order kinetic law, we can write:

$$r_{obs} = k_{1s} C_{is} = k_c (C_{ib} - C_{is}) \quad \text{(moles/time external surface)} \tag{6.30}$$

where C_{is} is the concentration of i on the surface of the catalyst particle and C_{ib} is the concentration of i inside the fluid bulk. Solving for C_{is} yields:

$$C_{is} = \frac{k_c C_{ib}}{k_{1s} + k_c} \tag{6.31}$$

Hence,

$$r_{obs} = \frac{k_{1s} k_c C_{ib}}{k_{1s} + k_c} = \frac{C_{ib}}{\frac{1}{k_{1s}} + \frac{1}{k_c}} = \frac{k_{1s} C_{ib}}{1 + \left(\frac{k_{1s}}{k_c}\right)} = \frac{k_{1s} C_{ib}}{1 + Da} = k_{obs} C_{ib} \tag{6.32}$$

where Da is the Damkohler number $= k_{1s}/k_c$, thus giving a comparison between reaction rate and diffusion rate. Considering Eq. (6.32), the first observation arises. We know that the kinetic constant, $k_{1s,}$ strongly depends on temperature according to the following relation:

$$k_{1s} = k_{1s}^0 \exp\left(-\frac{\Delta E}{RT}\right) \tag{6.33}$$

while the mass-transfer coefficient is relatively insensitive to this parameter. Neglecting any dependence on the temperature of k_c, we can write:

$$k_{obs} = \frac{1}{\frac{1}{k_{1s}} + \frac{1}{k_c}} = \frac{1}{\frac{1}{k_{1s}^0 \exp\left(-\frac{\Delta E}{RT}\right)} + \frac{1}{k_c}} = \frac{k_c k_{1s}^0 \exp\left(-\frac{\Delta E}{RT}\right)}{k_c + k_{1s}^0 \exp\left(-\frac{\Delta E}{RT}\right)} \tag{6.34}$$

From this relation, it is possible to deduce that at low temperature:

$$k_{obs} \approx k_{1s}^0 \exp\left(-\frac{\Delta E}{RT}\right) \tag{6.35}$$

while at high temperature $k_{obs} \approx k_c$. In conclusion, the external mass-transfer limitation can mask the effective dependence on the temperature of the occurring reaction rate. Another observation can be made always considering Eq. (6.32). In fact, it is possible to observe that if the reaction is very fast, $k_{1s} \gg k_c$, that is, the reaction is completely dominated by the external mass transfer, which in this case holds:

$$r_{obs} = k_c C_{ib} \tag{6.36}$$

This means that C_{is} is negligible with respect to C_{ib} and can be approximated to zero.

In contrast, if mass transfer is much faster than the reaction rate, $k_c \gg k_{1s}$, the chemical regime dominates, and we can write:

$$r_{obs} = k_{1s} C_{ib} \tag{6.37}$$

In conclusion, two limiting cases can be individuated, respectively, corresponding to a diffusional regime or to a chemical regime. Under intermediate conditions, Eq. (6.32) is valid.

Let us consider now what happens when the reaction order is different from 1. We can write an observed rate:

$$r_{obs} = k_{ns} C_{is}^n = k_c (C_{ib} - C_{is}) \tag{6.38}$$

Again, if the reaction rate is much larger than the mass-transfer rate, $C_{is} \approx 0$; therefore:

$$r_{obs} = k_c C_{ib} \tag{6.39}$$

that is, the reaction will appear of first order, and the true kinetics of nth order is masked as a consequence of the external mass-transfer limitation. To avoid the masking effect of the external mass-transfer limitation on the kinetics, operative conditions in laboratory are chosen in such a way to exclude this influence. For this purpose, some kinetic runs are performed at different flow rates conditions, and data are plotted as in Fig. 6.4. Clearly, kinetic runs are performed at flow rates at which the chemical regime is operative. Then, kinetic runs are also performed at different temperatures to individuate the range in which the mass-transfer limitation occurs. This can be performed by putting in an Arrhenius plot the ln r_{obs} as a function of $1/T$ as qualitatively shown in Fig. 6.5.

A treatment analogous to the one described for the mass transfer can be made for the heat transfer through the boundary layer. Starting from the Fourier law and assuming the reasonable approximation of a linear profile of the temperature inside the boundary layer, the heat flow can be expressed as:

$$q = h(T_b - T_s) = (\text{heat/time·surface area}) \tag{6.40}$$

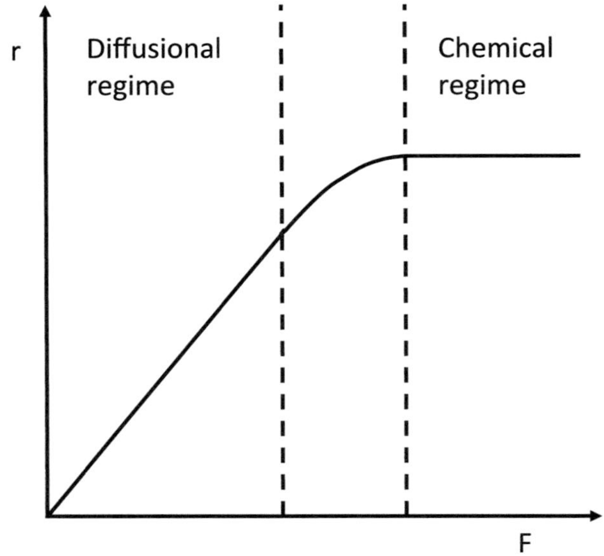

Fig. 6.4 Effect of flow rate on reaction rate in the presence of external mass-transfer limitation

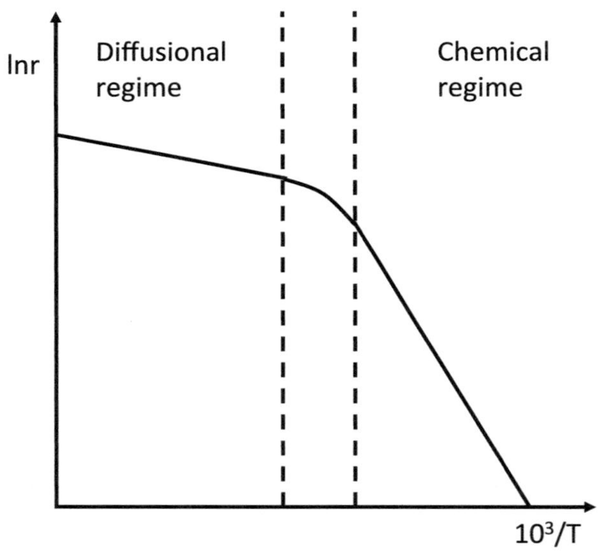

Fig. 6.5 Effect of external diffusional regime on apparent activation energy

where T_s is the temperature on the surface of the catalytic particle, T_b is the temperature inside the fluid bulk, and h is the heat-transfer coefficient related to the thermal conductivity of the fluid and to the size of the boundary layer, that is, $h = k_t/\delta$.

The thickness of the boundary layer depends on the flow conditions (laminar or turbulent). The mass-transfer coefficient depends not only on the average flow rate of the reacting mixture but also on both the most relevant physical properties of the fluid and the size of the catalyst particles. Dimensional analysis can be used to express the mentioned dependence in terms of dimensionless groups such as:

$$\text{Sherwood number} = \text{Sh} = \frac{\text{Mass transfer velocity}}{\text{Diffusion velocity}} = \frac{k_c d_p}{D_{i,m}} \quad (6.41)$$

$$\text{Schmidt number} = \text{Sc} = \frac{\text{Momentum diffusivity}}{\text{Mass diffusivity}} = \frac{\mu}{\rho D_{i,m}} = \frac{\nu}{D_{i,m}} \quad (6.42)$$

$$\text{Reynolds number} = \text{Re} = \frac{\text{Inertial forces}}{\text{Viscous forces}} = \frac{G_m d_p}{\mu} = \frac{u \rho d_p}{\mu} = \frac{u d_p}{\nu} \quad (6.43)$$

where d_p is the particle diameter (if the shape of the particle is different from a sphere, d_p can be assumed to be an equivalent diameter corresponding to the diameter of a sphere having the same external surface area), $D_{i,m}$ is the molecular diffusion coefficient i in the reacting mixture, μ is the viscosity of the fluid, ρ is the density of the fluid, ν is the kinematic viscosity $= \mu/\rho$, G_m is the mass velocity of the fluid per unit area of the empty cross section of the tubular reactor, and $u =$ linear fluid velocity. For the flow around a spherical particle, the Frössling (1938) correlation is valid (see also Fogler 1986):

$$\text{Sh} = 2 + 0.6\,\text{Re}^{1/2}\text{Sc}^{1/3} \quad (6.44)$$

Th Sh number becomes equal to 2 when the Re number is equal to 0, that is, when a spherical particle is immersed in a stagnant fluid. Under turbulent conditions, this value is negligible with respect to the second term, in particular whether the fluid is liquid. Neglecting term 2, we can write:

$$k_c = 0.6\left(\frac{D_{i,m}}{d_p}\right)\text{Re}^{1/2}\text{Sc}^{1/3} = 0.6\left(\frac{D_{i,m}^{2/3}}{\nu^{1/6}}\right)\left(\frac{u^{1/2}}{d_p^{1/2}}\right) \quad (6.45)$$

As can be seen for increasing k_c, it is necessary to decrease the particle size or increase the velocity of the fluid. In particular, if the fluid velocity is doubled, the mass-transfer rate increases by a factor of $\sqrt{2} = 1.41$.

According to Thoenes and Kramers (1958), k_c for the flow through a packed bed can be evaluated using the correlation:

$$\text{Sh}' = 1.0(\text{Re}')^{1/2}\text{Sc}^{1/3} \quad (6.46)$$

where:

$$\text{Sh}' = \frac{\text{Sh}\varepsilon_b}{(1-\varepsilon_b)\gamma} \quad \text{and} \quad \text{Re}' = \frac{\text{Re}}{(1-\varepsilon_b)\gamma} \quad (6.47)$$

ε_b is the void fraction of the packed bed, and γ is the shape factor = external surface area/πd_p^2.

This correlation is valid for $0.25 < \varepsilon_b < 0.5$; $40 < \mathrm{Re}' < 4000$; $1 < \mathrm{Sc} < 4000$.

A similar approach can be followed for the heat transfer, where the Nusselt dimensionless number (Nu) takes the place of the Sherwood number (Sh) and the Prandtl number (Pr) takes the place of the Schmidt number (Sc), that is:

$$\text{Nusselt number} = \mathrm{Nu} = \frac{\text{Heat transfer velocity}}{\text{Heat diffusion velocity}} = \frac{h d_p}{k_t} \tag{6.48}$$

$$\text{Prandtl number} = \mathrm{Pr} = \frac{\text{Diffusivity of momentum}}{\text{Diffusivity of energy}} = \frac{\mu C_p}{k_t} \tag{6.49}$$

The correlation is similar to the Frössling relation (Eq. 6.26) (see Ranz and Marshall 1952a, b):

$$\mathrm{Nu} = 2 + 0.6\ \mathrm{Re}^{1/2}\mathrm{Pr}^{1/3} \tag{6.50}$$

Experimental data of heat and mass transfer, as collected by a great number of researchers, often were correlated in terms of Colburn factors J_D for the mass transfer and J_H for the heat transfer (Colburn analogy 1934; Chilton and Colburn 1934; see also Satterfield and Sherwood 1963):

$$J_D = \frac{\mathrm{Sh}}{\mathrm{Re}\mathrm{Sc}^{1/3}} = \frac{k_c \rho}{G}\mathrm{Sc}^{2/3} = \frac{k_g P}{G_m}\mathrm{Sc}^{2/3} = f(\mathrm{Re}) \tag{6.51}$$

$$J_H = \frac{\mathrm{Nu}}{\mathrm{Re}\mathrm{Pr}^{1/3}} = \frac{h}{C_p G}\mathrm{Pr}^{2/3} = f(\mathrm{Re}) \tag{6.52}$$

where G is the mass velocity based on cross-sectional area of the empty reactor, P is the total pressure, G_m is the moles of the mixture per unit time per total bed cross-section, and Pr is the Prandtl number $\mathrm{Pr} = \mu\, C_p/k$ with C_p = specific heat of the fluid. f(Re) changes passing from liquid to gas. Dwydevi and Upadhay (1977), for example, for packed beds found the following correlation:

$$J_D = \frac{0.454}{\varepsilon_b}\mathrm{Re}^{-0.41} \quad \text{in the range} \quad 1 < \mathrm{Re} < 10{,}000 \tag{6.53}$$

Some other correlations of this type can be found in the literature, for example, for gaseous reactants whereby $3 < \mathrm{Re} < 2000$ with an interparticle void fraction $0.416 < \varepsilon < 0.788$, it holds:

$$J_D = \frac{0.357}{\varepsilon_b}\mathrm{Re}^{-0.359} \tag{6.54}$$

while for liquid reactants with $55 < \mathrm{Re} < 1500$ and $0.35 < \varepsilon < 0.75$

$$J_{\mathrm{D}} = \frac{0.25}{\varepsilon_{\mathrm{b}}} \mathrm{Re}^{-0.31} \tag{6.55}$$

and always for liquid reactant in the case of $0.0016 < \mathrm{Re} < 55$

$$J_{\mathrm{D}} = \frac{1.09}{\varepsilon_{\mathrm{b}}} \mathrm{Re}^{-0.67} \tag{6.56}$$

Initially, Chilton and Colburn (1934) assumed $J_{\mathrm{D}} = J_{\mathrm{H}}$, but according to many experimental observations collected and elaborated by different authors (see Carberry 1960; Bradshaw and Bennet 1961), it is possible to write that approximately $J_{\mathrm{H}} \simeq 1.08\, J_{\mathrm{D}}$ in agreement with the plot constructed by De Acetis and Thodos (1960) (see also Satterfield and Sherwood 1963) and reported in Fig. 6.6.

Re-elaborated from data of De Acetis and Thodos (1960), Copyright (1960) American Chemical Society

From this plot, if we know the Re number, it is possible to evaluate the corresponding values of J_{D} and J_{H}.

As has been seen, mass and heat transfer are strictly connected, and we also can write:

$$k_{\mathrm{c}}(C_{i\mathrm{b}} - C_{i\mathrm{s}})(-\Delta H) = h(T_{\mathrm{b}} - T_{\mathrm{s}}) \tag{6.57}$$

and hence,

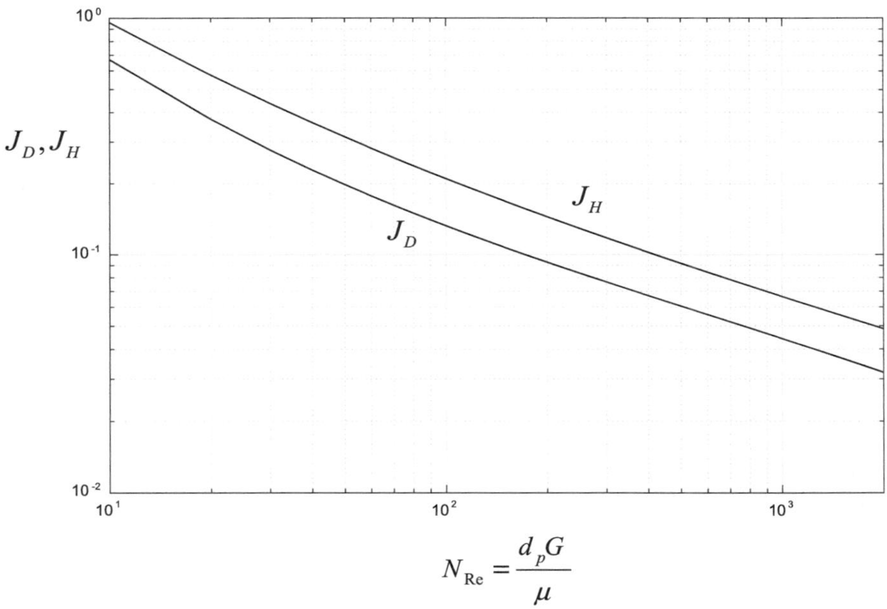

Fig. 6.6 Variation of J_{D} and J_{H} with the Reynolds number in a fixed bed

$$\Delta T = \Delta C(-\Delta H)\frac{k_c}{h} \quad (6.58)$$

that is, the temperature gradient is a function of the concentration gradient: Putting in (6.58) k_c and h determined through the empirical relations of J_D and J_H, we obtain:

$$\Delta T = \Delta C(-\Delta H)\frac{1}{\rho C_p}\left(\frac{C_p \mu / k_t}{\mu / \rho D_{i,m}}\right)\left(\frac{J_D}{J_H}\right) \quad (6.59)$$

$$\text{Le} = \text{Lewis number} = \frac{\text{Pr}}{\text{Sc}} = \frac{C_p \mu / k_t}{\mu / \rho D_{i,m}} \approx 1 \quad \text{and also} \quad \frac{J_D}{J_H} \approx 1 \quad (6.60)$$

Therefore,

$$\Delta T = \Delta C \frac{(-\Delta H)}{\rho C_p} \quad (6.61)$$

This relation shows that a significant temperature gradient is possible although the concentration gradient is very low. In fact, in the presence of a highly exothermic reaction sometimes the concentration gradient can be neglected, while this is impossible for the temperature gradient. For example, in the case of the reaction:

$$H_2 + \frac{1}{2}O_2 \overset{(pt)}{\rightarrow} H_2O \quad (6.62)$$

it resulted that for $\Delta C / C_b = 0.05$, $\Delta T = 115$ °C. At last, by assuming the limit condition for which $C_s \simeq 0$, we can easily calculate the maximum possible thermal gradient:

$$\Delta T_{max} = \frac{C_b(-\Delta H)}{\rho C_p} \quad (6.63)$$

This calculation can be used as a useful criterion for evaluating the importance of this gradient. The external temperature gradient for gas–solid reactions is normally more important than the internal one in terms of affecting the reaction rate because the gas is less conductive than the solid. In contrast, this gradient is less important for liquid–solid systems because of the high thermal conductivity of the liquids.

To conclude, to evaluate quantitatively the effects of the external mass and heat transfer, related to a single component, four equations are needed: two corresponding to the physical transport rates and two corresponding to the chemical rates:

$$\left.\begin{array}{l} J = k_m a_m (C_b - C_s) \\ Q = h a_m (T_b - T_s) \end{array}\right\} \quad \text{Physical transport rates} \quad (6.64)$$

$$\left.\begin{array}{l} r = f(C_s, T_s) \\ Q' = r(-\Delta H) \end{array}\right\} \quad \text{Chemical rates} \tag{6.65}$$

The term $k_m a_m$ is more conveniently used than k_c because it can be more easily measured. k_m is the mass-transfer coefficient expressed for example in cm/s, while a_m is the external surface area of the particles per unit mass of the pellet (e.g., cm^2/g). As a consequence, the mass-transfer rate is expressed in mol/g s, and the same units are used for expressing the reaction rate.

By introducing the steady-state approximation, the needed relations are reduced to two:

$$k_m a_m (C_b - C_s) = f(C_s, T_s) \tag{6.66}$$

$$h a_m (T_b - T_s) = f(C_s, T_s)(-\Delta H) \tag{6.67}$$

In the presence of a reaction, involving more components, the mass balance (Eq. 6.66) must be applied to any single component involved in the reaction, while just one equation for solving the heat balance is necessary. For a complex reaction scheme with different occurring reactions, the general approach developed in Chap. 5 is still valid, and the introduction of the mass- and heat-transfer relations has only the scope to evaluate the unknown C_s and T_s, thanks to Eqs. (6.66) and (6.67). It must be pointed out that it would be more correct to consider "activities" instead of "concentrations" in all of the mass and heat balance relations if we have the possibility to evaluate the related coefficients. However, as seen in Chap. 5, we normally have continuous stirred-tank reactors (CSTRs) or plug flow reactors (PFRs), and the mass and heat balance are characterized by an algebraic equation system in the former case and a system of differential equations in the latter.

Exercise 6.5
Estimation of Mass and Heat Transfer Coefficients in Gas–Solid Catalytic Reactors

Consider the same reaction of Exercise 6.3, that is:

$$CH_3CH_2OH \rightarrow CH_3CHO + H_2 \quad \Delta H = 16,811 \quad (\text{cal/mol})$$

Imagine making the reaction at 300 °C and atmospheric pressure in a tubular reactor of 4 cm diameter filled with 15 g of catalyst in pellets of cylindrical shape having a diameter = height = 0.5 cm. The density of the catalyst is 1.2 g/cm^3, and the void fraction $\varepsilon = 0.4$. The molar feed of the reactant ethanol is 8 mol/h, and the measured reaction rate for a conversion of 0.5 is 5.5×10^{-5} mol/g s. Consider the changes occurring in a range of conversion between 0 and 0.5.

1. Evaluate the mass- and heat-transfer coefficients.
2. Evaluate the external gradients of temperature and concentration.

Part 1

To evaluate the mass-transfer coefficient, we first must evaluate all of the numbers of Reynolds (Re) and Schmidt (Sc).

$$\mathrm{Re} = \frac{G d_{\mathrm{P}}}{\mu_{\mathrm{m}}}$$

At the reactor inlet, we have the feeding of only ethanol with a known flow rate, but along the reactor, the flow increases because from each reacted mole two moles of products are formed. Thus, we evaluate the Reynolds number at the reactor inlet as follows:

$$d_{\mathrm{peq}} = \sqrt{l d_{\mathrm{p}} + \frac{d_{\mathrm{p}}^2}{2}} = 0.61 \ \ \mathrm{cm}$$

$$G = \frac{F_{\mathrm{ET}}^{o} M_{\mathrm{ET}}}{\pi \varepsilon \left(\frac{d_{\mathrm{tube}}}{2}\right)^2} = \frac{8 \times 46}{3.14 \times 4 \times 0.4 \times 3600} = 2.034 \times 10^{-2} \left(\frac{\mathrm{g}}{\mathrm{s\ cm^2}}\right)$$

The viscosity of any single component and of the mixtures, at 300 °C, was already estimated Exercise 6.3 and resulted in similar for each component and for the mixtures. We can assume, therefore, that this value is a constant, independent from the conversion, and equal to approximately 2.53×10^{-4} Poise.

$$\mathrm{Re} = \frac{G d_{\mathrm{Peq}}}{\mu_{\mathrm{m}}} = \frac{0.61 \times 2.034 \times 10^{-2}}{2.53 \times 10^{-4}} = 49$$

Although the overall volumetric flow rate changes with the conversion because the reaction occurs with an increase of moles, the mass flow rate does not change; therefore, G remains constant, and consequently the Reynolds number also remains constant. Let us to calculate now the Schmidt number:

$$\mathrm{Sc} = \frac{\mu_{\mathrm{m}}}{\rho_{\mathrm{m}} D_{i,\mathrm{m}}}$$

$$M_{\mathrm{Av}} = \sum_i M_i y_i = \left(\frac{1-\lambda}{1+\lambda}\right) 46 + \left(\frac{\lambda}{1+\lambda}\right) 44 + \left(\frac{\lambda}{1+\lambda}\right) 2$$

Assuming a conversion $\lambda = 0.361$, it results in:

$$M_{\mathrm{Av}} = 0.468 \times 46 + 0.265 \times 44 + 0.265 \times 2 = 33.718$$

$$\rho_{\mathrm{m}} = \left(\frac{M_{\mathrm{Av}}}{22414}\right)\left(\frac{273}{573}\right) = 7.17 \times 10^{-4} \ \ (\mathrm{g/cm^3})$$

The value of $D_{1,m}$ was already calculated in Exercise 6.3 and holds 0.393 cm²/s; therefore,

$$\mathrm{Sc} = \frac{2.53 \times 10^{-4}}{7.17 \times 10^{-4} \times 0.397} = 0.888$$

The Sc number for conversion null is 0.658.

$J_\mathrm{D} = 1.66\mathrm{Re}^{-0.51} = 0.228$ (see Froment and Bischoff 1990);

$$J_\mathrm{D} = \frac{k_\mathrm{Gm}\mathrm{PM}_\mathrm{Av}}{G}\mathrm{Sc}^{2/3} = \frac{k_\mathrm{Gm} \times 1 \times 33.718}{2.04 \times 10^{-2}}(0.888)^{2/3} = 1527k_\mathrm{Gm}$$

$$k_\mathrm{Gm} = \frac{0.228}{1527} = 1.49 \times 10^{-4}\left(\frac{\mathrm{mols}}{\mathrm{s\ atm\ cm^2}}\right)$$

Another correlation reported by Smith (1981) gives the same result:

$$J_\mathrm{D} = \frac{0.458}{\varepsilon}\mathrm{Re}^{-0.407} = \frac{0.458}{0.4}(49)^{-0.407} = 0.234$$

A similar result can also be obtained from the plot of Fig. 6.6.

Then, it is known that $J_\mathrm{H} \approx 1.08\ J_\mathrm{D}$, that is, $J_\mathrm{H} = 0.246$. Therefore:

$$J_\mathrm{H} = \frac{h_\mathrm{m}}{GC_\mathrm{P}}\mathrm{Pr}^{2/3} \quad \mathrm{Pr} = \frac{\mu_\mathrm{m}C_\mathrm{Pm}}{k_\mathrm{m}} = \text{Prandtl Number}$$

$$h_\mathrm{m} = 0.246\frac{GC_\mathrm{Pm}}{\mathrm{Pr}^{2/3}}$$

$$C_{P-\mathrm{ET}} = 25.4\ \ (\mathrm{cal/mol\ K}) \quad C_{P-\mathrm{Ac}} = 19.41 \quad C_{P-\mathrm{H2}} = 7.00$$

$$C'_\mathrm{Pm} = \sum_i y_i C_{P-i} = 0.33 \times 25.4 + 0.33 \times 19.41 + 0.33 \times 7 = 17.10\ \ (\mathrm{cal/mol\ K})$$

$$C_\mathrm{Pm} = \frac{C'_\mathrm{Pm}}{M_\mathrm{Av}} = \frac{17.10}{33.718} = 0.507\ \ (\mathrm{cal/g\ \ K})$$

From Exercise 6.3, we have that $k_\mathrm{m} = 1.01 \times 10^{-4}$ (cal/cm s K)

$$\mathrm{Pr} = \frac{2.53 \times 10^{-4} \times 0.507}{1.01 \times 10^{-4}} = 1.279$$

$$h_\mathrm{m} = 0.246\frac{2.04 \times 10^{-2} \times 0.507}{1.279^{2/3}} = 2.166e - 3\ \left(\frac{\mathrm{cal}}{\mathrm{s\ \ K\ \ cm^2}}\right)$$

Part 2

The gradients of concentration and temperature can be evaluated by solving the two equations:

$$r_{obs} = k_{Gm} a (p_b - p_s) = k_{Gm} a P (y_b - y_s)$$

$$r_{obs}(-\Delta H) = h_m a (T_b - T_s)$$

$$a = \frac{2\pi r_p l + 2\pi r_p^2}{\pi r_p^2 l \rho_p} = \frac{2l + 2r_p}{r_p l \rho_p} = \frac{1 + 0.5}{0.25 \times 0.5 \times 1.2} = 10 \quad \left(\frac{\text{cm}^2}{\text{g}}\right)$$

$$\Delta y = \frac{r_{obs}}{k_{Gm} a P} = \frac{5.5 \times 10^{-5}}{1.49 \times 10^{-4} \times 10 \times 1} = 0.037$$

$$y_s = y_b - 0.037 = 0.33 - 0.037 = 0.293$$

$$\Delta T = \frac{r_{obs}(-\Delta H)}{h_m a} = \frac{-5.5 \times 10^{-5} \times 16,811}{0.0022 \times 10} = -42.68 \quad (\text{K})$$

$$T_s = T_b - (-42.68) = 573 - (-42.68) = 615.68\ \text{K} \quad \text{that is } 342.68\ ^\circ\text{C}$$

Exercise 6.6
Examples of Kinetic Behavior of Gas–Solid Reactions Dominated by External Mass-Transfer Limitation

Gas–solid external mass-transfer limitation normally occurs when reaction rate is very fast, in particular, when the catalyst is not porous. An example of this type of reaction is the oxidation of ammonia to nitrogen oxide, the first step in the production of nitric oxide. The main occurring reaction is:

$$NH_3 + 5/4O_2 \rightarrow NO + 3/2H_2O \quad (-\Delta H_{298K} = 54,250\ \text{cal/mol})$$

Different other side reactions can occur, in particular,

$$NH_3 + 0.75\,O_2 \rightarrow 0.5\,N_2 + 1.5\,H_2O \quad (-\Delta H_{298K} = 76,000\ \text{cal/mol})$$

The presence of a Pt-10% Rh alloy promotes the first reaction, giving place to a satisfactory selectivity. The catalyst is in the form of a stack of knitted or woven gauze. The employed screens are put one over the other for an overall height that would be sufficient to obtain the desired conversion. Ignition is initiated by directing a hydrogen torch to the center of the gauze. When the catalyzed reaction starts, the torch is extinguished, and the reaction spreads across the gauze sustained by the exothermic reaction. Therefore, after a transient period, steady-state conditions are reached also considering the high thermal conductivity of the metal. This reactor is a variation of the fixed-bed reactor because it constitutes an assembly of screens or gauze of the catalytic solid over which the reacting fluid flows. Assuming the system as a PFR, neglecting the contribution of all side reactions, and considering negligible the change in the number of moles—because ammonia is normally consistently diluted in air—, and on the basis of the data furnished in the text:

Part 1 Develop a model that can predict the overall height of the gauze to be employed for obtaining a desired conversion by feeding a mixture of ammonia and air containing a molar fraction of ammonia of 0.1 at an overall pressure of 7.807 atm.

Determine the number of screens necessary to obtain a conversion of 96% while preheating the inlet gas at 337 °C and keeping the reactor at approximately 900 °C with the aim to produce the NO necessary for obtaining 245 tons/day of HNO_3. Estimate the mass- and heat-transfer coefficients, the temperature increases in the reactor as a consequence of the reaction exothermicity and the temperature gradient between the gas and the solid surface at the end of the reactor.

Part 2 Evaluate using a MATLAB program how to change the conversion by changing the number of screens, mesh size, wire diameter, feed mass flow rate, and reactor radius.

Thermodynamic data (see Rase 1977). As has been mentioned, the reaction is extremely exothermic. Thus, it is possible to write:

$$(-\Delta H) = 54,250 - 0.4\ \ (T - 298)\ \ (\text{cal/mol of reacted NH}_3)$$
$$(-\Delta H) = 54,250 - 0.4\ \ (1173 - 298) = 54,250 - 350 = 53,900\ \ \text{cal/mol}$$

Some useful data are then:

$$C_{\text{pm}} = 6.45 + T\left(1.52 + 4.08\, y_{\text{NH}_3}\right)\ 10^{-3}(\text{calories/mol K})$$
$$y_{\text{NH}_3} = \text{molar fraction of NH}_3.$$
$$\text{Inlet } C_{\text{pm}-i} = 6.45 + 610 \times (1.52 + 4.08 \times 0.1) \times 10^{-3} = 7.62\ \ (\text{cal/mol K})xx$$
$$\text{Outlet } C_{\text{pm}-o} = 8.393\ \ (\text{cal/g K})$$

An average value calculated on the composition arising from the reactor at 1000 K (see Stull et al. 1969):

$$k_{\text{f}}\left(3.31 + 12.77 \times 10^{-3}\ \ \text{T}\right) \times 10^{-6}\ \ (\text{kcal/m s K})$$
$$k_{\text{f}}\left(3.31 + 12.77 \times 1173 \times 10^{-3}\right) \times 10^{-6}$$
$$= 1.828 \times 10^{-5}\ \ (\text{kcal/m s K})\ \ \text{or}\ \ 1.828 \times 10^{-4}\ \ (\text{cal/cm s K})$$
$$\mu_{\text{f}} = \left(12.5 + 29.20 \times 10^{-3}\ \ \text{T}\right) \times 10^{-5}(\text{g/cm s})$$
$$\mu_{\text{f}} = \left(12.5 + 29.20 \times 1173 \times 10^{-3}\right) \times 10^{-5} = 4.675 \times 10^{-4}(\text{g/cm s})$$

M_{f} = Average molecular weight of feed

$$M_{\text{f}} = 32 \times 0.21 \times \left(1 - y^{o}_{\text{NH}_3}\right) + 28.01 \times 0.79 \times \left(1 - y^{o}_{\text{NH}_3}\right) + 17.03 y^{o}_{\text{NH}_3}$$
$$= 28.85 + 11.82 y^{o}_{\text{NH}_3} = 27.66$$
$$\rho_{\text{f}} = \frac{M_{\text{f}} P}{\text{RT}} = \frac{27.66 \times 7.803}{82.058 \times 1173} = 2.434 \times 10^{-3}$$

Catalyst Structure

The catalyst is characterized by 80-mesh gauze with wires having a diameter of $d_{\text{w}} = 7.62 \times 10^{-3}$ cm Diameter of the gauze pad = 89.38 cm.

Cross-sectional area = 6272 cm^2.
Thickness of each screen = $2d_w$ = 1.524 × 10^{-2} cm.
Surface area per unit volume a = 101.75 cm^{-1}.
ε = void fraction = 0.806.

Operative Conditions

From the desired productivity of HNO_3, we can evaluate the amount of NH_3 that must be fed:

$$(245 \times 1000 \times 1000)/(24 \times 3600 \times 63(\text{Mol weight of } HNO_3))/0.96(NH_3 \ \text{conversion}) \\ = 46.886 \ \text{mol/s} = 797\,\text{g/s}$$

Inlet ammonia molar fraction $y_{NH_3} = 0.100$; the air molar fraction is consequently 0.9 in which 21% is oxygen; and 79% is nitrogen + argon (this last is approximately 1% and here is considered as nitrogen). We can now calculate the amount of oxygen and nitrogen that must be fed together with the ammonia:

Oxygen flow rate	2836 g/s	88.614 mol/s	$y_{O_2} = 0.189$
Nitrogen flow rate	9334 g/s	333.357 mol/s	$y_{N_2} = 0.711$
Total flow rate	12967 g/s	468.86 mol/s	

The mixture of gas is preheated at 610 K.
The pressure is P = 7.803 atm.
The reactor can be considered adiabatic.
G = mass gas flow rate = overall mass flow rate/cross-sectional area = 2.0673 g/s cm^2.
Gas flow velocity = G/ρ_f = 920.9 cm/s.

Solution

Part 1

Evaluate the temperature of the gas at the exit of the reactor by solving the thermal balance.

$$\text{Heat flux at the inlet} + \text{Generated heat by the reaction} = \text{Heat flux at the outlet (cal/s)}$$

$$F_{ov-i} C_{pm} T_i + F_{NH3} \lambda (-\Delta H) = F_{ov-out} C_{pm} T_{out}$$

where $F_{ov\text{-}i}$ and $F_{ov\text{-}out}$ are the molar flow rates in (mols/s) at the inlet and outlet, respectively; λ is the conversion; and T_i and T_{out} are the temperatures, respectively, at the inlet and outlet of the reactor.

$$T_{out} = \frac{F_{ov-i} C_{pm} T_i + F_{NH_3} \lambda (-\Delta H)}{F_{ov-out} C_{pm}} \\ = \frac{468.86 \times 7.626 \times 610 + 46.89 \times 0.96 \times 53,900}{480.11 \times 8.188} = 1142 \ \text{K}$$

which is in agreement with the value assumed as reaction temperature.

This reaction is completely dominated by the external diffusion; therefore, we can write:

$$(-r_{NH_3}) = k_g a(p_{NH_3} - p_{s-NH_3}) = k_g a\ p_{NH_3} = k_g aP\ y_{NH_3}$$

P_{s-NH_3} can be neglected because will be very low if compared with p_{NH_3}.

a = surface area/volume of the gauze
P = total pressure
y_{NH_3} = ammonia molar fraction
r_{NH_3} = reaction rate (converted mols of NH_3/cm^3 s)
k_g = mass-transfer coefficient (mols/cm^2 s atm)

Considering a plug flow regime, neglecting axial and radial diffusion, and assuming approximately constant the number of moles, for the low concentration of ammonia, we can write:

$$\frac{G}{M_f}\frac{dy_{NH_3}}{dz} = (-r_{NH_3}) = k_g a\ P\ y_{NH_3}$$

G = mass flow for surface unit (g/s cm^2)
M_f = average molecular weight of feed

$$M_f = 32 \times 0.21 \times \left(1 - y^o_{NH_3}\right) + 28.01 \times 0.79 \times \left(1 - y^o_{NH_3}\right) + 17.03 y^o_{NH_3}$$
$$= 28.85 + 11.82 y^o_{NH_3} = 27.66$$

z = length of the catalytic bed (cm)
Separating the variables and integrating:

$$\int_{y^o_{NH_3}}^{y_{NH_3}} \frac{dy_{NH_3}}{y_{NH_3}} = -\int_0^z \frac{M_f k_g aP}{G} dz = -\frac{M_f k_g aP}{G} z$$

$$\ln\left(\frac{y_{NH_3}}{y^o_{NH_3}}\right) = -\frac{M_f k_g aP}{G} z$$

But $y_{NH_3} = y^o_{NH_3}(1 - \lambda_{NH_3})$, and substituting y_{NH_3} in the previous equation:

$$\ln(1 - \lambda_{NH_3}) = -\frac{M_f k_g aP}{G} z$$

We imposed a value of the ammonia conversion of 0.96; therefore:

$$3.2188 = \frac{(28.85 + 11.82 \times 0.1) \times 101.75 \times 7.803\ k_g}{2.0673} z = 11533.916\ z\ k_g$$

To solve the problem, we must evaluate before k_g. For this purpose, Satterfield and Cortez (1970) suggested the use of a model based on the mass transfer to the surface of an infinite cylinder and assumed, as a characteristic dimension of this system, the wire diameter d_w, writing the Reynolds number as:

$$\mathrm{Re} = \frac{d_w G}{\mu_f \varepsilon} = \frac{7.62 \times 10^{-3} \times 2.0673}{0.806 \times 4.675 \times 10^{-4}} = 41.7922$$

That is, an interstitial velocity has been used instead of a superficial velocity to describe the Reynolds number. The infinite wire model is more suitable to describe mass transfer with respect to the packed-bed model, and Satterfield suggested the employment of the correlation:

$$J_D = 0.865\ \ \mathrm{Re}^{-0.648} = 0.07699$$

But we also can write:

$$J_D = \frac{k_g \varepsilon \mathrm{PM_f}}{G} \mathrm{Sc}^{2/3} = \frac{0.806 \times 27.66 \times 7.803 \times k_g}{2.0673} \mathrm{Sc}^{2/3} = 91.36\ k_g\ \ \mathrm{Sc}^{2/3}$$

To evaluate the Schmidt number, it is necessary to know the value of the molecular diffusion coefficient of ammonia in air, that is:

$$D_{\mathrm{NH_3-air}} = 0.198\left(\frac{T}{273}\right)^{3/2}\left(\frac{1}{P}\right)\left(\frac{\mathrm{cm}^2}{\mathrm{s}}\right) = 0.198\left(\frac{1173}{273}\right)^{3/2}\left(\frac{1}{7.803}\right)\left(\frac{\mathrm{cm}^2}{\mathrm{s}}\right)$$
$$= 0.226\left(\frac{\mathrm{cm}^2}{\mathrm{s}}\right)$$

$$\text{Hence : } \mathrm{Sc} = \frac{\mu_f}{\rho_f D_{\mathrm{NH_3-air}}} = \frac{4.675 \times 10^{-4}}{0.226 \times 2.242 \times 10^{-3}} = 0.922$$

$$k_g = \frac{0.07699}{91.36 \times 0.922^{2/3}} = 9.6512 \times 10^{-4}\ \left(\mathrm{mols/cm^2\ \ s\ \ atm}\right)$$

It derives that:

$$z = 0.316\ \ \mathrm{cm}$$

$$\text{Number of screens } n_S = 0.316/2d_w = 0.316/(2 \times 0.00762) = 20.7$$

For this system, Satterfield and Cortez (1970) also showed that $J_D \simeq J_H$; therefore:

$$J_{\rm D} \simeq J_{\rm H} \simeq 0.07699$$

$$\text{But}: \quad J_{\rm H} = \frac{ha}{GC_{\rm pm}} Pr^{2/3}$$

Pr is the Prandtl number: $\Pr = \frac{\mu_{\rm f} C_{\rm pm}}{k_{\rm f} M_{\rm f}} = \frac{8.188 \times 4.675 \times 10^{-4}}{27.66 \times 1.828 \times 10^{-4}} = 0.757$

$$h = \frac{J_{\rm H} GC_{\rm pm}}{M_{\rm f} Pr^{2/3}} = \frac{0.07699 \times 2.0673 \times 8.188}{27.66 \times 0.757^{2/3}} = \frac{1.303}{22.285}$$
$$= 5.67 \times 10^{-2} \ \left(\text{cal/s K cm}^2\right)$$

Part 2

1. On the basis of the available data, a MATLAB program was elaborated for evaluating the effect on the reaction of the number of screens, mesh size, wire diameter, feed composition, feed mass flow rate, and reactor radius. For all of these evaluations, only one equation must be solved giving the ammonia conversion as a function of the mentioned variables, that is:

$$\lambda = 1 - e^F \quad \text{with} \quad F = \frac{\left(5.817 \times 10^{-5}\right) n_{\rm S} f_{\rm w} T^{0.333} \left(28.85 + 11.82 y^o_{\rm NH_3}\right)^{0.667}}{\varepsilon^{0.352} d_{\rm w}^{0.648} G^{0.648} \mu_{\rm F}^{0.019}} \tag{6.68}$$

where $f_{\rm w} = 2ad_{\rm w}$ = wire area per gauze sectional area related to one gauze. Moreover,

$$l_{\rm w} = \sqrt{\left(\frac{1}{n_{\rm w}}\right)^2 + d_{\rm w}^2} \quad \text{where} \quad n_{\rm w} = \text{mesh number} \tag{6.69}$$

$$a = \pi l_{\rm w} n_{\rm w}^2 \quad \text{and} \quad \varepsilon = 1 - ad_{\rm w}/4 \tag{6.70}$$

Plots are constructed giving the conversion as a function of each mentioned variable by solving a MATLAB program.

The obtained results are reported in Fig. 6.7:

All of the described results were obtained using a MATLAB program available as Electronic Supplementary Material, Data 6.1.

6.2.2 *Mass and Heat Transfer Inside the Catalytic Particles*

As mentioned previously, the diffusion of reagents and products inside the pores of a solid catalyst occurs together with the reaction, that is, the two processes are

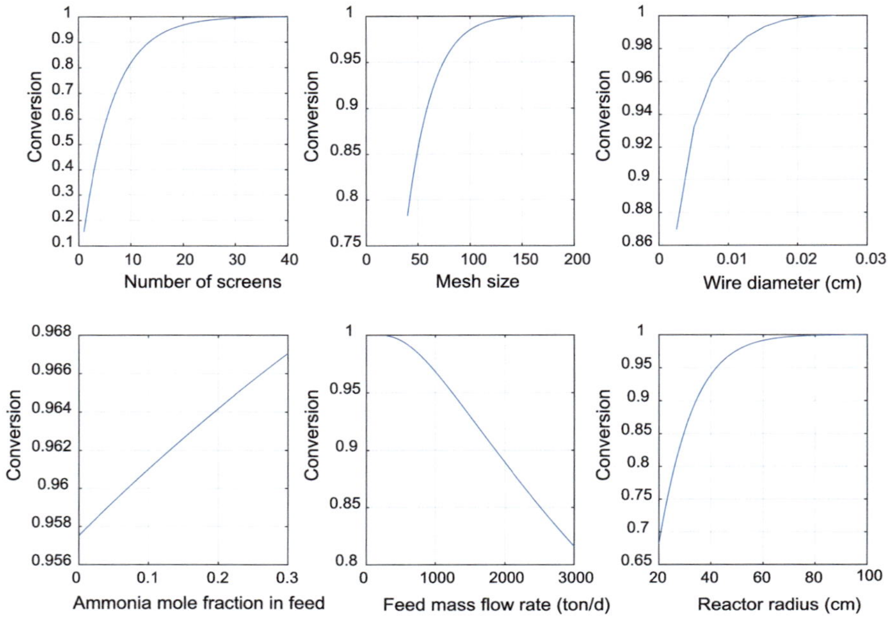

Fig. 6.7 Results of calculations for determining the evolution of the conversion with the number of screens, mesh size, wire diameter, ammonia molar fraction in feed, feed mass flow rate, and reactor radius

simultaneous and not consecutive as in the case of external diffusion. For this reason, the influence of the internal diffusion in limiting the reaction rate must be described introducing this limitation into the reaction-rate relationship. For a reaction of n order, for example, we can write:

$$r = \eta k_n C_{\mathrm{As}}^n \tag{6.71}$$

where η is the effectiveness factor corresponding to:

$$\eta = \frac{\text{observed reaction rate}}{\text{intrinsic chemical reaction rate}} \tag{6.72}$$

where η is a factor describing the effect of internal diffusion on the reaction rate. It is defined as the ratio between the observed reaction rate, which is affected by the internal diffusion, and the rate that would occur on the basis of the kinetic law and not limited by internal diffusion. C_{As} in Eq. (6.71) is the concentration of the reagent A on the catalyst surface. That is, to know the effect of the internal diffusion, we must also consider the effect of the external mass and heat transfer to evaluate both the concentration of all of the involved components and of temperature on the external surface of the catalytic particle. The rigorous determination of the

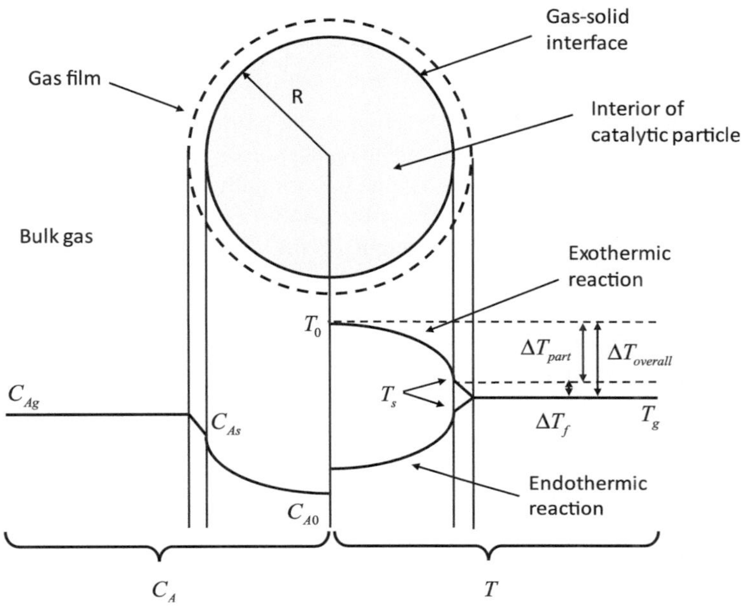

Fig. 6.8 Qualitative description of the gradients occurring in a catalyst particle as a consequence of a chemical reaction

effectiveness factor requires the definition of the profiles, inside the catalytic particle, of both concentration and temperature. An example of these profiles is qualitatively shown in Fig. 6.8.

The motion of the molecules inside the pores of the catalytic pellets occurs by the same mechanism seen for the external diffusion in the boundary layer, that is, by a slow process of molecular diffusion, and is regulated by D_{eff}, the effective diffusional coefficient. This term can be determined from two other terms, D_{be} and D_{ke}, acting as in series resistance to the motion of molecules; therefore, it is possible to write the relation:

$$\frac{1}{D_{\text{eff}}} = \frac{1}{D_{\text{be}}} + \frac{1}{D_{\text{ke}}} \tag{6.73}$$

where D_{be} is the bulk diffusion coefficient, that is, the diffusion coefficient of the fluid flowing in the macropores, and D_{ke} is the Knudsen diffusion coefficient corresponding to the diffusion in the micropores. The diffusion in micropores has a particular mechanism because micropores have diameters that are shorter than the free mean path of the molecules. However, to evaluate D_{eff}, we also can write:

$$D_{\text{be}} = \frac{D_{12}\theta}{\tau} \tag{6.74}$$

$$D_{\mathrm{ke}} = 1.94 \times 10^4 \frac{\theta^2}{\tau S_{\mathrm{g}} \rho_{\mathrm{p}}} \sqrt{\frac{T}{M}} \tag{6.75}$$

where θ is the solid porosity (void fraction in the porous mass), τ is the tortuosity factor (this is an empirical factor dependent on the porosity of the catalyst; its value is characteristic of the employed material and has normally a value falling in the range 0.3–10), S_{g} is the specific surface area of the catalyst (surface/mass), ρ_{p} the catalyst particle density, M the molecular weight of the diffusing molecule, and T the temperature in K. Considering the uncertainty of some parameters in the theoretical evaluation of D_{eff}, such as, for example, the tortuosity factor, D_{eff}, is often experimentally evaluated with an apparatus such as the one reported in Fig. 6.9.

In this apparatus, two different gases, kept at the same pressure, flow one over and the other under the catalytic particle, which is well-sealed inside the cell. Diffusion of one gas into the other occurs, thus passing through the catalytic particles. Under steady-state conditions, the concentration of one gas in the other will remain constant, and this concentration can be continuously determined with a thermal conductivity (TCD) sensor. Experimental data can be interpreted by applying the integrated Fick's law:

$$N = D_{\mathrm{eff}} \frac{\Delta C}{\text{Pellet lenght}} \tag{6.76}$$

Because the overall flows in the two branches are known and the flow of gas into the other also is known, it is possible to estimate the effective diffusion coefficient, D_{eff}.

Exercise 6.7
Estimation of the Molecular Diffusion Coefficient, D_{eff}, of a Gas in a Porous Catalyst

Imagine that the hydrogenation of benzene to cyclohexane, described in Exercises 6.1 and 6.2, occurs on a catalyst of ruthenium supported on alumina. The reaction conditions are the same as the previously mentioned exercises, that is, 250 °C and 5 atm. The catalyst, in cylindrical pellets, has a BET surface area of 200 m^2/g, a density of 1.5 g/cm^3, and a void volume fraction of 0.38. Evaluate the effective

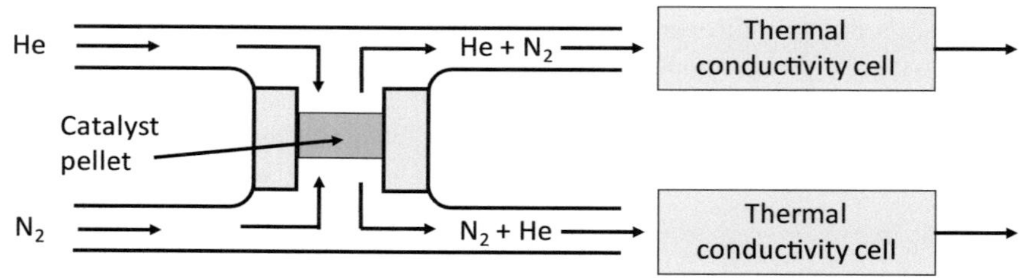

Fig. 6.9 Intrapellet diffusivity cell according to a device suggested by Carberry (1987)

molecular diffusion coefficient, D_{eff}, of benzene in hydrogen remembering that D_{12}, calculated in Exercise 6.1, is $= 0.22\ \text{cm}^2/\text{s}$. Assume a value of the tortuosity factor as $\tau = 6$.

Solution

$$\frac{1}{D_{\text{eff}}} = \frac{1}{D_{\text{b}}} + \frac{1}{D_{\text{K}}}$$

The bulk diffusion can be calculated as:

$$D_{\text{b}} = \frac{D_{12}\theta}{\tau} = \frac{0.22 \times 0.38}{6} = 1.39 \times 10^{-2} \left(\frac{\text{cm}^2}{\text{s}}\right)$$

$$D_{\text{K}} = 19,400 \frac{\theta^2}{\tau S_{\text{g}} \rho_{\text{p}}} \sqrt{\frac{T}{M}} = 19,400 \frac{0.38^2}{6 \times 2 \times 10^6 \times 1.5} \sqrt{\frac{523}{78}} = 4.029 \times 10^{-4} \left(\frac{\text{cm}^2}{\text{s}}\right)$$

$$\frac{1}{D_{\text{eff}}} = \frac{1}{1.39 \times 10^{-2}} + \frac{1}{4.029 \times 10^{-4}} = 71.7 + 2482 = 2553$$

$$D_{\text{eff}} = \frac{1}{2553} = 3.917 \times 10^{-4} \left(\frac{\text{cm}^2}{\text{s}}\right)$$

6.2.3 Mass and Heat Balance in a Catalytic Particle: Calculation of the Effectiveness Factor with the Traditional Approach

By considering a spherical particle, the mass balance related to a single component can be written by examining what happens in a spherical shell of a thickness dr and radius r (see Fig. 6.10):

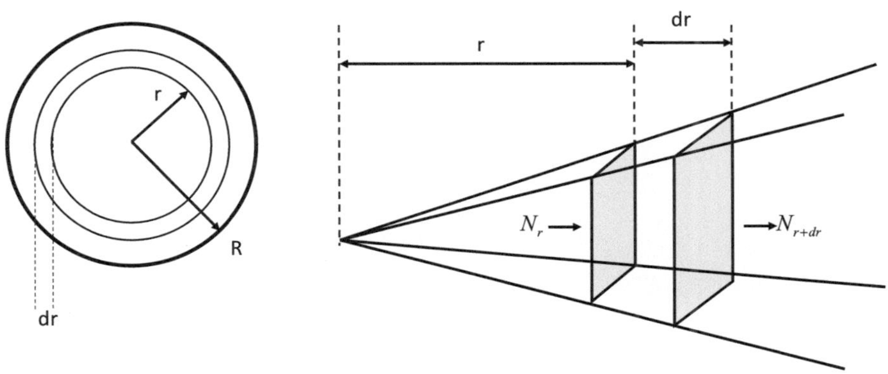

Fig. 6.10 Diffusion with reaction in a spherical particle

For this purpose, we can write:

$$\begin{bmatrix}\text{diffusion rate}\\ \text{inward at } r=r+\mathrm{d}r\end{bmatrix}-\begin{bmatrix}\text{diffusion rate}\\ \text{outward at } r=r\end{bmatrix}-\begin{bmatrix}\text{reaction rate}\\ \text{into the shell}\end{bmatrix}=[\text{accumulation}] \tag{6.77}$$

by assuming a steady-state condition the accumulation term is null, therefore:

$$4\pi(r+\Delta r)^2 N_{r+\Delta r}-4\pi r^2 N_r-4\pi r^2\Delta r v=0 \tag{6.78}$$

where ν is the reaction rate referred to the catalyst volume.

Dividing all members by $4\pi\Delta r$ and assuming a limit for $\Delta r\to 0$, it results in:

$$\lim_{\Delta r\to 0}\left(\frac{r^2 N_{r+\Delta r}-r^2 N_r}{\Delta r}\right)=r^2 v \tag{6.79}$$

Therefore:

$$\frac{\mathrm{d}(r^2 N)}{\mathrm{d}r}=r^2 v \tag{6.80}$$

By introducing Fick's law, $N=D_{\mathrm{eff}}\frac{\mathrm{d}C}{\mathrm{d}r}$, we obtain:

$$-\frac{\mathrm{d}}{\mathrm{d}r}\left(r^2\frac{\mathrm{d}C}{\mathrm{d}r}\right)+\frac{r^2 v}{D_{\mathrm{eff}}}=0 \tag{6.81}$$

By differentiating then the first term and dividing all terms by r^2, we obtain the equivalent expression:

$$\frac{2}{r}\frac{\mathrm{d}C}{\mathrm{d}r}+\frac{\mathrm{d}^2C}{\mathrm{d}r^2}=\frac{v}{D_{\mathrm{eff}}} \tag{6.82}$$

Assuming a rate law of n order, $v=S_{\mathrm{v}}k_{\mathrm{S}}C^n$, where S_{v} is the surface area for the volume unit of the solid and k_{S} is the kinetic constant related to the solid surface. Now we can write:

$$\frac{2}{r}\frac{\mathrm{d}C}{\mathrm{d}r}+\frac{\mathrm{d}^2C}{\mathrm{d}r^2}=\frac{S_{\mathrm{v}}k_{\mathrm{S}}C^n}{D_{\mathrm{eff}}} \tag{6.83}$$

with the following boundary conditions:

for $r = R_p$ $C = C_S$

for $r = 0$ $\frac{dC}{dr} = 0$ Postulating a symmetric profile.

Following a quite similar approach for the heat balance and introducing Fourier's law

$$q = -k_{eff}\frac{dT}{dr}$$

instead of the Fick's Law, the following equation can be derived for a single reaction:

$$-\frac{d}{dr}\left(r^2\frac{dT}{dr}\right) + \frac{r^2(-\Delta H)S_v k_S C^n}{k_{eff}} = 0 \tag{6.84}$$

and the equivalent:

$$\frac{2}{r}\frac{dT}{dr} + \frac{d^2T}{dr^2} = \frac{S_v k_S C^n}{k_{eff}} \tag{6.85}$$

With the boundary conditions:

for $r = R_p$ $T = T_S$

$r = 0$ $\frac{dT}{dr} = 0$ Considering a symmetric profile.

By eliminating the common terms of Eqs. (6.83) and (6.85), we obtain:

$$(T - T_S) = \frac{D_{eff}}{k_{eff}}(C - C_S)(-\Delta H) \tag{6.86}$$

that is, for any profile of C inside the particle, we have a corresponding profile of T that can be easily determined by solving the algebraic Eq. (6.86). The maximum temperature gradient, ΔT_{max}, is obtained when the concentration at the center of the particle is quite low, assuming in this case $\Delta C \approx C_S$; we obtain:

$$\Delta T_{max} = \frac{D_{eff}}{k_{eff}}C_S(-\Delta H) \tag{6.87}$$

ΔT_{max} can be also related to T_S, the temperature on the catalyst surface, thus obtaining the dimensionless Prater's number: $\beta = \Delta T_{max}/T_S$.

Catalysts are normally insulating materials, but their thermal conductivities are much higher than those of the gaseous reaction mixtures. Therefore, under steady-state conditions, internal temperature gradients are rarely important in practice.

However, the determination of the internal profiles of concentration requires the solution of Eq. (6.83) for each component of the reaction, while the internal profile of the temperature can be determined by solving Eq. (6.85) or (6.86).

Sometimes, it could be convenient to introduce dimensionless terms, such as:

$$\varepsilon = \frac{r}{R_P} \quad \gamma = \frac{C}{C_s} \quad \phi = R_P\sqrt{\frac{k_s S_v C_s^{n-1}}{D_{\text{eff}}}} \tag{6.88}$$

where ε (adimensional radius of the particle) is the ratio between the internal radius of a spherical particle and the radius of the particle, R_P; γ is the ratio of the concentration inside the particle in the point r and the concentration at the external surface C_s; ϕ is the Thiele modulus; and the square of this modulus is a measure of the ratio of a surface reaction rate to a rate of diffusion through the catalyst pellet. If the Thiele modulus is large, the reaction is limited by the internal mass transfer; on the contrary, if the Thiele modulus is low, a chemical regime is operative. When the catalyst particle is not spherical, an equivalent spherical radius is evaluated by considering a sphere having the same external surface as the particle with a different shape.

By introducing the dimensionless terms in Eq. (6.83), we obtain:

$$\frac{2}{\varepsilon}\frac{d\gamma}{d\varepsilon} + \frac{d^2\gamma}{d\varepsilon^2} - \gamma^n \phi_n^2 = 0 \tag{6.89}$$

with the following boundary conditions:

$$\text{for} \quad \varepsilon = 1 \quad \gamma = 1$$
$$\varepsilon = 0 \quad \frac{d\gamma}{d\varepsilon} = 0$$

It is interesting to observe that for a first-order reaction, that is, for $n = 1$, the Thiele modulus becomes independent of the concentration, and for this particular case, Eq. (6.83) or (6.89) can be solved analytically, and Eq. (6.89) becomes:

$$\frac{2}{\varepsilon}\frac{d\gamma}{d\varepsilon} + \frac{d^2\gamma}{d\varepsilon^2} - \gamma \phi_1^2 = 0 \tag{6.90}$$

with the same boundary conditions. The analytical solution of the previous differential equation is:

$$\gamma = \frac{C}{C_s} = \frac{1}{\varepsilon}\left(\frac{\sinh(\varepsilon\phi_1)}{\sinh(\phi_1)}\right) \tag{6.91}$$

When the reaction order is different from 1, currently it is more convenient to adopt a numerical solution approach as will be described in more detail later in the text. In any case, when the internal concentration profile is known, it is possible to evaluate η, the "effectiveness factor," with the following relation:

$$\eta = \frac{\nu_{\mathrm{obs}}}{\nu_{\max}} = \frac{\int_0^{R_{\mathrm{P}}} 4\pi r^2 \nu(C)\mathrm{d}r}{\frac{4}{3}\pi R_{\mathrm{P}} \nu(C_{\mathrm{s}})} = 3\int_0^1 \varepsilon^2 \psi(\gamma)\mathrm{d}\varepsilon \tag{6.92}$$

where $\Psi(\gamma) = \nu(C)/\nu(C_{\mathrm{s}})$.

As previously mentioned, η is a factor describing directly the effect of internal diffusion on the reaction rate, and for a reaction rate of a single reaction of nth order, we can simply write:

$$\nu = \eta k_{\mathrm{s}} S_{\mathrm{v}} C_{\mathrm{s}}^n = \eta k_{\mathrm{v}} C_{\mathrm{s}}^n \tag{6.93}$$

Then, by considering that, under steady-state conditions, the overall reaction rate in a particle equals the rate of external mass transfer, Eq. (6.92) can be rewritten as:

$$\eta = \frac{4\pi R_{\mathrm{P}}^2 D_{\mathrm{eff}} \left(\frac{dC}{dr}\right)_{R_P}}{\frac{4}{3}\pi R_P^3 \nu(C_s)} = \frac{-3D_{\mathrm{eff}} \left(\frac{dC}{dr}\right)_{R_P}}{R_P \nu(C_s)} \tag{6.94}$$

As previously mentioned for a first-order reaction, the concentration profile of the reactants can be analytically determined using Eq. (6.91). By differentiating those relations, evaluating the derivative at $r = R_{\mathrm{P}}$, and substituting this into Eq. (6.94), we obtain:

$$\eta = \frac{3}{\phi}\left[\frac{1}{\tanh\phi} - \frac{1}{\phi}\right] \tag{6.95}$$

The concentration profile and the effectiveness factor depend on the geometry of the catalyst particles. Aris (1957) defined a single Thiele modulus for any geometry:

$$\phi_L = L\sqrt{\frac{k_{\mathrm{s}} S_{\mathrm{v}} C_{\mathrm{s}}^{n-1}}{D_{\mathrm{eff}}}} \tag{6.96}$$

where L is a length parameter characteristic for each shape and normally is defined by the ratio:

$$L = \frac{V_{\mathrm{P}}}{S_{\mathrm{P}}} \tag{6.97}$$

where V_{P} is the volume of the particle and S_{P} is the external surface of the particle.

In the case of slab shape, for example, L is the thickness of the slab/2 if both the sides of the slab are exposed to the reactants. The effectiveness factor for a first-order reaction can be calculated, in this case, with the relation:

$$\eta = \frac{\tanh(\phi_L)}{\phi} \tag{6.98}$$

where L is equal to $R_P/3$ for a sphere and to $R_P/2$ for a cylinder.

Evaluation of the effect of the internal diffusion is complicated by the fact that the kinetic constant k_v is often unknown, while conversely experimental data of the reaction rate are available. In this case, it can be useful to define another dimensionless modulus, called the "Weisz modulus" (see Weisz and Prater 1954; Weisz and Hicks 1962), that is:

$$M_W = \frac{\nu L^2}{C_s D_{eff}} = \eta \phi^2 \tag{6.99}$$

This relation can be obtained, for a first-order reaction and a spherical geometry, simply by eliminating s_v from the two relations:

$$\nu = \eta k_v C_s \quad \text{and} \quad \phi = L\sqrt{\frac{k_v}{D_{eff}}} \tag{6.100}$$

M_W can be determined from "observable variables" and can be used as a criterion for estimating the role of the internal diffusion in affecting a reaction rate when experimental reaction-rate data are known.

Figures 6.11 and 6.12 report the plots of η versus ϕ and M_w, respectively, for spherical particles of the catalyst calculated for a first-order reaction. Observing the plots reported in these figures, two large asymptotic zones can be recognized: The first is characterized by the chemical regime ($\phi < 0.4$ or $M_w < 0.15$), and for the second one, in which internal diffusion limitations are predominant, ($\phi > 4$ and $M_w > 7$). In the last case, the effectiveness factor can be calculated in a simplified way as $\eta = 1/\phi = 1/M_w$. In the intermediate zone, a more rigorous calculation of η is necessary; however, if the asymptotic approximation is adopted, the error introduced in the calculation of η is <5%.

These two diagrams allow us to individuate the operating regime, that is, chemical or diffusional, and to estimate, with the asymptotic approach, an approximated value of the effectiveness factor. It is interesting to observe that the effect of the shape of the catalyst particles on the effectiveness factor is moderate as can be appreciated in Fig. 6.13. The effect of the reaction order is small, too, as it can be observed in Fig. 6.14.

The influence of internal diffusion on reaction rates also can be determined experimentally by performing two different types of experience, that is: (1) The reaction rates are determined in the presence of catalyst particles of different sizes,

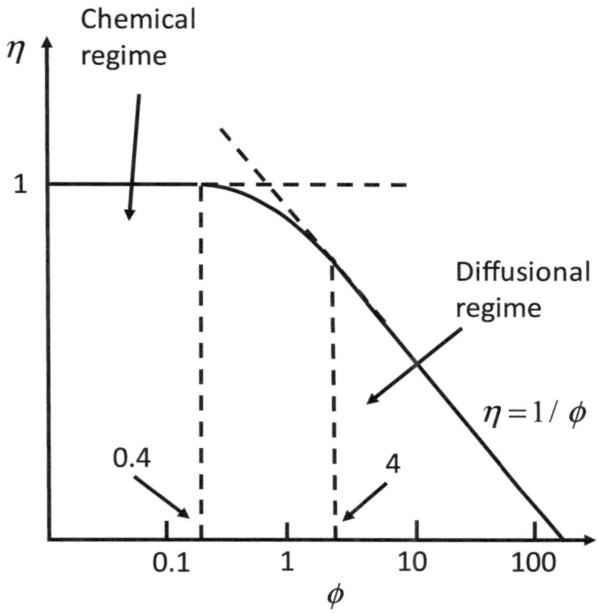

Fig. 6.11 Plot of the effectiveness factor as a function of the Thiele modulus

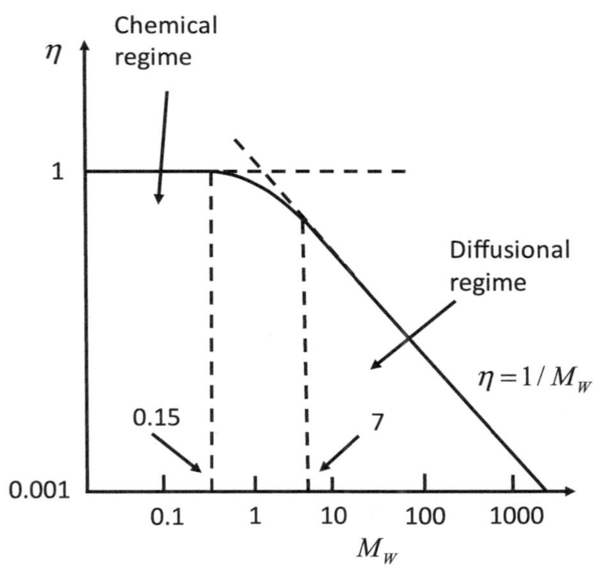

Fig. 6.12 Plot of the effectiveness factor as a function of the Weisz modulus

and (2) the reaction rates for a given size at different temperature are evaluated. In the first case, because the effectiveness factor is sensitive to particle sizes, the reaction rate strongly decreases by increasing the catalyst particle size as qualitatively shown

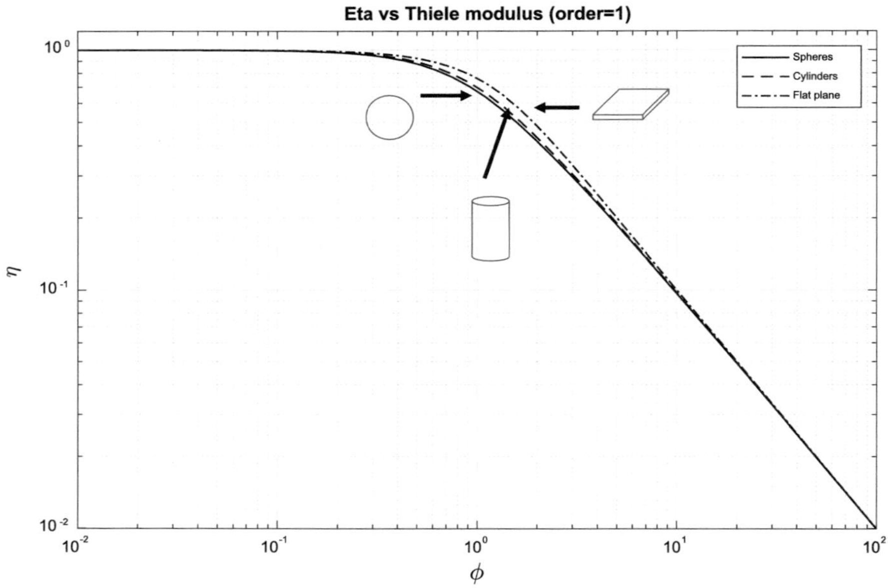

Fig. 6.13 Effect of the shape of the catalyst particles on the effectiveness factor in relation to the Thiele modulus

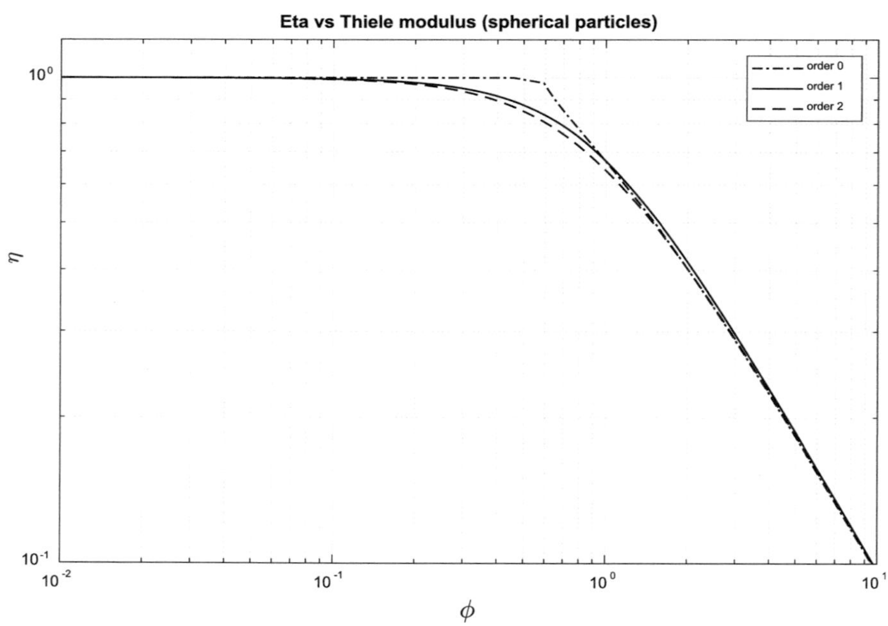

Fig. 6.14 Effect of the reaction order on the effectiveness factor in relation to the Thiele modulus

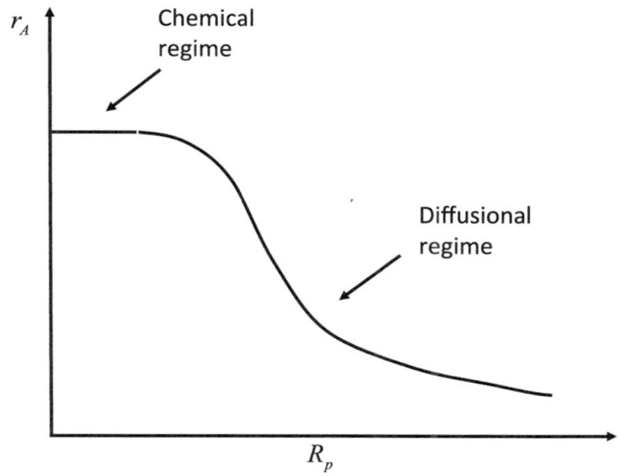

Fig. 6.15 Evolution of reaction rate with particle radius

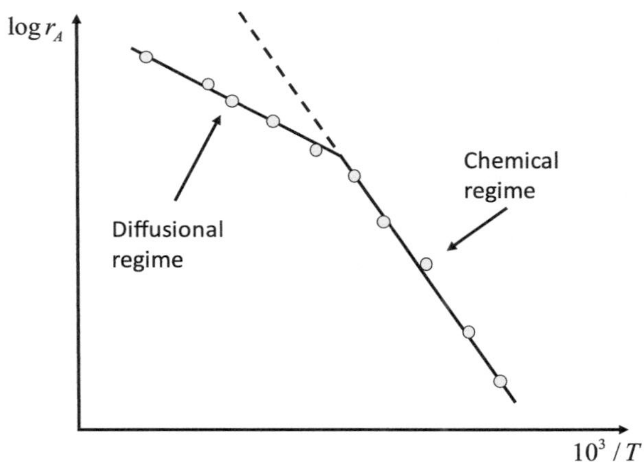

Fig. 6.16 Effect of internal diffusion on the apparent activation energy

in Fig. 6.15. In this case, the effectiveness factor η can be determined directly as the ratio of the reaction rates in a diffusional regime and a chemical regime, respectively.

In the second experimental approach, a catalyst is subject to kinetic runs at different temperatures, and the logarithm of the reaction rate is plotted as a function of the reciprocal of the temperature, as in an Arrhenius plot. When the internal diffusion is operative, a trend, such as the one qualitatively shown in Fig. 6.16, is obtained.

The apparent activation energy corresponding to the diffusional regime will be approximately half that in the chemical regime, that is, the mean between the activation energy of the chemical reaction, observed at a lower temperature, and that of the diffusional process that is not activated is quite low.

Let us consider now what happens in a non-isothermal particle. As already mentioned, catalysts are normally insulating materials; nevertheless, their thermal

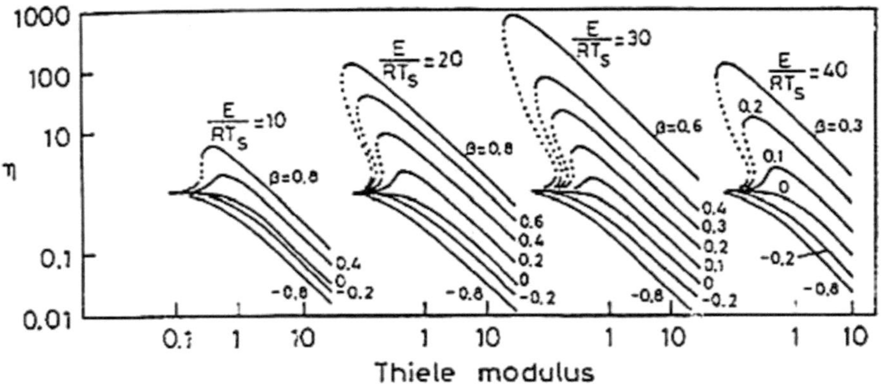

Fig. 6.17 Plots of the effectiveness factors against the Thiele modulus in the case of both exothermic and endothermic reactions with: an internal temperature gradient. Reproduced with permission from Santacesaria (1997), Copyright Elsevier (1997)

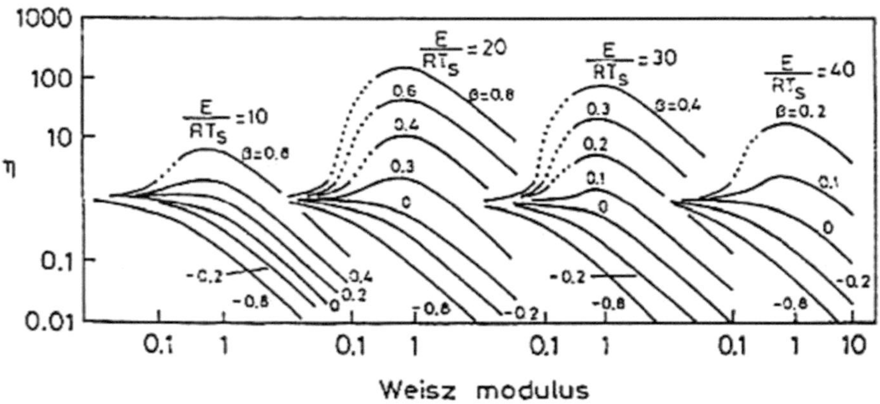

Fig. 6.18 Plots of the effectiveness factors against the Weisz modulus in the case of both exothermic and endothermic reactions with: an internal temperature gradient. Reproduced with the permission from Santacesaria (1997), Copyright Elsevier (1997)

conductivities are much higher than those of the reacting gases. For this reason, under steady-state conditions, the internal temperature gradients are seldom important in practice. However, when this gradient is present, the effectiveness factor can assume values >1 in the case of exothermic reactions as shown in Figs. 6.17 and 6.18, where plots for $\eta - \phi$ and $\eta - M_w$ are reported for different values of both the dimensionless parameters (E/RT_s) and the Prater number $\beta = \Delta T_{max}/T_s$.

The internal temperature gradients can be important in transient conditions and can determine thermal shock to the catalyst particles, followed by breaking and sintering. The temperature inside the particle can be experimentally measured inserting a thin thermocouple inside the catalytic particles.

6.2.4 *Effect of Diffusion on Selectivity*

Diffusion limitations influence the selectivity of solid catalysts in different ways according to the type of complex reactions involved. Let us consider three very simple examples, such as:

$$\begin{array}{ccc} A \xrightarrow{k_1} B \ \text{(desired)} & A \xrightarrow{k_1} B \ \text{(desired)} & A \xrightarrow{k_1} B \xrightarrow{k_2} C \\ R \xrightarrow{k_2} S & A \xrightarrow{k_2} C & \quad \text{(desired)} \\ \text{Independent reactions} & \text{Parallel reactions} & \text{Consecutive reactions} \\ (1) & (2) & (3) \end{array} \tag{6.101}$$

In all cases, for simplicity, we will consider only the first-order reaction.

Case no. 1
In the first case, we have, for each reaction, the overall reaction rate:

$$r = \frac{C_\text{b}}{\frac{1}{k_\text{m} a_\text{m}} + \frac{1}{\eta k}} \tag{6.102}$$

in which the contributions of external and internal diffusion are considered. The selectivity of the pellet can be expressed as the ratio between r_1 and r_2:

$$S = \frac{r_1}{r_2} = \frac{\left[1/(k_\text{m})_R a_\text{m} + 1/\eta_2 k_2\right] C_A}{\left[1/(k_\text{m})_A a_\text{m} + 1/\eta_1 k_1\right] C_R} \tag{6.103}$$

In the case of a chemical regime, selectivity becomes:

$$S = \frac{k_1 C_A}{k_2 C_R} \tag{6.104}$$

By comparing Eqs. (6.103) and (6.104), it can be seen that external and internal diffusion both decrease the selectivity to the desired product B. When internal diffusion gives an important contribution, it is possible to approximate $\eta \cong 1/\phi$; therefore,

$$r_1 = \frac{1}{\phi_1} k_1 C_A = \frac{3}{r} \sqrt{\frac{k_1 (D_A)_\text{e} C_A}{\rho_\text{p}}} \tag{6.105}$$

$$r_2 = \frac{1}{\phi_2} k_2 C_R = \frac{3}{r} \sqrt{\frac{k_2 (D_R)_\text{e} C_R}{\rho_\text{p}}} \tag{6.106}$$

The selectivity becomes:

$$S = \frac{r_1}{r_2} = \left(\frac{k_1}{k_2}\right)^{1/2} \frac{C_A}{C_R} \tag{6.107}$$

considering that $(D_A)_e \cong (D_R)_e$. By comparing Eqs. (6.104) and (6.107), it is noted that a strong internal diffusion reduces the selectivity by the square root of the kinetic constant ratio.

Case no. 2
For parallel reactions, diffusion limitations have no effect on the selectivity, with the exception of reactions that have different reaction orders. Because the reactant is the same, it is obvious that reduction in the concentration of A due to diffusion limitation has a greater effect for a reaction of the second order with respect to a reaction of the first order, and consequently the selectivity is affected, too.

Case no. 3
In the absence of diffusion resistance, selectivity can be written as:

$$S = \frac{B\text{ production}}{A\text{ consumption}} = \frac{k_1 C_A - k_2 C_B}{k_1 C_A} = 1 - \frac{k_2 C_B}{k_1 C_A} \tag{6.108}$$

In the case of significant internal diffusion resistance, we must calculate the concentration profile for both A and B. For strong diffusion resistance ($\eta < 0.2$) and equal effective diffusivities, the selectivity results are:

$$S = \frac{(k_1/k_2)^{1/2}}{1 + (k_1/k_2)^{1/2}} - (k_2/k_1)^{1/2} \frac{C_B}{C_A} \tag{6.109}$$

In addition, in this case, selectivity is consistently lowered by the intervention of the internal mass-transfer limitation.

6.3 Mass and Heat Balance in a Catalytic Particle: Calculation of the Effectiveness Factor with a Numerical Approach

6.3.1 *Isothermal Spherical Particle*

Consider an isothermal spherical catalyst particle, in which, for example, a second-order irreversible reaction occurs with the following stoichiometry:

$$A + B \rightarrow C \tag{6.110}$$

Assume that for this reaction, the intrinsic kinetic (not affected by diffusional phenomena) is known, and its expression is:

$$R = kC_A C_B \ \text{mol}/\left(\text{m}^3\ \text{s}\right) \tag{6.111}$$

If we consider a generic position along the radius R_p of the particle, r (see Fig. 6.19), the material balance of the specie A between r and r + dr is given by the following expression, which is valid under nonstationary conditions:

$$4\pi r^2 D_e \frac{\partial C_A}{\partial r} + \frac{\partial}{\partial r}\left[4\pi r^2 D_e \frac{\partial C_A}{\partial r}\right] dr - 4\pi r^2 D_e \frac{\partial C_A}{\partial r} - (R) 4\pi r^2 dr = \frac{\partial}{\partial t}\left(C_A 4\pi r^2 dr\right) \tag{6.112}$$

This expression contains the effective diffusivity of the species inside catalyst, D_e. Under steady-state conditions, the accumulation term at the right side of this relation is null, and then the material balance is reduced to:

$$4\pi r^2 D_e \frac{dC_A}{dr} + \frac{d}{dr}\left[4\pi r^2 D_e \frac{dC_A}{dr}\right] dr - 4\pi r^2 D_e \frac{dC_A}{dr} - (R) 4\pi r^2 dr = 0 \tag{6.113}$$

Because now C_A is function only of the position along the particle radius, the partial derivatives were replaced by total derivatives. On simplification, we obtain:

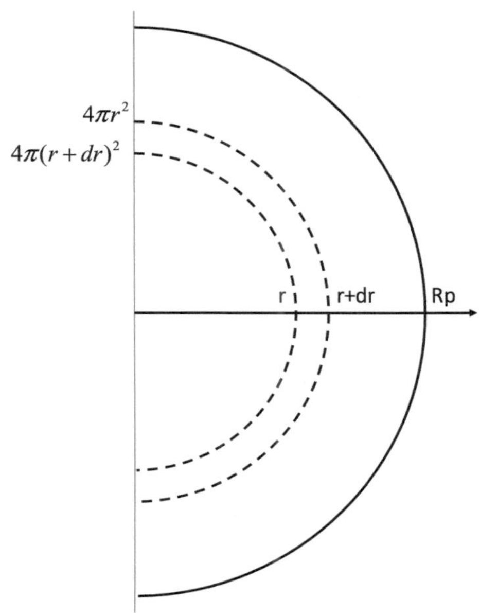

Fig. 6.19 Scheme of a spherical catalyst particle

$$\frac{\mathrm{d}}{\mathrm{d}r}\left[4\pi r^2 D_\mathrm{e}\frac{\mathrm{d}C_A}{\mathrm{d}r}\right]\mathrm{d}r-(R)4\pi r^2\mathrm{d}r=0 \tag{6.114}$$

By developing then the derivative:

$$4\pi D_\mathrm{e}\left[2r\frac{\mathrm{d}C_A}{\mathrm{d}r}+r^2\frac{\mathrm{d}^2C_A}{\mathrm{d}r^2}\right]\mathrm{d}r-(R)4\pi r^2\mathrm{d}r=0 \tag{6.115}$$

Dividing both terms by $4\pi r^2\mathrm{d}r$:

$$D_e\left[\frac{2}{r}\frac{\mathrm{d}C_A}{\mathrm{d}r}+\frac{\mathrm{d}^2C_A}{\mathrm{d}r^2}\right]-(R)=0 \tag{6.116}$$

In a similar way, the material balances for the other components, B and C, can be obtained, and by assuming the same effective diffusivity, we can write:

$$D_\mathrm{e}\left[\frac{2}{r}\frac{\mathrm{d}C_B}{\mathrm{d}r}+\frac{\mathrm{d}^2C_B}{\mathrm{d}r^2}\right]-(R)=0 \tag{6.117}$$

$$D_e\left[\frac{2}{r}\frac{\mathrm{d}C_C}{\mathrm{d}r}+\frac{\mathrm{d}^2C_C}{\mathrm{d}r^2}\right]+(R)=0 \tag{6.118}$$

Equations (6.116)–(6.118) represent a system of coupled second-order differential equations that must be solved for the profiles of C_A, C_B, and C_C inside the particle. For the solution of such system, suitable boundary conditions must be adopted.

The first boundary conditions at the particle surface $r = R_\mathrm{p}$ are:

$$C_A=C_A^\mathrm{S}\quad C_B=C_B^\mathrm{S}\quad C_C=C_C^\mathrm{S} \tag{6.119}$$

The second group of boundary conditions in correspondence of the particle center, $r = 0$, is:

$$\left.\frac{\mathrm{d}C_A}{\mathrm{d}r}\right|_{r=0}=0\quad \left.\frac{\mathrm{d}C_B}{\mathrm{d}r}\right|_{r=0}=0\quad \left.\frac{\mathrm{d}C_C}{\mathrm{d}r}\right|_{r=0}=0 \tag{6.120}$$

This last boundary conditions represents the symmetry of concentration profiles around the center of the particle.

For the numerical solution of the differential system represented by Eqs. (6.116)–(6.118) subject to the boundary conditions (Eqs. 6.119 and 6.120), a simple possibility is to transform the differential equations into algebraic nonlinear equations by approximating the derivatives with suitable finite difference formulae.

In this approach, a discretization of the particle coordinate (radius) is necessary, and a certain number of points (nodes) along the radius must be defined. The more

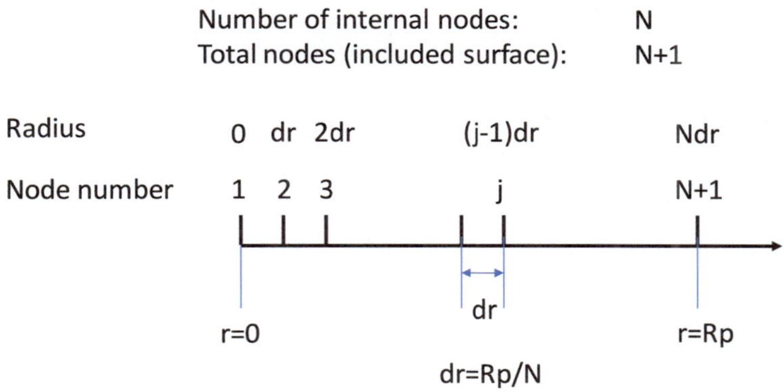

Fig. 6.20 Discretization nodes along particle radius

points are defined, the more accurate the obtained solution. In each node of the grid, radial derivatives can be calculated by finite difference formulas that differ in complexity and accuracy. Figure 6.20 shows discretization nodes along the particle radius.

In our development, we will use very simple formulas for the first and second derivatives with respect to radius as:

$$\frac{\mathrm{d}C_A^i}{\mathrm{d}r} = \frac{C_A^{i+1} - C_A^{i-1}}{2\Delta r} \tag{6.121}$$

$$\frac{\mathrm{d}^2 C_A^i}{\mathrm{d}r^2} = \frac{C_A^{i+1} + C_A^{i-1} - 2C_A^i}{(\Delta r)^2} \tag{6.122}$$

substituting these approximated expressions into Eq. (6.116), we have:

$$D_\mathrm{e}\left[\frac{2}{(i-1)\Delta r}\frac{C_A^{i+1} - C_A^{i-1}}{2\Delta r} + \frac{C_A^{i+1} + C_A^{i-1} - 2C_A^i}{(\Delta r)^2}\right] - (R_i) = 0 \tag{6.123}$$

In this equation, the term R_i is the reaction rate evaluated in correspondence of the concentrations at node ith. Relations similar to Eq. (6.123), which holds for component A, also can be written for components B and C. In this way, we transform a second-order differential equation into a system of nonlinear algebraic system with $C_A{}^i$ as unknowns. According to the radius discretization adopted, at the center of the particle, the zero derivative along the radius [symmetry (Eq. 6.120)] can be written for component A as:

$$\frac{C_A^2 - C_A^1}{\Delta r} = 0 \tag{6.124}$$

and, similarly, for B and C.

Some Remarks

If we have N discretization nodes and three components involved in the reaction, we have a system of 3 * N algebraic eqs. A good choice for N could be between 50 and 200 nodes: The equation system could become very large; however, the convergence is quite fast because the nonlinearity of the problem resides almost only in the term R_i of Eq. (6.123). However, in some cases, a higher number of nodes also could be necessary.

In the present development, the problem has a generality character because R can be represented by any kinetic expression.

This approach could be easily extended to a multiple reaction system by substituting R with a sum of all R in which the specific component is involved (see, e.g., the stoichiometric matrix method).

The method can easily be modified by introducing different effective diffusivity for each component by a slight modification of Eqs. (6.116)–(6.118).

The numerical solution proceeds through a well-known iterative algorithm and can be strongly improved by starting from an initial profile that is then refined to the desired tolerance.

The convergence of this method could be improved and the number of nodes reduced if more accurate finite difference formulae are employed.

In the material balance equations, the reactor rate is expressed as per unit of particle volume, while usually the reaction rate is referred to the mass of the catalyst. When the present approach is extended to a whole reactor, an opportune conversion factor should be introduced.

6.3.2 *Effectiveness Factor*

Once solved, the equation system for the unknown concentrations, the effectiveness factor of the catalyst can be calculated by its definition:

$$\eta = \frac{R_{\text{average}}}{R_{\text{surf}}} = \frac{\frac{1}{V_{\text{part}}}\int_{r=0}^{r=R_p} 4\pi r^2 R(r)\mathrm{d}r}{R_{\text{surf}}} = \frac{\int_{r=0}^{r=R_p} 4\pi r^2 R(r)\mathrm{d}r}{(4/3)\pi R_p^2 R_{\text{surf}}} \tag{6.125}$$

where R_{surf} is the reaction rate evaluated at the condition of the particle's external surface.

Exercise 6.8

Example of Calculation of the Effectiveness Factor for an Isothermal Particle Using the Numerical Approach

Consider the reaction written previously:

$$A + B \rightarrow C \tag{6.126}$$

As an example of the described calculation procedure, assume a reference case in which the numerical values of the parameters are as follows:

$C_s = [1\ 0.8\ 0]$	Concentration on surface for A, B and C (mol/m^3)
$R = 0.01$	Particle radius (m)
$K = 0.001$	Kinetic constant (m^3 mol/s)
$D = 1\text{e}{-9}$	Effective diffusivity (m^2/s)
NN = 100	Number of interior points (−)

In this case, the value of effectiveness factor is $\eta = 0.2318$.

Let us consider now the internal profile of, respectively, A, B and C of the studied reaction. Evaluate then the reaction rate changes with the particle radius; and evaluate how the effectiveness factor changes with the number of considered nodes, with the kinetic constant, with the particle radius of the catalyst, and with the effective diffusion coefficient. Plot the results of all of these calculations.

By applying the previously described concepts, an opportune MATLAB program can be developed. All of the obtained results are reported in Figs. 6.21, 6.22, 6.23, 6.24, 6.25 and 6.26. The MATLAB codes can be found presented as Electronic Supplementary Material, Data 6.1.

Fig. 6.21 Concentration profiles inside the particle for the reference case

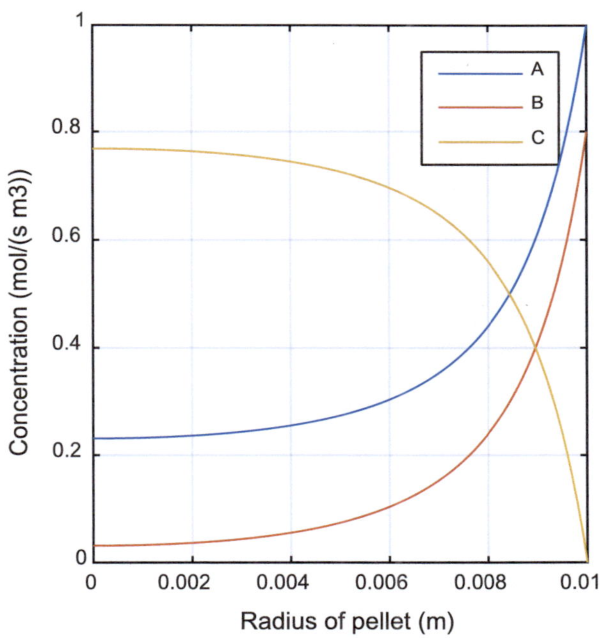

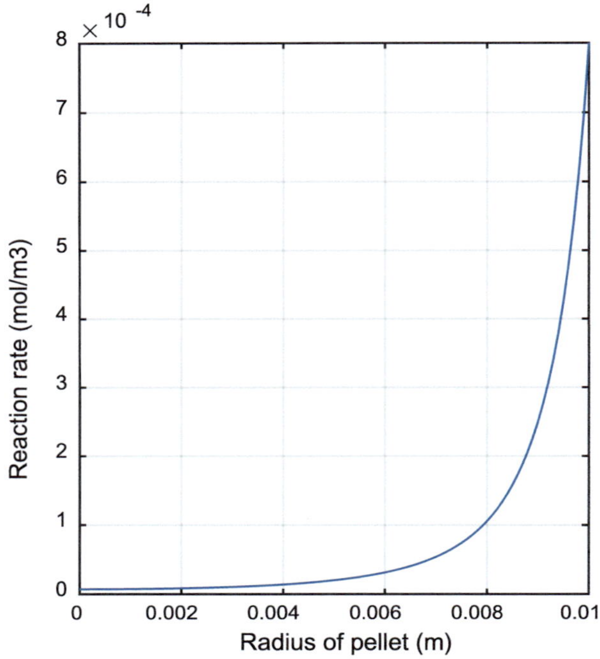

Fig. 6.22 Reaction-rate profile for the reference case

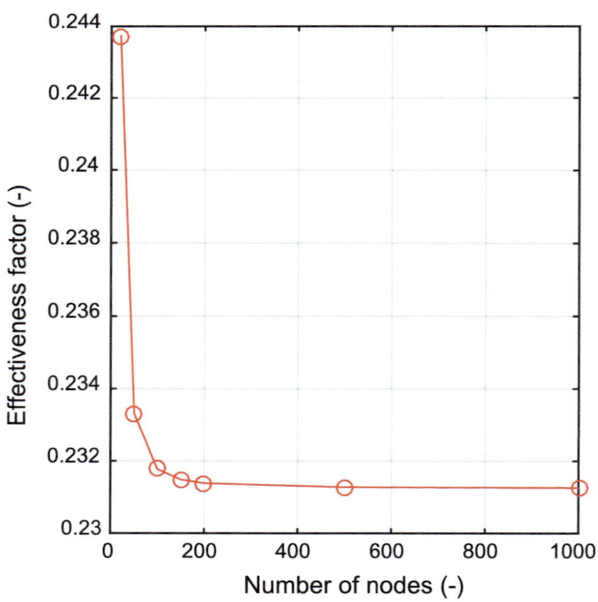

Fig. 6.23 Effect of the number of discretization nodes on the calculated values of the effectiveness factor. In this figure, it is evident that a number of discretization nodes equal to approximately ≥200 yield the same value of effectiveness factor

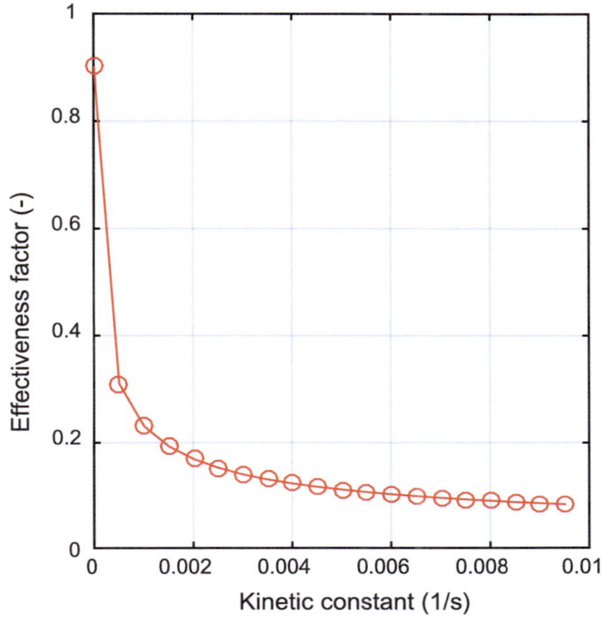

Fig. 6.24 Effect of the kinetic constant on the effectiveness factor. By increasing the value of the kinetic constant, the reaction becomes faster, and the effectiveness factor decreases to a low value as the effect of diffusion limitations become increasingly more important

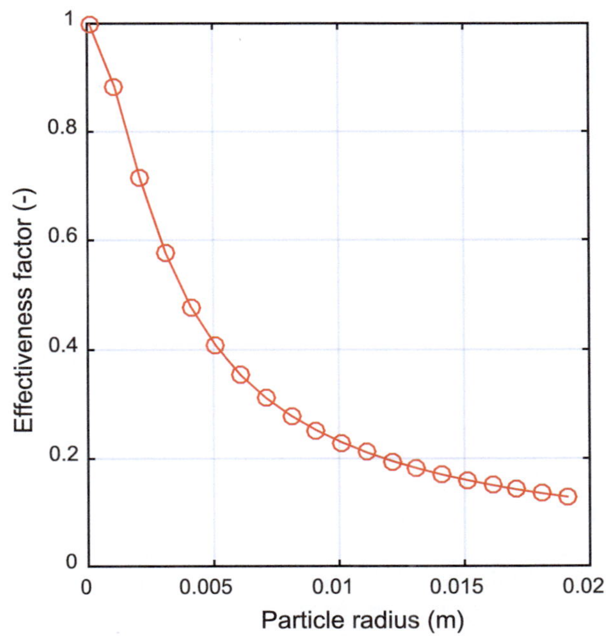

Fig. 6.25 Effect of the particle radius on the effectiveness factor. As expected, small particles are characterized by effectiveness factor near to unity, while as the particles becomes larger, the effectiveness factor decreases

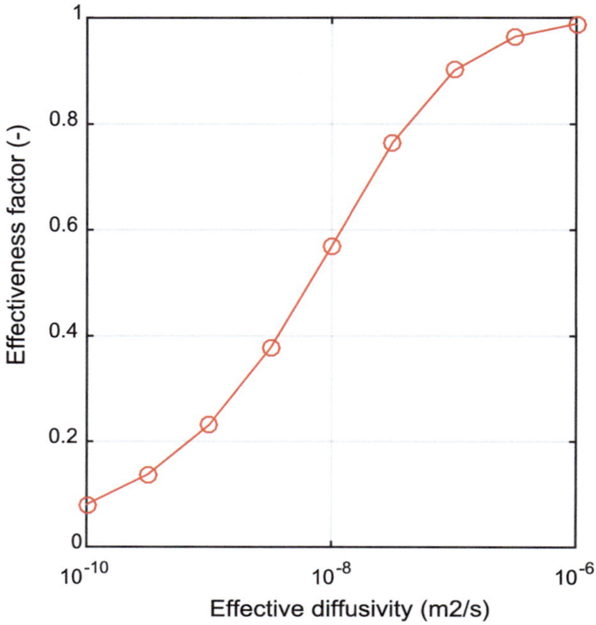

Fig. 6.26 Effect of effective diffusivity on effectiveness factor. High values of effective diffusivity (assumed equal for the three components) increase the effectiveness factor because the internal diffusion, in this case, not a limiting factor for the reaction. In contrast, when effective diffusivity assumes increasingly lower values, the effectiveness factor decreases

Exercise 6.9
Example of Calculation of the Effectiveness Factor for an Isothermal Particle with Multiple Reactions Using the Numerical Approach

It is interesting to observe that the approach presented for a single reaction can be extended quite easily to an isothermal particle in which multiple reactions occur. As an example, we can consider two reactions in series with the scheme:

$$A \rightarrow B \rightarrow C \tag{6.127}$$

The two reactions are of the first order with kinetic expressions and parameters:

$$r_1 = k_1 C_A \quad k_1 = 0.001$$
$$r_2 = k_2 C_B \quad k_2 = 0.002$$

A parametric study similar to the one performed for a single reaction was conducted in this different situation developing an opportune MATLAB program. The results were plotted in the same way as seen in the previous exercise and are reported in Figs. 6.27, 6.28, 6.29, 6.30, 6.31 and 6.32.

All of the described results were obtained using a MATLAB program available as Electronic Supplementary Material, Data 6.1.

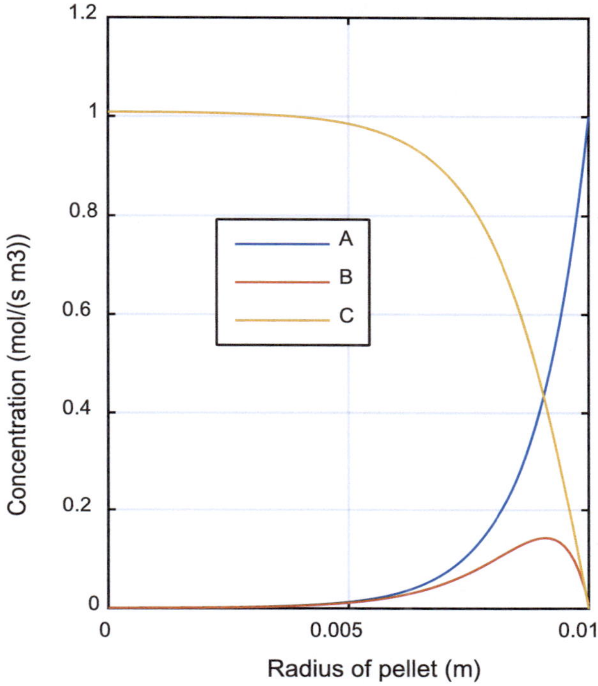

Fig. 6.27 Concentration profiles inside the particle. As expected, the concentration of *B* goes through a maximum and then decreases as it is converted to *C*

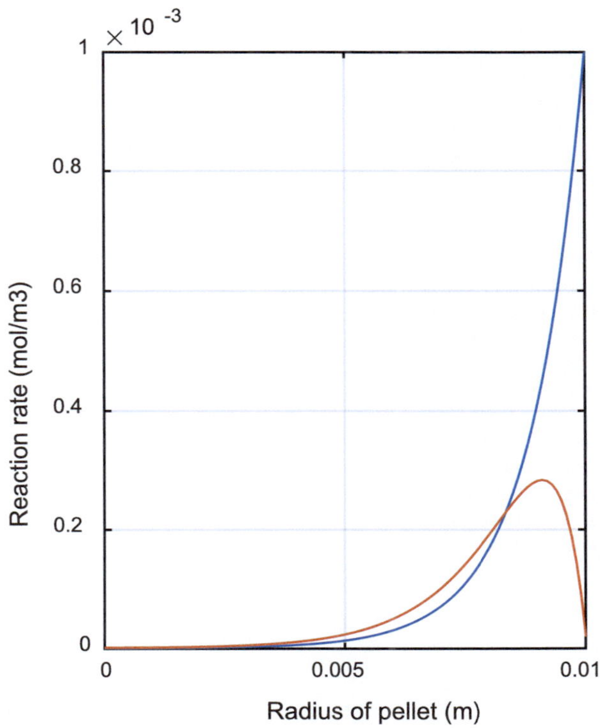

Fig. 6.28 Reaction-rate profiles from the particle surface toward the center. At the surface, the rate of the second reaction is practically zero because *B* is not initially present. When we move from the surface toward the center, the two reaction rates vanish because all *A* is converted to *B*, and all *B* is converted to *C*

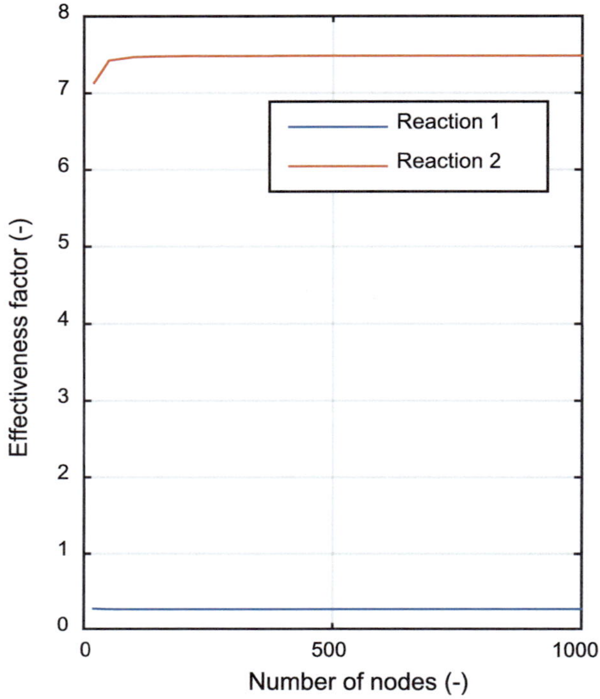

Fig. 6.29 The influence of number of discretization nodes is quite low because a number of points > 100 seems to be accurate enough

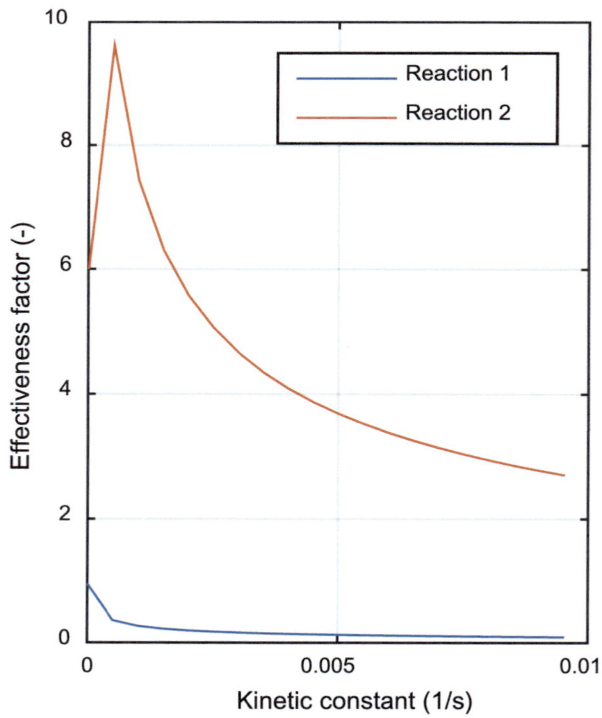

Fig. 6.30 In this plot, the effect of kinetic constants on the effectiveness factor is presented. The most interesting result is that, as in the case of not uniform temperature, an effectiveness factor well above unity was obtained for reaction no2. This can be explained by considering that, as stated previously, the rate of reaction no. 2 is very low at the surface, while inside the particle it assumes finite values. This means that the average reaction rate in the particle is greater than that on the surface, thus resulting in effectiveness factor > 1

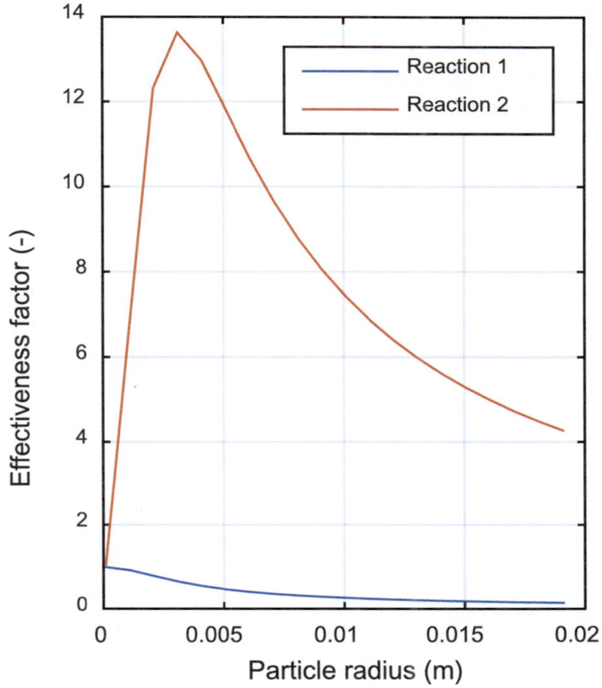

Fig. 6.31 In this plot, the effect of particle radius on the effectiveness factor is presented. A behavior similar to that reported in Fig. 6.30 has been obtained. Also in this case, an effectiveness factor well above unity was obtained for reaction no. 2

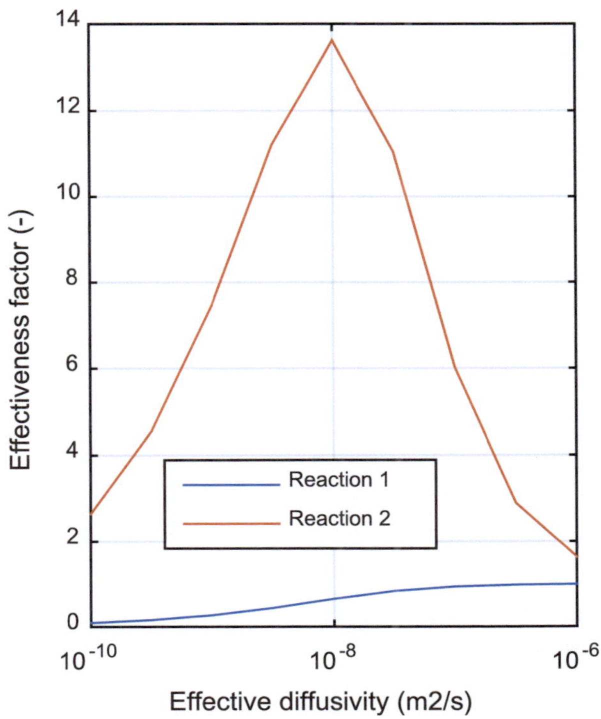

Fig. 6.32 In this figure, the effect of effective diffusivity on the effectiveness factor is presented. An effectiveness factor well above unity was also obtained for reaction no. 2. The reason is that the rate of reaction n. 2 is very low at the surface, and so the average reaction rate in the particle is greater than that on the surface, thus resulting in an effectiveness factor > 1

Exercise 6.10
Example of Calculation of the Effectiveness Factor for an Isothermal Particle with Single Equilibrium Reaction Using the Numerical Approach

This solved exercise is similar to the already presented Exercise 6.8, but now the reaction is reversible. The reaction scheme is simply the following:

$$A + B \rightleftarrows C$$

The behavior of the reversible reaction can be considered as the sum of two contribution from, respectively, the forward and the reverse reaction. The two reactions are assumed of the first order with respect to the reactants and the kinetic expressions and parameters are the following:

$$r_{\mathrm{forw}} = k_{\mathrm{forw}} C_A C_B \quad k_{\mathrm{forw}} = 0.001$$
$$r_{\mathrm{rev}} = k_{\mathrm{rev}} C_C \qquad k_{\mathrm{rev}} = 0.001$$

A parametric study similar to the one performed for a single irreversible reaction (Exercise 6.8) was conducted in this different situation developing an opportune MATLAB program. The results were plotted in the same way as seen in the previous exercise and are reported in Figs. 6.33.

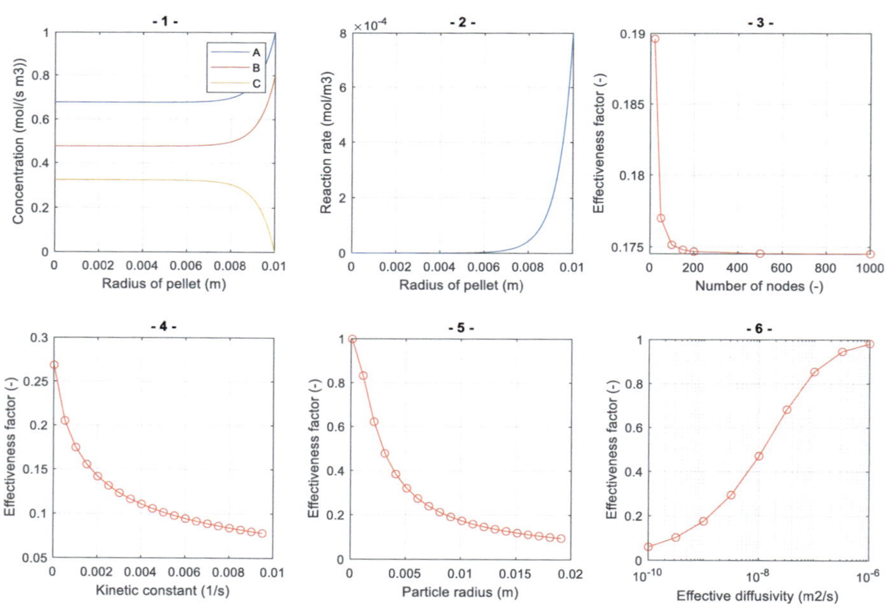

Fig. 6.33 Results for an isothermal particle in which a reversible reaction occurs. (1) Particle concentration profiles; (2) profile of reaction rate inside the particle; (3) effect of the number of nodes; (4) effect of kinetic constant; (5) effect of particle size; (6) effect of effective diffusivity

It is worth to note that in the plot (1) of Fig. 6.33, the internal concentration profiles, below a certain value of the radius, are flat as the system reaches the equilibrium. Moreover, also in the plot (2) of the same figure, the attainment of the equilibrium is evident as the net reaction rate is zero.

All of the described results were obtained using a MATLAB program available as Electronic Supplementary Material, Data 6.1.

6.3.3 Non-Isothermal Spherical Particle

When the heat of the reaction (released or absorbed) is high, the occurrence of a reaction inside catalytic pellets involves also thermal effects that cannot be neglected. An energy balance equation, describing the evolution of the temperature along the particle radius, must be coupled to the already described mass balance (Eqs. 6.116–6.118).

In analogy to what has been written for the mass balance, the energy balance differential equation is:

$$k_{\mathrm{T}}\left[\frac{2}{r}\frac{\mathrm{d}T}{\mathrm{d}r}+\frac{\mathrm{d}^2T}{\mathrm{d}r^2}\right]+(-\Delta H_R)(R)=0 \tag{6.128}$$

where k_{T} is the effective thermal conductivity of the pellet, ΔH_{R} is the reaction enthalpy, and T is the temperature. To solve this equation, two boundary conditions must be defined, as follows:

$$\begin{aligned} &T=T_{\mathrm{S}} && \text{at} \quad r=0 \\ &\left.\frac{\mathrm{d}T}{\mathrm{d}r}\right|_{r=0}=0 && \text{at} \quad r=R_{\mathrm{p}} \end{aligned} \tag{6.129}$$

Regarding the numerical solution, approximated temperature derivatives can be used as finite different formulas, as performed before for concentrations. The formulas for the first and second derivatives are then:

$$\frac{\mathrm{d}T^i}{\mathrm{d}r}=\frac{T^{i+1}-T^{i-1}}{2\Delta r} \tag{6.130}$$

$$\frac{\mathrm{d}^2T^i}{\mathrm{d}r^2}=\frac{T^{i+1}+T^{i-1}-2T^i}{(\Delta r)^2} \tag{6.131}$$

By substituting the finite difference approximations into the thermal balance (Eq. 6.128), we obtain:

$$k_T\left[\frac{2}{(i-1)\Delta r}\frac{T^{i+1}-T^{i-1}}{2\Delta r}+\frac{T^{i+1}+T^{i-1}-2T^i}{(\Delta r)^2}\right]+(-\Delta H_R)(R_i)=0 \quad (6.132)$$

At the center of the particle, the zero derivative along the radius (symmetry [Eq. 6.129]) can be written for the temperature as:

$$\frac{T_2-T_1}{\Delta r}=0 \quad (6.133)$$

Exercise 6.11

Example of Calculation of the Effectiveness Factor for a Non-Isothermal Particle with Single Reaction Using the Numerical Approach

Following an approach similar to the previous ones we considered, a simple non-isothermal reaction system characterized by the second-order reaction $A + B \rightarrow C$.

The input data are reported below:

$C_s = [1\ 0.8\ 0]$	Concentration on the surface for A, B, and C (mol/m^3)
$T_s = 350$	Surface temperature (K)
$r = 0.01$	Particle radius (m)
$k(1) = 1e6$	Kinetic constant pre-exp (m^3 mol/s)
$k(2) = 15{,}000$	Activation energy (J/mol)
$k(3) = -25{,}000$	Heat of reaction (J/mol)
$D = 1e-9$	Effective diffusivity (m^2/s)
$k_T = 1e-6$	Effective thermal conductivity (J/(m s K))
NN = 300	Number of interior points (–)

A MATLAB program, available as Electronic Supplementary Material, Data 6.1, was developed obtaining the results summarized in Figs. 6.34, 6.35, 6.36, 6.37, 6.38, 6.39, 6.40, and 6.41.

All of the described results were obtained using a MATLAB program available as Electronic Supplementary Material, Data 6.1.

6.4 Mass and Heat Transfer in Packed-Bed Reactors: Long-Range Gradients

As we have already seen, a reaction occurring inside a catalyst particle consuming reagents, giving products, and releasing or absorbing heat, according to the reaction, is exothermic or endothermic and generates gradients of both concentration and temperature. When the particles are put in a reactor, such as, for example, a fixed-bed reactor (see Fig. 6.42), long-range gradients (axial and radial) can be observed as a consequence of both the average reaction rate in any single particle and the regime of

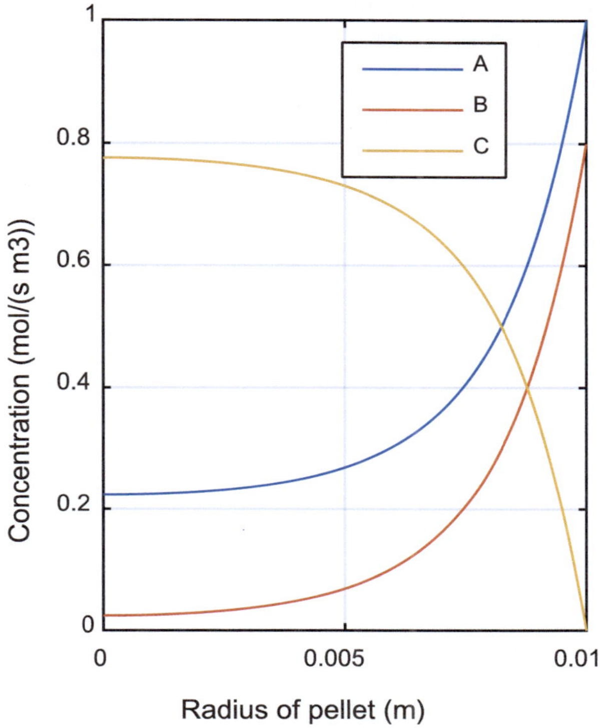

Fig. 6.34 Concentration profiles and reaction-rate profile inside particle

mass and heat flow adopted in the device. These long-range gradients can be minimized in the so-called "gradient-less" reactors, which are isothermal CSTRs, that are normally employed in laboratory kinetic studies, as seen in the previous chapter.

A rather large number of industrial catalytic processes are performed in fixed-bed reactors, which are usually large-capacity units reaching, as in the case of ammonia synthesis, capacities of more than half a million tons per year. Such reactors are not constituted only by a single tube packed with catalyst, but they are arranged in a complex scheme provided with all of the auxiliary equipment, such as feeding, compressing, heating, or cooling. The necessity of supplying or removing heat constitutes the main reason for which it is preferred to build up the reactor with multiple tubes, even in a number of thousands, especially for large-capacity reactors and for very exothermic reactions. The energy exchange with the surroundings is obtained by circulating fluid between the tubes to prevent excessive temperature increase of the flowing mixture, particularly in the case of strongly exothermic reactions.

The effort is to approach isothermal conditions; however, very frequently this ideal condition cannot be reached. In contrast, when the reactive system involves an equilibrium reaction, for example:

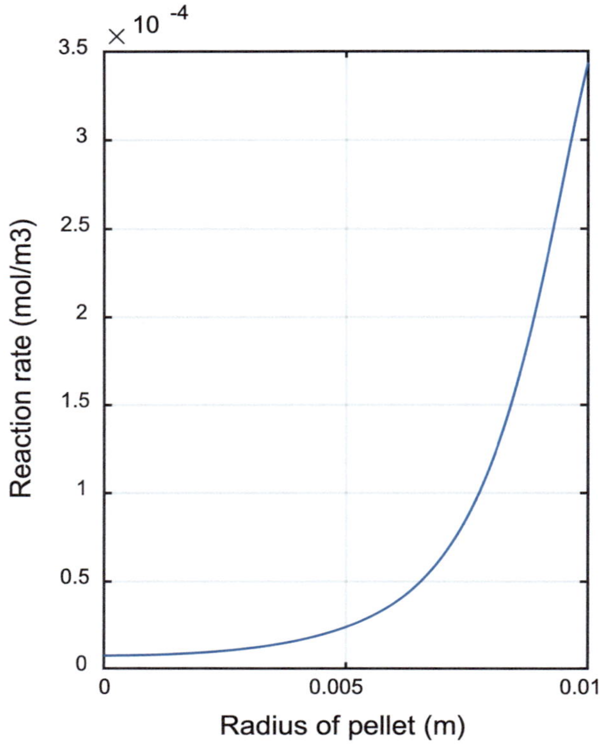

Fig. 6.35 Profile of reaction rate inside catalyst pellet

$$SO_2 + 1/2O_2 \rightleftarrows SO_3 \tag{6.134}$$

where a single large-diameter reactor, containing packed beds with different heights and operating under adiabatic conditions, is preferred because it is possible to control the overall conversion through the temperature of the outlet flow stream. The heat removal, in this case, is obtained by cooling the flow stream between two different successive stages. Therefore, regarding heat transfer, we have two ideal limit conditions: (1) the isothermal one, which occurs when the heat exchange at the reactor wall is very efficient, and (2) the adiabatic regime, where heat exchange is very poor. The intermediate condition, not isothermal and not adiabatic, is the most frequently encountered in common practice, and we will deal with this more complex situation considering isothermal and adiabatic conditions as particular cases.

Many other aspects play an important role in the design of a fixed-bed reactor: pressure drop, safe operation (runaway problems), operating temperature range, mode of catalyst packing, etc. A general approach to the design problem of fixed-bed catalytic reactors consists of setting up and solving conservation equations for mass and energy in a rigorous way. Such equations have analytical solutions only for a few particular and simple cases, but more generally, the solution can be achieved

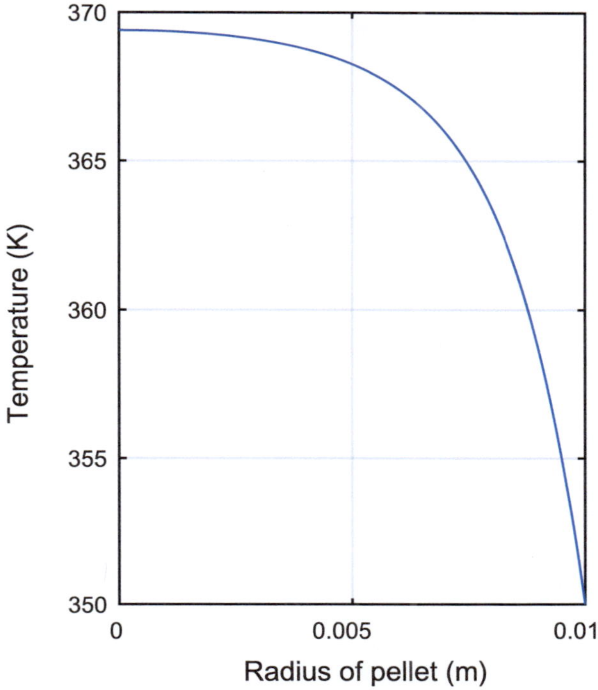

Fig. 6.36 Temperature profile. Assuming that the reaction is exothermic, the temperature increases from the surface to the center. A temperature gradient of approximately 20 K is calculated for the reference case

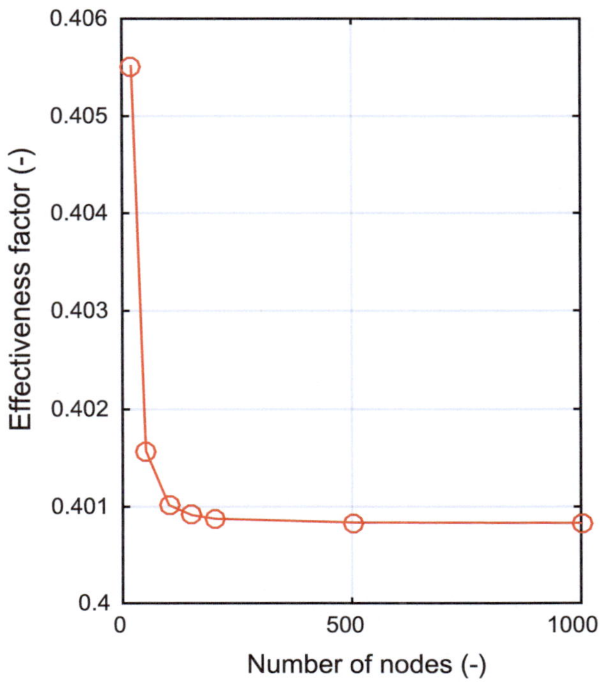

Fig. 6.37 Effect of discretization resolution. After 150–200 nodes, a stable value of the effectiveness factor is obtained

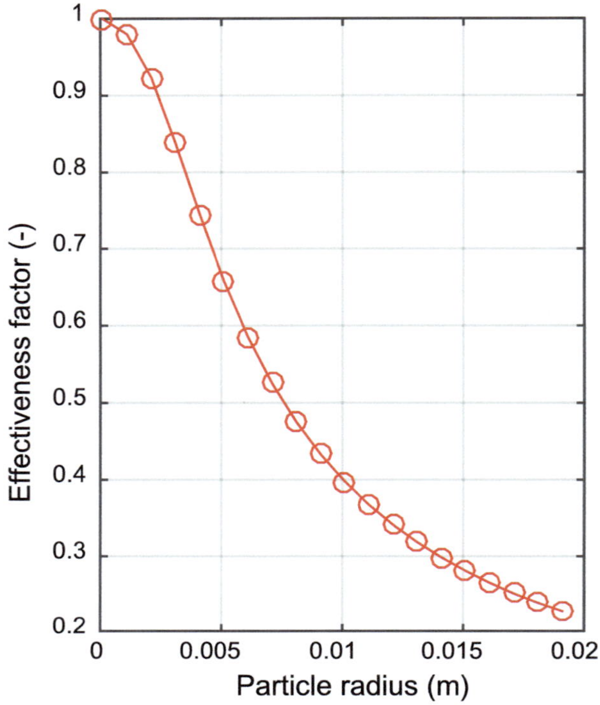

Fig. 6.38 The effects of particle radius on the effectiveness factor are as expected. By increasing the particle size, the effectiveness factor of the catalyst decreases due to the intervention of more severe internal diffusion limitations

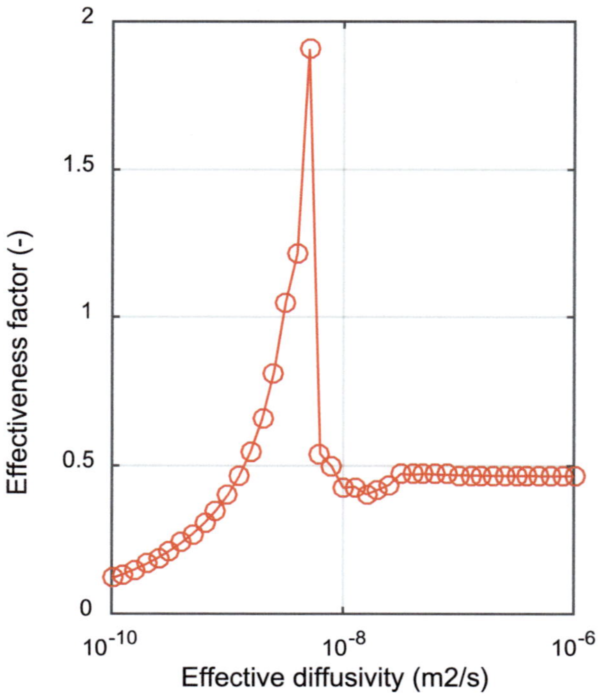

Fig. 6.39 The effect of effective diffusivity is quite complex, probably due to the simultaneous influence of various parameters. In a low range, an increase in effectiveness factor is observed until a maximum, and then a decrease occurs

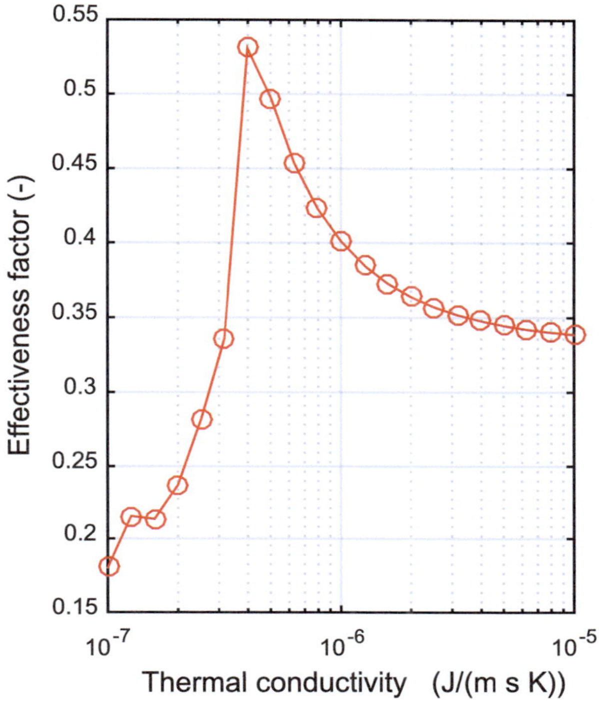

Fig. 6.40 The effect of particle thermal conductivity on the effectiveness factor is similar to that of effective diffusivity. Also, in this case, this is probably due to the simultaneous influence of various parameters. In a low range, an increase in effectiveness factor is observed until a maximum, and then a decrease occurs

only numerically, especially for systems characterized by a complex reaction scheme. The complexity of the problem can be better understood considering that the problem must be solved simultaneously both at a local level (e.g., obtaining particle profiles and effectiveness factor for the occurring reactions) and at the reactor level (e.g., reproducing the long-range profiles). This means that, virtually at each point of the catalytic bed, an effectiveness-factor calculation must be performed at the conditions valid at each point itself. Many books, papers, and reviews have been devoted to this subject; see, for example, Froment and Bishoff (1990), Smith (1981), Fogler (1986), Horak and Pasek (1978), Levenspiel (1984), Satterfield (1972), Holland and Anthony (1979), Carrà and Forni (1974), Winterbottom and King (1999), Missen et al. (1999), Carberry (1987), Bird et al. (1970), and Carrà et al. (1969).

6.4.1 Mass and Energy Balances in Fixed-Bed Reactors

In a system constituted of N_c components, in which N_r chemical reactions occur, the general conservation equation for the mass of the chemical species i can be written as it follows:

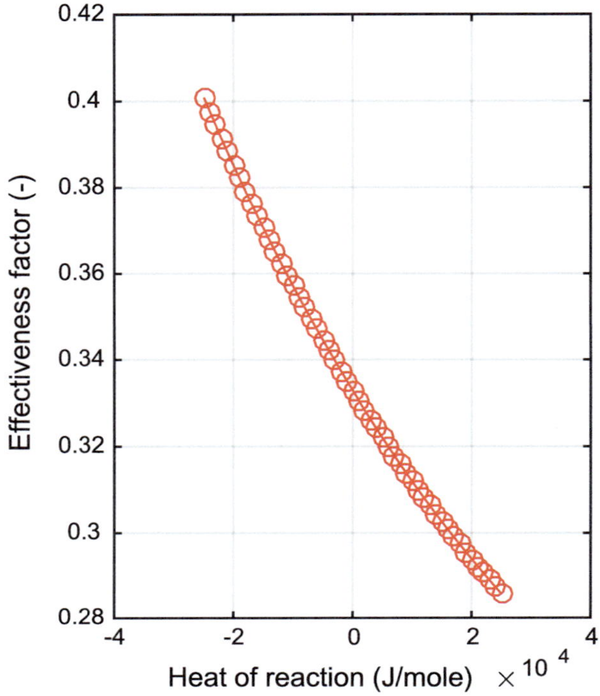

Fig. 6.41 In this plot, the effect of the heat of reaction is shown from exothermic to endothermic. The endothermic reaction leads to a lower effectiveness factor because the reaction inside the particle becomes lower than the one on the surface

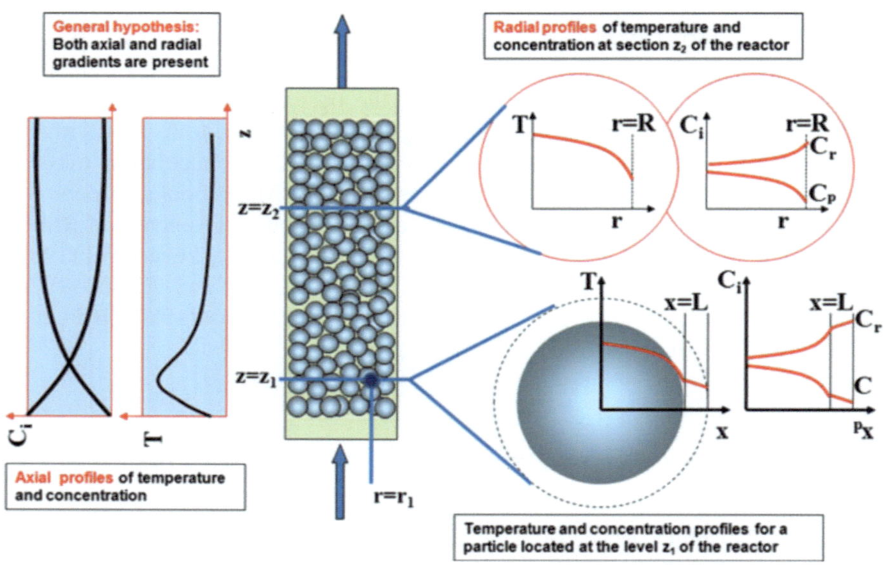

Fig. 6.42 Long-range gradients in a packed-bed tubular reactor

$$\frac{\partial C_i}{\partial t} = -\nabla(C_i u + J_i) + \sum_{j=1}^{N_r} \gamma_{i,j} R_j \tag{6.135}$$

where C_i is the concentration of the component i, u is the velocity vector, $\gamma_{i,j}$ is the stoichiometric coefficient of species i in reaction j, and R_j is the rate of reaction j expressed on the basis of fluid volume. J_i represents the molar flux of the component i due to the gradients of concentration, temperature, and pressure, and it is related to the effective dispersion coefficient D_i by Fick's law:

$$J_i = -D_i \nabla C_i \tag{6.136}$$

Equation (6.135) holds also for transient conditions and takes into account that the accumulation term results from the difference between the input and output terms plus the term due to all of the chemical reactions occurring in a defined control volume.

For a fixed-bed reactor in a cylindrical coordinate system, we can choose an appropriate annular control volume, around which to apply conservation concepts expressed by Eq. (6.135), as shown in Fig. 6.43. If only the velocity in the direction of flow ($u_z = v$) is considered, general Eqs. (6.135) and (6.136) can be combined to give:

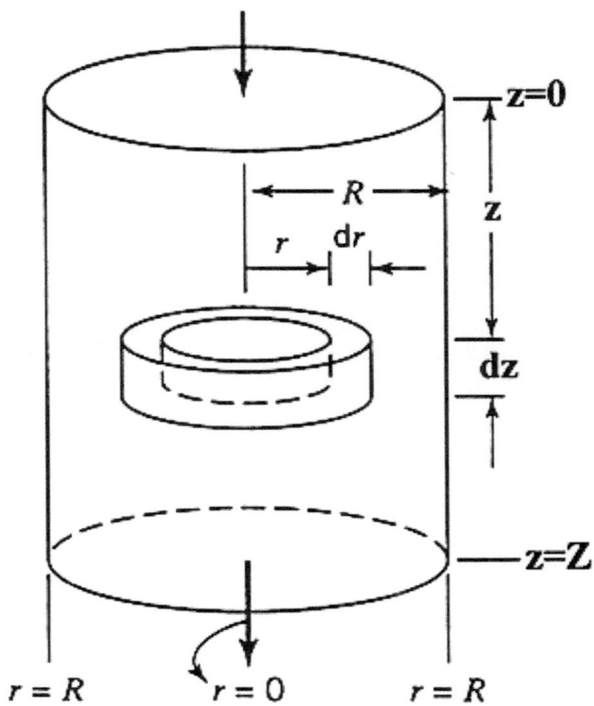

Fig. 6.43 Schematic representation of control volume for the development of conservation equations

$$\varepsilon_B \frac{\partial C_i}{\partial t} = -\frac{\partial}{\partial z}(vC_i) + \frac{\partial}{\partial z}\left[D_{a_i}\frac{\partial C_i}{\partial z}\right] + \frac{1}{r}\frac{\partial}{\partial r}\left[D_{r_i}\frac{\partial C_i}{\partial r}\right] + (1-\varepsilon_B)\sum_{j=1}^{N_R}\gamma_{i,j}R_{G_j} \tag{6.137}$$

where D_{a_i} and D_{r_i} are, respectively, the axial and radial effective dispersion coefficients for the component i (often called "diffusivities") based on the total area (void plus non-void) perpendicular to the direction of diffusion and v is the superficial velocity and ε_B is the bed-void fraction. The global rate R_{Gi} is multiplied by the factor $(1 - \varepsilon_B)$ because this rate is based on the pellet volume.

A first simplification can be operated on Eq. (6.137) by assuming constant the velocity along the z direction and diffusivities not sensitive to z and r. With these assumptions, Eq. (6.137) can be rewritten as follows:

$$\varepsilon_B \frac{\partial C_i}{\partial t} + v\frac{\partial C_i}{\partial z} - D_{a_i}\frac{\partial^2 C_i}{\partial z^2} - D_{r_i}\left[\frac{\partial^2 C_i}{\partial r^2} + \frac{1}{r}\frac{\partial C_i}{\partial r}\right] = +(1-\varepsilon_B)\sum_{j=1}^{N_R}\gamma_{i,j}R_{G_j} \tag{6.138}$$

In a similar way, we can write an equation for energy conservation by replacing, in Eq. (6.138), the concentration of the chemical species C_i with the term $\rho C_p T$, the diffusivities D with the effective thermal conductivities k, and the reaction term R_{Gj} with the reaction enthalpy term $(-\Delta H_j)\, R_{Gj}$:

$$\varepsilon_B \frac{\partial T}{\partial t} + v\frac{\partial T}{\partial z} - K_a\frac{\partial^2 T}{\partial z^2} - K_r\left[\frac{\partial^2 T}{\partial r^2} + \frac{1}{r}\frac{\partial T}{\partial r}\right] = \frac{(1-\varepsilon_B)}{\rho C_p}\sum_{j=1}^{N_R}(-\Delta H_j)R_{G_j} \tag{6.139}$$

where ρ and C_P, respectively, are the average density and the specific heat referred to the gas mixture.

If we consider a fixed-bed reactor, the temperature and concentration of the bulk can be regarded, from a general point of view, as functions of both the axial and radial coordinates, that is:

$$\begin{aligned} C_i^B &= f(z, r) \\ T_B &= g(z, r) \end{aligned} \tag{6.140}$$

With the previous assumptions, we can write the general conservation relations for mass and energy in a fixed-bed reactor, in which N_r chemical reactions—involving N_c components—occur as follows:

$$\varepsilon_B \frac{\partial C_i^B}{\partial t} + v \frac{\partial C_i^B}{\partial z} - D_{a_i} \frac{\partial^2 C_i^B}{\partial z^2} - D_{r_i} \left[\frac{\partial^2 C_i^B}{\partial r^2} + \frac{1}{r} \frac{\partial C_i^B}{\partial r} \right] = (1 - \varepsilon_B) \sum_{j=1}^{N_R} \gamma_{i,j} R_{G_j} \tag{6.141}$$

$$\varepsilon_B \frac{\partial T_B}{\partial t} + v \frac{\partial T_B}{\partial z} - K_a \frac{\partial^2 T_B}{\partial z^2} - K_r \left[\frac{\partial^2 T_B}{\partial r^2} + \frac{1}{r} \frac{\partial T_B}{\partial r} \right] = \frac{(1 - \varepsilon_B)}{\rho C_p} \sum_{j=1}^{N_R} (-\Delta H_j) R_{G_j} \tag{6.142}$$

where $i = 1, 2, \ldots N_c$ and $j = 1, 2, \ldots N_r$.

Equations (6.141) and (6.142) constitute a set of coupled partial differential equations that must be solved keeping in mind some suitable boundary conditions relative to variables and their derivatives. Usual boundary conditions can be written as follows:

$$\frac{\partial T_B}{\partial r} = \frac{\partial C_i^B}{\partial r} = 0 \quad \text{at the centerline of the reactor } (r = 0) \text{ for all } z \tag{6.143}$$

$$\frac{\partial C_i^B}{\partial r} = 0; \quad h_w (T_B - T_C) = -\rho C_p K_r \frac{\partial T_B}{\partial r} \quad \text{at the wall of reactor } (r = R) \text{ for all } z \tag{6.144}$$

The first boundary condition (Eq. 6.143) results from symmetry consideration around the centerline of a tubular reactor, while the second one (Eq. 6.144) is related to the fact that no reactant transport takes place across the reactor wall and that the heat transferred to the cooling medium, whose temperature is T_c, is equal to the heat conducted at the wall.

Regarding axial boundary conditions at the reactor inlet, we can write the following relations:

$$\left.\begin{aligned} (v C_i^B)_{in} &= \left(v C_i^B - D_{a_i} \frac{\partial C_i^B}{\partial z} \right)_{z=0} \\ (v T_B)_{in} &= \left(v T_B - K_a \frac{\partial T_B}{\partial z} \right)_{z=0} \end{aligned}\right\} \quad \text{at } z = 0 \tag{6.145}$$

and for the outlet:

$$\frac{\partial C_i^B}{\partial z} = \frac{\partial T_B}{\partial z} = 0 \quad \text{at} \quad z = Z \tag{6.146}$$

The previously mentioned boundary conditions are based on the continuity of the flux (mass or heat) across a boundary represented by the catalytic bed inlet and outlet.

6.4.2 External-Transport Resistance and Particle Gradients

To relate macroscopical, or "long-range," concentration and temperature gradients, involved in the conservation equations for the whole reactor, to the microscopic situation developed around and inside catalytic particles, we must write a relation between the global reaction rate and the intrinsic reaction rate. The global reaction rate, R_{Gi}, is equal to the rate at which mass is transferred across the interface between fluid and solid phase, which in turn is related to the flux at the catalyst surface:

$$\sum_{j=1}^{N_r} \gamma_{i,j} R_{G_j} = \frac{k_g}{L}\left(C_i^B - C_i^S\right) = \frac{D_{ei}}{L}\frac{\partial C_i^P}{\partial x}\bigg|_{x=L} = \sum_{j=1}^{N_r} \gamma_{i,j}\eta_j r_{C_j} \quad j = 1, 2, \ldots, N_r \tag{6.147}$$

with the following:

k_g	Film mass-transfer coefficient
L	Particle characteristic length (radius for spherical pellets)
C_i^S	Concentration of component i at the surface
C_i^P	Concentration of component i inside the particle
D_{ei}	Effective diffusivity of component i into the particle
x	Particle radial coordinate
η_j	Effectiveness factor for reaction j
r_{cj}	Intrinsic rate of reaction j

In a similar way, we can write a relation for the thermal flux:

$$\sum_{j=1}^{N_r} (-\Delta H_j) R_{G_j} = \frac{h}{L}(T_S - T_B) = -\frac{K_e}{L}\frac{\partial T_P}{\partial x}\bigg|_{x=L} \tag{6.148}$$

where:

h	Film heat-transfer coefficient
T_S	Temperature at the surface of the pellet
T_P	Temperature inside the pellet
K_e	Effective thermal conductivity of the catalytic particle

As can be seen from Eq. (6.147), the relation between the global rate and the intrinsic rate is expressed, for each reaction, by the effectiveness factor η or, equivalently, by means of the concentration gradients at the particle surface. This aspect introduces the necessity of solving mass- and energy-conservation equations relative to the catalytic particles to obtain local or microscopic concentration and temperature profile virtually at each position along the reactor.

Conservation equations for the particles can be summarized as follows:

$$\varepsilon_{\mathrm{P}} \frac{\partial C_i^{\mathrm{P}}}{\partial t} = D_{\mathrm{e}_i} \left[\frac{\partial^2 C_i^{\mathrm{P}}}{\partial x^2} + \frac{2}{x} \frac{\partial C_i^{\mathrm{P}}}{\partial x} \right] - \rho_{\mathrm{P}} \sum_{j=1}^{N_r} \gamma_{i,j} r_{c_j} \quad i = 1, 2, \ldots, N_{\mathrm{c}} \tag{6.149}$$

$$\varepsilon_{\mathrm{P}} \rho_{\mathrm{P}} C_{\mathrm{P}}^{\mathrm{P}} \frac{\partial T_{\mathrm{P}}}{\partial t} = K_{\mathrm{e}} \left[\frac{\partial^2 T_{\mathrm{P}}}{\partial x^2} + \frac{2}{x} \frac{\partial T_{\mathrm{P}}}{\partial x} \right] - \rho_{\mathrm{P}} \sum_{j=1}^{N_r} (-\Delta H_j) r_{c_j} \tag{6.150}$$

with the following meaning of symbols:

ε_{P}	Catalytic particle void fraction
ρ_{P}	Catalytic particle density
$C_{\mathrm{P}}^{\mathrm{P}}$	Catalytic particle specific heat

The simultaneous solution of Eqs. (6.149, 6.150) must be accomplished by the following boundary conditions being valid, respectively, at the center and at the external surface of the catalytic pellet and arising from symmetry and continuity considerations relative to concentration and temperature:

$$\begin{aligned} &\frac{\partial C_i^P}{\partial x} = 0 \quad \frac{\partial T_P}{\partial x} = 0 \quad \text{at} \quad r = 0 \ \text{(center)} \\ &C_i^{\mathrm{P}} = C_i^S \quad T_{\mathrm{P}} = T_{\mathrm{S}} \quad \text{at} \quad r = \mathrm{L} \ \text{(surface)} \end{aligned} \tag{6.151}$$

The explained problem consists of a set of nonlinear partial differential equations that must be solved both locally for the catalytic particle and long-range for the whole reactor. The solution of the problem in the full form, expressed by Eqs. (6.141)–(6.151), is not a simple task even from the point of view of a numerical solution, while an analytical solution is impossible for most of the practical cases. In the following paragraph, we shall introduce some useful simplifications that can be operated, in some cases, on the reported equations with: the purpose to convert the problem into a form that is easier to solve.

6.4.3 Conservation Equations in Dimensionless Form and Possible Simplification

For introducing the mentioned simplifications, it is convenient to rewrite conservation equations for the reactor in a dimensionless form, for both emphasizing some parameters of the reactor and implementing a more stable numerical solution procedure.

We can pose:

$$n = \frac{Z}{d_P} \quad m = \frac{R}{d_P} \quad A = \frac{Z}{R} \quad \theta = \frac{Z}{v} \quad \overline{C}_i = \frac{C_i^{B}}{C_i^{B(in)}}$$
$$\overline{T} = \frac{T_B}{T_{B_{(in)}}} \quad \overline{r} = \frac{r}{R} \quad \overline{z} = \frac{z}{Z} \quad \overline{t} = \frac{t}{\theta} \tag{6.152}$$

with:

d_P	particle diameter
R	fixed-bed reactor radius
Z	fixed-bed reactor length
$C_i^{B(in)}$	reactor-inlet concentration
$T_{B(in)}$	reactor-inlet temperature

Within these assumptions, the reactor conservation equations become:

$$\varepsilon_B \frac{\partial \overline{C_i}}{\partial t} + \frac{\partial \overline{C_i}}{\partial \overline{z}} - \frac{1}{nP_{ma}} \frac{\partial^2 \overline{C_i}}{\partial \overline{z}^2} - \frac{1}{mP_{mr}} \left[\frac{\partial^2 \overline{C_i}}{\partial \overline{r}^2} + \frac{1}{\overline{r}} \frac{\partial \overline{C_i}}{\partial \overline{r}} \right] = \frac{(1-\varepsilon_B)\theta}{C_i^{B(in)}} \sum_{j=1}^{N_R} \gamma_{i,j} R_{G_j} \tag{6.153}$$

$$\varepsilon_B \frac{\partial \overline{T}}{\partial \overline{t}} + \frac{\partial \overline{T}}{\partial \overline{z}} - \frac{1}{nP_{ha}} \frac{\partial^2 \overline{T}}{\partial \overline{z}^2} - \frac{A}{mP_{hr}} \left[\frac{\partial^2 \overline{T}}{\partial \overline{r}^2} + \frac{1}{\overline{r}} \frac{\partial \overline{T}}{\partial \overline{r}} \right] = \frac{(1-\varepsilon_B)\theta}{\rho C_p T_{B(in)}} \sum_{j=1}^{N_R} (-\Delta H_j) R_{G_j} \tag{6.154}$$

where $i = 1, 2, \ldots, N_c$ and $j = 1, 2, \ldots N_r$.

In Eqs. (6.153) and (6.154), some fundamental dimensionless groups related to mass dispersion can be recognized, in both the axial and radial directions, represented by Peclet's numbers expressed by the following relations:

$$P_{ma} = \frac{d_P v}{D_a} \text{ (axial)} \quad P_{mr} = \frac{d_P v}{D_r} \text{ (radial)} \tag{6.155}$$

and analogously for heat dispersion, we have:

$$P_{ha} = \frac{d_P v}{k_a} \text{ (axial)} \quad P_{hr} = \frac{d_P v}{k_r} \text{ (radial)} \tag{6.156}$$

The Peclet's numbers, together with reactor-to-particle size ratios—such as n, m, and A—can be adopted as suitable criteria for determining the extent to which the dispersion phenomena affect the global behavior of the reactor and, in many cases, can help to decide whether or not to introduce some simplifications. The reaction characteristics and operative conditions can suggest a variety of simplifications that can be applied to the written conservation equations. Among these, the first and more common is that of steady state, involving the elimination of time variable and all of the derivative with respect to it on the left-hand side of Eqs. (6.141), (6.142), (6.149), and (6.150). Another important role is played by the heat of reactions. When the heat involved in the reactions is negligible, the reactor can be run isothermally and, because the temperature is constant, all of the heat balance equations can be eliminated. When the reactor is operated as adiabatic, as many reactors are in practice, radial gradients could be negligible, and therefore, a one-dimensional treatment (only in an axial direction) of the conservation equations is sufficient for the description of the reactor itself. An intermediate situation, occurring between these two limit cases, is represented by non-isothermal and nonadiabatic reactors arising from very exothermic reactions for which external cooling is required to ensure the safety of the reactor and to preserve the integrity of the catalyst. In this case, a numerical solution of a full conservation equation appears to be the only feasible approach; however, some simplifications can be still applied even if the problem remains much more complex to solve than the two earlier cited limit cases. Usually, for a highly exothermic reaction, the packed beds have a relatively small diameter, thus favoring the heat removal, and then the radial temperature profile can be neglected, reconducting the problem to a one-dimensional one. In general, according to Carberry (1987), radial gradients, for practical purposes, can be neglected if the radial aspect ratio $m = R/d_p$ is <3 or 4. Furthermore, indications result from Peclet's numbers and reactor aspect ratios; for example, the axial aspect ratio $n = Z/d_P$ is usually large and considering that P_{ma} is approximately 2 for gases flowing through a packed bed for Reynold's number (based on d_P) > 10, then the term nP_{ma} is quite large, thus indicating that axial mass dispersion can be neglected. As a general guideline, Table 6.7 (see Lee 1984) lists the main simplifications that can be introduced into the conservation equations for a packed-bed reactor operating under steady-state conditions. In this table, we report the two limit cases represented by isothermal and adiabatic reactors and also for an intermediate situation, in which the reactor can be considered neither isothermal nor adiabatic.

6.4.4 Examples of Applications to Non-Isothermal and Nonadiabatic Conditions: Oxidation of Orto-Xylene to Phthalic Anhydride

The first example of a reaction performed in a packed-bed tubular reactor, operating under non-isothermal and nonadiabatic conditions, is the production of phthalic anhydride (PA) starting from o-xylene (OX) and oxidizing it with oxygen (O_2). A simplified scheme for this oxidation reaction can be expressed as it follows:

Table 6.7 Possible simplifications on the left-hand side of conservation equations under steady-state conditions

Reactor conditions	Aspect ratio criteria	Left-hand side of Eq. (6.149) and (6.150)
Isothermal		$v\frac{\partial C_B^i}{\partial z}$
Adiabatic	$\left(\frac{Z}{d_P}\right)\left(\frac{d_P}{D_a}\right) > 300 \quad R_e > 10$	$v\frac{\partial C_B^i}{\partial z}$ $\rho C_P v\frac{\partial T_B}{\partial z}$
	$\left(\frac{Z}{d_P}\right)\left(\frac{d_P}{D_a}\right) < 300 \quad R_e > 10$	$v\frac{\partial C_B^i}{\partial z}$ $\begin{cases} \rho C_P v\frac{\partial T_B}{\partial z} & \text{or, if necessary} \\ \rho C_P v\frac{\partial T_B}{\partial z} - K_a\frac{\partial^2 T_B}{\partial z^2} \end{cases}$
Non-isothermal and nonadiabatic	$\frac{R}{d_P} > 4$	$v\frac{\partial C_B^i}{\partial z} - D_r\left(\frac{\partial^2 C_B^i}{\partial r^2} + \frac{1}{r}\frac{\partial C_B^i}{\partial r}\right)$ $\rho C_P v\frac{\partial T_B}{\partial z} - K_r\left(\frac{\partial^2 T_B}{\partial r^2} + \frac{1}{r}\frac{\partial T_B}{\partial r}\right)$
	$\frac{R}{d_P} \leq 4 \quad R_e > 30$	$v\frac{\partial C_B^i}{\partial z}$ $\rho C_P v\frac{\partial T_B}{\partial z}$

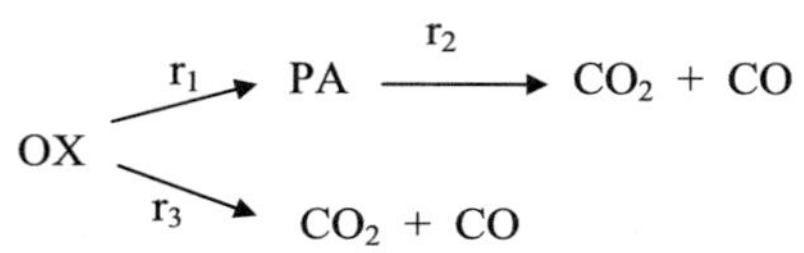

(6.157)

The reaction is catalyzed by V_2O_5, which is frequently supported on γ-alumina, and is highly exothermic (Gimeno et al. 2008). From this scheme, it is evident that the reaction can evolve, if not properly controlled from a thermal point of view, to CO_2 and CO production, giving a low yield of PA. It is relevant, therefore, for the reactor to have a simulation that takes into account the thermal effect due to the reaction and to the heating medium. The rate equations and kinetic parameters for the mentioned reactions (Eq. 6.157) are listed in Table 6.8, together with the characteristics of the reactor and the catalytic particles used in the calculations (see Froment 1967).

A peculiarity of this reactor is the catalyst dilution with an inert in the initial part of the reactor itself (0.75 m), which is implemented with the scope of a better temperature control.

The starting assumptions, related to the reactor adopted for the model development, are the following:

- No axial and radial dispersion.
- No radial temperature and concentration gradients in the reactor tube.

Table 6.8 Kinetic data for the conversion of o-xylene to phthalic anhydride

$r_1 = k_1 P_{OX} P_O$ (Kmol/kg cat h)	$\ln k_1 = -27{,}000/RT + 19.837$
$r_2 = k_2 P_{PA} P_O$ (Kmol/kg cat h)	$\ln k_2 = -31{,}000/RT + 20.860$
$r_3 = k_3 P_{OX} P_O$ (Kmol/kg cat h)	$\ln k_3 = -28{,}600/RT + 18.970$
$\Delta H_1 = -307$ kcal/mol	
$\Delta H_2 = -783$ kcal/mol	
$\Delta H_3 = -1090$ kcal/mol	
$U = 82.7$ kcal/ m^2 h °C	Overall heat-transfer coefficient
$D = 0.025$ m	Reactor diameter
$Z = 3$ m	Reactor length
$d_P = 0.003$ m	Particle diameter
$C_P = 0.25$ kcal/kg °C	Average specific heat
$\rho_B = 300$ kg/m^3	Bulk density of the bed
Feed composition	$y_{OX} = 0.0093$
	$y_O = 0.208$
Feed molar flow rate	$F = 0.779$ mol/h
Inert dilution of the catalyst	$m_I = 0.5$ for the first quarter
Inlet temperature	$T_0 = 370$ °C

- Plug flow through the reactor.
- No limitation related to diffusion inside catalytic pellets.

The assumptions relative to radial profiles can be supported by the criteria expressed in Table 6.7 for the radial aspect ratio $m = R/d_p$, which can be estimated as $m = 4.1$ and then slightly above the limit. For this system, and with the previous assumptions, we can derive a material balance equation directly from Eq. (6.141) in steady state:

$$\frac{dF_i}{dz} = F\frac{dy_i}{dz} = \rho_B \frac{\pi D^2}{4} \sum_{j=1}^{N_r} \gamma_{i,j} \frac{\overline{R_j}}{(1+m_I)} \qquad i = 1, 2, \ldots, N_c \tag{6.158}$$

assuming a constant molar flow rate, F, and with the following substitution:

$$v = \frac{Q}{A} \quad QC_i = F_i \quad A = \frac{\pi D^2}{4} \quad F_i = y_i F \tag{6.159}$$

where:

Q	volumetric overall flow rate
A	cross-section of the reactor tube
D	reactor diameter
F_i	component molar flow rate
y_i	mole fraction of component i
m_I	mass of inert per unit mass of catalyst (dilution ratio)
$\overline{R_j}$	reaction rate for reaction j based on catalyst mass

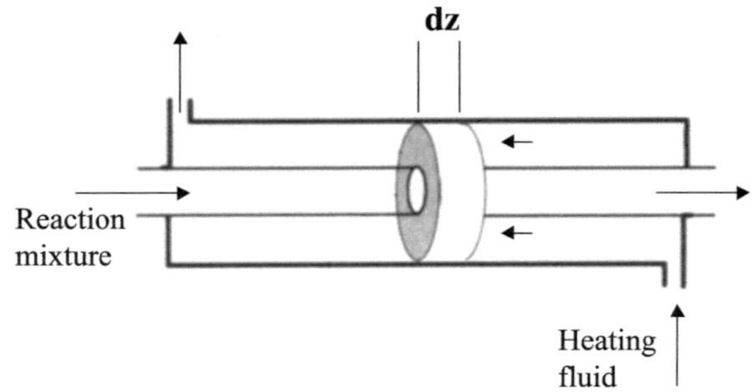

Fig. 6.44 Structure of the reactor similar to a double-pipe heat exchanger

The energy balance, as represented by Eq. (6.142), must be modified, as performed for the material balance, according to the assumed absence of radial profiles and to the presence of external cooling fluid. The thermal exchange with the surrounding now cannot be considered only as a boundary condition but as a term in the energy-conservation equation. In fact, if we assume for the reactor a behavior similar to a double-pipe heat exchanger (see Fig. 6.44), the heat transferred across the external surface per unit of reactor volume is defined as:

$$q = \frac{U(T_C - T)\pi D\mathrm{d}z}{A\mathrm{d}z} = \frac{U(T_C - T)\pi D}{A} = \frac{4U(T_C - T)}{D} \tag{6.160}$$

This term must be added algebraically, in the balance equation, to the heat associated with the reaction, thus yielding the following expression:

$$\frac{\mathrm{d}T}{\mathrm{d}z} = \frac{\rho_B}{\mathrm{GC_P}} \sum_{j=1}^{N_r} (-\Delta H_j) \frac{\overline{R}_j}{(1 + m_I)} + \frac{4U}{\mathrm{DGC_P}} (T_C - T) \tag{6.161}$$

with : $G = \frac{F \cdot M_F}{A}$ with the following meaning of symbols:

G	mass velocity
M_F	average molecular weight of mixture

The system of differential Eqs. (6.158) and (6.161) can be integrated along the axial reactor direction, *z*, thus obtaining the temperature and composition profiles. The temperature profile resulting from this mono-dimensional model is reported in Fig. 6.45, and in this diagram, we report, for comparison, the results of a more sophisticated two-dimensional model, in which radial profiles also are taken into account.

The two-dimensional model, as has been seen before, implies the solution of partial differential equations that, in the case of the present example, have been performed with the method of finite differences. The two models give results, in terms of axial temperature profiles, that are close enough to conclude that a one-dimensional model is sufficient for many practical situations. In the case of the two-dimensional model, however, a slightly higher conversion to carbon oxide is obtained due to the higher temperature attained along the reactor.

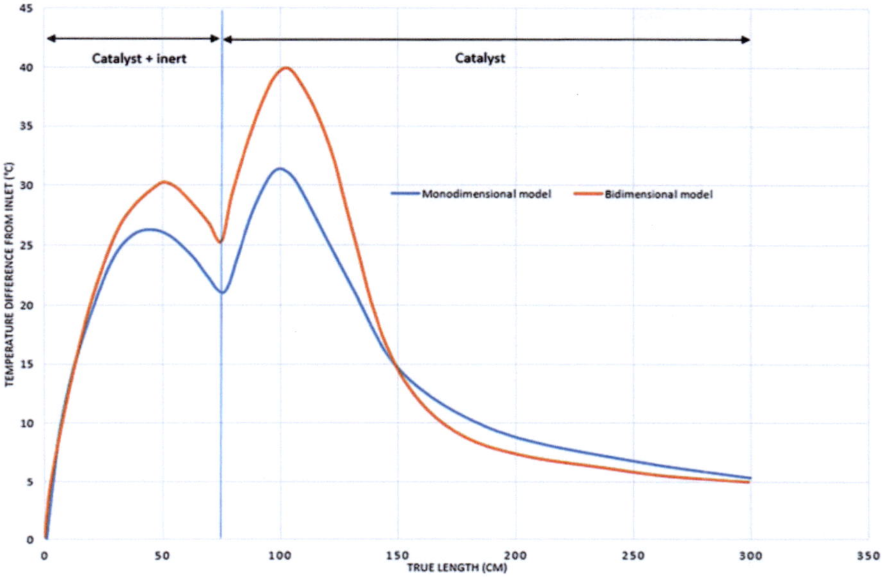

Fig. 6.45 A qualitative comparison of the results of the unidimensional and bidimensional models for reactor simulation (inspired from the work of from Froment 1967)

Oxidation of Methanol to Formaldehyde

As a further example of a system that can be considered neither isothermal nor adiabatic, we propose the catalytic conversion of methanol to formaldehyde catalyzed by iron molybdate. Two reactions occur:

$$\begin{array}{l} CH_3OH + 1/2O_2 \rightarrow CH_2O + H_2O \\ CH_2O + 1/2O_2 \rightarrow CO + H_2O \end{array} \tag{6.162}$$

The reaction was performed in a packed-bed tubular reactor filled with catalyst and surrounded by a heat-transfer fluid that has the function of controlling the temperature. Table 6.9 lists the characteristics of the reactor and the adopted operating conditions. By using these conditions and the kinetic data reported by Riggs (1988), we performed a simulation of the reactor behavior in terms of temperature and composition profiles along the axis. In this case, a further complication must be introduced in the framework of the model consisting of the calculation of the effectiveness factor along the reactor and taking into account diffusion limitations inside the particles.

The basic assumptions adopted here for the model layout can be summarized in the following points, which are similar to those of the example reported in the previous section:

- No axial and radial dispersion.
- No radial temperature and concentration gradients in the reactor tube.
- Plug flow through the reactor.

Table 6.9 Reactor characteristic and operating conditions

Inlet temperature	539 K
Total pressures	1.68 atm
Bulk density of the bed	0.88 kg/m^3
Overall heat transfer coefficient U	0.171 kJ/(m^2 s K)
Heating medium temperature	544 K
Reactor diameter	2.54×10^{-2} m
Particles diameter	3.5×10^{-3} m
Reactor length	0.35 m
Gas inlet composition (mol%) CH_3OH	9
O_2	10
CH_2O	0.5
H_2O	2
CO	1
N_2	77.5

Regarding the radial profiles, according to the criteria listed in Table 6.7, they can be considered negligible having a radial aspect ratio $m = R/d_P = 3.6$, which is below the limit value of 4. Considering these assumptions, it is evident that the resulting model is mono-dimensional because it takes into account only for axial, or longitudinal, profiles along the reactor. In each point of the axial coordinate, an effectiveness factor calculation is performed to obtain a reaction-rate operative in that point and also determine a profile also for the effectiveness factors. On the basis of the exposed assumptions and introducing the molar flow rate relative to each component, the material balance can be expressed by the following simplified relations:

$$\frac{\mathrm{d}F_i}{\mathrm{d}z} = \rho_B \frac{\pi D^2}{4} \sum_{j=1}^{N_r} \gamma_{i,j} \overline{R}_j \quad i = 1, 2, \ldots, N_c \tag{6.163}$$

which can be derived from the following substitution in Eq. (6.141):

$$v = \frac{Q}{A} \quad QC_i = F_i \quad A = \frac{\pi D^2}{4} \tag{6.164}$$

The energy balance, represented by Eq. (6.142), must be simplified, as performed for the material balance, according to the assumed absence of radial profiles and to the presence of cooling fluid, as reported in the previous example. The heat transferred across the external surface, per unit of reactor volume, can be defined as follows:

$$q = \frac{U(T_C - T)\pi D \mathrm{d}z}{A \mathrm{d}z} \tag{6.165}$$

This term must be added algebraically, in the balance equation, to the heat associated with the reaction, thus yielding the following expression:

$$\left(\sum_{i=1}^{N_c} F_i C_{P_i}\right)\frac{dT}{dz} = \frac{\pi D^2}{4}\rho_B \sum_{j=1}^{N_r}(-\Delta H_j)\overline{R}_j + \pi D U(T_C - T) \tag{6.166}$$

The set of $N_c + 1$ ordinary differential Eqs. (6.163) and (6.166) can be integrated along the z direction to obtain the desired profiles of temperature and concentration. At each integration step along the axis of the reactor, the effectiveness factor for each reaction must be determined employing the previously described procedure. The adopted algorithm for axial integration is that of Runge–Kutta with a variable z step size, which is inversely proportional to the quantity dT/dz, so that a smaller step size results when the temperature variation increases corresponding to a steeper profile. The main result for this calculation is the temperature profile along the reactor, as shown in Fig. 6.46, from which it is evident that when the reaction mixture enters the reactor, the gas temperature increases rapidly due to the exothermic nature of the reactive system. As the methanol is consumed (see composition profile reported in Fig. 6.47), the main reaction rate begins to decrease, and the same trend is shown by the temperature. Another interesting result of the simulation is the profile of the effectiveness factor, which is shown in Fig. 6.48 for both of the considered reactions.

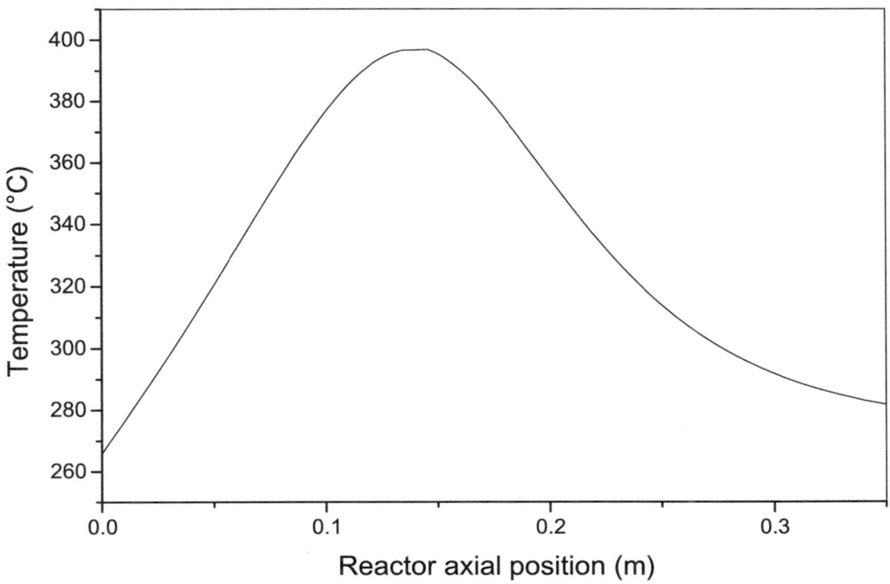

Fig. 6.46 Temperature profile for the reactor described in Table 6.9 and related to the oxidation of methanol to formaldehyde

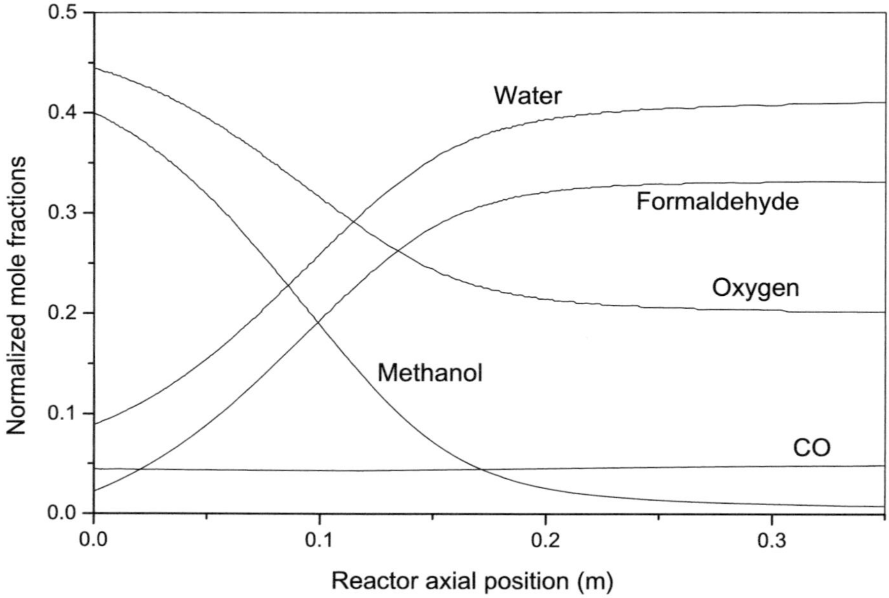

Fig. 6.47 Concentration profiles for the reactor described in Table 6.9 and related to the oxidation of methanol to formaldehyde

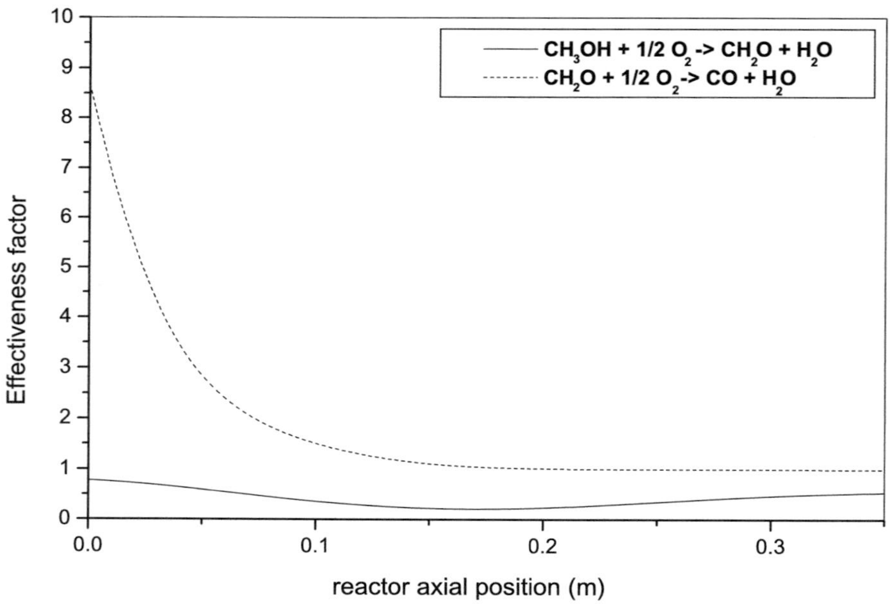

Fig. 6.48 Effectiveness-factor profile for the reactor described in Table 6.9 and for the two reactions occurring in the oxidation of methanol to formaldehyde

Appendices

Appendix 1: Lennard–Jones Force Constants Calculated From Viscosity Data

Compound	ϵ/k, °K	σ, A	Compound	ϵ/k, °K	σ, A
Acetylene	185	4.221	Hydrogen	33.3	2.968
Air	97	3.617	Hydrogen chloride	360	3.305
Argon	124	3.418	Hydrogen iodide	324	4.123
Arsine	281	4.06	Iodine	550	4.982
Benzene	440	5.270	Krypton	190	3.61
Bromine	520	4.268	Methane	136.5	3.822
i-Butane	313	5.341	Methanol	507	3.585
n-Butane	410	4.997	Methylene chloride	406	4.759
Carbon dioxide	190	3.996	Methyl chloride	855	3.375
Carbon disulfide	488	4.438	Mercuric iodide	691	5.625
Carbon monoxide	110	3.590	Mercury	851	2.898
Carbon tetrachloride	327	5.881	Neon	35.7	2.789
Carbonyl sulfide	335	4.13	Nitric oxide	119	3.470
Chlorine	357	4.115	Nitrogen	91.5	3.681
Chloroform	327	5.430	Nitrous oxide	220	3.879
Cyanogen	339	4.38	*n*-Nonane	240	8.448
Cyclohexane	324	6.093	*n*-Octane	320	7.451
Ethane	230	4.418	Oxygen	113	3.433
Ethanol	391	4.455	*n*-Pentane	345	5.769
Ethylene	205	4.232	Propane	254	5.061
Fluorine	112	3.653	Sulfur dioxide	252	4.290
Helium	10.22	2.576	Water	356	2.649
n-Heptane	282	8.88	Xenon	229	4.055
n-Hexane	413	5.909			

See Satterfield and Sherwood (1963), Hirschfelder et al. (1949, 1954), Rowlinson and Townley (1953)

Appendix 2: Collision Integrals Ω_μ and Ω_D as a Function of $T^* = K_B T/\varepsilon$ for Apolar Molecules: Lennard–Jones Approach

As explained in the text, the following correlation was used for interpolating both of the collision integrals:

$$\Omega_i = 10^{\left(ax^6+bx^5+cx^4+dx^3+ex^2+fx+g\right)} \quad \text{where} \quad x = \log_{10}(T^*)$$

The best-fitting coefficients for the polynomials related to Ω_D and Ω_μ are listed in the following table.

Integral	*a*	*b*	*c*	*d*	*e*	*f*	*g*
Ω_D	−0.0120	0.0877	−0.2146	0.1426	0.1948	−0.4848	0.1578
Ω_μ	−0.0165	0.1204	−0.3011	0.2360	0.1708	−0.4922	0.1997

Data to be interpolated are reported in the following table.

T^*	Ω_D	Ω_μ	T^*	Ω_D	Ω_μ	T^*	Ω_D	Ω_μ
0.30	2.6620	2.7850	1.65	1.1530	1.2640	4.00	0.8836	0.9700
0.35	2.4760	2.6280	1.70	1.1400	1.2480	4.10	0.8788	0.9649
0.40	2.3180	2.4920	1.75	1.1280	1.2340	4.20	0.8740	0.9600
0.45	2.1840	2.3680	1.80	1.1160	1.2210	4.30	0.8694	0.9553
0.50	2.0660	2.2570	1.85	1.1050	1.2090	4.40	0.8652	0.9507
0.55	1.9660	2.1560	1.90	1.0940	1.1970	4.50	0.8610	0.9464
0.60	1.8770	2.0650	1.95	1.0840	1.1860	4.60	0.8568	0.9422
0.65	1.7980	1.9820	2.00	1.0750	1.1750	4.70	0.8530	0.9382
0.70	1.7290	1.9080	2.10	1.0570	1.1560	4.80	0.8492	0.9343
0.75	1.6670	1.8410	2.20	1.0410	1.1380	4.90	0.8456	0.9305
0.80	1.6120	1.7800	2.30	1.0260	1.1220	5.00	0.8422	0.9269
0.85	1.5620	1.7250	2.40	1.0120	1.1070	6.00	0.8124	0.8963
0.90	1.5170	1.6750	2.50	0.9996	1.0930	7.00	0.7896	0.8727
0.95	1.4760	1.6290	2.60	0.9878	1.0810	8.00	0.7712	0.8538
1.00	1.4390	1.5870	2.70	0.9770	1.0690	9.00	0.7556	0.8379
1.05	1.4060	1.5490	2.80	0.9672	1.0580	10.00	0.7424	0.8242
1.10	1.3750	1.5140	2.90	0.9576	1.0480	20.00	0.6640	0.7432
1.15	1.3460	1.4820	3.00	0.9490	1.0390	30.00	0.6232	0.7005
1.20	1.3200	1.4520	3.10	0.9406	1.0300	40.00	0.5960	0.6718
1.25	1.2960	1.4240	3.20	0.9328	1.0220	50.00	0.5756	0.6504
1.30	1.2730	1.3990	3.30	0.9256	1.0140	60.00	0.5596	0.6335
1.35	1.2530	1.3750	3.40	0.9186	1.0070	70.00	0.5464	0.6194
1.40	1.2330	1.3530	3.50	0.9120	0.9999	80.00	0.5352	0.6076
1.45	1.2150	1.3330	3.60	0.9058	0.9932	90.00	0.5256	0.5973
1.50	1.1980	1.3140	3.70	0.8998	0.9870	100.00	0.5130	0.5882
1.55	1.1820	1.2960	3.80	0.8942	0.9811	200.00	0.4644	0.5320
1.60	1.1670	1.2790	3.90	0.8888	0.9755	300.00	0.4360	0.5016
						400.00	0.4170	0.4811

Data from Hirschfelder et al. (1954)

The coefficients were obtained using a MATLAB program importing all data from an Excel file. By applying the correlations found to the collision integrals, average absolute percent errors of 0.1396 and 0.2343% are obtained for, respectively, Ω_D and Ω_μ. Plots of the fittings obtained with the same program are reported in the following figures.

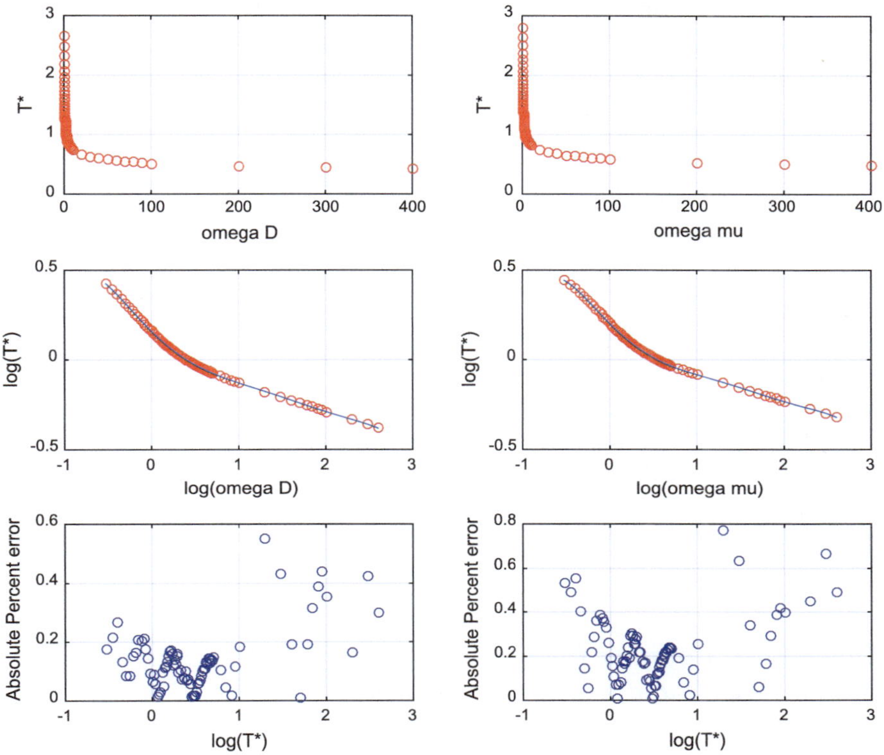

Figures related to Appendix 2. Fittings obtained by mathematical regression analysis on the data of Ω_D and Ω_μ available in the literature and plots of the errors.

These results can be obtained using a MATLAB program available as Electronic Supplementary Material.

Appendix 3: Parameters of the Stockmayer Equation for Some Polar Molecules

Substance	Dipol moment μ_d(D)	σ $(\AA)$	$\epsilon_o/k_B(k)$	$\delta = \frac{\mu_d^2}{2\epsilon_o\sigma^3}$
H_2O	1.85	2.52	775	1.0
NH_3	1.47	3.15	358	0.7
HCl	1.08	3.36	328	0.34
HBr	0.80	3.41	417	0.14
HI	0.42	4.13	313	0.029
SO_2	1.63	4.04	347	0.42
H_2S	0.92	3.49	343	0.21
NOCl	1.83	3.53	690	0.4

(continued)

Substance	Dipol moment μ_d(D)	σ (Å)	$\epsilon_o/k_B(k)$	$\delta = \frac{\mu_d^2}{2\epsilon_o\sigma^3}$
$CHCl_3$	1.013	5.31	355	0.07
CH_2Cl_2	1.57	4.52	483	0.2
CH_3Cl	1.87	3.94	414	0.5
CH_3Br	1.80	4.25	382	0.4
C_2H_5Cl	2.03	4.45	423	0.4
CH_3OH	1.70	3.69	417	0.5
C_2H_5OH	1.69	4.31	431	0.3
n-C_3H_7OH	1.69	4.71	495	0.2
i-C_3H_7OH	1.69	4.64	518	0.2
$(CH_3)_2O$	1.30	4.21	432	0.19
$(C_2H_5)_2O$	1.15	5.49	362	0.08
$(CH_3)_2CO$	1.20	3.82	428	1.3
CH_3COOCH_3	1.72	5.04	418	0.2
$CH_3COOC_2H_5$	1.78	5.24	499	0.16
CH_3NO_2	2.15	4.16	290	2.3

1 D (Debye) = 1×10^{-18} statcoulomb × cm = 1×10^{-18} dine $^{1/2}$ × cm^2

Appendix 4: Collision Integral Ω_μ as a Function of T = K_BT/ε and δ for Polar Molecules*

The following correlations were used for determining both the collision integrals:

$$f_1 = d_1 + d_2\frac{\delta}{T^*} + d_3\frac{\delta^2}{T^{*2}} + d_4\frac{\delta^3}{T^*} + d_5\delta^4$$

$$f_2 = \frac{f_1\delta^{1.5}}{K^{\log_{10}(T^*)}}$$

$$f_3 = a_1x^6 + a_2x^5 + a_3x^4 + a_4x^3 + a_5x^2 + a_6x + a_7$$

$$\Omega_i = 10^{f_2+f_3} \quad \text{where} \quad x = \log_{10}(T^*) \quad i = D \text{ or } \mu$$

The best-fitting coefficients for the functions related to Ω_D and Ω_μ are listed the following table.

	Collision integrals	
Parameter	Ω_D	Ω_μ
d_1	0.066225	0.067498
d_2	−0.002888	−0.002375
d_3	0.0000707	0.0000618
d_4	−0.0000626	−0.0000865
d_5	−0.0000785	−0.0001022
K	4.507	3.934

(continued)

Parameter	Collision integrals Ω_D	Ω_μ
a_1	0.010254	0.010948
a_2	−0.033249	−0.039147
a_3	−0.014026	−0.005066
a_4	0.096320	0.105559
a_5	0.068759	0.049660
a_6	−0.434055	−0.425989
a_7	0.163439	0.203885

These coefficients were determined by mathematical regression analysis made on the data reported by the literature and summarized in the following two tables.

Ω_D								
T^*	δ							
	0.00	0.25	0.50	0.75	1.00	1.50	2.00	2.50
0.10	4.00790	4.00200	4.65500	5.52100	6.45400	8.21300	9.52400	11.31000
0.20	3.13000	3.16400	3.35500	3.72100	4.19800	5.23000	6.22500	7.16000
0.30	2.64940	2.65700	2.77000	3.00200	3.31900	4.05400	4.78500	5.48300
0.40	2.31440	2.32000	2.40200	2.57200	2.81200	3.38600	3.97200	4.53900
0.50	2.06610	2.07300	2.14000	2.27800	2.47200	2.94600	3.43700	3.91800
0.60	1.87670	1.88500	1.94400	2.06000	2.22500	2.62800	3.05400	3.47400
0.70	1.72930	1.73800	1.79100	1.89300	2.03600	2.38800	2.76300	3.13700
0.80	1.62200	1.62200	1.67000	1.76000	1.88600	2.19800	2.53500	2.87200
0.90	1.51750	1.52700	1.57200	1.65300	1.76500	2.04400	2.34900	2.65700
1.00	1.43980	1.45000	1.49000	1.56400	1.66500	1.91700	2.19600	2.47800
1.20	1.32040	1.33000	1.36400	1.42500	1.50900	1.72000	1.95600	2.19900
1.40	1.23360	1.24200	1.27200	1.32400	1.39400	1.57300	1.77700	1.99000
1.60	1.16790	1.17600	1.20200	1.24600	1.30600	1.46100	1.63900	1.82700
1.80	1.11660	1.12400	1.14600	1.18500	1.23700	1.37200	1.53000	1.69800
2.00	1.07530	1.08200	1.10200	1.13500	1.18100	1.30000	1.44100	1.59200
2.50	1.00060	1.00500	1.02000	1.04600	1.18000	1.17000	1.27800	1.39700
3.00	0.95003	0.95380	0.96560	0.98520	1.01200	1.08200	1.16300	1.26500
3.50	0.91311	0.91620	0.92560	0.94130	0.96260	1.01900	1.09000	1.17000
4.00	0.88453	0.88710	0.89480	0.90760	0.92520	0.97210	1.03100	1.09800
5.00	0.84277	0.84460	0.85010	0.85920	0.87160	0.90530	0.94830	0.99840
6.00	0.81827	0.81420	0.81830	0.82510	0.83440	0.85980	0.89270	0.93160
7.00	0.78976	0.79080	0.79400	0.79930	0.80660	0.82650	0.85260	0.88360
8.00	0.77111	0.77200	0.77450	0.77880	0.78460	0.80070	0.82190	0.84740
9.00	0.75553	0.75620	0.75840	0.76190	0.76670	0.78000	0.79760	0.81890
10.00	0.74220	0.74280	0.74460	0.74750	0.75150	0.76270	0.77760	0.79570
12.00	0.72022	0.72060	0.72200	0.72410	0.72710	0.73540	0.74640	0.76000
14.00	0.70254	0.70290	0.70390	0.70550	0.70780	0.71420	0.72280	0.73340
16.00	0.68776	0.68800	0.68880	0.69010	0.69190	0.69700	0.70400	0.71250
18.00	0.67510	0.67530	0.67600	0.67700	0.67850	0.68270	0.68840	0.69550
20.00	0.66405	0.66420	0.66480	0.66570	0.66690	0.67040	0.67520	0.68110

(continued)

Ω_D								
25.00	0.64136	0.64150	0.64180	0.64250	0.64330	0.64570	0.64900	0.65310
30.00	0.62350	0.62360	0.62390	0.62430	0.62490	0.62670	0.62910	0.63210
35.00	0.60882	0.60890	0.60910	0.60940	0.60990	0.61120	0.61310	0.61540
40.00	0.59640	0.59640	0.59660	0.59690	0.59720	0.59830	0.59980	0.60170
50.00	0.57626	0.57630	0.57640	0.57660	0.57680	0.57750	0.57850	0.57980
75.00	0.54146	0.54150	0.54160	0.54160	0.54180	0.54210	0.54240	0.54290
100.00	0.51803	0.51810	0.51820	0.51840	0.51840	0.51850	0.51860	0.51870

Ω_μ							
T^*	δ						
	0.00	0.25	0.50	0.75	1.00	1.50	2.00
0.10	4.10050	4.26600	4.83300	5.74200	6.62900	8.62400	10.34000
0.20	3.26260	3.30500	3.51600	3.91400	4.43300	5.57000	6.63700
0.30	2.83990	2.83600	2.93600	3.16800	3.51100	4.32900	5.12600
0.40	2.53100	2.52200	2.58600	2.74900	3.00400	3.64000	4.28200
0.50	2.28370	2.27700	2.32900	2.46000	2.66500	3.18700	3.72700
0.60	2.08380	2.08100	2.13000	2.24300	2.41700	2.86200	3.32000
0.70	1.92200	1.92400	1.97000	2.07200	2.22500	2.61400	3.02800
0.80	1.79020	1.79500	1.84000	1.93400	2.07000	2.41700	2.78800
0.90	1.68230	1.68900	1.73300	1.82000	1.94400	2.25800	2.59600
1.00	1.59290	1.60100	1.64400	1.72500	1.83800	2.12400	2.43500
1.20	1.45510	1.46500	1.50400	1.57400	1.67000	1.91300	2.18100
1.40	1.35510	1.36500	1.40000	1.46100	1.54400	1.75400	1.98900
1.60	1.28000	1.28900	1.32100	1.37400	1.44700	1.63000	1.83800
1.80	1.22190	1.23100	1.25900	1.30600	1.37000	1.53200	1.71800
2.00	1.17570	1.18400	1.20900	1.25100	1.30700	1.45100	1.61800
2.50	1.09330	1.10000	1.11900	1.15000	1.19300	1.30400	1.43500
3.00	1.03880	1.04400	1.05900	1.08300	1.11700	1.20400	1.31000
3.50	0.99630	1.00400	1.01600	1.03500	1.06200	1.13300	1.22000
4.00	0.96988	0.97320	0.98300	0.99910	1.02100	1.07900	1.15300
5.00	0.92676	0.92910	0.93600	0.94730	0.96280	1.00500	1.05800
6.00	0.89616	0.89790	0.90300	0.91140	0.92300	0.95450	0.99550
7.00	0.87272	0.87410	0.87800	0.88450	0.89350	0.91810	0.95050
8.00	0.85379	0.85490	0.85800	0.86320	0.87030	0.89010	0.91640
9.00	0.83795	0.83880	0.84140	0.84560	0.85150	0.86780	0.88950
10.00	0.82435	0.82510	0.82730	0.83080	0.83560	0.84930	0.86760
12.00	0.80184	0.80240	0.80390	0.80650	0.81010	0.82010	0.83370
14.00	0.78363	0.78400	0.78520	0.78720	0.78990	0.79760	0.80810
16.00	0.76834	0.76870	0.76960	0.77120	0.77330	0.77940	0.78780
18.00	0.75518	0.75540	0.75620	0.75750	0.75920	0.76420	0.77110
20.00	0.74364	0.74380	0.74450	0.74550	0.74700	0.75120	0.75690
25.00	0.71982	0.72000	0.72040	0.72110	0.72210	0.72500	0.72890
30.00	0.70097	0.70110	0.70140	0.70190	0.70260	0.70470	0.70760
35.00	0.68545	0.68550	0.68580	0.68610	0.68670	0.68830	0.69050

(continued)

Ω_μ							
40.00	0.67232	0.67240	0.67260	0.67280	0.67330	0.67450	0.67620
50.00	0.65099	0.65100	0.65120	0.65130	0.65160	0.65240	0.65340
75.00	0.61397	0.61410	0.61430	0.61450	0.61470	0.61480	0.61480
100.00	0.58870	0.58890	0.58940	0.59000	0.59030	0.59010	0.58950

A mathematical regression analysis was performed using a MATLAB program available as Electronic Supplementary Material.

By applying the correlations found in the calculation of the collision integrals, average absolute percent errors of 1.62 and 1.85% are obtained for, respectively, Ω_D and Ω_μ. The obtained fittings can be appreciated in the plots reported in the following figures.

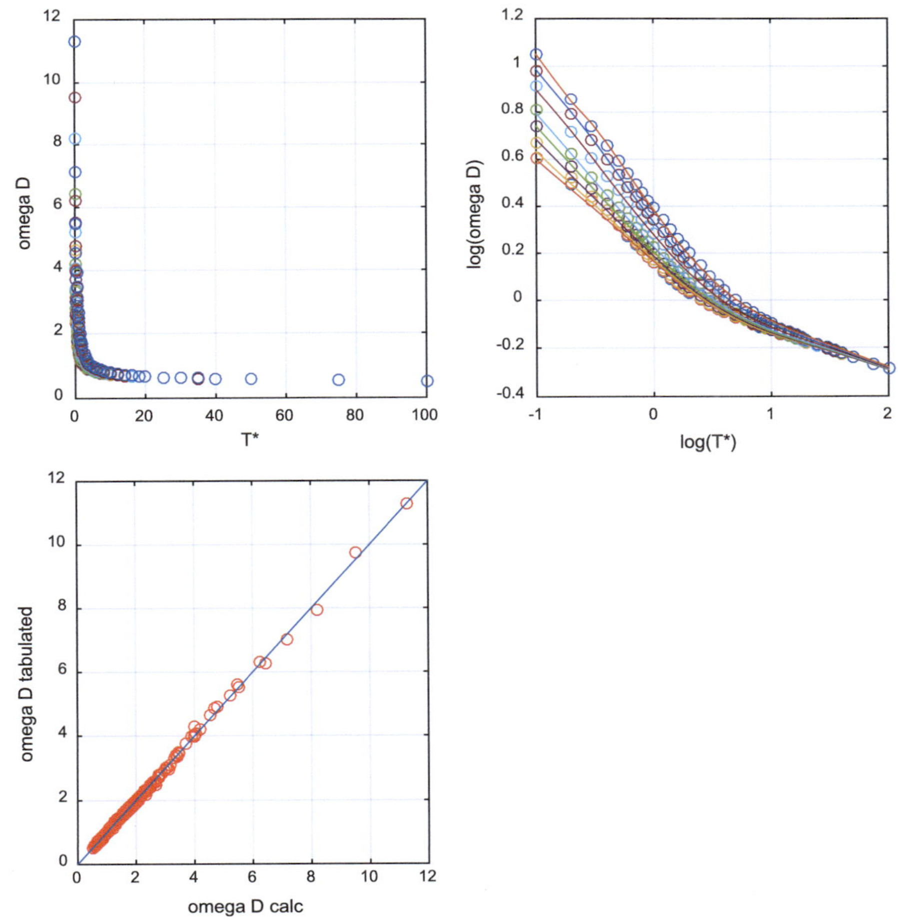

Figures 1 related to Appendix 4. Fittings obtained by mathematical regression analysis on the data of Ω_D available in the literature and parity plot.

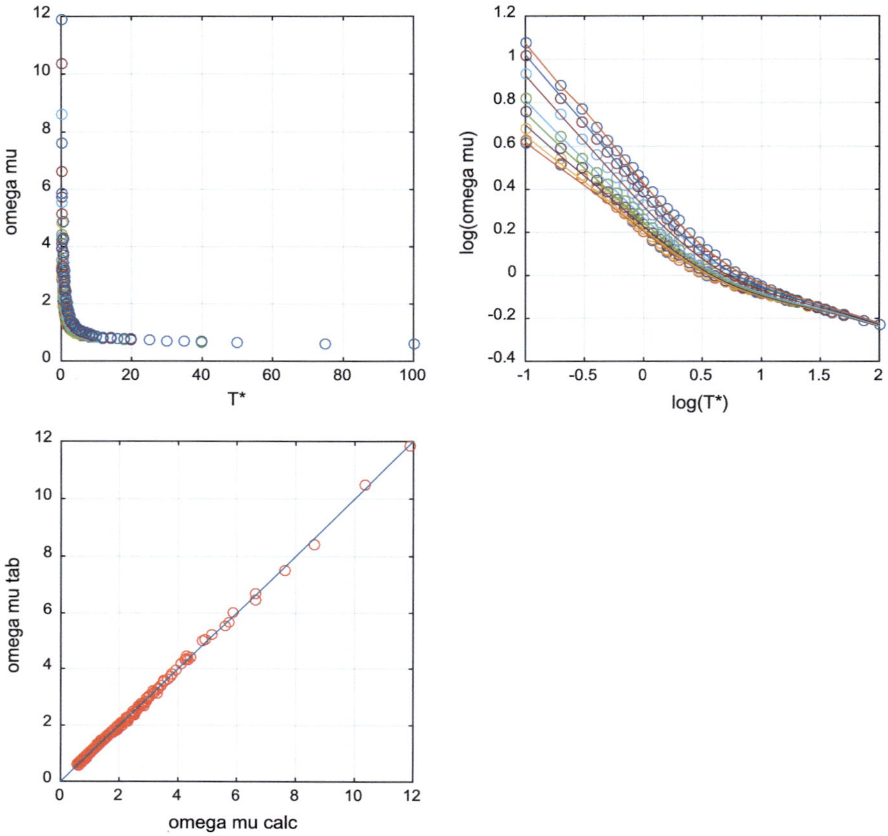

Figures 2 related to Appendix 4. Fittings obtained by mathematical regression analysis on the data of Ω_μ available in the literature and parity plot.

Appendix 5: Additive Volume Increments for the Estimation of the Molar Volume V_b *at Normal Boiling Point*

Substance	V_b increment, cm^3/g mol
Air	29.9
Ammonia	25
Bromine	27
Carbon	14.8
Chlorine, terminal, as R–CI	21.6
Medial, as R–CHC1–R	24.6
Fluorine	8.7

(continued)

Substance	V_b increment, cm^3/g mol
Helium	1.0
Hydrogen (in compound)	3.7
Hydrogen (molecular)	14.3
Mercury	15.7
Nitrogen	31.2
In primary amines	10.5
In secondary amines	12.0
Oxygen, molecular	14.8
Doubly bound	7.4
Methyl esters and ethers	9.1
Ethyl esters and ethers	9.9
Higher esters and ethers	11.0
Acids	12.0
Joined to S, P, or N	8.3
Phosphorus	27
Sulfur	25.6
Rings: 3-membered	−6
4-membered	−8.5
5-membered	−11.5
6-membered	−15
Naphthalene	−30
Anthracene	−47.5

References

Aris, R.: On shape factors for irregular particles—I: The steady state problem. Diffusion and reaction. Chem. Eng. Sci. **6**(6), 262–268 (1957)

Bird, R.B., Stewart, W.E., Lightfoot, E.N.: Transport Phenomena. Wiley (1960)

Bird, R.B., Stewart, W.E., Lightfoot, E.N.: Transport Phenomena Italian Edition Casa Editrice Ambrosiana (1970)

Bradshaw, R.D., Bennet, C.O.: Fluid-particle mass transfer in a packed bed. AICHE J. **7**(1), 48–52 (1961)

Bridgman, P.W.: The thermal conductivity of liquids. Proc. Natl. Acad. Sci. USA. **9**(10), 341–345 (1923)

Brush, G.: Kinetic Theory. In: The Nature of Gases and of Heat, vol. 1, Oxford (1965)

Carberry, J.J.: A boundary-layer model of fluid-particle mass transfer in fixed beds. A.I.Ch. E. J. **6**(3) (1960)

Carberry, J.J.: Physico-chemical aspects of mass and heat transfer in heterogeneous catalysis (Chap. 3). In: Anderson, J.R., Boudart, M. (eds.) Catalysis, vol. 8, pp. 131–171. Springer, Berlin (1987)

Carrà, S., Forni, L.: Aspetti Cinetici della Teoria del Reattore Chimico. Tamburini Ed. (1974)

Carrà, S., Ragaini, V., Zanderighi, L.: Operazioni di Trasferimento di Massa. Manfredi Editore, Milano (1969)

Chapman, S.: The kinetic theory of simple and composite monatomic gases: viscosity, thermal conduction, and diffusion. Proc. Roy. Soc. London A. **93**, 1–20 (1916)

Chapman, S., Cowling, T.G.: The Mathematical Theory of Non-Uniform Gases, 3rd edn. Cambridge University Press (1970)

Chilton, T.C., Colburn, A.P.: Mass transfer (absorption) coefficients prediction from data on heat transfer and fluid friction. Ind. Eng. Chem. **26**(11), 1183–1187 (1934)

De Acetis, J., Thodos, G.: Mass and heat transfer in flow of gases through spherical packings. Ind. Eng. Chem. **52**(12), 1003–1006 (1960)

Dwydevi, P.N., Upadhay, S.N.: Particle-fluid mass transfer in fixed and fluidized beds. Ind. Eng. Chem. Process. Des. Dev. **16**, 157 (1977)

Einstein, A.: Über die von der molekularkinetischen Theorie der Wärme geforderte Bewegung von in ruhenden Flüssigkeiten suspendierten Teilchen. Ann. J. Physik. **17**, 549–561 (1905)

Enskog, D.: Kinetische Theorie der Vorgänge in mässig verdiinten Gases (Almqvist and Wiksells, Uppsala, (1917); translation by S.G. Brush in Kinetic Theory Vol. 3, Pergamon, Oxford (1965)

Fairbanks, D.F., Wilke, C.R.: Diffusion coefficients in multicomponent gas mixtures. Ind. Eng. Chem. **42**(3), 471–475 (1950)

Fogler, H.S.: Elements of Chemical Reaction Engineering. Prentice Hall Int. Editions (1986)

Forni, L.: Fenomeni di Trasporto. Edizioni Cortina Milano (1979)

Froment, G.F.: Fixed bed catalytic reactors—current design status. Ind. Eng. Chem. **59**(2), 18–27 (1967)

Froment, G.F., Bischoff, K.B.: Chemical Reactor Analysis and Design. Wiley, New York (1990)

Frössling, N.: Über die Verdunstung fallender Tropfen. Gerlands Beitr. Geophys. **52**, 170–216 (1938)

Gimeno, M.P., Gascon, J., Tellez, C., Herguido, J., Menedez, M.: Selective oxidation of o-xylene to phthalic anhydride over V_2O_5/TiO_2: kinetic study in a fluidized bed reactor. Chem. Eng. Process. **47**(9–10), 1844–1852 (2008)

Hirschfelder, J.O., Bird, R.B., Spotz, E.L.: The transport properties of gases and gaseous mixtures. Chem. Revs. **44**(1), 205–231 (1949)

Hirschfelder, J.O., Curtiss, C.F., Bird, R.B.: Molecular theory of gases and liquids. Wiley, New York (1954)

Holland, C.D., Anthony, R.G.: Fundamentals of Chemical Reaction Engineering. Prentice-Hall, London (1979)

Horak, J., Pasek, J.: Design of Industrial Chemical Reactors from Laboratory Data. Heyden, London (1978)

Johnson, P.A., Babb, A.L.: Liquid diffusion of non-electrolytes. Chem. Rev. **56**, 387–453 (1956)

Lee, H.H.: Heterogeneous Reactor Design. Butterworth Pu (1984)

Levenspiel, O.: The Chemical Reactor Omnibook. OSU Book Store, Oregon (1984)

Missen, R.W., Mims, C.A., Saville, B.A.: Introduction to Chemical Reaction Engineering and Kinetics. Wiley, New York (1999)

Ranz, W.E., Marshall Jr., W.R.: Evaporation from drops. Chem. Eng. Prog. **48**(3), 141–146 (1952a)

Ranz, W.E., Marshall Jr., W.R.: Evaporation from drops part II. Chem. Eng. Prog. **48**(4), 173–180 (1952b)

Rase, H.F.: Chemical Reactor Design for Process Plant, Vol. 2: Case Study N. 109, pp. 115–122j. Wiley, New York (1977)

Riggs, J.B.: Introduction to numerical methods in chemical engineering. Texas Tech University Press (1988)

Rowlinson, J.S., Townley, J.R.: The application of the principle of corresponding states to the transport properties of gases. Trans. Faraday Soc. **49**, 20–27 (1953)

Santacesaria, E.: Kinetics and transport phenomena in heterogeneous gas-solid and gas-liquid-solid systems. Catal. Today. **34**(3–4), 411–420 (1997)

Satterfield, C.N., Sherwood, T.K.: The Role of Diffusion in Catalysis. Addison Wesley Pu. Co. Inc (1963)

Satterfield, C.N.: Heterogeneous Catalysis in Practice. Addison-Wesley (1972)

Satterfield, C.N., Cortez, D.H.: Mass transfer characteristics of woven-wire screen catalysts. Ind. Eng. Chem. Fundam. **9**(4), 613–620 (1970)

Smith, J.M.: Chemical Engineering Kinetics. Mc Graw-Hill Book Co., New York (1981)

Stull, D.R., Westrum, E.F., Sinke, G.C.: The Chemical Thermodynamics of Organic Compounds. Wiley, New York (1969)

Thoenes, D., Kramers, H.: Mass transfer from spheres in various regular packings to a flowing fluid. Chem. Eng. Sci. **8**(3–4), 271–283 (1958)

Treybal, R.E.: Mass Transfer Operations. Mc Graw-Hill Co., New York (1955)

Weisz, P.B., Hicks, J.S.: The behavior of porous catalyst particles in view of internal mass and heat diffusion effects. Chem. Eng. Sci. **17**, 265–275 (1962)

Weisz, P.B., Prater, C.D.: Interpretation of measurements in experimental catalysis. Adv. Catal. **6**, 143–196 (1954)

Wilke, C.R., Chang, P.: Correlation of diffusion coefficients in dilute solutions. AICHE J. **1**(2), 264–270 (1955)

Winterbottom, J.M., King, M.: Reactor Design for Chemical Engineers, 1st edn. CRC Press (1999)

Chapter 7
Kinetics and Transport Phenomena in Multiphase Reactors

Abstract This chapter is dedicated to the topic of multiphase reactors, characterized by reactions occurring in a liquid phase in which one or more of the reactants, coming from another gaseous or liquid phase, are dissolved. The catalyst promoting the reaction can be a component of the reacting mixture (e.g., gas–liquid reactors), a solid wetted by liquid phase (gas–liquid–solid reactors), or, more seldom, dissolved in another immiscible liquid (liquid–liquid or gas–liquid–liquid reactors). Modeling such systems is challenging because chemical, physical, and fluid dynamic factors are all involved, sometimes giving place to peculiar phenomena. The occurrence of a reaction in the liquid phase containing the catalyst determines the occurrence of gradients at the interfaces, followed by a mass-transfer flow across the interface. The main theories presented in this chapter are two-film theory, surface renewal theory, and penetration theory. Examples presented in this chapter include the oxidation of tetra-hydro-anthraquinone ($THEAQH_2$) and o-cresol alkylation. The MATLAB code associated with these examples is available online.

7.1 Introduction

Multiphase reactors are characterized by reactions occurring in a liquid phase in which one or more of the reactants, coming from another gaseous or liquid phase, are dissolved. The catalyst promoting the reaction can be a component of the reacting mixture (e.g., gas–liquid reactors), a solid wetted by the liquid phase (gas–liquid–solid reactors), or, more seldom, dissolved in another immiscible liquid (liquid–liquid or gas–liquid–liquid reactors).

Modeling multiphase reactors are a difficult problem because chemical, physical, and fluid dynamic factors are all involved, sometimes giving place to peculiar phenomena. The occurrence of the reaction in the liquid phase containing the catalyst determines the decrease of the reactant concentration and an increase of

Supplementary Information The online version contains supplementary material available at https://doi.org/10.1007/978-3-031-82502-6_7.

© The Author(s), under exclusive license to Springer Nature Switzerland AG 2025

R. Tesser, E. Santacesaria, *The Chemical Reactor from Laboratory to Industrial Plant*, https://doi.org/10.1007/978-3-031-82502-6_7

the reaction product, that is, the formation of gradients at the interfaces, followed by a mass-transfer flow across the interface. Mass-transfer flow occurs until thermodynamic equilibrium is reached, that is, when the concentration of the flowing substance at the interface corresponds to the solubility concentration of that substance. This means that, for correct modeling, we must also know all of the solubility parameters.

As mentioned in the previous chapter, interface thermal gradients are less important for the reactions occurring in the liquid phase compared with those occurring in the gaseous phase because of the high thermal conductivity of liquids.

However, we can classify the multiphase reactors as follows:

1. Gas–liquid reactors
2. Gas–liquid–solid reactors
3. Liquid-liquid phase reactors
4. Liquid–liquid–solid reactors

In this chapter, the kinetics and transport phenomena for the more frequently employed gas–liquid, and gas–liquid–reactors will be examined in detail, while some aspects of liquid–liquid, and liquid–liquid–solid reactors will be discussed.

7.2 Kinetics and Transport Phenomena in Gas–Liquid Reactors

Many different industrial processes are characterized by the presence of gas–liquid reactor units. Some examples are (1) homolytic liquid-phase oxidation using air as oxidant (see Table 4.22); (2) heterolytic oxidation of olefins to aldehydes or ketones, in particular, oxidation of ethylene to acetaldehyde [Wacker process (see Sect. 4. 4.12)]; (3) hydroformylations, such as the reactions of olefins with a mixture of CO and H_2; (4) carbonylation, that is, the reaction of the insertion of carbon monoxide in an organic molecule, such as methanol carbonylation, to obtain acetic acid; (5) hydrogenation using metal complexes; (6) hydrochlorination, such as the reaction of glycerol with hydrochloric acid, to obtain chlorohydrins in the first step and then epichlorohydrin; (7) chlorination; (8) polyethoxylation and polypropoxylation as well as other polymerization, where the monomer is gaseous; (9) bioreactors consuming oxygen of air; etc.

The absorption of a gas in a liquid occurs according to Fick's law until equilibrium solubility is reached. In the presence of a chemical reaction occurring in the liquid phase, the absorption rate is altered, and we can recognize two different limit conditions. For example, when a gas–liquid reaction is slow, the mass-transfer rate is limited by the chemical reaction rate, and this is well-described by the intrinsic kinetic law ("chemical regime"). In this case, the reaction occurs in the liquid bulk, and the concentration gradients at the interface are negligible. In contrast, in the case of a very fast reaction, the chemical reaction mainly occurs inside the hydrodynamic quiescent film at the liquid interface (boundary layer) together with the gas–liquid mass transfer. In this case, the reaction has the singular effect of enhancing the mass-transfer rate with respect to the physical absorption rate by a factor called the "enhancement factor." Intermediate conditions exist between these two limit conditions that require opportune mathematical treatment. Clearly, to describe adequately the mass-transfer rate in gas–liquid reactors, it is helpful to know the mass-transfer rate of absorption in the absence of the reaction and related parameters, that is, the mass-transfer coefficient and the solubility of the gas in the liquid.

To explain the experimentally observed phenomenon of mass transfer–rate enhancement, different theoretical models have been proposed in the literature, in particular, (1) the two-films theory proposed by Lewis and Whitman (1924); (2) the penetration theory proposed by Higbie (1935); and (3) the surface-renewal theory proposed by Danckwerts (1951a, b).

7.2.1 Two-Films Theory

The two-films theory is the oldest theory for describing gas–liquid mass transfer formulated by Lewis and Whitman (1924). Although the two-films theory is the less reliable of the mentioned theories, it remains the most popular because it is simple, intuitive, and works well in many cases. For this reason, this chapter will be considered as basic for describing gas–liquid reactors. Consider first, as a basic reference, the absorption of a gas in a liquid in the absence of any reaction. We can measure with an amperometric electrode how pure oxygen is absorbed by water or by another solvent that was gently stirred in a vessel without breaking the gas–liquid interface. Thus, we know the gas–liquid interface surface area. The electrode signal of the current is proportional to the oxygen concentration and can be relieved both in absorption and desorption as a function of time as shown in Fig. 7.1.

Data can be interpreted by applying Fick's law:

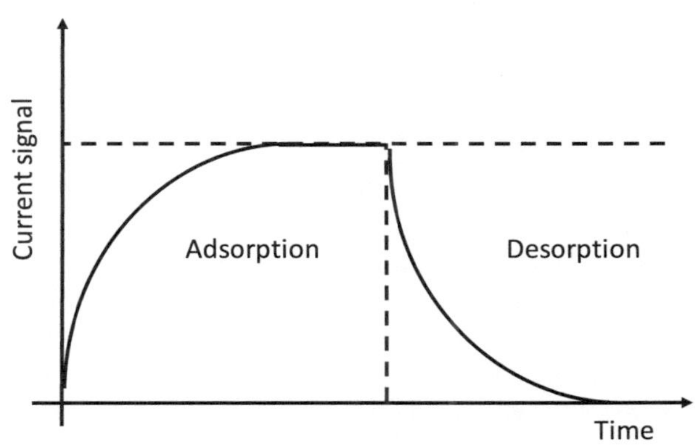

Fig. 7.1 Transient experiments for measuring the oxygen absorption in water using an amperometric electrode

Fig. 7.2 Concentration profiles in gas and liquid quiescent films at the interfaces

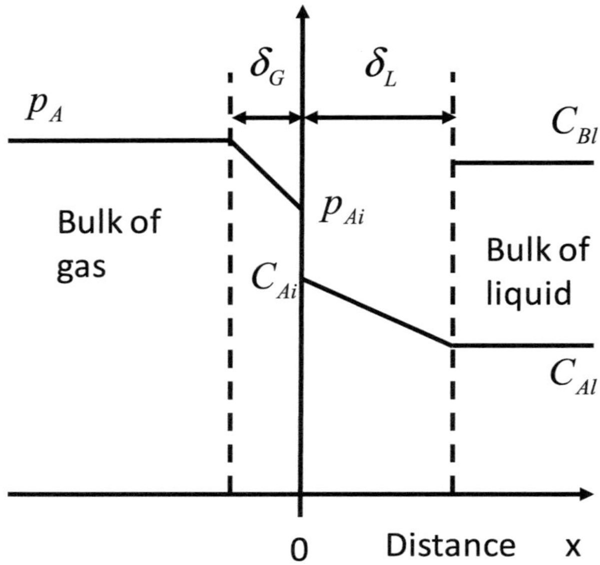

$$\pm\frac{\mathrm{d}C_\mathrm{A}}{\mathrm{d}t}=\pm\frac{\mathrm{d}i}{\mathrm{d}t}=k_\mathrm{L}a_\mathrm{L}\left(C_\mathrm{A}^{*}-C_\mathrm{A}\right)=k_\mathrm{L}a_\mathrm{L}\left(\frac{p_\mathrm{A}}{H_\mathrm{A}}-C_\mathrm{A}\right)=k_\mathrm{L}a_\mathrm{L}\left(i_\mathrm{sat}-i\right) \quad (7.1)$$

where C_A^{*} is the oxygen concentration at the interface that corresponds to the solubility equilibrium. For this reason, it has been substituted by $p_\mathrm{A}/H_\mathrm{A}$, where H_A is the Henry solubility parameter of oxygen in water $H_\mathrm{A}=p_\mathrm{A}/C_\mathrm{A}^{*}$. The solubility value can be better ascertained by the evaluation of the concentration corresponding to the value of the current obtained in the absorption run in correspondence with the final plateau. i, in this case, represents the measured current.

By integrating directly with respect to the current signal:

$$\ln\frac{1}{1-i/i_\mathrm{sat}}=k_\mathrm{L}a_\mathrm{L}t \quad (7.2)$$

Because in this case a_L is known, k_L can be determined by fitting the curves, such as the ones shown in Fig. 7.1.

In the two-film theory, we imagine that even if the two phases are well-mixed, two quiescent thick films are formed at the interface (boundary layers), a film of gas on the gas side and a liquid film in the liquid side. In these two films, diffusion occurs by a slow molecular diffusion mechanism that is relatively fast on the gas side and slow in the liquid one, considering a difference of three orders of magnitude in the diffusion coefficients passing from the gas to the liquid phase. Considering a second-order reaction of type A + zB → P, where A is the gaseous reagent, the occurrence of the reaction generates two gradients as shown in Fig. 7.2.

The two gradient profiles have been reasonably approximated to a linear trend, although it must be kept in mind that the actual concentration profiles are not linear. However, in such simplest case, occurring when the reaction rate is comparable with the mass overall mass-transfer rate, and assuming steady-state conditions, we can write, in agreement with Fick's law, two expressions for the mass-transfer rates, which are:

$$J_G = k_G a_{gl}(p_A - p_{Ai}) = \text{Mass transfer rate for gas-side interface} \tag{7.3}$$

$$J_L = k_L a_{gl}\left(\frac{p_{Ai}}{H} - C_{Al}\right) = \text{Mass transfer rate for liquid-side interface} \tag{7.4}$$

and an expression for the reaction rate, that is,

$$r = k_2 C_{Bl} C_{Al} \tag{7.5}$$

where k_G and k_L are the mass-transfer coefficients, the physical mean of which would be:

$$k_G = \frac{D_{AB\text{-}G}}{\delta_G} \qquad k_L = \frac{D_{AB\text{-}L}}{\delta_L} \tag{7.6}$$

where $D_{AB\text{-}G}$ is the molecular diffusion coefficient of A in the gaseous phase, δ_G is the gas-side film thickness, $D_{AB\text{-}L}$ is the molecular diffusion coefficient of A in the liquid phase, δ_L is the liquid-side film thickness, p_A is the partial pressure of A in the gas bulk, p_{Ai} is the value of the partial pressure at the interface, C_{Al} is the concentration of A in the liquid bulk, and H is the Henry solubility coefficient describing the partition of A at the equilibrium, that is, $H = \frac{p_{Ai}}{C_A^*}$. In other words, it has been assumed that at the interface A concentration $C_A{}^*$ instantaneously reaches equilibrium with A in the gaseous phase expressed by the partial pressure p_{Ai}.

Under steady-state conditions, we can equate the three rate expressions and write:

$$r = k_2 C_{Bl} C_{Al} = k_G a_{gl}(p_A - p_{Ai}) = k_L a_{gl}\left(\frac{p_{Ai}}{H} - C_{Al}\right) \tag{7.7}$$

This allows to eliminate p_{Ai}, thus obtaining a rate expression containing only measurable quantities. Another simple condition is when the reaction is slow and the system is in a chemical regime: in this case, the intrinsic kinetic law is valid because the mass transfer is rate limited by the chemical reaction rate, for example, in the described case $r = k_2 C_{Bl} C_{Al}$. However, as will be seen in the next paragraph, the situation is much more complicated when the reaction occurs partially or completely inside the liquid film instead of in the liquid bulk. In this case, we cannot assume a linear profile of the gradient in the liquid film and because the reaction and diffusion occur together inside the film, a more complicated model must be elaborated for describing the effective concentration profile in the film.

7.2.2 Penetration Theory

The "penetration theory" was proposed by Higbie (1935) (see also Danckwerts 1950, 1951a, b). Higbie assumed that each liquid element at the gas–liquid interface is exposed to the gas for a short time as shown in Fig. 7.3.

According to the Higbie theory: (1) mass transfer from the gas to a liquid element occurs according to an unsteady process when they are in contact; (2) each liquid element has the same contact time; and (3) equilibrium exists at the interface. Therefore, the Higbie penetration model, for its transient character, is a dynamic model and can be advantageously employed to describe a gas–liquid system operating under dynamic conditions.

The penetration theory describes the liquid-side mass-transfer coefficient as a function of the contact time τ and the molecular diffusivity D_{AB} of the gas in the liquid:

$$k_L = 2\sqrt{\left(\frac{D_{AB}}{\pi\tau}\right)} \tag{7.8}$$

As can be seen, in this theory, k_L is proportional to $\sqrt{D_{AB}}$, while in the two-films theory, it was proportional to D_{AB}.

7.2.3 Surface-Renewal Theory

The surface-renewal theory was developed by Danckwerts (1951a, b, 1970). In this theory, the liquid is considered divided into two portions: (1) a large well-mixed

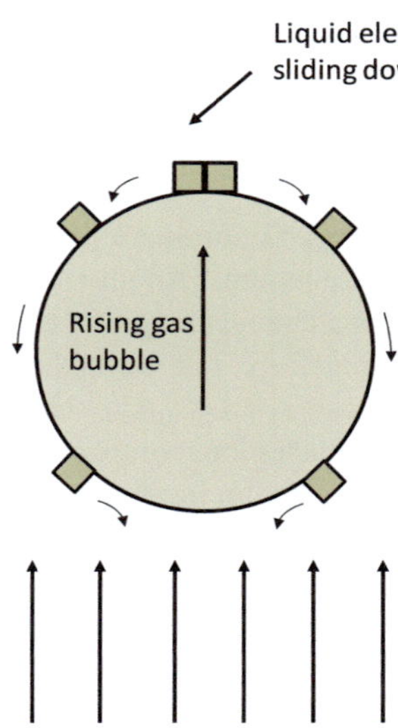

Fig. 7.3 Higbie model

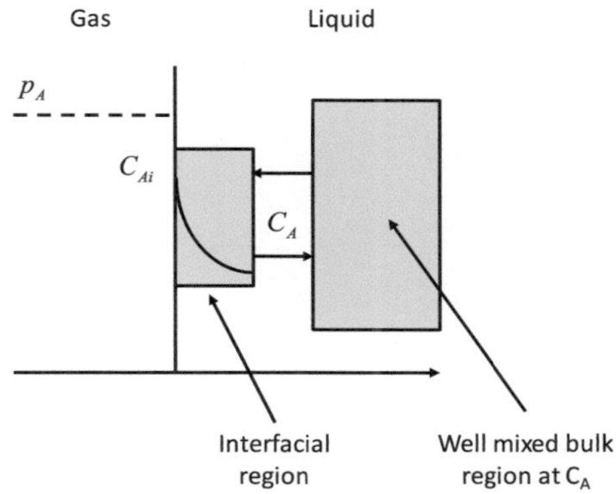

Fig. 7.4 Scheme of surface-renewal theory

portion, that is, the liquid bulk, and (2) an interfacial region of liquid elements corresponding to a thin film, which is rapidly renewed as shown in Fig. 7.4.

According to the theory, (1) the liquid elements at the interface are randomly substituted by fresh elements from the bulk; (2) each liquid element at the interface has the same probability of being substituted by a fresh element; and (3) mass transfer from the gas into the liquid element at the interface is a transient process. Instead of considering the contact time, as in the Higbie theory, Danckwerts introduced the concept of surface-renewal frequency (f) and expressed the liquid-side mass-transfer coefficient as a function of this parameter and of the molecular diffusivity of the gas in the liquid:

$$k_{\mathrm{L}} = \sqrt{fD_{\mathrm{AB}}} \tag{7.9}$$

Also, in this case k_{L} is proportional to $\sqrt{D_{\mathrm{AB}}}$.

As mentioned previously, both the penetration theory and the surface-renewal theory are more reliable than the two-films theory; nevertheless, this last theory is much more frequently used in the simulation of gas–liquid reactors for its simplicity and because in practice it works well. Therefore, we will use the two-films theory to examine the more complex cases in which the reaction occurs inside the liquid film together with the mass transfer.

7.2.4 *Application of the Two-Films Theory to the Elaboration of Kinetic and Mass-Transfer Data*

Let us consider again a second-order reaction of the type: A + zB → P. In Sect. 7.2.1, we described two possible situations, one in which the reaction was so slow that the rate can be expressed by the intrinsic kinetic law (chemical regime) and another one in which the reaction rate and mass-transfer rate are comparable and we can consider diffusion and reaction as consecutive separated steps. When the reaction rate is much

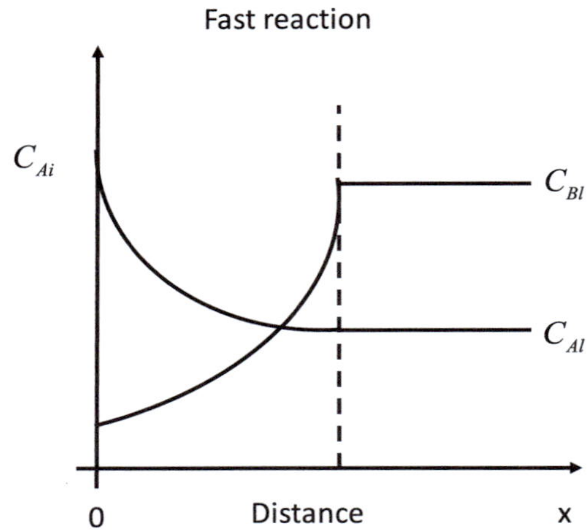

Fig. 7.5 Effect of a fast reaction on the concentration profile

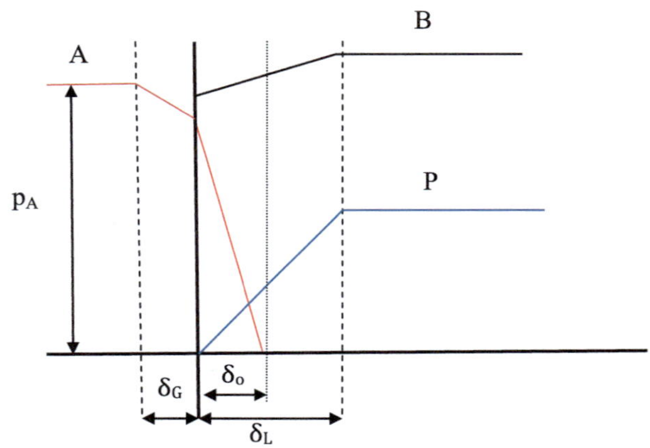

Fig. 7.6 Concentration gradients for a very fast second-order reaction

higher, part of the reaction occurs inside the liquid film and part in the liquid bulk, and the concentration profile cannot be considered to be linear (see Fig. 7.5).

Clearly, the extent of the reaction occurring in the liquid film increases by increasing the chemical reaction rate, and for extremely fast reactions, the reaction occurs completely inside the hydrodynamic film. This can determine a singular phenomenon that has been experimentally observed in some gas–liquid systems, that is, very fast reactions accelerate the mass transfer. This phenomenon can be well explained with the two-films theory by assuming that the diffused reactant is exhausted in the liquid film before reaching the limit of the boundary layer as shown in Fig. 7.6.

As can be seen, the concentration of the gaseous reactant A in the liquid phase is zeroed in correspondence with the distance δ_o, that is, before reaching the limit of the boundary layer corresponding to δ_L. Remembering that $k_L = (D_{AB}$/diffusion film thickness), we can conclude qualitatively that a very fast reaction determines the acceleration of the mass-transfer rate by increasing the value of k_L, which becomes $k_L = (D_{AB}/\delta_o)$ instead of $k_L = (D_{AB}/\delta_L)$.

If the reaction is extremely fast or instantaneous, the apparent mass-transfer coefficient also increases for one or two orders of magnitude. By comparing the observed mass-transfer rate in the presence of the reaction with the maximum rate observable in physical absorption, we can evaluate the so-called enhancement factor as:

$$E = \frac{\varphi_c}{\varphi_{max}} = \frac{\text{Gas absorption rate in the presence of the reaction}}{\text{Physical maximum absorption rate in the absence of the reaction } (C_{AI} = 0)} \tag{7.10}$$

This factor defines the effect of a fast reaction on the mass transfer.

To describe more rigorously and quantitatively the kinetic- and mass-transfer rate in gas–liquid reactors, we must evaluate the concentration profiles of the reactants and products in the liquid film. To do this, it is necessary to solve the mass-balance equations for a second-order reaction of the type:

$$\mathrm{A(gas)} + z\mathrm{B} \xrightarrow{k_2} \mathrm{P} \tag{7.11}$$

which corresponds to the equations reported below:

Mass balance

$$D_A\left(\frac{d^2 C_A}{dx^2}\right) - r_A = 0 \quad \text{with } r_A = k_2 C_A C_B \tag{7.12}$$

$$D_B\left(\frac{d^2 C_B}{dx^2}\right) - r_B = 0 \quad \text{with } r_B = z r_A \tag{7.13}$$

Boundary conditions

$$C_A = C_A^* = p_A/H \quad \text{(Henry law)} \quad x = 0 \tag{7.14}$$

$$\frac{dC_B}{dx} \quad x = 0 \tag{7.15}$$

$$C_B = C_{Bb} \quad x = \delta_L \tag{7.16}$$

$$-D_A\left(\frac{dC_A}{dx}\right)_{x=\delta_L} = k_2 C_A C_B \left\{\frac{(1-\varepsilon_g)}{a} - \delta_L\right\} \tag{7.17}$$

This last boundary condition (Eq. 7.17) takes into account the amount of A reacting in the liquid film. The analytical solution of these mass-balance equations is possible only for a few very simple cases. More generally, the system of equations must be solved using a numerical method. This can be made more easily by introducing three dimensionless parameters, which are:

$$1.\quad \text{Hatta number}\, H_a = \frac{(D_A k_2 C_{Bb})^{1/2}}{k_L} \tag{7.18}$$

$$2.\quad Z_D = \left(\frac{D_B}{zD_A}\right)\left(\frac{C_{Bb}}{C_A^*}\right) = E_1 - 1 \tag{7.19}$$

$$3.\quad M = \left(\frac{1-\varepsilon_g}{a}\right)\left(\frac{k_L}{D_A}\right) \tag{7.20}$$

The results can be expressed in terms of enhancement factor E or Z_D as a function of the Hatta number. A complete solution of these mass-balance equations was performed by Charpentier (1981), and the solution is shown in Fig. 7.7, where the

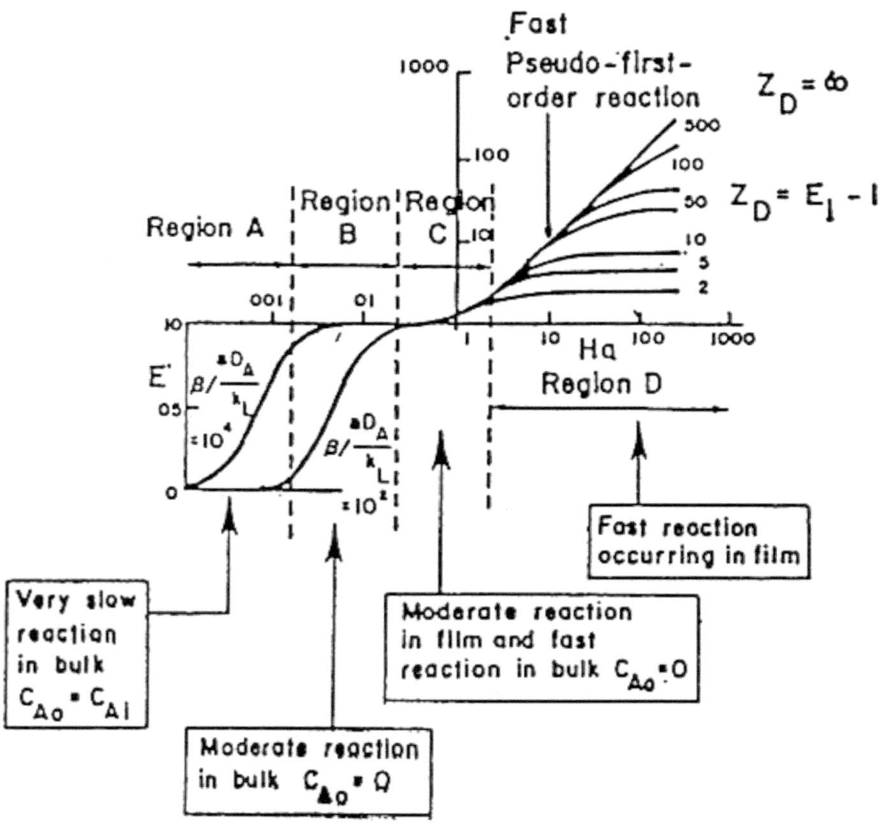

Fig. 7.7 Kinetic regimes for gas–liquid reactions: The enhancement factor as a function of the Hatta number. With the permission from Charpentier (1981)

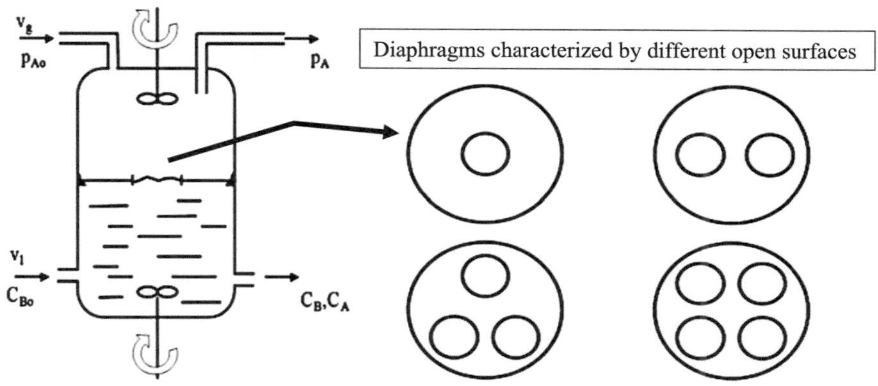

Fig. 7.8 Laboratory reactor proposed by Levenspiel for studying gas–liquid reactions

enhancement factor E was calculated as a function of the Hatta number H_a. In this plot, four regimes can be recognized as characterized by an increasing reaction rate.

Therefore, to study the kinetics and mass transfer of gas–liquid reactions, first, it is important to individuate the regime of rate characterizing the reaction between the four mentioned possibilities because the knowledge of the operative regime allows the best choice of the reactor to be used in the study as well as the mathematical approach to be followed. For this purpose, Levenspiel (1972), and Levenspiel and Godfrey (1974), for example, developed a reactor with a variable interface area and a strategy for individuating the operative kinetic regime. The reactor is a continuous stirred-tank reactor (CSTR) with a double-stirring system agitating well both gas and liquid phase as can be seen in Fig. 7.8.

A slight modification was introduced by Santacesaria et al. (1987), using floatings instead of diaphragms to avoid the formation of foams (see Fig. 7.9). Different holed diaphragms allow for change in the gas–liquid interface area.

In the Levenspiel reactor, it is possible to change:

1. The flow rate of gas and liquid
2. The concentration of A and B
3. The liquid volume in the reactor
4. The interface area
5. The gas pressure (in a stainless-steel reactor)

Considering again the second-order reaction A(gas) + zB → P, because the reactor, a CSTR for both gas and liquid phase, the consumption of gas can be directly determined by measuring the change of concentration of both A and B, obviously considering the stoichiometric coefficient z.

According to Levenspiel, four different situations can be envisaged:

1. The mass transfer from gas to liquid is slow with respect to the reaction rate. The rate increases by increasing the stirring rate of the fan ("extremely fast reaction").

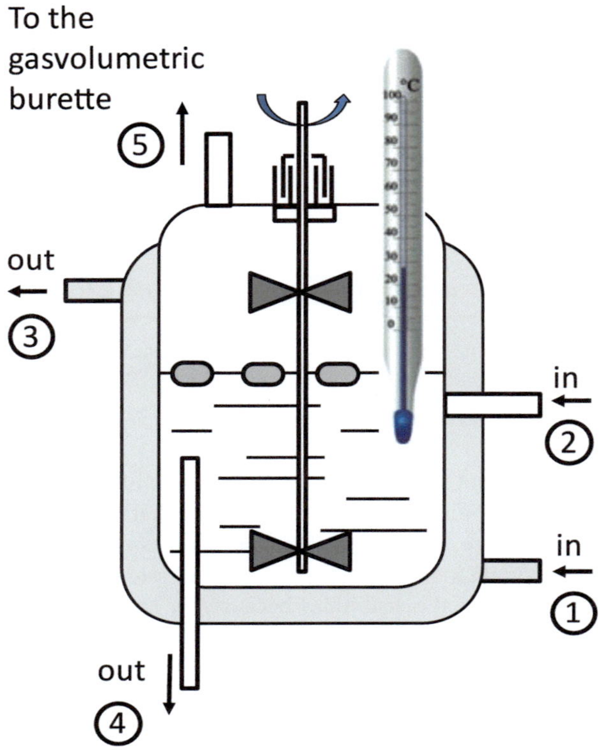

Fig. 7.9 Laboratory reactor proposed by Santacesaria et al. (1987) for studying gas–liquid reactions

2. The reaction is very fast and occurs mainly in the quiescent liquid film. The rate is affected by the gas–liquid interface area S but not by the liquid volume V_L ("very fast reaction").
3. The reaction rate and mass-transfer rate of the gaseous reagent are comparable. The rate depends on both the interface area S and V_L ("moderately fast reaction").
4. The reaction is very slow and is affected by V_L but not by S ("slow and very slow reactions").

Levenspiel then split these four possibilities into eight different regimes, considering also the effect of a high concentration of B with respect to A. For each regime, it is possible to give an expression of the reaction rate with the exclusion of regimes 5 and 6, requiring the numerical solution of the already reported mass-balance equations.

Regime 1
Extremely fast reaction with relatively low B concentration (see Fig. 7.10)

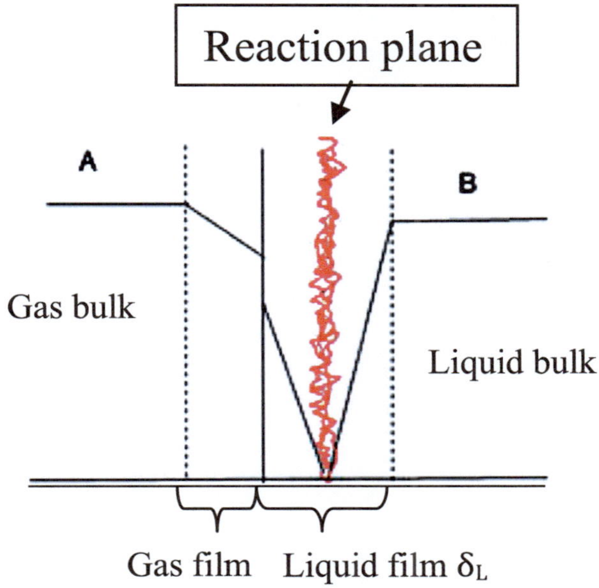

Fig. 7.10 Concentration profiles for regime 1

$$-r_A = -\frac{r_B}{z} = k_{Ag}(p_A - p_{Ai}) = k_{Al}(C_{Ai} - 0)\frac{\delta_L}{\delta_o} = \frac{k_{Bl}}{z}(C_B - 0)\frac{\delta_L}{\delta_L - \delta_o} \quad (7.21)$$

As previously seen, $p_{Ai} = H_A\ C_{Ai}$. Because the motion inside the liquid film occurs only by molecular diffusion in the case of the two-films theory, we can write:

$$\frac{k_{Al}}{k_{Bl}} = \frac{D_{Al}}{D_{Bl}} \quad (7.22)$$

By eliminating x, x_o, p_{Ai}, and C_{Ai} from the previous expressions, we obtain:

$$-r_A = -\frac{1}{S}\frac{dN_A}{dt} = \frac{\frac{D_{Bl}}{D_{Al}}\frac{C_B}{b} + \frac{p_A}{H_A}}{\frac{1}{H_A k_{A_g}} + \frac{1}{k_{Al}}}$$

Normally, the gas side resistance to the mass flow is negligible, that is, $k_{gA} \simeq \infty$; consequently, $p_A = p_{Ai}$. In this case, the previous relation becomes:

$$-r_A = k_{Al}C_{Ai}\left(1 + \frac{D_{Bl}C_B}{zD_{Al}C_{Ai}}\right) \quad (7.23)$$

By comparing this relation with the one of the maximum physical absorption rate, $-r_A = k_{Al}C_{ai}$, the term under the brackets corresponds to the enhancement factor E; thus, we can write:

Fig. 7.11 Concentration profiles for regime 2

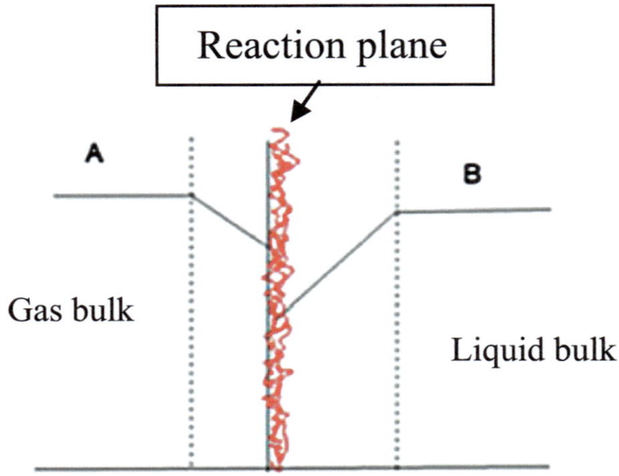

$$E = \text{Enhancement factor} = 1 + \frac{D_{Bl}C_B}{zD_{Al}C_{Ai}} \tag{7.24}$$

In conclusion, in this case, we can write:

$$-r_A = k_{Al}C_{Ai}E \tag{7.25}$$

Regime 2

Extremely fast reaction with high B concentration C_B

If the B concentration is high, the reaction plane is shifted toward the interface as shown in Fig. 7.11. In particular, if the following condition is valid:

$k_{Ag}p_A < \frac{k_{Bl}}{z}C_B$, only the gas-phase resistance to the mass transfer is operative, and we can write:

$$-r_A = -k_{Ag}p_A \tag{7.26}$$

This expression can be used when the following approximated condition is respected:

$$\frac{10^{-4}C_{Bl}}{z} > C_{Ag} = \frac{p_A}{RT} \tag{7.27}$$

This condition derives from the comparison of the diffusive coefficient, respectively, in the gas and the liquid film.

Regime 3

Relatively fast reaction

In this case, again the reaction occurs mainly inside the film as can be seen in Fig. 7.12. However, there is not a reaction plane because the reaction zone is

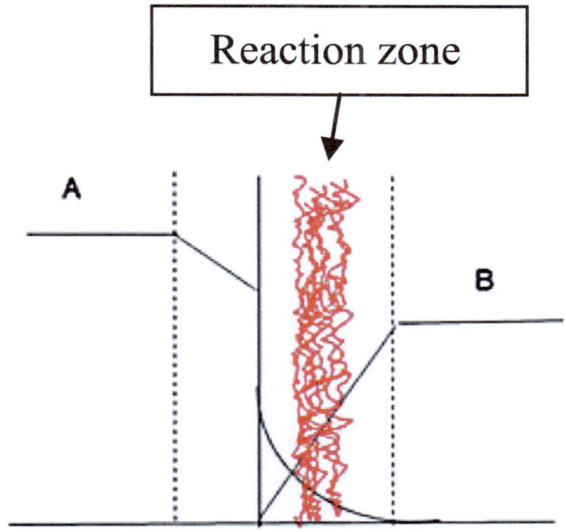

Fig. 7.12 Concentration profiles for regime 3

diffused. The mass transfer in the two films, both gaseous and liquid, can be written as:

$$-r_A = k_{Ag}(p_A - p_{Ai}) = k_{Al} C_{Ai} E \tag{7.28}$$

By eliminating C_{Ai} and p_{Ai}, we obtain:

$$-r_A = \frac{1}{\frac{1}{k_{Ag}} + \frac{H_A}{k_{Al}E}} p_A \tag{7.29}$$

E is a complex function of many variables, that is, $E = E(k_{Al}, k_2, z, C_B/C_{Ai})$, and no analytical expression has been found for calculating this parameter. An approximate solution was suggested by van Krevelen and Hoftijzer (1948). However, the alternative is the numerical solution of the mass-balance differential equations.

Regime 4

Relatively fast reaction with a high concentration of B

In this case, we can introduce the acceptable approximation that C_B compared with C_A remains relatively constant. The reaction zone is shifted toward the interface as can be seen in Fig. 7.13. However, considering the high C_B concentration, the reaction becomes of pseudo–first order, that is:

$$-r_A = k_2 C_A C_B = (k_2 C_B) C_A = k_1 C_A \tag{7.30}$$

This approximation allows the obtainment of an analytical expression for E, which is:

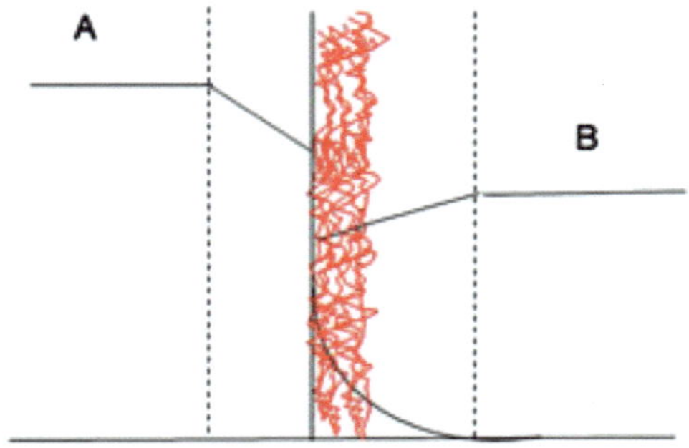

Fig. 7.13 Concentration profiles for regime 4

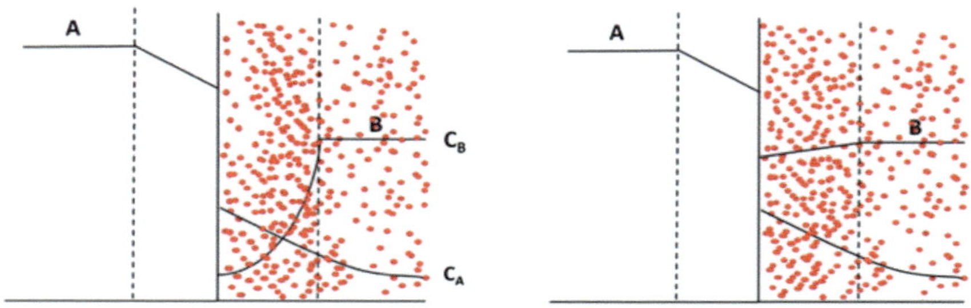

Fig. 7.14 Concentration profiles for regimes 5 and 6

$$E = \frac{\sqrt{D_{Al} k_1}}{k_{Al}} \tag{7.31}$$

By eliminating in the previous expressions C_{Ai} and p_{Ai} and combining them with the analytical expression of E, we obtain:

$$-r_A = \frac{1}{\frac{1}{k_{Ag}} + \frac{H_A}{\sqrt{D_{Al} k C_B}}} \tag{7.32}$$

Regimes 5 and 6

These two regimes are characteristic of moderately fast reactions. The reaction occurs both inside the liquid film, together with the mass transfer, and in the liquid bulk (see Fig. 7.14). Analytical kinetic expressions are not available for these two cases, and the differential mass-balance equations must be solved by numerical methods to determine the concentration profiles of reagents and products inside the film. A great concentration of B increases the extent of the reaction inside the film.

However, it is useful to point out that the absorption rate for these reactions depends on both the interface area and the volume of the liquid. It is useful, therefore, to define a specific surface area given by:

$$a_{\mathrm{i}} = \frac{S}{V_{\mathrm{l}}} \tag{7.33}$$

Regime 7

Slow reactions

In this case, we can write:

$$-\frac{1}{S}\frac{\mathrm{d}N_{\mathrm{A}}}{\mathrm{d}t} = k_{\mathrm{Ag}}(p_{\mathrm{A}} - p_{\mathrm{Ai}}) = k_{\mathrm{Al}}(C_{\mathrm{Ai}} - C_{\mathrm{A}}) \tag{7.34}$$

and also:

$$-\frac{1}{V_{\mathrm{l}}}\frac{\mathrm{d}N_{\mathrm{A}}}{\mathrm{d}t} = k_2 C_{\mathrm{A}} C_{\mathrm{B}} \tag{7.35}$$

By eliminating the intermediate variables that are not measurable, it is possible to obtain the expressions:

$$-r_{\mathrm{A}} = -\frac{1}{S}\frac{\mathrm{d}N_{\mathrm{A}}}{\mathrm{d}t} = \frac{p_{\mathrm{A}}}{\frac{1}{k_{\mathrm{Ag}}} + \frac{H_{\mathrm{A}}}{k_{\mathrm{Al}}} + \frac{H_{\mathrm{A}} a_{\mathrm{i}}}{k_2 C_{\mathrm{B}}}} \quad \text{or} \quad -r_{\mathrm{A}} = -\frac{1}{V_{\mathrm{l}}}\frac{\mathrm{d}N_{\mathrm{A}}}{\mathrm{d}t} = \frac{p_{\mathrm{A}}}{\frac{1}{k_{\mathrm{Ag}} a_{\mathrm{i}}} + \frac{H_{\mathrm{A}}}{k_{\mathrm{Al}} a_{\mathrm{i}}} + \frac{H_{\mathrm{A}}}{k_2 C_{\mathrm{B}}}} \tag{7.36}$$

In this case, the three occurring processes (the mass transfer of the gaseous reactant gas side, the mass-transfer of the liquid side, and the reaction) can be considered as consecutive steps. The specific interface surface area a_{i} referred to as the liquid volume has been introduced because the reaction rate is normally expressed in moles/time × liquid volume.

Regime 8

Very slow reactions

In this case, we have the chemical regime, and the absorption rate can be described with the kinetic law of the chemical reaction:

$$-r_{\mathrm{A}} = -\frac{1}{V_{\mathrm{l}}}\frac{\mathrm{d}N_{\mathrm{A}}}{\mathrm{d}t} = k_2 C_{\mathrm{A}} C_{\mathrm{B}} \tag{7.37}$$

The mass transfer does not affect the reaction rate.

Levenspiel and Godfrey (1974) also suggested a strategic approach for determining the operative kinetic regime by employing a continuous laboratory reactor similar to the one shown in Fig. 7.7. The scheme of this approach is shown in Fig. 7.15.

The approach suggested by Levenspiel and Godfrey (1974) is useful to identify the kinetic regime for an unknown system; however, because the interface area is quite low, the obtained conversions normally also are low. Therefore, having identified the regime with the described approach, it is opportune to employ a

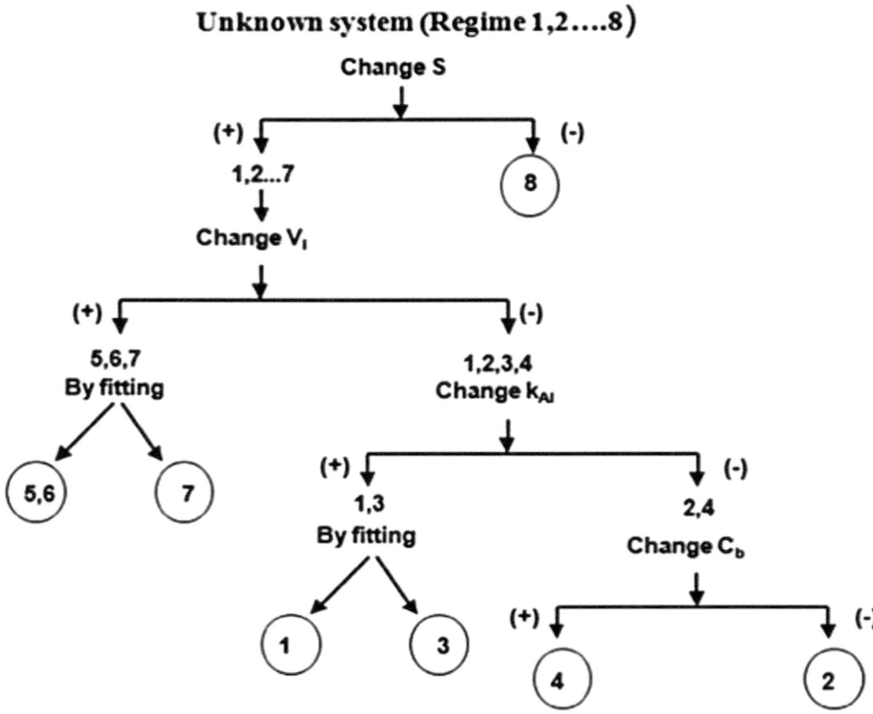

Fig. 7.15 Experimental approach to gas–liquid reaction suggested by Levenspiel and Godfrey. Reproduced with permission from Levenspiel and Godfrey (1974), Copyright Elsevier (1974)

different laboratory reactor, characterized by a high interface area, to study the kinetics and mass transfer. An example of a reactor suitable for the scope is shown in Fig. 7.16. This reactor is characterized by the development of a large interface surface area due to the use of a MagneDrive stirrer sucking the gas from the gaseous mixture and bubbling it inside the liquid. The stirrer can operate at different stirring rates from 500 to 2500 rpm. The same reactor also can be used under semi-batch conditions. However, a difficulty arises by using this reactor in the determination of, respectively, k_{Al} and a_i. In fact, by increasing the stirring rate, both of these parameters change, and it is difficult to separate their contribution to the determinable value of their product ($k_{Al}a_i$). The interface area can be determined using the sulfite method described by Charpentier (1981) and by Linek and Vacek (1981). This method consists of evaluating the absorption rate of oxygen, for different stirring rates, in an aqueous solution of sulfite containing an appropriate concentration of a cobalt salt acting as catalyst. The oxidation of sulfite is very fast, and the kinetic parameters of this reaction are well-known. The obtained values of the absorption rate are then compared with the rate obtained by stirring without producing bubbles, that is, under a condition of known interface area. An apparent interface

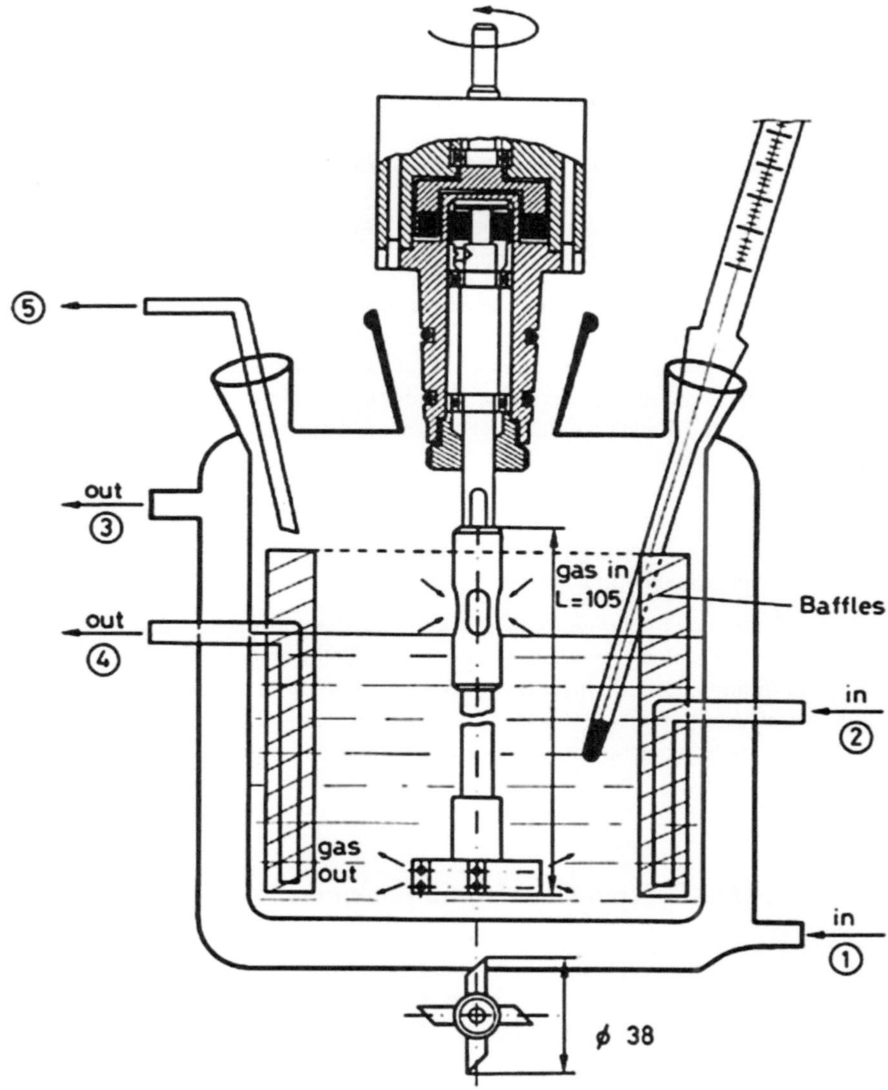

Fig. 7.16 A continuous laboratory reactor characterized by the development of a large interface surface area. Reproduced with permission from Santacesaria et al. (1987), copyright American Chemical Society (1987)

area can be obtained by dividing the absorption rate found for a definite stirring rate by the absorption under the reference condition of the known interface area. The obtained area is apparent because by increasing the stirring rate, the value of k_{Al} moderately increases. Another possibility is to employ the reaction of CO_2 with an alkaline solution. This reaction also is very fast, and its kinetic behavior has been well-studied in the literature. At last, the use of the already described amperometric electrode is also helpful for measuring the physical absorption of oxygen in water or other solvents (not reacting with oxygen) at different stirring rates.

In any case, although we can choose the fluid dynamic condition, the described measurements are often made in solvents that are different with respect to the one used in a specific reaction, and the extrapolation of the data collected requires some caution.

In conclusion, the study of the kinetics and mass transfer of gas–liquid reactions is performed through the following steps:

1. Define the kinetic regime using the Levenspiel and Godfrey (1974) reactor and related strategic approach.
2. Determine the mass-transfer parameters and the solubility of the gaseous reagents using independent techniques and in the absence of the reaction.
3. Characterize the fluid dynamic of the reactor working with a reactor characterized by a large interface area.
4. Perform kinetic runs (batch and/or continuous) in the same reactor used for the fluid-dynamics characterization.
5. Verify the kinetic models and the collected parameters in different types of reactors (for example, in a bubble-column reactor).

In particular, it is important to define the partition equilibria between the gas and liquid phases for both the reagents and the products. For this purpose, the solubility and vapor–liquid equilibria must be determined possibly in independent ways, that is, in the absence of the reaction. This aspect is particularly relevant when the reaction occurs at increased pressure or when the product is obtained in a gaseous phase.

7.2.5 Bubble-Column Gas–Liquid Reactors

The bubble column is a cylindrical vessel in which a liquid is fed on the top or on the bottom until filling an appropriate volume of the column, the level of which is regulated by the weight of the hydrostatic column. A reacting gas is sparged into the column in the form of bubbles that rise due to the effect of gravity. The traveling bubbles grow in volume from both the effect of pressure decreasing and the coalescence, but they decrease in volume due to the consumption of the gaseous reactant. The coalescence effect can be contrasted by putting perforated plates at different levels in the column for an efficient redistribution of the gas, inside the liquid, thus reducing the bubble mean size. A simplified scheme of a bubble column is shown in Fig. 7.17.

Two parameters are important to characterize the bubble columns—the gas and the liquid hold-up—that is, the volume of the column filled, respectively, by the gas and by the liquid. Clearly these volumes under steady-state conditions are constant, and different correlations have been proposed in the literature, such as:

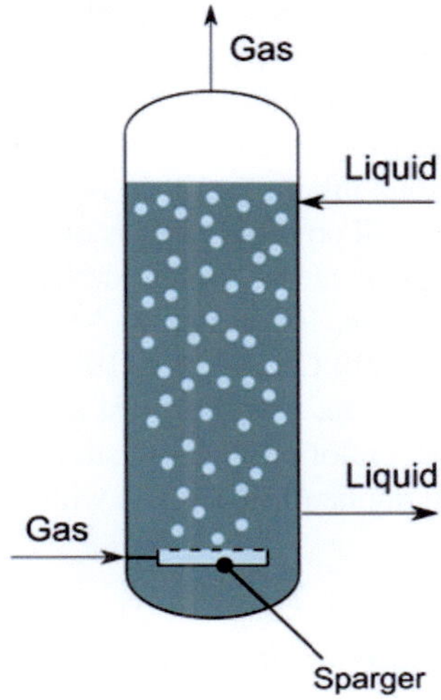

Fig. 7.17 Scheme of a bubble column

$$\frac{\varepsilon_{\mathrm{g}}}{(1-\varepsilon_{\mathrm{g}})^4}=0.20\left(\frac{gd_{\mathrm{c}}^2\rho_{\mathrm{l}}}{\sigma}\right)^{1/8}\left(\frac{gd_{\mathrm{c}}^3}{\nu_{\mathrm{l}}}\right)^{1/12}\left(u_{\mathrm{g}}\sqrt{gd_{\mathrm{c}}}\right) \tag{7.38}$$

which was suggested by Akita and Yoshida (1973) for estimating the gas holdup ε_{g}. In the relation, g is the gravity constant; d_{c} is the column diameter; ρ_{l} is the liquid density; σ is the surface tension; ν_{l} is the cinematic viscosity of the liquid phase; and u_{g} is the superficial velocity of the gas. In addition, the volumetric mass-transfer coefficient can be estimated from the physical properties of the liquid mixture and the dynamic conditions with relations of the type (see Akita and Yoshida 1973):

$$k_{\mathrm{L}}a=0.6\mathrm{Sc}_{\mathrm{l}}^{0.5}\left(\frac{gd_{\mathrm{c}}^2\rho_{\mathrm{l}}}{\sigma}\right)^{0.62}\left(\frac{gd_{\mathrm{c}}^3}{\nu_{\mathrm{l}}}\right)^{0.31}\varepsilon^{1.1}\left(\frac{D_i}{d_{\mathrm{c}}^2}\right) \tag{7.39}$$

where Sc is the Schmidt number $= \nu_{\mathrm{l}}/D_i$ and D_i is the diffusion coefficient of the i component.

The simulation of a bubble column is complicated by (1) the presence of a considerable effect of back-mixing in both phases and (2) the type of operative gas–liquid regime. The liquid phase, for example, can be well-mixed, and this has a great impact on the elaboration of the kinetic model. Back-mixing coefficient E_{L} can be estimated using relations of the type:

$$E_{\mathrm{L}}=0.678d_{\mathrm{c}}^{1.4}u_{\mathrm{g}}^{0.3}\quad E_{\mathrm{L}}=0.01d_{\mathrm{c}}^{1.5}u_{\mathrm{c}} \tag{7.40}$$

which was first proposed by Dechwer et al. (1974) and second by Joshi and Sharma (1976).

In contrast, in the presence of the previously seen gas–liquid regimes 5 and 6, we must solve by the numerical method the mass-balance system of differential equations to evaluate the concentration profiles inside the liquid film corresponding to any single bubble in a similar way described for the long-range gradients of gas–solid tubular reactors. A general description of the kinetic- and mass-transfer models for all possible cases is outside the scope of this book, and we suggest consulting the specialized literature devoted to the subject.

See, for example, Astarita (1967), Danckwerts (1970), Carrà and Santacesaria (1980), Charpentier (1981), and Carrà and Morbidelli (1987).

Here, we report just a specific example of a gas–liquid reaction studied from the laboratory to the industrial plant, that is, the oxidation of 2-ethyl-tetra-hydro-anthraquinone ($THEAQH_2$) with air to obtain hydrogen peroxide.

7.2.6 The Oxidation of $THEAQH_2$ with Air to Obtain Hydrogen Peroxide: An Example of Gas–Liquid Reaction Studied from the Laboratory to the Industrial Plant

(A) The Laboratory Approach

In the process called All-TETRA, for producing hydrogen peroxide, 2-ethyl-tetra-hydro-anthraquinone (THEAQ), is subject to cyclic reduction and oxidation according to the following scheme:

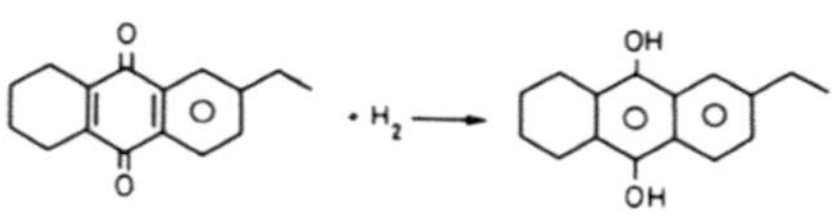

(7.41)

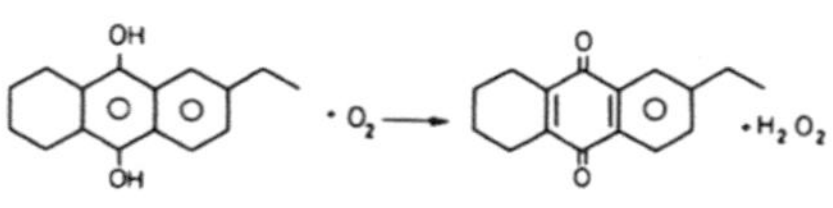

(7.42)

A mixture of 70% THEAQ and 30% 2-ethylanthraquinone (EAQ), both dissolved in an appropriate solvent, is normally employed in the industrial plant devoted to hydrogen peroxide production. EAQ such as THEAQ is also hydrogenated to $EAQH_2$, but it readily reacts, thus transferring the hydrogen to THEAQ due to the following fast reaction:

$$EAQH_2 + THEAQ \rightleftarrows THEAQH_2 + EAQ \tag{7.43}$$

Therefore, in the successive oxidation step only $THEAQH_2$ is the reactant. The kinetics and mass transfer of both the hydrogenation (see Santacesaria et al. 1988b) and oxidation (see Santacesaria et al. 1987) have been studied in laboratory reactors,

Table 7.1 Kinetic and thermodynamic parameters determined for the oxidation of $THEAQH_2$

Henry solubility (atm cm^3/mol)	$H_{20\ °C} = 112{,}100$ $H_{50\ °C} = 109{,}000$
Mass-transfer coefficient (cm/s)	k_L (20 °C) $= 5.5 \times 10^{-3}$ k_L (50 °C) $= 6.6 \times 10^{-3}$
Specific interface area (cm^2/cm^3) at 2000 rpm	$a_i = 42$
Kinetic constant at 50 °C (cm^3/mol s)	3830 ± 300
Activation energy (cal/mol)	13,000

and the obtained results have then been employed for simulating different industrial plants reactors. In particular, the oxidation reaction was industrially conducted in countercurrent bubble-column reactors.

In a work devoted to the study of $THEAQH_2$ oxidation, a reactor such as the one shown in Fig. 7.8 of known interface area was employed for evaluating first the mass-transfer rate of oxygen in a solution of already oxidized THEAQ, at 20 and 50 °C, using a Clark amperometric electrode for measuring the evolution with time of the oxygen concentration. The same reactor was then used, at 50 °C, to perform kinetic runs by changing the liquid flow rate, the partial pressure of oxygen, and the interface surface area. The solubility of oxygen in the working solution was evaluated by the method suggested by Nitta et al. (1983) (see Table 7.1). From these runs it was possible to achieve the following information:

1. The value of the mass-transfer coefficient, in the absence of the reaction, has been determined at 50 °C (see Table 7.1).
2. The reaction rate, under the adopted conditions, is very low because the resulting conversion also was low despite the long residence time (3% of conversion for 1 h of residence time). However, the rate increases by increasing the interface area. In conclusion, at a low interface area, this reaction is completely dominated by the oxygen mass-transfer rate.
3. Assuming that the gradient in the liquid film is the maximum possible, we can write:

$$r = k_L V_L a_i \frac{P_{O_2}}{H} \tag{7.44}$$

By comparing the k_L determined in the presence of the reaction (evaluated from this relation) with the value determined in the absence of the reaction, the ratio resulted in the range of 0.5–0.75. This means that the gradient is not maximum and that the enhancement factor can be assumed to be $E = 1$. In consequence, the mass transfer and reaction in this particular case can be considered independent consecutive steps.

As has been seen, the Levenspiel reactor allowed for individuation of the kinetic regime, but the conversions obtained were too low for a complete kinetic study. Therefore, other kinetic runs were performed, at 50 °C, in a reactor similar to the one shown in Fig. 7.16 that was able to develop a large interface area. First, the reactor was characterized for the fluid-dynamic aspects, and the interface area was determined for different stirring rates by using the already mentioned sulfite method. Then, different kinetic runs were performed at a constant stirring rate of approximately 2000 rpm, at which an apparent interface area of approximately $a_i = 42$ cm^2/cm^3 was developed. All of the kinetic runs were interpreted by assuming a second-order kinetic law and by solving the mass-balance equations:

$$\frac{R_{O_2}}{V_L} = k_L a_i \left(\frac{P_{O_2}}{H} - C_{O_2} \right) = k_2 C_R C_{O_2} + \frac{F}{V_L} \tag{7.45}$$

where C_R is the concentration of the organic reagent $THEAQH_2$ and F is the volumetric flow rate. Then, by eliminating C_{O_2} from the two relations the following expression of reaction rate corresponding to the oxygen absorption rate, we can obtain:

$$R_{O_2} = \frac{V_L \left(\frac{P_{O_2}}{H} \right)}{\frac{1}{k_2 C_R + \frac{F}{V_L}} + \frac{1}{k_L a_i}} \quad (\text{mol/s}) \tag{7.46}$$

The kinetic runs were performed at 50 °C by changing the liquid flow rate and the concentration of $THEAQH_2$. The obtained conversions were in the range of 78–92%, and an average value of the kinetic constant k_2 is listed in Table 7.1. Runs performed at different temperature allowed to evaluate an activation energy of 13,000 cal/mol. It is known, then, that the mass-transfer coefficient is proportional to $\sqrt{T}$.

Exercise 7.1. Oxidation of Hydrogenated $THEAQH_2$

The oxidation kinetic of hydrogenated $THEAQH_2$ was studied by Santacesaria et al. (1987). The authors studied this system in different reactors (batch and continuous) and under different conditions.

Part 1. Determination of the Gas–Liquid Mass-Transfer Coefficient k_L

Using the experimental data listed in Table 7.2, taken from Santacesaria et al. (1987) and related to the current intensity as a function of time for testing the physical

Table 7.2 Experimental data collected using an amperometric electrode measuring the concentration of absorbed oxygen

(nA)	Time (s)
11.27	20.14
115.15	432.11
189.71	887.01
235.06	1326.00
273.09	1718.35
296.52	2145.42
315.12	2871.17
326.87	3360.17
327.90	3852.91

Table 7.3 Some operative conditions of the performed runs

Run	Total volume added (cm^3)	Concentration of $THEAQH_2$ (g/L)
1	212	7.5
2	270	15.0
3	290	30.0

Table 7.4 List of parameters useful for interpreting the kinetic runs

Parameter	Value	Units
a_i	45	cm^2/cm^3
H	109,000	atm cm^3/mol
Rate	1.56	cm^3/s
V_{delay}	150	cm^3
k_L	6.6e−3	cm/s
k_c	3830	cm^3/(mol s)

absorption oxygen in a solution of THEAQ (already oxidized), evaluated by fitting the mass-transfer coefficient. Other conditions for this experiment are interfacial area $a_i = 0.16\ cm^2/cm^3$, $T = 20\ °C$, $P_{O_2} = 1$ atm, and $H = 112{,}100$ atm cm^3/mol.

Part 2. Kinetics in Fed-Batch Experiments

In the previously mentioned paper, the authors performed some kinetic experiments under fed-batch conditions, in which oxygen consumed is registered as a function of time. Different runs were performed by adding a solution of the organic reactant ($THEAQH_2$) with a certain flow rate and concentration to a reactor previously filled with pure oxygen. During the run, oxygen was also fed to the reactor to maintain its pressure constant at a fixed value. The conditions for this type of experiment in all cases were $T = 50\ °C$, $P = 1$ atm, and stirring speed of 1800 rpm. As stated by the authors, in these runs, the reaction starts when the liquid level in the reactor reaches the stirring speed that produces a high interfacial gas–liquid area. Table 7.3 lists data on the total volume of added working solution as well as the concentration of $THEAQH_2$. In addition, Table 7.4 lists some useful parameters related to these runs.

Using the data and parameters listed in Table 7.4, simulate the kinetic runs listed in Table 7.2 by developing a suitable model considering both mass-transfer and reaction rates.

Solution

Part 1

Evaluation of the gas–liquid mass-transfer coefficient k_L was performed using an electrode sensitive to oxygen concentration, for which the current intensity is proportional to the dissolved oxygen concentration according to the relation:

$$C_{O_2} = ib \tag{7.47}$$

where i is the current intensity and b is the proportionality factor relating the current and concentration. The dynamic material balance that describes the oxygen-transfer process is:

$$\frac{\mathrm{d}C_{O_2}}{\mathrm{d}t} = k_L a_i \left(\frac{P_{O_2}}{H} - C_{O_2} \right) \tag{7.48}$$

By substituting Eq. (7.47) into Eq. (7.48), we obtain:

$$\frac{\mathrm{d}i}{\mathrm{d}t} = \frac{k_L a_i}{b} \left(\frac{P_{O_2}}{H} - ib \right) \tag{7.49}$$

Equation (7.49) is suitable for describing the evolution in time of the current intensity. The unknown parameters in Eq. (7.49) can be evaluated by fitting the experimental data. Their values are, respectively, 6.9×10^{-3} cm/s and 2.69×10^{-8} mol/(cm^3 nA). The fitting of experimental data is shown in Fig. 7.18.

Part 2

In the fed-batch experiments, the volume of consumed oxygen is measured as a function of time. These experiments can be described by the following set of differential equations:

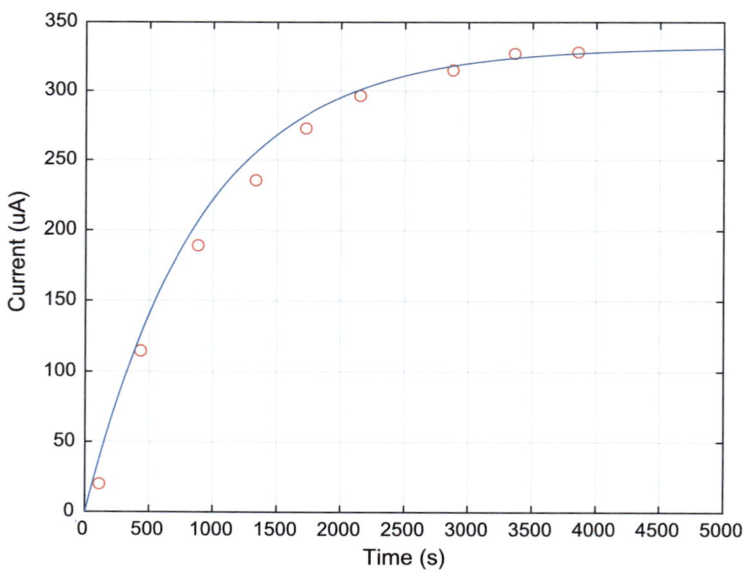

Fig. 7.18 Absorption of oxygen in the working solution determined by an amperometric electrode, the current signal of which is proportional to [O_2]

$$\frac{\mathrm{d}V_{\mathrm{O_2}}^{I}}{\mathrm{d}t} = V_{\mathrm{M}} V_{\mathrm{L}} k_{\mathrm{L}} a_{\mathrm{i}} \left(\frac{P_{\mathrm{O_2}}}{H} - C_{\mathrm{O_2}} \right) = \mathrm{O_T} \quad \text{oxygen transferred} \tag{7.50}$$

$$\frac{\mathrm{d}V_{\mathrm{O_2}}^{II}}{\mathrm{d}t} = V_{\mathrm{M}} V_{\mathrm{L}} k_{\mathrm{c}} C_{\mathrm{O_2}} C_{\mathrm{R}} = \mathrm{O_R} \quad \text{oxygen reacted} \tag{7.51}$$

$$\frac{\mathrm{d}C_{\mathrm{O_2}}}{\mathrm{d}t} = \frac{\mathrm{O_T} - \mathrm{O_R}}{V_{\mathrm{M}} V_{\mathrm{L}}} \tag{7.52}$$

$$\frac{\mathrm{d}V_{\mathrm{L}}}{\mathrm{d}t} = Q \tag{7.53}$$

$$\frac{\mathrm{d}C_{\mathrm{R}}}{\mathrm{d}t} = QC_{\mathrm{R}} - V_{\mathrm{L}} k_{\mathrm{c}} C_{\mathrm{O_2}} C_{\mathrm{R}} \tag{7.54}$$

By integrating this set of ordinary differential equations, the profiles shown in Fig. 7.19 can be obtained.

All of the described results were obtained by using a MATLAB program available as Electronic Supplementary Material.

(B) Example of Modeling an Industrial Bubble Column for the Oxidation of $\mathbf{THEAQH_2}$ with Air

Consider a countercurrent bubble column with the liquid fed from the top of the column, for abating the foam, and the gas distributed by a sparger located on the bottom. The mass balance has been reasonably simplified by considering the liquid in the column to be well mixed and the reacting gas as a plug flow. To verify the well-mixed condition, it is necessary to calculate the back-mixing coefficient E_{L} with the relation suggested by Dechwer et al. (1974) or by Joshi and Sharma (1976), while the plug-flow status of the gas phase can be verified by inoculating a pulse of a tracer and verifying the distribution concentration of the tracer at the outlet.

1. **Oxygen mass balance in the gas phase:**

$$-G^{\circ} \mathrm{d}Y = J \mathrm{d}V_{\mathrm{R}} \tag{7.55}$$

where:

$$Y = \frac{y}{1-y} = \frac{\text{molar fraction of oxygen in the gas phase}}{\text{molar fraction of nitrogen}} \tag{7.56}$$

V_{R}	reactor volume
G°	molar flow rate of nitrogen

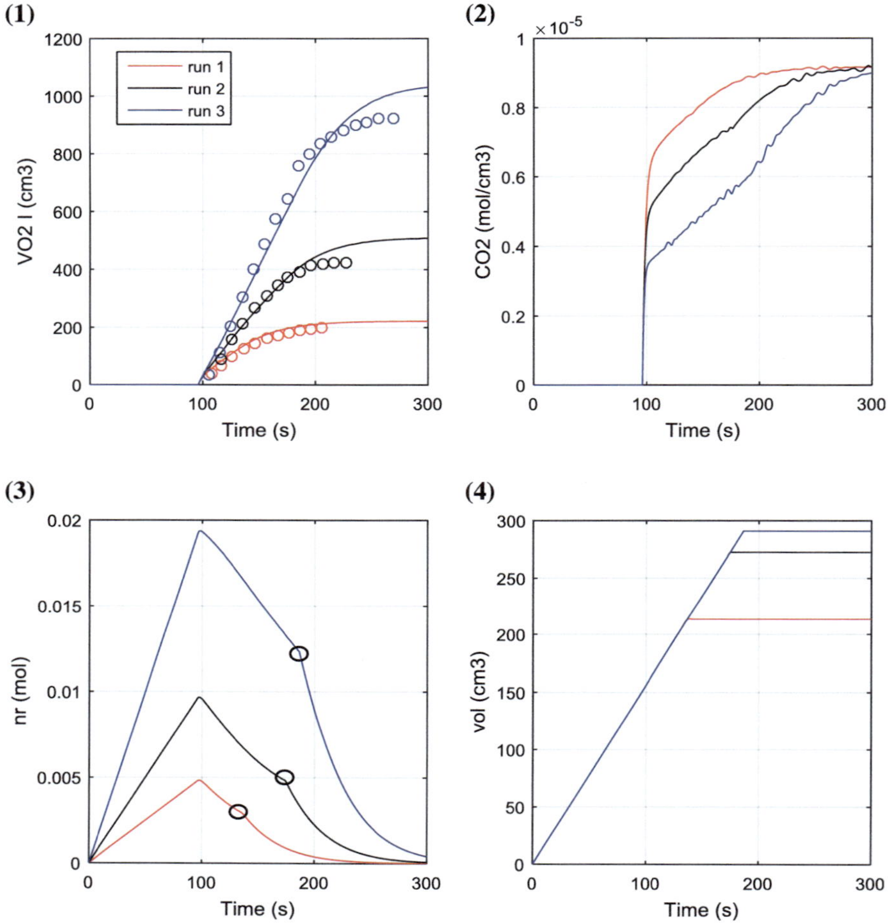

Fig. 7.19 Different plots obtained by simulation. (**Plot 1**) The first 100 s are necessary to fill the reactor until the stirrer is reached. At this time, the reaction starts and can be simulated with the developed model. The dots are experimental; the lines are calculated. (**Plot 2**) The oxygen concentration in the liquid phase has a steep increase when the stirrer is reached by the liquid because the interfacial area increases greatly. (**Plot 3**) The moles of $THEAQH_2$ increase with a linear trend up to 100 s because during this period the only phenomenon is the liquid feed. After this, a decrease in concentration is observed due to the simultaneous feeding and reaction. After the solution feeding is completed, the decrease is more pronounced because only an oxidation reaction occurs (circles). (**Plot 4**) The trend of the volume in the reactor is exactly the same, for the three runs, in the first phase of the reaction. Then, each run is stopped when all the volume has been fed to the reactor

$$J = \text{Oxygen mass} - \text{transfer rate} = K_G a_i P\left(y - \frac{S}{M_v}[O_2]\right) \tag{7.57}$$

S	Bunsen solubility = cm^3 of gas/cm^3 of liquid
M_v	molar volume of gas
K_G	overall mass-transfer coefficient. It holds:

$$\frac{1}{K_G P} = \frac{1}{k_G P} + \frac{S}{k_L M_v} \tag{7.58}$$

where k_G is the gas-side mass-transfer coefficient and k_L is the liquid-side mass-transfer coefficient.

It is then convenient to write that:

$$dV_R = A\,dz \text{ with } A = \text{column section and } z = \text{reactor height}$$

and

$$dY = \frac{dy}{(1-y)^2} \tag{7.59}$$

hence,

$$\frac{dy}{dz} = -\frac{A}{G^\circ}(1-y)^2 K_G a_i P\left(y - \frac{S}{M_v}[O_2]\right) \tag{7.60}$$

Eventually, this equation can be integrated analytically because the well-mixed liquid $[O_2]$ can be considered constant; thus, assuming $\alpha = \frac{S}{M_v}[O_2] = \text{costant}$, we can write:

$$\int_{y_{IN}}^{y_{OUT}} \frac{dy}{(1-y)^2(y-\alpha)} = -\frac{A K_G a_i P z}{G^\circ} \tag{7.61}$$

The integral is:

$$F(y) = \frac{1}{(1-\alpha)}\left[\frac{1}{1-y} + \frac{1}{(1-\alpha)}\log\left(\frac{y-\alpha}{1-y}\right)\right]_{y_{IN}}^{y_{OUT}} \tag{7.62}$$

2. **Oxygen mass balance in the liquid phase:**

$$V[O_2] = \overline{J} V_R - (1 - \varepsilon_g) V_R r \tag{7.63}$$

where:

V	liquid flow rate
$[O_2]$	oxygen concentration
ε_g	gas hold-up

$$r = \text{rate of } H_2O_2 \text{ formation} = k_2[\text{THEAQH}_2][O_2] \tag{7.64}$$

$$\overline{J} = \text{average mass transfer rate of oxygen} = K_G a_i P\left(\overline{y} - \frac{S}{M_v}[O_2]\right) = (\text{mol/h cm}^3) \tag{7.65}$$

with $\overline{y}$ = average molar fraction of oxygen. The rigorous calculation of this average is laborious.

3. **Overall mass balance of oxygen:**

$$G^\circ (Y_{IN} - Y_{OUT}) = V[O_2] + V\lambda[\text{THEAQH}_2]_{IN} \tag{7.66}$$

Or, alternatively,

$$G^\circ (Y_{IN} - Y_{OUT}) = V_R \overline{J} \tag{7.67}$$

where:

$$Y_{IN} = \text{moles of oxygen at the inlet/moles of nitrogen} = \frac{y_{IN}}{1 - y_{IN}} \tag{7.68}$$

$$Y_{OUT} = \text{Moles of oxygen at the outlet/moles of nitrogen} = \frac{y_{OUT}}{1 - y_{OUT}} \tag{7.69}$$

λ	conversion

The conversion can be calculated from the changes that occurred, respectively, in both the liquid phase and the gas phase:

$$\lambda = \frac{[\text{THEAQH}_2]_{\text{IN}} - [\text{THEAQH}_2]_{\text{OUT}}}{[\text{THEAQH}_2]_{\text{IN}}} = \frac{G^\circ (Y_{\text{IN}} - Y_{\text{OUT}}) - V[\text{O}_2]}{V[\text{THEAQH}_2]_{\text{IN}}} \quad (7.70)$$

where $[\text{THEAQH}_2]_{\text{IN}}$ = Initial concentration of the organic reagent; $[\text{THEAQH}_2]_{\text{OUT}}$ = Concentration of the organic reagent at the outlet

4. **Mass balance of the organic reagent:**

$$V\left([\text{THEAQH}_2]_{\text{IN}} - [\text{THEAQH}_2]_{\text{OUT}}\right) = V\lambda[\text{THEAQH}_2]_{\text{IN}} = \left(1-\varepsilon_g\right)V_R r \quad (7.71)$$

5. **Mass balance of the reaction product:**

$$V[\text{H}_2\text{O}_2] = V_R\left(1-\varepsilon_g\right)(r - r_d) \quad (7.72)$$

where $[H_2O_2]$ = Hydrogen peroxide concentration

$$r_d = \text{Hydrogen peroxide-decomposition rate} = k_d[\text{H}_2\text{O}_2] \quad (7.73)$$

The yields will be:

$$\eta = \frac{[\text{H}_2\text{O}_2]}{[\text{THEAQH}_2]_{\text{IN}}} \quad (7.74)$$

In this system, we know the following:

(a) The inlet conditions, that is, $[\text{THEAQH}_2]_{\text{IN}}$, Y_i, V, $G°$, and P_{In}
(b) The geometric characteristics of the reactor
(c) The kinetic, thermodynamic, and mass-transfer parameters, k_2, k_d, H, k_L, a_L, which are assigned or calculated

The unknowns that must be determined by solving the system of mass-balance equations include $[H_2O_2]$, $[\text{THEAQH}_2]_{\text{OUT}}$, $[O_2]$, and Y_{OUT}. From these, we also can calculate the conversion, the yield, and the gas-flow rate at the exit. In conclusion, we have four unknowns and five equations, that is, we can consider four mass balances to solve the problem, for example, the equations corresponding to the mass balances 1, 3, 4, and 5 being, as mentioned previously, the rigorous solution of equation related to the mass balance 2, which is laborious.

Exercise 7.2. Oxidation of THEAQH$_2$ with Air in a Bubble Column
An industrial bubble column for oxidating THEAQH$_2$, working in counter current, is fed from the top with a liquid flow rate of 206 m^3/h, while a flow of 9630 Nm^3 of air is sparged at the bottom. The diameter of the column is 3.6 m, and the height is 20 m. The initial THEAQH$_2$ concentration is $2.90\text{–}10^{-4}$ mol/cm^3. The liquid column height is 14 m, and the gas holdup is 0.35.

Do the following: (1) evaluate the conversion of THEAQH$_2$, the yield of H_2O_2, the concentration of oxygen in the gas at the reactor outlet, and the concentration of oxygen in the liquid phase [O_2]; (2) build a plot in which the oxygen mole fraction in gas phase (assumes as an ideal PFR) is reported against gas volume; and (3) evaluate the change in conversion and yield when the liquid feed flow rate is varied from 10 to 410 m^3/h.

Solution

The system to be solved is:

$$G(y_I - y_L) = Q[O_2] + Q\lambda[RH_2]^{\circ} \quad \text{overall oxygen balance} \tag{7.75}$$

$$Q\lambda = V_R(1-\varepsilon_G)k_1(1-\lambda)[O_2] \quad \text{balance on organic reactant} \tag{7.76}$$

$$Q[H_2O_2] = V_R(1-\varepsilon_G)\left[k_1[RH_2]^{\circ}[O_2](1-\lambda) - k_2[H_2O_2]\right] \quad \text{balance on product} \tag{7.77}$$

$$\frac{dG_{O_2}}{dV_G} = -k_G a\left(\frac{P}{H} - [O_2]\right)V_R(1-\varepsilon_G) \tag{7.78}$$

$$J = k_G a\left(\frac{P}{H} - [O_2]\right) \quad \text{mass transfer rate (gas} \rightarrow \text{liquid)}$$

In these relations, there are four unknowns: λ [O_2] y_L[H_2O_2]

The solution of the first part of the exercise involves a single solution of this equation system constituted by three algebraic equations and one ODE. In the second part, instead, the system can be solved inside a for-loop that performs the variation of liquid volumetric flow rate (Fig. 7.20).

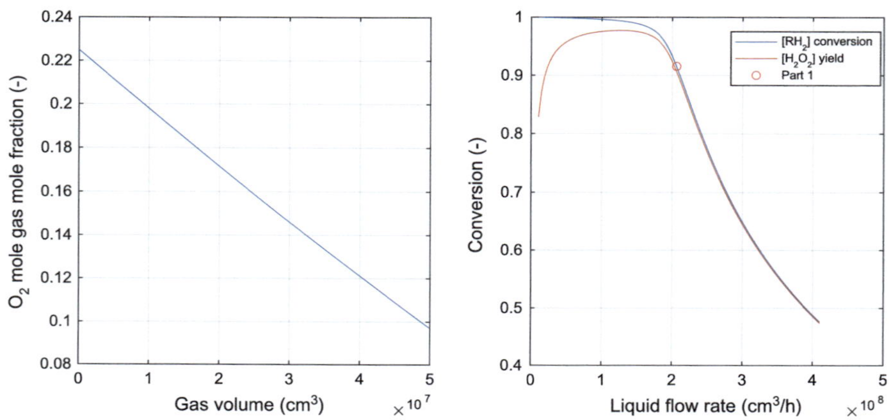

Fig. 7.20 Different plots obtained by simulation

Conversion results	0.9155	–
O_2 concentration	8.0344e−07	mol/cm^3
Y O_2 out	0.0973	–
Y O_2 in	0.2250	–
H_2O_2 concentration	2.6278e−04	mol/cm^3
$RH_2°$ concentration	2.9000e−04	mol/cm^3
RH_2 concentration	2.4498e−05	mol/cm^3
Liquid flow rate	2.0600e+08	cm^3/h
Gas flow rate	4.2964e+05	mol/h
Reactor volume	1.4250e+08	cm^3
Residence time	4.4964e−01	h
Average pressure	2.1497	atm
H_2O_2 yield	0.9062	–

All of the described results were obtained by using a MATLAB program available as Electronic Supplementary Material.

7.2.7 *Multistage Operation: Distillation with Reaction*

When the reaction product is more volatile than both the reactants and the solvent, it could be convenient to perform the reaction in a distillation column, thus coupling reaction and separation. The multistage operation is more convenient than a single-stage one because it approaches the performance of a plug flow reactor, while the single stage can be assimilated to that of a CSTR. Examples of reactions conveniently performed in a distillation column are chlorohydrin dehydrochlorination for obtaining epichlorohydrin or propylene oxide, respectively, obtained from the following reactions:

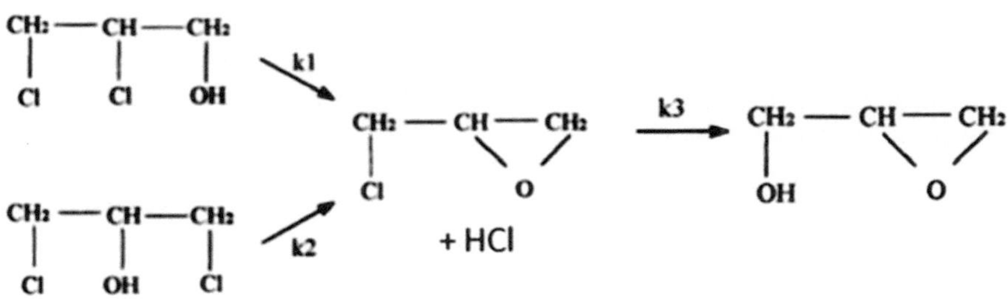

(7.79)

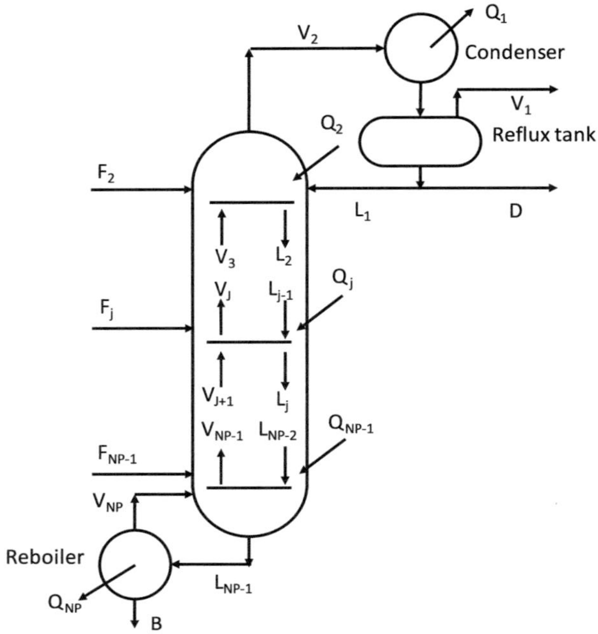

Fig. 7.21 Scheme of the multistage unit

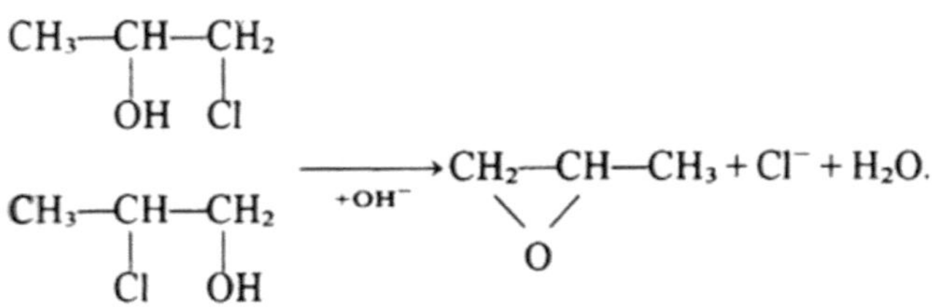

(7.80)

The kinetics of the first set of reactions were studied in detail by Carrà et al. (1979a), while the kinetics of the second group of reactions were studied in detail by Carrà et al. (1979b).

In both cases, chlorohydrins are put in contact in countercurrent with an aqueous suspension of lime ($Ca(OH)_2$) or a solution of NaOH in a multistage unit as the one shown in Fig. 7.21.

The alkaline environment catalyzes the dehydrochlorination reaction, but in the meantime, it is neutralized by the developed hydrochloric acid. The mathematical model of the multistage reaction with a distillation unit, similar in the two cases, was developed by Carrà et al. (1979c, d).

The developed model is based on the following equations referred to any single stage:

1. Material balance of the component i on plate j (number of equations = number of components × number of plates):

$$F_j z_{ij} + V_{j+1} y_{i,j+1} - V_j y_{i,j} + L_{j-1} y_{i,j-1} - L_j x_{i,j} - r_{i,j} \mathrm{Vol}_j = 0 \tag{7.81}$$

2. Enthalpy balances (number of equations = number of plates):

$$F_j h_j + V_{j+1} H_{j+1} - V_j H_j - L_j h_j + L_{j-1} h_{j-1} - \mathrm{Vol}_j \left(\sum_{k=1}^{NR} r_{k,j} \Delta H_{r,k} \right) - Q_j = 0 \tag{7.82}$$

3. Equilibrium relationships (number of equations = number of components × number of plates):

$$y_{i,j} - \eta_j K_{i,j} x_{i,j} - (1 - \eta_j) y_{i,j+1} = 0 \tag{7.83}$$

where η_j is the Murphee efficiency in the vapor phase on plate j:

$$\eta_j = \frac{y_{i,j} - y_{i,j+1}}{y_{i,j} - K_{i,j+1} x_{i,j+1}} \tag{7.84}$$

4. Stoichiometric equations (number of equations = number of plates):

$$\sum_i (x_{i,j} - y_{i,j}) = 0 \tag{7.85}$$

5. Total balance equations (number of equations = number of plates):

$$L_j - V_{j+1} - \sum_{k=1}^{j} F_k + D + V_1 = 0 \tag{7.86}$$

The input data of the model include:

(a) Feeding-flow characteristics
 F_j = feed rate on the j-th plate
 z_{ij} = mole fraction of component i in the plate feed
 h_j, H_j = liquid and vapor molar enthalpy on the j-th plate
(b) Number and volume of plates
 NP = number of plates, Vol_j = volume of the j-th plate.
(c) Profile of pressure along the column P_j
(d) Withdraws
 V_1 = flow rate of vapor distillate
 D = flow rate of condensed distillate
(e) Reflux ratio $R = L_1/D$

Unknown to be determined:

(a) Compositions:

$x_{i,j}$ and $y_{i,j}$ = liquid and vapor molar fraction of component i on the j-th plate

(b) Flow rates:

L_j and V_j = liquid and vapor molar flow rate from the j-th plate

(c) Plate temperature: T_j

(d) Exchanged heat: Q_1 = heat exchanged on the condenser, Q_j = flow of heat exchanged on the j-th plate

7.2.8 Others Gas–Liquid Reactors: Spray-Tower Loop Reactor, Venturi Tube Loop Reactor, Gas–Liquid Film Reactor, and Membrane Gas–Liquid Reactor

(A) **Spray-Tower and Venturi Loop Reactors**

The previous paragraphs have described the behavior of relatively slow gas–liquid reactions (e.g., the oxidation with air of $THEAQH_2$), which was characterized by an enhancing factor $E \approx 1$ and by a reaction mainly occurring in the liquid bulk but subject to mass-transfer limitation when the interface surface area is small. It has been seen in this case that it is opportune to use a reactor, in which it is possible to increase as much as possible both the liquid holdup and the interface area. The bubble column is suitable for such a scope. Some other reactions have the same requisite but are extremely exothermic and, for safety purposes, must conducted in special reactors. An example is the polyethoxylation of fatty alcohols (RXH) reacting with ethylene oxide (EO), the enthalpy change of which is approximately 22 kcal/mol. Runaway is dangerous for this reaction because ethylene oxide at high temperatures can be involved in other explosive reactions.

The reaction is catalyzed by an alkaline catalyst (KOH, NaOH, or related alkoxides) for producing surfactants. The reaction occurs according to the following scheme:

1. In situ catalyst formation

$$B^+OH^- + RXH \rightleftharpoons RX^-B^+ + H_2O \uparrow \tag{7.87}$$

2. Initiation reaction

$$RX^-B^+ + EO \xrightarrow{k_i} RX(EO)^-B^+ \tag{7.88}$$

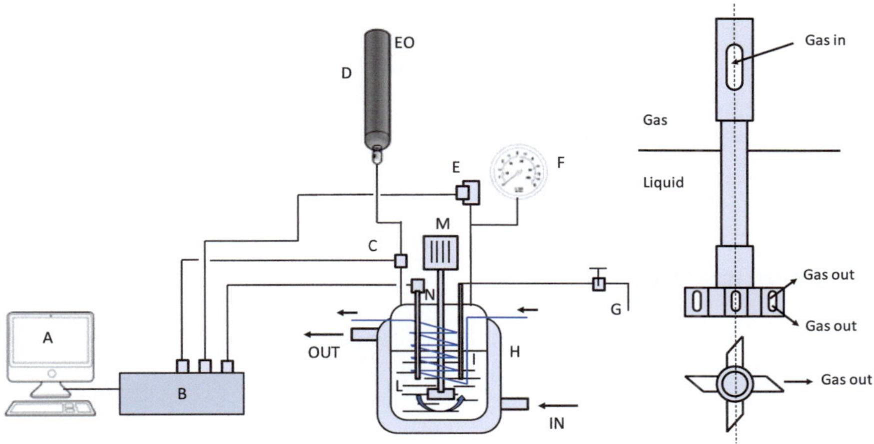

Fig. 7.22 Scheme of a laboratory plant used for studying fatty alcohol polyethoxylation. (A) computer; (B) computer interface; (C) on–off solenoid valve; (D) EO bottle; (E) pressure transducer; (F) manometer; (G) exit for withdrawal; (H) jacketed reactor; (I) freezing coil; (L) holed stirrer; (M) MagneDrive stirrer; and (N) thermocouple. Re-elaborated with permission from Di Serio et al. (1994), Copyright American Chemical Society (1994)

3. Propagation reactions

$$\mathrm{RX(EO)_J^-\,B^+ + EO \xrightarrow{k_p} RX(EO)_{J+1}^-B^+} \tag{7.89}$$

4. Proton-transfer equilibrium reactions

$$\mathrm{RX(EO)_iH + RX^-B^+ \overset{Ke_i}{\rightleftharpoons} RX(EO)_i^-\,B^+ + RXH} \tag{7.90}$$

This is a living polymer, and the reaction starts again every time ethylene oxide is added. To stop the reaction, the catalyst must be neutralized with an acid. The kinetics of this reaction was studied by Santacesaria et al. (1992) by employing a laboratory reactor as shown in Fig. 7.22.

As can be seen, a gas-entrainment impeller has been used for generating a high interface surface area inside the reactor. On the basis of the reaction mechanism, the following kinetic model can be used for interpreting laboratory kinetic runs:

Substrate consumption

$$\frac{\mathrm{d[RXH]}}{\mathrm{d}t} = -r_0 \tag{7.91}$$

Oligomer formation or consumption

$$\frac{\mathrm{d}\left[\mathrm{RX(EO)}_i\mathrm{H}\right]}{\mathrm{d}t} = r_{i-1} - r_i \quad i = 1, \ldots, n \tag{7.92}$$

Overall ethylene oxide consumption

$$\frac{\mathrm{d}n_{\mathrm{EO}}}{\mathrm{d}t} = V_1 \sum_{i=0}^{n} r_i = J \tag{7.93}$$

Eventual mass-transfer limitation

$$J = k_{\mathrm{L}} a_1 \cdot \left([\mathrm{EO}]_i - [\mathrm{EO}]_b\right) = \text{mass transfer rate} \tag{7.94}$$

Reactor heat balance

$$\frac{\mathrm{d}(\rho \cdot c_{\mathrm{P}} \cdot T)}{\mathrm{d}t} = \frac{1}{V_1} \cdot \frac{\mathrm{d}n_{\mathrm{EO}}}{\mathrm{d}t} \cdot \Delta H - q \tag{7.95}$$

The reaction rate law is all of the second order, that is:
Initiation rate

$$r_0 = k_0 \left[\mathrm{RX}^- \mathrm{M}^+\right][\mathrm{EO}] \tag{7.96}$$

Propagation rate

$$r_i = k_i \left[\mathrm{RX(EO)}_i^- \mathrm{M}^+\right][\mathrm{EO}] \quad i = 1, \ \ldots, \ n \tag{7.97}$$

The proton-transfer equilibria can be written as:

$$K_{\mathrm{e}_i} = \frac{[\mathrm{RXH}]\left[\mathrm{RX(EO)}_i^- \mathrm{M}^+\right]}{\left[\mathrm{RX}^- \mathrm{M}^+\right]\left[\mathrm{RX(EO)}_i\mathrm{H}\right]} \tag{7.98}$$

For dodecanol, it has been found that $k_0 = k_i = 6.55 \times 10^8 \exp(-13{,}280 \pm 228/RT)$ ($\mathrm{cm^3/mol\ s}$); K_{ei} = constant with temperature and number of adducts = 4.8.

To solve this model correctly also requires knowledge of the following:

1. One must know the density of the substrate and of all the oligomer mixtures and their related dependence on temperature. Empirical relations, such as the following, which is valid for dodecanol ethoxylation until an average number of 15 EO polymerized units is reached, can be usefully employed:

$$
\begin{aligned}
d = 0.86 + 2.50\times10^{-2}\left(\frac{\mathrm{EO_{units}}}{\mathrm{RXH}^{\circ}}\right) - 4.76\times10^{-4}\left(\frac{\mathrm{EO_{units}}}{\mathrm{RXH}^{\circ}}\right)^2 \\
- 2.69\times10^{-5}\left(\frac{\mathrm{EO_{units}}}{\mathrm{RXH}^{\circ}}\right)^3 - 7.7\times10^{-4}T
\end{aligned}
\tag{7.99}
$$

It is necessary to evaluate density data with independent experiments.

2. The solubility of ethylene oxide in the substrate and in all the oligomer mixtures obtained during the reaction and the related dependence of these solubilities on temperature. Ethylene oxide solubility in the reaction mixture can be considered a pseudo two-component system, with one component being ethylene oxide and the other being a more or less ethoxylated substrate with a ratio of a given number of EO units/RXH°.

   The vapor phase can normally be considered ideal because the adopted pressures are normally relatively low (2–10 atm), but the liquid phase is usually characterized by high nonideality; therefore, the equilibrium alkylene oxide molar fraction is:

$$
x_{\mathrm{AO}} = \frac{y_{\mathrm{EO}} \cdot P}{P^{\circ}_{\mathrm{EO}} \cdot \gamma_{\mathrm{EO}}}
\tag{7.100}
$$

where the ethylene oxide vapor pressure P°_{EO} can be determined using the Antoine equation:

$$
P^{\circ}_{\mathrm{EO}} = \frac{\exp\left[16.74 - \frac{2568}{(T - 29.01)}\right]}{760}
\tag{7.101}
$$

For the most common substrates, the Wilson and NRTL methods give the best performances. In particular for dodecanol ethoxylation, Di Serio et al. (1995) experimentally evaluated the Wilson parameters, which are useful for calculating the activity coefficients with the relations:

$$
\ln\gamma_1 = -\ln(x_1 + x_2\Lambda_{12}) + x_2\left[\frac{\Lambda_{12}}{(x_1 + x_2\Lambda_{12})} - \frac{\Lambda_{21}}{(x_1\Lambda_{21} + x_2)}\right]
\tag{7.102}
$$

$$
\ln\gamma_2 = -\ln(x_2 + x_1\Lambda_{21}) - x_1\left[\frac{\Lambda_{12}}{(x_1 + x_2\Lambda_{12})} - \frac{\Lambda_{21}}{(x_1\Lambda_{21} + x_2)}\right]
\tag{7.103}
$$

The binary interaction parameters Λ_{12} and Λ_{21} depend on the number of ethylene oxide adducts as follows:

$$
\Lambda_{12} = \Lambda_{12} + B_{12}n_{\mathrm{EO}} + C_{12}n_{\mathrm{EO}}^2
\tag{7.104}
$$

$$\Lambda_{21} = \Lambda_{21} + B_{21} n_{\mathrm{EO}} + C_{21} n_{\mathrm{EO}}^2 \tag{7.105}$$

The dependency of these parameters on the temperature was found negligible. The parameters found for these two equations are listed in Table 7.5.

However, in the absence of experimental solubility data, the UNIFAC predictive method can be used to estimate the activity coefficients.

Let us now consider industrial polyethoxylation reactors. Well-stirred semi-batch reactors are used in industry for small productions. The reactors differ in terms of modality for removing the reaction heat as shown in Fig. 7.23.

The best-performing reactor is the first one, which is characterized by liquid recirculation and by the presence of an external heat exchanger for removing the reaction heat. This reactor is shown in more detail in Fig. 7.24.

In contrast, reactors that are more commonly used for large productions of polyethoxylated fatty alcohols are of two different types: the Venturi loop reactor (see Fig. 7.25) and the spray-tower loop reactor (see Fig. 7.26).

The Venturi loop reactor is characterized by both a large gas–liquid interface area and high turbulence of the liquid phase induced by the Venturi tube. The advantage of the Venturi loop reactor is that mixing is obtained without the presence of moving parts inside the reactor, and this is important due to ethylene oxide instability. The kinetic model, as described for interpreting the kinetic runs performed in the well-stirred laboratory reactor, can be employed for describing the Venturi loop reactor as well as all of the semi-batch well-stirred industrial reactors.

Another reactor employed for large productions of polyethoxylated fatty alcohol is the spray-tower loop reactor (see Fig. 7.26). In this case, the liquid reactant is sprayed in an atmosphere of ethylene oxide, that is, the liquid is the dispersed phase. Small drops, flying for a few milliseconds, dissolve ethylene oxide until reaching saturation, which corresponds to the EO solubility, and then fall on the top of the liquid pool. Here, the reaction starts, and the conversion of the dissolved ethylene oxide and of the starter increases from the top to the bottom of the liquid column; the temperature also increases for the reaction exothermicity. In other words, it is possible to identify the presence of two zones of the reactor—the mass-transfer zone and the reaction zone—as shown in Fig. 7.27.

According to Santacesaria et al. (2005b), a good spray nozzle produces drops that are internally well stirred (see Fig. 7.27), and the mass-transfer coefficient can be determined with the relations suggested by Srinivasan and Aiken (1988). The approach of Srinivasan and Aiken is based on the Levich (1962) theory, which considers two different zones inside the drop: one near the free surface of small thickness, δ_1, in which the mass transfer occurs only by molecular diffusion, and another larger zone of thickness, δ_2, in which the mass transfer occurs by the whirling motion as shown in Fig. 7.28.

$$\mathrm{Sh} = 0.16 \cdot (\mathrm{Sc})^{\frac{1}{2}} \cdot (\mathrm{We})^{\frac{1}{2}} \cdot (\mathrm{Re})^{\frac{5}{16}} \tag{7.106}$$

where Sh = Sherwood dimensionless number

Table 7.5 Solubility parameters of the Wilson model for dodecanol ethoxylated mixtures

Substrate	A_{12}	B_{12}	C_{12}	A_{21}	B_{21}	C_{21}
Dodecanol + n EO	13.00	9.611×10^{-1}	-1.967×10^{-2}	-4.069×10^{-1}	4.714×10^{-2}	-1.340×10^{-3}

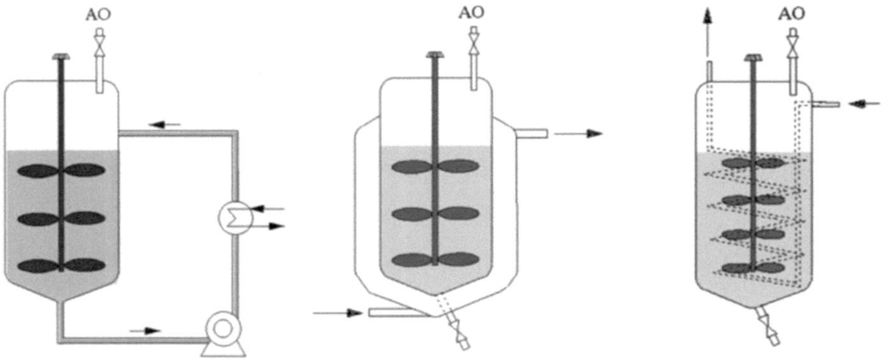

Fig. 7.23 Scheme of industrial reactors used in industry for small production of polyethoxylated fatty alcohols

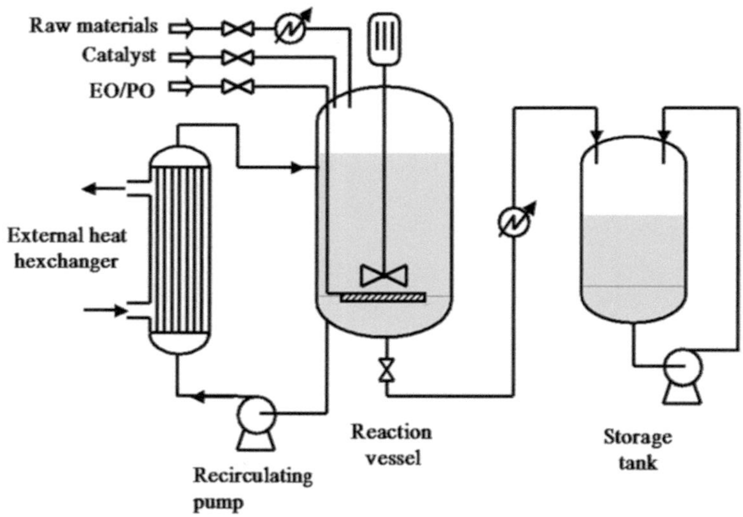

Fig. 7.24 Scheme of a semi-batch reactor employed in small production of polyethoxylated fatty alcohols

$$\mathrm{Sh} = \frac{k_{\mathrm{L}}}{\rho_{\mathrm{L}} D_{\mathrm{EO}}} \tag{7.107}$$

Sc = Schmidt dimensionless number

$$\mathrm{Sc} = \frac{\mu_{\mathrm{L}}}{\rho_{\mathrm{L}} D_{\mathrm{EO}}} \tag{7.108}$$

We = Weber dimensionless number

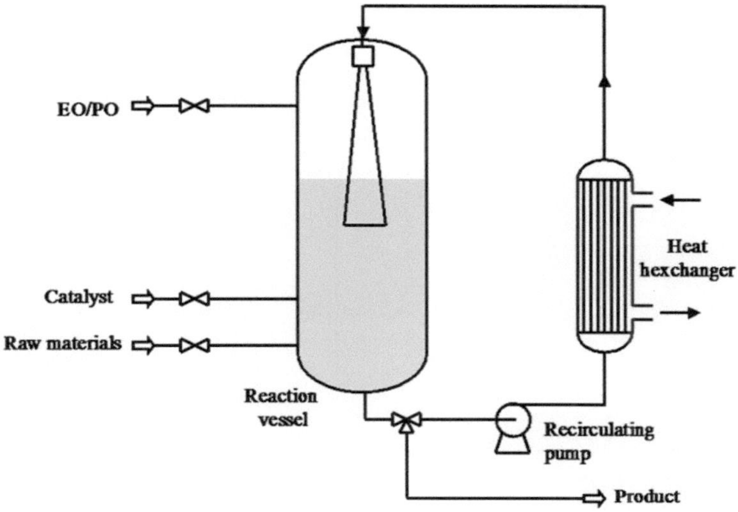

Fig. 7.25 Scheme of the Venturi loop reactor

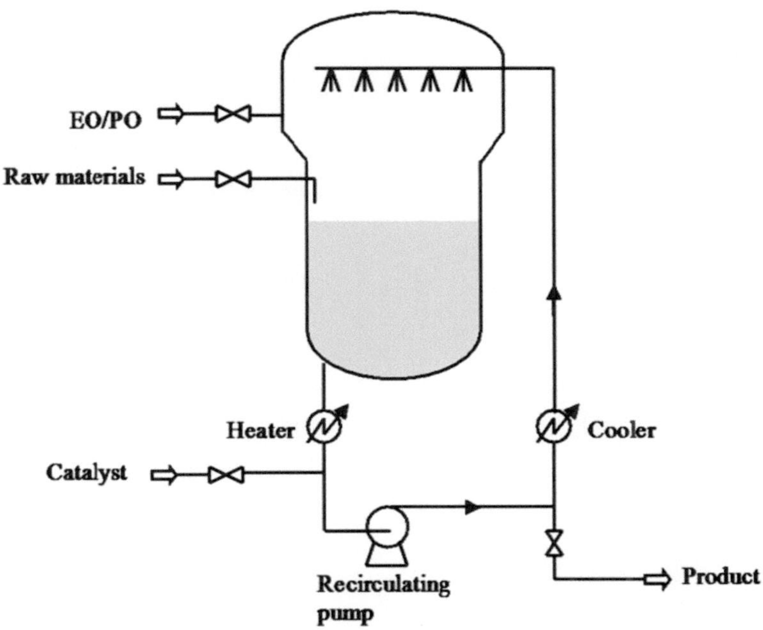

Fig. 7.26 Scheme of a spray-tower loop reactor

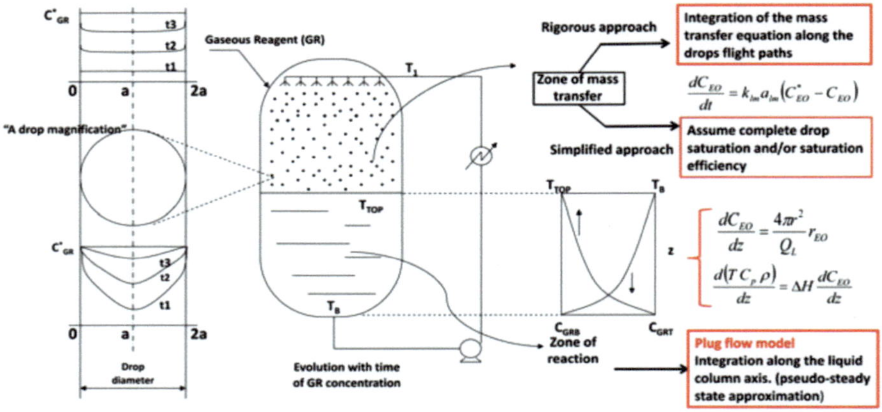

Fig. 7.27 Mass transfer and polyethoxylation reaction in a spray-tower loop reactor. Re-elaborated with permission from Santacesaria et al. (2005b), copyright American Chemical Society (2005)

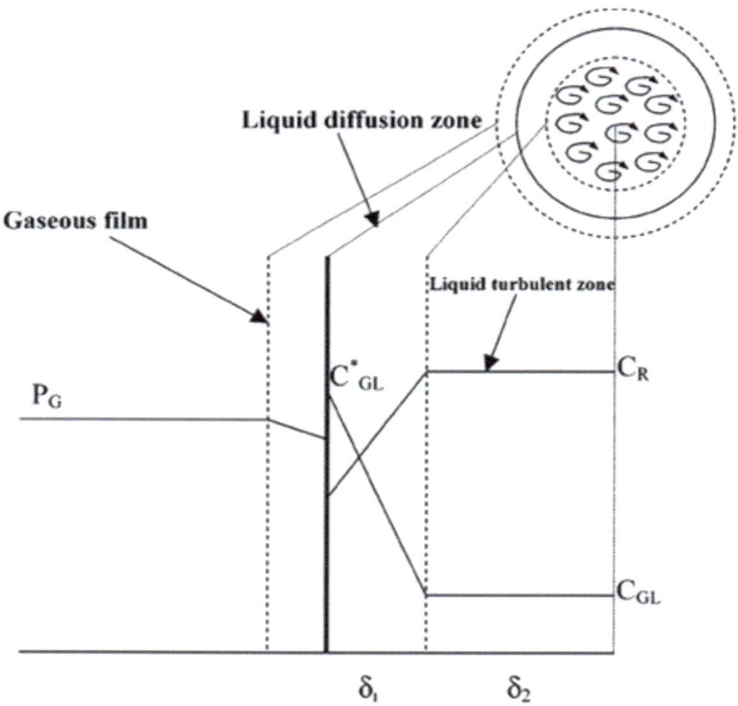

Fig. 7.28 Picture of internally well-mixed droplets formed at the outlet of a spray nozzle according to the model suggested by Srinivasan and Aiken (1988)

$$\mathrm{We} = \frac{v^2 \rho_L D_{32}}{\sigma_L} \tag{7.109}$$

Re = Reynolds dimensionless number

$$\mathrm{Re} = \frac{v \rho_L D_{32}}{\mu_L} \tag{7.110}$$

from which we can write:

$$k_L = \frac{D_{EO} \times 0.16}{D_{32}} \cdot \left(\frac{\mu_L}{\rho_L \cdot D_{EO}}\right)^{\frac{1}{2}} \cdot \left(\frac{v^2 \cdot \rho_L \cdot D_{32}}{\sigma_L}\right)^{\frac{1}{2}} \cdot \left(\frac{D_{32} \cdot v \cdot \rho_L}{\mu_L}\right)^{\frac{5}{16}} \tag{7.111}$$

where D_{EO} is the diffusion coefficient of EO in the liquid phase (cm^2/s), D_{32} is the Sauter diameter of drops, μ_L is the liquid viscosity (g/cm s); ρ_L is the density (g/cm^3), v is the drop speed (cm/s), and σ_L is the surface tension (g/cm^2).

It is possible to evaluate experimentally the Sauter mean diameter of the drops as:

$$D_{32} = \frac{\sum_i n_i d_i^3}{\sum_i n_i d_i^2} \tag{7.112}$$

by measuring the drop-size distribution with a laser-scattering technique (see Figs. 7.29 and 7.30) using water as the fluid of reference and then correcting the value for the reacting liquid by the relation taken from Perry and Green (1984):

$$\frac{D_{32}}{(D_{32})_{\mathrm{water}}} = \left(\frac{\sigma_L}{73}\right)^{0.5} \cdot \left(\frac{\mu_L}{0.1}\right)^{0.2} \cdot \left(\frac{1.0}{\rho_L}\right)^{0.3} \tag{7.113}$$

From the Sauter diameter, it is possible to evaluate the interface surface area and the overall surface area of the flying drops:

$$a_{lm} = \frac{\text{Surface area of drops}}{\text{Volume of drops}} = \frac{\pi \sum_i n_i \cdot d_i^2}{\frac{\pi}{6} \sum_i n_i \cdot d_i^3} = \frac{6}{D_{32}} \qquad a_G = \frac{6 Q t_{\mathrm{flight}}}{D_{32}} \tag{7.114}$$

where Q is the recirculating liquid flow rate (cm^3/s) and t_{flight} is the average drop flight time (s).

It has been shown in the mentioned work that drops are almost completely saturated during their flight inside the ethylene oxide atmosphere. It is also important in this case, therefore, to know the ethylene oxide solubility in both the starter fatty alcohol and the ethoxylated product measured by an independent way or estimated

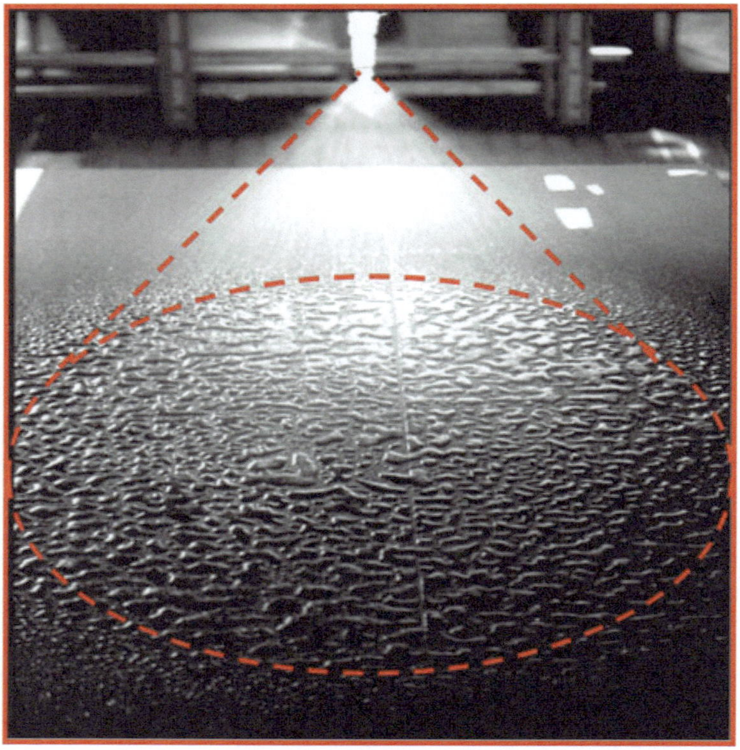

Fig. 7.29 Measure of the drop-size distribution. Reproduced with permission from Dimiccoli (2000), copyright American Chemical Society (2000)

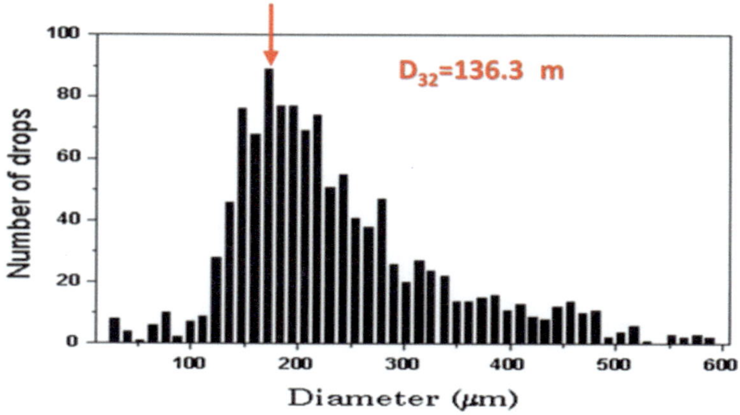

Fig. 7.30 Example of measured drop-size distribution. Reproduced with permission from Dimiccoli (2000), copyright American Chemical Society (2000)

with predictive methods, such as UNIFAC. Dissolved ethylene oxide then reacts along the column with a plug-flow behavior, and the internal profiles can be determined by integrating the mass- and heat-balance equations shown in Fig. 7.27. The reaction temperature is continuously restored by the external heat exchanger.

(B) **Gas–Liquid Film Reactors**

A gas–liquid film reactor is often used in laboratory devices to evaluate mass- and heat-transfer coefficients. Normally, the reactor consists of a falling liquid film column with a gas stream fed in the counter current. Gilliland and Sherwood (1934) (see also Chilton and Colburn 1934) found, in their pioneering work, that for the vaporization of pure liquids falling in a tube in contact with a flowing stream of air, the following relation characterizing the gas-side mass-transfer rate is valid:

$$\frac{d}{x} = \frac{k_G d M_m P_{BM}}{\rho D} = 0.023\ \mathrm{Re}^{0.83}\mathrm{Sc}^{0.44} = 0.023\left(\frac{dG}{\mu}\right)^{0.83}\left(\frac{\mu}{\rho D}\right)^{0.44} \tag{7.115}$$

where k_G is the gas-side mass-transfer coefficient, d is the tube diameter, D is the molecular-diffusion coefficient, x is the film thickness, M_m is the average molecular weight, P_{BM} is the log mean partial pressure of inert component in film, ρ is the density, μ is the viscosity, and G is the mass velocity. In the presence of a reaction, the mass-transfer coefficient can be enhanced if the reaction is very fast, but it is important to have the mass-transfer coefficient in the absence of the reaction as a reference for the enhancement factor evaluation. The relation suggested by Gilliland and Sherwood has been improved, for example, by Duduković and Pjanović (1999), who proposed slightly different exponents for the Re and Sc numbers. However, some industrial realizations using this type of reactor are known. In the case of an extremely fast reaction, for example, we have seen that the reaction occurs together with the mass transfer inside a thin liquid film at the boundary of the gas–liquid interface. In this case, it is not opportune to have a high liquid holdup in the reactor; rather, it is preferable to have just a reacting thin film in contact with the gaseous reagent, thus favoring the specific interface area. Such a type of reactor is used, for example, in the sulfonation of alkylbenzenes with diluted gaseous $S0_3$. The structure of a liquid-falling film reactor is simply a tube, the wall of which is wetted by the liquid as shown in Fig. 7.31. In the same figure, the gradients arising as a consequence of the reaction also are shown. The gas-side mass-transfer coefficient can be estimated with the correlation reported in the specific literature.

An example of a multi-tube gas–liquid film industrial reactor, used for the mentioned sulfonation reaction, is shown in Fig. 7.32.

In this specific case, mass transfer is surely enhanced by the extremely fast reaction, and the equations suggested by Levenspiel for the Regime 1 or 2 must be used in modeling the reactor.

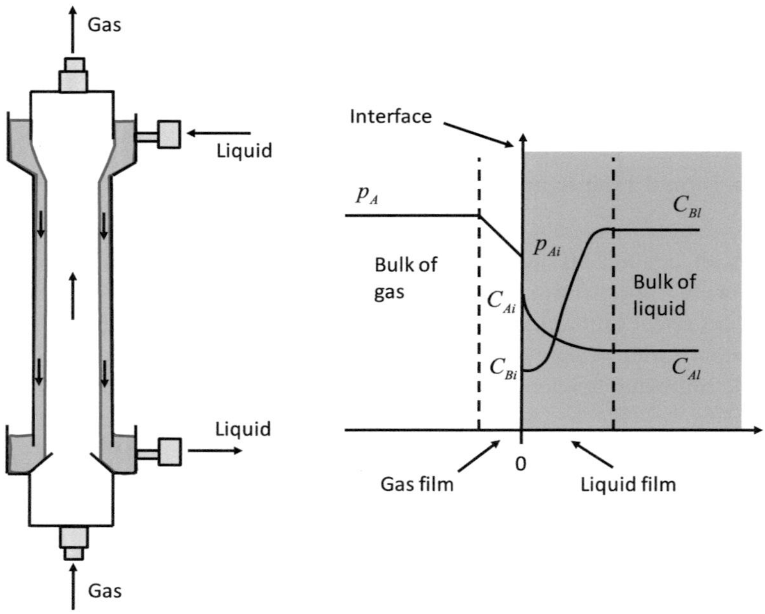

Fig. 7.31 Scheme of a liquid-falling film column

Exercise 7.3. Evaluate the Profiles in Liquid Film for Gas–Liquid Reactions
It is important sometime to define rigorously what happens when the reaction is so fast that mainly occurs inside the liquid film. At this purpose, let us consider generic gas–liquid reaction as in relation (7.11) in which A is the first gaseous reactant and B is the second one, present in the liquid phase with negligible volatility. This solved example examines how the liquid film profiles, for both A and B, are influenced by kinetic and physicochemical parameters adopted in the simulation. The objective is to solve the mass-balance PDE equations (7.12) and (7.13) obtaining concentration profiles for A and B in the liquid film in stationary conditions. The different numerical simulations must be developed with parameter sets as reported in Table 7.6.

Solution
The mass balances around the liquid-side stagnant film are represented by relations (7.12) and (7.13), two coupled PDEs whose solution requires proper boundary conditions (BC). These last correspond to equations from (7.14) to (7.17) in the text of the present chapter. The adopted numerical strategy for the solution consists of the spatial discretization of the film coordinate (distance from the gas–liquid interface) and transform, in this way, the PDEs system in a system of nonlinear algebraic equations to be solved iteratively. In fact, at each discretization point along x direction two mass balances, related to the two components A and B, can be written

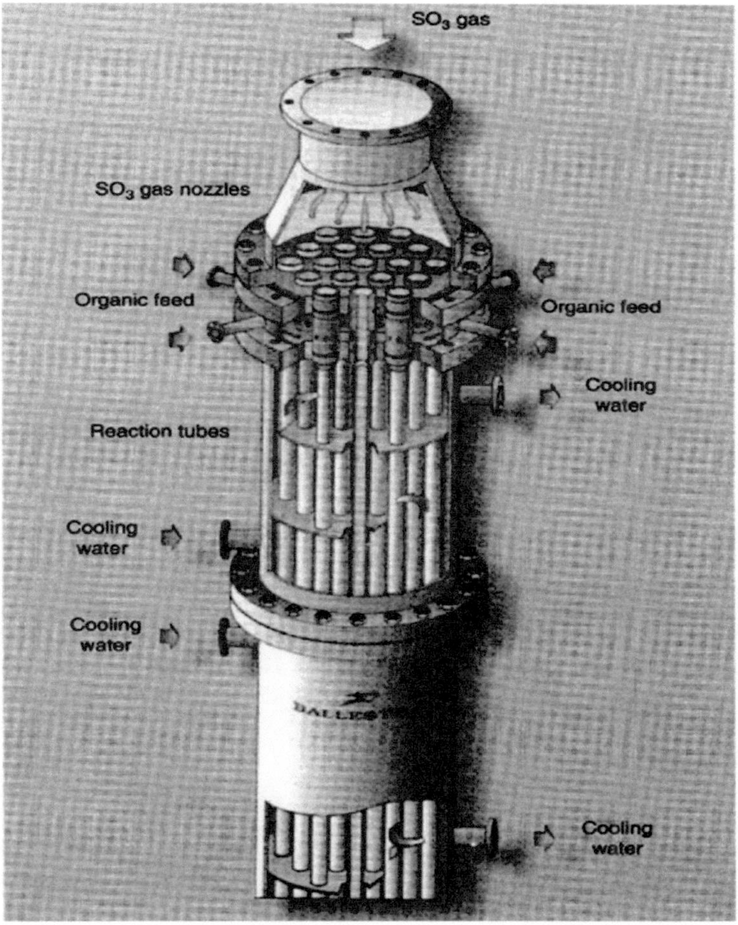

Fig. 7.32 Example of a gas–liquid film reactor. Reproduced with permission of Desmet-Ballestra Co.

in terms of algebraic equation in which the concentrations in adjacent points also appear. For example, for the component A, such equation is of the form:

$$D_{\mathrm{A}}\left[\frac{C_{\mathrm{A}}^{j+1}+C_{\mathrm{A}}^{j-1}-2C_{\mathrm{A}}^{j}}{\mathrm{d}x^{2}}\right]-kC_{\mathrm{A}}^{j}C_{\mathrm{B}}^{j}=0 \tag{7.116}$$

Obviously, as a high number of discretization points can be necessary for good calculation accuracy, the resulting nonlinear equations system is of high dimensionality.

Table 7.6 Simulation parameters for the base simulation (case 1) and others

Parameter	Units	Case 1 (ref.)	Case 2	Case 3	Case 4	Case 5
Film thickness	(m)	0.1e–3	0.1e–3	0.1e–3	0.1e–3	0.1e–3
Diffusivity of A in liquid	(m^2/s)	1e–9	1e–9	1e–9	1e–9	1e–9
Diffusivity of B in liquid	(m^2/s)	1e–9	1e–9	1e–9	1e–9	1e–9
Concentration of A at saturation	(mol/m^3)	0.5	0.5	0.5	0.5	0.5
Concentration of B in bulk liquid	(mol/m^3)	0.7	0.7	0.7	0.7	0.7
Specific interface area	(m^2/m^3)	0.5	0.5	0.5	0.5	0.5
Liquid holdup	(m^3/m^3)	0.8	0.8	0.8	0.8	0.8
Kinetic constant	(m^3/(mol/s))	1.8	100	300	0.01	1e–5
Number of discretization points	(–)	100	100	250	100	100
Stoichiometric coefficient of B	(–)	1	1	1	1	1
Mass-transfer coefficient	(m/s)	1e–5	1e–5	1e–5	1e–5	1e–5
Hatta number	(–)	3.550	26.458	37.417	0.265	0.008

Results

Case 1

The reference calculation has been done with parameters and conditions as reported in column 1 of Table 7.6. This case corresponds to a moderately fast reaction (Ha $>$ 3) for which concentration of A is completely depleted inside the liquid film and reaches a value around zero inside the liquid bulk. The reaction occurs practically throughout the whole film as can be seen in Fig. 7.33.

Cases 2 and 3

Cases 2 and 3 are related to a very fast reaction. As can be observed from both the plots of Fig. 7.34, as the reaction becomes faster and faster, the reaction zone inside the film is progressively reduced. In plot a, the reaction zone is between 0.2e–4 and 0.6e–4 mm inside the film as only in this region are both A and B components simultaneously present. In plot b, the reaction zone is located approximately between 0.3e–4 and 0.5e–4 mm as the kinetic constant was increased to 300 m^3/(mol/s). A more precise evaluation of the reaction zone inside the film can be done by examining the reaction rate profile. In plot c, the reaction rate profile inside the liquid film is reported. From this plot, related to the conditions of case 5 (plot b), it is more evident that the chemical reaction occurs only in a portion of the liquid film, while in the liquid bulk the reaction is absent. Moreover, two layers of the liquid film, marked in gray, are free from the reaction

Cases 4 and 5

Cases 4 and 5 are related to a very slow reaction. As can be observed from both the plots of Fig. 7.35, as the reaction becomes slower and slower, the liquid bulk concentration of A changes from a negligible to a finite value. This means that the

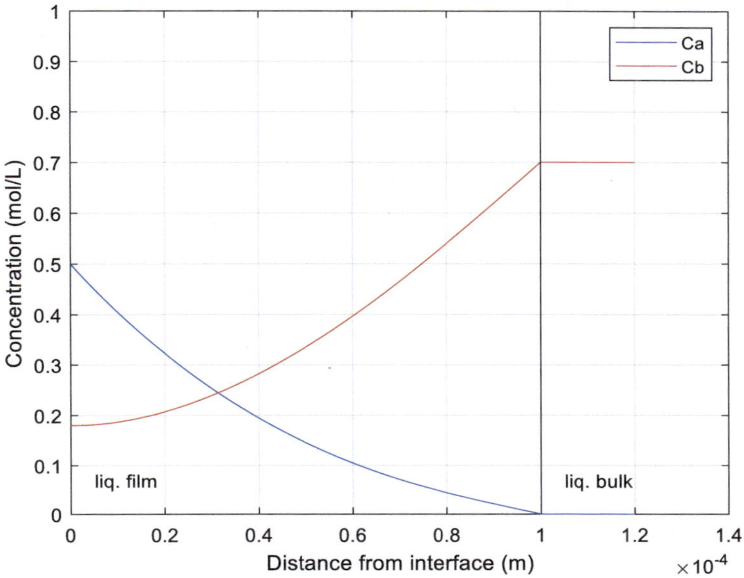

Fig. 7.33 Liquid film profiles related to reference case 1

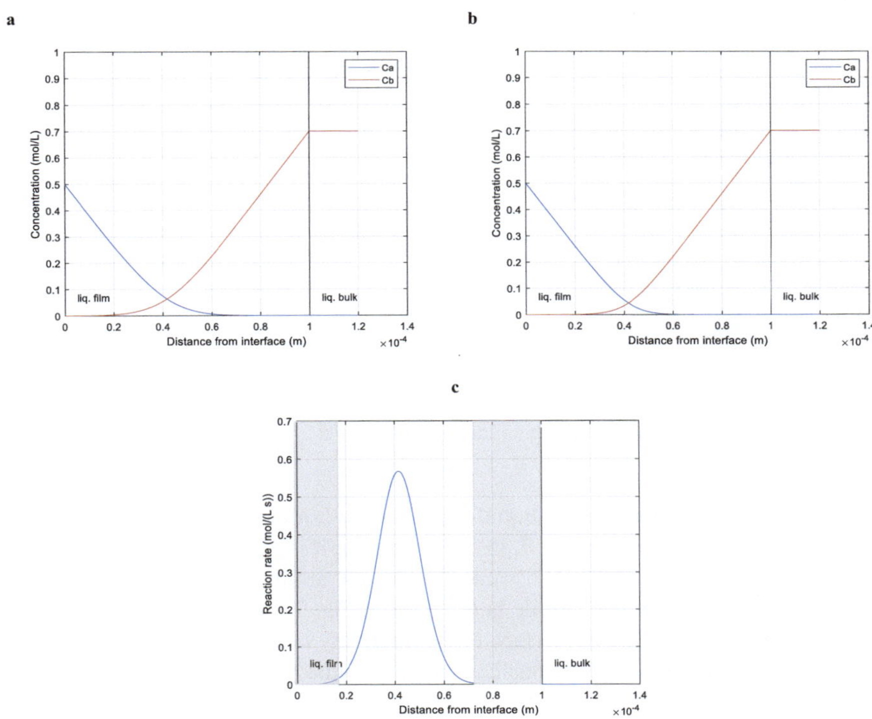

Fig. 7.34 Liquid film profiles related to cases 2 and 3

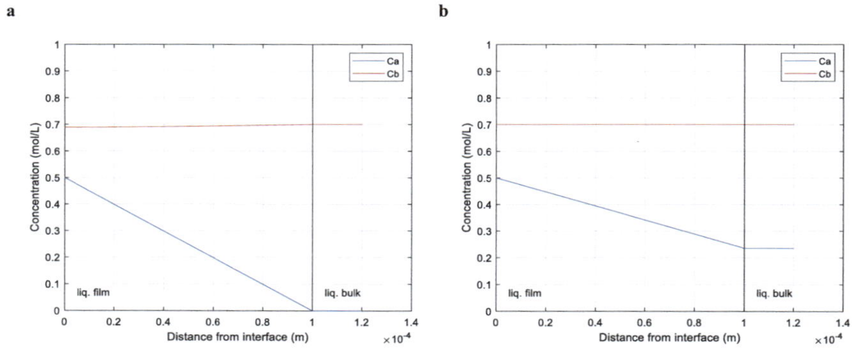

Fig. 7.35 Liquid film profiles related to cases 4 and 5

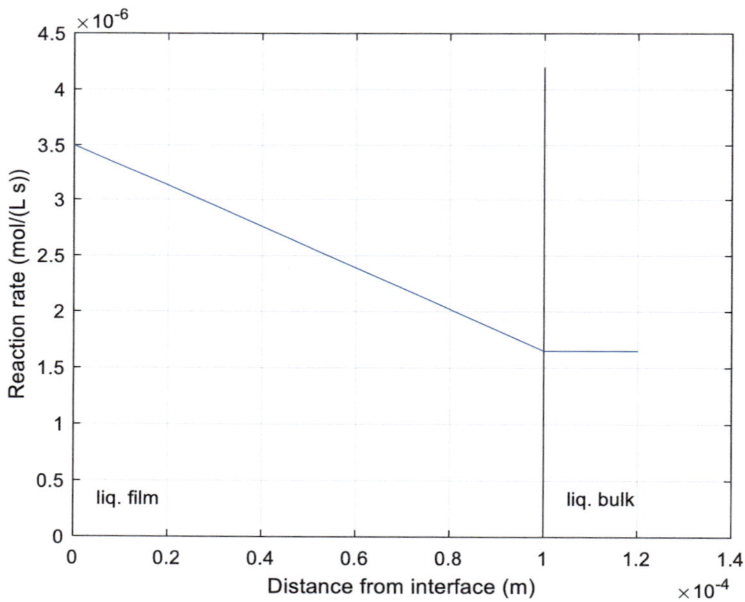

Fig. 7.36 Reaction rate profile in liquid film related to case 5

reaction propagates back to the liquid bulk as now both A and B are present there. This situation is depicted in Fig. 7.35, plot b, where it is possible to observe that the concentration of A is around 2.5 mol/L and not around zero as in plot a. The corresponding reaction rate profile is reported in Fig. 7.36. Reaction rate decreases linearly from the gas–liquid interface to the boundary between liquid film and liquid bulk and then, in the bulk, remains constant at a finite value.

7.3 Notes About Liquid–Liquid Reactions

Some reactions occur in the presence of two immiscible liquids, one polar and the other one apolar. In this case, one liquid is characterized by a continuous phase, and the other is dispersed in the form of small drops. An example of a reaction of this type is the transesterification of vegetable oils with methanol to obtain biodiesel. The catalysts for this reaction are normally NaOH, KOH, or their related alkoxides dissolved in the polar phase of methanol. At the end of the reaction, methyl esters (biodiesel) are formed that are apolar and glycerol, as a by-product, migrate in the residual methanol (polar phase). Another example of liquid–liquid reaction is the epoxidation of the double bonds of vegetable oils, in particular soybean oil, with hydrogen peroxide in the presence of formic acid and a mineral acid as catalyst in the presence of water as solvent. Formic acid is oxidized to performic acid, which is moderately soluble in the oil phase and gives place to the epoxidation reaction, thus restoring formic acid. Formic acid, diffusing from the oil phase to water, is ready for another reaction cycle. Other examples are nitration and sulfonation of aromatic compounds. Normally, the reaction occurs only in one phase, and problems of mass transfer arise when the interface area is kept low. Batch and continuous reactors can be used for industrial production. The problems of liquid–liquid reaction systems are not completely developed, considering that the dispersion of a liquid in another one can be affected by many different factors, in particular by the presence of tenside agents. The formation of emulsions or microemulsions can give place to a surprising increase in the reaction rate.

7.4 Gas–Liquid–Solid Reactors

Some examples of industrial processes characterized by the presence of gas–liquid–solid reactors include (1) many hydrogenation reactions, in particular, the hydrogenations of vegetable oils promoted by Ni-based catalysts and the hydrogenation of $THEAQH_2$, an important step in the production of hydrogen peroxide—the latter is promoted by supported palladium catalysts—and (2) reactions catalyzed by acid-exchange ion resins.

For gas–liquid–solid reactors, all of the observations reported in the sections devoted to gas–solid systems are valid with the exclusion of the following:

(a) In the gas–liquid–solid reactor, thermal gradients normally can be neglected for the high thermal conductivity of the liquids compared with those of the gases.
(b) The resistance to diffusion within the catalyst pores is higher than that observed in a gas–solid system because the molecular diffusion in the liquid is much slower.
(c) Transfer phenomena at the gas–liquid interface could yield reaction-rate limitations, that is, an additional interface must be considered.
(d) Kinetic runs are more often performed in batch reactors instead of continuous tubular reactors as in the gas–solid system.

7.4.1 Slurry Reactors

The most commonly used gas–liquid–solid reactors are "slurry reactors," which work with powdered catalysts. In a slurry reactor, we can identify different resistances to the mass transfer of the gaseous reagent as shown in Fig. 7.37. As can be seen, we have mass transfer from the gas to the liquid phase through both the gas and liquid quiescent films. Then, the gaseous reagent is transferred from the liquid to the external surface of the solid particles; last, it is transferred from the solid surface inside the pores where the reaction occurs. The reaction products, which are formed inside the solid particles, make the inverse itinerary. Although in slurry reactors there are more diffusional steps than in gas–liquid reactors, the interpretation of the kinetic data normally is simpler because the diffusional steps are all independent of the reaction excluding the internal diffusion occurring simultaneously with the reaction.

Let us consider a reaction of the type:

$$\mathrm{A(gas)} + z\mathrm{B(liquid)} \rightarrow \mathrm{P}$$

and suppose first order with respect to the gaseous reagent. By applying the steady-state approximation to all of the possible steps occurring before and after the reaction and considering all the possible gradients depicted in Fig. 7.38, we can write:

Overall reaction rate $= r_{\mathrm{A}}$

Fig. 7.37 Gradients at the gas–liquid and liquid–solid interphases in slurry reactors

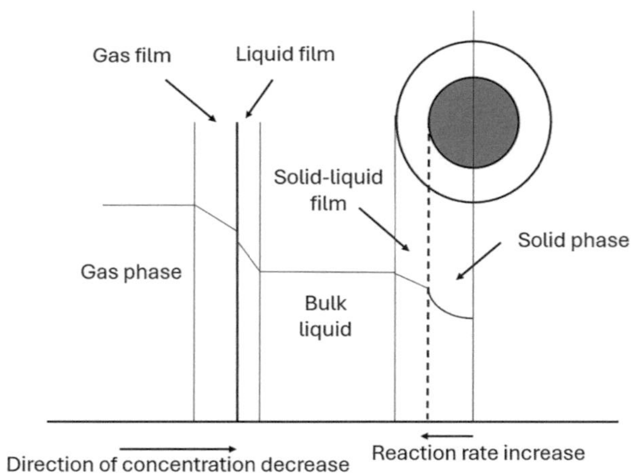

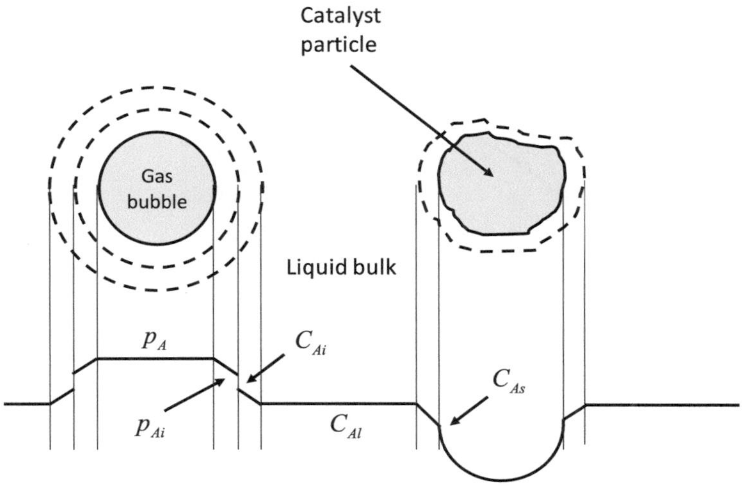

Fig. 7.38 Gradients that could be possible in a slurry reactor

$$r_A = k_g a_L (p_A - p_{Ai}) = \text{Gas} - \text{liquid mass transfer (gas side)} \tag{7.117}$$

$$r_A = k_L a_L \left(\frac{p_{Ai}}{H} - C_{AL}\right) = \text{Gas} - \text{liquid mass transfer(liquid side)} \tag{7.118}$$

$$r_A = k_s a_s (C_{AL} - C_{As}) = \text{Liquid} - \text{solid mass transfer} \tag{7.119}$$

$$r_A = \eta k_1 a_s C_{as} \quad \text{Internal diffusion + chemical-reaction rate} \tag{7.120}$$

where a_L is the gas–liquid interphase area, a_s the liquid–solid interphase area, and H is the Henry solubility constant for A. Combining the different relations with the elimination of the interphase concentrations, we obtain:

$$r_A = \frac{1}{k_g a_L} + \frac{H}{k_L a_L} + \frac{H}{k_s a_s} + \frac{H}{\eta k_I a_s} \tag{7.121}$$

Normally, $1/k_g\ a_L$ can be neglected mainly when a pure gas reagent is used. Because $a_s = 6\ \text{m}/\rho_p d_p$, where m is the catalyst hold-up, ρ_p is the density of the catalyst particle, and d_p is the particle diameter, we can obtain the following by substituting a_s in Eq. (7.121):

$$\frac{p_A}{H r_A} = \frac{1}{k_L a_L} + \frac{\rho_p d_p}{6m}\left(\frac{1}{k_S} + \frac{1}{\eta k_1}\right) \tag{7.122}$$

that is, for a first-order reaction, a linear correlation exists between p_A/Hr_A and $1/m$. When the reaction order is different from one, this linear correlation is not valid (see Fig. 7.39).

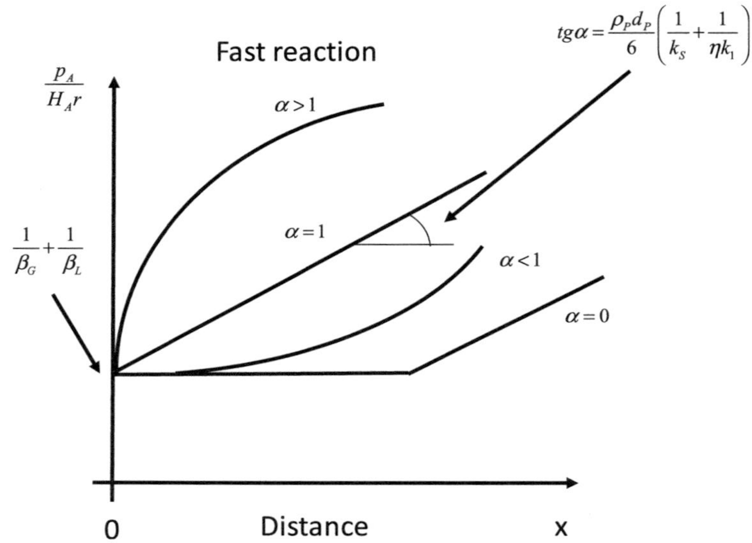

Fig. 7.39 Trend of p_A/Hr_A and $1/m$ for different values of α = reaction order in a conventional slurry reactor

The rigorous description of the kinetic behavior of reactions with orders different from one requires the determination of the concentration profiles of the gaseous reagent inside the catalyst particles. The parameters of the kinetic model, which are useful for describing gas–liquid–solid reactors, are k_L, a_L, k_s, a_s, H, k_n, and η. Some of them would be determined following independent routes with respect to the kinetic study of the reaction in order to avoid statistical correlation. The effectiveness factor η, for a first-order reaction, is a well-known function of the Thiele modulus:

$$\eta = \frac{3}{\phi}\left(\frac{1}{\tanh \phi} - \frac{1}{\phi}\right) = \frac{R}{R_{\max}} \tag{7.123}$$

$$\phi = R_p\left(\frac{S_V k_s}{D_{\text{eff}}}\right)^{1/2} \tag{7.124}$$

This approach, in some cases, is not valid because if the powdered catalyst is constituted by very thin particles with a diameter less than that of the film layer and the reaction is very fast, an enhancement factor can be observed similar to the one observed for the gas–liquid systems; however, this is less relevant because it is limited by the size of the particles. This phenomenon is due to the alteration of the concentration profile inside the boundary layer as shown in Fig. 7.40. The reaction can occur mainly inside the liquid film, and mass transfer is enhanced as a consequence.

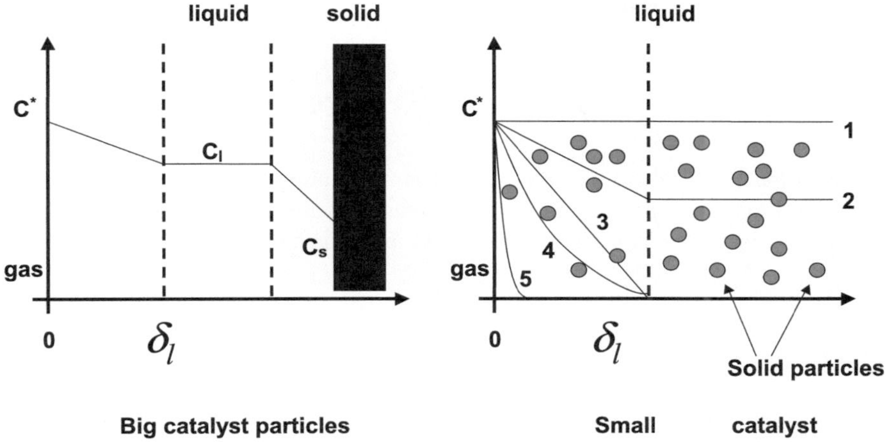

Fig. 7.40 Gas–liquid gradients for catalyst particles with diameters, respectively, larger and smaller than that of the boundary layer

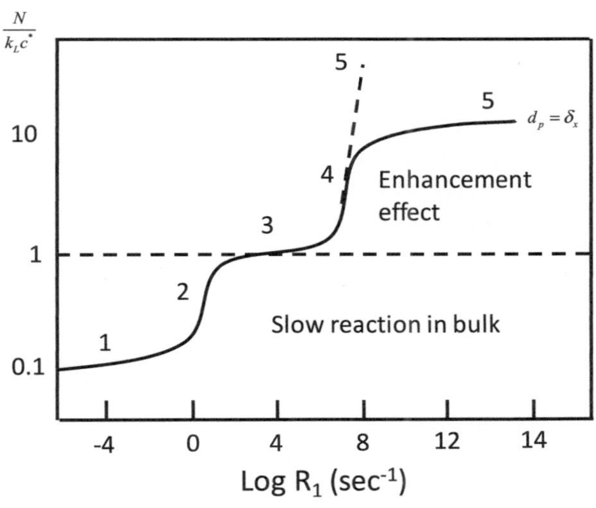

Fig. 7.41 Enhancement factor in a slurry reactor according to Alper et al. (1980) for the oxidation of glucose in the presence of active coal particles

An example of the enhancement effect was reported by Alper et al. (1980) for the oxidation of glucose in the presence of active coal as the catalyst. As can be seen in Fig. 7.41, the enhancement effect is limited by the size of the catalyst particles.

Determination of the Thiele modulus ϕ and of the effectiveness factor η becomes more complicated when the kinetic expressions have reaction orders different from one, in particular, when the reaction is not isotherm and the system is not stationary. The approach is the same already described for gas–solid reactions except that the mass-transfer rates in the liquid phase are much slower. The most complex and general case requires a numerical solution of the following differential equations:

$$\frac{\partial Ci}{\partial t} = D_{\text{eff}} \quad \nabla^2 C_i + R_i \tag{7.125}$$

$$C_{\text{p}} \frac{\partial T}{\partial t} = \lambda_{\text{eff}} \quad \nabla^2 T + Q_{\text{R}} \tag{7.126}$$

given, respectively, the concentration and temperature profiles inside the catalyst particles. From these profiles, it is possible then to evaluate the effectiveness factor.

Under steady-state conditions referring to only one component, we have:

$$D_{\text{eff}} \nabla^2 C + R = 0 \tag{7.127}$$

$$\lambda_{\text{eff}} \nabla^2 T - R\Delta H = 0 \tag{7.128}$$

with the boundary conditions:

$$C = C_{\text{s}} \quad T = T_{\text{s}} \quad r = R_{\text{p}} \tag{7.129}$$

$$\nabla T = \nabla C = 0 \quad r = 0 \tag{7.130}$$

It must be pointed out that in passing from a gas–solid to a liquid–solid system, the diffusion coefficient decreases by 3–4 orders of magnitude; consequently, to obtain the same value of the Thiele modulus the catalyst particles in contact with the liquid must be at least 1/100 smaller in diameter than the particles in contact with gaseous reagents. This means that in a liquid–solid system, for a moderately fast reaction, the effectiveness factor is always <1 for particles >50–100 μm.

In conclusion, the study of the kinetics of gas–liquid reactions is performed through the following steps:

1. Determine the physical parameters in an independent way, that is, in the absence of the reaction (solubilities, vapor–liquid equilibria, mass-transfer coefficients).
2. Characterize the fluid dynamic of the reactor (determine the gas–liquid and liquid–solid interface area).
3. Determine the slow steps determining the observed reaction rate.
4. Determine the intrinsic kinetic laws and related parameters.
5. Verify the kinetic model in different types of reactors.

As in the case of gas–liquid reactors, in slurry reactors, the partition of reactants and products between the gas and liquid phases is fundamental for a correct kinetic approach.

Exercise 7.4. Hydrogenation of THEAQH$_2$ in the Presence of Palladium Catalyst

Section 7.2.6 elaborates on the reactions occurring in the all-TETRA process for producing hydrogen peroxide, and the successive example describes the oxidation step in more detail. Santacesaria et al. (1988a, b) studied the kinetics of the hydrogenation step in a well-stirred semi-batch slurry reactor using a palladium-supported catalyst. The studied reaction is:

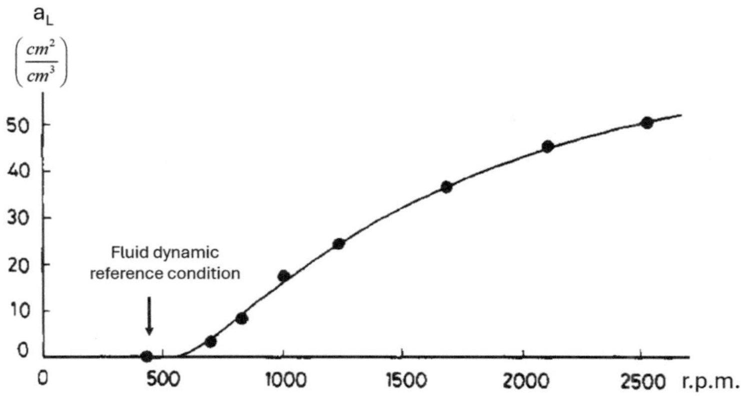

Fig. 7.42 Apparent specific interface area for different stirring speeds evaluated for the reactor shown in Fig. 7.16. Reproduced with permission from Santacesaria et al. (1988a), copyright American Chemical Society (1988)

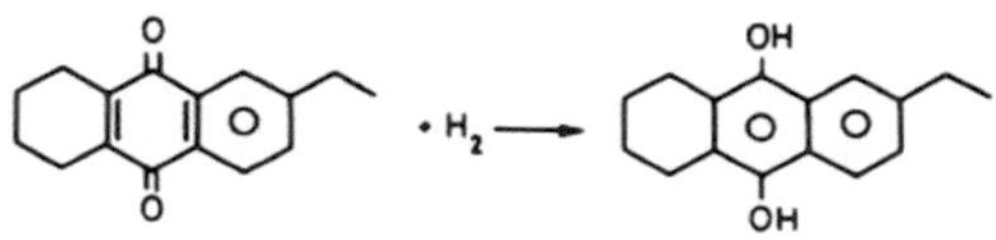

(7.131)

For studying the kinetics of this reaction, the authors employed a laboratory reactor similar to the one shown in Fig. 7.16.

Determination of the Gas–Liquid Mass-Transfer Parameter $\beta_L = k_L a_L$
First, fluid-dynamic characterization of the reactor was performed by measuring with the sulfite method the evolution of the apparent specific interface area with the stirring speed. The specific interface area is apparent because it also contains the contribution of the change of the mass-transfer coefficient with the stirrer rotating speed. The results obtained are shown in Fig. 7.42.

The fluid-dynamic reference condition, shown in Fig. 7.42, is characterized by a well-stirred solution but without producing cavitation. In this condition, the interface surface area is known, and measuring the hydrogen absorption rate in the absence of the reaction under reference conditions allows the determination of k_L, the mass-transfer coefficient. This was done using an amperometric electrode, giving a signal of current proportional to the hydrogen concentration. Two types of runs were performed: one evaluating the hydrogen-absorption rate in the working solution not containing the catalyst until reaching the saturation corresponding to the hydrogen solubility and another one by following the desorption rate. Different runs were performed, respectively, at 30, 40, and 50 °C for a given specific interface area $a_L = 0.16\ \text{cm}^{-1}$. Data were interpreted in the same way described in Exercise 7.1, and a value of k_L was determined for this reactor equal to:

$$k_L = 0.612\ \exp(-1135.3/T) \qquad (7.132)$$

The product of $\beta_L = k_L a_L$ is the true mass-transfer parameter to be used in the interpretation of the kinetic runs.

Determination of the Liquid–Solid Mass-Transfer Parameter k_s

Liquid–solid mass-transfer coefficients can be estimated with the correlation suggested by Sano et al. (1974) (see also Chaudhari and Ramachandran 1980).

$$\frac{k_s d_p}{DF_c} = 2 + 0.4\left(\frac{P_{ow} d_p^4 \rho_L^3}{\mu_L^3}\right)^{1/4}\left(\frac{\mu_L}{D\rho_L}\right)^{1/3} \qquad (7.133)$$

where $P_{ow} = \frac{8\omega^3 d_I^5 \psi}{L d_t^2}$, ω is the stirring rate (in rps), d_I is the impeller diameter (3.8 cm), d_t is the reactor diameter (9 cm), ψ is the correction factor (see Chaudhari and Ramachandran 1980), D is the diffusion coefficient (in cm^2/s), F_c is the catalyst shape factor (1), μ_L is the liquid viscosity (in g/cm s), d_p is the average catalyst particle diameter (0.013 cm), and ρ_L is the liquid density (in g/cm^3).

Changes of μ_L and ρ_L with temperature were measured, while changes of D were calculated as suggested by Berglin and Shoon (1981). The evolution with temperature of the mentioned parameters is shown in Fig. 7.43. The same plot also shows the values of the Henry constant describing the solubility of hydrogen in the working solution for different temperatures.

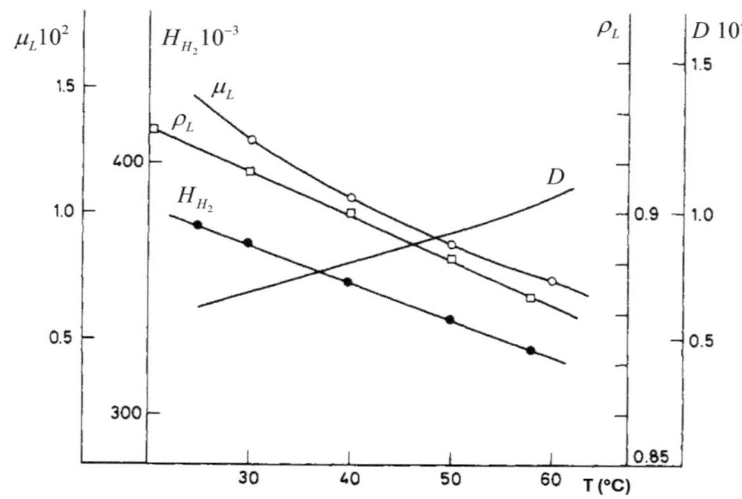

Fig. 7.43 Hydrogen solubility, H_{H2}, in the working solution at different temperatures: Behavior of the viscosity, density, and hydrogen diffusion coefficient by changing temperature. Reproduced with permission from Santacesaria et al. (1988a), copyright American Chemical Society (1988)

Table 7.7 Mass-transfer parameters obtained, respectively, by kinetic runs and by independent approaches

Parameters obtained in the presence of the reaction				Parameters obtained by physical measurements (β_L) or by calculations (k_s)		
Rotation speed (rpm)	700	1000	2000	700	1000	2000
$\beta_L = k_L a_L$	0.1	0.28	0.78	0.11	0.27	0.75
k_s	0.050	0.066	0.076	0.068	0.080	0.13

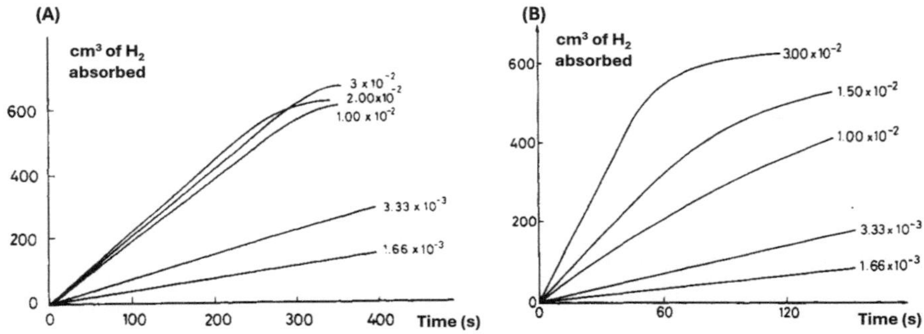

Fig. 7.44 Kinetic runs performed at (**a**) 50 °C and 700 rpm in the presence of different catalyst concentrations (g/cm^3) and (**b**) at 50 °C and 2000 rpm in the presence of different catalyst concentrations (g/cm^3). The anthraquinone concentration is 20.5 g/L in both cases. Reproduced with permission from Santacesaria et al. (1988a), copyright American Chemical Society (1988)

The authors demonstrated that k_s is poorly affected by the catalyst particle diameter; in contrast, it depends almost linearly on the stirring rate. The mass-transfer parameters, respectively obtained in the presence and the absence of the reaction, are listed in Table 7.7.

Many different kinetic runs were performed measuring the volume of hydrogen adsorbed with time and changing the stirring rate, the temperature, and the catalyst hold-up.

Figure 7.44 shows two examples of kinetic runs performed, respectively, at 700 and 2000 rpm, at 50 °C, by changing the catalyst hold-up.

Figure 7.45 shows experimental runs performed at different temperatures by changing the rotating speed.

By observing the trends of the kinetic runs, it can be noted that:

1. The measured reaction rates are almost independent of the organic reagent because all of the curves are linear for $>80\%$ of the reaction extent.
2. If the reaction rates that can be obtained by the slopes of the linear part of the curves are arranged in an Arrhenius-type plot, very low apparent activation energies are obtained: 1.5 kcal/mol for the runs performed at 700 rpm and 4.0 kcal/mol at 2000 rpm. This means that the reaction rate is strongly limited by the gas–liquid mass transfer, and it is reasonable to assume that the gas–liquid

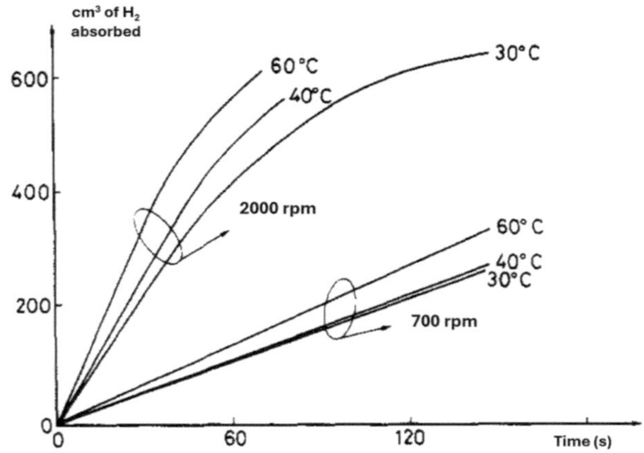

Fig. 7.45 Kinetic runs performed at different temperatures and at, respectively, 700 and 2000 rpm with a fixed catalyst concentration of 3.0×10^{-2} g/cm^3. The anthraquinone concentration is 20.5 g/L in both cases. Reproduced with permission from Santacesaria et al. (1988a), copyright American Chemical Society (1988)

mass transfer prevails at low stirring rates, while liquid–solid mass transfer is controlled at a higher stirring rate.

Therefore, for interpreting the kinetic runs, we chose to adopt the classical approach suggested by Satterfield and Sherwood (1963) by constructing the plot of the reciprocal of the observed reaction rates ($1/H_{H2}r$) against the reciprocal of catalyst concentrations ($1/m$) for two rotating speeds of, respectively, 700 and 2000 rpm. For this purpose, we consider the following relation:

$$\frac{p_{H_2}}{H_{H_2}r} = \frac{1}{k_L a_L} + \frac{1}{m}\frac{\rho_p d_p}{6k_s} \tag{7.134}$$

To construct the two plots, assume that the temperature is 50 °C, the pressure p_{H2} is 1 atm, the catalyst density is 2.25 g/cm^3, and the average d_p is 0.013 cm. Perform the calculations considering the parameters listed in Table 7.7. Compare the results with the experimental data shown in Fig. 7.46.

Solution

By observing the plot in Fig. 7.46, we see that the values of $1/HR$ at $1/m = 0$ are the values of $\beta_L = k_L a_L$ listed in Table 7.7 in the presence of the reaction. At high catalyst loading, the values of $1/HR$ are nearly constant, while decreasing the catalyst content, a linear trend between $1/HR$ and $1/m$ was observed.

The following plot (Fig. 7.47) was obtained, for the three sets of experimental data at different stirring rates (700, 1000, and 2000 rpm), using the values of the parameters listed in Table 7.7.

All of the described results were obtained using a MATLAB program available as Electronic Supplementary Material.

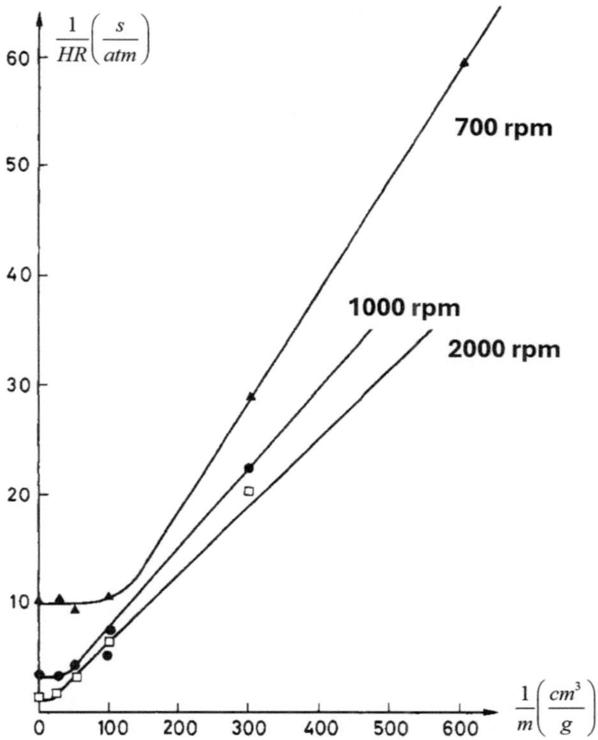

Fig. 7.46 Trend of $1/Hr$ against $1/m$ at different stirring rates. The values of $1/Hr$ at $1/m = 0$ correspond to the gas–liquid mass-transfer coefficients. Reproduced with permission from Santacesaria et al. (1988a), copyright American Chemical Society (1988)

Exercise 7.5. Alkylation of Cresol to BHT

Cresol alkylation reaction with isobutene has been extensively studied in the works of Santacesaria et al. (1988a, 2005a). The reaction occurs in the presence of an acid ion-exchange resin. The occurring reaction scheme is reported below.

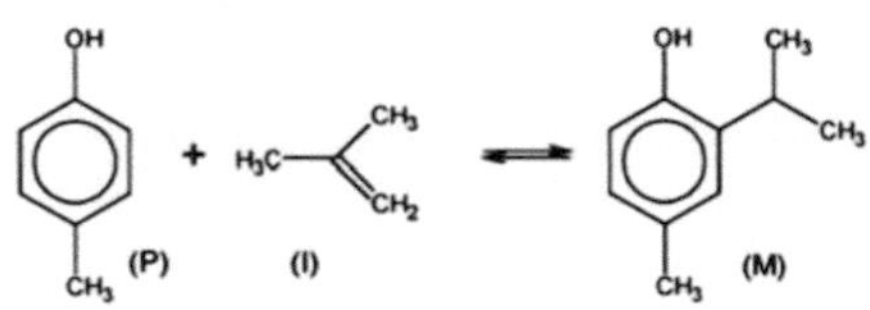

(7.135)

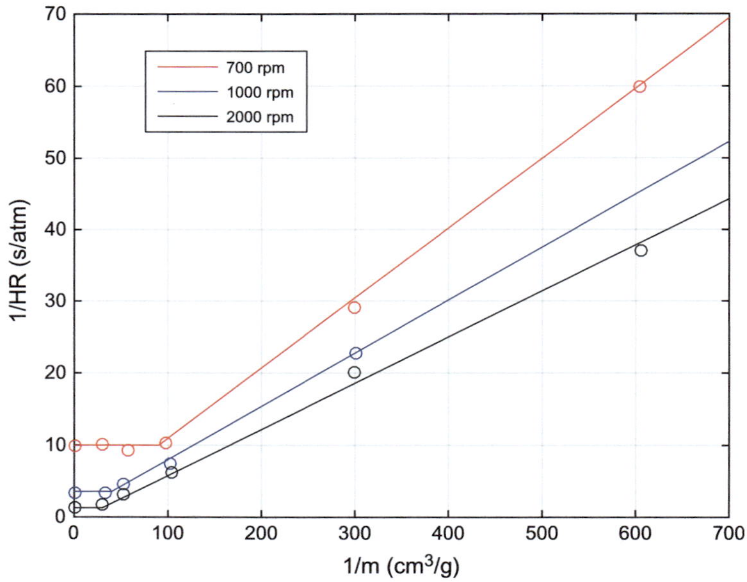

Fig. 7.47 Calculated trend of $1/Hr$ against $1/m$ at different stirring rates

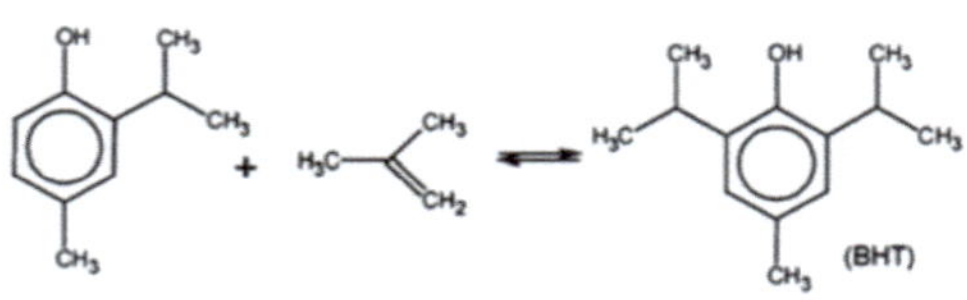

(7.136)

Isobutene oligomerization also occurs in the system with the following side reactions:

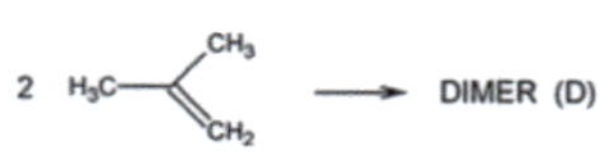

(7.137)

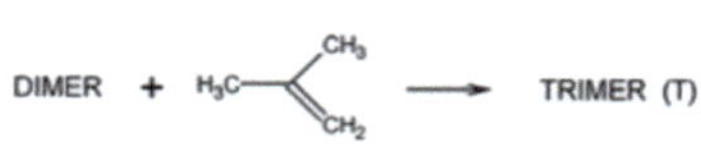

(7.138)

Both main and side reactions can be written in a compact form as:

$$1.\quad P + I \rightleftarrows M \quad \text{monoalkylation} \quad r_1 \tag{7.139}$$

$$2.\quad M + I \rightleftarrows BHT \quad \text{dialkylation} \quad r_2 \tag{7.140}$$

$$3.\quad 2I \rightarrow D \quad \text{dimerization} \quad r_3 \tag{7.141}$$

$$4.\quad D + I \rightarrow T \quad \text{trimerization} \quad r_4 \tag{7.142}$$

Table 7.8 Experimental data related of a run performed under conditions of $T = 50$ °C, catalyst = 10 g, and initial cresol = 140 g

Time (min)	Cresol (mol)	Time (min)	Monoalk. (mol)	Time (min)	BHT (mol)	Time (min)	Dimer (mol)	Time (min)	Trimer (mol)
0	1.296	6	0.154	6	0.009	12	0.002	40	0.005
6	1.132	12	0.283	12	0.038	25	0.016	60	0.009
12	0.974	25	0.483	25	0.112	40	0.018	90	0.011
25	0.699	40	0.654	40	0.199	60	0.022	120	0.015
40	0.441	60	0.809	60	0.340	90	0.028	160	0.024
60	0.145	90	0.836	90	0.458	120	0.029	220	0.031
		120	0.750	120	0.545	160	0.031	300	0.053
		160	0.705	160	0.589	220	0.038		
		220	0.600	220	0.694	300	0.044		
		300	0.498	300	0.797				

Table 7.9 Parameters useful for the calculations

Parameter	Value	Units
Isobutene gas phase mole fr.	1	–
Temperature	50	°C
Pressure	1	atm
Initial moles of cresol	1.296	mol
k_L	2.2e−3	cm/s
k_S	1.2e−2	cm/s
D_{eff}	1e–6	cm^2/s
R_p	0.035	cm
Specific area	500,000	cm^2/g
Mass of catalyst	10	g
Specific gas–liquid area	20	cm^2/cm^3
External catalyst surface	87	cm^2/g
Stirring speed	1100	rpm

From Santacesaria et al. (2005a, b), the experimental data listed in Table 7.8 were extracted:

Develop a kinetic and mass-transfer model and a related MATLAB code to optimize kinetic parameters starting from the values reported in the work of Santacesaria et al. (1988a).

Solution

From the previously mentioned works, the following parameters and data listed in Table 7.9 can be used in the model.

1. **Mass Transfer**

Isobutene mass transfer from the gas phase to the liquid phase and then to the catalyst solid surface involves the evaluation of three different concentrations related to this component: the gas–liquid interface equilibrium concentration, the bulk liquid-phase concentration, and the solid surface concentration. These concentrations can be calculated dynamically by solving the two following differential equations coupled with material balances (Sect. 7.3) and with thermodynamic equilibrium relation (Sect. 7.2). These differential equations have the form:

$$\frac{\mathrm{d}C_{\mathrm{I}}^{\mathrm{L}}}{\mathrm{d}t} = k_{\mathrm{L}} a_{\mathrm{L}} \left(C_{\mathrm{I}} - C_{\mathrm{I}}^{\mathrm{L}}\right) - k_{\mathrm{S}} a_{\mathrm{S}} \left(C_{\mathrm{I}}^{\mathrm{L}} - C_{\mathrm{I}}^{\mathrm{S}}\right) \tag{7.143}$$

$$\frac{\mathrm{d}C_{\mathrm{I}}^{\mathrm{S}}}{\mathrm{d}t} = k_{\mathrm{S}} a_{\mathrm{S}} \left(C_{\mathrm{I}}^{\mathrm{L}} - C_{\mathrm{I}}^{\mathrm{S}}\right) - \sum_{i=1}^{\mathrm{Nr}} \nu_{i\mathrm{I}} r_i \tag{7.144}$$

where

k_{L}	gas–liquid mass-transfer coefficient
a_{L}	gas–liquid specific area
C_{I}	isobutene equilibrium concentration
$C_{\mathrm{I}}^{\mathrm{L}}$	isobutene bulk liquid concentration
k_{S}	liquid–solid mass-transfer coefficient
a_{S}	gas–solid specific area
$C_{\mathrm{I}}^{\mathrm{S}}$	isobutene surface concentration
$\nu_{i\mathrm{I}}$	stoichiometric coefficient of isobutene in reaction *i*-th
r_i	reaction rate for reaction *i*-th
Nr	number of reactions

The solution of these two ordinary differential equations can be found using usual numerical techniques, or a pseudo–steady-state approximation can be introduced as:

$$\frac{\mathrm{d}C_{\mathrm{I}}^{\mathrm{L}}}{\mathrm{d}t} = k_{\mathrm{L}} a_{\mathrm{L}} \left(C_{\mathrm{I}} - C_{\mathrm{I}}^{\mathrm{L}}\right) - k_{\mathrm{S}} a_{\mathrm{S}} \left(C_{\mathrm{I}}^{\mathrm{L}} - C_{\mathrm{I}}^{\mathrm{S}}\right) \approx 0 \tag{7.145}$$

$$\frac{\mathrm{d}C_{\mathrm{I}}^{\mathrm{S}}}{\mathrm{d}t} = k_{\mathrm{S}} a_{\mathrm{S}} \left(C_{\mathrm{I}}^{\mathrm{L}} - C_{\mathrm{I}}^{\mathrm{S}}\right) - \sum_{i=1}^{\mathrm{Nr}} \nu_{i\mathrm{I}} r_i \approx 0 \tag{7.146}$$

In this case, the two resulting algebraic equations can be solved, at each step in time, and the values of isobutene concentration in bulk liquid and on the solid surface can be obtained.

The final step requires to also take into account the contribution of the internal diffusion. The catalyst effectiveness factor can be calculated, for the two main reactions, as follows:

Table 7.10 Activity coefficients at infinite dilution at 50 °C

Component	γ_{Ii}^{∞}
Cresol	3.33
Monoalkylate	1.15
BHT	0.81

$$\phi_i = r_p[C_c k_1 a_S / D_{eff}]^{0.5} \tag{7.147}$$

$$\eta_i = \frac{3}{\phi_i}\left[\frac{1}{\tanh \phi_i} - \frac{1}{\phi_i}\right] \quad i = 1, 2 \tag{7.148}$$

For the effectiveness factor of the two oligomerization reactions, unitary values are assumed because these reactions are very slow and occur with very moderate conversion.

2. **Gas–Liquid Equilibrium**

The nonideal behavior of the system can be accounted for by using the approach of infinite-dilution activity coefficients as reported by the abovementioned authors. The infinite-dilution activity coefficient for isobutene can be calculated as:

$$\ln \gamma_I^{\infty} = \sum x_i \ln \gamma_{Ii}^{\infty} \tag{7.149}$$

and from this, the activity coefficient of isobutene in the mixture is:

$$\ln \gamma_I = (1 - x_I)^2 \ln \gamma_I^{\infty} \tag{7.150}$$

The infinite-dilution activity coefficients of isobutene in the other components can be calculated according to Santacesaria et al. (1988a), and their values are listed in Table 7.10.

The isobutene molar fraction in the liquid phase must be calculated iteratively because the activity coefficient itself is a function of the composition. The value of this mole fraction is obtained by solving the following equilibrium relation:

$$f(x_I) = x_I - \frac{P y_I}{P_I^0 \gamma_I(x_I)} = 0 \tag{7.151}$$

For this type of calculation, the presence of isobutene dimer and trimer was neglected because these components are present in low amounts.

3. **Kinetics and Mass Balances**

The differential equations related to material balances for a slurry reactor are expressed in terms of concentration variation along time, that is:

$$\frac{dC_P}{dt} = -r_1 \quad \text{p-cresol} \tag{7.152}$$

$$\frac{dC_M}{dt} = r_1 - r_2 \quad \text{monoalkylate} \tag{7.153}$$

$$\frac{dC_{BHT}}{dt} = r_2 \quad \text{dialkylate} \tag{7.154}$$

$$\frac{dC_D}{dt} = r_3 - r_4 \quad \text{isobutene dimer} \tag{7.155}$$

$$\frac{dC_T}{dt} = r_4 \quad \text{isobutene trimer} \tag{7.156}$$

$$-\frac{1}{V_L}\frac{dn_I}{dt} = r_1 + r_2 + 2r_3 + r_4 \quad \text{isobutene consumption} \tag{7.157}$$

This set of differential equations must be integrated in time and must be coupled with the mass-transfer and equilibrium relations reported in Sects. 7.1 and 7.2. Kinetic parameters must be evaluated through the fitting of the experimental data. The only information that we need for completing the model is the expressions of the reaction rate. In the previously mentioned papers, the following relations are adopted:

$$r_1 = \eta_1 k_1 \left(C_{Ps} C_{Is} - \frac{C_{Ms}}{K_{e1}} \right) \tag{7.158}$$

$$r_2 = \eta_2 k_2 \left(C_{Ms} C_{Is} - \frac{C_{BHTs}}{K_{e2}} \right) \tag{7.159}$$

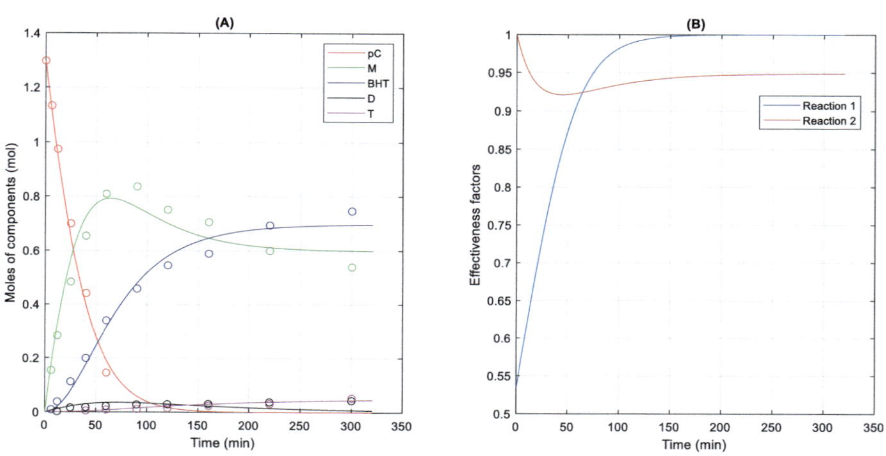

Fig. 7.48 (**a**) Mole number profiles. (**b**) Effectiveness factor for the two reactions

Table 7.11 Parameters comparison at $T = 50$ °C

Parameter	Optimized parameters	Data from Santacesaria et al. (1988a, b)
k_1	3.1049e−07	8.6616e−07
k_2	6.4220e−08	8.3389e−08
k_3	9.8173e−08	1.9270e−08
k_4	3.3954e−08	3.7613e−07
K_{e1}	2.7343e+13	2.7343e+13
K_{e2}	2.8992e+24	2.8992e+24

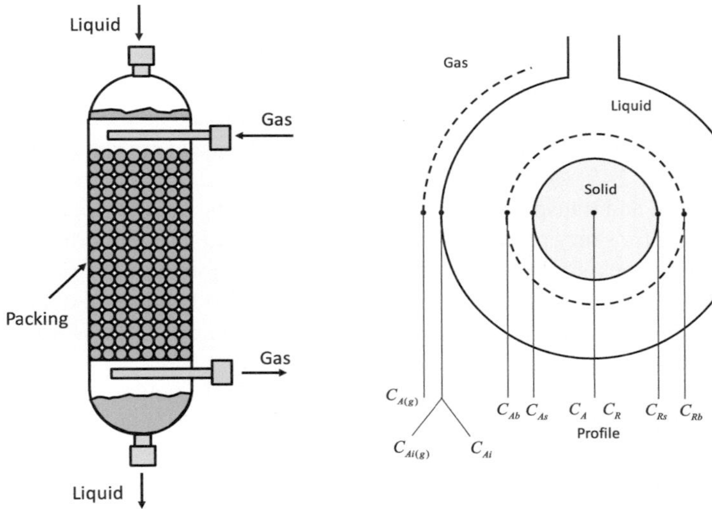

Fig. 7.49 Scheme of gas–liquid cocurrent trickle-bed reactor

$$r_3 = k_3 C_{\text{Is}}^2 \tag{7.160}$$

$$r_4 = k_4 C_{\text{Is}} C_{\text{Ds}} \tag{7.161}$$

As evident from these relations, all the component concentrations are referred to as catalysts' surface values. All of the kinetic and equilibrium constants are functions of temperature through the well-known Arrhenius and Van't Hoff equations. In our case, the objective of the calculation is the description of a single constant temperature run so k and K_e are really constants.

Results

In Fig. 7.48, the agreement between calculated and experimental values for component moles number is reported (plot a). In the plot (b), the evolution in time of the calculated effectiveness factors for the two reactions is depicted.

By observing the values in Table 7.11, the agreement between parameters calculated by optimization in this exercise and those taken from the literature is quite good, also considering that the values reported in the reference paper are calculated on several experimental runs at different temperatures and not on a single run at a fixed temperature as in the case of the present exercise.

All of the described results have been obtained using a MATLAB program available as Electronic Supplementary Material.

7.4.2 Trickle-Bed Reactors

Trickle-bed reactors are gas–liquid–solid reactors characterized by a fixed bed of catalytic particles continuously wetted by a liquid falling film, containing one or more reactants, with a cocurrent or countercurrent stream of reacting gas flowing across the reactor as shown in Fig. 7.49. By considering a single catalytic particle, the concentrations of the reacting gas and of the liquid reagent are always distributed as shown in Fig. 7.49, assuming a porous catalyst.

The reaction and transport steps in trickle-bed reactors and related equations are similar to those for slurry reactors. The main differences are just the correlations used to determine the mass-transfer coefficients that can be found in the rich literature devoted to the subject (see, e.g., Ranade Vivek et al. 2011; Gianetto and Specchia 1992; Ramachandran and Chaudhari 1983; Satterfield et al. 1969; Westerterp and Wammes 1992).

A difficulty in modeling these reactors arises when the catalytic particles are not uniformly wetted by the liquid film as a consequence of liquid maldistribution. Moreover, trickle-bed reactors are often applied to perform strong exothermic reactions and because trickle-bed reactors have poor capability to eliminate the heat involved with reactions, hot spots may be created. Hot spots can cause undesirable side reactions, runaway, and damage to the catalyst.

Trickle beds are used in many different industrial processes, such as the hydrodesulfurization of heavy oil stocks, the hydro-treating of lubricating oils, and many different hydrogenation processes.

References

Akita, K., Yoshida, F.: Gas holdup and volumetric mass transfer coefficient in bubble columns. Effects of liquid properties. Ind. Eng. Chem. Process Des. Dev. **12**(1), 76–80 (1973)

Alper, E., Wichtendahl, B., Deckwer, W.D.: Gas absorption mechanism in catalytic slurry reactors. Chem. Eng. Sci. **35**(1–2), 217–222 (1980)

Astarita, G.: Mass Transfer with Chemical Reaction. Elsevier, Amsterdam (1967)

Berglin, T., Shoon, N.H.: Kinetic and mass transfer aspects of the hydrogenation stage of the anthraquinone process for hydrogen peroxide production. Ind. Eng. Chem. Process Des. Dev. **20**(4), 615 (1981)

Carrà, S., Morbidelli, M.: Gas-liquid reactors. In: Carberry, J.J., Varma, A. (eds.) Chemical Reaction and Reactor Engineering. Marcel Dekker, New York (1987)

Carrà, S., Santacesaria, E.: Engineering aspects of gas-liquid catalytic reactions. Catal. Rev. Sci. Eng. **22**(1), 75–140 (1980)

Carrà, S., Morbidelli, M., Santacesaria, E., Buzzi, G.: Synthesis of propylene oxide from propylene chlorohydrins—II: Modeling of the distillation with chemical reaction unit. Chem. Eng. Sci. **34**(9), 1133–1140 (1979a)

Carrà, S., Santacesaria, E., Morbidelli, M., Cavalli, L.: Synthesis of propylene oxide from propylene chlorohydrins—I: kinetic aspects of the process. Chem. Eng. Sci. **34**(9), 1123–1132 (1979b)

Carrà, S., Santacesaria, E., Morbidelli, M., Schwarz, P., Divo, C.: Synthesis of epichlorohydrin by elimination of hydrogen chloride from chlorohydrins. 1. Kinetic aspects of the process. Ind. Eng. Chem. Process Des. Dev. **18**(3), 424–427 (1979c)

Carrà, S., Santacesaria, E., Morbidelli, M., Schwarz, P., Divo, C.: Synthesis of epichlorohydrin by elimination of hydrogen chloride from chlorohydrins. 2. Simulation of the reaction unit. Ind. Eng. Chem. Process Des. Dev. **18**(3), 428–433 (1979d)

Charpentier, J.C.: Mass transfer rates in gas-liquid absorbers. In: Drew, T.B. (ed.) Advances in Chemical Engineering, pp. 2–133. Elsevier, New York (1981)

Chaudhari, R.V., Ramachandran, P.A.: Three phase slurry reactors. AIChE J. **26**(2), 177–201 (1980)

Chilton, T.C., Colburn, A.P.: Mass transfer (absorption) coefficients prediction from data on heat transfer and fluid friction. Ind. Eng. Chem. **26**(11), 1183–1187 (1934)

Danckwerts, P.V.: Absorption by simultaneous diffusion and chemical reaction. Trans. Faraday Soc. **46**, 300–305 (1950)

Danckwerts, P.V.: Significance of liquid-film coefficients in gas absorption. Ind. Eng. Chem. **43**(6), 1460–1467 (1951a)

Danckwerts, P.V.: Absorption by simultaneous diffusion and chemical reaction into particles of various shapes and into falling drops. Trans. Faraday Soc. **47**, 1014–1022 (1951b)

Danckwerts, P.V.: Gas-Liquid Reactions. McGraw-Hill, New York (1970)

Dechwer, W.D., Burchart, R., Zoll, G.: Mixing and mass transfer in tall bubble columns. Chem. Eng. Sci. **29**(11), 2177–2188 (1974)

Dimiccoli, A., Di Serio, M., Santacesaria, E.: Mass transfer and kinetics in spray-tower-loop absorbers and reactors. Ind. Eng. Chem. Res. **39**(11), 4082–4093 (2000)

Di Serio, M., Di Martino, S., Santacesaria, E.: Kinetics of fatty acids polyethoxylation. Ind. Eng. Chem. Res. **33**(3), 509–514 (1994)

Di Serio, M., Tesser, R., Felippone, F., Santacesaria, E.: Ethylene oxide solubility and ethoxylation kinetics in the synthesis of nonionic surfactants. Ind. Eng. Chem. Res. **34**(11), 4092–4098 (1995)

Duduković, A., Pjanović, R.: Effect of turbulent Schmidt number on mass-transfer rates to falling liquid films. Ind. Eng. Chem. Res. **38**(6), 2503–2504 (1999)

Gianetto, A., Specchia, V.: Trickle-bed reactors: state of art and perspectives. Chem. Eng. Sci. **47**(13–14), 3197–3213 (1992)

Gilliland, E.R., Sherwood, T.K.: Diffusion of vapors into air streams. Ind. Eng. Chem. **26**(5), 516–523 (1934)

Higbie, R.: The rate of absorption of a pure gas into a still liquid during short periods of exposure. Trans. Am. Inst. Chem. Eng. **31**, 365 (1935)

Joshi, J.B., Sharma, M.M.: Mass transfer characteristics of horizontal sparged contactors. Trans. Inst. Chem. Eng. **54**, 42 (1976)

Levenspiel, O.: Chemical Reaction Engineering. Wiley, New York (1972)

Levenspiel, O., Godfrey, J.H.: A gradientless contactor for experimental study of interphase mass transfer with/without reaction. Chem. Eng. Sci. **29**(8), 1723–1730 (1974)

Levich, V.G.: Physicochemical Hydrodynamics. Prentice-Hall, Englewood Cliffs, NJ (1962)

Lewis, W.K., Whitman, W.G.: Principles of gas absorption. Ind. Eng. Chem. **16**(12), 1215–1220 (1924)

Linek, V., Vacek, V.: Chemical engineering use of catalyzed sulfite oxidation kinetics for the determination of mass transfer characteristics of gas-liquid contactors. Chem. Eng. Sci. **36**(11), 1747–1768 (1981)

Nitta, T., Akimoto, T., Matsui, A., Katoyama, T.: An apparatus for precise measurement of gas solubility and vapor pressure of mixed solvents. J. Chem. Eng. Jpn. **16**(5), 352–356 (1983)

Perry, R.H., Green, D.W.: Chemical Engineer's Handbook, 6th edn. McGraw Hill, New York (1984)

Ramachandran, P.A., Chaudhari, R.V.: Three-Phase Catalytic Reactors. Gordon and Breach, New York (1983)

Ranade Vivek, V., Chaudhari, R., Gunjal, P.R.: Trickle Bed Reactors: Reactor Engineering and Applications. Elsevier, Oxford (2011)

Sano, Y., Yamaguchi, N., Adachi, T.: Mass transfer coefficients for suspended particles in agitated vessels and bubble columns. J. Chem. Eng. Jpn. **7**(4), 255–261 (1974)

Santacesaria, E., Ferro, R., Ricci, S., Carrà, S.: Kinetic aspects in the oxidation of hydrogenated 2-ethyltetrahydroanthraquinone. Ind. Eng. Chem. Res. **26**(1), 155–159 (1987)

Santacesaria, E., Silvani, R., Wilkinson, P., Carrà, S.: Alkylation of p-cresol with isobutene catalyzed by cation-exchange resins: a kinetic study. Ind. Eng. Chem. Res. **27**(4), 541–548 (1988a)

Santacesaria, E., Wilkinson, P., Babini, P., Carrà, S.: Hydrogenation of 2-ethyl-tetrahydro-anthraquinone in the presence of palladium catalyst. Ind. Eng. Chem. Res. **27**(5), 780–784 (1988b)

Santacesaria, E., Di Serio, M., Garaffa, R., Addino, G.: Kinetics and mechanisms of fatty alcohol polyethoxylation. 1. The reaction catalyzed by potassium hydroxide. Ind. Eng. Chem. Res. **31**(11), 2413–2418 (1992)

Santacesaria, E., Di Serio, M., Tesser, R., Cammarota, F.: Comparison between the performances of a well-stirred slurry reactor and a spray loop reactor for the alkylation of *p*-cresol with isobutene. Ind. Eng. Chem. Res. **44**(25), 9473–9481 (2005a)

Santacesaria, E., Di Serio, M., Tesser, R.: Gas–liquid and gas–liquid–solid reactions performed in spray tower loop reactors. Ind. Eng. Chem Res. **44**(25), 9461–9472 (2005b)

Satterfield, C.N., Sherwood, T.K.: The Role of Diffusion in Catalysis. Addison Wesley, Reading, MA (1963)

Satterfield, C.N., Pelossof, A.A., Sherwood, T.K.: Mass transfer limitations in a trickle bed reactor. AIChE J. **15**, 226 (1969)

Srinivasan, V., Aiken, R.C.: Mass transfer to droplets formed by the controlled breakup of a cylindrical jet—physical absorption. Chem. Eng. Sci. **43**(12), 3141–3150 (1988)

van Krevelen, D.W., Hoftijzer, P.J.: Kinetics of gas-liquid reactions part I. General theory. Recueil des Travaux Chimiques des Pays-Bas. **67**(7), 563–586 (1948)

Westerterp, K.R., Wammes, W.J.A.: Three-phase trickle-bed reactors. In: Ullmann's Encyclopedia of Industrial Chemistry, pp. 309–320. VCH, Weinheim (1992)

Chapter 8
Polymerization Reactions and Reactors

Abstract After the examination of the historical evolution of the industrial production of polymers, different polymer classifications are described. Then, the polymers' average molecular weight and molecular weight distribution are defined together with the description of the techniques commonly employed for their experimental determination. The different mechanisms of polymerization and the kinetics of polymerization are described as a base for the polymerization reactor modeling. For this purpose, polycondensations, polyadditions, radical polymerizations, and stereospecific polymerizations are considered in detail for both the mechanism and kinetics. A paragraph devoted to polymerization reactor modeling concludes the chapter. Different solved examples and exercises are developed such as the following: kinetics of polycondensation, conversion from molecular weight distribution, kinetics of polyethoxylation, the method of moments applied to kinetics of polymerization, stepwise polymerization in different reactors, batch, plug flow, CSTR, and styrene radical polymerization in a batch reactor.

Keywords Polymers · Polymerization reactions · Polymerization reactors · Kinetics of polymerization · Polymerization mechanisms

8.1 Definitions, Historical Notes, and Classifications

Synthetic polymers can be obtained from simple molecules (monomers) assembled with each other to form macromolecules containing from a few tens of monomer units (oligomers) to several thousand ones (polymers). These macromolecules are characterized by an average molecular weight ($\overline{MW}$) and a molecular weight distribution (MWD). These two parameters are strictly related to the physical properties of the polymers that are fundamental for their final use. Product specifications of interest are, for example, the following: the mechanical strength, the

---

Supplementary Information The online version contains supplementary material available at https://doi.org/10.1007/978-3-031-82502-6_8.

© The Author(s), under exclusive license to Springer Nature Switzerland AG 2025
R. Tesser, E. Santacesaria, *The Chemical Reactor from Laboratory to Industrial Plant*, https://doi.org/10.1007/978-3-031-82502-6_8

electrical properties, the solvent solubility, the corrosion resistance, the melting point, the softening point, etc. Polymerization reactions are a combination of series and simultaneous reactions characterized by a scheme of complex chemical reactions. Therefore, the concepts developed in a previous chapter of this book for the complex reaction schemes can be applied also to the design of polymerization reactors, but with an important difference in the objective to be pursued. In a usual complex chemical reaction scheme, the goal is to maximize the obtainment of a specific compound through the steps of reaction, separation, and purification. In the case of polymers, the objective is to obtain a product with specific properties that are related, through a more or less complicated correlation, with both the $\overline{MW}$ and the MWD. Therefore, it is important to define the most appropriated range of $\overline{MW}$ and MWD to be achieved for each type of polymer and the correlated properties. The obtainment of the desired polymeric properties becomes much more difficult when polymerization involves more complex reaction systems, like chain branching or copolymerization, because those systems do not depend uniquely on the MWD. The production of synthetic polymers in the world and the growth over the years are impressive, as it can be appreciated from the plot reported in Fig. 8.1 taken from UNEP (2021). In the same plot is also reported the overall accumulation of plastics in the world determining serious environmental problems. The industry of polymers production is very recent if compared with that of other materials, for example, the production of metals. For this purpose, in Fig. 8.2, the milestones of the most important polymers production are reported (reproduced from Wikipedia 2024). Polymers can be classified in different ways. The first type of classification is simply based on the chemical starting compounds. Examples are:

- **Styrene** polymerizes giving polystyrene.
- **Ethylene** gives polyethylene.
- **Propylene** gives polypropylene.
- **Vinyl chloride** gives poly-vinyl-chloride.
- **Phenol and formaldehyde** give phenolic resins.
- Etc.

Another possible classification is based on the type of occurring reaction:

- **Addition polymers**: in this case, the structural unit of the polymer is identical to the monomer.
- **Condensation polymers**: in this case, the polymerization occurs by a reaction between two or more molecules with the elimination of smaller molecules. An example is the synthesis of Nylon 66 obtained by a reaction between hexamethylene diamine and adipic acid occurring with the elimination of water.

Another classification considers the polymer's molecular structure in relation with the application field. Based on this classification, we can have, plastomers, elastomers, and fibers.

- **Plastomers** are polymers that can be molded by heating in a reversible way (thermoplastic polymers) or in an irreversible process (thermosetting polymers).

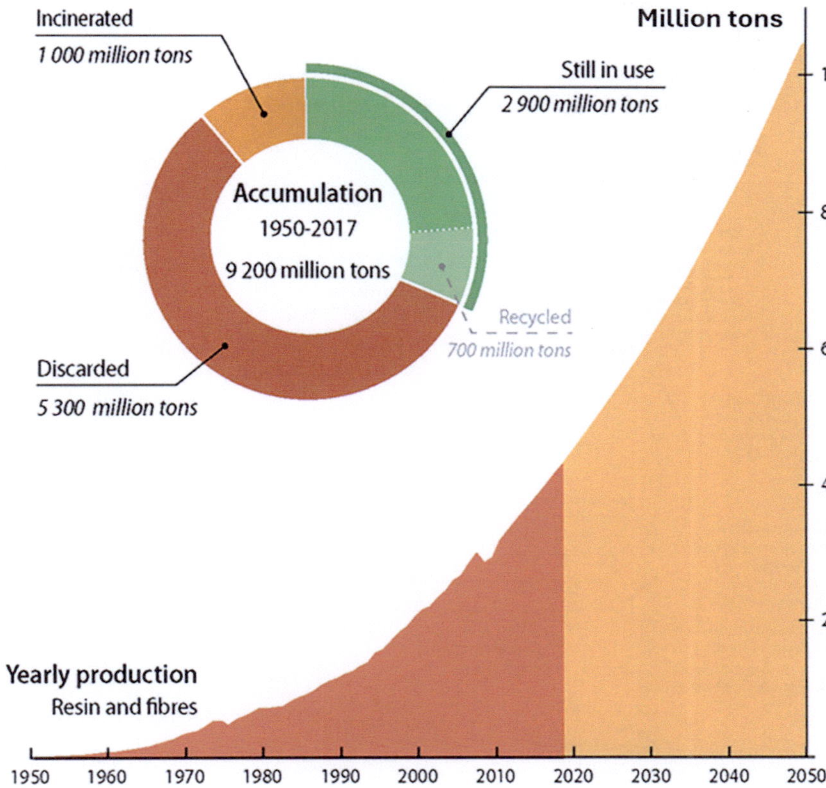

Fig. 8.1 Evolution of polymer production along the time in million metric tons and production trend for the future. The environmental problems due to the plastics accumulation. Plot taken from UNEP (United Nations Environment Programme) (2021) illustrated by GRID-Arendal

- **Elastomers** are polymers characterized by a considerable elasticity (rubber) due to the bridges existing between the linear polymeric chains.
- **Fibers** are polymers with a structure ordered along one dimension. The stretching of fibers favors a correct orientation of crystallites along the desired direction.

From this classification, it is possible to argue that there is a strict correlation between the molecular structure and the properties of a polymer. For this purpose, we can have polymers characterized by **linear** or **branched** chains, but the chains can also interact with each other giving place to secondary bonds between electrically charged atoms or hydrogen bonds. A certain degree of order can be obtained, and for very regular structures, crystalline polymers can be obtained. In any case, molecular weight and molecular weight distribution are always of paramount importance in determining the final properties. Polymers can be derived not only from a single monomer (**homopolymer**) but also from the simultaneous reaction of two different monomers (**random copolymers**). If the monomers are fed in sequence, it

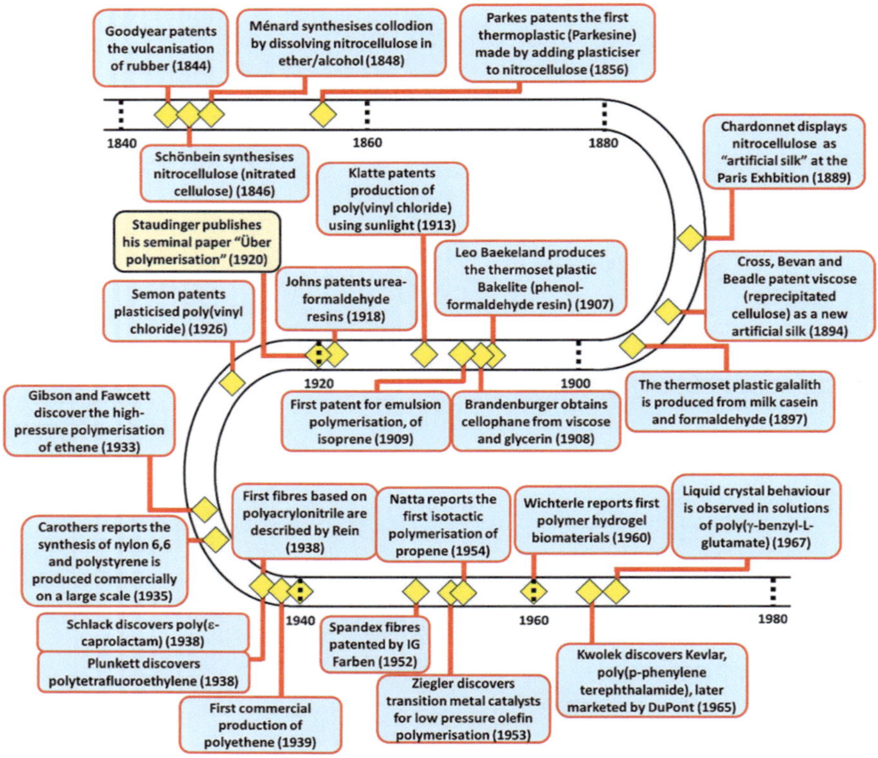

Fig. 8.2 Milestones along the time in the production of the polymers (Wikipedia 2024)

is possible to obtain **block copolymers**. Another possibility is the obtainment of **graft copolymers**. A scheme of all the mentioned possibilities is reported in Fig. 8.3.

Another universally accepted classification considers the behavior of the polymer with the temperature, which is important in polymers processing. It is possible to distinguish:

(a) **Thermoplastic polymers**—A thermoplastic polymer, when heated, softens, becoming liquid, and hardens when cooled with a reversible process. Normally, polymers constituted by linear chains are thermoplastic.
(b) **Thermosetting polymers**—A thermosetting polymer, when heated initially melts but then irreversibly hardens, becoming solid, thermostable, and insoluble in solvents, because of reticulations reactions, favored by the temperature.

Each polymerization system requires a particular type of polymerization reactor able to dissipate the large amount of reaction heat (13–21 kcal/mol, see Holland and Anthony 1979). **The mechanism of polymerization** is the most important element of classification because it allows the definition of the kinetic models of polymerization. Therefore, this aspect will be considered in detail in the next sections.

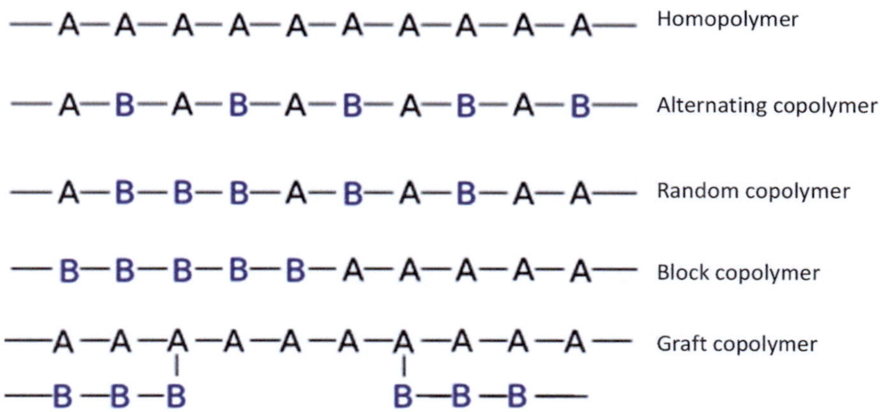

Fig. 8.3 Some different structures of copolymers

8.2 Polymers Average Molecular Weight ($\overline{MW}$) and Molecular Weight Distribution (MWD)

In the previous section, it has been stressed the importance of both the average molecular weight ($\overline{MW}$) and the molecular weight distribution (MWD) in affecting the physical properties of the polymers. There are some different ways to calculate $\overline{MW}$ and related MWD based on the method adopted for the measurement. The polymers $\overline{MW}$ can be determined with many different techniques, such as, for example:

(a) Analytical determination of the terminal groups
(b) Cryoscopy, ebullioscopy, or vapor pressure measurements
(c) Osmometry
(d) Light scattering
(e) Sedimentation measurements (ultracentrifugation)
(f) Viscosimetry

The results obtained by each of the mentioned experimental techniques are different and the following average $\overline{MW}$ definitions are the most employed.

- **Numerical Average Molecular Weight $\overline{MW_N}$**

When the average $\overline{MW}$ is measured through the evaluation of osmotic pressure and other colligative properties, or by the analysis of the terminal groups, being these measurements independent of the molecular size, but just dependent on the number of molecules, the obtained result is the numerical average molecular weight $\overline{MW_N}$, defined as:

$$\overline{\mathrm{MW_N}} = \frac{n_1M_1 + n_2M_2 + n_3M_3 + \cdots n_nM_n}{n_1 + n_2 + n_3 + \cdots n_n} = \frac{\sum_i^n n_iM_i}{\sum_i^n n_i} \tag{8.1}$$

n_i is the number of molecules of molecular weight M_i.

- **Weighted Average Molecular Weight $\overline{\mathbf{MW_W}}$**

Measuring the average molecular by light diffraction or by sedimentation rate methods, a different result is obtained because, in these cases, the average depends not only on the number of molecules but also on their size. The ponderal average molecular weight $\overline{\mathbf{MW_W}}$ can be defined as:

$$\overline{\mathrm{MW_W}} = \frac{m_1M_1 + m_2M_2 + mM_3 + \cdots m_nM_n}{m_1 + m_2 + m_3 + \cdots m_n} = \frac{\sum_i^n m_iM_i}{\sum_i^n m_i} = \frac{\sum_i^n n_iM_i^2}{\sum_i^n n_iM_i} \tag{8.2}$$

m_i is the overall mass of the specie i-th, that is, $m_i = n_iM_i$.

- **Viscosimetric Average Molecular Weight $\overline{\mathbf{MW_V}}$**

The relation between the viscosity of a polymer solution and the molecular weight, according to Mark (1938)-Houwink (1940), can be expressed as:

$$[\eta] = KM^a \tag{8.3}$$

where $[\eta]$ is the intrinsic viscosity of the solution, M is the molecular weight, K and a are constants characteristic of the polymer-solvent. If the system is polydisperse any i-th specie has its own intrinsic viscosity:

$$[\eta_i] = KM_i^a \tag{8.4}$$

This average molecular weight clearly will depend on the system solvent-polymer. For this purpose, we can write:

$$\overline{\mathrm{MW_V}} = \left[\frac{\sum_i^n n_iM_i^{a+1}}{\sum_i^n n_iM_i}\right]^{\frac{1}{a}} \tag{8.5}$$

n_i is the number of individual molecules having the molecular mass M_i. The exponent a, between 0.6 and 0.8 for many polymer/solvent systems, is best evaluated from measured viscosities of dilute solutions of narrow-molecular-mass fractions of polymers, determining the intrinsic viscosity $[\eta]$ for each fraction, then fitting Eq. (8.4). For $a = 1$, $\overline{MW_W} = \overline{MW_V}$ otherwise $\overline{MW_W} > \overline{MW_V}$.

- **Average Molecular Weight From Sedimentation Equilibrium $\overline{\mathbf{MW_Z}}$**

More seldom, the average molecular weight is estimated through the sedimentation equilibrium and is expressed as:

$$\overline{MW_Z} = \left[\frac{\sum_i^n n_i M_i^3}{\sum_i^n n_i M_i^2} \right] \tag{8.6}$$

It is possible to foresee that for a polydisperse polymer $\overline{\mathbf{MW_N}} < \overline{\mathbf{MW_V}} \le \overline{\mathbf{MW_W}} < \overline{\mathbf{MW_Z}}$.

- **Index of Polydispersity**

The ratio $\overline{\mathbf{MW_W}}/\overline{\mathbf{MW_N}}$ is the **index of polydispersity** of the polymer. This ratio is 1 when the polymer has a uniform molecular weight but becomes greater by increasing the polydispersity. The properties of the polymers are strongly affected by the degree of polydispersity, and the presence of short chains, normally, has a negative effect on the quality of the polymers.

- **Molecular Weight Distribution (MWD) and the Degree of Polymerization**

The **molecular weight distribution (MWD)** can be represented with distribution curves of integral or differential type (Bamford et al. 1958). In the curves of integral type, the ratio: $W/W_0 = f(x)$ is a function of x, the **degree of polymerization**. x is the number of structural units of monomers for a given polymeric chain, W is the weight of the polymer having polymerization degree equal or inferior to x, and W_0 is the total weight of the polymer. An integral curve is like the one reported in Fig. 8.4.

The mass differential distribution can be obtained by derivation of the integral curve as:

$$W_x = \frac{1}{W_0} \frac{dW}{dx} = \frac{xN_x}{\sum_x xN_x} \tag{8.7}$$

In Fig. 8.5, examples of differential molecular weight distribution of a linear polymer considering, respectively, the number molecular weight distribution (MWD_N) and the weight molecular weight distribution (MWD_W).

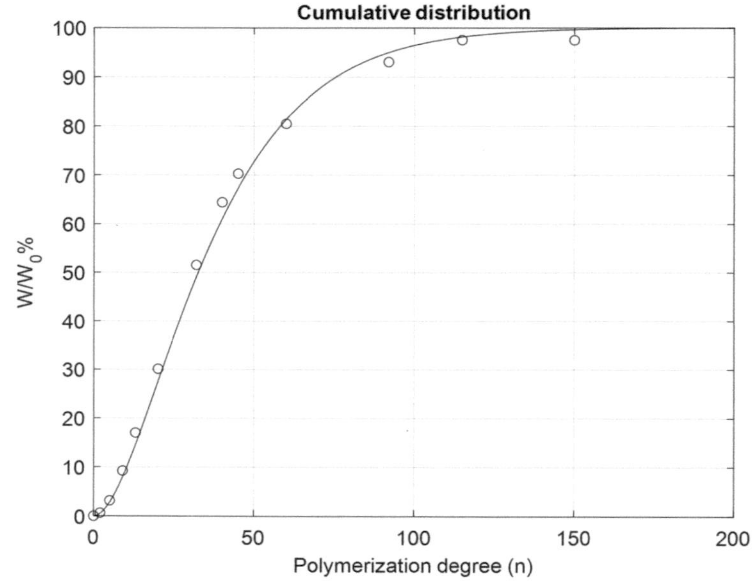

Fig. 8.4 An example of integral molecular weight distribution (polystyrene) (Natta et al. 1978)

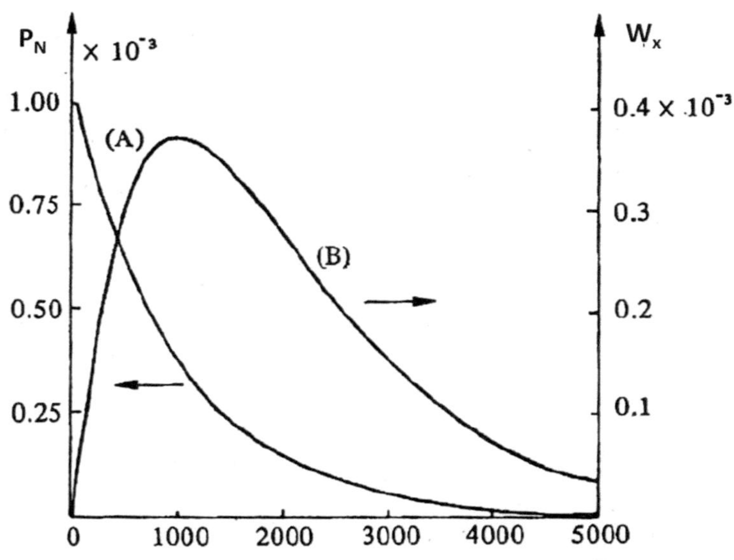

Fig. 8.5 Examples of differential molecular weight referred to (A) number molecular weight distribution (MWD_N) and (B) weight molecular weight distribution (MWD_W) (Natta et al. 1978)

8.3 Reaction Mechanisms and Kinetics of Polymerization as a Base for the Reactor Modeling

In general, the assembly of monomers for obtaining a polymer can occur through two main mechanisms that can be recognized as polycondensation or polyaddition reactions.

8.3.1 Polycondensation

Three types of polycondensation are worth to be mentioned:

1. **Polycondensation of two different bifunctional molecules.** It occurs by elimination of a small molecule such as water or a low-molecular-weight alcohol. Polyamides (Nylon 6.6, Nylon 6.10, and Kevlar), polyesters (PET-polyethylene terephthalate), and polycarbonates (obtained by condensation of bisphenol and phosgene) are obtained in this way.

   Polyester is reported in the following scheme as an example:

$$\mathrm{HO{-}(CH_2)_n{-}OH + HOOC{-}(CH_2)_m{-}COOH \rightarrow HO{-}(CH_2)_n{-}O{-}CO{-}(CH_2)_m{-}COOH + H_2O} \tag{8.8}$$

   Esterification proceeds in the same way in the successive steps.

2. **Polycondensation by elimination of chlorine from a bifunctional molecule.** Poly-hydrocarbons can be obtained in this way:

$$\mathrm{Cl{-}CH_2{-}(CH_2)_n{-}CH_2{-}Cl \rightarrow \left({-}CH_2{-}(CH_2)_n{-}CH_2{-}{-}CH_2{-}(CH_2)_n{-}CH_2{-}\right) + Cl_2} \tag{8.9}$$

3. **Polysulfide polymers. Are obtained from a dichloride molecule and an alkaline polysulfide:**

$$\mathrm{Cl{-}CH_2{-}(CH_2)_n{-}CH_2{-}Cl + Na_2S_4} \\ \mathrm{\rightarrow Cl{-}CH_2{-}(CH_2)_n{-}CH_2{-}S4{-}CH_2{-}(CH_2)_n{-}CH_2{-}Cl + 2\,NaCl} \tag{8.10}$$

Other products obtainable by polycondensation are as follows: polyurethanes, polyphenylesters, and polysiloxanes. In the three examples reported above, the monomers are all bifunctional and the polymers that can be obtained have linear chains and are therefore thermoplastic, but it is also possible to start from monomers having functionality greater than 2, obtaining, in this case, branched thermosetting polymers. Examples of branched-chain polymers, obtained by polycondensation, are phenolic, ureic, and epoxy resins. Some of the reactions occurring in the production of phenolic resins are reported in the following scheme as an example:

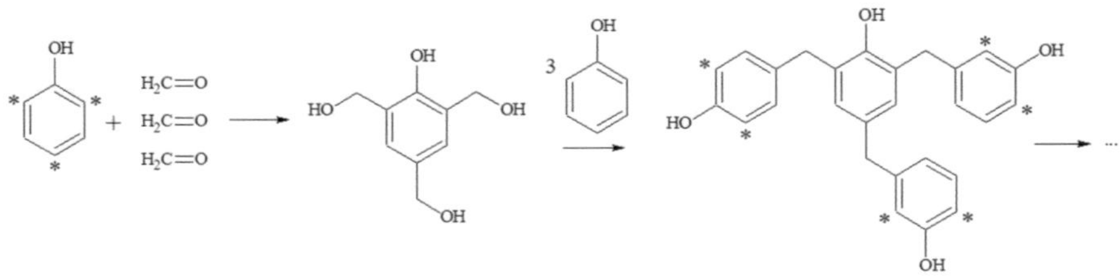

(8.11)

8.3.2 *Kinetics of Polycondensation*

Let us consider now the simplest cases of polycondensation, that is, the ones involving only bifunctional molecules. There are two different molecular systems that can polymerize by linear polycondensation:

1. The monomer is a single bifunctional molecule having two different reactive terminal groups. An example is that of the amino acid molecule $H_2N–(CH_2)_n–COOH$. It is possible to simplify the structure as aAb.
2. The monomers are two different bifunctional molecules, each having the same reactive terminal groups. An example is the reaction of a dicarboxylic acid $HOOC–(CH_2)_n–COOH$ with a diamine $H_2N–(CH_2)_n–NH_2$. In this case, it is possible to simplify by assuming the reaction of aAa with bBb.

The kinetic approach of both these systems is relatively simple, because, in these cases, we have a conventional complex reaction scheme characterized by consecutive reactions gradually consuming the monomer. High molecular weights can be obtained for very high monomer conversions. Moreover, the reaction can proceed also after the complete consumption of the monomer and can restart if further monomer is added. For this reason, these polymers have been named by Szwarc (1968) **living polymers**. Even if the set of consecutive reactions is normally equilibrium reactions, often the reverse reactions are neglected because the small molecule "ab," formed during the reaction, is removed by vaporization shifting to the right of the equilibrium. To obtain a regular polymer, in the case of using two different bifunctional molecules aAa and bBb, the monomers must be loaded respecting the 1:1 stoichiometry ratio. The kinetic approach is normally simplified by assuming that the reactivity of the functional terminal groups is independent of the chain length to which they are attached. In the case of polycondensation, starting from a single monomer, it is possible to write the following reaction scheme:

$$aAb + aAb \rightleftarrows aA_2b + ab$$

$$aA_2b + aAb \rightleftarrows aA_3b + ab$$

$$aA_2b + aA_2b \rightleftarrows aA_4b + ab \quad aA_3b + aAb \leftrightarrows aA_4b + ab \tag{8.12}$$

$$aA_3b + aA_2b \rightleftarrows aA_5b + ab \quad aAb + aA_4b \leftrightarrows aA_5b + ab$$

$$aA_3b + aA_3b \rightleftarrows aA_6b + ab \quad \text{etc.}$$

More generally, it is possible to write:

$$aA_pb + aA_qb \rightleftarrows aA_{p+q}b + ab \tag{8.13}$$

As the reactivity has been postulated the same for all the terminal functional groups, independently of the chain length, the reaction rate expression, in batch condition, will be:

$$r = -\frac{da}{dt} = -\frac{db}{dt} = k\,[a][b] \tag{8.14}$$

as a and b are terminal groups of the same molecule, their concentration is the same, hence:

$$r = -\frac{da}{dt} = k\,[a]^2 \qquad -\int_{a_0}^{a_t} \frac{da}{a^2} = kt \qquad \left(\frac{1}{a_t} - \frac{1}{a_0}\right) = kt \tag{8.15}$$

Then, the conversion of "a" can be defined as:

$$\lambda_a = \frac{a_0 - a_t}{a_0} = 1 - \frac{a_t}{a_0} \quad \text{therefore} \quad \frac{\lambda}{a_0(1-\lambda)} = kt \quad \text{or} \quad \frac{1}{1-\lambda} = 1 + ka_0t \tag{8.16}$$

By putting $(1 - \lambda)^{-1}$ or $\lambda(1 - \lambda)^{-1}$ as a function of t a straight line must be obtained.

Let us consider now a polymerization reaction of the type aAa + bBb. In this case, we have the following reaction scheme:

$$aAa + bBb \rightleftarrows aABb + ab$$

$$aAa + aABb \rightleftarrows aABAa + ab$$

$$bBb + aABb \rightleftarrows bBABb + ab \tag{8.17}$$

$$aAa + aABb \rightleftarrows aABAa + ab$$

$$\text{etc.}$$

Again, the reactivity of the terminal groups is postulated the same independently of the chain length. The difference with respect to the previous reaction system involving a single aAb molecule is that as the monomers are two different molecules aAa and bBb, their initial concentration could be different, and this aspect must be considered in the kinetic approach. However, when the stoichiometric ratio between the molecules is $R = 1$, the kinetic approach is the same reported before with relations (8.14–8.16). In the case in which the stoichiometric ratio $R < 1$, we have:

$$R = \frac{a_0}{b_0} \tag{8.18}$$

a_0 = initial concentration of functional group a and b_0 = initial concentration of functional group b. Remembering then that $b = b_0 -$ reacted functional groups $= b_0 - \lambda a_0$, it holds:

$$r = -\frac{\mathrm{d}a}{\mathrm{d}t} = kab = a_0\frac{\mathrm{d}\lambda}{\mathrm{d}t} = ka_0^2(1-\lambda)\left(\frac{1}{R} - \lambda\right) \tag{8.19}$$

By integrating dλ/dt, it results:

$$\ln\left(\frac{\lambda - \frac{1}{R}}{\lambda - 1}\right) = ka_0\left(\frac{1}{R} - 1\right)t \quad \text{Valid for } R < 1 \tag{8.20}$$

For any value of "a" or "b" terminal group conversion, we have an average molecular weight of the growing polymer and a molecular weight distribution. It is opportune to define, first of all, an average degree of polymerization $\overline{D_n}$.

$$\overline{D_n} = \frac{\text{Number of polymerized monomers}}{\text{Overall number of polymer molecules}} \tag{8.21}$$

$$\text{Number of polymerized monomers} = a_0\frac{\left\{1 + \frac{1}{R}\right\}}{2} \tag{8.22}$$

$$\begin{aligned} &\text{Overall number of polymer molecules} \\ &= \frac{\text{Number of terminal functional groups}}{2} = \frac{a_0(1-\lambda) + a_0\left(\frac{1}{R} - \lambda\right)}{2} \\ &= \frac{a_0\left(1 - 2\lambda + \frac{1}{R}\right)}{2} = \frac{a_0(R - 2\lambda R + 1)}{2R} \end{aligned} \tag{8.23}$$

Introducing Eqs. (8.22) and (8.23) into Eq. (8.21), it results:

Table 8.1 Trend of the average polymerization degree with the conversion

λ	$\overline{D_n}$
0	1
0.5	2
0.90	10
0.98	50
0.99	100
0.999	1000

$$\overline{D_n} = \frac{1+R}{1+R-2\lambda R} \quad \text{for } R=1 \quad \overline{D_n} = \frac{1}{1-\lambda} \tag{8.24}$$

In Table 8.1, the values of $\overline{D_n}$ for $R = 1$ as a function of the conversion are reported. As it can be seen, high molecular weights can be obtained only at very high monomers conversions.

The molecular weight distribution (MWD) of the linear polymers, obtained from a single bifunctional monomer or by the combination of two different monomers, can be obtained by solving the system of differential equations describing the evolution with time of any single species involved in the previously reported reactions schemes. MWDs can be more easily calculated by the Flory-Schulz statistical approach (Flory 1953). The parameter λ, that is, the conversion of the functional group a at time t, is also the probability that each functional group has had to react in the time interval $0 - t$. The difference $(1 - \lambda)$, on the contrary, is the probability that each functional group has had the possibility of not reacting. Defining x, the polymerization degree at time t, the probability that in the interval of time $0 - t$ polymeric molecules of polymerization degree x are formed, is:

$$P_x = \lambda^{x-1}(1-\lambda) = \frac{N_x}{N} = \frac{N_x}{N_0(1-\lambda)} \tag{8.25}$$

where N_x is the number of polymeric molecules of polymeric degree x, while N = total number of molecules in the system $N = N_0(1 - \lambda)$. Equation (8.25) corresponds to the numeral molecular weight distribution (MWD_N). It is then possible to calculate the ponderal molecular weight distribution (MWD_W) as:

$$W_x = \frac{N_x M_x}{N_0 M_0} \approx \frac{x N_x M_0}{N_0 M_0} = \frac{x N_x}{N_0} \quad \text{Combining this with relation (8.25)} \quad W_x = x\lambda^{x-1}(1-\lambda)^2 \tag{8.26}$$

In Figs. 8.6 and 8.7, some examples of differential numeral and ponderal molecular weights distribution curves as a function of the polymerization degree x for different monomer conversions are reported. Calculations have been made with the Flory-Schultz statistical relationships.

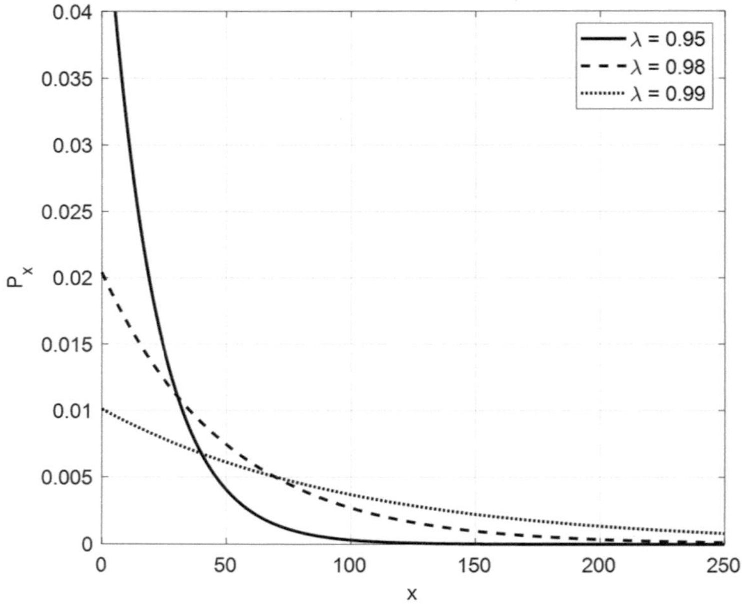

Fig. 8.6 Numeral molecular weight differential distribution (MWD_N) as a function of the polymerization degree x for different values of the conversion degree

Exercise 8.1. Polycondensation Kinetics

For a polycondensation reaction, the monomer conversion has been measured as a function of time. The resulting experimental data are reported in Table 8.2.

By using the experimental data reported in Table 8.2, determine the kinetic constant k in the two following ways:

1. Linearization of Eq. (8.20)
2. Fitting of the differential model represented by Eq. (8.19)

Solution—Part 1

Define, first of all, the two expressions:

$$A = \ln\left(\frac{\lambda - 1/R}{\lambda - 1}\right) \quad B = a_0\left(\frac{1}{R} - 1\right) \tag{8.27}$$

The linearization of Eq. (8.20) consists in a linear plot, in which A is reported as a function of t in correspondence of the experimental values. The fitting of these data (A vs. t) gives the values of slope and intercept, respectively. From the slope, k can be calculated as k = slope/B. In this case, the obtained k value was 0.0122 (L/(mol min)).

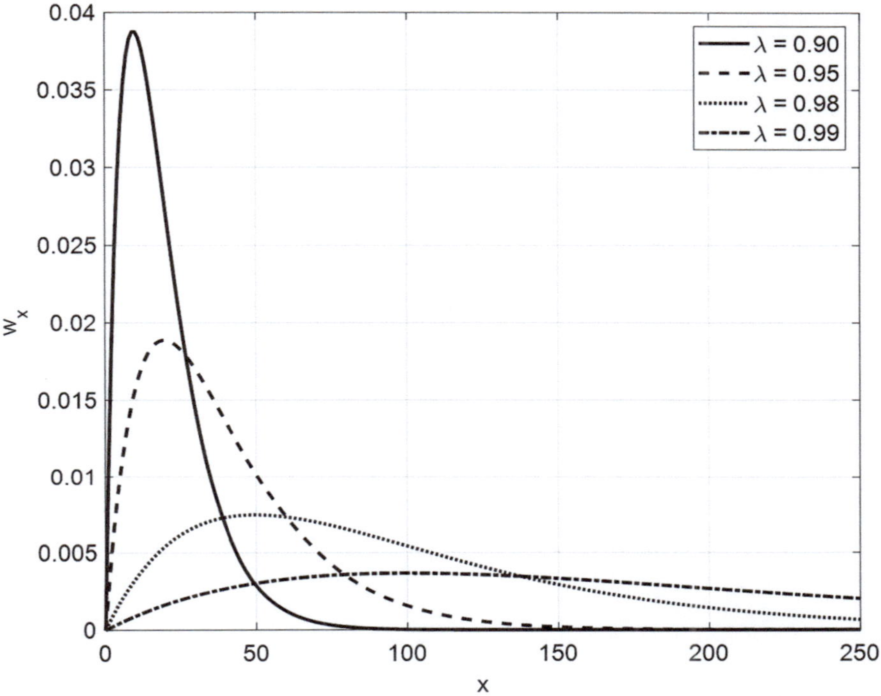

Fig. 8.7 Ponderal molecular weight differential distribution (MWD_W) as a function of the polymerization degree x for different values of the conversion degree

Table 8.2 Data of conversion as a function of time

Time (min)	0	2	4	7	11	20	40	60	110
$\lambda(-)$	0	0.18	0.34	0.51	0.63	0.79	0.90	0.96	0.99

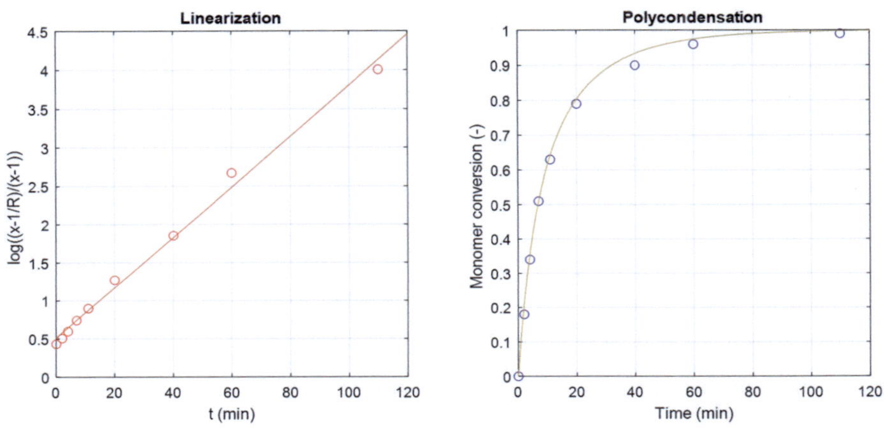

Fig. 8.8 Example of kinetics of polycondensation result

Table 8.3 Data of ponderal distribution as a function of polymerization degree

Dp (units)	0	2	5	9	13	20	32	40	45
Wx (–)	0	0.0048	0.0103	0.0153	0.0175	0.0187	0.0163	0.0130	0.0114
Dp (units)	60	92	115	150					
Wx (–)	0.0073	0.0022	0.0008	0.0001					

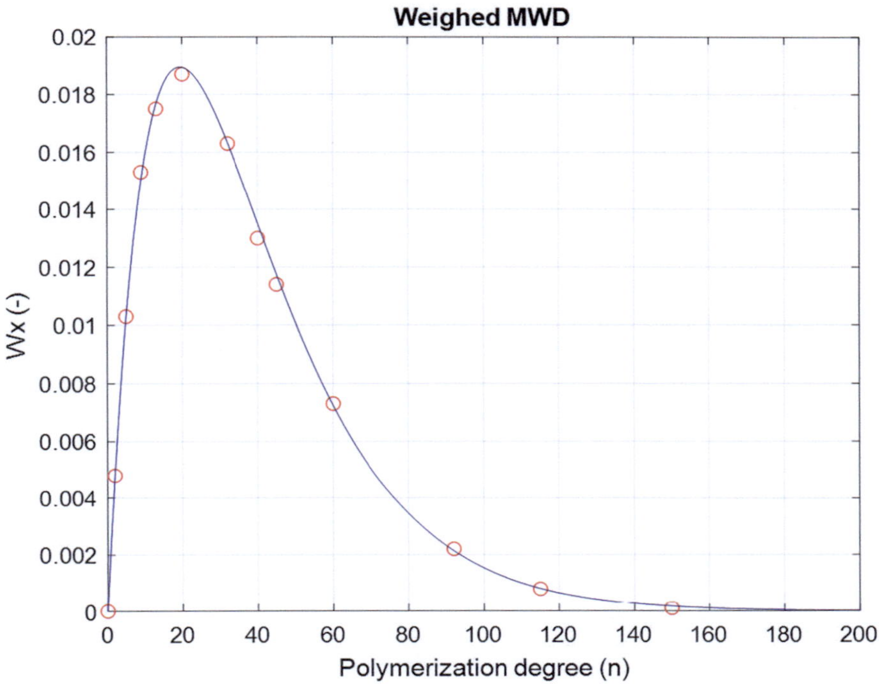

Fig. 8.9 Example of calculated MWD

Solution—Part 2

A more rigorous approach consists of the direct integration of Eq. (8.19) and by mathematical regression (least squares) determine the constant *k*. The value of *k*, found in this second approach, was 0.0165 (L/(mol min)). All the results are reported in Fig. 8.8.

The described results were obtained using a MATLAB program available as Electronic Supplementary Material.

Exercise 8.2. Conversion from Molecular Weight Distribution

For a polycondensation reaction, the ponderal MWD was measured, and a set of experimental data is reported in Table 8.3 (Dp is the polymerization degree expressed in monomeric units, and Wx is the ponderal distribution).

By using the experimental data reported in Table 8.3, it is possible to evaluate the monomer conversion to which the distribution corresponds, by using Eq. (8.26) and by numerical fitting.

Solution

By using mathematical fitting of the nonlinear function (8.26), a value of 0.9498 was found for the monomer fractional conversion. The graphical representation of the calculation is reported in Fig. 8.9.

The described results were obtained using a MATLAB program available as Electronic Supplementary Material.

Exercise 8.3. Cumulative Molecular Weight Distribution

Considering the same experimental data reported in Exercise 8.2 and the monomer conversion equal to 0.9498, obtained as results of the same exercise, build a plot in which the integral (cumulative) MWD is reported. Plot in the mentioned diagram both the experimental data and the theoretical distribution represented by the integral of Eq. (8.26).

Solution

The cumulative MWD can be calculated as the following integral:

$$W_{xc}(x) = \int_0^x W_x(p)\mathrm{d}p = \int_0^x p\lambda^{p-1}(1-\lambda)^2\mathrm{d}p \tag{8.28}$$

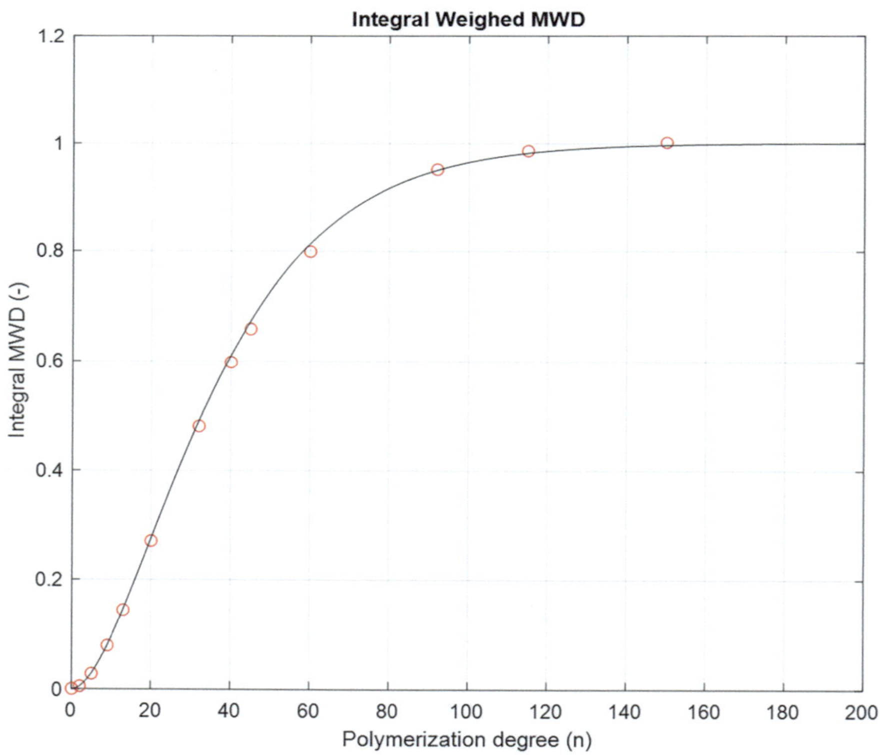

Fig. 8.10 Cumulative MWD as a function of the polymerization degree

The calculation can be easily performed in MATLAB by using two numerical solvers as depicted in the following points:

- Trapz: used for the calculation of integral based on experimental data
- Integral: used for the calculation of integral based on mathematical function representing the MWD

In Fig. 8.10, the cumulative MWD is reported as a function of the polymerization degree. Note that the asymptotic value of the function is equal to one; this means that all the polymeric chains are included in the range of about 0–150 monomeric units.

The described results were obtained using a MATLAB program available as Electronic Supplementary Material.

8.3.3 Polyadditions

It is possible to recognize at least four different types of polyaddition.

1. **Polyvinylic addition and dienes polymerization**. The monomer is a small molecule containing one or more double bonds. The polymerization occurs by opening the double bond as in the following example:

$$\begin{aligned}&H_2C=CHX+H_2C=CHX\rightarrow -H_2C\text{–}CHX\text{–}H_2C\text{–}CHX\text{–}+H_2C=CHX\rightarrow\\&-\ H_2C\text{–}CHX\text{–}H_2C\text{–}CHX\text{–}H_2C\text{–}CHX\text{–}\quad \text{etc.}\end{aligned} \tag{8.29}$$

This mechanism is common to a large number of polymers, like polyethylene, polypropylene, polystyrene, polybutadiene, polyisoprene, polyacrylates, poly-methyl-methacrylate, polyvinylchloride (PVC), polyvinyl acetate, polyacrylonitrile, etc. Butadiene and isoprene are polymerized in the production of rubber.

2. **Polyalkylation**. The monomer contains an epoxide ring. Polymerization occurs by opening the ring as in the following example related to ethylene oxide:

(8.30)

3. **Polyaddition of α-amino-carboxy-anhydride**. The anhydride ring is opened, and polymerization occurs with the elimination of CO_2.

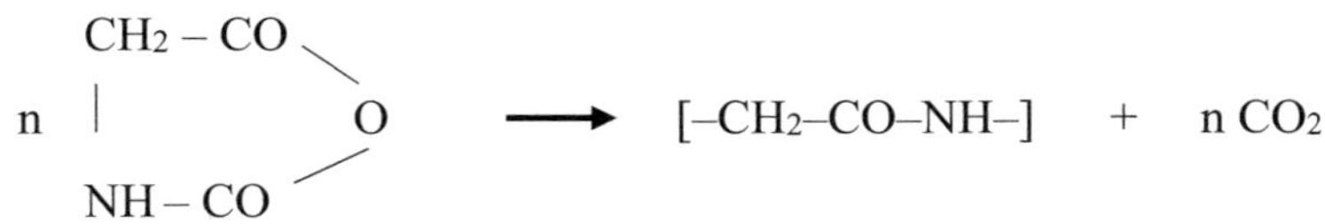

(8.31)

4. **Polyaddition between two different molecules by opening unstable bonds**. With this method, polyamides and polyurethanes are obtained according to the following reaction example:

$$\begin{aligned}&\mathrm{HOCH_2CH_2OH{+}O{=}C{=}N{-}(CH_2)_6{-}N{=}C{=}O{\rightarrow}HO{-}CH_2{-}CH_2O{-}CO{-}NH{-}(CH_2)_6{-}N{=}C{=}O}\\&\mathrm{HO{-}CH_2{-}CH_2O{-}CO{-}NH{-}(CH_2)_6{-}N{=}C{=}O{+}HOCH_2CH_2OH{\rightarrow}}\\&\mathrm{{\rightarrow}HO{-}CH_2{-}CH_2O{-}CO{-}NH{-}(CH_2)_6{-}NH{-}CO{-}OCH_2CH_2OH}\ldots\ldots\ldots\text{etc.}\end{aligned} \tag{8.32}$$

The polyadditions of type (4) have a behavior very similar to the polycondensation. Reactions (8.32), to produce polyurethanes, can be considered like a polycondensation of the type aAa + bBb with the only difference that there is not the elimination of a molecule of low-molecular-weight "ab." Moreover, the reactions in sequence are relatively slow, as in the case of polycondensation. Therefore, the kinetic approach and the determination of the average MW and of the MWD, for linear polymers of this type, are very similar to the one described for polycondensation.

The other polyadditions can be classified according to two different mechanisms: (a) step-growth mechanism and (b) chain-growth mechanism. The rates of the step-growth reactions are relatively slow and the obtainment of high MWs requires a long reaction time. Again, the characteristics are like that of polycondensation reactions. A general scheme for these reactions can be:

$$\begin{aligned}&\mathrm{I + M \rightarrow IM1} && \text{Initiation}\\&\mathrm{IM1 + M \rightarrow IM2} && \text{Propagation steps}\\&\mathrm{IM2 + M \rightarrow IM3}\\&\ldots\ldots\ldots\ldots\ldots\text{etc.}\end{aligned} \tag{8.33}$$

Again, we have living polymers that can growth indefinitely because there is not a termination reaction. The polyadditions of type 1 can be classified, on the contrary, for their chain-growth reaction mechanism. These reactions are characterized by the formation of very unstable reactive intermediate species. The reaction rate is very high, and polymers of high MW are obtained also for very low monomer conversions. Polymerization must be promoted by different types of initiators that can be: radical or ionic (cationic or anionic).

8.3.4 *Kinetics of Polyaddition*

Neglecting the polyadditions that have a kinetic behavior similar to polycondensation, only two mechanisms must be considered for developing the kinetic approach of the other polyaddition reactions: the step-growth mechanism and the chain-growth mechanism.

8.3.4.1 Kinetics of Step-Growth Polymerization

In the most general case, a reaction scheme (8.33) can be considered. In this case, it is easy to define, for a batch reactor, a set of ordinary differential equations whose integration gives the evolution with time of all the formed oligomers. A second-order reaction could be a preliminary choice, and the assumption that the reactivity of all the propagation steps is the same and the kinetic constant would be independent of the chain length is a reasonable approximation. Otherwise, the kinetic constant of each successive step must be estimated from the experimental determination of the MWDs at different reaction times. Some complications arise when the successive reactions of the scheme (8.33) are equilibria and/or chain transfer reactions. In the case of polyethoxylation (Santacesaria et al. 2018), for example, the reaction scheme is:

$$\mathrm{ROH} + \mathrm{KOH} \rightleftarrows \mathrm{RO^-K^+} + \mathrm{H_2O}\uparrow \quad \text{In situ catalyst generation} \tag{8.34}$$

$$\mathrm{RO^-K^+} + \mathrm{EO} \rightarrow \mathrm{REO^-K^+} \quad (r_0) \quad \text{Initiation} \tag{8.35}$$

$$\mathrm{RO(EO)}_i{}^-\mathrm{K^+} + \mathrm{EO} \rightarrow \mathrm{RX(EO)}_{i+1}{}^-\mathrm{K^+} \quad (r_i) \quad \text{Propagation} \tag{8.36}$$

$$\mathrm{RO(EO)}_i\mathrm{H} + \mathrm{RO^-K^+} \rightleftarrows \mathrm{ROH} + \mathrm{RO(EO)}_i{}^-\mathrm{K^+} \quad \text{Proton transfer} \tag{8.37}$$

Based on the mechanism reported above, the following kinetic expressions can be written:

$$\text{Initiation}: \quad r_0 = k_0[\mathrm{RO^-K^+}]\,[\mathrm{EO}] \tag{8.38}$$

$$\text{Propagation}: \quad r_i = k_i\left[\mathrm{RO(EO)}_i^-\,K^+\right]\,[\mathrm{EO}] \tag{8.39}$$

$$\text{Proton transfer equilibrium}: \quad K_{\mathrm{ei}} = \frac{[\mathrm{ROH}]\left[\mathrm{RX(EO)}_i^-\,\mathrm{K^+}\right]}{[\mathrm{RO^-K^+}]\left[\mathrm{RO(EO)}_i\mathrm{H}\right]} \tag{8.40}$$

Some useful approximations are represented by the following relations:

$$[\mathrm{ROH}] >> [\mathrm{RO^-K^+}] \quad \text{and} \quad \left[\mathrm{RO(EO)}_i\mathrm{H}\right] >> \left[\mathrm{RO(EO)}_i^-\,\mathrm{K^+}\right] \tag{8.41}$$

We can write consequently:

$$\left[\mathrm{RO(EO)}_i^- \mathrm{K}^+\right] = \frac{K_{\mathrm{ei}}[\mathrm{RO}^-\mathrm{K}^+]\left[\mathrm{RO(EO)}_i\mathrm{H}\right]}{[\mathrm{ROH}]} \tag{8.42}$$

Moreover, the initial catalyst concentration is:

$$B^0 = [\mathrm{RO}^-\mathrm{K}^+] + \sum_{i=1}^{n} \left[\mathrm{RO(EO)}_i^- \mathrm{K}^+\right] \tag{8.43}$$

That is:

$$B^0 = [\mathrm{RO}^-\mathrm{K}^+] + \sum_{i=1}^{n} \frac{K_{\mathrm{ei}}[\mathrm{RO}^-\mathrm{K}^+]\left[\mathrm{RO(EO)}_i\mathrm{H}\right]}{[\mathrm{ROH}]} \tag{8.44}$$

and also:

$$B^0 = [\mathrm{RO}^-\mathrm{K}^+]\left(1 + \sum_{i=1}^{n} \frac{K_{\mathrm{ei}}\left[\mathrm{RO\,(EO)}_i\mathrm{H}\right]}{[\mathrm{ROH}]}\right) \tag{8.45}$$

It is now possible to calculate the concentration of all the ionic species that appear in the reaction rates (8.38) and (8.39):

$$[\mathrm{RO}^-\mathrm{K}^+] = \frac{B^0}{\left(1 + \frac{1}{[\mathrm{ROH}]}\sum_{i=1}^{n} K_{\mathrm{ei}}\left[\mathrm{RO\,(EO)}_i\mathrm{H}\right]\right)} = \frac{B^0[\mathrm{ROH}]}{[\mathrm{ROH}] + \sum_{i=1}^{n} K_{\mathrm{ei}}\left[\mathrm{RO\,(EO)}_i\mathrm{H}\right]} \tag{8.46}$$

$$\begin{aligned}\left[\mathrm{RO(EO)}_i^- \mathrm{K}^+\right] &= \frac{K_{\mathrm{ei}}\left[\mathrm{RO\,(EO)}_i\mathrm{H}\right]}{[\mathrm{ROH}]} \frac{B^0[\mathrm{ROH}]}{[\mathrm{ROH}] + \sum_{i=1}^{n} K_{\mathrm{ei}}\left[\mathrm{RO\,(EO)}_i\mathrm{H}\right]} \\ &= \frac{B^0 K_{\mathrm{ei}}\left[\mathrm{RO\,(EO)}_i\mathrm{H}\right]}{[\mathrm{ROH}] + \sum_{i=1}^{n} K_{\mathrm{ei}}\left[\mathrm{RO\,(EO)}_i\mathrm{H}\right]}\end{aligned} \tag{8.47}$$

The reacted substrate has given place to different ethoxylated oligomers therefore, the following substrate balance equation can be written:

$$\sum_{i=1}^{n} \left[\mathrm{RO\,(EO)}_i\mathrm{H}\right] = [\mathrm{ROH}]^0 - [\mathrm{ROH}] \tag{8.48}$$

From which, it follows that:

$$r_0 = k_0[\mathrm{RO^-K^+}][\mathrm{EO}] = \frac{k_0 B^0[\mathrm{ROH}][\mathrm{EO}]}{[\mathrm{ROH}] + \sum_{i=1}^{n} K_{\mathrm{ei}}\left[\mathrm{RO\,(EO)}_i\mathrm{H}\right]} \tag{8.49}$$

$$r_i = k_i\left[\mathrm{RO\,(EO)}_i^-\,\mathrm{K^+}\right][\mathrm{EO}] = \frac{k_i K_{\mathrm{ei}} B^0\left[\mathrm{RO\,(EO)}_i\mathrm{H}\right][\mathrm{EO}]}{[\mathrm{ROH}] + \sum_{i=1}^{n} K_{\mathrm{ei}}\left[\mathrm{RO\,(EO)}_i\mathrm{H}\right]} \tag{8.50}$$

These rate equations are function only of measurable quantities and it is not required the introduction of the ionic species concentrations that are difficult or impossible to measure. Moreover, as often it occurs, $K_{\mathrm{e1}} = K_{\mathrm{e2},}$ etc. = Ke; moreover, a further simplification can be obtained introducing λ the conversion of the substrate ROH.

$$r_0 = \frac{k_0 B^0[\mathrm{ROH}][\mathrm{EO}]}{[\mathrm{ROH}] + \sum_{i=1}^{n} K_{\mathrm{ei}}\left[\mathrm{RO\,(EO)}_i\mathrm{H}\right]} = \frac{k_0 B^0[\mathrm{ROH}][\mathrm{EO}]}{[\mathrm{ROH}] + \mathrm{Ke}\left([\mathrm{ROH}]^0 - [\mathrm{ROH}]\right)}$$
$$= \frac{k_0 B^0[\mathrm{EO}]}{1 + \mathrm{Ke}/[(1/\lambda) - 1]} \tag{8.51}$$

$$r_i = \frac{k_i K_{\mathrm{ei}} B^0\left[\mathrm{RO\,(EO)}_i\mathrm{H}\right][\mathrm{EO}]}{[\mathrm{ROH}] + \sum_{i=1}^{n} K_{\mathrm{ei}}\left[\mathrm{RO\,(EO)}_i\mathrm{H}\right]} = \frac{k_i K_{\mathrm{e}} B^0\left[\mathrm{RO\,(EO)}_i\mathrm{H}\right][\mathrm{EO}]}{[\mathrm{ROH}] + \mathrm{Ke}\left([\mathrm{ROH}]^0 - [\mathrm{ROH}]\right)}$$
$$= \frac{k_i \mathrm{Ke}\, B^0\left[\mathrm{RO(EO)}_i\mathrm{H}\right][\mathrm{EO}]}{[\mathrm{ROH}]^0[(1-\lambda) + \mathrm{Ke}\lambda]} \tag{8.52}$$

Finally, the polyethoxylation kinetic model is the following one:

(a) Substrate consumption $\frac{d[\mathrm{ROH}]}{\mathrm{d}t} = -r_0$ (8.53)

(b) Oligomers formation and consumption $\frac{\mathrm{d}\left[\mathrm{RO(EO)}_i\mathrm{H}\right]}{\mathrm{d}t} = r_{i-1}$

$-r_i \quad i = 1 \ldots n$ (8.54)

(c) Overall ethylene oxide consumption $\frac{\mathrm{d}n_{\mathrm{EO}}}{\mathrm{d}t} = V_{\mathrm{L}} \sum_{i=0}^{n} r_i$ (8.55)

The liquid volume V_{L} increases during the reaction due to the continuous addition of EO molecules to the substrate or oligomers and is adjusted at any integration step of time knowing the density of the reaction mixture. The knowledge of the EO solubility in the substrate, at different temperatures, is necessary to solve this ODE system. At last, two other aspects must be considered: ethylene oxide, at the reaction

Table 8.4 Kinetic parameters and operative conditions for the polyethoxylation run performed at constant EO concentration

Parameter	Value	Units
Temperature	130	°C
Pre-exp. factor	6e8	mol/(min cm^3)
Activation energy/*R*	6640	–
Equilibrium constant	4.1	–
Catalyst concentration	2E–4	mol/cm^3
Final time	60	min
Maximum pol. degree	15	EO units

temperature, is a gaseous compound, and in some conditions the gas-liquid mass transfer rate could affect the chemical reaction rate. Moreover, in order to complete the kinetic model, with the scope of modeling a chemical reactor, also the heat balance must be considered.

Exercise 8.4. Polyethoxylation at Constant EO Concentration
Considering an ethoxylation reaction conducted by using dodecanol as substrate. Assuming that the liquid phase ethylene oxide concentration is dependent only from volume increase due to the addition of ethylene oxide, simulate the behavior of a semi-batch ethoxylation reactor with the purpose of synthesizing a product with EO/S = 4 as average ethylene oxide/substrate molar ratio.

The kinetic parameters and the conditions for the reaction are summarized in Table 8.4.

Develop a MATLAB code that builds the following three plots:

- Moles of substrate as a function of time
- Moles of each polymeric specie as a function of time
- Final distribution of the polymeric chains

Solution
The calculation can be developed by considering that the only variation of EO and catalyst concentration is due to the volume increase along the time for the addition, in the system, of ethylene oxide in fed batch modality. The volume of the reacting phase can be calculated by using the relation (8.56), expressing the dependence of liquid mixture density from ethoxylation degree (EO/S) and temperature.

$$\rho_{\text{mix}} = 0.86 - 2.5e-2\left(\frac{\text{EO}}{\text{S}}\right) - 4.76e-4\left(\frac{\text{EO}}{\text{S}}\right)^2 - 2.69e-5\left(\frac{\text{EO}}{\text{S}}\right)^3 - 7.74e-4(T+273) \tag{8.56}$$

The concentration of EO and catalyst can be updated at each instant of time with the following relations:

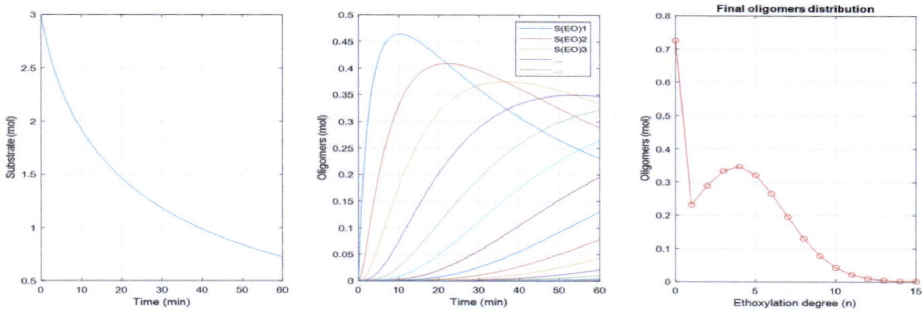

Fig. 8.11 Results of the polyethoxylation kinetics at constant EO concentration

Table 8.5 Kinetic parameters and operative conditions for the polyethoxylation run performed with variable EO concentration

Parameter	Value	Units
Temperature	160	°C
Pre-exp. factor	6e8	mol/(min cm^3)
Activation energy/*R*	6640	–
Equilibrium constant	4.1	–
Catalyst concentration	2E–4	mol/cm^3
Final time	60	min
Maximum pol. degree	15	EO units
Substrate initial moles	3	mol
Initial pressure	10	atm
EO feed	0.005	mol/s
Reactor volume	2000	cm^3

$$C_{\mathrm{EO}} = C^0_{\mathrm{EO}} \frac{V^0}{V} \quad B_0 = B^0_0 \frac{V^0}{V} \tag{8.57}$$

The simulation of the reactor behavior can be accomplished by solving a differential equations system (ODEs) with the initial conditions represented by the initial charge of the reactor. The assumption of constant concentration of EO means that the pressure is maintained at a fixed value and that a variation of liquid mixture composition does not affect solubility. Obviously, this simplifying assumption can be removed in a more sophisticated model.

The reaction rates (initiation and propagation) are represented by Eqs. (8.49) and (8.50), while the material balance equations are reported in the text as Eqs. (8.53–8.55).

The code implementing the described calculations produces the following output (Fig. 8.11):

On the left side, the substrate evolution in time is reported. After 60 min of reaction, the consumption of substrate is not yet completed, as can be appreciated also from the plot on the right side.

In the plot at the center, the evolution in time of the different substrate EO adducts is reported. The behavior is typical of a reaction in series.

The plot on the right side of the figure reports the final oligomer distribution. It is evident that an average polymerization degree of around 4 is obtained. In these conditions, the choice of a maximum polymerization degree of 15 is revealed correctly, as after this value, oligomer concentration is negligible.

The described results were obtained using a MATLAB program available as Electronic Supplementary Material.

Exercise 8.5. Polyethoxylation with Variable EO Concentration
Consider an ethoxylation reaction conducted by using dodecanol as substrate, like in the previous Exercise 8.4. Assuming that the liquid phase ethylene oxide concentration is dependent on pressure in the reactor and on the ethoxylation degree, simulate the behavior of a semi-batch ethoxylation reactor with the purpose to synthesize a product with EO/S $=$ 4 as average ethylene oxide/substrate molar ratio. Assume that at time 20 min, the feed of EO is stopped.

The kinetic parameters and the conditions for the reaction are summarized in Table 8.5.

Develop a MATLAB code that builds the following three plots:

- Moles of substrate as a function of time
- Moles of each polymeric specie as a function of time
- Final distribution of the polymeric chains
- Moles of EO in liquid and in gas phase
- Ethoxylation degree as a function of time

Solution
The calculation can be developed by considering that the EO solubility is a function of the pressure in the system and of the liquid phase composition (ethoxylation degree). Moreover, the catalyst concentration is due to the volume increase along the time for the addition, in the system, of ethylene oxide in fed-batch modality. The volume of the reacting phase can be calculated by using a relation (8.58) expressing the dependence of liquid mixture density from ethoxylation degree (EO/S) and temperature. The employed relation is:

$$\begin{aligned}\rho_{\text{mix}} = {} & 0.86 - 2.5e-2\left(\frac{\text{EO}}{\text{S}}\right) - 4.76e-4\left(\frac{\text{EO}}{\text{S}}\right)^2 - 2.69e-5\left(\frac{\text{EO}}{\text{S}}\right)^3 \\ & - 7.74e-4(T+273)\end{aligned} \tag{8.58}$$

The concentration of EO and catalyst can be updated at each instant of time with the following relations:

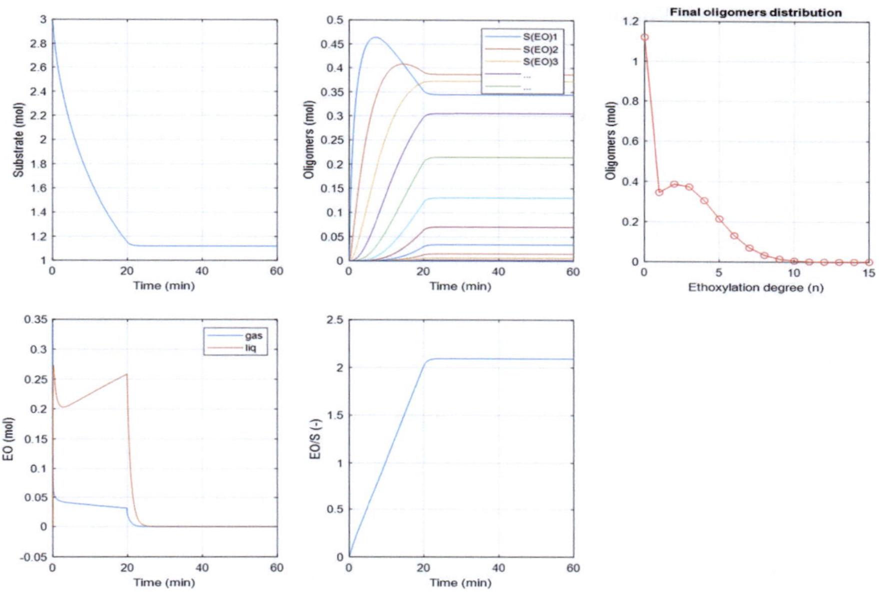

Fig. 8.12 Results of the polyethoxylation kinetics with variable EO concentration

$$C_{\mathrm{EO}} = C^0_{\mathrm{EO}} \frac{V^0}{V} \quad B_0 = B^0_0 \frac{V^0}{V} \tag{8.59}$$

For what concerns the EO solubility, the following expression can be used:

$$H_{\mathrm{EO}} = e^{(A+B)} \tag{8.60}$$

$$A = H_1 + \left(\frac{\mathrm{EO}}{\mathrm{S}}\right) H_2 \tag{8.61}$$

$$B = \left[H_3 + \left(\frac{\mathrm{EO}}{\mathrm{S}}\right) H_4\right] / (T + 273) \tag{8.62}$$

$$C_{\mathrm{EO}} = P^* H_{\mathrm{EO}} \tag{8.63}$$

The values of the constants H_1–H_4 are, respectively, as follows: –15.259, 0.029, 3079.6, and 4.83.

The simulation of the reactor behavior can be accomplished by solving a differential equations system (ODEs) with the initial conditions represented by the initial charge of the reactor. The reaction rates (initiation and propagation) are represented by Eqs. (8.49) and (8.50), while the material balance equations are reported in the text as Eqs. (8.53–8.55).

The code implementing the described calculations produces the following output (Fig. 8.12):

Description of the Subplots (from Left to Right)
The first plot reproduces the trend in substrate moles. At time of 20 min, the reaction is stopped.

The second plot reports the oligomers number of moles. In this case, also, after the stopping of feed, the system is stationary.

The third plot represents the oligomers' molar distribution at final time.

The successive plot (the first in the second row) shows the trend of ethylene oxide mole number in gas and in the liquid phase. Note that after the feed was stopped, the reaction continues for some minutes, during which the EO consumption proceeds to a completion.

The last plot shows the trend of ethoxylation degree as a function of time. Note that after 20 min, a value of 2 (EO/S) is reached, and then the EO feed is stopped. The further slight increase of the EO/S is due to the EO present in the system that will be then consumed (in the gas and in the liquid phase).

The described results were obtained using a MATLAB program available as Electronic Supplementary Material.

8.3.4.2 Kinetics of Step-Growth Copolymerization

In some cases, the polymerization of ethylene oxide is performed together with that of propylene oxide. Obviously, each monomer reacts with its own reactivity, giving place to an EO-PO copolymer. Two different types of copolymers can be produced according to the modality of adding the monomers to the growing chains. It is possible to obtain:

(a) Random copolymers, when the two monomers are fed together to the reacting mixture
(b) Block copolymers, when the monomers are fed alone in sequence

The properties of the copolymers will depend, in these cases, not only on both the average MW and the MWD but also on the sequence of monomers EO and PO composing the polymeric chains. In order to evaluate the evolution of the molecular structure along the reaction time, it is opportune to adopt a statistical kinetic Monte Carlo approach (Tesser and Santacesaria. 2021a, b). In step addition polymerization

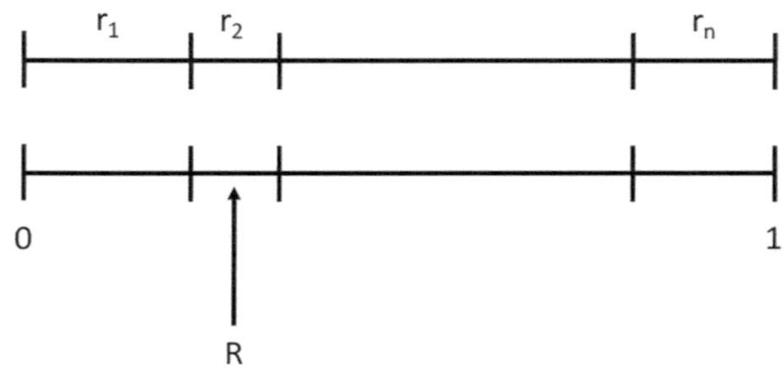

Fig. 8.13 Scheme of reaction rate selection

like ethoxylation, propoxylation, and the corresponding copolymerization, molecule addition to the starter is a discrete phenomenon, and molecular weight growth is discrete and not continuous as considered in the deterministic models. Therefore, the chain's growth can be described as the occurrence of the most probable additions of an alkoxide molecule to a growing chain in a process completely dominated by the statistical probability. If the kinetic is based on the overall alkoxide consumption, we can obtain the most probable oligomer distribution without solving a system of differential mass balance equations, one for each chemical species, like in the deterministic approach. Another advantage of this approach is that any single reaction event can be easily stored, representing the microstructure of the obtained polymer or copolymer. The basis of the classical kinetic Monte Carlo method consists of random number generation, according to which a chemical reaction occurrence is selected. Suppose that n distinct chemical reactions occur simultaneously in a system of p components and that their rate laws are known in terms of functionality and kinetic parameters. The first step in the simulation is the choice of a microscopic simulation volume, in which X_i is the number of molecules of component i-th that are initially present (initial state). Then, the reaction rates are calculated and organized as in the scheme of Fig. 8.13 to randomly select one of the reactions. The sum of reaction rates is scaled in a way that the cumulative sum varies from 0 to 1, allowing the selection among them by generating a random number with a uniform distribution in [0, 1].

In fact, with this scheme, generating a random number R in the range 0–1 with a uniform distribution, the random selection of the occurring reaction is straightforward. This strategy corresponds to the assumption that only one reaction at a time occurs. If the reaction k-th is selected by this procedure, the numbers of molecules in the microscopic simulation volume are updated with the following recursive relation:

$$X_i = X_i - \nu_{i,k} \quad i = 1 \ldots p \tag{8.64}$$

where $\nu_{i,k}$ is the stoichiometric coefficient of component i in the reaction k. The simulation time is then updated by advancing in time of a quantity τ as:

$$t = t + \tau \tag{8.65}$$

where τ is defined rigorously as:

$$\tau = \frac{\ln\left(\frac{1}{R_1}\right)}{\sum_{j=1}^{n} r_i} \tag{8.66}$$

where R_1 is another random number, generated with uniform distribution, between 0 and 1. The described procedure is repeated iteratively until the simulation time reaches the final time chosen to stop the simulation. A key point is the reasonable

OH on the left	OH on the right
SABAAAABBBABAAABABAB	SBBBBBBBABAABBABABABA
SBABBBABABBBA	SAAABBBBAAAAAAAAABBA
SABAAAABABABBBABBB	SABBABBABABAAAABABAB
SBABAAABABABAABAABB	SAAABABAAAABAAABB
SAABBABAAAAAAABAA	SBABABAAAAAAA
SAAAAABAAABABBABBBBAA	SBBBBAAABBABABBBAAA
SBBBAAAAAABABBBAB	SAABAAABBABBAAAAAA
SAABBBBBABA	SBABBBAABABAABAABABB
SAAAABBBAAAABAB	SBBABBAABAAAABAAAAAAAAABAABAAABB
SBBABBBBBBABBBAB	SAAAAABABBBAAABABB
SAAAAABAAABAAAB	SAABBAABBAAAB
SBAAABABBABB	SBABABAABBBBABBB
SBBAAAABBBB	SBBABAAAAABBB
SAAAAAAAABBAABAABB	SAAABABAAAABABAAABABAB
SAAABAAABBABAA	SBBAABBBAAAAABBABAAB
SAAABAAAABAB	SBBAAAABBABABBA
SAABAABBAAAAB	SAAABABAABBBAABA
SAAAAAABAAABBAAAABA	SBABAAAABAAAAABBA
SAABAABAAAAAAABB	SBAAAABBABAAAABBA
SBAAABABABAAABAAABBB	SBAAAAABBBABBB
SBBABAAABAAABBABBABBA	SABAAAABBAABBB
SBAABBBABBAABB	SAAAABAAABB
SBBABBBBABAAAAAA	SABBAABAAAAABAAAAB

Fig. 8.14 Examples of structural conformation determined with the Monte Carlo model at the end of the EG-PO-EO random copolymerization. The conformation is given for all the molecules (about 5000) present in the microscopic volume considered for the stochastic calculations. A are EO units and B are PO units (published with the permission of Tesser, E. Santacesaria; American Chemical Society: Ind. Eng. Chem. Res. 60, 10638–10653 (2021))

choice of the size of the microscopic simulation volume, which must be big enough to include a sufficiently high number of molecules for statistical significance. Generally, a choice in the range of 10^{-17}–10^{-19} cm^3 was found correct for many simulation purposes. The stochastic simulation algorithm (SSA) method requires the evaluation, at each step in time, of the reaction rates for which kinetic parameters are necessary. The relation between macroscopic constants and microscopic ones is obtained through the following scaling relation:

$$k_{\text{micro}} = \frac{k}{(n_A V_{\text{micro}})^{q-1}} \tag{8.67}$$

where k_{micro} is the microscopic scale parameter, k is the actual kinetic parameter, n_A is the Avogadro's number $= 6.023 \times 10^{23}$, V_{micro} is the microscopic simulation volume, and q is the reaction order (equal to 2 in our cases). With this method, only the reaction rates have to be defined, and there is no need to write material balances as the necessary information are stored in the stoichiometric matrix. This represents

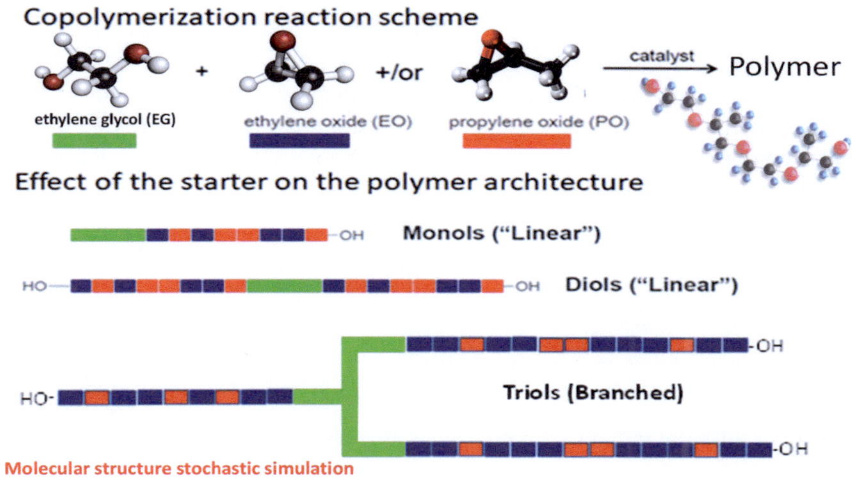

Fig. 8.15 Microstructures of different EO-PO copolymers (published with the permission of Tesser, E. Santacesaria; American Chemical Society: Ind. Eng. Chem. Res. 60, 10638–10653 (2021))

an obvious advantage as no differential equations are present and there is no need for an ODE solver. The kinetic parameter k is the same of the deterministic model. The simulation results satisfactory coincident to the deterministic output, by considering a volume containing more than 10,000 molecules. The advantage of the kinetic stochastic approach is that of allowing to define the microstructure of the copolymers' chains. In Fig. 8.14, an example of microstructures obtained by Monte Carlo simulation of the system ethylene glycol, ethylene oxide, and propylene oxide is reported as an example.

The same simulation can be made by considering different types of starters, respectively, monofunctional, bifunctional, or trifunctional as represented in Fig. 8.15.

8.3.4.3 Kinetics of Chain Growth Polymerization: Radical Polymerization

Chain growth polymerizations are normally characterized by highly reactive monomers that need an initiator promoting the first reaction step. The initiation can be promoted by a radical (radical polymerization) or by an ionic compound (cationic or Anionic polyaddition). The radical mechanism can be schemed as it follows:

$$\mathrm{I} \rightarrow 2\mathrm{R}_* \qquad \textbf{Initiator decomposition} \qquad r_\mathrm{d} = -\frac{\mathrm{d}I}{\mathrm{d}t} = k_\mathrm{d} I \tag{8.68}$$

Initiators of polymerization are unstable molecules that decompose, for a thermal or photochemical activation. Examples of initiators are organic peroxide or some diazo compounds, such as benzoyl peroxide or azo-bis-isobutyro-nitrile, that decompose as in the following examples:

$$\mathrm{C_6H_5{-}CO{-}O{-}O{-}CO{-}C_6H_5} \longrightarrow 2\,\mathrm{C_6H_5{-}CO{-}O_*} \longrightarrow 2\,\mathrm{CO_2} + 2\,\mathrm{C_6H_{5*}} \tag{8.69}$$

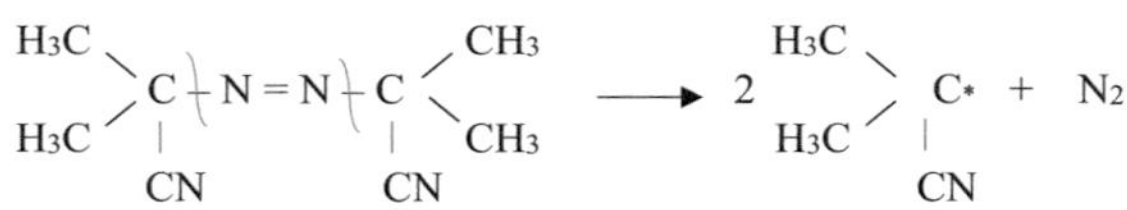

(8.70)

A lot of different initiators are nowadays available. The reaction then proceeds as it follows:

$$\mathrm{R}_* + \mathrm{M} \rightarrow \mathrm{M}_{1*} \qquad \textbf{Chain initiation} \qquad r_\mathrm{i} = \left(-\frac{\mathrm{d}R*}{\mathrm{d}t}\right)_\mathrm{i} = k_\mathrm{i} M R^* - 2 f k_\mathrm{d} I \tag{8.71}$$

where f is the initiator efficiency, that is, the fraction of radical formed by decomposition of I. Clearly, the maximum value for "f" is $f = 1$, but the typical values range in practice is 0.3–0.8.

As radicals react as soon as they are formed, a steady-state condition can be postulated, and their concentration can be assumed to be constant, that is, $(\mathrm{d}R_*/\mathrm{d}t) \simeq 0$, and consequently, it holds $r_\mathrm{i} = r_D$. Then, the propagation steps occur:

Propagation Steps

$$\begin{aligned} &\mathrm{M}_{1*} + \mathrm{M} \rightarrow \mathrm{M}_{2*} \qquad r_1 = k_\mathrm{p} M M_1^* \\ &\mathrm{M}_{2*} + \mathrm{M} \rightarrow \mathrm{M}_{3*} \\ &\ldots\ldots\ldots\ldots\ldots \\ &\ldots\ldots\ldots\ldots\ldots \end{aligned} \tag{8.72}$$

$$\mathrm{M}_{n-1*} + \mathrm{M} \rightarrow \mathrm{M}_{n*} \qquad r_{n-1} = k_\mathrm{p} M M_{n-1}^* \tag{8.73}$$

The propagation kinetic constant will reasonably be considered independent of the chain length. Therefore, the propagation rate would be:

$$r_\mathrm{p} = k_\mathrm{p} M \sum_{i=1}^{\infty} M_i^* \tag{8.74}$$

However, the radical concentration cannot be experimentally evaluated; as a consequence, this relation cannot be employed directly for evaluating the propagation rate. The radical concentration must be expressed in terms of known or measurable parameters. As the reactions involving radicals are extremely fast, it is possible to introduce again the steady-state approximation, that is, radicals are formed and destroyed at equal rates, and hence $r_{\mathrm{i}} = r_{\mathrm{t}}$, that is, the initiation rate is approximately equal to the termination rate.

Termination Reactions

$$\mathrm{M}_{i*} + \mathrm{M}_{n*} \rightarrow \mathrm{P}_{i+n} \quad \text{by radical coupling} \quad r_{\mathrm{tc}} = 2k_{\mathrm{tc}} M_i^* M_n^* \tag{8.75}$$

$$\mathrm{M}_{i*} + \mathrm{M}_{n*} \rightarrow \mathrm{P}_i + \mathrm{P}_n \quad \text{by disproportionation} \quad r_{\mathrm{td}} = 2k_{\mathrm{td}} M_i^* M_n^* \tag{8.76}$$

$$\mathrm{M}_{i*} + \mathrm{S} \rightarrow \mathrm{P}_i + \mathrm{S}_* \quad \text{by solvent chain transfer} \quad r_{\mathrm{tr}} = k_{\mathrm{cts}} M_n^* S \tag{8.77}$$

$$\mathrm{M}_{i*} + \mathrm{M} \rightarrow \mathrm{P}_i + \mathrm{M}_* \quad \text{by monomer chain transfer} \quad r_{\mathrm{tr}} = k_{\mathrm{cts}} M_n^* M \tag{8.78}$$

The number 2, for the rate laws of reactions (8.75) and (8.76), has been introduced because two radicals disappear in these reactions. As can be seen, the radicals are eliminated only by the reactions of coupling and disproportionation, not by chain transfer. It must be pointed out that if long chains are desired, the initiator concentration must be kept low; on the contrary, many short chains are obtained as the monomer units are distributed among a higher number of growing centers. The termination reactions (8.75–8.78) follow different chemical patterns. Radical coupling is characterized by the association of two growing radical chains, giving place to a sigma bond and doubling the molecular weight. The radical disproportionation occurs because a hydrogen atom is abstracted from the end of a chain to another, producing a polymer with a terminal unsaturated group and another with a saturated terminal group. The molecular weight, in this last case, remains the same. Chain transfer reactions destroy one radical, but another one is formed. The effect of chain transfer is that of decreasing the polymer chain length, and if the rate of chain transfer is much greater than the rate of propagation, very short chains are obtained with few monomeric units (telomerization).

It has been already mentioned that the life of the radical chains is very short. This means that some polymer chains are formed at the beginning, while others at the end of the process in conditions that are deeply changed for the concentration of the reagents, viscosity, and temperature. In conclusion, with conventional radical polymerization, the final product is not uniform but highly polydisperse, and the MWD is broad. A lot of publications and books have been published on the mechanism and kinetics of radical polymerization such as the following: Bamford et al. (1958), Moad and Solomon (1995), Matyjaszewski and Davis (2002), Odian (2004), Barner-Kowollik et al. (2002), and Coates and Sawamoto (2013). As before mentioned, the quality of a polydisperse polymer for some applications is low; efficient block copolymerization is not possible with conventional radical polymerization. To obtain narrower MWDs and polymers with controlled architecture several procedures have been developed and largely described in the literature (Fischer 2001; Rizzardo and Solomon 2012). These approaches are generally known by different terminologies as living radical polymerization (LRP), controlled radical polymerization (CRP), or reversible deactivation radical polymerization (RDRP). With these techniques, premature termination is minimized, and molecular weight proceeds linearly with time until all monomer molecules are consumed. Controlled radical polymerization can be made through three different fundamental methods that are:

1. Nitroxide-mediated polymerization (NMP)
2. Atom transfer radical polymerization (ATRP)
3. Reversible addition/fragmentation chain transfer polymerization (RAFT)

Nitroxide-Mediated Polymerization

The technique known as nitroxide-mediated polymerization (NMP) has been the first method of living radical polymerization developed by Dave Solomon group of CSIRO for controlling the average MW and for obtaining narrower MWD (Solomon et al. 1986). The NMP method is based on the addition of an alkoxyamine to the reacting system. Some alkoxyamines have the property to give place to reversible thermal homolysis (Kovtun et al. 1974), such as in the following scheme:

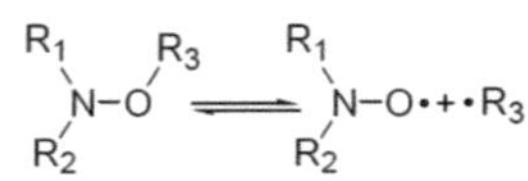

(8.79)

The nitroxide radical is a stable radical, unable to promote propagation It gives terminal reversible reaction with a growing radical chain deactivating it in the form of a dormant polymer. The dormant polymers, bonded to the nitroxide groups, are subjected to homolytic cleavage coming back to propagating radical chains and stable nitroxide radicals. This cycle of activation/deactivation is responsible for the reduction of irreversible termination and allows to obtain the desired MW and narrower MWD. Two different approaches have been developed. In the first developed technique, an initiator (benzoyl peroxide or 2,2′ azo-bis-isobutyronitrile) produces radicals that, in the presence of a monomer, start with the propagation reaction. Propagating radicals give then a reversible termination with nitroxide radicals, such as, for example, 2,2,6,6-tetramethylpiperidinyl-1-oxy (TEMPO), according to the scheme (8.80). The initial ratio nitroxide/initiator is of paramount importance for instauring the NMP process, because an excess of nitroxide radical shifts the activation/deactivation equilibrium toward the formation of dormant polymeric species. In the meantime, the polymerization rate is decreased. However, as the efficiency of the initiator in producing active radicals cannot be predicted, the kinetics of polymerization is not reproducible. For this reason, a successive approach avoided the use of the initiator. For this purpose, Hawker (1994) proposed the employment of a thermolabile alkoxyamine.

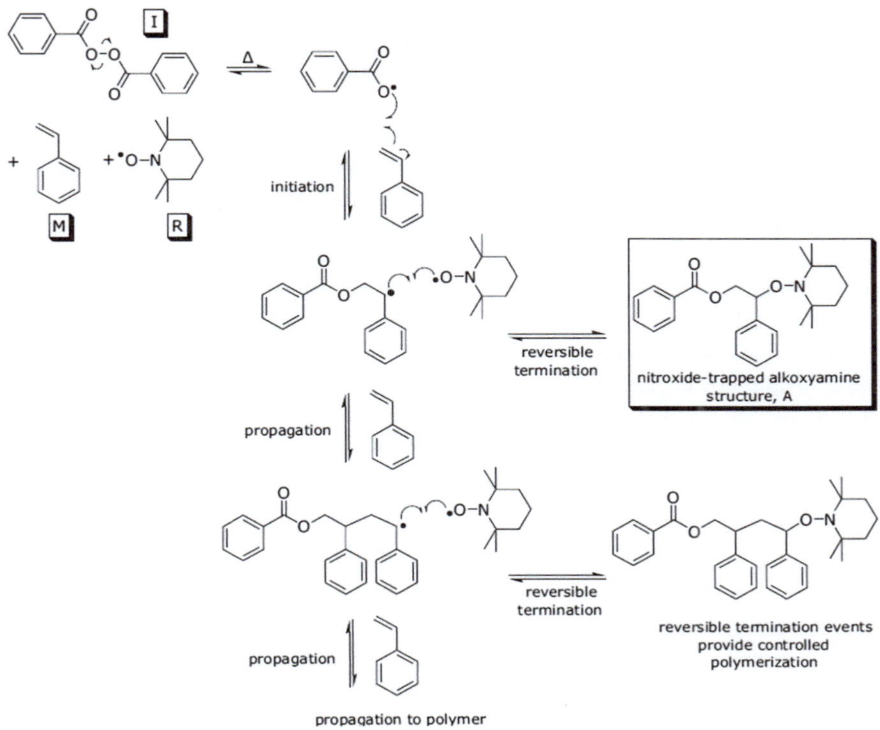

(8.80)

Thermolabile alkoxyamines (Lamontagne and Lessard 2020), easily decompose by heating giving both a reactive initiating radical and the stable mediating nitroxide radical, so allowing a simplified procedure shown in the following scheme:

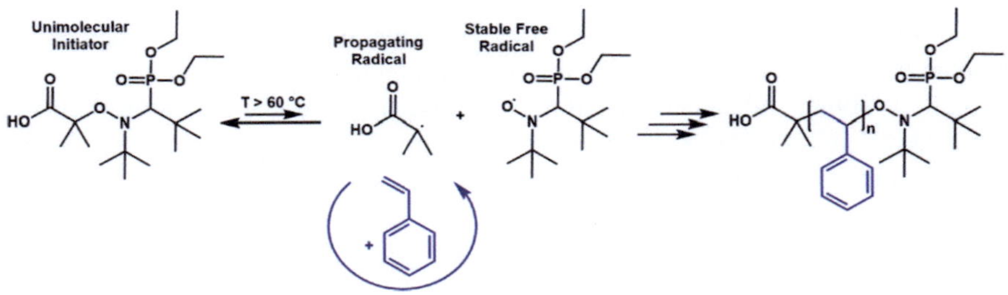

(8.81)

Different alkoxyamines able to promote the NMP process are nowadays commercially available. In this case, the initial ratio nitroxide/initiator is always equal to the stoichiometric value of 1 and the MWD control is optimal. For a kinetic analysis, it is possible to simplify the reaction scheme as follows:

$$\mathrm{R-Y}\underset{k_c}{\overset{k_d}{\rightleftarrows}}\mathrm{R^*+Y^*} \tag{8.82}$$

$$\mathrm{R^*+R^*}\xrightarrow{k_t}\mathrm{R-R} \tag{8.83}$$

R – Y is the original alkoxyamine that decomposes, giving a transient radical R* and a persistent nitroxide radical Y*. Then, only two reactions can occur consuming R*, one is the irreversible self-termination and another one is the reverse of the initial reaction. In the beginning, R* linearly increases with time and Y* increases, too, with the same rate $r_D = k_D\,[R - Y]$. The concentration of both R* and Y* increases until the two bimolecular reactions of the previous scheme take place. The disappearance of R* due to the irreversible reaction determines the lowering of its concentration and the relative increase of the persistent radical Y*. Therefore, the recombination of R* and Y* becomes more and more favorable with respect to the irreversible termination. For long reaction times, the R* concentration becomes null and the Y* concentration increases until to equate the initial $[\mathrm{R-Y}]_0$ concentration. According to Nicolas et al. (2013), the following unusual relations hold:

$$[Y^*]=\left(\frac{3k_tk_d^2[R-Y]_0^2}{k_c^2}\right)^{\frac{1}{3}}t^{1/3} \tag{8.84}$$

$$[R^*]=\left(\frac{k_d[R-Y]_0}{3k_tk_c}\right)^{\frac{1}{3}}t^{-1/3} \tag{8.85}$$

The combination of these two equations brings to:

$$k_c[R^*][Y^*]=k_d[R-Y]_0 \tag{8.86}$$

The equilibrium $k_d/k_c = K$ exists if the following condition is respected $K<\frac{k_c[R-Y]_0}{4k_t}$. If this condition is not satisfied, the cross-reaction between R* and Y* is not favored, and the system follows the conventional radical polymerization, reaching a steady state for R*. In the optimal conditions, termination events are strongly lowered, and an accumulation of the living macro-alkoxyamines occurs at the end of polymerization. Neglecting the propagation steps occurring before reaching the equilibrium the evolution with time of the monomer concentration is:

$$\ln\frac{[M]_0}{[M]}=\frac{3k_p}{2}\left(\frac{k_d[R-Y]_0}{3k_tk_c}\right)^{\frac{1}{3}}t^{2/3} \tag{8.87}$$

The polydispersity PDI of the dormant chains can be determined as:

$$\mathrm{PDI} = 1 + \frac{1}{\mathrm{DP_n}} + \frac{[M]_0^2}{\mathrm{DP_n^2}[R-Y]_0^2}\left(\frac{\pi k_\mathrm{p}^3[R-Y]_0}{k_\mathrm{d}k_\mathrm{c}k_\mathrm{t}}\right)^{\frac{1}{2}}\mathrm{erf}(z) \tag{8.88}$$

Being $\mathrm{DP_n}$ the number average degree of polymerization and erf(z) the error function:

$$\mathrm{erf}(u) = \frac{2}{\sqrt{\pi}}\int_0^u e^{-t^2}\mathrm{d}t \qquad z = (3k_\mathrm{p})^{1/2}\left(\frac{K[R-Y]_0}{3k_\mathrm{t}}\right)^{\frac{1}{6}} t^{1/3} \tag{8.89}$$

PDI can be approximated for the initial reaction and after long time:

$$\mathrm{PDI_0} = 1 + \frac{1}{\mathrm{DP_n}} + \frac{8}{3}(k_\mathrm{d}t)^{-1} \qquad \mathrm{PDI_\infty} = 1 + \frac{[R-Y]_0}{[M]_0} + \left(\frac{\pi k_\mathrm{p}^3[R-Y]_0}{k_\mathrm{d}k_\mathrm{c}k_\mathrm{t}}\right)^{\frac{1}{2}} \tag{8.90}$$

Using the previously reported equations, it is possible to evaluate analytically the final PDI value, the monomer conversion, and the concentration of dormant polymers if all the kinetic rate constants are known.

Atom Transfer Radical Polymerization (ATRP)

ATPR is a living radical polymerization method proposed by Wang and Matyjaszewski (1995a, b). A detailed review related to this technique with the improvements developed over time has been published by Krys and Matyjaszewski (2017). ATPR is a catalytic process promoted by a metal complex. Most commonly, a copper complex is preferred, and $\mathrm{Cu^I/L}$ (L is a ligand) reacts with an alkyl halide initiator RX to yield a radical R$*$ and the corresponding copper complex X-$\mathrm{Cu^{II}/L}$, at a higher oxidation state, with coordinated halide anion. In practice, this is the "activation step." Then, R$*$ can react with the monomer M, giving place to a radical polymeric chain R–P_{n*} that is subjected to the following fast activation-deactivation equilibrium:

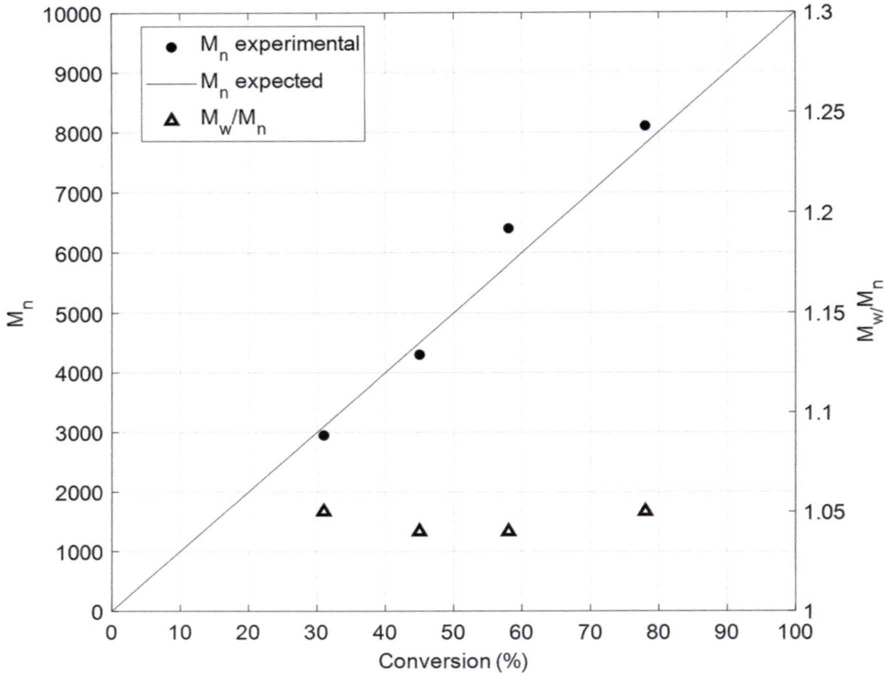

Fig. 8.16 Number average molecular weight as a function of monomer conversion and experimental trend of polydispersity index for bulk ATPR polymerization of styrene at 110 °C. 1-Phenylethyl bromide has been used as an initiator with a mole ratio styrene/initiator = 100-di-; two equivalents of the ligand 4,4′-di-(5-nonyl)-heptyl-2,2′-bipyridine were used. $[CuBr]_o$ = $[1\text{-}PEBr]_o$ = $[dNbipy]_o/2$ = 0.087 M (published with the permission of K. Matyjaszewski et al. (1997))

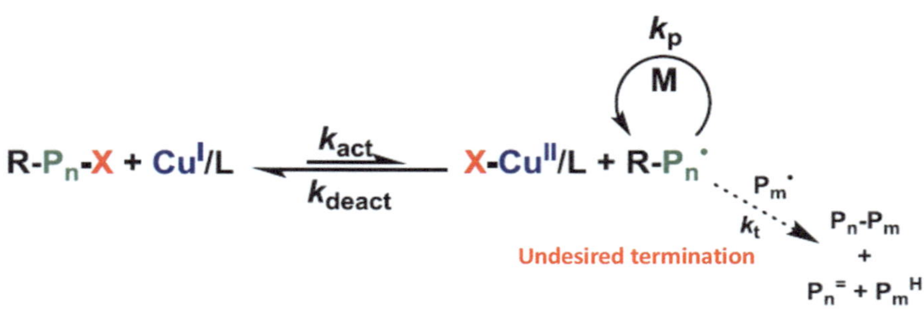

(8.91)

After the activation step, the polymer grows as a radical but is partially deactivated giving place to a temporarily inactive polymeric chain (dormant). Deactivation must be fast and efficient to obtain the desired MW with a narrow

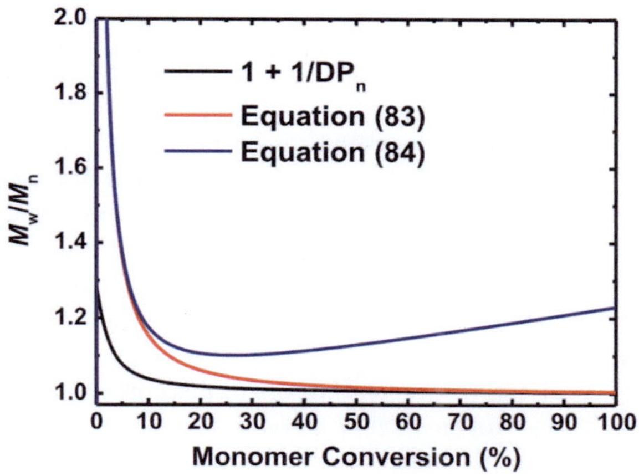

Fig. 8.17 Polydispersity as a function of the monomer conversion for ATPR. Used conditions: [M] = 250, [R-X] = 1, $[Cu^I/L] = 0.9$, $[X\text{-}Cu^{II}/L] = 0.1$, $[M]_o = 5M$, $k_p = 1000\ M^{-1}s^{-1}$, $k_t = 10^8\ M^{-1}s^{-1}$, $k_a = 1.0\ M^{-1}s^{-1}$, $k_d = 10^6\ M^{-1}s^{-1}$ (published with the permission of P. Krys and K. Matyjaszewski (2017))

MWD. Termination and disproportionation have a small contribution as the radicals' concentration is kept always very low in the system. Thanks to the equilibrium (8.91), the dormant polymeric chains can be reactivated to form any type of block copolymers. Many attempts have been made to improve the ATPR process by changing the metal of the complex, the ligand (firstly used 2,2′-bipyridine), and the alkyl halide or by adding some components, giving positive interference with the scheme of reaction (8.91). Azo-bis-isobutyronitrile is added, for example, in some cases, at the scope to reduce copper from Cu^{II} to Cu^I by generating radicals. Other reducing agents have been tested, too.

The rate of polymerization in the ATPR system is:

$$R_P = -\frac{dM}{dt} = k_P M \left[\sum_{n=1}^{n} P_n^* \right] \tag{8.92}$$

The ATPR equilibrium can be written as:

$$K_{ATPR} = \frac{\left[X - Cu^{II}/L\right] \left[\sum_{n=1}^{n} P_n^* \right]}{\left[Cu^I/L\right] \left[\sum_{n=1}^{n} RP_n X \right]} \tag{8.93}$$

By eliminating the overall radical concentration, we obtain:

$$R_{\mathrm{P}} = -\frac{\mathrm{d}M}{\mathrm{d}t} = k_{\mathrm{P}} K_{\mathrm{ATPR}} M \left[\sum_{n=1}^{n} RP_n X \right] \frac{[\mathrm{Cu^I/L}]}{[\mathrm{X - Cu^{II}/L}]} \tag{8.94}$$

This equation shows that the overall rate of ATPR polymerization depends on the $[\mathrm{Cu^I}]/[\mathrm{Cu^{II}}]$ ratio and not on the total amount of loaded copper.

The advantage of using the ATPR method in radical polymerization can be appreciated in Fig. 8.16, where the number average molecular weight $\overline{\mathrm{MW_N}}$, experimentally obtained in ASTR styrene polymerization, is reported as a function of the monomer conversion. In the same plot is also reported the index of polydispersity $\overline{\mathrm{MW_W}}/\overline{\mathrm{MW_N}}$.

As it can be seen in Fig. 8.16, the number average molecular weight linearly increases with the monomer conversion, and polydispersity is near to the ideal value of 1 corresponding to a monodisperse system. ASTR polymerization allows to obtain predetermined MWs maintaining narrow MWDs. The desired average MW can be obtained by establishing the molar ratio between the initial concentration of the monomer and the alkyl halide initiator, $[\mathrm{M}]_\mathrm{o}/[\mathrm{RX}]_\mathrm{o}$ = overall degree of polymerization. Polymers with MW_n from 1000 to 1,000,000 have been obtained with this technique. Polydispersity PDI in ATPR polymerization changes with the monomer conversion and can be predicted with relations of the type:

$$\mathrm{PDI} = \frac{\overline{M_w}}{\overline{M_n}} = 1 + \frac{1}{\mathrm{DP_n}} + \left(\frac{k_\mathrm{p}[\mathrm{RX}]_\mathrm{o}}{k_\mathrm{d}\left[\mathrm{Cu^{II}/L}\right]} \right) \left(\frac{2}{\lambda} - 1 \right) \tag{8.95}$$

$\mathrm{DP_n}$ is the number average degree of polymerization, k_p rate constant of polymerization and k_d is $\mathrm{Cu^{II}}$ deactivation constant, and λ is the monomer conversion. This equation predicts that PDI decreases monotonically with the monomer conversion, but the trend experimentally observed is different. To obtain a better prediction, it is necessary to consider also the contribution of termination reactions, and a more accurate relation has been proposed:

$$\mathrm{PDI} = \frac{\overline{M_w}}{\overline{M_n}} = 1 + \frac{1}{\mathrm{DP_n}} + \left(\frac{k_\mathrm{p}[\mathrm{RX}]_\mathrm{o}}{k_\mathrm{d}\left[\mathrm{Cu^{II}/L}\right]} \right) \left(\frac{2}{\lambda} - 1 \right) + \left(\frac{k_\mathrm{t} k_\mathrm{a}\left[\mathrm{Cu^{II}/L}\right]_\mathrm{o}}{4\, k_\mathrm{p} k_\mathrm{d}\left[\mathrm{Cu^{II}/L}\right]_\mathrm{o}} \right) \lambda \tag{8.96}$$

k_a is $\mathrm{Cu^I}$ activation constant and k_t is the termination constant.

In Fig. 8.17, the evolution of the polydispersity with the monomer conversion, calculated with, respectively, relation (8.95) and (8.96), is reported for comparison. As it can be seen, at low conversions, the trends are similar but are quite different at high conversion.

Reversible Addition/Fragmentation Chain Transfer (RAFT) Polymerization
RAFT is another efficient technique to provide living characteristics to radical polymerization. The technique has been developed by the Australian CSIRO group, the same that proposed also the NMP methodology. The term RAFT

polymerization was coined by Chiefari et al. (1998). The historical development of both the NMP and RAFT techniques is described in detail by Rizzardo and Solomon (2012). Many publications and reviews have been devoted to the improvement of the RAFT technique (Barner-Kowollik 2007). The RAFT method is based on the addition of a thiocarbonylthio compound to the reactants, and the polymerization reaction occurs as in the following scheme (8.97):

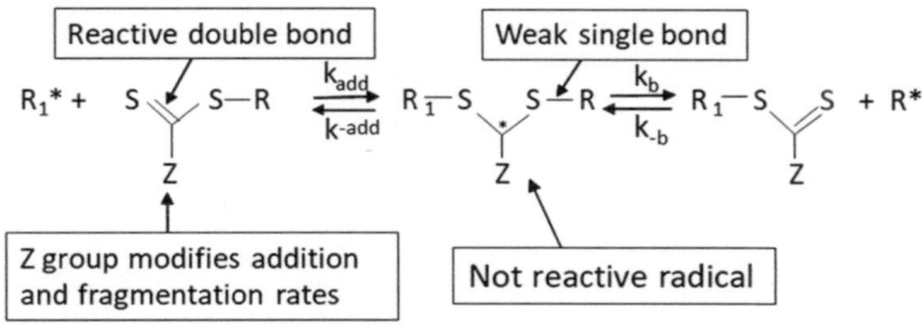

(8.97)

Z = Ph, CH_3, etc. As it can be seen, the unsaturated compound can act as a chain transfer agent (RAFT agent). The most common RAFT agents are dithioesters, dithiocarbamates, and xanthates. The radical intermediate, formed by R_1* radical addition, has very low activity in propagation for steric hindrance and for the Z group influence but easily undergoes scission to form a new radical R* that reinitiates polymerization or, alternatively, comes back with the addition equilibrium. The use of a suitable chain transfer agent allows to obtain polymers with a high degree of functionality and narrow MWD. The mechanism (8.97) can be simplified by defining X the thioester capping agent: X = –SC=S and X–P the compound subjected to addition; therefore, it is possible to write:

$$\text{Addition:}\quad P_A^* + X\text{-}P_B \underset{k_{-add}}{\overset{k_{add}}{\rightleftarrows}} P_A\text{-}X^*\text{-}P_B \underset{k_{-b}}{\overset{k_{b}}{\rightleftarrows}} P_A\text{-}X + P_B^* \quad\text{:Fragmentation} \tag{8.98}$$

A propagating radical P_A* attacks the RAFT agent ($X–P_B$), giving place to another radical chain P_B* plus a **dormant polymeric** P_A–X chain. P_A–X can then be involved in a new addition reaction so starting for a new cycle. k_{add} and k_b of relation (8.98) are the kinetic constants of the addition reactions, while k_{-add} and k_{-b} are the kinetic constants of the fragmentation reactions.

A further simplification can be introduced by considering this a "degenerative chain transfer mechanism," where:

$$P_A{}^* + X - P_B \underset{k_{BA}}{\overset{k_{AB}}{\rightleftarrows}} P_A - X + P_B{}^* \tag{8.99}$$

being k_{AB} and k_{BA} the chain transfer rate constants. k_{AB} is related to the kinetic constant of the reaction (8.86) through the following relation (Moad et al. 2008):

$$k_{AB} = k_{add}\phi = k_{add}\frac{k_b}{k_{add} + k_b} \tag{8.100}$$

being ϕ a partition coefficient of the adduct between products and starting material.

The most complete reaction mechanism of RAFT polymerization according to Zapata-Gonzalez et al. (2011) consists of five stages that are:

1. **The initiation step**, In the initiation step a suitable initiator generate radicals. These last polymerize very rapidly giving place to more or less long polymer chains.

$$\mathrm{I} \xrightarrow{k_d} 2\mathrm{I}_* \tag{8.101}$$

$$\mathrm{I}_* + \mathrm{M} \xrightarrow{k_i} \mathrm{P}_1^* \tag{8.102}$$

2. **Pre-equilibrium reversible chain transfer** corresponds to the already described reactions of addition and fragmentation.

$$\mathrm{P}_r^* + \mathrm{X} - \mathrm{R} \underset{k_{f1}}{\overset{k_{a1}}{\rightleftarrows}} \mathrm{P}_i - \mathrm{X}^* - \mathrm{R} \underset{k_{a1}}{\overset{k_{f1}}{\rightleftarrows}} \mathrm{P}_n - \mathrm{X} + \mathrm{R}^* \tag{8.103}$$

3. **Reinitiation and propagation**, due to the new radical formed in the previous stage of chain transfer.

$$\mathrm{R}^* + \mathrm{M} \xrightarrow{k_p} \mathrm{P}_1 \tag{8.104}$$

$$\mathrm{P}_r^* + \mathrm{M} \xrightarrow{k_p} \mathrm{P}_{r+1}^* \tag{8.105}$$

4. **Chain equilibration or main addition-fragmentation equilibrium**

$$\mathrm{P}_r^* + \mathrm{X} - \mathrm{P}_s \underset{k_f}{\overset{k_a}{\rightleftarrows}} \mathrm{P}_r - \mathrm{X}^* - \mathrm{P}_s \underset{k_f}{\overset{k_a}{\leftrightarrows}} \mathrm{P}_r - \mathrm{X} - \mathrm{P}_s^*$$

5. **Termination**

$$\mathrm{P}_r^* + \mathrm{P}_s^* \xrightarrow{k_{\mathrm{tc}}} \mathrm{P}_{r+s} \tag{8.106}$$

$$\mathrm{P}_r^* + \mathrm{P}_s^* \xrightarrow{k_{\mathrm{td}}} \mathrm{P}_r + \mathrm{P}_s \tag{8.107}$$

Modeling and simulation of the full molecular weight dispersion (FMWD) of RAFT polymerization is still controversial, because often induction period and rate retardation are observed. For describing these experimentally observed aspects, three different interpretations have been proposed, giving place to different models and corresponding kinetic parameters (Zapata-Gonzalez et al. 2011; Pintos et al. 2016). The different models are named:

Slow fragmentation (SF): The hypothesis assumed is that the intermediate radicals are relatively stable and give place to a slow fragmentation reaction.

Intermediate radical termination (IRT): In this model, it has been supposed that the intermediate radicals are unstable and can give place to cross-termination, giving a structure with three polymeric arms.

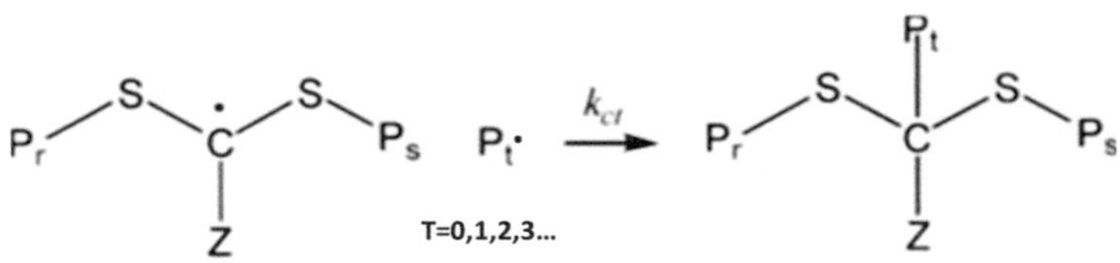

(8.108)

Intermediate radical termination with oligomers (IRTO): In this model, the cross-termination occurs only with oligomeric living radicals containing 0, 1 or 2 monomeric units.

By adopting the SF model, the fragmentation constant results of about $k_{\mathrm{f}} = 10^{-2}$ s^{-1} and the corresponding equilibrium constant is $K = 10^7\ \mathrm{M}^{-1}$, while, for the IRT model, k_{f} is more than six orders of magnitude greater $k_{\mathrm{f}} \approx 10^3$–$10^5\ \mathrm{s}^{-1}$ and $K = 50$–$55\ \mathrm{M}^{-1}$. Clearly, this is an unsolved problem, and it is not easy to discriminate what is the correct interpretation. At last, IRTO model is a compromise between the two hypotheses. In any case, different techniques have been used in the literature for modeling and simulating RAFT polymerization. A method based on a weighted residual (Galekin h-p method) that is commercially available (PREDICI) has been largely employed to evaluate the MWD. The method resulted powerful but incomplete. Monte Carlo technique has also been employed although heavy for the computational aspects. The most popular is the method of moments for calculating $\overline{M_n}, \overline{M_w}$, MWD and the polydispersity index (PDI). More recently, Zapata-Gonzalez et al. (2011) proposed a method based on the direct integration of the balance equations of all the involved components. The mass balances of all the involved species require the integration of many thousands or millions ODE equations and considering the large difference in the lifetime of those species the system is always characterized by stiffness. Stiffness can be removed by applying, when possible, the quasi-steady-state approximation (QSSA). This QSSA condition is valid when the rate of formation and consumption of species can be considered equal. In this case, by equating the rate expressions of radical formation and consumption, an algebraic expression is obtained that must be updated at any integration step. Therefore, the large ODE system is coupled with a large algebraic equations (AE) system. For example, if the rate of radical addition to the RAFT chain transfer agent is comparable with the fragmentation rate, based on the scheme (8.98), it is possible to write:

$$\frac{d[P_A - X^* - P_B]}{dt} \simeq 0 \tag{8.109}$$

Therefore, the addition rates can be equated to the fragmentation rates:

$$k_{add}\left(\left[P_A^*\right]\ [X - P_B]\right) + k_b\left(\left[P_B^*\right]\ [P_A - X]\right) = (k_{-add} + k_{-b})[P_A - X^* - P_B] \tag{8.110}$$

$$[P_A - X^* - P_B] = \frac{k_{add}\left[P_A^*\right][X - P_B] + k_b\left[P_B^*\right][P_A - X]}{k_{-add} + k_{-b}} \tag{8.111}$$

Moreover, when P_A^* and P_B^* radicals are kinetically identical, relation (8.100) becomes:

$$[P_A - X^* - P_B] = \frac{(k_{add} + k_b)[P - X][P^*]}{(k_{-add} + k_{-b})} = \frac{k_{add}[P - X][P^*]}{k_{frag}} = K_{RAPT}[P - X][P^*] \tag{8.112}$$

By observing the reaction scheme (8.101–8.107), four long-chain populations can be identified having each one its own MWD: living radicals, intermediate radicals with two arms of different length, dormant polymer, and dead polymer.

The three models considered (SF, IRT, IRTO) have the same ODE equations for the Initiator and the Monomer:

$$\frac{dI}{dt} = -2fk_d[I] \tag{8.113}$$

$$\frac{d[M]}{dt} = -k_p[M]\sum_{r=0}^{\infty}\left[P_r^*\right] \tag{8.114}$$

The other equations are related to species having a specific chain length and are reported in detail in the work published by Zapata-Gonzalez et al. (2011) together with the corresponding equations of the moments that had been used to verify the correctness of the results of the calculations made by direct integration.

8.3.4.4 Cationic Polyaddition

Polyaddition can also be promoted by an ionic mechanism. The reaction starts with an ionic attack to a double bond that could be cationic or anionic. In the case of a cationic polyaddition, the first step is the formation of a carbocation. The polymerization of a vinylic monomer can be promoted by a strong Bronsted (H_2SO_4, $HClO_4$, H_3PO_4 etc.) or Lewis ($AlCl_3$, BF_3, $SnCl_4$, $FeCl_3$ etc.) acid according to the following schemes (Matyjaszewski 1996; Odian 2004):

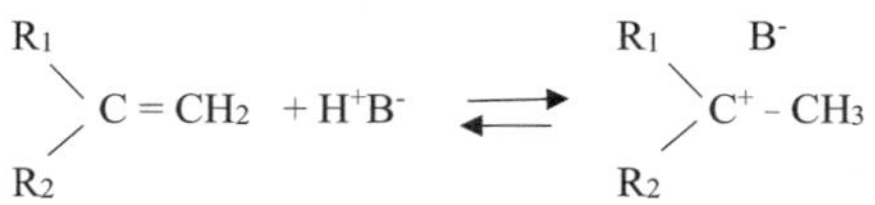

(8.115)

Initiation $$H^+B^- + M \leftrightarrows HM^+B^- \quad \text{Fast reaction} \tag{8.116}$$

Propagation $$HM^+B^- + M \rightarrow HM_2{}^+B^- \quad \text{Slow reactions}$$

$$\ldots\ldots\ldots\ldots\ldots\ldots\ldots \tag{8.117}$$

$$\ldots\ldots\ldots\ldots\ldots\ldots\ldots$$

$$HM_n{}^+B^- + M \rightarrow HM_{n+1}{}^+B^- \tag{8.118}$$

Termination by chain transfer $$HM_n{}^+B^- + M \rightarrow P_n + HM^+B^- \tag{8.119}$$

It is also possible the reaction:

$$HM_n{}^+B^- \rightarrow P_n + H^+B^- \tag{8.120}$$

P_n is a polymer chain terminating with a double bond. Effective termination occurs only if the acid is neutralized.

In the case of a Lewis acid, a cocatalyst is necessary, for example:

$$BF_3 + H_2O \leftrightarrows \text{Adduct} \leftrightarrows [BF_3OH]^-H^+ \tag{8.121}$$

Then, the proton attacks the double bond of a monomer M giving a carbocation and the mechanism proceeds in the same way seen for the Bronsted acids. It is important to point out that the chemical environment of polymerization is often apolar, therefore, a tight ionic couple is normally operative.

The cationic polymerization, promoted by a Lewis acid catalyst is the most complicated system of ionic polymerization. As seen, in this case we have a catalyst forming an adduct with a cocatalyst and two equilibria must be considered:

$$K_{e1} = \frac{[\text{Adduct}]}{[\text{Cat}]\,[\text{Cocat}]} \qquad K_{e2} = \frac{[I^+][A^-]}{[\text{Adduct}]} \tag{8.122}$$

$$[I^+] = [A^-] = \sqrt{K_{e2}[\text{Adduct}]} = \sqrt{K_{e1}K_{e2}\,[\text{Cat}][\text{Cocat}]} \tag{8.123}$$

$$\textbf{Initiation rate} \quad r_i = k_i[I^+][M] \tag{8.124}$$

$$\textbf{Propagation rate} \quad r_p = k_p\left[M_n^+\right][M] \tag{8.125}$$

$$\textbf{Termination via chain transfer} \quad r_{ct1} = k_{ct1}\left[M_n^+\right][M] \tag{8.126}$$

$$r_{ct2} = k_{ct2}\left[M_n^+\right] \tag{8.127}$$

It is possible to assume, as a first approximation, that k_p is the same independently of the chain length and the same approximation can be made for the chain transfer termination constants. A rigorous kinetic approach requires to write the mass balances for all the components involved in the reaction as it will be seen in a next section.

8.3.4.5 Anionic Polyaddition

Polymerization of vinylic compounds can be promoted also by an anionic attack in particular when the monomers contain electron-attracting groups. Anionic catalysts can be, alkaline metals, metal alkyls, metal alkoxides, amides etc. In this case, the initiation reaction is:

$$CH_2 = CHX + A^+B^- \rightleftarrows B - CH_2 - CHX^- A^+ \tag{8.128}$$

Propagation is:

$$B - CH_2 - CHX^- + n\,CH_2 = CHX \;\rightarrow B - (CH_2 - CHX)_n CH_2CHX^- \tag{8.129}$$

Termination occurs, also in this case, by chain transfer but adding monomer the reaction can go again (leaving polymer). The reaction can be stopped only by neutralizing the basic catalyst.

8.3.4.6 Stereospecific Polymerization, Ziegler-Natta Polymerization

In this type of polymerization, the monomer is coordinated on a metal catalyst and the polymerization occurs by insertion of the monomer molecules on an active center. In some conditions, the metal imposes a particular orientation to the reacting monomer giving place to a stereoregular polymer of high crystallinity. Catalysts of this type are classified as Ziegler-Natta catalysts (Ziegler et al. 1955; Natta 1955). Ziegler-Natta catalysts are constituted by a mixture of metal-organic compounds

Table 8.6 Performances of the different generations of Ziegler-Natta catalysts (Syahrudin and Herlambang 2020)

Catalyst generations	Composition	Productivity Kg PP/g Ti	Isotactic index %
I	$TiCl_3$-$Al(C_2H_5)_2Cl$	3–5	88–91
II	Well-dispersed $TiCl_3$-$Al(C_2H_5)_2Cl$	12–20	94–97
III	$TiCl_3$-ethyl benzoate-$MgCl_2$-$Al(C_2H_5)_2Cl$-methyl-p-toluate	300	90–95
IV	$TiCl_3$-di-iso-butyl phtalate-$MgCl_2$-Al $(C_2H_5)_2Cl$-$RSi(OCH_3)_2$	1000	97–99
V	$TiCl_3$-2,2′dimethoxypropane-$MgCl_2$-Al $(C_2H_5)_2Cl$-$RSi(OCH_3)_2$	2000	97–99
VI	$TiCl_3$-dialkylsuccinate-$MgCl_2$-$Al(C_2H_5)_2Cl$-$RSi(OCH_3)_2$	1500	97–99

(alkyls, aryls, alkyl-halides) of I, II and III groups (Li, Be, Mg, Al etc.) and transition metals compounds of the IV, V, VI and VIII group (Ti, V, Cr, Mo, Rh, Co, Ni, etc.). have shown that, starting from propylene as monomer, using different catalysts, different stereoregular structures (Natta 1960) can be obtained, such as:

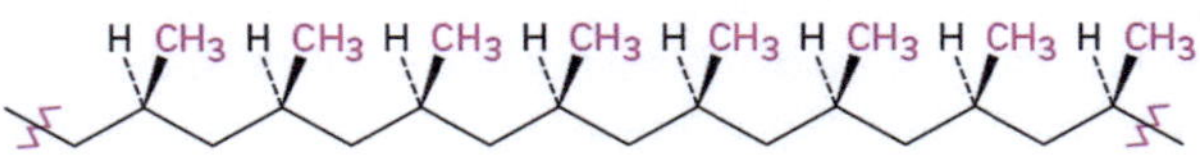

Isotactic (same side)

Catalyst $TiCl_3$-$Al(Et)_2Cl$

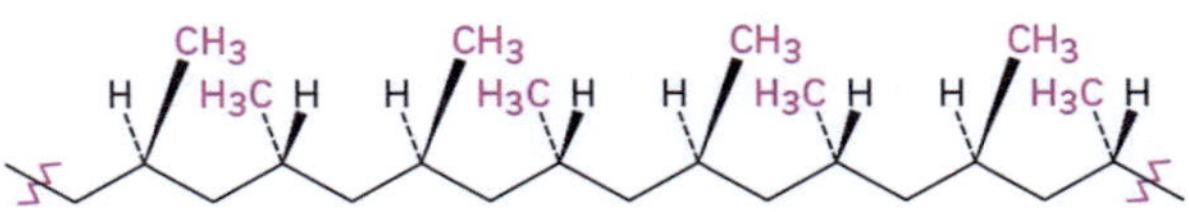

Syndiotactic (alternating sides)

Catalyst VCl_4-$Al(iBu)_2Cl$

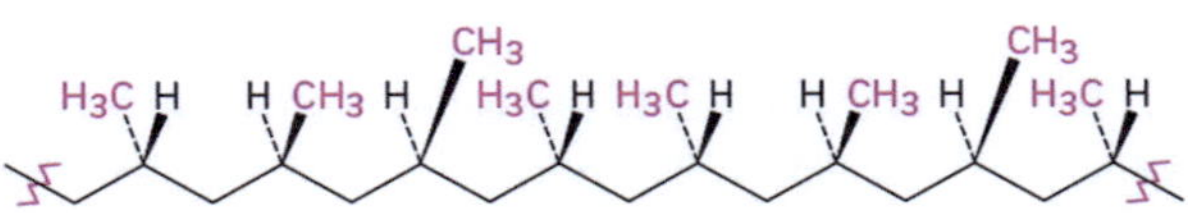

Atactic (random)

Radical polymerization

Stereoregular polymers as isotactic and syndiotactic have a crystalline structure. Crystalline polymers melt at temperatures higher than the corresponding amorphous ones and have better physical properties. For these reasons, Ziegler-Natta catalysts have been intensively studied and continuously improved during the years to produce polymers with higher and higher productivity and crystallinity by using simpler processes, reducing as much as possible the costs. Six generation catalysts have been developed (Vittoria et al. 2017) and the research on the topic is still active for further improvements. The first-generation Ziegler-Natta catalyst was the couple

$TiCl_3$-$Al(C_2H_5)_2Cl$ with $TiCl_3$ in the form of very small crystals dispersed in a hydrocarbon solvent. The active centers of polymerization were located in particular points of the $TiCl_3$ crystals, as a consequence of the reactions occurring by the contact with the cocatalyst $Al(C_2H_5)_3$ or $Al(C_2H_5)_2Cl$ such as:

$$TiCl_3 + Al(C_2H_5)_3 \rightleftarrows TiCl_3 \cdot Al(C_2H_5)_3 \rightleftarrows C_2H_5TiCl_2 \cdot Al(C_2H_5)_2Cl \tag{8.130}$$

An active bimetallic complex was formed with structure of the type:

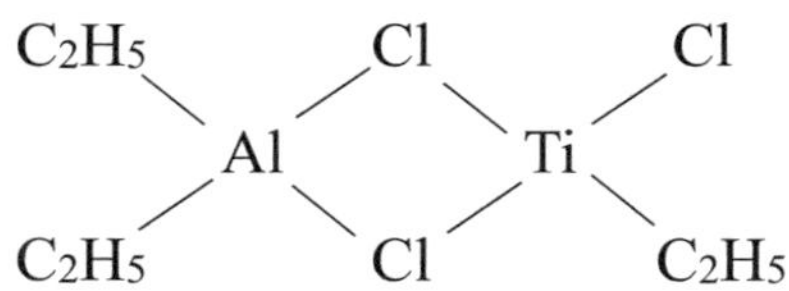

(8.131)

The double bond of the monomer is previously coordinated as ligand on the complex and then inserted in a growing chain. The titanium atom, having an incomplete electronic structure, can coordinate both the monomer and a growing chain, while the presence of alkyl aluminum favors the stereospecific growth. The rate of polymerization is proportional to both the propylene concentration and the amount of $TiCl_3$ present in the system, being, on the contrary, independent of the alkyl aluminum quantity. As this catalytic system is heterogeneous, the dispersion of $TiCl_3$ in very small solid particles resulted of paramount importance for the productivity. This aspect inspired the development of the second-generation Ziegler-Natta catalyst characterized by smaller $TiCl_3$ crystallites of γ and δ form instead of β and by the washing procedure to eliminate the catalyst poison $Al(C_2H_5)Cl_2$. Productivity increased by fivefold, corresponding to an analogous increase of the catalyst specific surface area. The isotactic index increased from 90 to 95%, as it can be seen in Table 8.6. The third-generation catalyst had been a breakthrough in the field, because the titanium chloride precursor was supported on a carrier like $MgCl_2$, rendering easier the catalyst dispersion. The obtainment of good productivity with powerful stereoselectivity enhancement required the addition of two Lewis bases acting as donor compounds. One is added during the pre-catalyst preparation step (internal donor, ID) and another in combination with the AlR_3 activator (external donor, ED). In the fourth-generation catalyst, ethyl benzoate was the ID and methyl-para-toluate was the ED. The fourth-generation catalyst corresponds to an exceptional improvement because it allows a radical simplification of the process (spheripol process). Titanium chloride was again supported on $MgCl_2$, but instead of adding esters of aromatic monocarboxylic acids, esters of phthalic acids (di-iso-butyl-phthalate) and polyalkoxy silanes (i.e., $RSi(OCH_3)_2$) were added. With this catalyst, isotactic index reached 98% and productivity became very high. However, the restriction on the use of carcinogenic phthalates determined the advent of the fifth and sixth generations, in which this ID Lewis base has been substituted by other ones.

In 1991, a new class of Ziegler-Natta catalysts was discovered, constituted by metallocene complexes with a structure of Cp_2MCl_2 (being M = Ti, Zr or Hf) and a cocatalyst of MAO (methylaluminoxane). In some cases, the two cyclopentadienyls (Cp) are linked with chemical bridges, such as -CH_2-CH_2-, and, according to the type of cyclopentadienyl ligands, the catalyst produces isotactic or syndiotactic polymers (Huang and Rempel 1995). These catalysts are homogeneous, because they soluble in the reaction medium and are the only examples of stereospecific homogeneous catalysts. Moreover, for these catalysts, the reaction mechanism has been completely elucidated:

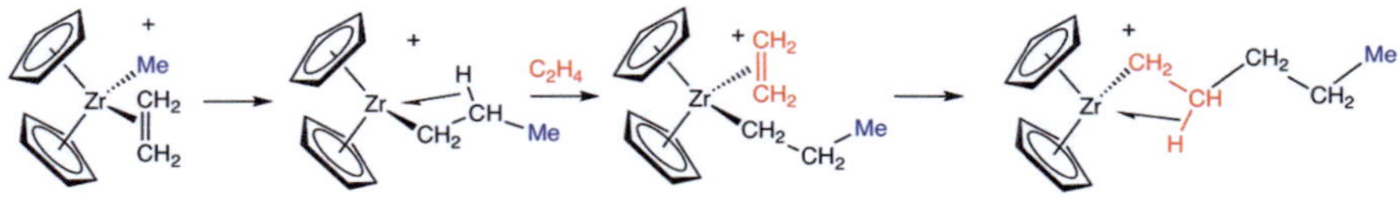

(8.132)

The metallocene complex alone is inactive. The metallocene halide reacts with MAO, giving a metallocenium ion $Cp_2Zr^+CH_3$ that becomes the true catalyst. The polymer grows after the alkene coordination on a single site of the metal atom, followed by the insertion of the C=C bond into the M-C bond of the active center.

8.3.4.7 Kinetics of Polymerization by the Method of Moments

The method of moments applied to polymer science was first proposed by Bamford and Tompa (1953). After that time, the method has been extensively employed for a simplified determination of the average molecular weight and of the molecular weight distribution (MWD), so defining the physicochemical properties of a polymer. A detailed and complete kinetic approach to radical polymerization requires, as previously mentioned, the solution of a set of a great number of differential equations to follow the evolution along the time of all the molecular species involved in the polymerization. However, the lifetime of the involved components is quite different and this determines numerical problems of stiffness in the calculations. The method of moments greatly simplifies the numerical approach drastically reducing the number of differential equations to be integrated. A detailed description of the method of moments applied to polymerization reaction has been published by Mastan and Zhu (2015) and by Zapata-Gonzalez and Saldivar-Guerra (2023) It is possible to define the moments related to, respectively, the population balance equations (PBE) of the polymeric chain species P_r and D_r. P_r is the population balance of the polymer live radicals, while D_r, that of the dead polymer chains. The moments can be defined as:

$$\mu_K = \sum_{r=0}^{\infty} r^K P_{\mathrm{r}}^*; \qquad K = 0, 1, 2, \ldots \qquad \lambda_K = \sum_{r=0}^{\infty} r^K D_{\mathrm{r}}; \qquad K = 0, 1, 2, \ldots \tag{8.133}$$

The zero moment $K = 0$ corresponds to the total concentration of each polymer species, while the first moment $K = 1$ is the average molar monomer concentration contained in the polymer chain.

In order to calculate the moments, we have to define:

1. A detailed reaction scheme.
2. Two population balance equations, respectively, related to the living radical polymers and dead polymer chains.
3. Substitute Eq. (8.133) with $K = 0$ in the PBE mentioned in point 2.

Reaction Scheme

Suppose that the following reaction scheme occurs:

$$\text{Initiation} \qquad \mathrm{I} \overset{2fk_{\mathrm{d}}}{\rightarrow} 2\mathrm{P}_0^* \tag{8.134}$$

$$\text{Propagation} \qquad \mathrm{P}_r^* + \mathrm{M} \overset{k_{\mathrm{p}}}{\rightarrow} \mathrm{P}_{r+1}^* \tag{8.135}$$

$$\text{Termination} \qquad \mathrm{P}_r^* + \mathrm{P}_s^* \overset{k_{\mathrm{td}}}{\rightarrow} \mathrm{P}_r + \mathrm{P}_s \quad \text{Disproportionation} \tag{8.136}$$

$$\mathrm{P}_r^* + \mathrm{P}_s^* \overset{k_{\mathrm{ts}}}{\rightarrow} \mathrm{P}_{r+s} \quad \text{Combination} \tag{8.137}$$

The subscripts r and s are the numbers of monomeric units in the polymeric chains (radical and dead chains). On the basis of the reaction scheme (8.134–8.137), it is possible to write the following system of mole balance equations:

$$\frac{\mathrm{d}[I]}{\mathrm{d}t} = -k_{\mathrm{d}}[I] \tag{8.138}$$

$$\frac{\mathrm{d}[M]}{\mathrm{d}t} = -k_{\mathrm{p}}[M] \sum_{r=0}^{\infty} [P_r^*] \tag{8.139}$$

$$\frac{\mathrm{d}[P_0^*]}{\mathrm{d}t} = 2fk_{\mathrm{d}}[I] - k_{\mathrm{p}}[M][P_0^*] - (k_{\mathrm{tc}} + k_{\mathrm{td}})[P_0^*] \sum_{s=0}^{\infty} [P_s^*] \tag{8.140}$$

$$\frac{\mathrm{d}[P_r^*]}{\mathrm{d}t} = k_{\mathrm{p}}[M]\,[P_{r-1}^*] - k_{\mathrm{p}}[M]\,[P_r^*] - (k_{\mathrm{tc}} + k_{\mathrm{td}})[P_r^*] \sum_{s=0}^{\infty} [P_s^*] \quad r \geq 1 \tag{8.141}$$

Table 8.7 Solution of the double summation of Eq. (8.142)

$[P_s^*][P_{r-s}^*]$	$s=0$	$s=1$	$s=2$	$s=3$	$s=\ldots$	Addition
$r=0$	$[P_0^*][P_0^*]$					↓
$r=1$	$[P_0^*][P_1^*]$	$+\ [P_1^*][P_0^*]$				
$r=2$	$[P_0^*][P_2^*]$	$+\ [P_1^*][P_1^*]$	$+\ [P_2^*][P_0^*]$			
$r=3$	$[P_0^*][P_3^*]$	$+\ [P_1^*][P_2^*]$	$+\ [P_2^*][P_1^*]$	$+\ [P_3^*][P_0^*]$		
$r=4$	$[P_0^*][P_4^*]$ ⋯⋯	$+\ [P_1^*][P_2^*]$	$+\ [P_2^*][P_2^*]$	$+\ [P_3^*][P_1^*]$		
$r=\ldots$	…	…				
Addition	$[P_0^*]\mu_0$	$+\ [P_1^*]\mu_0$	$+\ [P_2^*]\mu_0$	$+\ [P_3^*]\mu_0\ +$		$=\mu_0\mu_0$

$$\frac{d[P_r]}{dt} = k_{td}[P_r^*]\sum_{s=0}^{\infty}[P_s^*] + \frac{k_{tc}}{2}\sum_{s=0}^{r}[P_s^*]\,[P_{r-s}^*] \quad r \geq 0 \tag{8.142}$$

Equations (8.138–8.142) can be simplified and rendered solvable by introducing the zeroth, first and second moment balances. The equations related to the polymer radical chains are:

$$\frac{d\mu_0}{dt} = 2fk_d[I] + \sum_{r=0}^{\infty}\left(k_p[M]\,[P_{r-1}^*] - k_p[M]\,[P_r^*] - (k_{tc}+k_{td})\,[P_r^*]\mu_0\right) =$$
$$= 2fk_d[I] + k_p[M]\sum_{r=0}^{\infty}[P_{r-1}^*] - k_p[M]\sum_{r=0}^{\infty}[P_r^*] - (k_{tc}+k_{td})\mu_0\sum_{r=0}^{\infty}[P_r^*] \tag{8.143}$$

$$\frac{\mathbf{d}\boldsymbol{\mu}_\mathbf{0}}{\mathbf{d}\boldsymbol{t}} = 2fk_d[I] - \mu_0^2(k_{tc}+k_{td}) \tag{8.144}$$

$$\frac{d\mu_1}{dt} = \sum_{r=0}^{\infty} r\left(k_p[M]\,[P_{r-1}^*] - k_p[M]\,[P_r^*] - (k_{tc}+k_{td})\,[P_r^*]\mu_0\right) =$$
$$= k_p[M]\sum_{r=0}^{\infty}[(r-1)+1][P_{r-1}^*] - k_p[M]\sum_{r=0}^{\infty} r[P_r^*] - (k_{tc}+k_{td})\,\mu_0\sum_{r=0}^{\infty} r[P_r^*] \tag{8.145}$$

$$\frac{\mathbf{d}\boldsymbol{\mu}_\mathbf{1}}{\mathbf{d}\boldsymbol{t}} = k_p[M]\mu_0 - (k_{tc}+k_{td})\,\mu_0\mu_1 \tag{8.146}$$

$$\frac{d\mu_2}{dt} = \sum_{r=0}^{\infty} r^2 \left(k_p[M]\left[P^*_{r-1}\right] - k_p[M]\left[P^*_r\right] - (k_{tc} + k_{td})\left[P^*_r\right]\mu_0\right) =$$
$$= k_p[M]\sum_{r=0}^{\infty}[(r-1)+1]^2\left[P^*_{r-1}\right] - k_p[M]\sum_{r=0}^{\infty} r^2\left[P^*_r\right] - (k_{tc} + k_{td})\,\mu_0\sum_{r=0}^{\infty} r^2\left[P^*_r\right] \quad (8.147)$$

$$\frac{d\mu_2}{dt} = k_p[M]\mu_0 + 2k_p[M]\mu_1 - (k_{tc} + k_{td})\,\mu_0\mu_2 \quad (8.148)$$

Less simple is the derivation of the moment balances for the polymer dead chains. The zero moment contains a double summation that is apparently hard to solve by expressing it in terms of moments:

$$\frac{d\lambda_0}{dt} = \sum_{r=0}^{\infty}\left(k_{td}\left[P^*_r\right]\mu_0 + \frac{k_{tc}}{2}\sum_{s=0}^{r}\left[P^*_s\right]\left[P^*_{r-s}\right]\right) = k_{td}\mu_0^2 + \frac{k_{tc}}{2}\sum_{r=0}^{\infty}\sum_{s=0}^{r}\left[P^*_s\right]\left[P^*_{r-s}\right] \quad (8.149)$$

It is possible to demonstrate by considering just the first 4–5 terms of the double summation for $r = 0,1,2,3,4\ldots$ in correspondence of $s = 0,1,2,3,4$ that the double summation corresponds to $\mu_0{}^2$ (see Table 8.7).

$$\frac{\mathbf{d}\boldsymbol{\lambda}_0}{\mathbf{d}t} = \left(k_{td} + \frac{k_{tc}}{2}\right)\mu_0^2 \quad (8.150)$$

The moments related to polymer chain radicals are much smaller than the ones of dead chains polymers, that is, $\mu_i << \lambda_i$; hence, $\lambda_i + \mu_i \approx \lambda_i$. This approximation corresponds to the quasi-steady-state assumption for the radicals. This allows us to approximate the dead polymer chains as:

$$\frac{\mathbf{d}\boldsymbol{\lambda}_0}{\mathbf{d}t} = \frac{\mathbf{d}(\boldsymbol{\lambda}_0 + \boldsymbol{\mu}_0)}{\mathbf{d}t} = 2fk_d[I] - \frac{k_{tc}}{2}\mu_0^2 \quad (8.151)$$

This equation contains the generation and consumption rate terms of polymer chains. A method similar can be employed to define the first- and second-moment balances of the dead polymer chains considering, respectively, the terms:

$$r\left[P^*_s\right]\left[P^*_{r-s}\right] = [s + (r-s)]\left[P^*_s\right]\left[P^*_{r-s}\right] \text{ and } r^2\left[P^*_s\right]\left[P^*_{r-s}\right] = [s + (r-s)]^2\left[P^*_s\right]\left[P^*_{r-s}\right] \quad (8.152)$$

The first-moment balance can be written as:

Table 8.8 Parameters and data for calculating the kinetic of polymerization with the method of moments

Parameter	Value	Units
Initiator decomposition constant	1.0E–5	h^{-1}
Propagation constant	1600	L/(h mol)
Efficiency factor for initiator	0.6	–
Termination by coupling constant	2.0e8	L/(h mol)
Termination by disproportionation constant	1.0e8	L/(h mol)
Initial monomer concentration	5	mol/L
Initial initiator concentration	0.1	mol/L

$$\frac{d\lambda_1}{dt}=\sum_{r=0}^{\infty} r\left(k_{td}\left[P_r^*\right]\mu_0+\frac{k_{tc}}{2}\sum_{s=0}^{r}\left[P_s^*\right]\left[P_{r-s}^*\right]\right)=k_{td}\mu_0\sum_{r=0}^{\infty} r\left[P_r^*\right]+\frac{k_{tc}}{2}\sum_{r=0}^{\infty} r\sum_{s=0}^{r}\left[P_s^*\right]\left[P_{r-s}^*\right] \tag{8.153}$$

Repeating the determination of the first 4–5 terms of the double summation for $r = 0,1,2,3,4\ldots$ in correspondence of $s = 0,1,2,3,4$, it is possible to simplify the expression as:

$$\frac{\mathbf{d}\boldsymbol{\lambda}_1}{\mathbf{d}\boldsymbol{t}} = (k_{tc} + k_{td})\,\mu_0\mu_1 \tag{8.154}$$

More details about the procedure can be found in the article published by Mastan and Zhu (2015).

Again, assuming the steady-state conditions for the radicals, we can also write:

$$\frac{\mathbf{d}\boldsymbol{\lambda}_1}{\mathbf{d}\boldsymbol{t}} \approx k_p[M]\mu_0 \tag{8.155}$$

Finally, the second moment balance can be written as:

$$\begin{aligned}\frac{d\lambda_2}{dt} &= \sum_{r=0}^{\infty} r^2\left(k_{td}\left[P_r^*\right]\mu_0+\frac{k_{tc}}{2}\sum_{s=0}^{r}\left[P_s^*\right]\left[P_{r-s}^*\right]\right)\\ &= k_{td}\mu_0\sum_{r=0}^{\infty} r^2\left[P_r^*\right]+\frac{k_{tc}}{2}\sum_{r=0}^{\infty} r^2\sum_{s=0}^{r}\left[P_s^*\right]\left[P_{r-s}^*\right]\end{aligned} \tag{8.156}$$

As previously made by determining the first 4–5 terms of the double summation for $r = 0,1,2,3,4\ldots$ in correspondence of $s = 0,1,2,3,4$, it is possible to simplify the expression as:

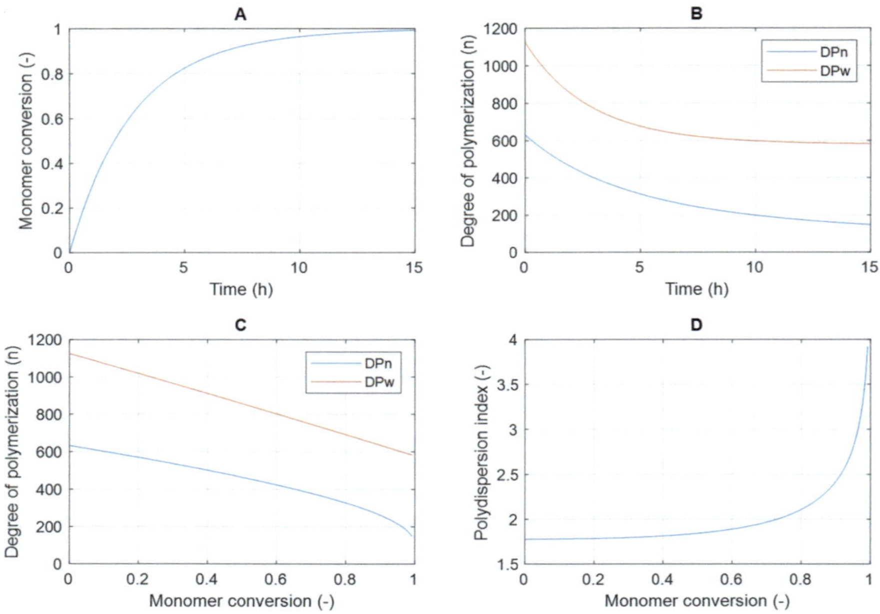

Fig. 8.18 An example of kinetic of polymerization faced with the method of moments

$$\frac{\mathbf{d}\lambda_2}{\mathbf{d}t} = (k_{\mathrm{tc}} + k_{\mathrm{td}})\,\mu_0\mu_2 + k_{\mathrm{tc}}\mu_1^2 \tag{8.157}$$

After the radical steady-state assumption:

$$\frac{\mathbf{d}\lambda_2}{\mathbf{d}t} \approx k_{\mathrm{p}}[M]\mu_0 + 2k_{\mathrm{p}}[M]\mu_1 + k_{\mathrm{tc}}\mu_1^2 \tag{8.158}$$

The first-moment balance is equal to the monomer consumption rate and holds $\mu_1 = [M_0] - [M]$

That is, the amount of monomer units of the polymeric chains is equal to the monomer concentration consumed in the polymerization.

To solve the problem of polymerization, modeling needs to solve an ODE system of six equations that are:

1. The balance equation related to the initiator I (Eq. 8.138)
2. The balance equation related to the monomer M (Eq. 8.139)
3. The balance equation related to the zero moment of the radical chain polymers μ_0 (Eq. 8.144)
4. The balance equation related to the first moment of the radical chain polymers μ_1 (Eq. 8.146)
5. The balance equation related to the zero moment of the dead chain polymers λ_0 (Eq. 8.151)

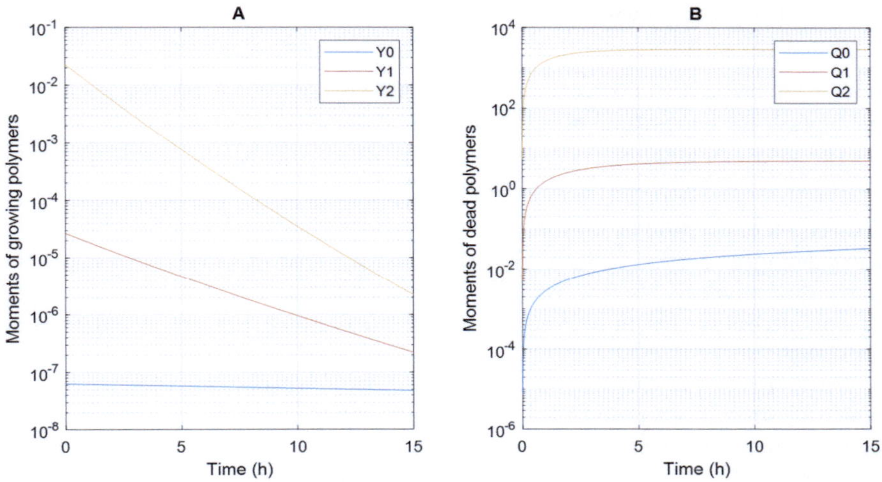

Fig. 8.19 Trends of the moments of order 0, 1, and 2 of growing chains polymers (**a**) and of dead chains polymers (**b**) as a function of time

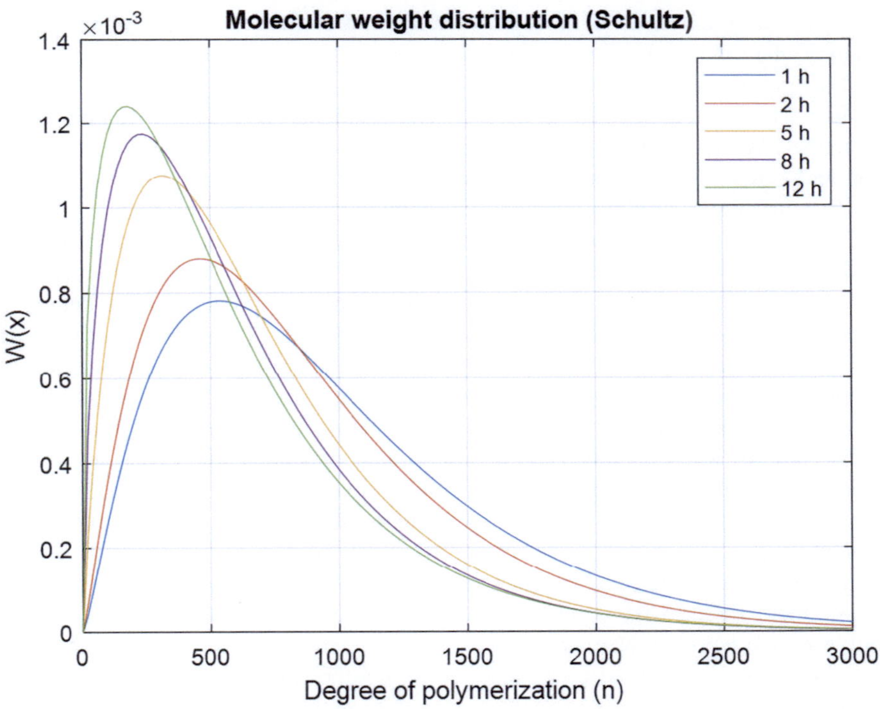

Fig. 8.20 Change of the MWD with the degree of polymerization

6. The balance equation related to the second moment of the dead chain polymers λ_2 (Eq. 8.157 or 8.158)

It is not necessary to solve the balance of the first moment λ_1 because, as already mentioned, the corresponding equation is equal to the one of monomer balance.

The initial conditions for all the moments are obviously zero, while the initial values of the initiator and of monomer concentration are known data.

Exercise 8.6. Method of Moments for Polymerization Kinetic
Considering a reaction scheme as represented by Eqs. (8.134–8.148), apply the method of moments, as is reported by Mastan and Zhu (2015), for the simulation of polymerization reaction. Parameters and other useful data are summarized in Table 8.8:

Develop a MATLAB code that solves the material balances for a batch reactor in which the mentioned reaction network occurs. Produce, then, a plot reporting the conversion of monomer along the time.

Draw also other plots that are representative of the considered reacting system (e.g., molecular weight distribution).

Solution
The calculation can be developed by considering that a general definition of moments of various order is the relation (8.133). According to the text of the present chapter, the monomer and initiator mass balance is expressed by Eqs. (8.139) and (8.138), respectively.

For what concerns living polymers, the moments of order 0, 1, and 2 are described by Eqs. (8.144), (8.146) and (8.148), while the corresponding moments for dead polymers are expressed by relations (8.150), (8.155) and (8.157).

As a first group of plots, consider the following reported in Fig. 8.18.

In the plot (a), the monomer conversion as a function of time is reported; after 15 h of reaction, the conversion is almost complete. In the plot (b), the numerical and weighed average molecular weights, expressed in terms of polymerization degree in monomer units, also as a function of time are reported. (c) In this plot, the same quantities as in the previous one, molecular weights, are reported as a function of monomer conversion. As we can observe, the higher is the monomer conversion, the lower are the molecular weights. (d) in the las plot of the figure, the polydispersity index (ratio between weighed and numeric average molecular weights) is reported as a function of monomer conversion. At high conversion, the polydispersity index increases very steeply due to the rapid decrease of numerical averaged molecular weight.

In Fig. 8.19, the trends of, respectively, (a) moments of 0, 1, and 2 orders related to polymeric radical (living polymers) and (b) moments of 0, 1, and 2 related to dead polymers are reported. Moments of living polymers decreases along the time as the corresponding group of species decrease too. On the contrary, the moments calculated for the dead polymers increase very steeply in the initial stage of the reaction, while at longer times, quite constant values are reached.

In Fig. 8.20, we can appreciate how the molecular weight distribution evolves as the reaction proceed. The initial distribution (after 1 h) is quite narrow, and the average molecular weight is relatively high. As the reaction proceed, the distribution becomes more and more narrow and shifted toward lower molecular weights.

The described results were obtained using a MATLAB program available as Electronic Supplementary Material.

8.4 Polymerization Reactor Modeling

As already mentioned, polymerization can be considered a system of complex chemical reactions (i.e., a combination of series and simultaneous reactions), and the design of the polymerization reactors can be made by following the same concepts developed in the previous chapters for those systems, but different are the objectives to be pursued in the optimization. In the case of a conventional complex chemical reaction system, the main objective is to obtain a single specific compound with a high yield and purity. In the case of a polymerization reaction, the obtainable product is a mixture of molecules of different molecular weights containing a different number of monomeric units. This product would have certain specifications related to the following: mechanical strength, elasticity, electrical properties, corrosion resistance, performance in extrusion equipment, etc. These properties depend on both the average molecular weight ($\overline{MW}$) and the molecular weight distribution (MWD) in a complex and often unpredictable way. Therefore, polymerization reactor modeling has normally the scope to obtain the $\overline{MW}$ and MWD more suitable for the required specifications. As the correlation between $\overline{MW}$ and MWD with the mentioned specifications is often unknown, the best values are usually obtained by an extensive experimental approach. This problem becomes more and more complicated in the case of copolymerization or when branching occurs during the polymerization reaction. However, as for the conventional complex chemical systems, a valuable kinetic model, based on a reliable reaction mechanism, is basilar for the polymerization reactor design. First of all, it must be considered that polymerization can be homogeneous or heterogeneous. Mass and heat transfer must obviously be considered in the model when necessary. Mixing and removing the heat of reaction (13–21 kcal/mol) are also very important in the reactor design. Moreover, the reaction can be performed in batch, semi-batch, or continuous reactors, and the kinetic model must be adapted to the reactor configuration. Finally, another important aspect is related to the polymerization modalities. We can have for example:

- **Bulk or mass polymerization** involves a monomer and an initiator as the main components, without a solvent. The monomer is almost completely converted. The polymer is in the form of a melted mass becoming solid by cooling.

- **Polymerization in solution:** the monomer is dissolved in a solvent, while the polymer can be soluble or insoluble in the same solvent. In the second case, the polymer is separated by precipitation, while, in the first one, by solvent evaporation.
- **Suspension polymerization:** the monomer is dispersed in a liquid in which it is insoluble. Small liquid drops of spherical shape are formed that become solid after polymerization. Surfactants are normally added to keep stable the dispersion.
- **Emulsion polymerization:** with the addition of opportune surfactants, the monomer is emulsified in water. The polymerization reaction occurs inside the micelles.

As it has been seen, a lot of different possibilities exist, each corresponding to a different kinetic model, that is, a general unified approach to the polymerization reactor modeling is not possible, and we can just suggest the pathways to properly set the approach to be followed for some examples of polymerization as reported in the following scheme.

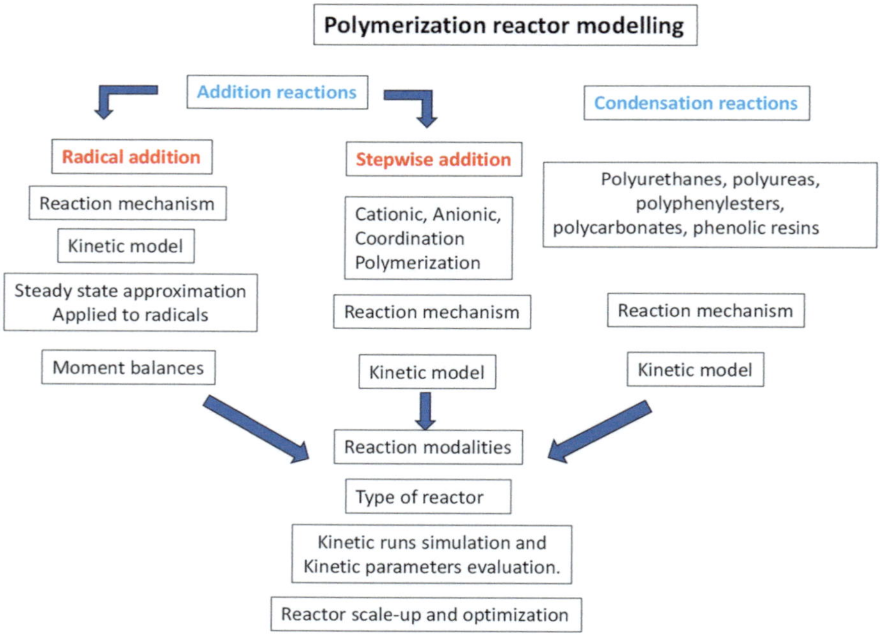

8.4.1 Selected Examples of Polymerization Reactor Modeling: Stepwise Addition Polymerization

As mentioned in some previous sections, stepwise addition polymerization gives place to **living polymers**, that is, in the simplest case, the reaction scheme is:

$$\mathrm{I} + \mathrm{M} \xrightarrow{k_\mathrm{i}} \mathrm{P}_1 \qquad \text{Initiation} \tag{8.159}$$

$$\mathrm{P}_n + \mathrm{M} \xrightarrow{k_\mathrm{p}} \mathrm{P}_{n+1} \qquad \text{Propagation} \tag{8.160}$$

No chain transfer or termination reactions occur. Therefore, the reaction can continue indefinitely if further monomer is added. The mechanism is similar for cationic, anionic, and coordinated polymerization. The initiation reaction will be:

$$r_\mathrm{i} = k_\mathrm{i}[I][M] \tag{8.161}$$

Considering as a first approximation k_p independent of the chain length, the propagation rate of any single propagation step will be:

$$r_{P_1} = k_\mathrm{p}[P_1][M] \tag{8.162}$$

$$\begin{aligned} r_{P_2} &= k_\mathrm{p}[P_2][M] \\ &\ldots\ldots\ldots\ldots\ldots \end{aligned} \tag{8.163}$$

$$r_{P_n} = k_\mathrm{p}[P_n][M] \tag{8.164}$$

From this scheme, it is possible to define the mass balance related to any involved chemical specie. Considering a batch reactor, we have, for example:

Kinetic Model to Describe the Behaviour of a Stepwise Addition Polymerization Occurring in a Batch Reactor

$$\frac{dI}{dt} = -k_i[I][M] \tag{8.165}$$

$$\frac{dM}{dt} = -r_i - \sum_{n=1}^{\infty} r_{P_n} = -M\left(k_i[I] + k_P \sum_{n=1}^{\infty} [P_n]\right) \tag{8.166}$$

$$\frac{dP_1}{dt} = k_i[I][M] - k_p[P_1][M] \tag{8.167}$$

$$\frac{dP_2}{dt} = k_p[P_1][M] - k_p[P_2][M] \tag{8.168}$$

....

$$\frac{dP_n}{dt} = k_p[P_{n-1}][M] - k_p[P_n][M] \tag{8.169}$$

Initial conditions: For $t = 0$ it holds $[I] = [I]_0$, $[M] = [M]_0$, $[P_1] = [P_2] = \cdots\cdots [P_n] = 0$

If the monomer is continuously fed to the reactor, the model is slightly modified introducing the monomer molar flow rate into the monomer balance expression (8.166) that becomes:

Table 8.9 Parameters and data useful for the calculations

Parameters and operative data	Value	Units
Initiation constant	2.0E–04	L/(h mol)
Propagation constant	3.5E–03	L/(h mol)
Maximum polymerization degree	150	–
Final time	10	H
Initial monomer concentration	5	mol/L
Initial initiator concentration	0.1	mol/L
Initial reaction volume	1	L
Monomer feed	0.001	mol/s
Monomer molecular weight	100	g/mol
Mixture density	1000	g/L

Adjustment of the Batch Kinetic Model to Describe a Semi-batch Reactor

$$\frac{\mathrm{d}M}{\mathrm{d}t} = \frac{F_\mathrm{M}}{V_\mathrm{R}} - r_\mathrm{i} - \sum_{n=1}^{\infty} r_{P_n} = \frac{F_\mathrm{M}}{V_\mathrm{R}} - M\left(k_\mathrm{i}[I] + k_P \sum_{n=1}^{\infty}[P_n]\right) \tag{8.170}$$

The volume of the reacting mixture must be adjusted after each integration step.

Initial conditions could be the same of batch reactor with the option of $[M] = [M]_0$ or $[M] = 0$

Finally, considering a continuous reactor, there are two different ideal conditions: CSTR (continuous stirred tank reactor) or PFR (plug flow reactor).

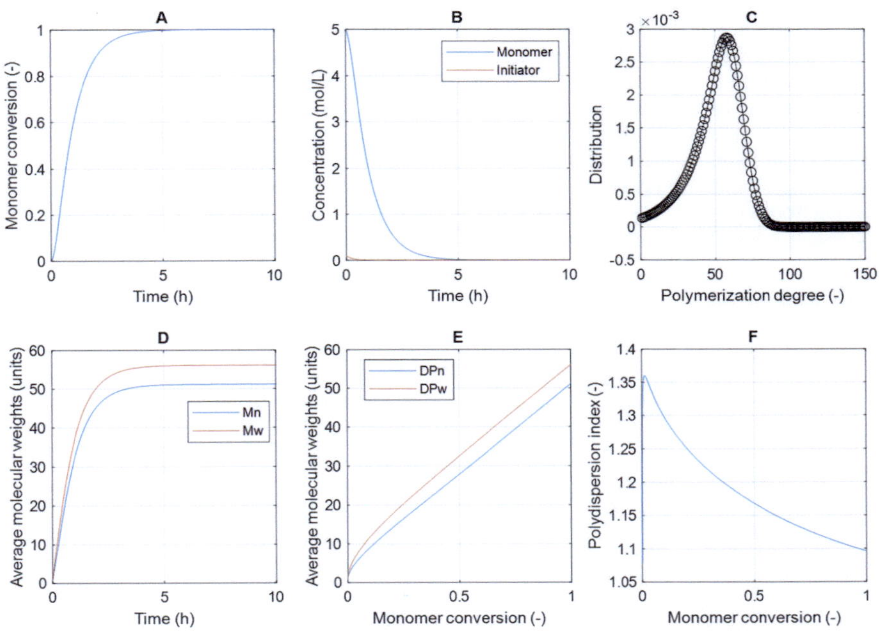

Fig. 8.21 Results of calculations. Plot (**a**), monomer conversion as a function of time; plot (**b**), concentration of monomer and initiator; plot (**c**), final molecular weight distribution; plot (**d**), numerical and weighed average molecular weights, expressed in terms of monomeric units (polymerization degree); plot (**e**), numerical and weighed average molecular weights as a function of monomer conversion; plot (**f**), polydispersity index as a function of monomer conversion

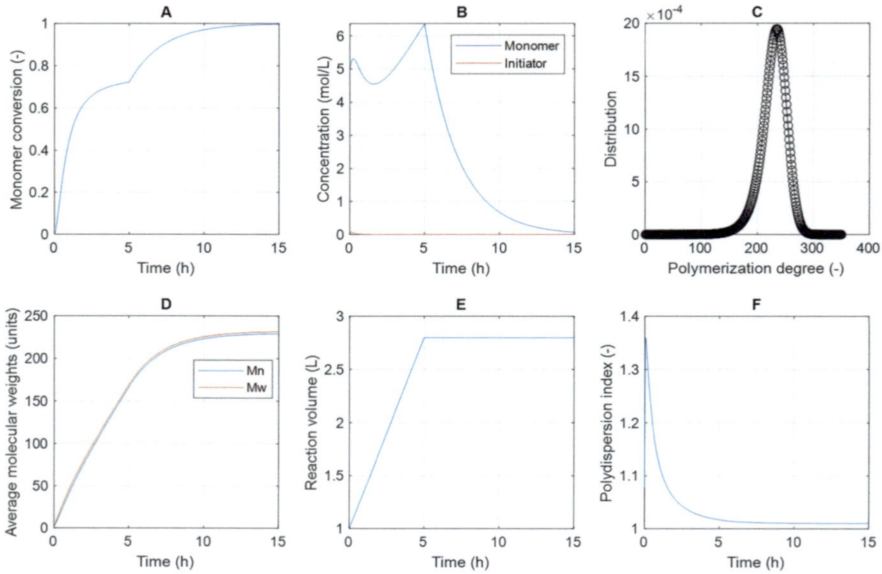

Fig. 8.22 Results of the calculations: Plot (**a**), monomer conversion as a function of time. From this plot is evident that, in the first 5 h of reaction, the system is not able to consume all the monomer fed to the reactor. When the feed is stopped, the conversion goes to completion in the following time. Plot (**b**): concentration of monomer and initiator as function of time. The concentration of monomer is quite fluctuating due to the balance between the monomer consumed and monomer fed. Plot (**c**): final molecular weight distribution. Plot (**d**): numerical and weighed average molecular weights, expressed in terms of monomeric units (polymerization degree) as a function of time. Plot (**e**): reaction volume as a function of time. In the first 5 h, the volume increases linearly with time, starting from the initial value. Obviously, when the monomer feed is stopped, the volume maintains a constant value. Plot (**f**): polydispersity index as a function of time

Table 8.10 List of parameters and data useful for the calculation

Parameter	Value	Units
Initiation constant	2.0E–04	L/(h mol)
Propagation constant	3.5E–03	L/(h mol)
Maximum polymerization degree	150	–
Volumetric flow rate	10	L/h
Reactor volume	80	L
Inlet monomer concentration	5	mol/L
Inlet initiator concentration	0.1	mol/L

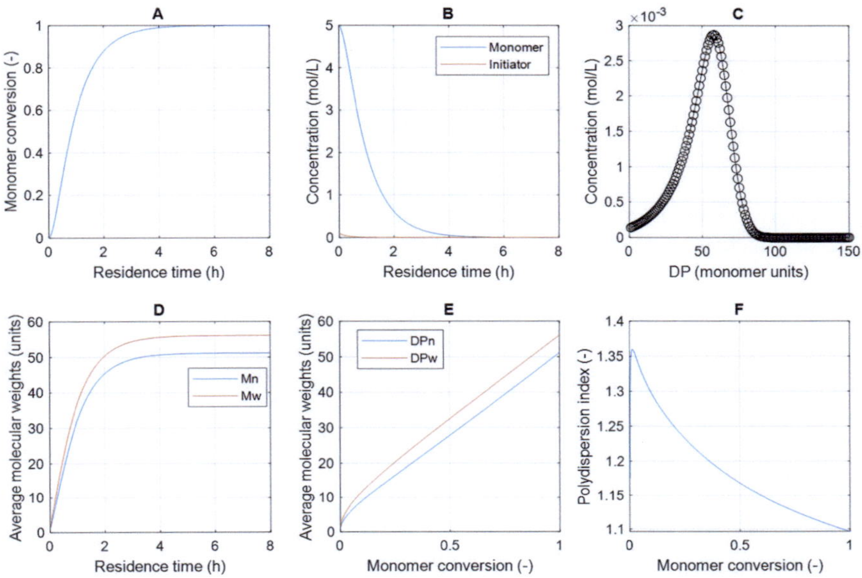

Fig. 8.23 Results of the calculations: Plot (**a**): monomer conversion as a function of residence time. Plot (**b**): concentration of monomer and initiator as function of residence time. Plot (**c**): final molecular weight distribution. Plot (**d**): numerical and weighed average molecular weights, expressed in terms of monomeric units (polymerization degree) as function of residence time. Plot (**e**): numerical and weighed average molecular weights as a function of monomer conversion. Plot (**f**): polydispersity index as a function of monomer conversion

Kinetic Model to Describe the Behavior of a Stepwise Addition Polymerization Occurring in a PFR Reactor

$$\frac{\mathrm{d}F_{\mathrm{I}}}{\mathrm{d}V_{\mathrm{R}}} = -k_{\mathrm{i}}[I][M] \tag{8.171}$$

$$\frac{\mathrm{d}F_{\mathrm{M}}}{\mathrm{d}V_{\mathrm{R}}} = -r_{\mathrm{i}} - \sum_{n=1}^{\infty} r_{P_n} = -M\left(k_{\mathrm{i}}[I] + k_P \sum_{n=1}^{\infty} [P_n]\right) \tag{8.172}$$

$$\frac{\mathrm{d}F_{P_1}}{\mathrm{d}V_R} = k_{\mathrm{i}}[I][M] - k_{\mathrm{p}}[P_1][M] \tag{8.173}$$

$$\frac{\mathrm{d}F_{P_2}}{\mathrm{d}V_{\mathrm{R}}} = k_{\mathrm{p}}[P_1][M] - k_{\mathrm{p}}[P_2][M] \tag{8.174}$$

....

(continued)

$$\frac{dF_{P_n}}{dV_R} = k_p[P_{n-1}][M] - k_p[P_n][M] \tag{8.175}$$

Initial conditions: For $V_R = 0$, it holds $[I] = [I]_0$, $[M] = [M]_0$, $[P_1] = [P_2] = \cdots\cdots [P_n] = 0$

Kinetic Model to Describe the Behavior of a Stepwise Addition Polymerization Occurring in a CSTR Reactor

$$\frac{Q}{V_R}\left([I]_0 - [I]\right) = k_i[I][M] \tag{8.176}$$

(continued)

Table 8.11 List of parameters and data useful for the calculation

Parameter	Value	Units
Initiation constant	2.0E–04	L/(h mol)
Propagation constant	3.5E–03	L/(h mol)
Maximum polymerization degree	500	–
Volumetric flow rate	12.5	L/h
Reactor volume	100	L
Inlet monomer concentration	5	mol/L
Inlet initiator concentration	0.1	mol/L

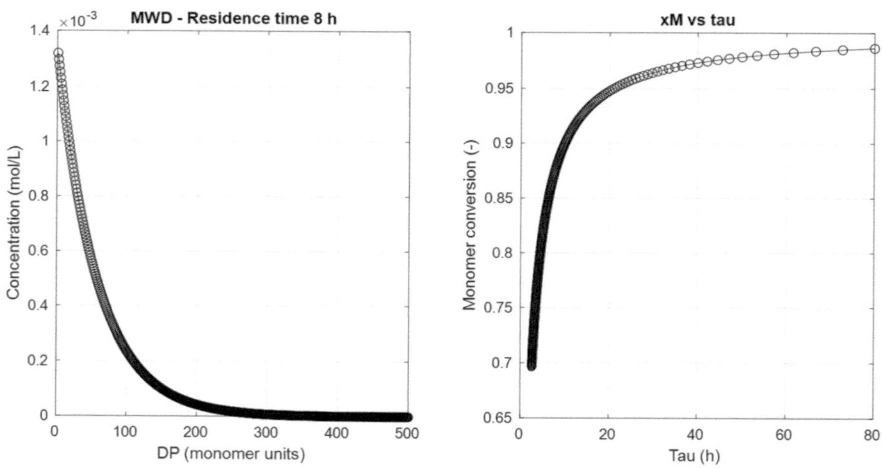

Fig. 8.24 Results of the calculations. In the plot on the left, the MWD distribution is reported, while, in the plot on the right, the trend of monomer conversion is reported against the residence time

$$\frac{Q}{V_{\mathrm{R}}}([M]_0 - [M]) = M\left(k_{\mathrm{i}}[I] + k_P\sum_{n=1}^{\infty}[P_n]\right) \tag{8.177}$$

$$\frac{Q}{V_{\mathrm{R}}}([P_1]_0 - [P_1]) = -k_{\mathrm{i}}[I][M] + k_{\mathrm{p}}[P_1][M] \tag{8.178}$$

$$\frac{Q}{V_{\mathrm{R}}}([P_2]_0 - [P_2]) = -k_{\mathrm{p}}[P_1][M] + k_{\mathrm{p}}[P_2][M] \tag{8.179}$$

....

$$\frac{Q}{V_{\mathrm{R}}}([P_n]_0 - [P_n]) = -k_{\mathrm{p}}[P_{n-1}][M] + k_{\mathrm{p}}[P_n][M] \tag{8.180}$$

Composition fed flow: $[I] = [I]_0$, $[M] = [M]_0$, $[P_1] = [P_2] = \cdots\cdots [P_n] = 0$ or $[P_n]_0$

Exercise 8.7. Stepwise Polymerization in Batch and Fed Batch Reactors
Considering a reaction scheme as represented by Eqs. (8.159 and 8.160), apply the reaction rates (Eqs. 8.161–8.164) and material balance for a batch reactor (Eqs. 8.165–8.169) and related initial conditions for the simulation of a batch reactor, in which a stepwise polymerization is conducted. Parameters and other useful data are summarized in Table 8.9:

Part 1—Develop a MATLAB code that solves the material balances for a batch reactor, in which the mentioned reaction network occurs. Produce a group of plots in which the main variables of the system are plotted against the time.

Part 2—Develop a further MATLAB code, similar to that of Part 1 but related to a fed-batch reactor in which the monomer is continuously fed at a fixed molar flow rate (see Table 8.9) for the first 5 h of operation. After this time, the feed of monomer is stopped.

Solution
Part 1

The calculations can be developed by considering that Eqs. (8.165–8.169) are ordinary differential equation that can be integrated starting from opportune initial conditions. If DPmax is the maximum polymerization degree, then the dimension of the ode system id DPmax+2.

The MATLAB code produces the following figures (Fig. 8.21).

Part 2

The fed-batch reactor can be modeled by using mass balance equations similar to that of batch reactor with the difference in the monomer balance (Eq. 8.170). In this equation, the monomer feed must be considered. Another difference is that the reaction volume increases along the time due to the feed.

By considering the described differences, a MATLAB code was developed that produced the group of plots of Fig. 8.22.

The described results were obtained using a MATLAB program available as Electronic Supplementary Material.

Exercise 8.8. Stepwise Polymerization in Plug Flow Reactor
Considering a reaction scheme as represented by Eqs. (8.159 and 8.160), apply the reaction rates (Eqs. 8.161–8.164) and material balance for a PFR reactor (Eqs. 8.171–8.175) and related initial conditions (reactor inlet) for the simulation of a continuous PFR reactor in which a stepwise polymerization is conducted. Parameters and other useful data are summarized in Table 8.10:

Develop a MATLAB code that solves the material balances for a PFR reactor, in which the mentioned reaction network occurs. Produce a group of plots in which the main variables of the system are plotted against the time.

Solution
The calculations can be developed by considering that Eqs. (8.171–8.175) are ordinary differential equations (ODE) that can be integrated starting from opportune initial conditions (reactor inlet conditions). If DPmax is the maximum polymerization degree, then the dimension of the ode system id DPmax+2.

The MATLAB code produced the results reported in Fig. 8.23.

If these results are compared to those obtained for the batch reactor, a correspondence is evident between the two, and the same performance is obtained in two types of reactors, providing the same residence time.

The described results were obtained using a MATLAB program available as Electronic Supplementary Material.

Exercise 8.9. Stepwise Polymerization in CSTR Reactor
Considering a reaction scheme as represented by Eqs. (8.159 and 8.160), apply the reaction rates (Eqs. 8.161–8.164) using the material balance for a CSTR reactor Eqs. (8.176–8.180) and the condition at the reactor feed for the simulation of a continuous CSTR reactor in which a stepwise polymerization is conducted. Parameters and other useful data are summarized in Table 8.11:

Develop a MATLAB code that solves the material balances for a CSTR reactor in which the mentioned reaction network occurs.

Produce the following two plots:

1. A first plot in which the final molecular weight distribution is constructed.
2. A second plot in which the monomer conversion is plotted against the volumetric feed flow rate between 1.25 and 37.5 L/h.

Solution
The calculations can be developed by considering that Eqs. (8.176–8.180) are nonlinear algebraic equations (NLE) that can be simultaneously solved. If DPmax is the maximum polymerization degree, then the dimension of the NLE system id DPmax+2.

The MATLAB code produce the following figures (Fig. 8.24).

As expected, at low residence time, lower conversion is obtained, while, by increasing the residence time, the monomer conversion approaches the completion.

Table 8.12 Reactor's conditions and kinetic constants (taken from Hui and Hamielec 1972)

Reactor's conditions					
T(K)	P(atm)	Time (min)	M_0 (mol/L)	I_0 (mol/L)	S_0 (mol/L)
373.15	1.0	80	4.8824	6.8794×10^{-3}	3.5233
Kinetic constants					
Initiator dissociation		k_d (1/min)		$1.0272 \times 10^{17} \exp\left(\frac{-15{,}924}{T}\right)$	
Constant of propagation		k_p (L/mol/min)		$6.306 \times 10^{8} \exp\left(\frac{-35{,}774}{T}\right)$	
Constant of termination		k_{tc} (L/mol/min)		$7.53 \times 10^{10} \exp\left(\frac{-844}{T}\right)$	
Monomer chain transfer constant		k_{ctm} (L/mol/min)		$1.2 \times k_p \times 10^{-5}$	
Monomer chain transfer constant		k_{cts} (L/mol/min)		$7.0 \times k_p \times 10^{-5}$	

The described results were obtained using a MATLAB program available as Electronic Supplementary Material.

8.4.2 Selected Examples of Polymerization Reactor Modeling: Radical Polymerization

It has already been seen that is possible to compute both the $\overline{\mathrm{MW}}$ and MWD of a polymer avoiding the solution of an ODE system containing thousands of equations. Nonetheless, in this case, a leak of information is unavoidable, and the rigorous computation is today feasible thanks to the availability of high-speed computers. Therefore, we will apply this approach to the polymerization in a batch reactor of styrene dissolved in benzene. The reaction mechanism and related kinetic equations for any single step are the same reported in Sect. 8.3.4.3. Therefore, we can write the mass balance equations for each component involved in the reaction.

Kinetic Model to Describe the Behavior of a Radical Polymerization Occurring in an Isothermal Batch Reactor

$$\text{Initiator decomposition}: \quad \frac{\mathrm{d}I}{\mathrm{d}t} = -k_d I \tag{8.181}$$

$$\text{Radical chain initiation}: \quad \frac{\mathrm{d}R^*}{\mathrm{d}t} = 2fk_d I - k_i R^* M \tag{8.182}$$

Monomer's radical:

(continued)

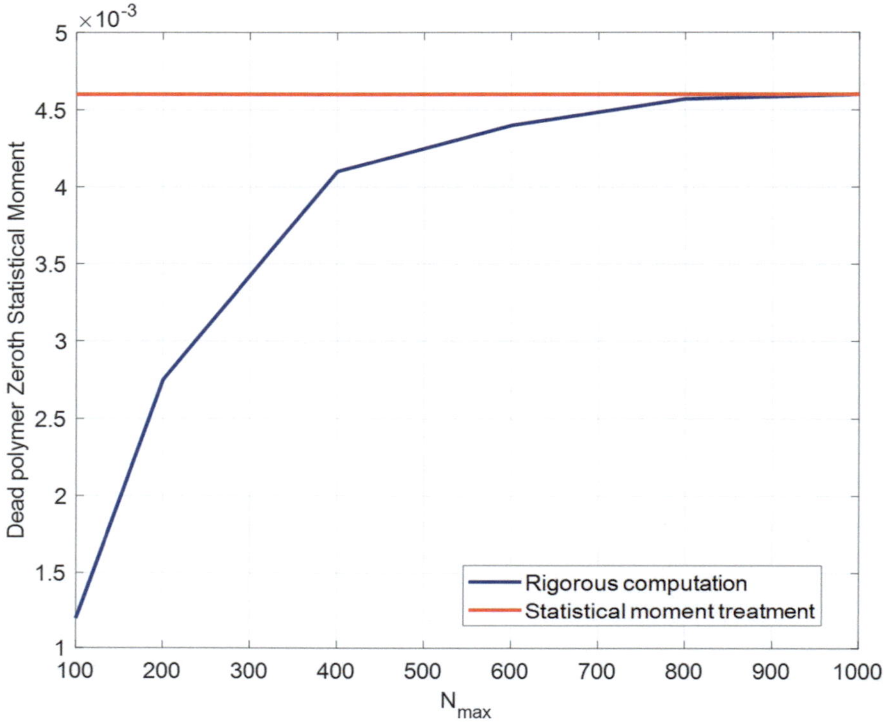

Fig. 8.25 Comparison of the concentration of the dead polymers calculated, respectively, with a rigorous solution of the ODE system and with the method of moments

$$\frac{dM_1^*}{dt}=k_iR^*M-k_pM_1^*M-k_{tc}M_1^*\sum_{i=1}^{\infty}M_i^*-k_{tc}MM_1^*\sum_{i=1}^{\infty}M_i^*+(k_{cts}S+k_{ctm}M)\sum_{i=1}^{\infty}M_i^* \tag{8.183}$$

Polymer's radical:

$$\frac{dM_n^*}{dt}=k_pM_{n-1}^*M-k_pM_n^*M-k_{tc}M_n^*\sum_{i=1}^{\infty}M_i^*-(k_{cts}S+k_{ctm}M)\sum_{i=1}^{\infty}M_i^* \tag{8.184}$$

(continued)

$$\text{Monomer}: \quad \frac{\mathrm{d}M}{\mathrm{d}t} = -k_{\mathrm{i}}R^{*}M - k_{\mathrm{p}}M\sum_{i=1}^{\infty}M_{i}^{*} - (k_{\mathrm{cts}}S + k_{\mathrm{ctm}}M)\sum_{i=1}^{\infty}M_{i}^{*} \tag{8.185}$$

$$\text{Dead polymer}: \quad \frac{\mathrm{d}P_{n}}{\mathrm{d}t} = \frac{1}{2}k_{\mathrm{tc}}\sum_{i=1}^{\infty}M_{i}^{*}M_{n-i}^{*} + (k_{\mathrm{cts}}S + k_{\mathrm{ctm}}M)\sum_{i=1}^{\infty}M_{i}^{*} \tag{8.186}$$

For styrene, termination by disproportionation is negligible.

$$\text{Solvent}: \quad \frac{\mathrm{d}S}{\mathrm{d}t} = -k_{\mathrm{cts}}S\sum_{i=1}^{\infty}M_{i}^{*} \tag{8.187}$$

The initial conditions are as follows: $I = I_0$, $M = M_0$, $S = S_0$, $R^* = 0$, M_1^* to $M_i^* = 0$, P_1 to $P_n = 0$

There are two problems in solving this system, one is related to the stiffness of some equations and the other is how to render reasonable the computation time. The first problem can be solved by adopting a suitable numerical integration solver and by assuming the QSSA (quasi-steady-state approximation) for the radicals, for example, it is possible to write

$$\frac{\mathrm{d}R^{*}}{\mathrm{d}t} = 2fk_{\mathrm{d}}I - k_{\mathrm{i}}R^{*}M = 0 \tag{8.188}$$

The second problem is to define when to stop the calculation being impossible to solve an infinite number of ODEs. It is necessary to estimate an opportune value of the maximum chain length. An arbitrary truncation of the calculation does not guarantee to obtain a chain length corresponding to a real value. The problem can be solved by calculating the zero moment of the dead polymers:

$$\mu_{0} = \sum_{n=1}^{\infty}P_{n} \tag{8.189}$$

which is the overall concentration of the dead polymer in the solution. The rigorous calculation of the ODE's system can be truncated when the calculated overall polymer concentration is equal to μ_0. For this purpose, the following ODE's system must be solved:

$$\frac{\mathrm{d}I}{\mathrm{d}t} = -k_{\mathrm{d}}I \tag{8.190}$$

$$\frac{dM}{dt} = -k_p M \lambda_0 - 2f k_d I - k_{ctm} M \lambda_0 \tag{8.191}$$

$$\frac{d\lambda_0}{dt} = 2f k_d I - k_{tc} \lambda_0^2 \tag{8.192}$$

$$\frac{d\mu_0}{dt} = -k_{ctm} M \lambda_0 + k_{cts} S \lambda_0 + \frac{k_{tc}}{2} \lambda_0^2 \tag{8.193}$$

$$\frac{dS}{dt} = -k_{cts} S \lambda_0 \tag{8.194}$$

This system of only five ODEs can be easily and promptly solved by assuming the following initial conditions: $I = I_0$, $M = M_0$, and $S = S_0$ and the reactor's conditions and kinetic parameters reported in Table 8.12. A value of $\mu_0 = 0.046$ (mol/L) is obtained as the solution that must be compared with the increasing values of the rigorous calculation as in the plot of Fig. 8.25. As it can be seen, truncation of the integration can be made after the addition of about 1000 monomer units.

It is then opportune to mention that as the polymerization proceeds, the viscosity of the system increases and the termination rate decreases being, affected by diffusion. This effect known as Norrish-Trommsdorff or gel effect determines an auto-acceleration of the polymerization reaction at high monomer conversions, because the propagation rate becomes predominant. The behavior is different for different monomers, being very important for the polymerization of methyl-methacrylate and less for styrene, The phenomenon is mitigated by the presence of a solvent. According to some different authors (Schmidt et al. 2003), the termination rate can be considered constant until a certain critical value and then progressively decreases. It is possible, according to Mahabadi (1985), to correlate the termination kinetic constant with the length of the interacting molecules:

$$k_{tc}(i,j) = \frac{k_{tc0}}{(ij)^{\alpha}} \tag{8.195}$$

k_{tc0} is the rate constant related to the termination of two monomeric radicals (Barner-Kowollik and Russel 2009), i and j are the chain lengths of two interacting molecules, and α is an adjustable parameter, normally is a value with an order of magnitude 0.1.

Other correlations have also been proposed in the literature such as:

Table 8.13 Parameters and data useful for the calculations

Parameter	Value	Units
Temperature	100	°C
Efficiency factor for the initiator	0.71	–
Final time	80	min
Initial monomer concentration	4.8824	mol/L
Initial initiator concentration	6.8794E–3	mol/L
Initial solvent concentration	3.5233	mol/L

$$k_{\mathrm{tc}}(i,j) = \frac{1}{2}\,k_{\mathrm{tc}0}\left(i^{-\alpha} + j^{-\alpha}\right) \tag{8.196}$$

$$k_{\mathrm{tc}}(i,j) = k_{\mathrm{tc}0}\left(\frac{2ij}{i+j}\right)^{-\alpha} \tag{8.197}$$

The α parameter has a strong influence on the MWD, and it must be expected a wider MWD and greater chain lengths if α increases. The introduction of a variable termination kinetic constant modifies the mathematical model to be solved and consistently increases the computation time, because, in correspondence of k_{tc}, all the binary combinations of the terminating radicals would be considered.

Plug Flow Reactor

The equation describing the profiles in a tubular reactor of all the reaction components of a radical polymerization is the same as seen for the batch reactor, provided that the following conditions are satisfied: (a) The flow rate through the reactor satisfies the conditions of plug flow. (b) Isothermal and steady-state conditions are valid. (c) The volumetric flow rate of the reacting mixture remains constant along the reactor. To obtain the plug flow reactor model, the equations stated above for modeling a batch reactor must be modified, changing concentrations and time that are, respectively, replaced by the corresponding molar flow rates, divided by the total volumetric flow rate, and plug flow reactor volume, divided by the total volumetric flow rate, for example:

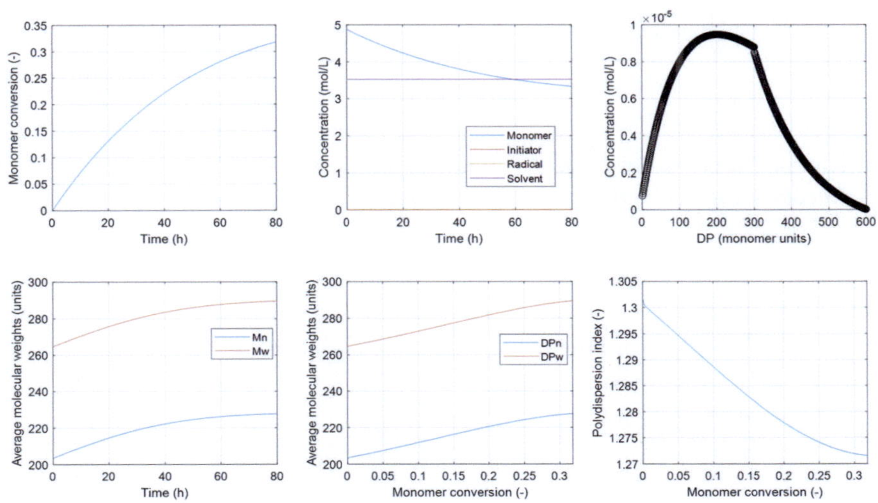

Fig. 8.26 Results obtained with the rigorous calculation approach

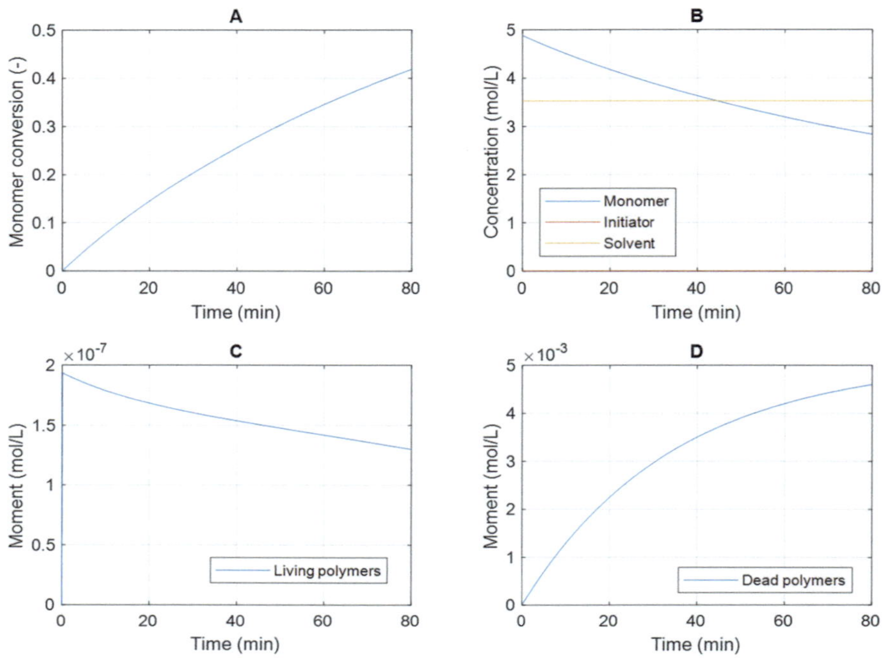

Fig. 8.27 Results obtained with the method of moments. The plots in the figure reported the following trends. Plot (**a**): monomer conversion as a function of time. Plot (**b**): concentration of monomer, initiator, and solvent as a function of time. Plot (**c**): zero-th moment of living polymers. Plot (**d**): zero-th moment of dead polymers

$$M = \frac{F_{\mathrm{M}}(\text{Monomer molar flow rate})}{Q\ (\text{Total volumetric flow rate})} \qquad t = \frac{V_{\mathrm{R}}(\text{Volume of the plug flow reactor})}{Q(\text{Total volumetric flow rate})} \tag{8.198}$$

Kinetic Model to Describe the Behavior of a Radical Polymerization Occurring in an Isothermal CSTR Reactor

$$\text{Initiator decomposition}: \quad \frac{Q}{V_{\mathrm{R}}}\left([I]_0 - [I]\right) = -k_{\mathrm{d}} I \tag{8.199}$$

$$\text{Radical chain initiation}: \quad \frac{Q}{V_{\mathrm{R}}}[R^*] = 2fk_{\mathrm{d}} I - k_{\mathrm{i}} R^* M \tag{8.200}$$

Monomer's radical:

(continued)

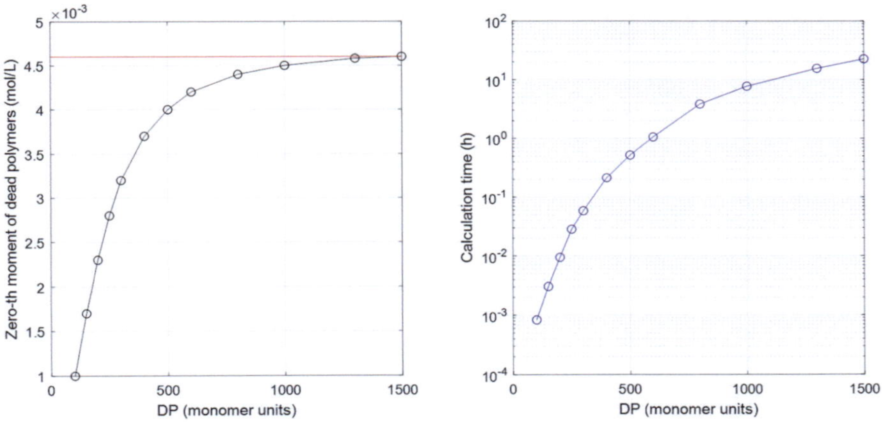

Fig. 8.28 A comparison between the two used methods and increase of the time calculation with DPmax when the rigorous method is adopted

$$\frac{Q}{V_{\mathrm{R}}}[M_1^*]=k_{\mathrm{i}}R^*M-k_{\mathrm{p}}M_1^*M-k_{\mathrm{tc}}M_1^*\sum_{i=1}^{\infty}M_i^*-k_{\mathrm{tc}}MM_1^*\sum_{i=1}^{\infty}M_i^*+(k_{\mathrm{cts}}S+k_{\mathrm{ctm}}M)\sum_{i=1}^{\infty}M_i^* \tag{8.201}$$

Polymer's radical:

$$\frac{Q}{V_{\mathrm{R}}}[M_n^*]=k_{\mathrm{p}}M_{n-1}^*M-k_{\mathrm{p}}M_n^*M-k_{\mathrm{tc}}M_n^*\sum_{i=1}^{\infty}M_i^*-(k_{\mathrm{cts}}S+k_{\mathrm{ctm}}M)\sum_{i=1}^{\infty}M_i^* \tag{8.202}$$

$$\text{Monomer}: \frac{Q}{V_{\mathrm{R}}}([M]_0-[M])=-k_{\mathrm{i}}R^*M-k_{\mathrm{p}}M\sum_{i=1}^{\infty}M_i^*-(k_{\mathrm{cts}}S+k_{\mathrm{ctm}}M)\sum_{i=1}^{\infty}M_i^* \tag{8.203}$$

$$\text{Dead polymer}: \quad \frac{Q}{V_{\mathrm{R}}}[P_n]=\frac{1}{2}k_{\mathrm{tc}}\sum_{i=1}^{\infty}M_i^*M_{n-i}^*+(k_{\mathrm{cts}}S+k_{\mathrm{ctm}}M)\sum_{i=1}^{\infty}M_i^* \tag{8.204}$$

$$\text{Solvent}: \frac{Q}{V_{\mathrm{R}}}([S]_0-[S])=-k_{\mathrm{cts}}S\sum_{i=1}^{\infty}M_i^* \tag{8.205}$$

Composition fed flow: $I=I_0$, $M=M_0$, $S=S_0$, $R^*=0$, $M_1{}^*$ to $M_i{}^*=0$, P_1 to $P_n=0$

Finally, the mass balances for a continuous CSTR isothermal reactor would be

Exercise 8.10. Radical Polymerization of Styrene in a Batch Reactor
Considering a simplified reaction scheme for radical polymerization of styrene as follow:

$$\begin{aligned}
&\mathrm{I} \overset{k_{\mathrm{d}}}{\rightarrow} 2\mathrm{R} \\
&\mathrm{R} + \mathrm{M} \overset{k_{\mathrm{i}}}{\rightarrow} \mathrm{P}_1 \\
&\mathrm{P}_i + \mathrm{M} \overset{k_{\mathrm{p}}}{\rightarrow} \mathrm{P}_{i+1} \\
&\mathrm{P}_i + \mathrm{S} \overset{k_{\mathrm{i}}}{\rightarrow} \mathrm{D}_i + \mathrm{S}^* \\
&\mathrm{S}^* + \mathrm{M} \overset{k_{\mathrm{cts}}}{\rightarrow} \mathrm{P}_1 + \mathrm{S} \\
&\mathrm{P}_i + \mathrm{M} \overset{k_{\mathrm{ctm}}}{\rightarrow} \mathrm{P}_1 + \mathrm{D}_i \\
&\mathrm{P}_i + \mathrm{P}_k \overset{k_{\mathrm{tc}}}{\rightarrow} \mathrm{P}_{i+k}
\end{aligned} \tag{8.206}$$

In this scheme, P are living polymers, while D are dead polymers. Solvent is indicated by S, monomer by M, and radicals by R. The task of this exercise is the simulation of such type of reactions conducted in a batch reactor with the conditions reported in Table 8.13.

The temperature dependence of the kinetic constants is reported in Table 8.12 of the present chapter. The material balance for individual species is represented by Eqs. (8.181–8.187), while the moments for both dead and living polymers are represented by Eqs. (8.190–8.194).

Develop MATLAB codes that compare the solution obtained by solving a rigorous ODE system related to each specie with that obtained by adopting the method of moments. In particular, vary the DPmax (maximum polymerization degree) in order to establish a value of this parameter for which the two approaches are substantially equivalent.

Solution

Rigorous Approach

The rigorous approach for the solution of the exercise consists in solving the material balance related to each of the chemical species involved. The system of ODE Eqs. (8.181–8.187) has a size that depends on the maximum polymerization degree that is assumed. If DPmax is the maximum allowable polymerization degree, the size of ODE system is 3*DPmax+3. It is then evident that for a polymerization degree in the range of 1000–5000, a system of 3000–15,000 ODE must be solved with the obvious implication in computing time issue. On the contrary, the method of moments (see the following section) involves only 5 ODE, and computing time is negligible. For the comparison (see the last section of the exercise), the calculation with rigorous approach must be repeated at different DPmax values, and the results must be compared with those obtained with the method of moments. For example, in correspondence of a DPmax = 300, the following plot can be produced (Fig. 8.26):

Method of Moments

By adopting the method of moments, only five differential equations need to be solved. The first three are the differential material balance on, respectively, initiator, solvent, and monomer. Coupled with these ODE, two additional ODEs must be solved that are related to the first moment of living polymers and of dead polymers. The system of 5 ODE, even if these have stiff character, is easily solved in MATLAB with a computing time of the order of few seconds or even less.

The developed code implementing this approach gives the following figure as output (Fig. 8.27).

The final value of the zero-th moment of dead polymers is, in this case, 0.0046 mol/L; this value will be used for the comparison with the rigorous approach (see the corresponding section of this exercise).

Comparison

When the code implementing the rigorous approach is executed with increasing values of DPmax, different values of the zero-th moment of dead polymers are obtained. The obtained values are approaching to that obtained in the method of moments when a sufficiently high DPmax is used, but this involved higher and higher computing times. A comparison is made in the first plot of Fig. 8.28.

The described results were obtained using a MATLAB program available as Electronic Supplementary Material.

References

Bamford, C.H., Tompa, H.: On the calculation of molecular weight distributions from kinetic schemes. J. Polym. Sci. **X**, 3, 345 (1953)

Bamford, C.H., Barb, W.G., Jenkins, A.D., Onyon, P.F.: Kinetics of Vinyl Polymerization by Radical Mechanisms. Butterworth, London (1958)

Barner-Kowollik, C.: Handbook of RAFT Polymerization. Wiley-VCH, Weinheim (2007)

Barner-Kowollik, C., Russel, G.T.: Chain-length dependent termination in radical polymerization: Subtle revolution in tackling a long-standing challenge. Prog. Polym. Sci. **34**, 1211–1259 (2009)

Barner-Kowollik, C., Vana, P., Davis, T.P.: The kinetics of free radical polymerization, Chapter 4. In: Matyjaszewski, K., Davis, T.P. (eds.) Handbook of Radical Polymerization. Wiley, New York (2002)

Chiefari, J., Chong, Y.K., Ercole, F., Krstina, J., Jefferi, J., Le, T.P.T., Mayadunne, R.T.A., Meijs, G.F., Moad, C.L., Moad, G., Rizzardo, E., Thang, S.H.: Living free-radical polymerization by reversible addition—Fragmentation chain transfer: The RAFT process. Macromolecules. **31**, 5559–5562 (1998)

Coates, G.W., Sawamoto, M.: Introduction and overview. In: Matyjaszewski, K., Moller, H. (eds.) Polymer Science, a Comprehensive Reference. Vol. 3: Chain Polymerization of Vynil Monomers. Elsevier, Amsterdam (2013)

Fischer, H.: The persistent radical effect: a principle for selective radical reactions and living radical polymerization. Chem. Rev. **101**, 3581–3610 (2001)

Flory, P.J.: Principles of Polymers Chemistry. Cornell University Press, Ithaca (1953)

Hawker, C.J.: Molecular-weight control by living free-radical polymerization process. J. Am. Chem. Soc. **116**, 11185 (1994)

Holland, C.D., Anthony, R.G.: Fundamentals of Chemical Reaction Engineering. Prentice Hall, Englewood Cliffs (1979)
Houwink, R.: Zusammenhang zwischen viscosimetriscH und osmotisch bestimmten Polymerisationsgraden bei Hochpolymeren. J. Prakt. Chem. **157**, 15 (1940)
Huang, J., Rempel, G.L.: Ziegler-Natta catalysts for olefin polymerization: mechanistic insights from metallocene systems. Prog. Polym. Sci. **20**, 459–526 (1995)
Hui, A.W., Hamielec, A.E.: Thermal polymerization of styrene at high conversion and temperatures. An experimental study. J. Appl. Polym. Sci. **16**(3), 749–769 (1972)
Kovtun, G.A., Aleksandrov, A.L., Golubev, V.A.: Interaction of peroxide radicals with esters of hydroxylamines. Russ. Chem. Bull. **23**, 2115–2121 (1974)
Krys, P., Matyjaszewski, K.: Kinetics of atom transfer radical polymerization. Eur. Polym. J. **89**, 482–523 (2017)
Lamontagne, H.R., Lessard, B.H.: Nitroxide-mediated polymerization: a versatile tool for the engineering of next generation materials. ACS Appl. Polym. Mater. **2**, 5327–5344 (2020)
Mahabadi, H.K.: Effect of chain length dependence of termination rate constants on the kinetics of free radical polymerization. Macromolecules. **18**(6), 1319–1324 (1985)
Mark, H.: Über die entstehung und eigenschaften hochpolymerer festkörper. In: Sänger, R. (ed.) Der feste Körper, pp. 65–104. Hirzel, Leipzigin (1938)
Mastan, E., Zhu, S.: Method of moments: A versatile tool for deterministic modelling of polymerization kinetics. Eur. Polym. J. **68**, 139–160 (2015)
Matyjaszewski, K.: Cationic Polymerization: Mechanisms, Synthesis and Applications. Marcell Dekker, New York (1996)
Matyjaszewski, K., Davis, T.P.: Handbook of Radical Polymerization. Wiley, New York (2002)
Matyjaszewski, K., Patten, T.E., Xia, J.: Controlled/"living" radical polymerization. Kinetics of the homogeneous atom transfer radical polymerization of styrene. J. Am. Chem. Soc. **119**, 674–680 (1997)
Moad, G., Solomon, D.H.: The Chemistry of Free Radical Polymerization. Pergamon, Oxford (1995)
Moad, G., Rizzardo, E., Thang, S.H.: Radical addition-fragmentation chemistry in polymer synthesis. Polymer. **49**, 1079–1131 (2008)
Natta, G.: Une Nouvelle Classe de Polymeres Alpha-Olefines Ayant Une Regularitè Exceptionelle. J. Polym. Sci. **16**, 82, 143–154 (1955)
Natta, G.: Stereospecific polymerizations. J. Polym. Sci. **XLVIII**, 219–239 (1960)
Natta, G., Pasquon, I., Centola, P.: Principi di Chimica Industriale, vol. 2. CLUP, Milano (1978)
Nicolas, J., Guillaneuf, Y., Lefay, C., Bertin, D., Gigmes, D., Charleux, B.: Nitroxide-mediated polymerization. Prog. Polym. Sci. **38**, 63–235 (2013)
Odian, G.: Principles of Polymerization. Wiley-Interscience, New York (2004)
Pintos, E., Sarmoria, C., Brandolin, A., Asteasuain, M.: Modeling RAFT polymerization processes using an efficient Monte Carlo algorithm in Julia. Ind. Eng. Chem. Res. **55**, 8534–8547 (2016)
Rizzardo, E., Solomon, D.H.: On the origins of nitroxide mediated polymerization (NMP) and reversible addition-fragmentation chain transfer (RAFT). Aust. J. Chem. **65**, 945–969 (2012)
Santacesaria, E., Tesser, R., Di Serio, M.: Polyethoxylation and polypropoxylation reactions: Kinetics, mass transfer and industrial reactor design. Chin. J. Chem. Eng. **26**, 1235–1251 (2018)
Schmidt, A.D., Russel, G.T., Heuts, J.P.A.: Termination in dilute solution free radical polymerization: a composite model. Macromol. Theory Simul. **12**, 299–314 (2003)
Solomon, D.H., Rizzardo, E., Cacioli, P.: Polymerization process and polymers produced thereby. US Patent 4,581,429 (1986)
Syahrudin, R.I., Herlambang, B.A.: Sixth-generation of Ziegler-Natta catalyst for homopolymer polypropylene polymerization: Effect to product properties and regulation. In: 4th International Symposium on Current Progress in Functional Materials (2019); IOP Conference Series: Materials Science and Engineering 902 (2020)
Szwarc, M.: Carbanions, Living Polymers and Electron Transfer Processes. Wiley-Interscience, New York (1968)

Tesser, R., Santacesaria, E.: Monte Carlo simulation of glycerol poly-alkoxylation: a novel application of a stochastic algorithm. Ind. Eng. Chem. Res. **60**, 16619–16634 (2021a)

Tesser, R., Santacesaria, E.: Monte Carlo approach to the simulation of ethylene oxide and propylene oxide polymerization and copolymerization. Ind. Eng. Chem. Res. **60**, 10638–10653 (2021b)

UNEP (United Nations Environment Programme): Pollution to solution: A global assessment of marine litter, illustrated by GRID-Arendal (2021)

Vittoria, A., Meppelder, A., Friederichs, N., Busico, V., Cipullo, R.: Demystifying Ziegler-Natta catalysts: the origin of stereoselectivity. ACS Catal. **7**, 4509–4518 (2017)

Wang, J., Matyjaszewski, K.: Controlled/"Living" radical polymerization. Atom transfer radical polymerization in the presence of transition-metal complexes. J. Am. Chem. Soc. **117**, 5614–5615 (1995a)

Wang, J., Matyjaszewski, K.: "Living"/Controlled radical polymerization. transition-metal-catalyzed atom transfer radical polymerization in the presence of a conventional radical initiator. Macromolecules. **28**, 7572–7573 (1995b)

Wikipedia: Polymer. https://en.wikipedia.org/wiki/Polymer (2024)

Zapata-Gonzalez, I., Saldivar-Guerra, E.: The method of moments used in polymerization reaction engineering for 70 years: An overview, tutorial, and minilibrary. Can. J. Chem. Eng. **101**, 5324–5356 (2023)

Zapata-Gonzalez, I., Saldivar-Guerra, E., Ortiz-Cisneros, J.: Full molecular weight distribution in RAFT polymerization. New mechanistic insight by direct integration of the equations. Macromol. Theory Simul. **20**, 370–388 (2011)

Ziegler, K., Holzkamp, E., Breil, H., Martin, H.: Polymerisation Von Athylen Und Anderen Olefinen. Angew. Chem. Int. Ed. **16**, 67, 426 (1955)

Chapter 9
Bioreactors

Abstract A list of the microorganisms that can be cultivated for obtaining chemical products or biomass is reported in the introduction of this chapter. Fermentation processes and microorganism kinetic growth are then described. The historical evolution of the kinetic models, for the interpretation of the fermentation processes, is successively reported, considering both unstructured and structured models with their respective variants of segregated and unsegregated models. Starting from the unstructured Monod kinetic model and related modifications the compartmental structured models are then considered, in particular metabolic and cybernetic model. These models have been applied to the production of baker yeast from *Saccharomyces cerevisiae*, as a useful example of microorganism kinetic growth. It has also been considered the production of bioethanol, always employing the same microorganism but in anaerobic conditions. At last, a brief review of the nonmechanistic models such as neural network mathematical models and fuzzy models is reported at the end of the chapter. Different examples and solved exercises, related to both unstructured and structured models, described in the chapter are reported along the text.

Keywords Bioreactors · Growth kinetics of microorganism · Microbial kinetics · Unstructured and structured kinetic model · Baker yeast production · *Saccharomyces cerevisiae* · Fuzzy models

9.1 Introduction

Biotechnology has the scope of using different types of microorganisms to manufacture useful products. Biotechnology requires the involvement of various scientific disciplines like microbiology, biochemistry, and engineering. The products of biotechnology are obtained from two types of processes, that is, products derived from

Supplementary Information The online version contains supplementary material available at https://doi.org/10.1007/978-3-031-82502-6_9.

© The Author(s), under exclusive license to Springer Nature Switzerland AG 2025

R. Tesser, E. Santacesaria, *The Chemical Reactor from Laboratory to Industrial Plant*, https://doi.org/10.1007/978-3-031-82502-6_9

living processes and processes using biocatalysts (enzymes) obtained from cellular materials. The processes, based on the use of enzymes, have already been described in a previous chapter and the reactors used in this case are like those used for the conventional catalytic system. Biological reactions can be induced by different microorganisms such as: viruses, bacteria, fungi, animal cells, and plant cells (Emery 1999).

- **Viruses** are submicroscopic organisms with a size inferior to 0.1 μm. Viruses can exist and reproduce only by infecting other living organisms acting as hosts. Their use in biotechnology is limited to the production of vaccines and some insecticides. Viruses are a problem for bacterial cultures being often responsible for undesired infections.
- **Bacteria** are small microorganisms ($\simeq$1 μm) that can be distinguished by their need for oxygen. There are **Aerobes**, **Anaerobes**, and **Facultative** bacteria. Aerobes need oxygen for their life and reproduction, anaerobes need not, facultative can live both in the presence and absence of oxygen. Clearly, the metabolisms of these microorganisms are different, and different is their behavior in a bioreactor changing both their growth conditions and replication rates. However, as the bacteria growth rate can be very rapid, doubling the population in less than 20 min, and reaching very high concentration (more than 100 kg/m^3), the design and control of the bioreactor is a challenging task, also considering the necessity to feed nutrients and eventually oxygen enough to support their exponential growth. Moreover, the bacterial growth occurs in a triphasic system and mass transfer limitations at the interface could be operative.
- **Fungi** can be classified in two different categories that are **yeasts** and **mycelial**. Yeasts are microorganisms of size larger than bacteria (2–12 μm), that grow more slowly (duplication of population occurs in 1–2 h) and, with respect to oxygen, are often facultative. A high concentration of the microorganism can be reached (100 kg/m^3).

  Mycelial fungi are filamentous and grow forming a branched network of microtubes (**hyphae**) called **mycelium**. The growth rate is much slower with a slow duplication requiring more than 4 h. Concentration must be kept low (<15 kg/m^3) to avoid an excessive viscosity of the culture.
- **Animal cells** are extracted from animal tissue and slowly replicated with the main scope to produce a host for the viral infection so producing vaccines. This topic, although very interesting, is outside the scope of this text.
- **Plant cells**. The culture of plant cells, in particular microalgae, has problems similar to that of animal cells, is slow growing, and is difficult to avoid contamination. The most relevant difference is that plant cells can use light energy. This topic also must be considered outside the scope of this book.

In conclusion, in this chapter, the fermentation processes and related bioreactors will be described in detail, and the microorganisms considered will be bacteria and yeasts. However, it is opportune to mention that bioengineering like chemical engineering includes kinetics, transport phenomena, reactor design, and unit operations (Moulijn et al. 2001).

9.2 Fermentation Processes

There are two distinct phases in the fermentation processes: the phase of production occurring in a bioreactor (**fermentation procedure**) and the phase of product recovery (**downstream processing**). In the first phase of production, the microorganism is the center of attention. The choice of the more suitable strain is fundamental for an optimal productivity. The best available strains are often the result of genetic engineering modifications made with the objective of improving the quality of the microorganism for a specific process. Then, it is opportune the provision of adequate supply of nutrients and oxygen that are necessary to sustain the growth. Obviously, all the nutrients must be maintained and fed in sterile conditions to avoid any contamination. The nutrients must contain a "carbon" source (the **Substrate**) and carbohydrates are traditionally used at the scope. Glucose, sucrose, molasses, malt extract, starch, and dextrin are commonly employed as both carbon and energy sources. Then, it is also necessary a source of nitrogen to build the protein material of the cells. Many large-scale processes utilize at this scope: ammonium salts, urea, gaseous ammonia, and sources of amino acids, such as corn steep liquor, yeast extracts, and soy meals. At last, some minor components are necessary, too, such as vitamins, phosphates, sulphates, chlorides minerals, and traces of some key elements, like K, Na, Mg, Ca, Fe, Co, and Zn. Such components are, normally, contained in some of the mentioned substrates, otherwise must be added in a sufficient amount. Fermentation starts by inoculating a certain amount of a microorganism in a solution containing an opportune concentration of all the mentioned nutrients dissolved in water. The suspension is gently stirred and when necessary, an adequate amount of air is bubbled into the reactor. As the cell's population grows, it necessitates an increasing amount of nutrients, oxygen, and reaction volume. Therefore, a common practice is to use the first reactor as inoculum for a bigger second one and so on until reaching a satisfactory production size. The plant is constituted by a battery of fed batch reactors of increasing capacity as shown in the scheme of Fig. 9.1. It is clear the difference with respect to the traditional chemical reactors. In this case, the production starts in a very small reactor (first inoculum) and proceeds with a series of scaling up adapting the feed of nutrients and oxygen to the needs of the microorganism increasing population. In alternative, a continuous bioreactor can be adopted. In any case, it is important not to limit the population growth due to the scarcity of one or more nutrients.

The products of fermentation can be some specific metabolites, such as alcohol, organic acids, polysaccharides, antibiotics etc., or the production of cell biomass, and the fermentation procedure must be tuned to optimize the desired production. At the end of fermentation starts the product recovery phase. The purification steps depend on the type of product to be recovered, its overall amount, its concentration and the required degree of purification, the presence of other side products, and the stability of the employed microorganism. Many techniques are available for the separation and purification of the product, such as the following: centrifugation, filtration, extraction, distillation, crystallization, dialysis, drying, preparative

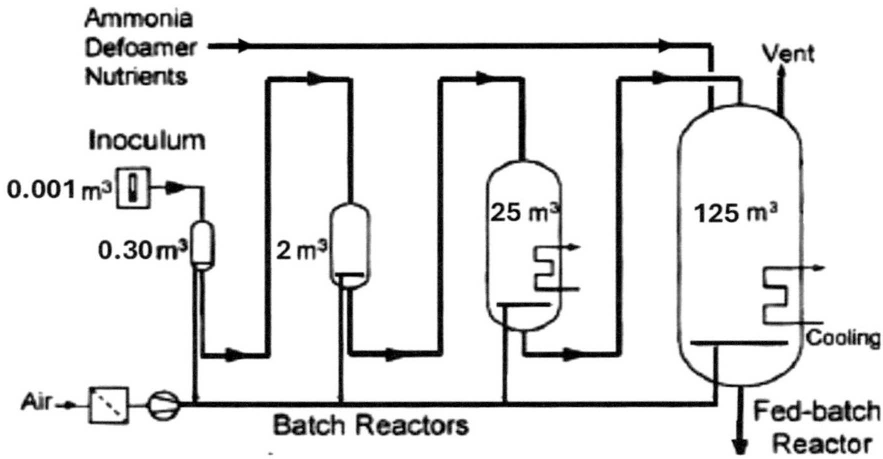

Fig. 9.1 Simplified scheme of a battery of fed-batch reactors

chromatography, etc. Any type of production requires a defined protocol of separation and purification integrated with the specific bioreactor production. Therefore, it is not possible a unified treatment of the bioreactors downstream processing.

9.2.1 Growth Kinetics of Microorganisms

The knowledge of microorganisms' growth kinetics is fundamental for optimizing any process of fermentation. Fermentation can be conducted in batch, semi-batch, or continuous bioreactor. Batch reactors are normally employed for laboratory investigation purposes with the aim of collecting kinetic data, elaborating a kinetic model, and evaluating the most reliable kinetic parameters. On the contrary, semi-batch and continuous reactors are preferred for the industrial production. In fed-batch bioreactors, nutrients and oxygen are progressively added to the growing biomass with a rate that is regulated in a way to satisfy the needs of the microorganisms. At last, continuous bioreactors can be of CSTR or PFR type, provided that the residence time is sufficient to reach a stationary phase of cells growth.

9.2.1.1 Batch Bioreactors

An example of a batch bioreactor, suitable also for fed-batch operation, is reported in Fig. 9.2.

A sterilized solution, containing the nutrients, is loaded into the reactor. Some predetermined physiological conditions, such as temperature, pH, and stirring rate, are established. An antifoam agent could be useful sometimes. At time $t = 0$, a

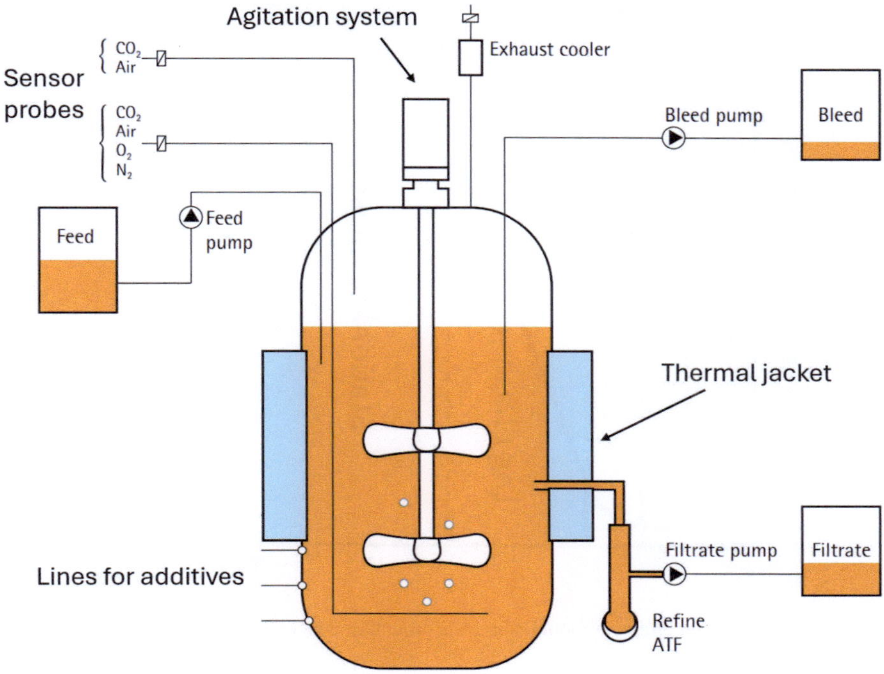

Fig. 9.2 An example of a batch bioreactor suitable also for fed-batch operation

known amount of microorganism is inoculated into the solution and the fermentation starts. During the fermentation, nothing is added to the reaction mixture with the exception of oxygen, if the microorganism is aerobic, and an acid or a base to maintain the optimal pH. The cells' population increases and the nutrients' concentration decreases. The population growth occurs in four different phases represented in Fig. 9.3 (Crueger and Crueger 1989).

This behavior can be explained as it follows:

1. **The lag phase**. The cells contained in the inoculum need some time to adapt themselves to the new environment and related conditions. The length of the lag phase depends on the characteristics of the strain, its physiological conditions, and its concentration. It is not known whether the length of the lag phase can influence the successive product formation, and this aspect must be ascertained on an empirical basis with laboratory experiences.
2. **The log phase**. After the adaption phase, the cells start to replicate growing with an exponential trend. In this phase, it is possible to evaluate quantitatively what is the growth rate expressed as doubling cell number per unit time. In a semilogarithmic plot, the logarithm of the cell number as a function of time is a straight line. In this phase, the growth is accompanied by both a consistent increase of

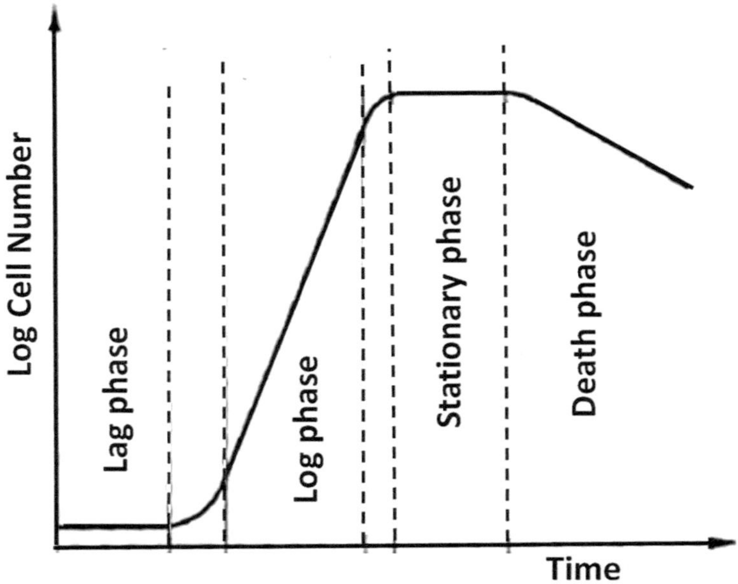

Fig. 9.3 The different phases of population cells growth in a batch reactor

nutrients consumption and a release of metabolic products. In the presence of an excess of substrate concentration, the growth rate prosecutes independently along the time.

3. **The stationary phase.** The exponential growth of the microorganisms decreases the concentration of the substrate and determines the accumulation of toxic agents. Moreover, some cells after a certain number of duplications dies. The probability of dead cells accumulation gradually increases. Dead cells are subjected to lysis releasing substances that constitute a new source of nutrients for the living cells. In the stationary phase, we can imagine that the number of newborn cells is approximately equal to the dead ones and the overall growth appears to be stopped. The length of this phase depends on both the type of microorganism and the biological adopted conditions.
4. **The death phase.** As the energy reserves available to the cells are exhausted the death of the cells predominates and living cells decrease exponentially. In industrial plants, fermentation is stopped before starting the death phase.

From this brief description of the behavior of the microorganisms in a batch bioreactor, it arises that a reliable mathematical model, describing all the four mentioned phases and the consumption along the time of the nutrients, would be very useful for modeling the fermentation performed in bioreactors of whatever size, from laboratory to industrial size and of any type, that is, batch, fed-batch, and continuous. The elaboration of such a model is a very difficult task because the behavior of a living microorganism depends on both the internal metabolism and the interaction with the external environment, this last changing along the time. However, the interpretation with an adequate model of runs performed in small laboratory reactors, describing the overall growth history and the consumption of the nutrients as a function of time, allows the estimation of the kinetic parameters of the model.

9.2.1.2 Kinetic Models for the Interpretation of the Fermentation Processes

The kinetics of cells growth can be affected by many different factors as it can be appreciated in Fig. 9.4 (Bailey and Ollis 1986).

Clearly, it is not possible the formulation of a kinetic model including all the factors reported in Fig. 9.4. It is necessary to introduce some approximations. It is possible, for example, to use a substrate, in which all the components are in excess with the exclusion of one which concentration becomes limiting. In this case, the growth rate is affected only by the limiting nutrient and, in that model, we can consider just the effect of the limiting component. Then, some other parameters such as pH, temperature, and oxygen concentration can be kept constant.

A classification of the kinetic models is based on their description of multi-component systems (Structured) or single component (Unstructured) and model considering cells as heterogeneous, discrete units (Segregated) or considering the average properties of the cells (Unsegregated) (Bailey and Ollis 1986; Buchanan and Whiting 1993; Dunn et al. 2003). In Fig. 9.5, a scheme of these different possibilities is reported.

A structured model attempts to describe intracellular processes offering a more realistic representation of a cell. Structured models are single-cell models and often are not applicable to the description of the overall process. For this reason, the majority of the models proposed in the literature can be classified as unstructured. Unstructured models describe the cell growth and the properties of the culture neglecting the intracellular processes. Therefore, the most realistic but complicated models are structured and segregated.

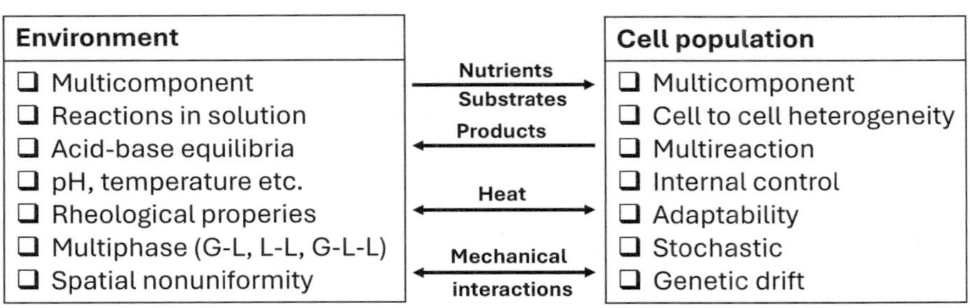

Fig. 9.4 Factors affecting the cell growth kinetics

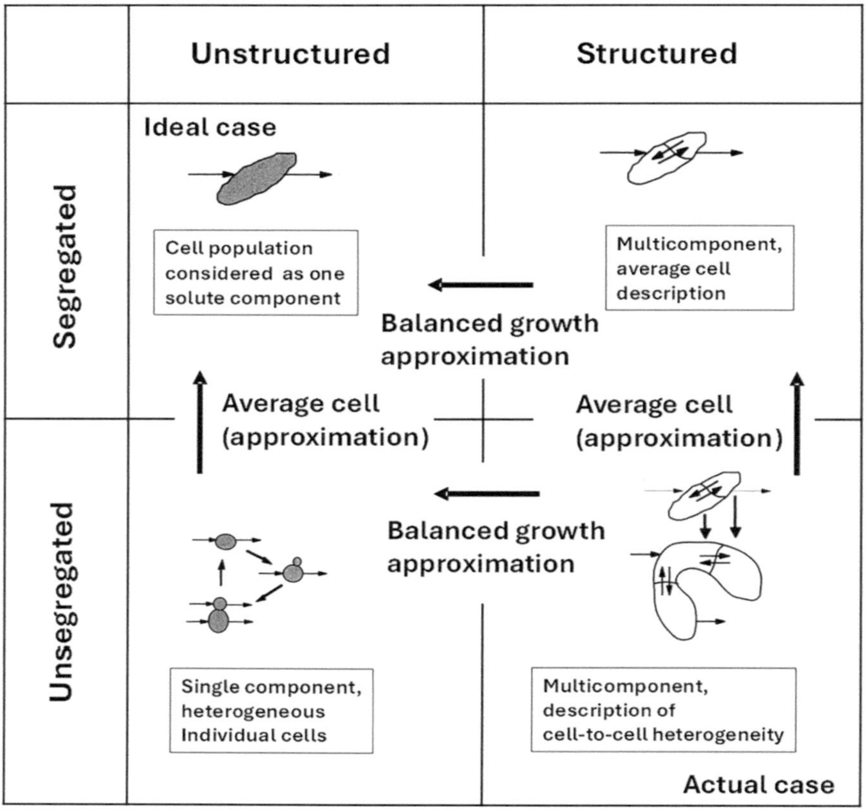

Fig. 9.5 Classification of the cell growth kinetic models

The Unstructured Models

Simple Microbial Kinetics: Primary Models

The primary models describe the kinetics of the growth process using as few parameters as possible.

- **Malthusian Model**

Malthusian model is the simplest cell growth model (Malthus 1830) known also as the exponential law model, because it has been experimentally observed that microorganisms composing biomass increase exponentially along the time. This occurs because all the cells have the same probability of multiplying. Therefore, the rate of biomass production is proportional to the biomass itself, that is:

$$r_X = \mu X \tag{9.1}$$

being r_X the rate growth of the cells (kg cells/m^3 h), X is the cell concentration (kg cells/m^3), and μ is the kinetic growth constant (1/h), also called specific growth rate. In batch conditions, it is possible to write:

$$\frac{dX}{dt} = \mu X \tag{9.2}$$

This equation can easily be analytically solved as:

$$X(t) = X_o \exp(\mu t) \tag{9.3}$$

where X_o is the cell concentration at time $t = 0$. The Malthusian model neglects that cells need resources to grow and that often these resources are limited.

- **Logistic Model**

The logistic model was firstly proposed by Pierre Francois Verhulst (1838) and is based on the following equation rate:

$$\frac{dX}{dt} = \mu X\left(1 - \frac{X}{K}\right) \tag{9.4}$$

As seen, a new parameter K (kg cells/m^3) has been introduced. K is the maximum concentration of cells supported by the environment. Unfortunately, K can be measured just by making an experimental run and observing when the growing stops. The analytical integration is:

$$X(t) = \frac{K}{1 + \left(\frac{1}{X_o} - \frac{1}{K}\right) K \exp(-\mu t)} \tag{9.5}$$

The curve of growth, in this case, is a sigmoid instead of the exponential curve of the Malthusian model. An empirical modification of the logistic model, largely employed in the literature (see McKellar and Lu 2004), is based on the following relation:

$$\log X(t) = A + \frac{C}{1 + \exp[-B(t - M)]} \tag{9.6}$$

A, B, and C are empirical parameters.

Exercise 9.1
In this exercise, a simulation of the cell growth over time according to the logistic model is considered. The adopted kinetic parameters are $X_o = 3 \times 10^4$ (cells/cm^3),

$\mu = 0.08$ (1/h) and $K = 2 \times 10^6$ (cells/cm^3). Then, the effect of μ on the growth rate is considered by changing this value from 0.06 to 0.15 (1/h).

Solution

Part 1

By integrating Eq. (9.4) with the mentioned parameters, the plot in Fig. 9.6 is obtained:

Part 2

The integration is then repeated by changing the value of μ obtaining different curves as reported in the plot of Fig. 9.7.

The described results were obtained by using a MATLAB program that is available to the readers as Electronic Supplementary Material of this book.

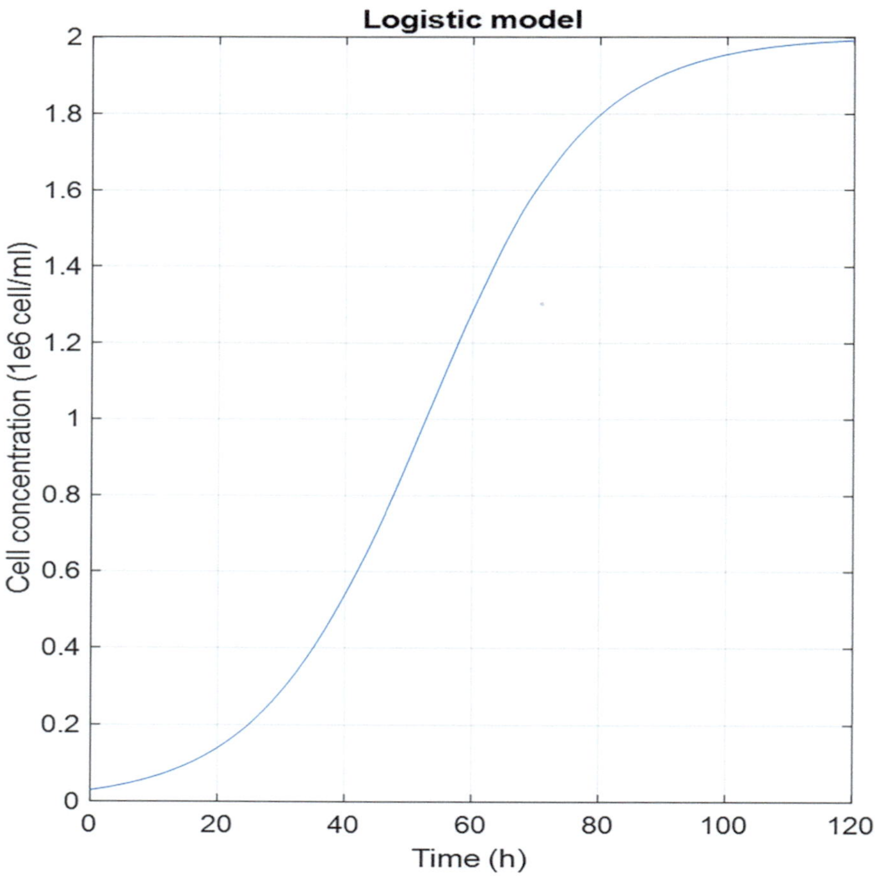

Fig. 9.6 Evolution of cell growth over time according to the logistic model

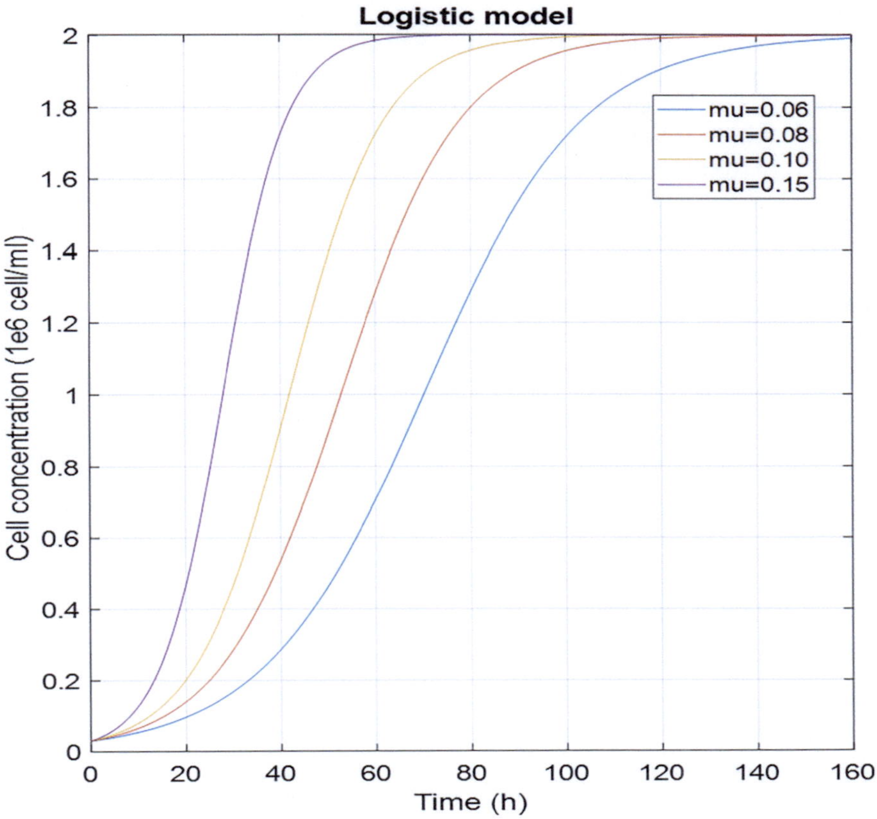

Fig. 9.7 Effect of changing the μ parameter on the cells growth rate

- **Gompertz Model**

The Gompertz (1825) model is similar in trend to the logistic model. The difference is that in the logistic model, the asymptotes of the curve are symmetric, while in the Gompertz model, the final asymptote is approached more gradually. The expression of the Gompertz model is:

$$X(t) = a\ \exp[-b\ \exp(-ct)] \tag{9.7}$$

Parameter "*a*" regulates the upper asymptotic value; parameter "*b*" defines the *X* displacement, and parameter "*c*" sets the growth rate. The initial cell concentration is $X_o = a\ \exp\ (-b)$. Gompertz model has been modified in an empirical form by assuming a relationship of the type (see Zwietering et al. 1992):

$$\log X(t) = A + C \exp(-\ \exp[-B(t - M)]) \tag{9.8}$$

being *A*, *B*, *C*, and *M* empirical parameters of the model.

Exercise 9.2

In this exercise, a simulation of the cell growth over time according to the Gompertz model is considered. The adopted kinetic parameters are as follows: $a = 2 \times 10^6$ (cells/cm^3), $b = 10$, and $c = 0.05$ (1/h), while the same parameters used for the logistic model have been taken, that is, $X_o = 3 \times 10^4$ (cells/cm^3), $\mu = 0.08$ (1/h), and $K = 2 \times 10^6$ (cells/cm^3).

Solution

Part 1

A simulation of the cells growth along the time using the Gompertz method has been made using the previously reported parameters. The obtained plot is reported in Fig. 9.8.

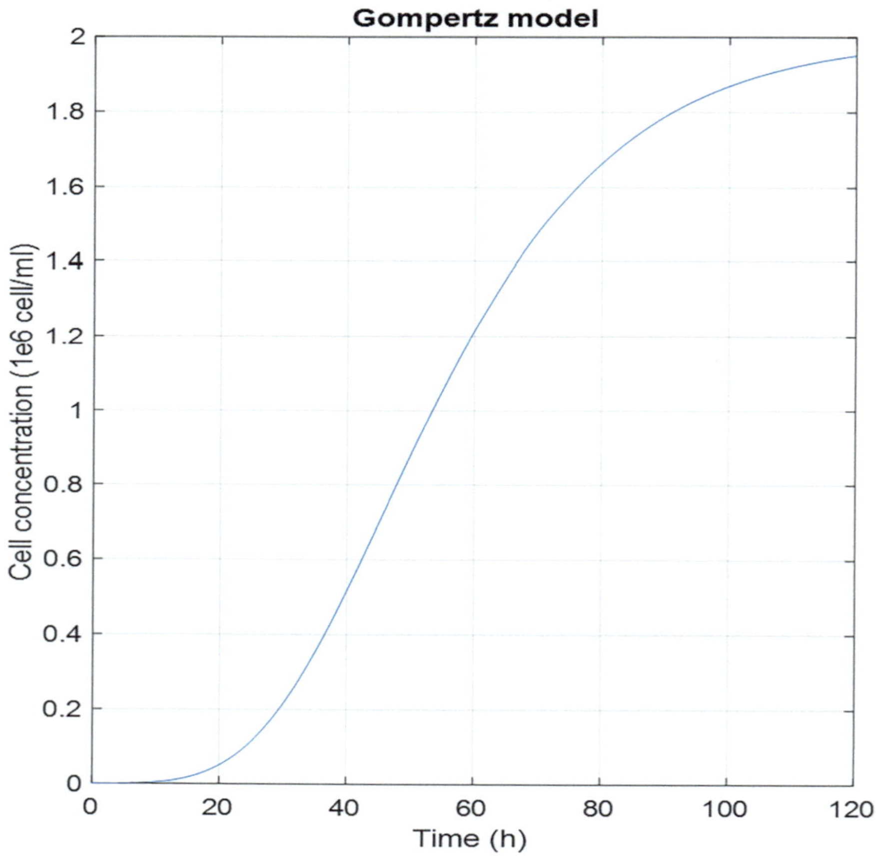

Fig. 9.8 Example of cell growth simulation interpreted with the Gompertz model

Part 2

A comparison is made between the logistic and the Gompertz model by using the same X_o, μ, and K parameters. The obtained results are reported in Fig. 9.9.

The described results were obtained by using a MATLAB program that is available to the readers as Electronic Supplementary Material of this book.

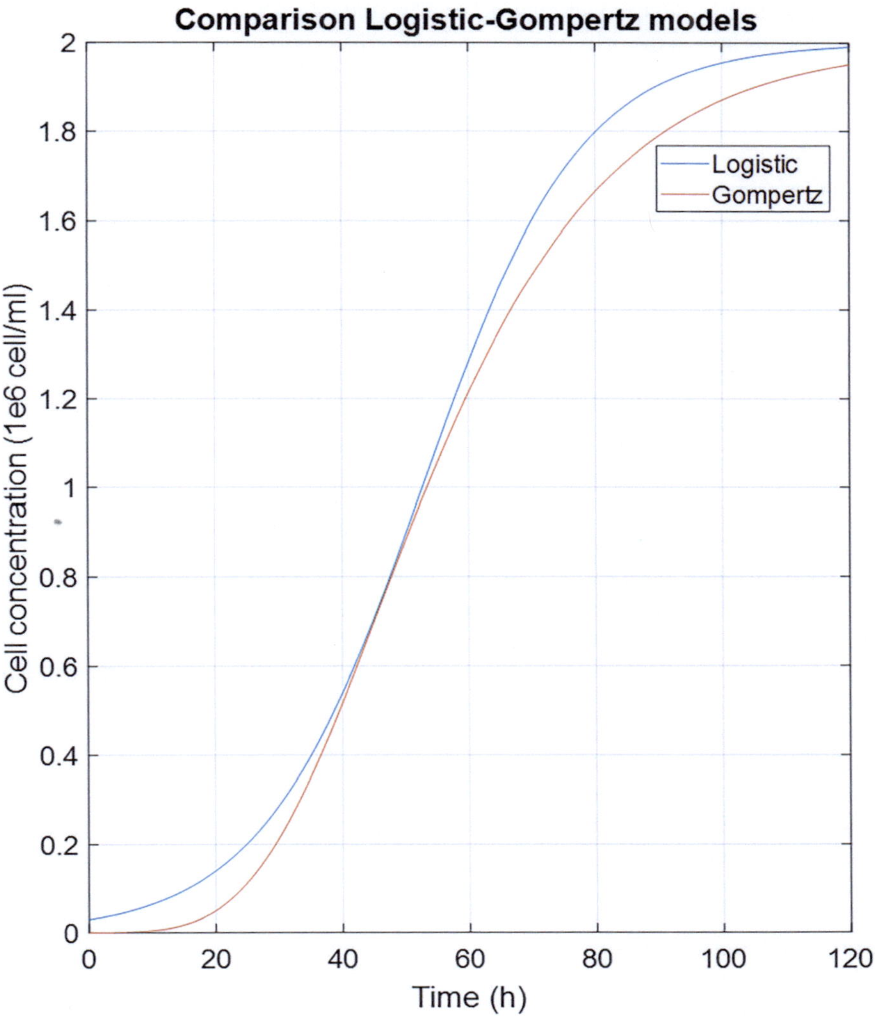

Fig. 9.9 A comparison between the logistic and the Gompertz models

Simple Microbial Kinetics: Secondary Models

Secondary models are based on the knowledge of the effect of the environmental conditions on the behavior of the cells. In these models, relationships between cell growth, substrate concentration, and temperature are introduced.

- **The Monod Kinetic Model**

The simplest kinetic model, describing the microorganism growth rate, is the semiempirical Monod model (Monod 1942). According to this model, the growth rate depends just on the cells concentration M (g/L) or N (number of cells/L). That is:

$$\frac{\mathrm{d}M}{\mathrm{d}t} = \mu M \quad \frac{\mathrm{d}N}{\mathrm{d}t} = \mu N \tag{9.9}$$

being μ the specific growth rate (h^{-1}) of the cells. μ can be determined by the Monod equation:

$$\mu = \mu_{\max} \frac{S}{S + K_S} \tag{9.10}$$

Therefore, it holds:

$$\frac{\mathrm{d}M}{\mathrm{d}t} = \mu_{\max} \frac{S}{S + K_S} M(t) \tag{9.11}$$

$\mu_{\max}$ is the maximum growth rate, S is the concentration of the **limiting nutrient** and K_S, called half saturation constant or affinity constant, is a constant like the Michaelis-Menten constant used in the kinetics of reactions catalyzed by enzymes. If S is very high, the growth rate becomes independent of the substrate, while, on the contrary, if S is very low, the rate is limited by the nutrient concentration. K_S is an empirical constant, dependent on many different factors that are characteristic of any couple cell substrate. If $K_S = S(t)$, it holds $\mu = 0.5\ \mu_{\max}$. $\mu_{\max}$ is a characteristic of a specific microorganism defining its ability in reproducing itself in the optimal nutritional conditions without the presence of limiting nutrients. The log phase is crucial to produce both the cellular biomass and the desired metabolites and the **maximum specific growth rate** $\mu_{\max}$ is very important for the industrial production. In Table 9.1, some data of $\mu_{\max}$ related to different microorganisms (Anderson et al. 1975) are reported.

The link between growth rate and substrate utilization was proposed by Monod who linearly related the specific biomass growth rate to the specific rate of substrate consumption through the yield coefficient Y_{MS} (conversion efficiency of substrate into cell material). As seen, according to the equation of Monod, μ shows a linear trend with S for a low value of this variable, while when $S >> K_S$, it becomes independent of S and $\mu \simeq \mu_{\max}$.

A complete Monod model is a system of two differential equations with two parameters that must be evaluated:

Table 9.1 Maximum specific growth rate μ_{max} for different fungi using glucose as substrate

Microorganism	T (°C)	μ_{max} (h^{-1})	Doubling time (h)
Aspergillus niger	30	0.20	3.46
Aspergillus nidulans	30	0.215	3.23
Penicillium chrysogenum	25	0.123	5.65
Mucor hiemalis	25	0.17	4.10
Fusarium avenaceum	25	0.18	3.80
Fusarium graminearum	30	0.28	2.48
Verticillium agaricinum	25	0.24	2.90
Geotrichum candidum	25	0.41	1.70
Neurospora sitophila	30	0.40	1.73

$$\frac{dX}{dt} = \mu_{max} \frac{S}{S + K_S} M(t) \tag{9.12}$$

$$\frac{dS}{dt} = -\frac{1}{Y_{MS}} \mu_{max} \frac{S}{S + K_S} M(t) \quad \text{where} \quad Y_{MS} = \frac{dX}{dS} = \text{yield factor} \tag{9.13}$$

The system can be solved if the initial conditions X_o and S_o are known. The advantage of the nonstructured models, like the Monod model, is their simplicity, but such models are not able to describe the lag phase and give no information about the variables affecting the cells growth and the metabolism of the microorganism.

Exercise 9.3

The Monod model is employed to describe the cell growth and the substrate consumption along the time, by solving the system of differential equations (9.12 and 9.13). The adopted parameters were as follows: $X_o = 0.3 \times 10^5$ (cells/cm^3), $S_o = 0.2$ (g/L), $\mu_{max} = 0.15$ (1/h), $K_S = 0.1$ (g/L), and $Y_{xs} = 0.5 \times 10^8$.

Solution

The obtained results are reported in Fig. 9.10.

The described results were obtained by using a MATLAB program that is available to the readers as Electronic Supplementary Material of this book.

- **Modifications of the Monod Model**

The Monod model has been opportunely modified in many cases to allow the description of more complex situations. In the presence of two different substrates, for example, the Monod equation can be written as:

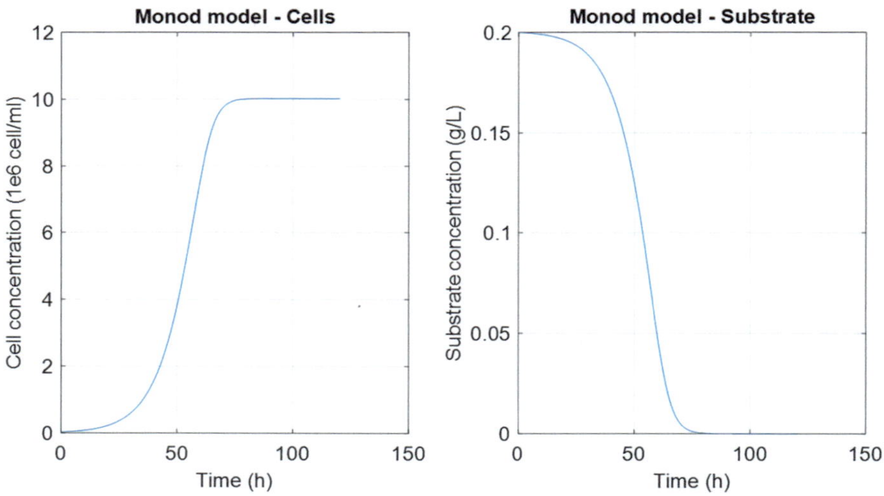

Fig. 9.10 The performance of the model of Monod in describing the cell growth and substrate consumption

$$\mu = \mu_{\max}\left(\frac{S_1}{S_1 + K_{S_1}}\right)\left(\frac{S_2}{S_2 + K_{S_2}}\right) \tag{9.14}$$

In this case, both the substrates may be limiting when the others are in excess.

Exercise 9.4

An example of calculation using the modified Monod model (Eq. 9.14) to describe the behavior of the microorganisms in the presence of two different substrates. It will be calculated the evolution of both the cells concentration and that of the two substrates along the time. The used parameters are the following:

$X_0 = 0.3 \times 10^5$ (cell/ml), $S_1° = 0.5$ (g/L), $S_2° = 0.05$ (g/L), $\mu_{\max} = 0.1$ (1/h), $k_{s1} = 0.035$ (g/L), $k_{s2} = 0.01$ (g/L), Y_{XS} 0.3×10^8, $c_1 = 5$, $c_2 = 1$.

Solution

The obtained results are reported in the plots in Fig. 9.11.

The described results were obtained by using a MATLAB program that is available to the readers as Electronic Supplementary Material of this book.

Another case of modification of the Monod model is the one in which the substrates are consumed in parallel but with different rates. In this case, it holds:

$$\mu = \mu_{\max}\left(\frac{1}{k_1 + k_2}\right)\left(k_1\frac{S_1}{S_1 + K_{S_1}}\right)\left(k_2\frac{S_2}{S_2 + K_{S_2}}\right) \tag{9.15}$$

The specific growth rate may be negatively affected also by the biomass accumulation, the substrate accumulation, or the fermentation product. The anaerobic

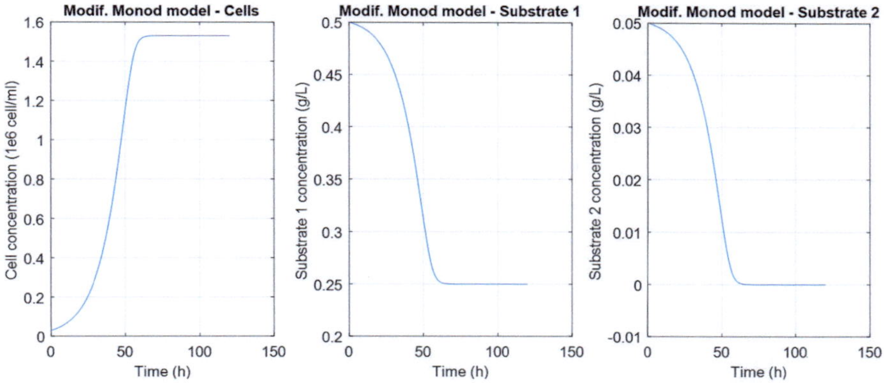

Fig. 9.11 Example of simulation of the cells' growth in the presence of two different substrates performed with a modified Monod model

fermentation of glucose, promoted by *Saccharomyces cerevisiae*, for example, is inhibited by the product ethanol. The inhibiting effect of the product can be described, according to Aiba et al. (1968), by the term $K_P/(K_P + P)$ that must be coupled with the nutrients limiting terms of the previously seen expressions.

$$\mu = \mu_{max} \frac{S}{S + K_S} \frac{K_P}{K_P + P} \tag{9.16}$$

K_P is an empirical constant and P is the product concentration.

Ruggeri et al. (1988), always studying the *Saccharomyces* behavior in glucose fermentation to ethanol, introduced two inhibiting terms, one related to the substrate ($\exp(-S/K_I)$) and another one related to the product ($1 - P/K_P$), that is:

$$\mu = \mu_{max} \frac{S}{S + K_S} \exp\left(-\frac{S}{K_I}\right)\left(1 - \frac{E}{K_E}\right) \tag{9.17}$$

Exercise 9.5

Ruggeri et al. (1988), for interpreting the low-concentration glucose fermentation to ethanol, suggested the solution of the following system of differential equations:

$$\frac{dX}{dt} = \frac{\mu_{max} SX}{K_S + S}\left(-\frac{S}{K_I}\right)\left(1 - \frac{E}{K_E}\right) \tag{9.18}$$

$$\frac{dS}{dt} = -A\frac{dX}{dt} - BX \tag{9.19}$$

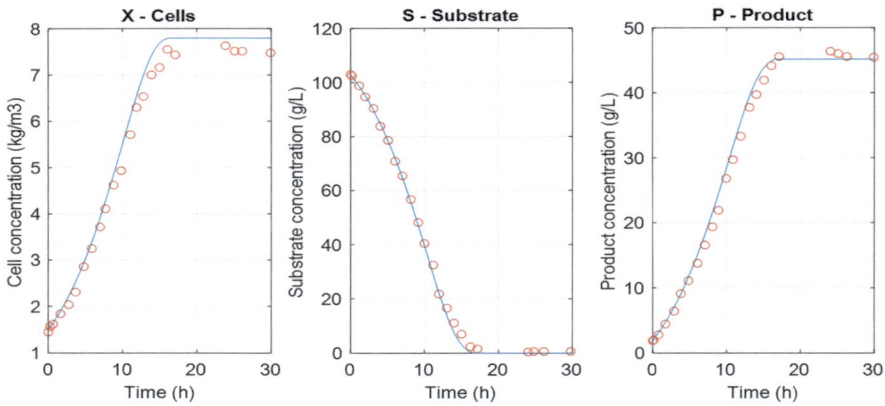

Fig. 9.12 Example of simulation of the cells' growth (*Saccharomyces cerevisiae*), substrate consumption (glucose), and product accumulation (ethanol) with the model suggested by Ruggeri et al. (1988)

$$\frac{dE}{dt} = \alpha_E \frac{dX}{dt} + \beta_E X \tag{9.20}$$

where X, S, and E represent the concentrations of biomass, substrate, and ethanol, respectively (kg/m^3). K_E is the ethanol concentration that shuts down the biomass growth rate (kg/m^3). K_I is the substrate inhibition constant (kg/m^3). A = substrate consumption, constant associated to growth (dimensionless), B = substrate consumption not associated to growth (1/h), α_E = production growth associated constant (dimensionless), β_E = production nongrowth associated (1/h), and μ_{max} = maximum specific growth rate (1/h).

Solution

The obtained results are reported in the plots in Fig. 9.12.

The described results were obtained by using a MATLAB program that is available to the readers as Electronic Supplementary Material of this book.

Rivera et al. (2006) proposed a more complete relationship containing also a term limiting the biomass growth $(1 - X/X_{max})$:

$$\mu = \mu_{max} \frac{S}{S + K_S} \exp(-K_I S) \left(1 - \frac{X}{X_{max}}\right)^m \left(1 - \frac{P}{P_{max}}\right)^n \tag{9.21}$$

For *Saccharomyces cerevisiae*, it resulted $m = 1$ and $n = 1.5$. In this work, many different runs have been performed at different temperatures in the range of 28–40 ° C so determining also the effect of the temperature on the kinetic parameters of the model.

The Monod equation can be adapted also to describe the diauxic growth. This occurs when two different substrates are present, in which one is first utilized as an

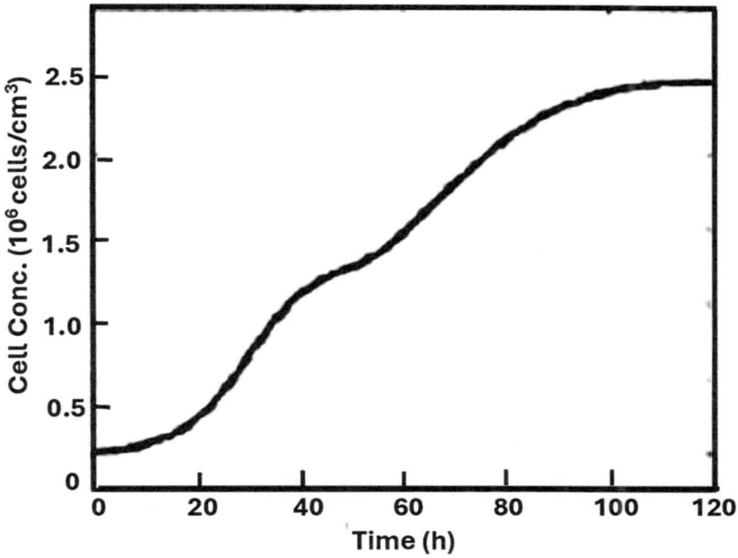

Fig. 9.13 Example of diauxic simulation interpreted with a modified Monod model. $M_0 = 0.3 \times 10^5$ (cells/cm^3)

energy source until its exhaustion and then the second substrate is used by the microorganism in sequence. In this case, we have:

$$\mu = \mu_{\max 1}\left(\frac{S_1}{S_1 + K_{S_1}}\right) + \mu_{\max 2}\left(\frac{S_2}{S_2 + K_2 + \frac{S_1^2}{K_1}}\right) \tag{9.22}$$

K_I is the inhibition constant. In this case, the consumption of the substrate S_2 is inhibited until S_1 is almost completely consumed, and then the growth restarts with another rate as reported in Fig. 9.13.

The original Monod equation does not consider the possibility that cells can consume substrate and give product when are not growing. As a matter of fact, cells, normally, require energy to survive by replacing degraded proteins also when they are not growing. To take into account this aspect, the Monod equation has been modified by Herbert (1958) by introducing a factor m = maintenance rate (kg substrate/kg cell × h) by assuming that the maintenance energy per unit mass remains constant.

By introducing the maintenance substrate uptake factor in the original Monod model (Eqs. 9.9 and 9.10), we obtain:

$$\frac{dM}{dt} = \mu_{max} \frac{S}{S + K_S} M(t) - mY_{MS} M(t) \tag{9.23}$$

$$\frac{dS}{dt} = -\frac{1}{Y_{MS}} \mu_{max} \frac{S}{S + K_S} M(t) \tag{9.24}$$

It has been demonstrated by Xing et al. (2010) that the maintenance energy is negligible during the growth phase but becomes relevant when growth is inhibited.

In the described models, the effect of temperature on growth rate is normally described by an Arrhenius-type relationship applied to μ, that is:

$$\mu = A \exp\left(\frac{\Delta E}{RT}\right) \tag{9.25}$$

being A the pre-exponential constant, ΔE the activation energy of the process, T the absolute temperature, and R the general constant of gases. Obviously, this equation is valid just in the narrow range of temperature in which the microorganisms are active.

Finally, Zwietering et al. (1992) introduced the concept of dimensionless growth factors, according to which any factor affecting the growth rate can be expressed as a fraction of the maximum growth rate. Each microorganism, in optimal conditions, shows a reproducible maximum growth rate. Any inhibitory effect due to the employed substrate can be defined by a dimensionless growth factor having a value falling between 0 and 1.

Structured Kinetic Models of Cell Growth

Structured kinetic models consider in some detail processes and reactions occurring in a single cell. This brings to a more realistic description of the cell behavior. The main metabolic pathways determining any possible activity of the microorganism are defined and considered in the elaboration of the kinetic models. A limitation of this approach is the difficulty of determining reliable parameters of the reactions occurring inside the cells.

- **Structured Compartmental Model**

The simplest structured model is the compartmental model developed by Williams (1967). The assumptions of this model are:

1. The cell is assumed to be structured in two compartments, one synthetic (S) and another one structural/genetic (N).

2. The compartment S is fed with an external nutrient (C) while the compartment N is fed by S.
3. The cell division occurs only if the part N doubles its initial size.

The assumption made for the single cell can be extended also to the entire population, where the overall biomass includes both the compartments S and N, that is, $M = N + S$. The number of cells is proportional to N, and the ratio M/N is the average size of the cells. Referring to the entire cells' population, in a batch condition, we can write:

$$\frac{dC}{dt} = -k_1 CM \tag{9.26}$$

$$\frac{dS}{dt} = k_1 CM - k_2 SN \tag{9.27}$$

$$\frac{dN}{dt} = k_2 SN \tag{9.28}$$

$$\frac{dM}{dt} = k_1 CM \tag{9.29}$$

considering C_0 and M_0 the initial quantities of nutrient and biomass, respectively. Hence,

$C = M_0 + C_0 - M$. As a consequence:

$$\frac{dM}{dt} = k_1 (C_0 + M_0) M - k_1 M^2 \tag{9.30}$$

and

$$M = \frac{C_0 + M_0}{1 + \frac{C_0}{M_0} \exp\left(\frac{-k_1 t}{C_0 + M_0}\right)} \tag{9.31}$$

Williams successfully tested the model describing both batch and chemostat conditions reproducing, in particular the following: (1) the lag phase during, which cells grow without division; (2) the exponential phase, with a fast increase of both cell numbers and total biomass; (3) a stationary phase, in which the nutrient C has been completely consumed and the cells don't grow and divide and their average size return to the initial value. In Fig. 9.14, the typical behavior of a microorganism in a batch reactor, according to the Williams model, is reported.

As seen, the performance of this model is satisfactory, but with this model, no information about the metabolic pathways occurring inside the microorganism is furnished. Harder and Roels (1982) developed also a three-compartment model.

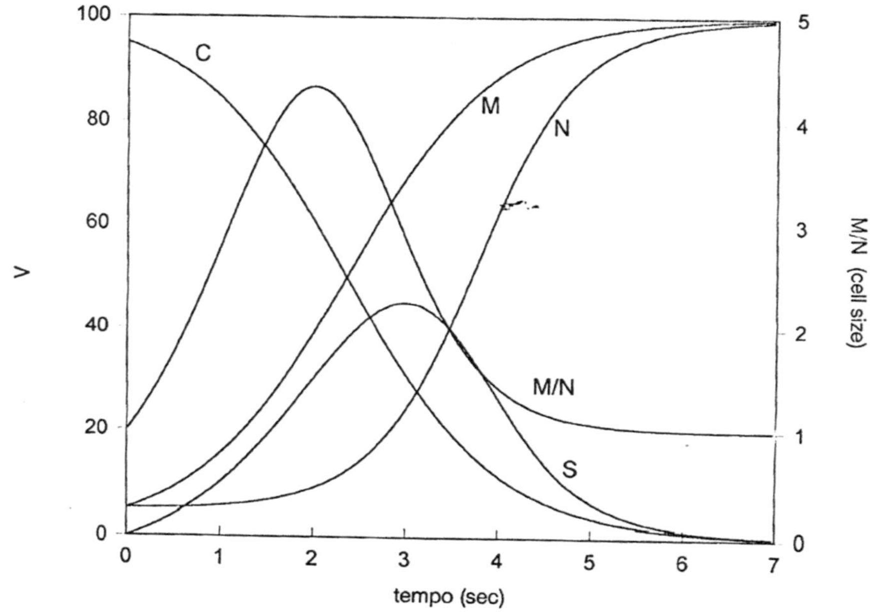

Fig. 9.14 Simulation of a batch cells growth according to the compartmental William model. The constants k_1 and k_2 have been assumed $k_1 = 0.0125\ (\mathrm{s})^{-1}$ and $k_2 = 0.025\ (\mathrm{s})^{-1}$. Reported with the permission of J. Theoret. Biol. (Williams 1967)

- **Structured Metabolic Models**

These structured kinetic models include some aspects of cellular metabolism. Clearly, the inclusion of more details about what happens inside the microorganism renders the model specific for that microorganism. An example of a metabolic model is the one proposed by Bijkerk and Hall (1977) to describe the aerobic growth of the microorganism *Saccharomyces cerevisiae*. The aerobic growth of *Saccharomyces cerevisiae* is industrially practiced in the production of bakers' yeast. Many different works have been published on this subject observing a complex interaction between the specific growth rate, the availability of sugar and oxygen, cell yield, cell morphology, and intracellular enzyme levels. The Bijkerk and Hall model starts from the observation of the experimental peculiarities of the growth process and proposes a novel mechanism of the control of the growth processes. According to Bijkerk and Hall, from the experimental observations, it can be deduced that the microorganism *Saccharomyces cerevisiae* grows following two different metabolic routes that are fermentation and respiration. In the presence of an excess of oxygen and of the nutrient glucose as a source of energy, the fermentation occurs with a growth of the biomass and production of ethanol and of CO_2. In this phase, the crab tree effect occurs with the inhibition of the respiratory enzymes, and oxygen is not consumed. When the nutrient glucose has been totally consumed, the biomass continues to grow consuming the ethanol accumulated in the first fermentation phase as a new source of energy. In this phase of respiration, oxygen is consumed to sustain the process. The alternance of fermentation and respiration is well described by the metabolic model developed by Bijkerk and Hall. This model is based on the following assumptions:

(i) The model is segregated, that is, the single cellular units are not considered.
(ii) The limiting nutrient S is the source of carbon and energy.
(iii) The biomass is divided in two parts: mass A consuming the substrate and producing energy and mass B being responsible of the cellular division.
(iv) The cellular division occurs in a fixed time and is described by the conversion of the mass B in the mass A.
(v) Ethanol E, produced during the fermentation phase is consumed in the respiration phase.

The occurring reactions are:

$$\text{Fermentation}: \quad A_w + a_1 S \xrightarrow{r_A} 2B_w + a_2 E + CO_2 \tag{9.32}$$

$$\text{Respiration}: \quad A_w + a_3 E + O_2 \xrightarrow{r_B} 2B_w + CO_2 \tag{9.33}$$

$$\text{Cellular division}: \quad B_w \xrightarrow{r_C} A_w \tag{9.34}$$

A_w = mass of A, B_w = mass of B and X = overall biomass = $A_w + B_w$

The kinetic expressions are:

$$r_A = \frac{k_1 A_w S}{K_S + S} \tag{9.35}$$

$$r_B = \frac{k_2 A_w E}{K_E + E} \tag{9.36}$$

$$r_C = K\,B_w \tag{9.37}$$

where k_1 is the maximum rate of energy production during the fermentation process while k_2 is the maximum rate of energy production during the respiration phase. The constants K_E and K_S are the concentrations at which the substrate consumption rates are limited by the low substrate concentration. Then, it is possible to write the following mass balance equations:

$$\frac{dA_w}{dt} = -r_A - r_B + r_C \tag{9.38}$$

$$\frac{dB_w}{dt} = 2r_A + 2r_B - r_C \tag{9.39}$$

$$\frac{dE}{dt} = a_2 r_A - a_3 r_B \tag{9.40}$$

$$\frac{dS}{dt} = -a_1 r_A \tag{9.41}$$

a_1, a_2, and a_3 are stoichiometric coefficients. a_1 and a_3 have been determined from the yields related, respectively, to glucose or ethanol. Yield on glucose in fermentation = 0.15 g of cell/g of glucose = $1/a_1$, hence $a_1 = 6.67$. Yield on ethanol in respiration = 0.45 g of cell/g of glucose = $1/a_3$, hence $a_3 = 2.22$.

a_2 has been assumed equal to 2.8. In conclusion:

$$\text{Fermentation :} \quad A_w + 6.67\,S \rightarrow 2\,B_w + 2.8\,E + CO_2 \tag{9.42}$$

$$\text{Respiration :} \quad A_w + 2.2\,E + O_2 \rightarrow 2\,B_w + CO_2 \tag{9.43}$$

$$\text{Cellular division :} \quad B_w \rightarrow A_w \tag{9.44}$$

The specific rate of CO_2 production (Q_{CO2}) can be determined by the carbon mass balance:

Carbon in CO_2 = carbon in glucose–(carbon in ethanol + carbon in cells)

The carbon content in mass has been calculated 0.48 g/g cell mass; then, it has been assumed that ethanol and CO_2 are produced in equimolecular quantities. Hence, we can write:

$$Q_{CO_2} = (\text{mols/g h}) = (0.40\,Q_S - 0.52\,Q_E - 0.48\,\mu)/12 \tag{9.45}$$

Q_S, Q_E, μ must be calculated as:

$$Q_S = \frac{dS}{dt}/X = \frac{a_1 r_A}{X} \tag{9.46}$$

$$Q_E = \frac{dE}{dt}/X = \frac{a_2(r_A - r_B)}{X} \tag{9.47}$$

$$\mu = \frac{dX}{dt}/X \tag{9.48}$$

The oxygen consumption is considered proportional to the respiration growth rate:

$$\frac{dO_2}{dt} = -k_{O_2} r_B \quad \text{and} \quad Q_{O_2} = k_{O_2} r_B/X \tag{9.49}$$

Starting from the observation that the complete oxidation of ethanol to CO_2 and water requires three moles of oxygen for every two of CO_2 produced, we can suppose that the overall oxidation maintains the same stoichiometry. Therefore, it would be considered a theoretical value of 1.53 g of oxygen required for each gram of dry cell formed during the respiration process:

$$Q_{O_2}\ (\text{mols/g h}) = \frac{1.53}{32} r_B/X = 0.048\ r_B/X \tag{9.50}$$

The best fitting has been obtained by assuming $k_1 = 5.0\ \text{h}^{-1}$, $k_2 = 0.51\ \text{h}^{-1}$ and $K = 0.38\ \text{h}^{-1}$ for reproducing the growth rates of the experimental run reported in Figs. 9.15 and 9.16.

The metabolic model of Bijkerk and Hall simulates the presence of a diauxic growth occurring in the transition from the fermentation to the respiration phase, but the model is not able to describe the diauxic lag phase, that is, the time occurring to the cells to pass from the fermentation to the respiration phase as it can be appreciated in Fig. 9.17.

This time is necessary for the cells to activate the enzymes promoting the respiration process. Pamment et al. (1978) improved the metabolic model proposed by Bijkerk and Hall, introducing the effect of the enzymes necessary, respectively, to the fermentation and the respiration on the growth of the microorganism population. In this model, in addition to the assumptions made by Bijkerk and Hall that remain still valid, four new assumptions have been formulated that are:

(vi) All the enzymes promoting the fermentation process are grouped and named E_f, while the enzymes responsible for the respiration process are grouped, too, and indicated as E_r.

(vii) Two different systems of control influence the production of each enzymatic system. First, enzymes are produced at a rate proportional to the metabolic flux through that enzymatic system. Second, there is an additional production of each enzymatic system that is proportional to the difference between the existing concentration of the enzyme and a target value E_F for the fermentation and E_R for the respiration. The target values are proportional to the flux occurring if the enzyme concentration is not limiting.

(viii) The rates of fermentation and respiration depend linearly on E_f/E_F and E_r/E_R, respectively.

(ix) Glycolytic enzymes are produced as a consequence of the flow through the respiration process.

The following stoichiometry of growth has been assumed:

$$\text{Fermentation}: \quad A_w + a_1 S + (E_f) \rightarrow 2B_w + a_2 E + (E_f) + (CO_2) \tag{9.51}$$

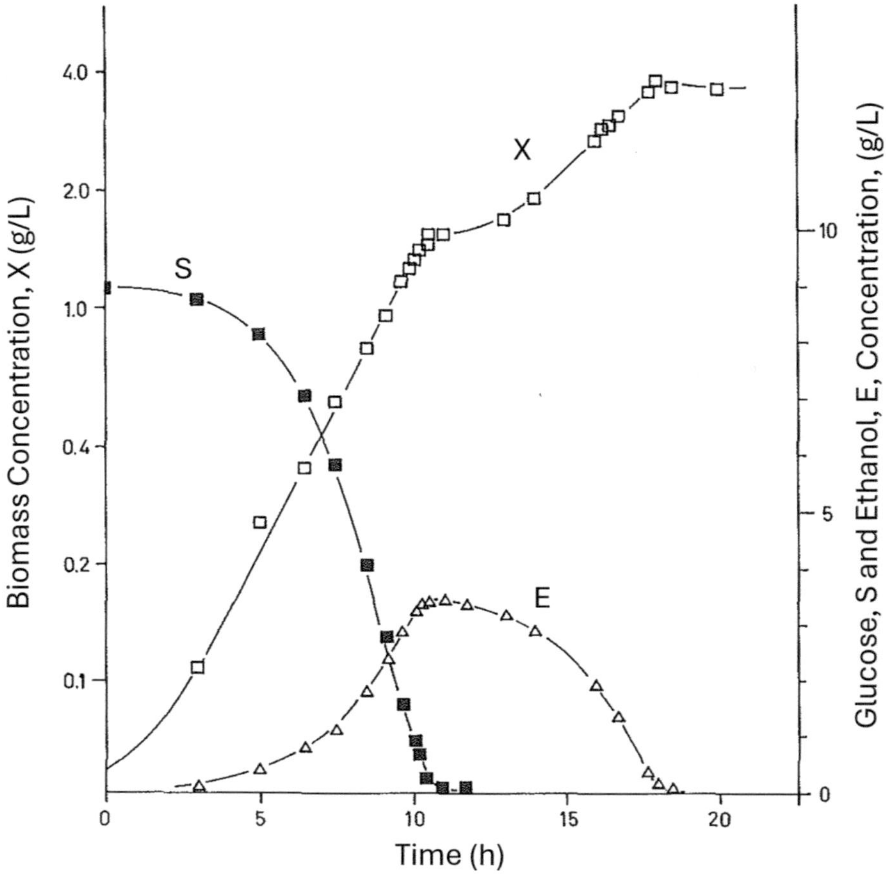

Fig. 9.15 Behavior of *Saccharomyces cerevisiae* in a batch bioreactor (runs performed by Von Meyenburg 1969) open squares: cell dry mass × g/L; open triangles: glucose S (g/L); open circles: ethanol E (g/L)

$$\text{Respiration}: \quad A_{\mathrm{w}} + a_3E + (\mathrm{O}_2) + (E_{\mathrm{f}}) + (E_{\mathrm{r}}) \rightarrow 2B_{\mathrm{w}} + (E_{\mathrm{r}}) + (E_{\mathrm{f}}) + (\mathrm{CO}_2) \tag{9.52}$$

$$\text{Division}: \quad B_{\mathrm{w}} \rightarrow A_{\mathrm{w}} \tag{9.53}$$

The equations of rates are:

$$r_A = \frac{k_1 A_{\mathrm{w}} S}{K_{\mathrm{S}} + S}\left(\frac{E_{\mathrm{f}}}{E_{\mathrm{F}}}\right) \tag{9.54}$$

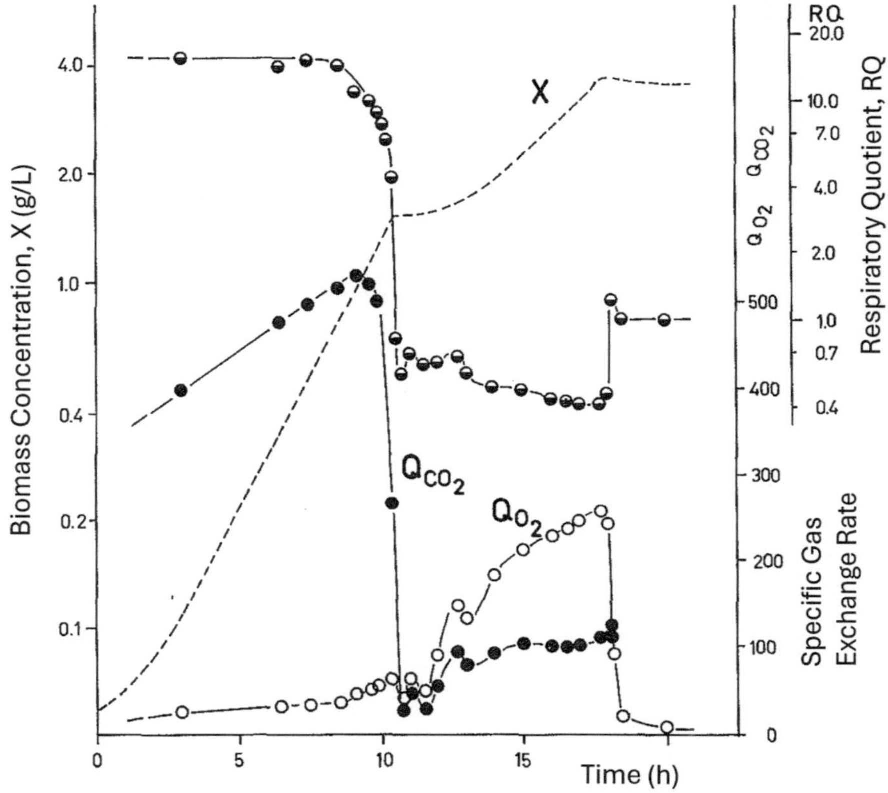

Fig. 9.16 Gas exchange during batch cultivation of *Saccharomyces cerevisiae* (runs performed by Von Meyenburg 1969) open triangles: Q_{O_2} = (mL/(g dry wt h) = specific oxygen uptake; filled squares: Q_{CO_2} = (mL/(g dry wt h) = specific carbon dioxide production; open circles: respiratory quotient (RQ)

$$r_B = \frac{k_2 A_w E}{K_E + E}\left(\frac{E_r}{E_R}\right) \tag{9.55}$$

$$r_C = K\, B_w \tag{9.56}$$

where:

$$E_F = \frac{k_f k_1 A_w S}{K_S + S} + \frac{k_7 k_2 A_w E}{K_E + E} \qquad E_R = \frac{k_h k_2 A_w E}{K_E + E} \tag{9.57}$$

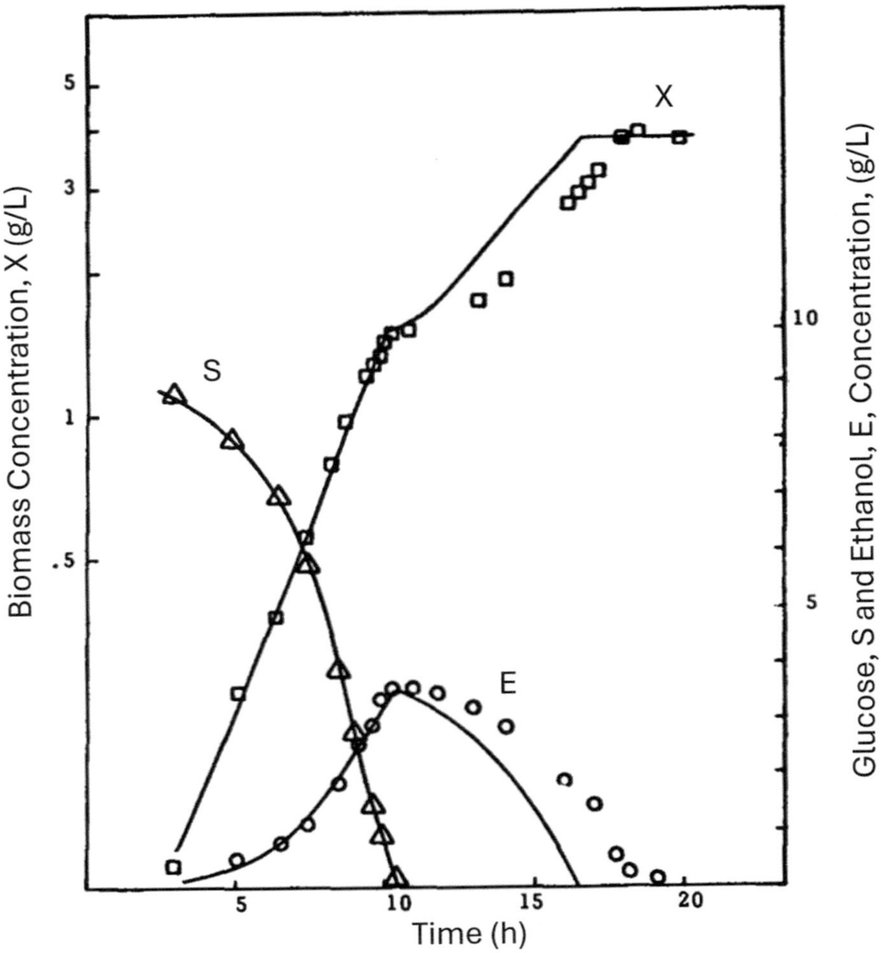

Fig. 9.17 Comparison of the model predictions (solid lines) and experimental observations (points): (runs performed by Von Meyenburg 1969) open squares: cell dry mass × g/L; open triangles: glucose S (g/L); open circles: ethanol E (g/L)

$$k_f = \frac{1}{k_1}\frac{\lambda + 1}{\lambda} \quad k_h = \frac{1}{k_2}\frac{\lambda' + 1}{\lambda'} \quad \lambda = \frac{-(k_1 - K) + \left[(k_1 - K)^2 + 8\,k_1 K\right]^{1/2}}{4k_1}$$

$$\lambda' = \frac{-(k_2 - K) + \left[(k_2 - K)^2 + 8\,k_2 K\right]^{1/2}}{4k_2} \tag{9.58}$$

The mass balance equations are:

$$\frac{dA_w}{dt} = -r_A - r_B + r_C \tag{9.59}$$

$$\frac{dB_w}{dt} = 2r_A + 2r_B - r_C \tag{9.60}$$

$$\frac{dE}{dt} = a_2 r_A - a_3 r_B \tag{9.61}$$

$$\frac{dS}{dt} = -a_1 r_A \tag{9.62}$$

$$\frac{dE_f}{dt} = r_A + k_5\, r_C + k_4(E_F - E_f) \tag{9.63}$$

$$\frac{dE_r}{dt} = r_B + k_6(E_R - E_r) \tag{9.64}$$

In Fig. 9.18, the experimental points related to the growth of *Saccharomyces cerevisiae* and the simulation with the two described metabolic models of the run performed are reported.

As it can be seen, the model developed by Pamment et al. would be more adherent to the experimental behavior.

9.3 The Cybernetic Model: Application to the Industrial Production of Baker's Yeast

In this and the next sections, the behavior in different conditions of the microorganism *Saccharomyces cerevisiae* will be described in detail as a useful didactic example. This is because *Saccharomyces cerevisiae* behavior has been intensively studied in the past, and several publications are reported in the literature on the subject. Several million tons of fresh baker's yeast are produced yearly for human food use. *Saccharomyces cerevisiae* is the microorganism employed for this production, and molasses, ammonia or ammonium salts, phosphates, and vitamins dissolved in water are used as nutrients. The microorganism growth can be performed in batch, fed-batch, or continuous bioreactors. Normally, the batch reactor is used with the main scope of testing the growth peculiarity of the available strain and nutrients. However, baker's yeast is usually produced starting from a small quantity of yeast added to a solution containing the nutrients at a suitable pH (5–6) and temperature (about 30 °C) (**the inoculum**). Once the cell population is grown enough, it is transferred into a larger bioreactor for a new growth stage. This operation is repeated for 3–4 further stages passing from an initial volume of 1–10 L to 100–150 m^3 as already described in the scheme of Fig. 9.1. The smallest bioreactor used for the initial stage normally operates in batch and anaerobic conditions, while, in the successive stages, an increasing flow of oxygen is supplied to the reactor and nutrients are fed at a variable optimal rate. Sugar is the main energy

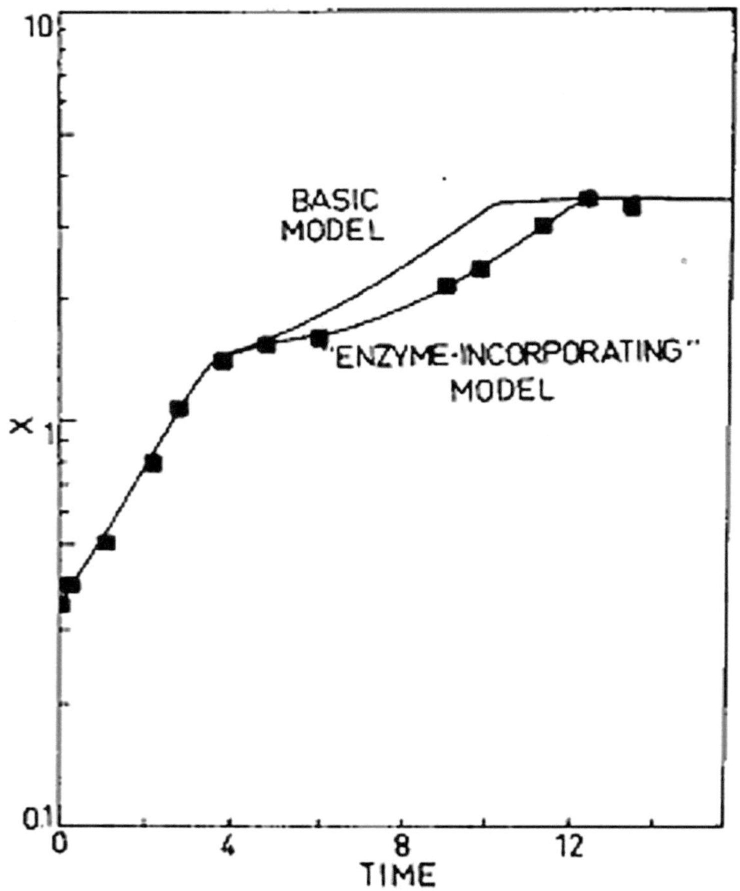

Fig. 9.18 Simulation of the diauxic growth of *Saccharomyces cerevisiae* with, respectively, the basic metabolic model of Bijkerk and Hall (1977) and the model of Pamment et al. (1978)

source, and it is metabolized via two different pathways, "fermentation" and "oxidation," depending on its concentration level. At high concentrations, oxidation is suppressed occurring mainly fermentation (crab tree effect). Oxidation is predominant when sugar concentration falls below 0.10–0.15 g/L. Ethanol is produced when the oxygen concentration is low, but it is consumed by the microorganism, as an alternative energy source, when sugar concentration becomes very low. The optimization of the entire process requires the definition of the following: (1) the best initial concentration of the inoculum, (2) the optimal bioreactor size, (3) the duration of each stage, (4) the nutrient flow rates modalities at each stage, and (5) the oxygen flow rate at each stage. However, as the yeast growth rate depends on both the employed strain and the composition of the molasses, the overall process optimization can be made by using a kinetic model that is able to describe a batch and a fed-batch preliminary runs having the scope to determine the parameters characteristics of the couple "strain substrate." If the model is valuable in describing those runs, it reasonably becomes predictive for all the scale-up of the different mentioned stages or for interpreting a continuous bioreactor.

As already mentioned, many different models have been proposed in the literature to describe the *Saccharomyces cerevisiae* growth rate in batch, fed-batch, and continuous laboratory bioreactors, but few have been tested on industrial data of baker's yeast production. However, the cybernetic approach proposed by Kompala et al. (1986), improved by Jones and Kompala (1999) and by Di Serio et al. (2001), resulted performant in correctly describing both laboratory, pilot, and industrial bioreactors resulting suitable for the overall process optimization.

Moreover, *Saccharomyces cerevisiae* is also employed for the industrial production of ethanol, mainly for food purposes (wine, beer), but recently ethanol acquired importance also as a renewable, alternative energy vector, and this production is destined to increase exponentially. Therefore, the mathematical-kinetic models, employed to optimize the baker's yeast production, could be adapted to the optimization of ethanol production, too. In the production of yeast, it is important to maximize both the yield in comparison with the substrate conversion (g of cells/g of sugar) and the productiveness (g of cells/L h). While the effects of variables, such as pH and temperature, are well-known and their optimal values can easily be defined, the yield and productiveness are largely affected by the concentration of biomass, sugar, and oxygen, and ethanol formation. The optimal conditions change along the time with the growth of the population cells. As a consequence, the feeding rate of nutrients must be changed, too, in an opportune way. The problem can be solved by adopting a structured unsegregated model able to individuate the prevailing metabolic pathways, the corresponding growth rate, and the influence of the operative conditions on the growth. The "cybernetic model" resulted suitable for this purpose.

Saccharomyces cerevisiae has, very probably, internal regulatory mechanisms directing the microorganism toward the most convenient metabolic pathway, able to optimize the use of the available resources. Therefore, the cybernetic modeling framework is based on the hypothesis that microorganisms optimize the utilization of the available substrates always to maximize their growth rate. The model is based on the reasonable assumption that the growth can occur following three different metabolic pathways that are: sugar fermentation, sugar respiration, and ethanol respiration. Three different sets of enzymes are specialized in promoting the mentioned metabolic pathways. The model needs to define an initial distribution of the three sets of enzymes in the inoculum, the value of biomass yields, and the values of the Monod kinetic parameters for any single metabolic pathway growth. Two cybernetic variables u_i and v_i are introduced into the model with the scope to optimize the strategy of enzymes in promoting, respectively, the synthesis and the activity. The value of u_i can be defined assuming that the cell resources will be allocated in such a way as to obtain the maximum biomass growth rate:

$$u_i = \frac{r_i}{\sum_j r_j} \tag{9.65}$$

The variable controlling the eventual inhibition/activation mechanism of the enzyme e_i (v_i) is determined considering null the inhibition effect when the microorganism grows on the substrate accelerating at maximum the biomass growth rate, whereas the inhibition effect progressively increases at a decreasing growth rate (Kompala et al. 1986). Therefore:

$$v_i = \frac{r_i}{\max_j(r_j)} \tag{9.66}$$

The specific growth rate corresponding to the three different metabolic pathways is defined according to modified Monod rate equations (Di Serio et al. 2001). The modification consists in the fact that each growth rate is assumed proportional to the ratio $e_i/e_{i,\max}$, that is, the relative promoting enzyme concentration:

$$r_1 = \mu_{1,\max}\left(\frac{e_1}{e_{1,\max}}\right)\frac{S_1}{K_1 V_L + S_1} = \text{sugar fermentation} \tag{9.67}$$

$$r_2 = \mu_{2,\max}\left(\frac{e_2}{e_{2,\max}}\right)\frac{S_2}{K_2 V_L + S_2}\ \frac{O_x}{K_{Ox} + O_x} = \text{etanol oxidation} \tag{9.68}$$

$$r_3 = 3\left(\frac{e_2}{e_{2,\max}}\right)\frac{S_1}{K_3 V_L + S_1}\ \frac{O_x}{K_{Ox} + O_x} = \text{sugar oxidation} \tag{9.69}$$

The ratio ($e_i/e_{i,\max}$) can change just in the interval 0–1. S_1 and S_2 are, respectively, the quantity of sugar and ethanol in the bioreactor. O_x is the oxygen concentration in the liquid phase, and V_L is the liquid volume in the reactor. K_i is the saturation constant for the substrates of each metabolic pathway, and K_{Ox} is the saturation constant of the dissolved oxygen. This last has been considered independent of any single oxidative metabolic pathway. The mass balance equations for batch ($F_{in} = 0$) and fed-batch ($F_{in} \neq 0$) bioreactors can be written as:

$$\frac{dX}{dt} = \left(\sum_i (r_i v_i)\right)X \quad \text{Balance on biomass} \tag{9.70}$$

$$\frac{dS_1}{dt} = F_{in}S_1^0 - \left(\frac{r_1 v_1}{Y_1} + \frac{r_3 v_3}{Y_3}\right)X \quad \text{Balance on sugar} \tag{9.71}$$

$$\frac{dS_2}{dt} = \left(\Phi_1 \frac{r_1 v_1}{Y_1} + \frac{r_2 v_2}{Y_2}\right)X \quad \text{Balance on ethanol} \tag{9.72}$$

$$\frac{dV_L}{dt} = F_{in} \quad \text{Balance on liquid volume} \tag{9.73}$$

Balance on key enzymes promoting fermentation:

$$\frac{d\left(\frac{e_1}{e_{1,\max}}\right)}{dt} = (\mu_{1,\max} + \beta)\left(1 - \varepsilon + \varepsilon u_i \frac{S_1}{K_1 V_L + S_1}\right) - \left(\sum_{j=1,3}(r_j v_j) + \beta\right)\left(\frac{e_1}{e_{1,\max}}\right) \tag{9.74}$$

Balance on key enzymes promoting ethanol oxidation:

$$\frac{d\left(\frac{e_2}{e_{2,\max}}\right)}{dt} = (\mu_{2,\max} + \beta)\left(1 - \varepsilon + \varepsilon u_2 \frac{S_2}{K_2 V_L + S_2}\frac{O_x}{K_{Ox} + O_x}\right) - \left(\sum_{j=1,3}(r_j v_j) + \beta\right)\left(\frac{e_2}{e_{2,\max}}\right) \tag{9.75}$$

Balance on key enzymes promoting sugar oxidation:

$$\frac{d\left(\frac{e_3}{e_{3,\max}}\right)}{dt} = (\mu_{3,\max} + \beta)\left(1 - \varepsilon + \varepsilon u_3 \frac{S_1}{K_3 V_L + S_1}\frac{O_x}{K_{Ox} + O_x}\right) - \left(\sum_{j=1,3}(r_j v_j) + \beta\right)\left(\frac{e_3}{e_{3,\max}}\right) \tag{9.76}$$

Balance on oxygen dissolved in the liquid phase:

$$\frac{dO_x}{dt} = k_L a\,(O_x^* - O_x) - \left(\Phi_2 \frac{r_2 v_2}{Y_2} + \Phi_3 \frac{r_3 v_3}{Y_3}\right)\frac{X}{V_L} \tag{9.77}$$

X = biomass amount in the reactor; F_{in}= feed stream; $S_1°$ = sugar concentration in the feed; $k_L a$ = gas-liquid mass transfer coefficient; O_x^* oxygen concentration at the interface, corresponding to oxygen solubility, β enzyme decay coefficient; and $\varepsilon = \alpha/(\alpha - \alpha^*)$, where α is the enzyme synthesis rate constant and α^* is a small constitutive synthesis term for all the enzymes (important in predicting the induction time of enzymes which have been repressed for long period of time). Finally, Y_i are the yields and Φ_i the stoichiometric coefficients for the different metabolic pathways, respectively.

Then, we can calculate the respiratory quotient (RQ), that is, the ratio of CO_2 moles produced on the oxygen moles consumed, as:

$$RQ = -\frac{\left(\frac{\Phi_1}{46}\frac{r_1 v_1}{Y_1} + \frac{\Phi_2}{32}\frac{r_2 v_2}{Y_2}\frac{2}{3} + \frac{\Phi_3}{32}\frac{r_3 v_3}{Y_3}\right)}{\frac{\Phi_2}{32}\frac{r_2 v_2}{Y_2} + \frac{\Phi_3}{32}\frac{r_3 v_3}{Y_3}} \tag{9.78}$$

Table 9.2 Kinetic parameters of the cybernetic model applied to *Saccharomyces cerevisiae*

E	0.909	Y_1	0.15	K_3 (g/dm^3)	0.001
β (h^{-1})	0.2	Y_2	0.74	Φ_1	0.41
μ_1 (h^{-1})	0.45	Y_3	0.5	Φ_2	1.067
μ_2 (h^{-1})	0.2	K_1 (g/dm^3)	1	Φ_3	0.52
μ_3 (h^{-1})	0.33	K_2 (g/dm^3)	0.08	K_{Ox} (g/dm^3)	4.6×10^{-5}

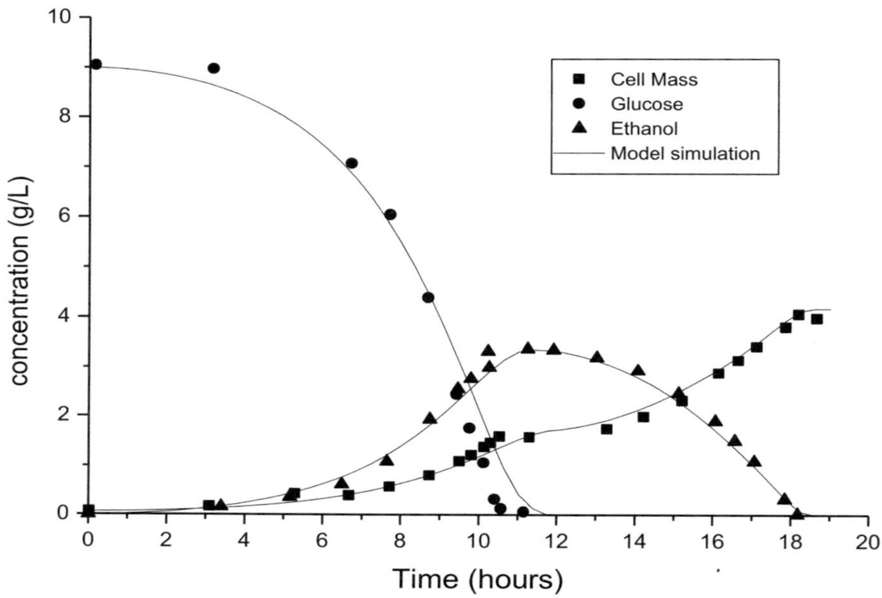

Fig. 9.19 Simulation (lines) of the experimental data (points) collected by Von Meyenburg (1969) for the growth of *Saccharomyces cerevisiae* microorganism in aerobic batch culture (Di Serio et al. 2001)

Different kinetic runs have been simulated in batch and fed-batch conditions (Di Serio et al. 2001) using the parameters reported in Table 9.2.

Some of the obtained results for the batch bioreactor can be appreciated in Figs. 9.19 and 9.20.

The same model has been tested also on the runs performed by Wank et al. (1979) in a fed-batch reactor using molasses as sugar source. In Fig. 9.21, the results of the simulation of the fed-batch kinetic run are reported.

In this run, to ensure full aerobic conditions, both air flow and stirring rate were varied to keep the oxygen concentration greater of a critical value corresponding to 15% of the saturation. Although molasses has been employed in this case, instead of glucose, the parameters of Table 9.3 resulted still valid. It is interesting to observe that no initial lag phase is present in Fig. 9.21. This means that a previous history of the employed inoculum affected the initial values of the $e_i/e_{i,\max}$. The initial values

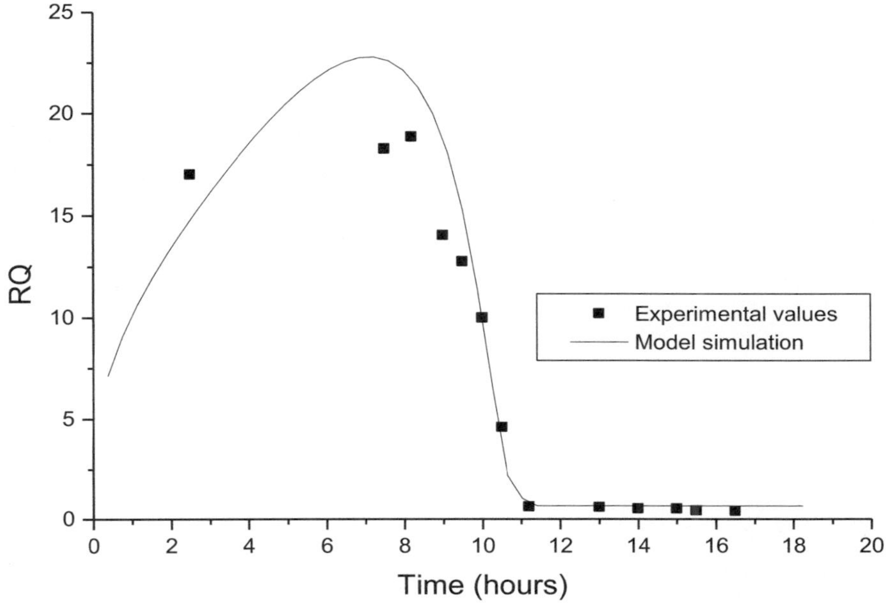

Fig. 9.20 Simulation (lines) of the respiratory quotient data (points) collected by Von Meyenburg (1969) related to the same run of Fig. 9.13 (Di Serio et al. 2001).

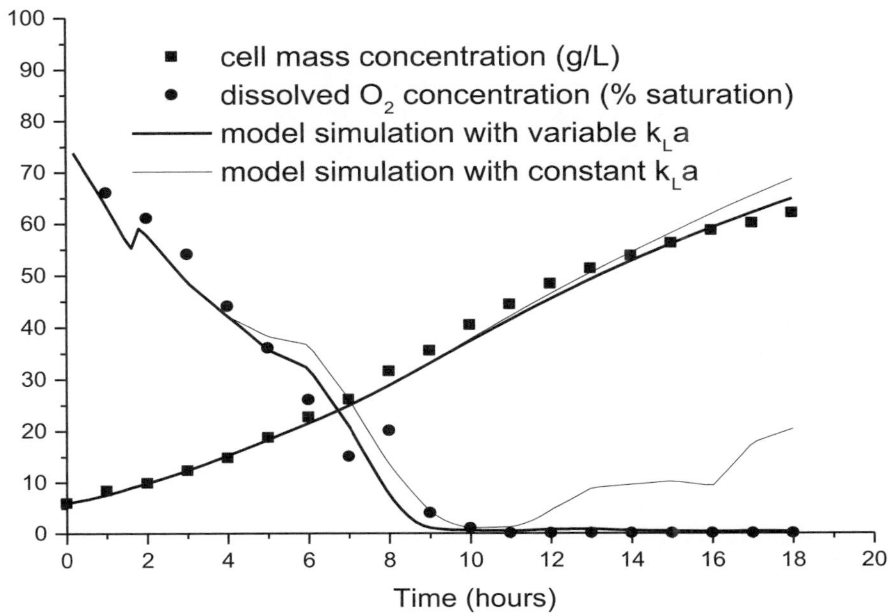

Fig. 9.21 Model simulation of the experimental data collected by Wank et al. (1979) in a fed-batch bioreactor (Di Serio et al. 2001)

Table 9.3 Initial conditions for different experimental runs performed by Gonzàlez-Hernàdez et al. (2022) in both anaerobic and aerobic conditions

Run	S_{Glc} (g/L)	S_{O_2} (mg/L)	S_E (g/L)	S_{Gly} (g/L)	X_Y (g/L)	T (°C)	Condition
A	13.75	–	52.13	0.26	2.43	26.10	Anaerobic
B	15.37	–	10.60	0.12	0.30	25.20	Anaerobic
C	31.06	–	10.99	0.15	0.81	26.50	Anaerobic
D	10.24	6.08	49.44	0.16	0.81	24.60	Aerobic
E	15.93	5.84	9.11	0.10	0.50	26.10	Aerobic
F	30.79	2.86	9.84	0.25	1.31	25.20	Aerobic
G	12.41	–	1.27	0.15	0.40	23.80	–

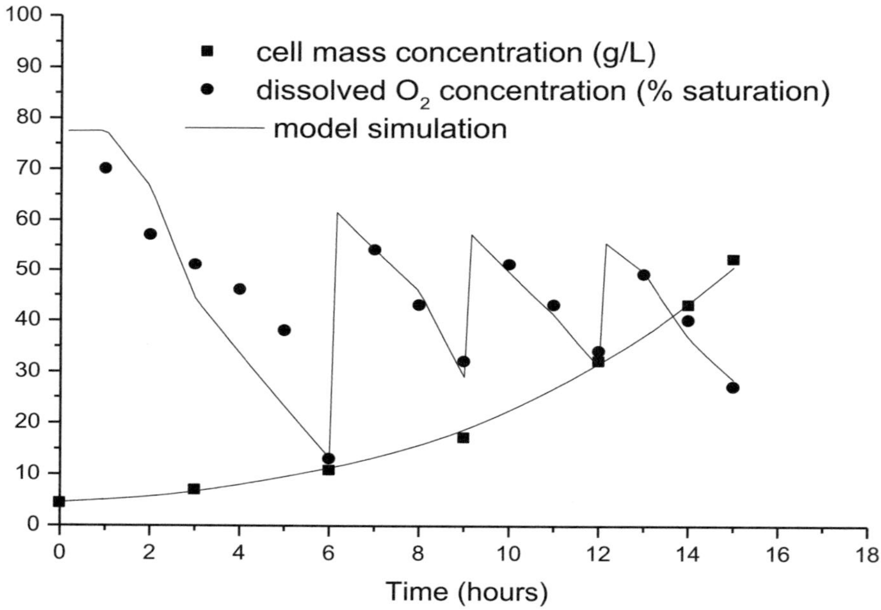

Fig. 9.22 Dynamic simulation of both the cell mass concentration and oxygen decay along the time. The oxygen never falls under conditions limiting the productivity (Di Serio et al. 2001)

used for the simulation, determined by trial-and-error procedure, are $e_1/e_{1,\max} = 0.2$, $e_2/e_{2\max} = 0.2$, and $e_3/e_{3,\max} = 0.7$. From Fig. 9.21, it can be argued that a correct simulation requires to change the mass transfer coefficient k_La that decreases at high biomass concentration. A correct estimation of this parameter can be achieved by measuring the oxygen concentration decay in dynamic runs like the one reported in Fig. 9.22. As it can be seen, also this type of run can be correctly simulated with the developed cybernetic model.

Finally, the same model has been employed to simulate the biomass growth in an industrial bioreactor. The reactor was a fed-batch bubble column with the inoculum

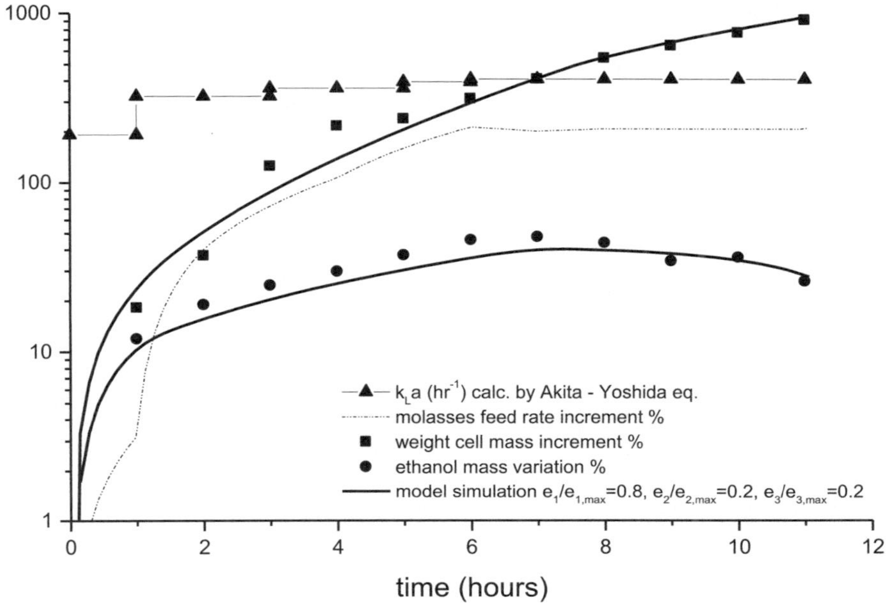

Fig. 9.23 Percent increase of molasses range in comparison with the initial feeding rate. Model simulation and experimental data for biomass and ethanol percentage variation. Values of k_La calculated using the Akita-Yoshida (1973) equation used in the simulation. Data taken from Di Serio et al. (2001)

coming from a previous batch stage. The liquid and gas phases were considered well mixed because of the high flow rate of air in the system. Available data were biomass concentration, ethanol concentration, air flow rate, and molasses addition rates at different times. Simulations have been made still using the parameters of Table 9.2, assuming for the initial enzyme concentrations of those of the previous inoculum: $e_1/e_{1,max} = 0.8$, $e_2/e_{2max} = 0.2$, and $e_3/e_{3,max} = 0.2$. The mass transfer coefficient has been estimated with the Akita-Yoshida (1973, 1974) relationships. In Fig. 9.23, the results obtained in the simulation of the industrial baker's yeast bioreactor with the cybernetic model are reported.

As the yeast growth is affected not only by the operative conditions, like temperature, pH, and sugar concentration, but also by intrinsic properties of the investigated system, that is, the type of strain, the culture medium, and the physiological state of the inoculum, Di Serio et al. (2003) have made several experimental runs with the aim to evaluate quantitatively also the effect of the intrinsic properties. The runs have been carried out by using different types of strain, culture medium, and physiological state of the inoculum. All the experimental runs performed in batch and fed-batch bioreactors have been interpreted with the already described cybernetic model determining, by mathematical regression analysis on the experimental data, all the kinetic parameters governing the effect of the intrinsic properties on the cell population growth. The parameters related to the different physiological states of the inoculum resulted independent from the strain and culture medium employed. This means that one batch and one fed-batch run, performed in well-defined operative conditions, are enough to determine the whole set of model parameters to be used in a predictive scale-up. The parameters of Table 9.2 can be used as the first approximation in the mathematical regression analysis.

Exercise 9.6
The cybernetic model, described in the previous section, has been applied, assuming data reported by Di Serio et al. 2001 and related parameters, to simulate the run reported in Fig. 9.19.

Solution
Experimental data, collected by Von Meyenburg (1969) for the growth of *Saccharomyces cerevisiae* microorganism in aerobic batch culture, have been perfectly simulated with the parameters of Table 9.2. The obtained results are reported in Fig. 9.24.

The described results were obtained by using a MATLAB program that is available to the readers as Electronic Supplementary Material of this book.

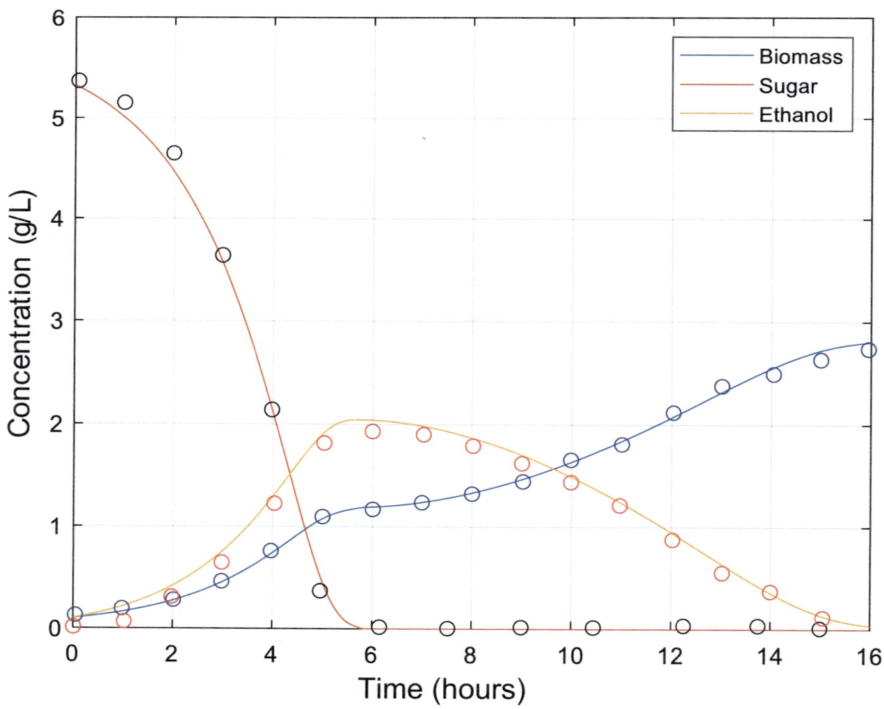

Fig. 9.24 Performance of the cybernetic model in the simulation of *Saccharomyces cerevisiae* behavior in a batch aerobic condition

9.4 The Production of Bioethanol as an Alternative Energy Vector

The world production of bioethanol increased very much in the last decades (from 31 billion Liters in 1996 (Nissen et al. 2000) to 111 billion in 2023 (MarketsandMarkets.com) with the scope of using ethanol as a renewable source of liquid fuel. Ethanol is produced almost entirely by anaerobic fermentation of various sugar sources by *Saccharomyces cerevisiae*. This reaction has been extensively studied, and the occurring metabolism is well-known, as it can be seen in the reaction scheme of Fig. 9.25 (Maicas 2020). As it can be seen, in anaerobic conditions, sugar is fermented, giving ethanol and CO_2 with the following stoichiometry:

$$C_6H_{12}O_6\,(\text{glucose}) \rightarrow 2\,C_2H_5OH\,(\text{ethanol}) + 2\,CO_2\,(\text{Carbon dioxide}) \qquad (9.79)$$

Normally, a small amount (about 4%) of glycerol can be obtained as by-product (Nissen et al. (2000), and metabolic engineering can be employed with the aim to reduce glycerol and other by-products. At this purpose, Nissen et al. (2000), for example, have performed anaerobic batch cultivations using different strains comparing, in particular, the results obtained by a wild type of strain TN1 with others submitted to genetic manipulation. The genetic manipulations made have given place to less active microorganisms and only one reduced glycerol formation maintaining an activity similar to the wild strain.

Rivera et al. (2006) used the ethanol fermentation process as a case study to test the validity of a hybrid neural model and a balance-based model in describing experimental data in different operative conditions. They focused the attention, in particular, on the kinetic effect of the temperature and to the methodology of reestimating the kinetic parameters when the substrate or the strain are changed.

It must be pointed out that the biomass yield of glucose in a respiratory culture is 0.5 g of biomass/g of glucose, while biomass yield of an anaerobic fermentation culture is fivefold lower, that is, 0.1 g of biomass/g of glucose (Bakker et al. 2001). As the baker's yeast production and the production of ethanol use the same microorganism *Saccharomyces cerevisiae*, the first in aerobic conditions and the second in anaerobic ones, it seems that it would be possible to employ a unique kinetic model to describe both the processes. For this purpose, some recent works deals with the description of the switch from anaerobic to aerobic conditions (Otterstedt et al. 2004; Gonzàlez-Hernàdez et al. 2022) of the *Saccharomyces cerevisiae* activity. The work of Gonzàlez-Hernandez proposes, in particular, a kinetic model that seems an advancement of the already described cybernetic model, because it considers four different metabolic pathways instead of three, that is, the sugar fermentation in anaerobic conditions is considered a pathway different with respect to the fermentation in aerobic conditions. The following stoichiometric reactions are suggested by the authors:

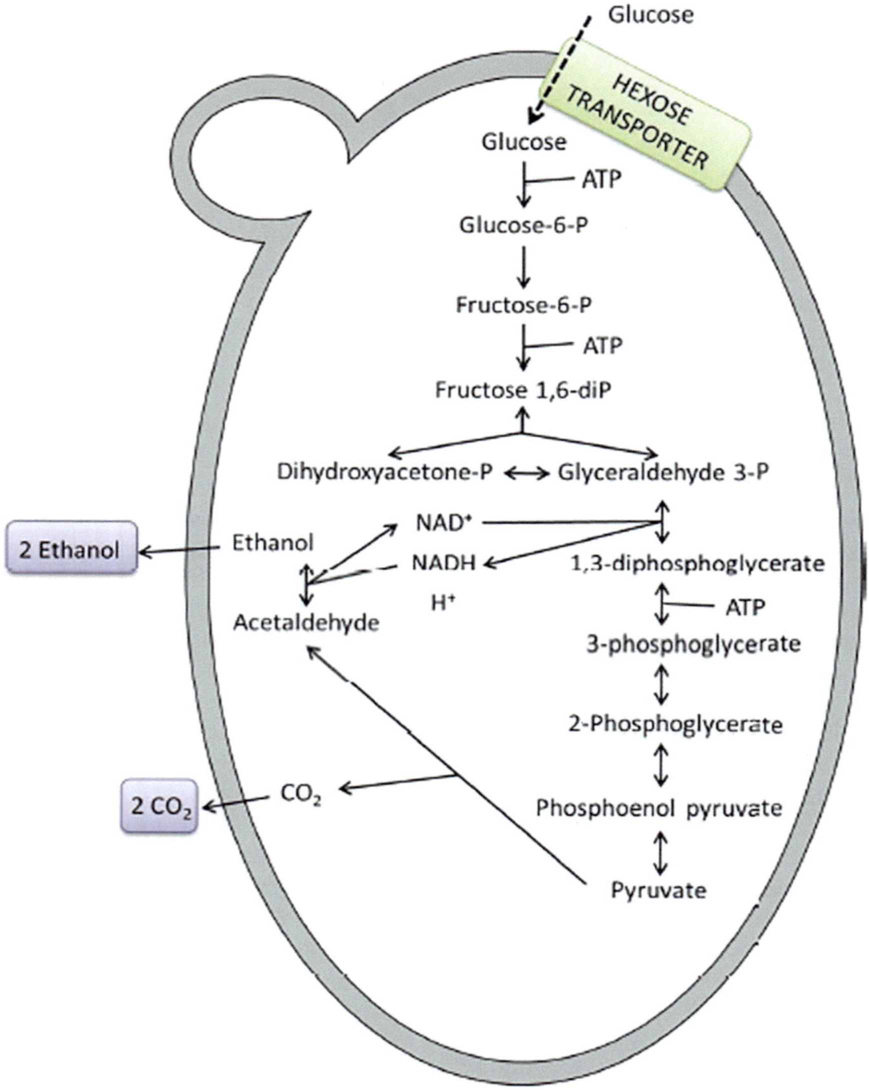

Fig. 9.25 Metabolism of yeast fermentation (Maicas 2020)

1. Anaerobic fermentation:

$$C_6H_{12}O_6 + \gamma c_1 NH_3 \rightarrow d_1 C_\alpha H_\beta N_\gamma O_\omega + e_1 C_2H_6O + f_1 C_3H_8O_3 + g_1 CO_2 + h_1 H_2O \quad (9.80)$$

As it can be seen, some glycerol is produced during this fermentation path. The following rate expression has been proposed.

$$R_1 = \mu_{\max 1} \frac{S_{\mathrm{Glc}}}{K_{\mathrm{Glc,Ax}} + S_{\mathrm{Glc}}} \frac{K_{\mathrm{O_2}}}{K_{\mathrm{O_2}} + S_{\mathrm{O_2}}} \frac{K_{\mathrm{E,Ax}}^{\mathrm{inh}}}{K_{\mathrm{E,Ax}}^{\mathrm{inh}} + S_{\mathrm{E}}} X_{\mathrm{Y}} \tag{9.81}$$

being S_{Glc} = glucose concentration (g/L), $K_{\mathrm{Glc,Ax}}$ = half saturation coefficient for glucose under anaerobic conditions (g/L), $K_{\mathrm{O_2}}$ = half saturation coefficient for oxygen (mg/L), $S_{\mathrm{O_2}}$ = dissolved oxygen concentration (g/L), S_{E} = ethanol concentration (g/L), $K_{\mathrm{E,Ax}}^{\mathrm{inh}}$ = half saturation coefficient for ethanol inhibition under anaerobic conditions (g/L), and X_{Y} = yeast concentration (g/L).

According to Bakker et al. (2001) a typical chemical composition of a yeast biomass $C_{\alpha}H_{\beta}N_{\gamma}O_{\omega}$ is $C_{3.75}H_{6.6}N_{0.63}O_{2.10}$, approximately in agreement with Rieger et al. (1983) that reported $C_{1.0}H_{1.79}N_{0.15}O_{0.57}$.

2. Aerobic fermentation:

$$C_6H_{12}O_6 + b_2O_2 + \gamma c_2NH_3 \rightarrow d_2C_{\alpha}H_{\beta}N_{\gamma}O_{\omega} + e_2C_2H_6O + g_2CO_2 + h_2H_2O \tag{9.82}$$

Fermentation occurs, in aerobic conditions, when glucose concentration exceeds 0.10–0.15 g/L giving ethanol and carbon dioxide as products. Anaerobic fermentation can be operative also in aerobic conditions, when the amount of supplied oxygen is not sufficient to satisfy the demand from the cell population. In this case, some amount of glycerol can be obtained as by-product as evidence of the contribution of this alternative pathway. The rate expression given for the aerobic fermentation is:

$$R_2 = \mu_{\max 2} \lambda_r \frac{S_{\mathrm{Glc}}}{K_{\mathrm{Glc,Ox}} + S_{\mathrm{Glc}}} \frac{S_{\mathrm{O_2}}}{K_{\mathrm{O_2}} + S_{\mathrm{O_2}}} \frac{K_{\mathrm{E,Ox}}^{\mathrm{inh}}}{K_{\mathrm{E,Ox}}^{\mathrm{inh}} + S_{\mathrm{E}}} X_{\mathrm{Y}} \tag{9.83}$$

being $K_{\mathrm{Glc,Ox}}$ = half saturation coefficient for glucose under aerobic conditions (g/L) and $K_{\mathrm{E,Ox}}^{\mathrm{inh}}$ = half saturation coefficient for ethanol inhibition under aerobic conditions (g/L).

3. Glucose respiration:

$$C_6H_{12}O_6 + b_3O_2 + \gamma c_3NH_3 \rightarrow d_3C_{\alpha}H_{\beta}N_{\gamma}O_{\omega} + g_3CO_2 + h_3H_2O \tag{9.84}$$

When glucose concentration is below the threshold value of 0.10–0.15 g/L, the metabolism rapidly switches from aerobic fermentation to respiration, and the expression rate becomes:

$$R_3 = \mu_{\max 3}(1-\lambda_r)\frac{S_{\mathrm{Glc}}}{K_{\mathrm{Glc,Ox}}+S_{\mathrm{Glc}}}\frac{S_{\mathrm{O_2}}}{K_{\mathrm{O_2}}+S_{\mathrm{O_2}}}\frac{K^{\mathrm{inh}}_{\mathrm{E,Ox}}}{K^{\mathrm{inh}}_{\mathrm{E,Ox}}+S_{\mathrm{E}}}X_{\mathrm{Y}} \tag{9.85}$$

4. Ethanol respiration:

$$\mathrm{C_2H_6O + b_4O_2 + \gamma c_4NH_3 \rightarrow d_4C_\alpha H_\beta N_\gamma O_\omega + g_4CO_2 + h_4H_2O} \tag{9.86}$$

When glucose is depleted, ethanol becomes the energy source and ethanol respiration starts with the following rate equation.

$$R_4 = \mu_{\max 4}\frac{\delta K_{\mathrm{Glc,Ox}}}{\delta K_{\mathrm{Glc,Ox}}+S_{\mathrm{Glc}}}\frac{S_{\mathrm{E}}}{K_{\mathrm{E}}+S_{\mathrm{E}}}\frac{S_{\mathrm{O_2}}}{K_{\mathrm{O_2}}+S_{\mathrm{O_2}}}\frac{K^{\mathrm{inh}}_{\mathrm{E,Ox}}}{K^{\mathrm{inh}}_{\mathrm{E,Ox}}+S_{\mathrm{E}}}X_{\mathrm{Y}} \tag{9.87}$$

being δ = safety factor of switching function for the ethanol respiration process activation, K_{E} = half saturation coefficient for ethanol (g/L), S_{E} = ethanol concentration (g/L), and $\mu_{\max(i)}$ = maximum growth rate for the process I ($\mathrm{h^{-1}}$).

Then, λ_r is a stepwise function that has the scope to warrant a smooth passage in the shift between fermentation and respiration.

$$\lambda_r = \frac{1+\tanh\left[\alpha\left(\frac{S_{\mathrm{Glc}}}{S_{\mathrm{Glc,c}}}-1\right)\right]}{2} \tag{9.88}$$

This function is defined by two parameters: the shift value $S_{\mathrm{Glc,c}}$, that is, the glucose concentration value at which the transition occurs ($S_{\mathrm{Glc,c}} = 0.125$ g/L) and the parameter α defining the sharpness of the transition. Assuming $\alpha = 150$ means to have a transition between 0.10 and 0.15 g/L of glucose concentration.

The expression of mass transfer rate is:

$$R_5 = k_{\mathrm{L}}a\left(M_{\mathrm{O_2}}H_{(T)}P_{\mathrm{O_2}} - S_{\mathrm{O_2}}\right) \tag{9.89}$$

where k_{L} = mass transfer coefficient, a = specific interface area, $k_{\mathrm{L}}a$ = overall oxygen transfer rate ($\mathrm{h^{-1}}$), $M_{\mathrm{O_2}}$ = oxygen molar mass (g/mol), $H_{(T)}$ = Henry constant a T temperature (mol/L/atm), $P_{\mathrm{O_2}}$ = oxygen partial pressure in gaseous phase (atm), and $S_{\mathrm{O_2}}$ = dissolved oxygen concentration (g/L).

The system of differential equations to be integrated in matrix form is:

$$\frac{\mathrm{d}C_{(j)}}{\mathrm{d}t} = \sum_{i=1}^{5}\sum_{J=1}^{5}\nu_{(i,j)}R_i \tag{9.90}$$

Gonzàlez-Hernàdez et al. (2022) have interpreted seven experimental runs, respectively, performed either in anaerobic or aerobic conditions. In Table 9.3, the initial conditions of the mentioned runs are reported. Runs 1–3 were made in

Table 9.4 Stoichiometric and kinetic parameters of the Gonzàlez-Hernadez model for the anaerobic process

Parameters	Values	Units
d_1	1.05	Mol
e_1	1.70	mol
f_1	0.10	mol
$K_{\text{Glc, Ax}}$	1.72	g/L
$K_{\text{E,Ax}}^{\text{inh}}$	202.83	g/L
$\mu_{\max 1}$	0.31	h^{-1}

Table 9.5 Stoichiometric and kinetic parameters of the Gonzàlez-Hernadez model for the aerobic process

Parameters	Values	Units
b_2	0.37	mol
d_2	1.40	mol
e_2	1.26	mol
b_3	2.50	mol
d_3	3.33	mol
b_4	1.61	mol
d_4	1.17	mol
K_{O_2}	0.10	mg/L
$K_{\text{Glc, Ox}}$	1×10^{-3}	g/L
K_{E}	0.10	g/L
$K_{\text{E,Ox}}^{\text{inh}}$	19.70	g/L
$\mu_{\max 2}$	0.51	h^{-1}
$\mu_{\max 3}$	0.20	h^{-1}
$\mu_{\max 4}$	0.11	h^{-1}
$k_L a$	27.27	h^{-1}

anaerobic conditions, runs 4–6 were in aerobic conditions, and run 7 was used for testing the predictivity of the model. This last run has been made by keeping a constant atmosphere of oxygen (1 atm) on the gas headspace of the bioreactor, so aerobic and anaerobic metabolic pathways take place simultaneously, and the model reproduces this behavior that is characterized by a modest production of glycerol.

A good agreement was obtained by using the kinetic parameters reported in Table 9.4 for the anaerobic runs and in Table 9.5 for the aerobic ones. Some kinetic parameters have been taken from the literature, and some others have been determined from the experimental data

In conclusion, in this chapter, we have described in detail the behavior of the microorganism *Saccharomyces cerevisiae* because it has been extensively studied in the literature for the very important industrial production of baker's yeast and bioethanol. The methods described for studying the behavior of this microorganism can be applied also to other microorganisms taking into account that any microorganism is a living being having its own peculiarity that needs to be accurately investigated before modeling its behavior.

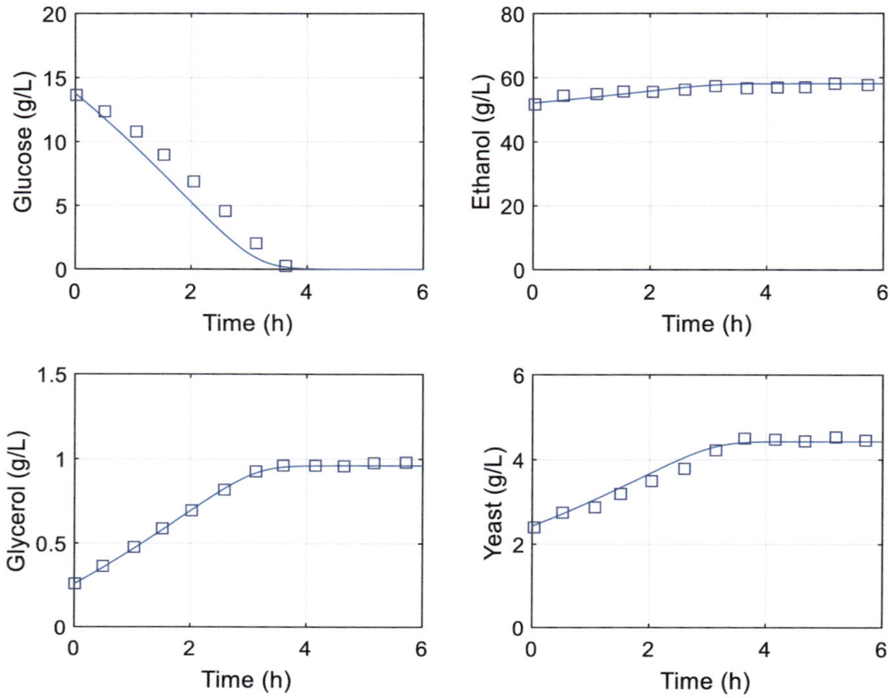

Fig. 9.26 Simulation of the run A of Table 9.3

Exercise 9.7

As seen, Gonzàlez-Hernàdez et al. (2022) have made different kinetic runs in anaerobic conditions defining a set of differential equations to be integrated for simulating the performed runs. Initial conditions and kinetic parameters are reported in Tables 9.3 and 9.4.

Solution

The results obtained for the simulation of the runs A, B, and C of Table 9.3 are reported in the plots of Figs. 9.26, 9.27, and 9.28.

The described results were obtained by using a MATLAB program that is available to the readers as Electronic Supplementary Material of this book.

Exercise 9.8

Gonzàlez-Hernàdez et al. (2022) have made also different kinetic runs in aerobic conditions, defining a set of differential equations to be integrated for simulating the performed runs. Initial conditions and kinetic parameters are reported in Tables 9.3 and 9.5. In the exercise, run E of Table 9.3 has been simulated as an example.

Solution

The results obtained for the simulation of run E of Table 9.3 are reported in the plots of Fig. 9.29.

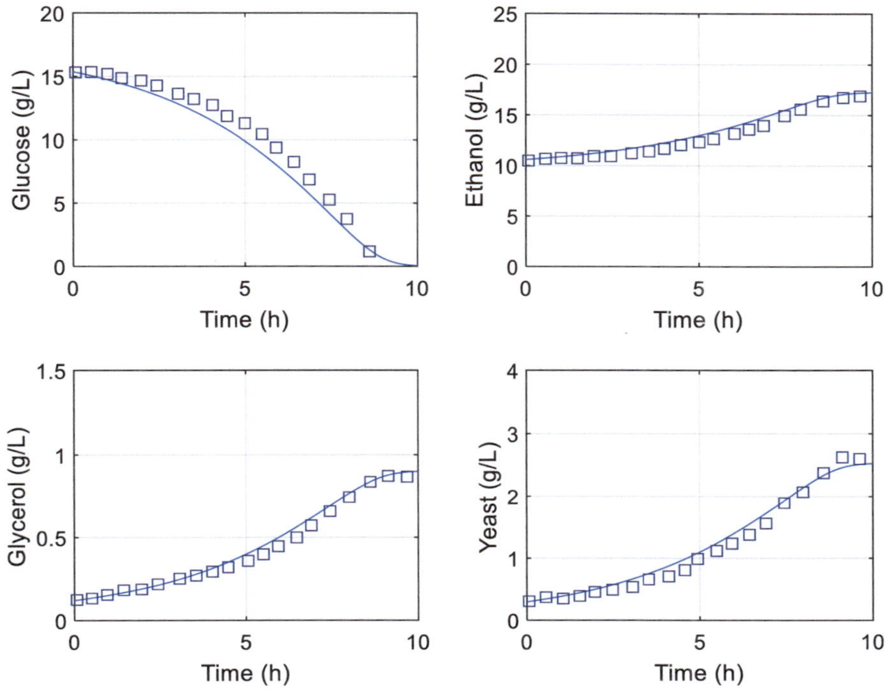

Fig. 9.27 Simulation of the run B of Table 9.3

The described results were obtained by using a MATLAB program that is available to the readers as Electronic Supplementary Material of this book.

Exercise 9.9

Veloso et al. (2019), using an industrial strain of *Saccharomyces cerevisiae*, studied batch and fed-batch ethanol fermentation with the aim to obtain highly concentrated alcoholic wine. Different runs have been made by the authors, at different temperatures between 28 and 34 °C. The kinetic results have been interpreted with the Andrews-Levenspiel hybrid kinetic model, and the kinetic parameters collected have been used for a successive step of optimization. The model is based on the following system of differential equations:

$$\frac{\mathrm{d}C_{\mathrm{VX}}}{\mathrm{d}t} = \left(\mu - \frac{F}{V}\right) C_{\mathrm{VX}} \tag{9.91}$$

$$\frac{\mathrm{d}C_{\mathrm{S}}}{\mathrm{d}t} = (C_{\mathrm{SM}} - C_{\mathrm{S}}) \frac{F}{V} \frac{1}{Y_{\mathrm{VX/S}}} \mu C_{\mathrm{VX}} \tag{9.92}$$

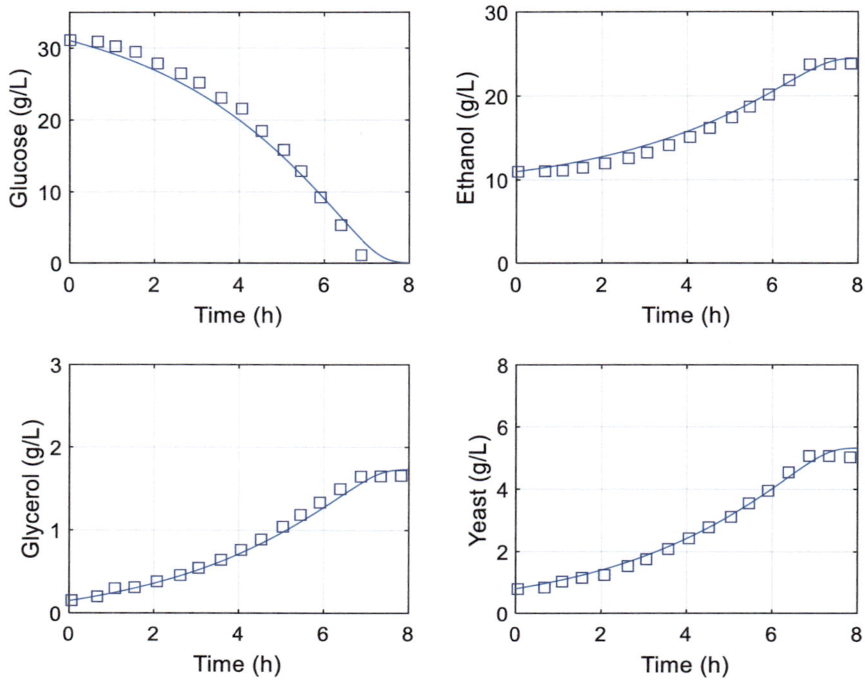

Fig. 9.28 Simulation of the run C of Table 9.3

$$\frac{dC_E}{dt} = \frac{Y_{E/S}}{Y_{VX/S}}\mu C_{VX} - \frac{F}{V}C_E \tag{9.93}$$

$$\frac{dV}{dt} = F \tag{9.94}$$

where μ is the specific cell growth rate (h^{-1}), C_{VX} is the viable cells concentration (g/L), F is the feed flow rate (L/h), obviously for the batch runs $F = 0$, V is the reaction volume (L), C_S is the limiting substrate concentration (g/L), C_{SM} is the substrate concentration in the feeding flow (g/L), C_E is the ethanol concentration (g/L), $Y_{VX/S}$ is the cell's yield coefficient (g_{VX}/g_S), and $Y_{E/S}$ is the ethanol yield coefficient (g_E/g_S). As mentioned, the Andrews-Levenspiel (Andrews 1968; Levenspiel 1980) hybrid kinetic model has been adopted by the authors for interpreting the experimental data. In this model, the cell growth inhibition by the substrate and product is described by the relationship:

$$\mu = \mu_{max}\frac{C_S}{K_S + C_S + \frac{C_S^2}{K_{IS}}}\left(1 - \frac{C_E}{C_{Emax}}\right)^n \tag{9.95}$$

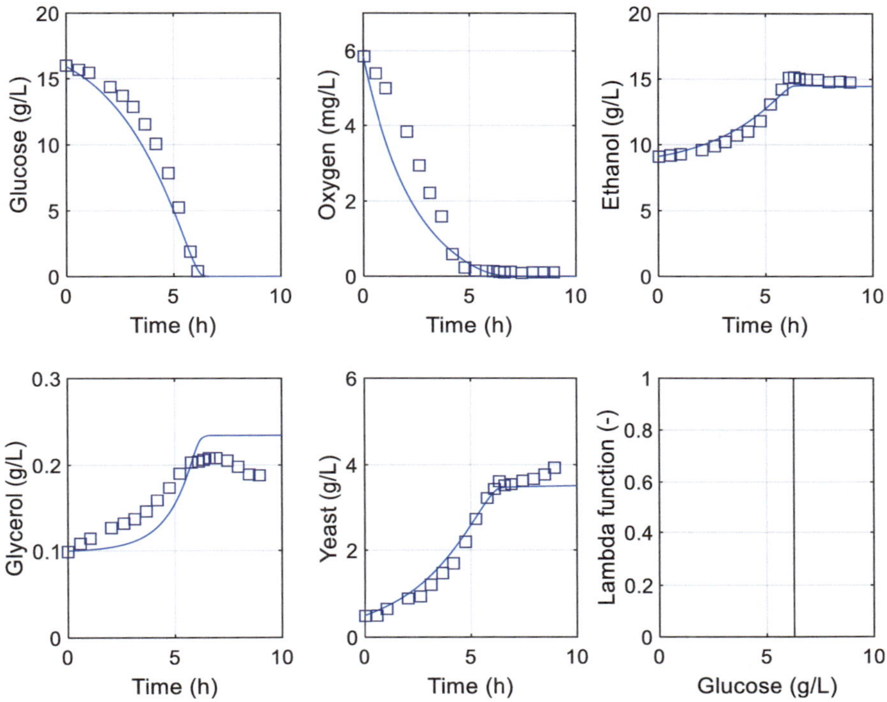

Fig. 9.29 Simulation of the run E of Table 9.3

μ_{max} is the maximum specific cell growth rate (h^{-1}), K_S is the saturation constant (g/L), K_{IS} is the substrate inhibition constant (g/L), C_{Emax} is the maximum ethanol concentration after which cell growth stopped (g/L), and n is a dimensionless constant describing toxic effects.

For batch fermentation, the yield coefficients $Y_{VX/S}$ and $Y_{E/S}$ can be determined by the following equations:

$$Y_{VX/S} = \frac{C_{VXf} - C_{VX0}}{C_{S0} - C_{Sf}} \tag{9.96}$$

$$Y_{E/S} = \frac{C_{Ef} - C_{E0}}{C_{S0} - C_{Sf}} \tag{9.97}$$

where "0" and "f" are related to initial and final times. In the case of fed-batch fermentation, the yield coefficients are determined with the following relationships:

$$Y_{VX/S} = \frac{V_f C_{VXf} - V_0 C_{VX0}}{(V_f - V_0) C_{SM} - V_f\, C_{Sf}} \tag{9.98}$$

Table 9.6 Kinetic parameters determined by Veloso et al. (2019) for interpreting ethanol fermentation batch runs and initial conditions of a run simulated as an example

Parameters	Values	Initial conditions	Values
$Y_{VX/S}$	0.0622	C_{VX0}	12.2
$Y_{E/S}$	0.446	C_{E0}	2.40
μ_{max}	0.157	C_{S0}	168
K_S	19.98		
K_{IS}	170		
C_{Emax}	139.7		
N	0.125		

$$Y_{E/S} = \frac{V_f C_{Ef}}{(V_f - V_0)C_{SM} - V_f\, C_{Sf}} \tag{9.99}$$

Part 1: A Batch Run Simulation

Veloso et al. have made four runs in batch conditions in the same conditions, changing just the temperature. As an example, in this exercise, one of these runs, performed at 28 °C, has been simulated.

In Table 9.6 are reported the kinetic parameters and the initial conditions related to the mentioned run.

Solution

The obtained results in the simulation of the run performed by Veloso et al. at 28 °C are reported in Fig. 9.30.

Part 2: A Fed-Batch Run Simulation

By using the same strain of *Saccharomyces cerevisiae*, Veloso et al. (2019) have made also eight fed-batch runs that have been simulated by the authors using the already described kinetic model. In this exercise, a fed batch, performed at the same temperature of 28 °C of the previously considered batch run, has been simulated. The kinetic parameters of Table 9.6 have been used with the exception of the ones reported in Table 9.7. In the same table are also reported the initial conditions.

Solution

The obtained results in the simulation of the fed-batch run performed by Veloso et al. at 28 °C are reported in Fig. 9.31.

The described results were obtained by using a MATLAB program that is available to the readers as Electronic Supplementary Material of this book.

Exercise 9.10

Rivera et al. (2013) have studied the ethanol fermentation in a fed-batch reactor, using a strain of *Saccharomyces cerevisiae* normally employed in industrial plants of ethanol production. The kinetics of the cell growth, adopted by the authors, was:

$$r_X = \mu_{max} \frac{S}{S + K_S} \exp(-K_I S)\left(1 - \frac{X}{X_{max}}\right)^m \left(1 - \frac{P}{P_{max}}\right)^n X \tag{9.100}$$

The ethanol formation rate was assumed:

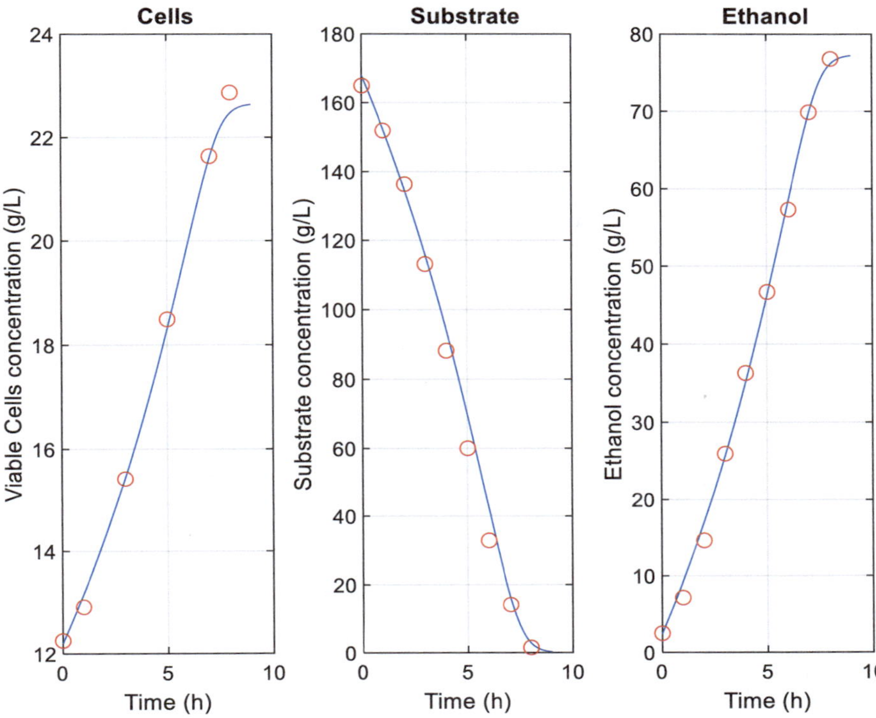

Fig. 9.30 Results of the simulation of a batch run of ethanol fermentation, performed at 28 °C. The simulation has been made by using the Andrews-Levenspiel kinetic model

Table 9.7 Kinetic parameters determined by Veloso et al. (2019) for interpreting ethanol fermentation fed-batch runs and initial conditions of a run simulated as an example

Parameters	Values	Initial conditions	Values
$Y_{VX/S}$	0.0598	C_{VX0}	30
$Y_{E/S}$	0.455	C_{E0}	0.0
F	179	C_{S0}	0.0
F_t (filling time)	1.4/F	V_0	0.6
C_{SF}	180 × 1.45		
$C_{sm} = C_{SF}$	180 × 1.45		

$$r_P = Y_{P/X} r_X \frac{\beta_{mP} S}{K_{\beta s1} + S} X \tag{9.101}$$

The substrate consumption rate has been assumed:

$$r_S = Y_{P/X}(r_X/Y_X) \frac{\beta_{mS} S}{K_{\beta s2} + S} X \tag{9.102}$$

μ_{max} is the maximum specific rate (h^{-1}), X is the cell concentration (kg/m^3), K_S is the substrate saturation constant (kg/m^3), K_I is the substrate inhibition parameter

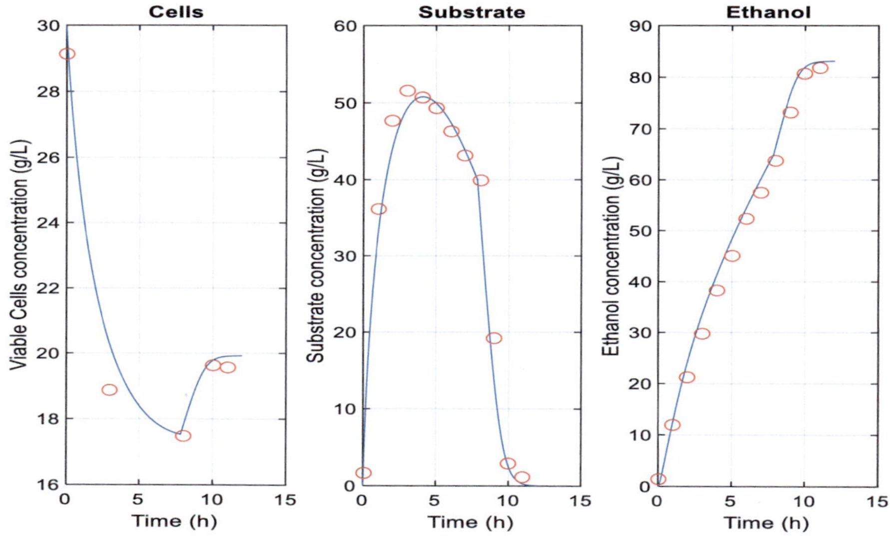

Fig. 9.31 Results of the simulation of a fed-batch run of ethanol fermentation, performed at 28 °C. The simulation has been made by using the Andrews-Levenspiel kinetic model

(kg/m^3), X_{max} the cell concentration where the growth stops (kg/m^3), P_{max} the ethanol concentration at which the cell growth ceases (kg/m^3), and m and n are empirical parameters. $Y_{P/X}$ (kg/kg) is the product yield based on cell growth, Y_X (kg/kg) denote the limit cellular yield, β_{mP} and β_{mS} (kg/kg h) are parameters describing maintenance, and $K_{\beta s1}$ and $K_{\beta s2}$ (kg/m^3) are saturation parameters.

The fed-batch model is represented by the following set of ordinary differential equations:

$$\frac{dX}{dt} = r_X - \frac{F_A X}{V} \quad \text{Balance on cells} \tag{9.103}$$

$$\frac{dS}{dt} = \frac{F_A(S_A - S)}{V} - r_s \quad \text{Balance on the substrate} \tag{9.104}$$

$$\frac{dP}{dt} = r_P - \frac{F_A P}{V} \quad \text{Balance on ethanol} \tag{9.105}$$

$$\frac{dV}{dt} = F_A \quad \text{Volume} \tag{9.106}$$

Simulations of the kinetic runs have been performed with the kinetic parameters and initial values reported in Table 9.8.

Solution

The obtained results in the simulation of the fed-batch runs performed in the conditions reported in Table 9.8 are reported in Fig. 9.32.

Table 9.8 Kinetic parameters determined by Rivera et al. (2013) for interpreting ethanol fermentation fed-batch runs and initial conditions of a run simulated as an example

Parameters	Values	Initial values	Values (kg/m^3)	Conditions	Values
μ_{max}	0.33	X_0	40.55	V_i (m^3)	0.5
P_{max}	85.7	S_0	17.95	S_A (kg/m^3)	171.7
X_{max}	56.2	P_0	9.84	F_A (m^3/h)	0.43
$Y_{P/X}$	3.56				
Y_X	0.0909				
β_{mP}	0.229				
$K_{\beta s1}$	0.001				
β_{mS}	0.201				
$K_{\beta s2}$	0.001				
m	1				
n	1.5				
K_S	4.1				
K_I	0.002				

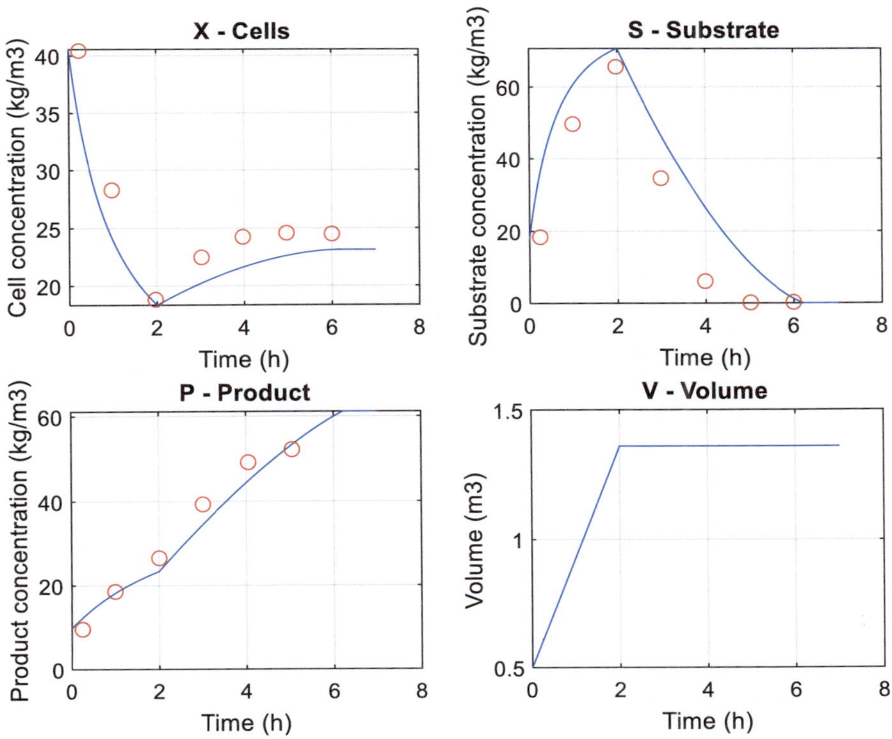

Fig. 9.32 Results of the simulation of a fed-batch run of ethanol fermentation, performed by Rivera et al. (2013). The simulation has been made by adopting the kinetic model suggested by the authors with the related parameters

The described results were obtained by using a MATLAB program that is available to the readers as Electronic Supplementary Material of this book.

9.5 Nonmechanistic Models

We have described in the previous sections some of the most popular mechanistic kinetic models, recognizing them through the classification of nonstructured, structured, nonsegregated, and segregated. More recently, different nonmechanistic kinetic models have been developed and proposed for the description of the cell's growth (Gonzàlez-Figueredo et al. (2018). These models consider the cells system as a "black box" and face the problem with a statistical approach or recurring to the artificial intelligence tools. The statistical approach considers the following: the experimental design, the response surface analysis, and data analysis. The approach based on the artificial analysis tools considers data mining, artificial network, and fuzzy logic as summarized in the following scheme.

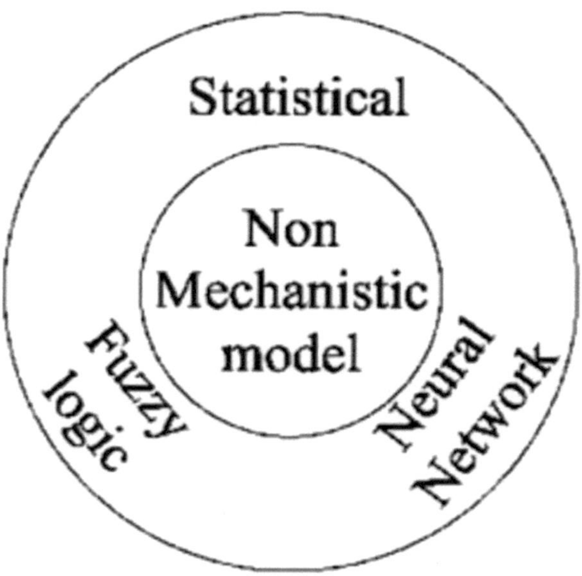

The advantage of these models is that they have a flexible structure and can be dependent on both qualitative features and quantitative values of the state variables. Attempts have also been made to combine the mechanistic and nonmechanistic approaches to developing **hybrid models** that are also known as **gray box models** (Psichogios and Ungar 1992; Bernard et al. 1999). The gray box models are more accurate, give better results in extrapolation, and furnish more detailed information about the occurring phenomenology and, therefore, resulted more useful in bioreactor modeling and optimization (Acuna et al. 2003; Zhang et al. 2019). Nowadays, several simulation programs, including neural networks and fuzzy logic, are also commercially available (see a list reported by Julien and Whitford 2007).

- **Artificial Neural Network (ANN)**

The use of the artificial neural network as a new approach to predict process variables from historical cultivation data was proposed in the 1990s (Noll and Henkel 2020; Thibault et al. 1990; Saner and Stephanopoulos 1992). Neural networks are mathematical models that imitate the way to interact of the brain neurons applied to the biological process. It is possible to imagine connections like the ones reported in the following scheme (Julien and Whitford 2007).

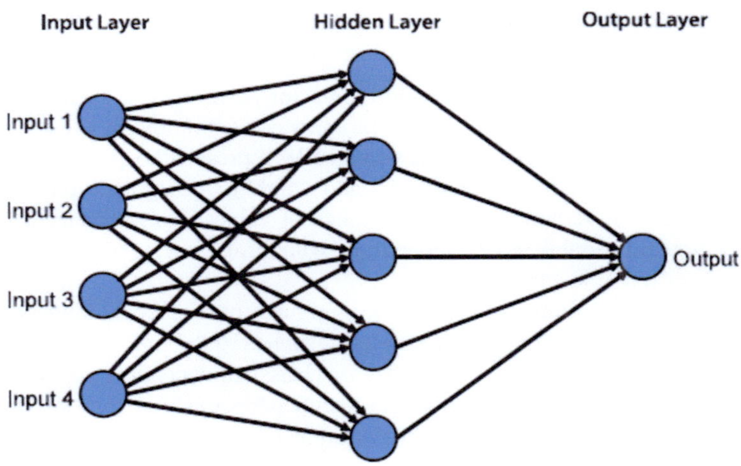

Defined the ANN structure, it is opportune to develop the memory form experience (i.e., the introduction of the experimental data). Then, a training algorithm must be introduced, fitting the process associated with each neuron in a way that the actions introduced (input signals) converge to the reactions produced (output signal). ANN does not furnish a physical interpretation of the phenomenon occurring in the system but can approximate quite well the dynamic behavior of the system. As seen in the scheme, ANNs are defined by three characteristics: the architecture, the activation functions, and the training algorithm. The architecture defining the interconnections is like the one represented in the scheme, the activation functions correspond to the dynamic characteristics of the neuron transfer functions, and the training algorithm is related to the parameter fitting procedure that is the basis of the learning ability.

- **Fuzzy Models**

Many uncertainties characterize the experimental data collected in bioreactors, and these uncertainties are clearly transmitted to the mechanistic models employed in their simulation (Ruggeri et al. 1994). Microorganisms of the same type, in comparable situations, often don't show the same behavior. This is typical of living systems that are susceptible to the external environment. Another source of uncertainty is determined by the arbitrary level of precision of the experimental investigation. At last, uncertainty can be related also to the phenomena occurring across the biological membrane. Fuzzy models are mathematical tools that have the scope to handle known precise details (bioreactor dimensions, mixing power, oxidation flow rate, etc.) together with imprecise details having both a role in the biocatalytic process. In other words, these models accommodate the inherent variability and uncertainty in biological processes by allowing for degrees of truth rather than strict binary outcomes. Fuzzy models, in the context of bioreactors, use the "fuzzy logic" to represent and analyze the behavior of biological systems within the reactor. Fuzzy logic enables the incorporation of qualitative and quantitative data, providing a more flexible and robust framework for describing and controlling bioreactor dynamics. Constructing a fuzzy framework involves several steps that are:

1. **Identify the key variables** affecting bioreactor performance, such as temperature, pH, substrate concentration, and biomass concentration.
2. **Define linguistic variables**. Linguistic variables represent the qualitative states of each variable (i.e., "low," "medium," and "high"). A linguistic term must be assigned to each variable to define its range of values. For example, you might have linguistic variable like "temperature" with terms: low, medium, and high to describe different levels.
3. **Membership functions**. Define membership functions for each linguistic term to specify how each value of the variable corresponds to the linguistic terms. These functions can be triangular, trapezoidal, or Gaussian, depending on the shape of the data distribution. For instance, the membership function for "low temperature" may be a triangular function that assigns higher degrees of membership to temperatures closer to the lower bound of the "low" range.
4. **Fuzzy rules**. Establish fuzzy rules to describe the relationships between the input variables and the output (e.g., biomass growth rate). These rules typically take the form of "if-then" statements, where the antecedents are fuzzy sets of input variables and the consequents are fuzzy sets of output variables. For example, a fuzzy rule might state: "If the temperature is low and the pH is high, then the biomass growth rate is medium." These rules capture the qualitative knowledge about how variables interact.
5. **Fuzzy inference system**. Implement a fuzzy inference system to process the fuzzy rules and compute the output based on the input variables. This involves fuzzification, rule evaluation, aggregation, and defuzzification. This operation involves several steps:

   (a) **Fuzzification**. Convert crisp input values into fuzzy sets using the membership functions.
   (b) **Rule evaluation**. Evaluate the degree to which each rule is satisfied based on the degree of membership of the input variables in antecedents.

(c) **Aggregation**. Combine the outputs of individual rules to obtain a comprehensive output fuzzy set.
(d) **Defuzzification**. Convert the aggregated fuzzy set back into crisp output values using methods like centroid, mean of maximum, or weighted average.

6. **Parameter tuning**. Fine-tune the parameters of the fuzzy model, such as membership function shapes and rule weights, using experimental data or expert knowledge to improve its accuracy and performance. This can be made by adjusting parameters such as the shapes of membership functions and the weights of fuzzy rules.
7. **Validation**. Validate the fuzzy model by comparing its predictions with experimental data or simulation results. Adjust the model as necessary for a more accurate simulation of bioreactor behavior.

By employing these mathematical techniques within a fuzzy framework, linguistic variables are transformed into computational operations that enable the modeling of bioreactors.

References

Acuna, G., Cubillos, F., Thibault, J., Latrille, E.: Comparison of methods for training grey-box neural network models. Process Engineering (2003)

Aiba, S., Shoda, M., Nagatani, M.: Kinetics of product inhibition in alcohol fermentation. Biotechnol. Bioeng. **10**, 845–864 (1968)

Akita, K., Yoshida, F.: Gas hold up and volumetric mass transfer coefficient in bubble columns. Ind. Eng. Chem. Process Des. Dev. **12**(1), 76–80 (1973)

Akita, K., Yoshida, F.: Bubble size, interfacial area and liquid-phase mass transfer coefficient in bubble columns. Ind. Eng. Chem. Process Des. Dev. **13**(1), 84–91 (1974)

Anderson, C.J., Longton, C., Maddix, G.W., Solomons, G.L.: The growth of microfungi on carbohydrates. In: Tannenbaum, S.R., Wang, D.I.C. (eds.) Single Cell Protein, vol. II, pp. 314–329. MIT Press, Cambridge, MA (1975)

Andrews, J.F.: A mathematical model for the continuous culture of microorganisms utilizing inhibitory substrate. Biotechnol. Bioeng. **10**, 707–723 (1968)

Bailey, J.E., Ollis, D.F.: Biochemical Engineering Fundamentals, 2nd edn. McGraw-Hill, New Delhi (1986)

Bakker, B.M., Overkamp, K.M., Van Maris, A.J.A., Kotter, P., Luttik, M.A.H., van Dijken, J.P., Pronk, J.T.: Stoichiometry and compartmentation of NADH metabolism in *Saccharomyces cerevisiae*. FEMS Microbiol. Rev. **25**, 15–37 (2001)

Bernard, O., Bastin, G., Angelov, P.: Hybrid modelling of biotechnological processes using neural network. In: 14th World Congress of IFAC, pp. 7573–7577 (1999)

Bijkerk, A.H.E., Hall, R.J.: A mechanistic model of the aerobic growth of *Saccharomyces cerevisiae*. Biotechnol Bioeng. **XIX**, 267–296 (1977)

Buchanan, R.L., Whiting, R.C.: A classification of model for predictive biotechnology. Food Microbiol. **10**, 175–177 (1993)

Crueger, W., Crueger, A.: A textbook of industrial microbiology, 2nd edn. Sinauer, Sunderland, MA (1989)

Di Serio, M., Tesser, R., Santacesaria, E.: A kinetic and mass transfer model to simulate the growth of baker's yeast in industrial bioreactors. Chem. Eng. J. **82**, 347–354 (2001)

Di Serio, M., Aramo, P., De Alteriis, E., Tesser, R., Santacesaria, E.: Quantitative analysis of the key factors affecting yeast growth. Ind. Eng. Chem. Res. **42**, 5109–5116 (2003)

Dunn, I.J., Heinzle, E., Ingham, J., Prenosil, J.E.: Biological Reaction Engineering. Wiley-VCH, Weinheim (2003)

Emery, A.N.: Chapter 10. Bioreactors. In: Winterbottom, J.M., King, M.B. (eds.) Reactor Design for Chemical Engineering, pp. 376–426. Stanley Thornes, Cheltenham (1999)

Gompertz, B.: On the nature of the function expressive of the law of human mortality and on a new mode of determining the value of life contingencies. Philos. Trans. R. Soc. Lond. **115**, 513–585 (1825)

Gonzàlez-Figueredo, C., Flores Estrella, R.A., Rojas-Rejòn, O.A.: Fermentation: metabolism, kinetic models, and bioprocessing. In: Current Topics in Biochemical Engineering, pp. 1–17. Intech Open, London (2018)

Gonzàlez-Hernàdez, Y., Michiels, E., Perré, P.: A comprehensive mechanistic yeast model able to switch metabolism according to growth conditions. Fermentation. **8**, 710 (2022)

Harder, A., Roels, J.A.: Application of simple structured models in bioengineering. In: Fiechter, A. (ed.) Advances in Biochemical Engineering, vol. 21. Springer, Berlin (1982)

Herbert, D.: Continuous culture of microorganisms; some theoretical aspects. In: Malek, I. (ed.) Continuous Cultivation of Microorganisms, a Symposium. Czech Academy of Science, pp. 45–52 (1958)

Jones, K.D., Kompala, D.S.: Cybernetic model of the growth dynamics of Saccharomyces Cerevisiae in batch and continuous cultures. J. Biotechnol. **71**, 105 (1999)

Julien, C., Whitford, W.: Bioreactor monitoring, modeling and simulation. BioProcess Int. **5**(01), S10–S17 (2007)

Kompala, D.S., Ramkrishna, D., Jansen, N.B., Tsao, G.T.: Investigation of bacterial growth on mixed substrates: experimental evaluation of cybernetic models. J. Biotechnol. Bioeng. **28**, 1044–1055 (1986)

Levenspiel, O.: The Monod equation: a revisit and a generalization to product inhibition situations. Biotechnol. Bioeng. **22**, 1671–1687 (1980)

Maicas, S.: Microorganisms (MPDI) **8**, 1142 (2020)

Malthus, T.: An Essay on the Principles of Population (1798 First Edition) with a Summary View and Introduction by A. Flew. Penguin Classics, London (1830)

McKellar, R.C., Lu, X.: Modeling Microbial Responses in Food. CRC Series in Contemporary Food Science. CRC, Boca Raton (2004)

Monod, J.: Recherches sur la croissance des cultures bacteriennes. Ed. Hermann and Cie, Paris (1942)

Moulijn, J.A., Makkee, M., Van Diepen, A.: Chemical Process Technology, pp. 353–379. Wiley, New York, Chapter 12 (2001)

Nissen, T.L., Kielland-Brandt, M.C., Nielsen, J., Villadsen, J.: Optimization of ethanol production in *Saccharomyces cerevisiae* by metabolic engineering of the ammonium assimilation. Metab. Eng. **2**, 69–77 (2000)

Noll, P., Henkel, M.: History and evolution of modeling in biotechnology: modeling & simulation, application and hardware performance. Comput. Struct. Biotechnol. J. **18**, 3309–3323 (2020)

Otterstedt, K., Larsson, C., Bill, M.R., Stahlberg, A., Boles, E., Hohmann, S., Gustafsson, L.: Switching the mode of metabolism in the yeast *Saccharomyces cerevisiae*. EMBO Rep. **5**, 5 (2004)

Pamment, N.B., Hall, R.J., Barford, J.P.: Mathematical modeling of lag phases in microbial growth. Biotechnol. Bioeng. **XX**, 349–381 (1978)

Psichogios, D.C., Ungar, L.H.: A hybrid neural network-first principles approach to process modelling. AIChE J. **38**, 1499–1511 (1992)

Rieger, M., Kappeli, O., Fiechter, A.: The role of limited respiration in the incomplete oxidation of glucose by Saccharomyces cerevisiae. Microbiology. **129**, 653–661 (1983)

Rivera, E.C., Costa, A., Atala, D.I.P., Maugeri, F., Maciel, M.R.W., Filho, R.M.: Evaluation of optimization techniques for parameters estimation: Application to ethanol fermentation considering the effect of temperature. Process Biochem. **41**, 1682–1687 (2006)

Rivera, E.C., Yamakawa, C.K., Garcia, M.H., Geraldo, V.C., Rossell, C.E.V., Filho, R.M., Bonomi, A.: A procedure for estimation of fermentation kinetic parameters in fed-batch bioethanol production process with cell recycle. Chem. Eng. Trans. **32**, 1369–1374 (2013)

Ruggeri, B., Sassi, G.: Acquisition of fuzzy rules by data analysis of biokinetics. In: Proceedings of 3rd IEEE International Conference on Fuzzy Systems IEEE World Congress on Computational Intelligence, Orlando, 26 June–2 July, vol. 3, pp. 2088–2094 (1994)

Ruggeri, B., Specchia, V., Sassi, G., Gianetto, A.: Numerical estimation of biokinetics parameters. Chem. Eng. J. **39**, 17–24 (1988)

Saner, U., Stephanopoulos, G.: Application of pattern recognition techniques to fermentation data analysis. IFAC Symp. Ser. **25**, 123–128 (1992)

Thibault, J., Van Breusegem, V., Chéruy, A.: On-line prediction of fermentation variables using neural networks. Biotechnol. Bioeng. **36**, 1041–1048 (1990)

Veloso, I.I.K., Rodrigues, K.C.S., Sonego, J.L.S., Cruz, A.J.G., Badino, A.C.: Fed-batch ethanol fermentation at low temperature as a way to obtain highly concentrated alcoholic wines: Modeling and optimization. Biochem. Eng. J. **141**, 60–70 (2019)

Verhulst, P.F.: Notice sur lam loi que la population pursuit dans son accroissement. Correspondance mathematique et physique. **10**, 113–121 (1838)

von Meyenburg, H.K.: Katabolit-Repression under Sprossungszklus von Saccharomyces cere6isiae. Dissertation No. 4279. Eidg. Techn. Hochschule, Zurich (1969)

Wank, H.Y., Cooney, C.L., Wang, D.I.C.: Computer control of baker's yeast production. Biothechnol. Bioeng. **21**, 975–995 (1979)

Williams, F.M.: A model of cell growth dynamic. J. Theor. Biol. **15**, 190–207 (1967)

Xing, Z., Bishop, N., Leister, K., Li, Z.J.: Modeling kinetics of a large-scale fed-batch CHO cell culture by Markov Chain Monte Carlo method. Biotechnol. Prog. **26**, 208–219 (2010)

Zhang, D., Del Rio-Chanona, E.A., Petsagkourakis, P., Wagner, J.: Hybrid physics-based and data-driven modelling for bioprocess online simulation and optimization. Biotechnol. Bioeng. **116**, 2919–2930 (2019)

Zwietering, M.H., Wijtzes, T., De Wit, J.C., Van't Riet, K.: A decision support system for prediction of the microbial spoilage in food. J. Food Prot. **55**, 973–979 (1992)

MIX
Papier aus verantwortungsvollen Quellen
Paper from responsible sources
FSC® C105338

If you have any concerns about our products,
you can contact us on
ProductSafety@springernature.com

In case Publisher is established outside the EU,
the EU authorized representative is:
Springer Nature Customer Service Center GmbH
Europaplatz 3, 69115 Heidelberg, Germany

Printed by Libri Plureos GmbH
in Hamburg, Germany